MATHEMATICS • PHYSICS
ASTRONOMY • BIOLOGY
CHEMISTRY • GEOLOGY • MEDICINE

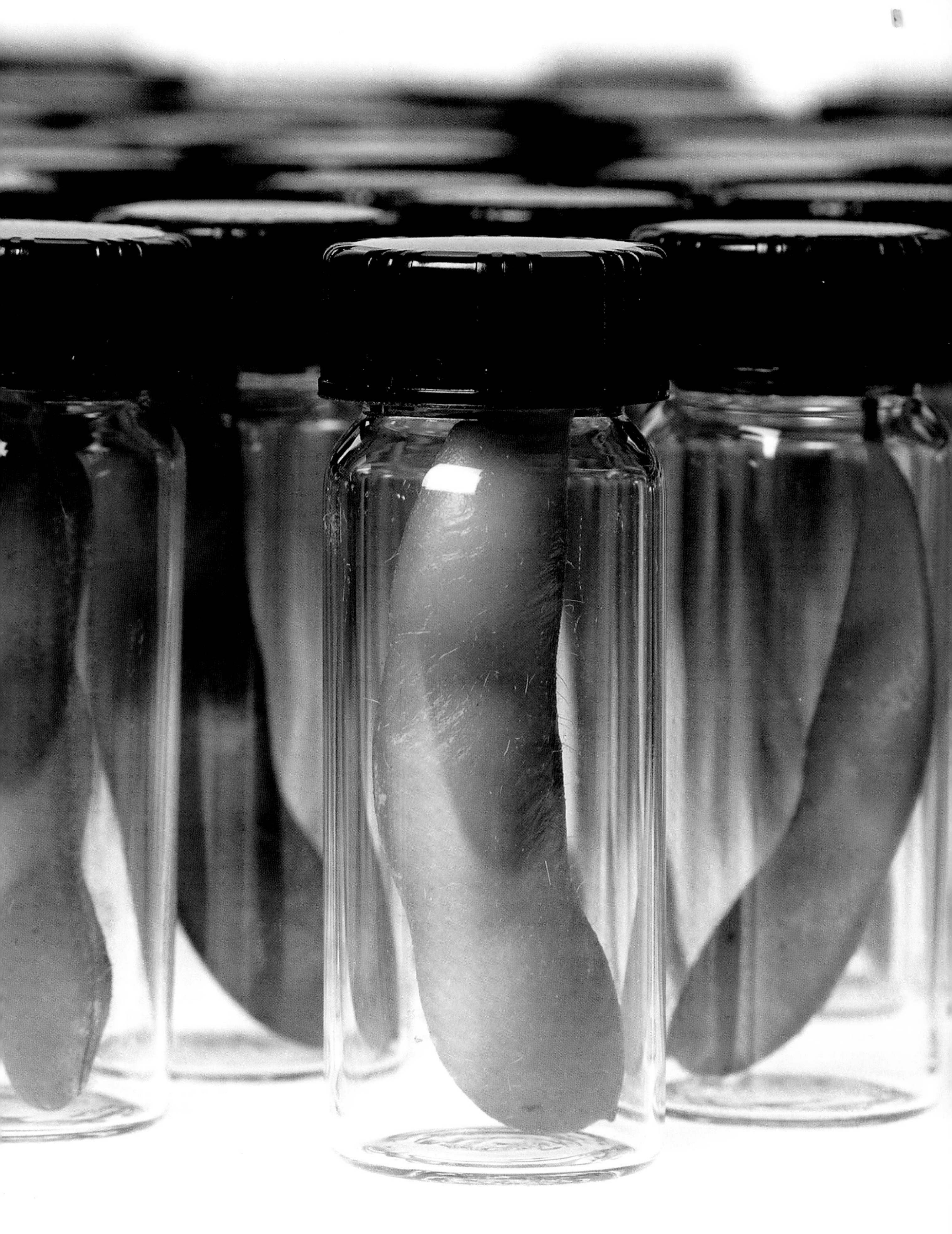

MATHEMATICS • PHYSICS
ASTRONOMY • BIOLOGY
CHEMISTRY • GEOLOGY • MEDICINE

Chief Consultant
Associate Professor Allan R. Glanville

Foreword
David Ellyard

MILLENNIUM HOUSE

First published in 2008 as *Scientifica* by
Millennium House Pty Ltd
52 Bolwarra Rd, Elanora Heights, NSW, 2101, Australia
Ph: 612 9970 6850
Fax: 612 9913 3500
Email: rightsmanager@millenniumhouse.com.au
Website: www.millenniumhouse.com.au

Reprinted 2009, 2010

ISBN: 978-1-921209-67-3

Authors: Millennium House would be pleased to receive submissions from authors. Please send brief submissions to editor@millenniumhouse.com.au
Photographers and illustrators: Millennium House would be pleased to receive submissions from photographers or illustrators. Please send submissions to editor@millenniumhouse.com.au

Color separation by Pica Digital Pte Ltd, Singapore
Printed in China

Photographs on cover and preliminary pages:

Front cover: A digital composite image of an X-ray of a spark between the fingers of two people.

Back cover (from top to bottom):
Soap bubble bursting: The colors produced just before the bubbles burst are due to interference between light rays reflected from the front and back of the thin film of water making the bubble.

Antique iron scale.

Fiber optic strands are thin, transparent fibers made of very specialized glass or plastic that transmit light along the length of their axes.

The wave rock formation reflected in water at Coyote Buttes North, Paria Canyon-Vermilion Cliffs Wilderness Area, Utah, USA.

A pair of harlequin crabs (*Lissocarcinus orbicularis*) on a sea cucumber.

The Alnitak region in Orion (Flame Nebula NGC 2024, Horsehead Nebula IC434).

X-ray effect image of a male skeleton, close-up.

Page 1: Blue glass globe filled with bright plasma lines.

Pages 2–3: Green beans in separate jars.

Page 6–7: Colored smoke rises in a vortex caused by the air flow from an agricultural plane landing. Testing such a wake vortex helps NASA and the US Federal Aviation Administration set standards for the distance aircraft stay from each other.

Pages 8–9: Storm researchers launch a weather balloon into a tornadic supercell thunderstorm.

Pages 10–11: Close-up of numbers on a protractor.

Pages 12–13: A scientist examines yeast growth in plant samples.

Pages 14–15: *The Alchemist* by William Fettes Douglas painted in 1853.

Publisher Gordon Cheers

Associate publisher Janet Parker

Art director Stan Lamond

Project manager Loretta Barnard

Chief consultant Associate Professor Allan R. Glanville

Consultants Dr Robert Coenraads, Paul Deans, David Ellyard, Dr Ron Haines

Contributors Dr Martin Anderson, Youna Angevin-Castro, Dr Phillip Arena, Yael Augarten, Brenton Banham, Dr Penny Bishop, Dr Luca Bombelli, Professor Richard Boyd, Colin Burgess, Heather Catchpole, Dr Stephen Clarke, Dr Steven Clay, Dr Robert Coenraads, Professor Christina Coughlan, Les Dalrymple, Paul Deans, Stefan Dieters, Emma Donnelly, Kerry Dougherty, Sean Elliott, David Ellyard, Margaret Etherton, Professor Anthony Fairall, Professor Jack Feinberg, Francis French, Melina Georgousakis, Dr Joel Gilmore, Andrew Grunseit, Dr Ron Haines, Nicole Harvey, Dr Susan Hawes, Catherine Healy, Margot Hislop, James Inglis, Imogen Jubb, Dr Peter Kappen, Andrew Ko, Henry Ko, John I. Koivula, Dr Kath Kovac, Dr Kathy Kramer, Dr Gary Lee, Sarah Lee, Tim Leslie, Nick Lomb, Mike McRae, Dr Jennifer Manyweathers, Jurgita Mikelenaite, Sir Patrick Moore, Jonathan Nally, Eleanor Neal, Dr L.E. Ohman, Tom O'Leary, Dr Armstrong Osborne, Professor Alfio Parisi, Tanya Patrick, Karen Pearce, Silvia Piviali, Dr Noel Roberts, Diane Robinson, Phil Rodwell, Christophe Rothmund, Philippa Rowlands, Dr Wayne Rowlands, Dr Andrew Sakko, Professor Jonathon Scott, Barry Stone, Robyn Stutchbury, Melissa Trudinger, Nicole Vanderkroef, Dr Anica Vasic, Dr Magda Wajrak, Alan Whitman

Editors Loretta Barnard, Helen Borger, Jennifer Coombs, Chris Edwards, Heather Jackson, Melody Lord, Nina Paine

Cover design Stan Lamond

Senior designer Jacqueline Richards

Designer Ingo Voss

Picture research Loretta Barnard, Casey Golden, Melody Lord

Illustrators Andrew Davies, Glen Vause

Index Tricia Waters

Production Simone Russell

Production assistants Bernard Roberts, Scott Quin

Chief Consultant

Associate Professor Allan R. Glanville MBBS, MD, FRACP, trained in Sydney, gaining his FRACP in 1985 before undertaking further education at the Brompton Hospital in London and the Stanford University Medical Center in California, USA. At Stanford University, he performed studies in the new discipline of human heart–lung transplantation, which led to the award of MD in 1990. Since returning to Australia, Dr Glanville has worked as a specialist in Lung Transplantation at St Vincent's Hospital, Sydney, where he is Director of the Department of Thoracic Medicine and Medical Director of Lung Transplantation. Dr Glanville is also an Associate Professor in the Faculty of Medicine at the University of New South Wales. He is actively involved with international trials into new immunosuppressive and antifibroproliferative agents and is Chair of the European and Australian Investigators in Lung Transplantation.

Dr Glanville is a foundation Director of LARA, Chair of the LAM Treatment Alliance Trials Group, and represents Australia on the LAM Foundation. He has been a member of the Editorial Board of the *American Journal of Respiratory and Critical Care Medicine* for the last four years. He is also Immediate Past President of the Pulmonary Council of the International Society for Heart and Lung Transplantation and is author of over 110 publications. Recent works have included novel approaches to therapeutic drug monitoring after lung transplantation and a reappraisal of the significance of subclinical acute lung rejection with emphasis on lymphocytic bronchiolitis as a determinant of long-term outcome after lung transplant. He is senior author of the International Guidelines for Lung Transplantation, a consensus document reflecting opinions of five international societies in the field. He is a foundation director of ShareLife Australia and considers it a privilege to work with like-minded Australians who give of their time freely to better the outcomes for all Australians in need of transplantation services.

Consultants

Dr Robert Coenraads is a consultant geoscientist, and author of four books and over 30 scientific publications. He has led archaeology, natural history, and geology field trips to various corners of the globe, including the magnificent Olmec, Maya, and Aztec sites of Mexico, active volcanoes of the Pacific region, and a number of fabulous gem mines. During his 30-year exploration career, travel to some of the world's poorest regions has sparked a strong humanitarian interest. Dr Coenraads is currently President of FreeSchools World Literacy Australia, and has established a support network to provide free education for underprivileged children in Bihar State, India. It is his firm belief that a solid education for all, not just a privileged few, is the key to solving the world's major problems such as overpopulation, hunger, and poverty. The privilege of helping produce quality educational books, such as *Scientifica*, is a step in this direction.

For many years **Paul Deans** wrote, produced, and directed multimedia shows for star theaters in Edmonton, Toronto, and Vancouver in Canada. In 2000 he switched careers and became an editor at Sky Publishing (Cambridge, MA, USA). Here he was an associate editor for *Sky & Telescope* and *Night Sky* magazines and the editor of *SkyWatch* magazine before ending up as Sky Publishing's book editor. Now based in Edmonton, Canada, Paul is currently a freelance science writer/editor and the editor of two digital magazines: *Mercury* and *Travel Quest*. He also specializes in science-themed travel, and particularly enjoys his annual aurora-watching trips to Iceland and his journeys to out of the way locations to witness total solar eclipses.

Trained in both science and education, **David Ellyard** has been involved in science and technology all his working life, as a researcher (including a year in Antarctica), teacher, radio and television broadcaster, writer, public speaker, and government policy and program officer. He worked on the pioneering Australian Broadcasting Corporation TV science programs *Towards 2000*, *Quantum*, and *Skywatch*, and was a weekend weather presenter on ABC TV for nearly 20 years. He has published more than a dozen books, including the award-winning *Oliphant* and *Who Discovered What When*. Many of his books combine his love of history with his concern for the impact of technology. His Starwheel map of the night sky has been in print for 20 years and has sold more than 100,000 copies. David is a former Governor of the University of New England and a former President of the Australian Science Communicators. He holds the New South Wales Director-General's award for services to public education.

Dr Ron Haines gained his BSc (with first class honors and university medal) and PhD from the University of New South Wales, Sydney, Australia. After a year of postdoctoral work using lasers to study the transfer of energy between molecules, he took up a tutorship with responsibility for planning and teaching for courses in computer applications in chemistry. After being appointed a lecturer, Ron's teaching widened to cover most areas of physical chemistry from first year to honors classes. Dr Haines maintains his interest in computers and their applications in chemistry by writing software for chemical education and by collaborating with computer scientists. In addition, he maintains the UNSW School of Chemistry website and recently became coordinator for first year laboratory classes in Chemistry at UNSW. He keeps fit by running and competes in races such as the Sydney half-marathon and marathon.

Contributors

Youna Angevin-Castro specializes in communicating science and innovation. Combining her natural curiosity with a love of the written word, Youna has published articles on everything from computational fluid dynamics to new technologies in food production.

Dr Phillip Arena is a member of the Australian Science Communicators. A herpetologist, he has published key works on animal welfare. His conservation efforts have achieved international recognition and as a lecturer at Murdoch University, Western Australia, he is acknowledged as one of Australia's foremost online teachers.

Yael Augarten has a physics degree with honors from the University of New South Wales. Currently a PhD student in Photovoltaics, Yael's research involves lasers to characterize solar cells, in order to make them cheaper. She is also a tutor in undergraduate physics.

Brenton Banham is a science communicator in chemistry, physics, and earth sciences with Flinders University, South Australia. He has extensive experience as a senior secondary science teacher both in Australia and in Papua New Guinea, Oman, Brunei, and the USA.

Dr Penny Bishop has extensive research experience in microbiology and molecular biology, and has spent the last 20 years teaching in the areas of microbiology and infection control at the University of Sydney.

Dr Luca Bombelli is an Italian physicist currently living in the United States with his wife Viki and their two sons. His main research area is quantum gravity and the structure of space–time, and he teaches at the University of Mississippi.

Professor Richard Boyd is the Director of Monash Immunology and Stem Cell Laboratories where he heads the Immune Regeneration Laboratory. He is also Chief Scientific Officer of the biotechnology company, Norwood Immunology, which has supported the translation of his laboratory's research to clinical trials in Australia and the USA. **Anne Fletcher** and **Jarrod Dudakov** are senior PhD students supervised by Professor Boyd, who completed their studies in 2008.

Heather Catchpole is a freelance science writer and author of three children's books in the popular "It's True" series (Allen & Unwin), as well as a geology recipe book for children. She writes regularly for the Australian Broadcasting Corporations's science website and COSMOS magazine.

Dr Stephen Clarke has taught innovation and chemistry at Flinders University, South Australia, since mid-2002, and also undertakes biofuels, desalination, nanotechnology, polymers, dendrimers, and organosilicon research. In 2008 he accepted an academic research-only position, leading the Materials and BioEnergy Group at Flinders.

Dr Steven Clay attended the Kirksville College of Osteopathic Medicine in Missouri and was residency trained in Youngstown, Ohio, USA. He is certified in family medicine, addiction medicine, and geriatric medicine. After eight years of private practice, he joined the Ohio University College of Osteopathic Medicine where he develops curricula, teaches medical students and residents, performs research, and treats patients.

Dr Christina Coughlan, Assistant Professor of Neuropharmacology at the University of Denver, received her degrees from the Pharmacology Department, UCD, Ireland. Her postdoctoral experience was gained at Ninewells Hospital Dundee, the University of Pennsylvania, and the University of Pittsburgh. Her areas of research expertise are Alzheimer's disease and dementia in diabetes.

Emma Donnelly has always had a keen interest in science. Since graduating from the University of Western Australia in 2000, she has worked in science communication and education. Through this she has been able to share her passion for science with school students and adults.

Sean Elliott is a Melbourne-based science communicator and part-time writer. He has previously worked at the Scienceworks Museum as a science show presenter and has written programs for family and school audiences. He now works for CSIRO Education as part of their science outreach program.

Margaret Etherton is a teacher and writer who actively champions an understanding of mathematics and computing. She has written widely on nature and environment in magazines, and educational resources for teachers.

Professor Jack Feinberg teaches physics at the University of Southern California. He studies the fundamental properties of optical materials and won a 1995 Discover Award for his use of temporal holography to see through human tissue using light instead of X-rays.

Melina Georgousakis is a PhD scholar researching the development of novel mucosal vaccines against the human pathogen, Group A streptococcus. A dedicated science communicator, she was recently acknowledged for her efforts as a 2008 Queensland state finalist for Young Australian of the Year.

Dr Joel Gilmore received his PhD in Physics from the University of Queensland in 2007. He currently works there as a science communicator, running the highly acclaimed Physics Demo Troupe that promotes science to schools around Australia.

Andrew Grunseit is Head Teacher of Mathematics at Barrenjoey High School, Sydney, Australia. A keen traveler, surfer, and skier, Andrew lives on the Northern Beaches of Sydney with his wife and two daughters.

Nicole Harvey is a science writer particularly interested in community health. She has studied and worked in several fields including virology, zoology, and medical writing. Nicole currently works at The Royal Australian College of General Practitioners.

Dr Susan Hawes is a research scientist and freelance writer, who is interested in how embryonic cells, like a clean slate, can be given instructions to generate complex cell types that behave like our body's cells. Susan writes about science and medicine.

Catherine Healy gained her qualifications from Monash University, Victoria, Australia, in 1999. After completing a teaching degree and working abroad, she was an education officer at Commonwealth Scientific and Industrial Research Organisation Education (CSIRO). Catherine currently works as a freelance writer.

Margot Hislop is a freelance science writer and communicator based in Canberra, Australia. Having studied marine biology, she writes about the marine environment and other areas of science, and has written for a number of Australian science publications. www.mangoh.com.au

Following her studies, **Imogen Jubb** worked as an ecologist, before branching out into science communication. She has traveled extensively delivering educational science programs, and has been a presenter for the Australian Broadcasting Corporation's science and technology program *Nexus*.

Dr Peter Kappen is a synchrotron scientist by training and also at heart. He did his PhD in Physics at the Hamburg Synchrotron Laboratory HASYLAB in Germany. Peter runs synchrotron projects in various fields including industrial research and environmental sciences. He lives in Melbourne, Australia.

Andrew Ko is an Australian-born Chinese whose academic areas of interest include biochemistry and microbiology. He intends to become a pharmacist after completing his master of pharmacy degree from the University of Sydney.

Henry Ko is a biomedical engineering researcher. He is involved in tissue engineering and regenerative medicine research, exploring socio-techno issues surrounding emerging biotechnologies, and in developing systematic reviews for clinical therapies and evidence-based medicine.

Dr Kath Kovac is a freelance science writer and editor based in Australia. She has a PhD in plant molecular genetics and is the former editor of *The Helix*, CSIRO's science magazine for teenagers.

Dr Kathy Kramer is a doctor, lecturer, writer, and editor. She lives with her husband, three small children, and an assortment of animals on an olive farm near Bellingen in New South Wales, Australia.

Dr Gary Lee is a senior lecturer at the University of Sydney. He has a PhD in microbiology and immunology. Dr Lee is the author of an acclaimed university textbook on microbiology and infection control.

Sarah Lee is a newspaper, magazine, and internet journalist. She has a Bachelor of Arts degree with an English major and a Certificate III in News Media acquired at TAFE NSW.

Tim Leslie is a PhD student at the University of New South Wales, studying Antarctic astronomy. His research covers areas including automated experiment design for hostile conditions, software development methodologies, data analysis, and modeling.

Mike McRae has worked as a medical scientist in a pathology laboratory and as a science educator in both Australia and the UK. He currently lives in Canberra where he writes for the CSIRO and works on a weekly radio science show, *Fuzzy Logic*.

Dr Jennifer Manyweathers is a veterinary surgeon and freelance science writer, currently living in Japan. She teaches science communication at Tsukuba University and writes for the Japanese Science and Technology (JST) website for children. She is married with two children.

Jurgita Mikelenaite graduated from the University of Klaipeda, Lithuania, with a master's degree in ecology and environmental science. She is currently researching the growth and distribution of juvenile brown trout in Lithuanian rivers.

Eleanor Neal has tutored at the University of Melbourne where she earned her three degrees. She has volunteered at the Australian Academy of Technological Sciences and Engineering (ATSE) Clunies Ross Foundation Extreme Science Experience, explaining scientific concepts to non-scientists.

Dr L.E. Ohman is a freelance medical writer. She completed a PhD in physiological psychology and began her career as a biomedical researcher. She has worked as a journalist and public relations officer in the area of health sciences.

Tom O'Leary is an Australian freelance writer, currently based in Japan. He writes for print and web publications on topics ranging from science and health to business and personal development. Tom specializes in communicating important ideas to important people.

Dr Alfio Parisi is a physicist and Associate Professor in the Faculty of Sciences at the University of Southern Queensland. He has over 20 years experience in lecturing science and has established a research unit in solar ultraviolet radiation with significant research publications.

Tanya Patrick is currently editor of CSIRO's popular science magazine *Scientriffic*. Based in Canberra, Australia, Tanya was recently awarded one of three annual Australian Antarctic Division Arts fellowships. The resulting photographs and articles have been published around the world.

Karen Pearce is a science communicator who has worked for the Commonwealth Scientific and Industrial Research Organisation, the Australian Bureau of Meteorology, and the Australian Greenhouse Office. She has also written on a range of science topics for print and online publications.

Silvia Piviali emigrated from Italy to Western Australia at the age of six. Her passion for science led to a career as a medical scientist. She also follows other pursuits, such as freelance writing.

Dr Noel Roberts, formerly Associate Professor of Chemistry at the University of Tasmania, is author of *From Piltdown Man to Point Omega: The Evolutionary Theory of Teilhard de Chardin* (New York: Peter Lang, 2001). Along with his science degrees, Noel has degrees in philosophy and theology, and is literate in seven languages.

Philippa Rowlands is a professional writer who has publications in fields as diverse as science and technology, travel, education, parenting, self-help, and popular culture. The effective and clear communication of science and technology is a strong feature of her career.

Dr Wayne Rowlands is a physicist and lecturer at Swinburne University of Technology, Melbourne, Australia, where he specializes in laser cooling of atoms and molecules. He is passionate and enthusiastic about the accessible communication of science.

Dr Andrew Sakko received his Bachelor of Biotechnology and PhD from Flinders University, Australia. He then spent several years researching cancer at the Karolinska Institute in Sweden and continues to research cancer at the University of Adelaide, Australia.

Jonathan Scott became Foundation Professor of Electronic Engineering at the University of Waikato, New Zealand, in 2006. Prior to that he worked for Agilent Technologies' Microwave Technology Center and Hewlett-Packard in California, as a senior lecturer at the University of Sydney, and also for a variety of electronics companies in Australia.

Melissa Trudinger has worked as a science writer since 2002, notably for Australia's first dedicated publication for the biotechnology industry, *Australian Biotechnology News*. Before that, she worked in research and development at a number of biotechnology companies in the San Francisco Bay Area, USA.

Nicole Vanderkroef has extensive media experience and her writing has appeared in a variety of publications, including COSMOS magazine, *Medical Observer*, and several CSIRO publications. Recently she branched into industrial relations journalism, where she won an award for her work.

Dr Anica Vasic is a specialist anesthetist and pain management consultant, and Director of the Pain Management Unit, St George Hospital, Sydney, Australia. Her clinical interests include optimization of patients for anesthesia through perioperative medical assessment processes and analgesic techniques for day stay patients.

Dr Magdalena Wajrak is currently a lecturer in chemistry at Edith Cowan University, Western Australia. Her background is in physical chemistry, applied physics, and theoretical chemistry. She has worked as a science adviser and also as a research chemist.

Contents

Foreword

Scientifica has taken on a daunting task—to summarize the current state of scientific knowledge across seven of the major disciplines, namely astronomy, physics, chemistry, medicine, biology, geology, and mathematics.

It is daunting for two reasons. Scientific understanding is not only immense but continues to grow exponentially. By one reckoning, the sum total of scientific knowledge doubles every decade. A book of this kind must to some extent be out of date even before it is printed. That of course does not detract from its value. Much of the most valuable and fundamental scientific principles have been in place for decades or even centuries.

The second daunting aspect of this task is summarized thus—we have sought to present what we have come to know through science in a way that is readily accessible and useful to people who are not themselves scientists. Many parts of science seem obscure, and its practitioners often use strange words unintelligible to the rest of us.

Yet we all need to understand what science has taught us about the operation of the natural world in which we are embedded and on which we so depend. No human development has influenced the state of the world today as has knowledge gained through science and its application through technology and invention. It is very hard to make sense of many things happening almost daily unless we have some understanding of science.

Indeed we can argue we have a right to know. Many great scientific discoveries have been paid for from the public purse, so in a very real sense, we own this knowledge.

Science is a human activity. What we know about the workings of the world has not been let down from heaven on a golden string. It is rather the product of human curiosity, ingenuity, and persistence. Many of the most brilliant minds and outstanding personalities in the history of the human race have devoted their lives to science, believing that nothing matters more than well-founded scientific insight.

What they have discovered has not only changed our lives but fired our imaginations. We have learned from them that the universe from the smallest particle to the largest galaxy is an amazing place, full of wonders. Many of those wonders are in this book.

David Ellyard

INTRODUCTION

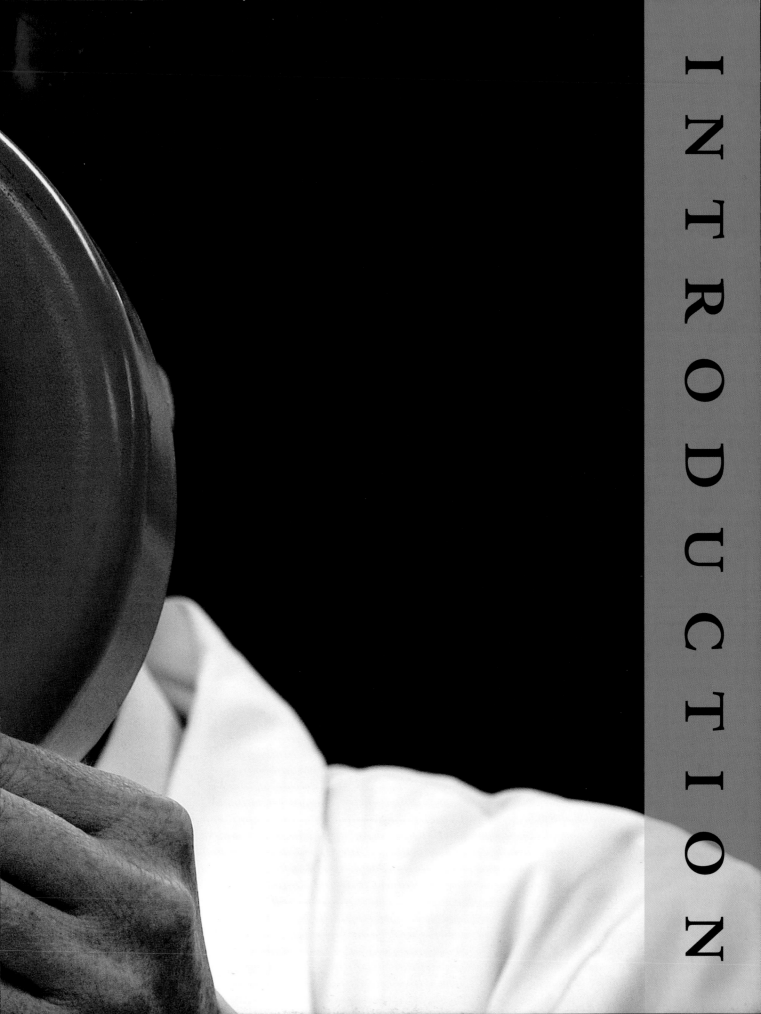

An enquiring yet critical mind is the origin of all thought and knowledge, so let us begin at the beginning and ask the most simple question, why Scientifica *or indeed why science? In fact, what is science after all? Science may be defined variably as the intellectual and practical activity encompassing those branches of study that apply objective methodologies to the phenomena of the physical universe, or as the state or art of knowing, or perhaps as a theoretical perception of truth contrasted with moral conscience. More commonly, science is defined as knowledge acquired by study leading to mastery over a department of learning.*

Introduction

Indeed, what is science but the quest for knowledge about the human condition and how we relate to all around us? In every field, science informs us, enchants us, and holds us in its thrall, promising a reward of information which helps make sense of natural phenomena. *Scientifica* contains the stuff of science, or knowledge if you will, which aims to translate esoteric disciplines so we can all understand them. How else to explain the intricacies of the natural laws that govern the form and function of all things? In past millennia great thinkers pondered on the nature of things and sought to understand why it was so. Now we know. Well, we know some of what is to be known, and with that knowledge comes the beginning of understanding that all is not known or perhaps even knowable.

Books such as *Scientifica* hold an important place in building our international vision of the natural world because they interpret and challenge at the same time. No one can read the content or view the illustrations without some sense of awe at how the scientific world provides a sense of order within seeming chaos, even if one must embrace chaos theory to understand this!

Above: Astronomer, physician, and philosopher Abu'l-Walid Ibn Rushd, better known as Averroes (1126–1198), was one of the most influential Arab scholars of all time. He wrote on medicine, physics, and logic.

So let us dissect (as this is a most appropriate term) the areas covered by *Scientifica* and commence with an overview of what, why, and how. This is no simple task as the mere enumeration of physics, chemistry, biology, astronomy, geology, mathematics, and medicine is enough to remind many of us of courses taken and examined formally during our schooling. However, in the greater school of life, education transcends formality, and the approach of *Scientifica* is to accept the challenge of making interesting former areas of struggle by showing the essential linkages between the way we live now and what has been discovered, analyzed, and developed. Technology has become our great servant (and at times master) but without development strategies the great ideas would remain just that—great ideas—and we would not have electricity, mass communication, telephones, the World Wide Web, or computers. Each

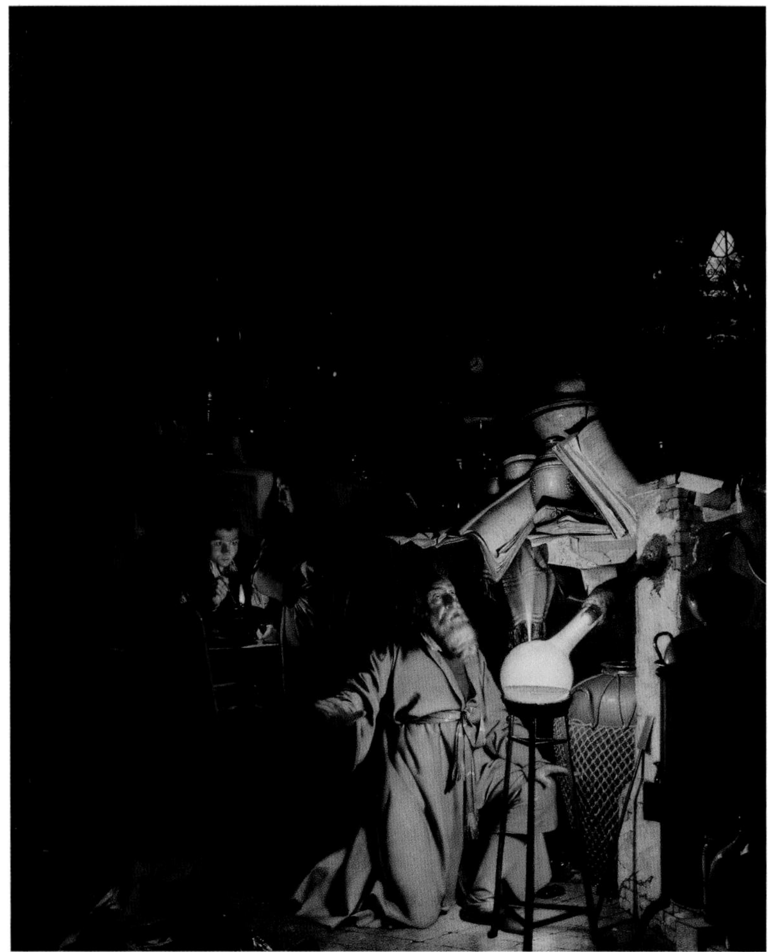

Left: Alchemy had its roots in Arab philosophy, encompassing elements of chemistry, physics, astrology, and metallurgy. Most alchemists, such as this one painted by Joseph Wright of Derby in 1771, tried to turn base metals into gold.

Above: A laboratory technician wears a hazmat suit to protect herself against possible chemical or biological contamination. Hazmat is short for "hazardous materials." Such garments allow biological research to be carried out safely.

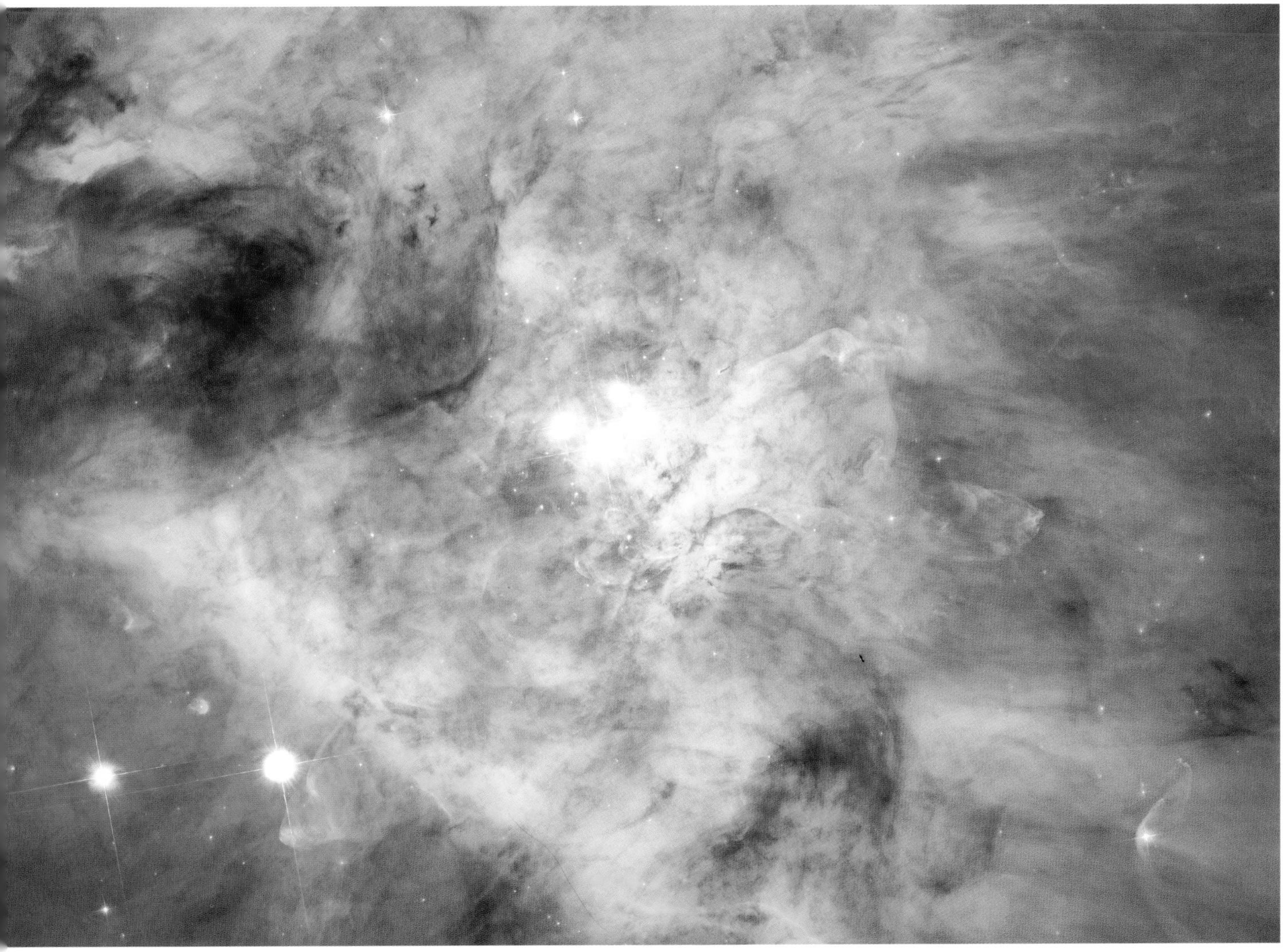

depends on science and discovery and translation of those great ideas into practicalities. The following paragraphs give a brief introduction and overview of just some of the intriguing information found in the text of *Scientifica*, showing how no field of scientific endeavor stands alone. They are all entwined.

Have you ever wondered why, as Isaac Newton did, that an apple falls to the ground rather than arbitrarily into the air? The answer of course is gravity. Gravity refers to the force of attraction that occurs between two masses and the bigger the mass of an object, the stronger the force of attraction. Consequently, because Earth's mass is so much greater than our own, the attraction is great enough to keep us grounded (at least physically!). Gravity acts locally and throughout the universe and a power greater than gravity is the likely explanation for an outstanding observation. The cosmos is actually getting larger as you are reading this! Astronomers have discovered that stars that exploded many many years ago were not as bright as conventional estimates predicted, which suggests that they are farther away than originally thought. Hence the rate of expansion of the universe must be accelerating and the reason for this is thought to be the repulsive force of dark energy. Dark energy began about nine billion years ago and between five to six billion years ago its repulsive force became sufficient to overcome the force of gravity, which led to the expansion of the universe accelerating.

While we are pondering the universe and the laws that govern it, let us consider the heavenly bodies. That unruly star that wakes us each morning and sets the pattern of our days is in fact much larger than Earth. The Sun is a massive 865 million miles (1.4 million km) in diameter while Earth measures only 7,920 miles (12,746 km) wide. The Sun has about 330,000 times more mass than Earth but is not a solid body, like a rocky planet, but rather a huge globe of intensely hot gas. Did you know that the Sun does not rotate as a solid mass? It rotates faster at the equator—around 25 days for a full rotation—than it does at its polar regions where it takes 35 days.

Gravity, mass, and time are all fundamentals of the universe but so too is distance which has been a dominant constraining factor in any attempt to realize extraterrestrial travel. So has the simple necessity of accuracy in measurement, which needs to be absolutely precise or you simply miss your target! In particular, travel to the Moon has been an abiding fascination of popular fiction but the truth is perhaps much more interesting. The planned landing site of the *Eagle* on the surface of the Moon was the Sea of Tranquillity, but as the landing began, it was discovered that the Lunar Module was further along its descent trajectory than planned and would have to land some distance west of the intended site. Manual control of the Lunar Module was necessary to guide the spacecraft to a landing on July 20, 1969, with only seconds' worth of fuel left.

Above: The Orion Nebula's biggest stars. Packed into the center of this region are the bright lights of the Trapezium stars, the four heftiest stars in the Orion Nebula. Ultraviolet light unleashed by these stars is carving a cavity in the nebula and disrupting the growth of hundreds of smaller stars.

Left: Swedish chemist Alfred Nobel (1833–1896) is remembered as the inventor of dynamite and the man who established the Nobel Prize for excellence in physics, chemistry, medicine, literature, and work for peace.

Above and left: Botany has long been an area of biological study. Above is an eighteenth century painting of the rose *Rosa gallica* by Redoute, where *Rosa* is the genus name and *gallica* is the species name. At left is an illustration of some botanical plants from *The Wonders of the Creation and the Curiosities of Existence* by Zakariya-ibn Muhammed al-Qazwini, dating from the thirteenth century.

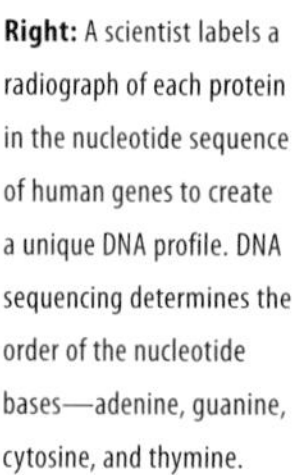

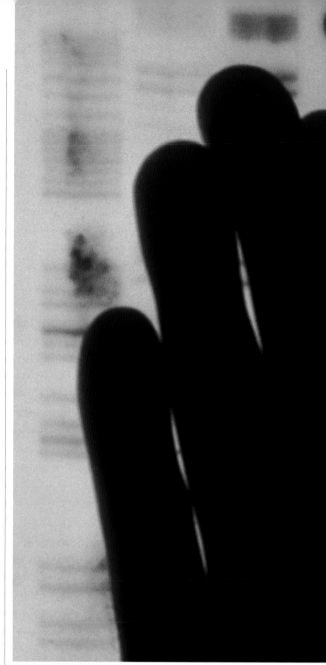

Right: A scientist labels a radiograph of each protein in the nucleotide sequence of human genes to create a unique DNA profile. DNA sequencing determines the order of the nucleotide bases—adenine, guanine, cytosine, and thymine.

At such a distance, even the smallest error in calculation would have resulted in a major catastrophe.

If the space race embodied mankind's quest for exploration, it is the science of biology that embodies the quest for understanding. Have you ever wondered why offspring look like their parents? Is it a random mechanism? As it happens, understanding the mystery of inheritance took centuries, beginning with the realization that both the male and the female of the species contribute to the offspring's characteristics. Biology has also helped us by putting a sense of order to life's diversity. Humans have been classifying the life around them since Aristotle's time. Prior to the mid-eighteenth century however, there were no clear rules for naming life, and attempts to classify organisms were sometimes confusing. This changed when Carl Linnaeus, a Swedish botanist and physician, proposed a system that assigns all living things a two-part name. The first part, the genus, identifies the group to which the organism belongs, while the second part of the name, the species name, distinguishes the organism from other organisms in that group.

One significant group of organisms are viruses, typified by the virus that causes the "flu." In 1918 an influenza epidemic swept the world causing the deaths of around 40 million people. We now know that the influenza A virus has two "spikes" on its outer envelope—hemagglutinin (H) which is involved in the attachment of the virus to its target cell, and neuraminidase (N) which allows the release of new viral particles from the infected cell. Genetic changes (mutation) produce new virus types with variations in the spikes which alter the infectivity of the virus. Different virus types are designated by numbers, for example avian flu is H5N1. Thankfully, in its present form avian flu is not very infectious to humans but there is concern that if it mutates it might cause a pandemic. Certainly the mortality rate in infected individuals is high (about 40 percent). Not surprisingly, the World Health Organization constantly monitors cases of H5N1 flu and other communicable diseases.

Bacteria are another type of organism. Thankfully, few bacteria cause disease in humans, but those that do can cause serious illness, even death. Among the most serious diseases caused by bacteria are tuberculosis, whooping cough, diphtheria, gastrointestinal diseases, pneumonia, meningococcal disease, sexually transmitted infections, and wound infections. Diseases such as botulism, tetanus, gangrene, and anthrax are caused by bacteria that form resistant endospores. Bacteria affect the body by production of enzymes and toxins that damage various organs. Before the discovery of antibiotics many of these diseases were often fatal. Antibiotics have altered the level of risk, but there is growing concern that many bacteria have become resistant to antibiotics due to antibiotic selection pressure from over-use. Responsible prescribing habits are necessary to prevent overuse of antibiotics in situations where they are contra-indicated, such as pure viral infections.

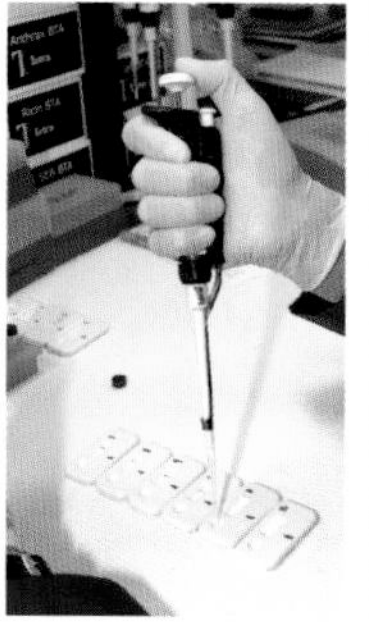

Above: An anthrax test kit provides on-site screening of biological threats such as anthrax, which is caused by the bacterium *Bacillus anthracis*. Anthrax is a serious disease of cattle and can be dangerous to humans. It is acknowledged as a possible biological weapon.

The building block for all life is DNA, which is found in every cell of a living organism. The way that the two strands of DNA meet and join is unusual, and was discovered in 1953 by James Watson, Francis Crick, and Maurice Wilkins. They found that two strands of DNA intertwine around each other forming a structure that is known as the double helix. This discovery was pivotal in understanding our genetic makeup. By June 26, 2000, the completion of the first draft sequence of the entire human genome was announced, which was in effect, an accounting of the three billion base pairs that make up the DNA sequence that spells out much of who we are. This project had started in earnest in 1990 and was considered by its critics an impossible dream. However, with the development of large automated sequencing machines and new techniques for generating data, the initial draft was completed two years ahead of schedule.

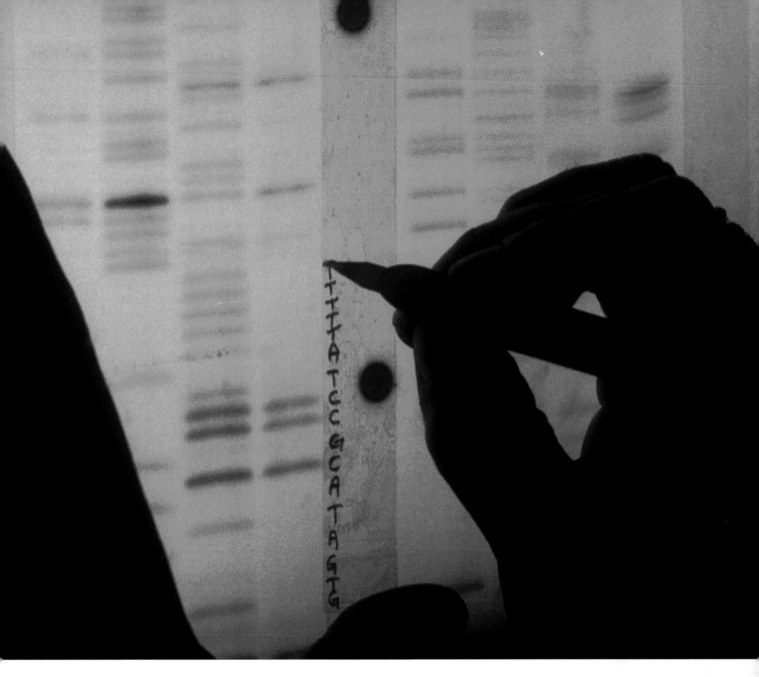

While we are thinking about DNA and genes let us ponder scale and measurement. Just how small can we think? How small can we know, act, and control? Well, a rather new science of small things has been developed. Nanotechnology refers to science and technology, nominally on a scale of 1 nm to 100 nm (nanometers) where a nanometer is one-billionth of a meter (10^{-9} meters). Although nanotechnology is considered to be a very new science, certain aspects of nanotechnology have been around for many years. It is the unique, multidisciplinary linking of chemistry, physics, and biology that characterized the growth of nanotechnology from the mid-1980s onward.

Conversely, on a massive scale, there are billions of cubic miles of water on Earth, 97.5 percent of which is undrinkable seawater. One method in which we could lower atmospheric carbon dioxide and reduce global warming might be to seize the opportunity to produce usable water by desalination which would then allow us to produce vast quantities of low-cost feedstock to manufacture renewable biofuels. Coupled with negative population growth the possibility of balancing sustainable resources may become a reality.

Clearly scale is a major issue and to grasp these arguments we require a basic feel for numbers. Like it or loathe it, we could not

Right: MEMS micromachined gears, magnified 700 times. MEMS stands for Micro-Electro-Mechanical Systems, which is technology at the nanoscale. MEMS devices might be no bigger than a grain of pollen. MEMS technology has many applications, including in the automotive industry, in science, and in medicine.

function in the modern world without mathematics. Numbers are tools to help us solve problems and to understand the world. Even more fundamental to our survival, we use numbers every day in countless ways. Numbers are the basis for all formulas but simple mistakes can happen if the correct units are not used. Formulas are only as good as the information put into them. There have been some spectacular examples where formulas have appeared to fail but where the disaster can actually be traced back to the data used. In 1998, NASA launched its Mars Climate Orbiter, designed to gather information about the climate of the red planet. Shortly before it was due to begin its orbit around Mars, it disappeared and was never found. The reason for the failure? While the formulas were found to be correct and appropriate, it was discovered that some scientists had entered inches and pounds into their formulas and others had used meters and kilograms!

Algebra is a very old science but is integral to modern life, underpinning everyday situations, and is used in scientific fields ranging from medicine to physics. In engineering, structural formulas and the principles of electronic circuits depend on algebra. Banking and financial matters like accounting and superannuation use formulas for calculating compound interest, mortgages, and payouts when people retire. In addition, the techniques of algebra can be applied to probability in areas such as health statistics, genetics, and games of chance. Cooks, accountants, builders, and engineers all use algebra. Not only is algebra fundamental to our world, the mental exercise of solving problems helps to develop logical thinking.

Logic underpins the computer that we now take for granted. Computers would not work if it were not for sensible programming. Programming is the set of rules that have been given to a computer, that describe what is to be done to any information entered into it. Early computer programming was achieved by physically changing the hardware of the computer. Computer programming is currently carried out in one of 2,000 languages that dictate what a computer will do with information that it is given. One application is in the statistical analysis of data. In fact statistics is the science of data. It is easy to forget that when we see a statistic, it usually relates to something physical in the real world. Road accident statistics describe the collisions of vehicles containing real people. In our daily lives, plans and actions are often dictated by subconscious estimations of chance. Even a young child understands the statement, "there is a very good chance it might rain today." But what is "a good chance"? Probability is the application of values to such likelihoods.

Computers are now most commonly used to convey information around the world via the internet. Some people reading this book grew up before the internet and the "web" existed, as strange as that may seem to contemporary behavior! In fact the World Wide Web was only invented by British scientist Tim Berners-Lee in 1989 and while the words "web" and "internet" are often used interchangeably they are not the same thing. The web is a collection of resources and documents connected by hypertext links, whereas the internet is the physical connection of computer networks. The development of the internet became possible from the 1960s by the implementation of packet-switching networks, the means through which the internet delivers information from one computer to another. At the source computer, information is divided into small blocks of data called packets, which are defined by a language known as the standard Internet Protocol (IP). Packets can be compressed and encrypted and are transmitted individually, often using different routes to make the transfer of information more efficient. The packets are transported using physical connections made of copper wire or fiber optics. At the destination computer, the packets are recompiled.

Left: These windmills in California, USA, are part of an extensive wind farm, an alternative source of electrical power. Scientists around the world are researching new ways to power our lives so that we cease to depend on fossil fuels.

Above: We might not realize it, but mathematics plays a very important role in our everyday lives. For example, we use formulas to assess body weight and bank interest, to determine how much food to buy, to predict the weather, and to work our computers.

By now you may wish to scream—so much information, so much to learn, and still so much yet to know. Even a scream has special and different properties depending on whether you are moving towards or away from the scream. When something noisy, such as a car with a siren, is standing still, the waves spread out evenly in all directions and it sounds the same no matter where you are standing. If the car is moving, however, the waves will be different depending on whether you are in front of or behind the car. The waves at the front will end up pushed closer together—the car is catching up with the sound waves it has already sent out—while the sound waves at the back will be stretched out. Standing in front of an oncoming car you will hear a higher pitched sound, while standing behind it you will hear a lower pitched sound. This shift in frequency is known as the Doppler effect and is used in medical diagnostic imaging to detect the flow of blood through a blood vessel.

Advances in medical science have depended to a great degree on advances in technology as described above. Every time you see your doctor or undergo a simple non-invasive test such as an X-ray or an ultrasound, the same principles are being applied and our knowledge of what to do when and how is informed by access to the most up-to-date information available on the "web." While the universe is demonstrably expanding so too is the capacity to obtain accurate diagnosis from which springs the possibility of a mutually satisfactory outcome. This is just one way that science impacts our daily life.

Seen in a challenging world of exploding information, *Scientifica* should act as a stimulus to encourage deeper study in chosen disciplines. The next important discovery for mankind is awaiting a new scientist of the twenty-first century. That could be you!

Allan Glanville

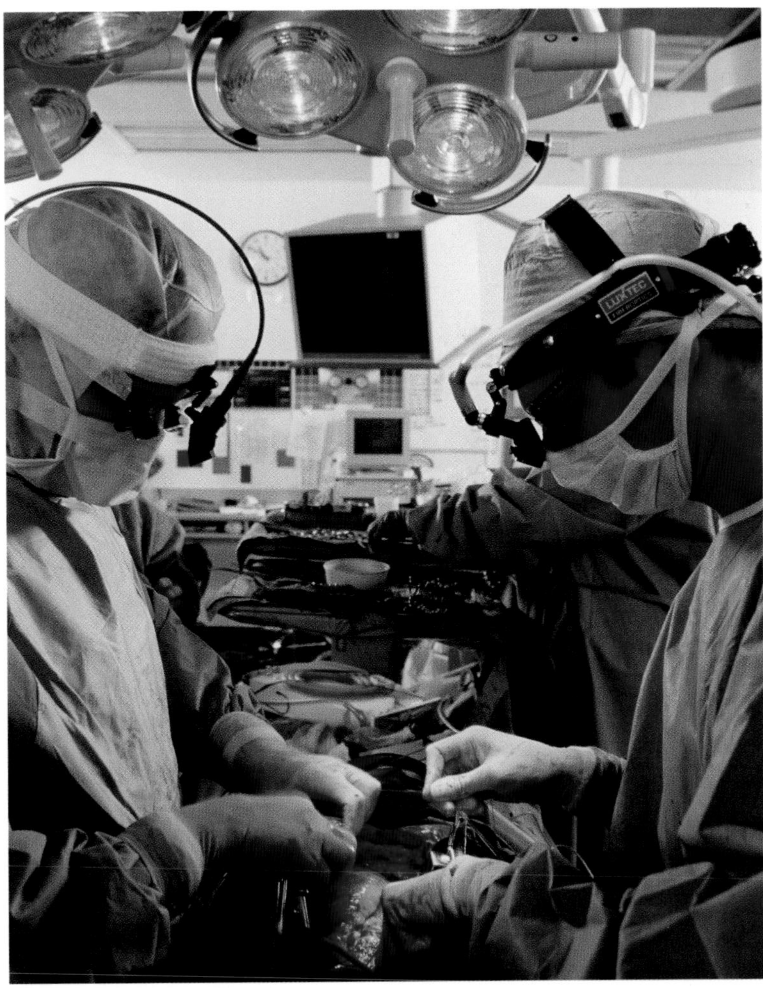

Right: Surgeons sew a donor heart into a patient. Since the first human heart transplant in 1967, many problems associated with organ rejection have largely been overcome through the use of immunosuppressive drugs.

PHYSICS

Physics describes the relationship between matter and energy, how they interact, and indeed, how they are interrelated. That relationship is important in all sciences. The interpenetration is profound. In the increasingly complex subdivision of the sciences, we now commonly talk about biophysics, chemical physics, physical chemistry, geophysics, medical physics, mathematical physics, and astrophysics.

Introduction

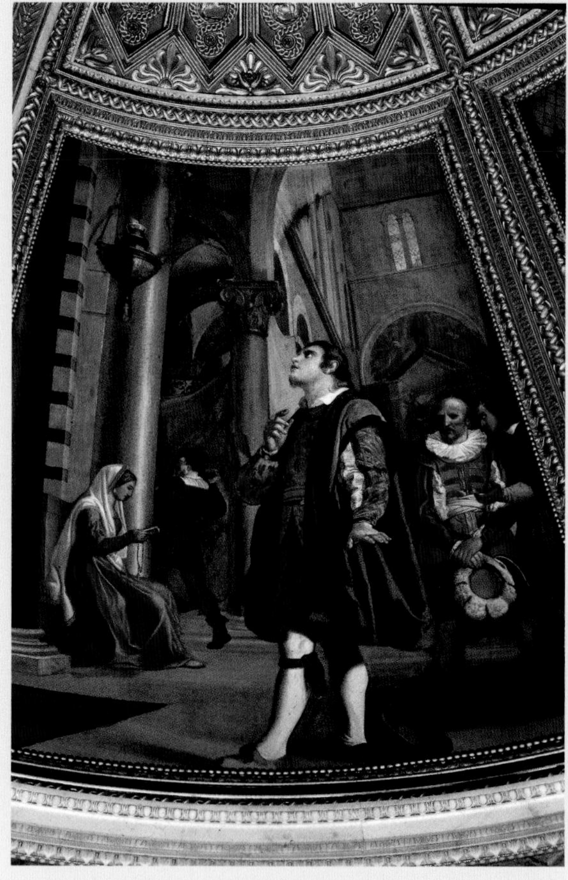

Right: Galileo (1564–1642) made a series of observations of the pendulum in Pisa Cathedral. Timing each swing against his pulse, he saw that the time it took to swing back and forth was independent of the arc of the swing. He then came up with the idea of using a pendulum to control the speed of a clock, thus making timekeeping more accurate.

The first glimpses of modern understanding in physics emerged in Europe around the turn of the seventeenth century. Until then, most of our ideas about matter and energy came from the ancient Greeks, in particular from Aristotle. Yet, as brilliant as he was, Aristotle was mostly wrong about physics.

Aristotle made much of "natural motion," the tendency of objects to move when left alone. He believed that all objects are composed of the four elements—earth, air, fire, and water. Two, air and fire, he asserted, were naturally light. They had "levity" and wanted to rise as they sought their place in nature. Earth and water, on the other hand, possessed "gravity." Left to themselves they wanted to fall toward the center of the universe, then thought to be coincident with Earth's center.

Because everyday objects contained varying quantities of the four elements, they would naturally rise and fall at different speeds. Galileo is credited with disproving this venerable assertion. This reputedly showed that two balls of very different weights hit the ground together when simultaneously dropped. From there, Isaac Newton devised his concept of universal gravity, which states that all objects pull on one another with a force that depends on their masses and on the distance between them.

In the seventeenth century we started looking at the world directly through experiment, rather than simply relying upon the words of long-dead sages. We also came to realize, as Galileo remarked, that "the book of the universe is written in the language of mathematics." The more precise the measurements, the more likely we were to uncover nature's secrets.

In the 1600s, Evangelista Torricelli found that air had weight, which led to atmospheric and steam engines, transforming industry. William Gilbert explored "lodestones" and "electrics," beginning the drive to understand electricity and magnetism and ultimately, electromagnetism. Debate began regarding the nature of light. Isaac Newton argued it was a stream of particles; others claimed light was a form of wave motion.

In the nineteenth century we tackled the nature of energy, and our capacity to transmute it into different forms. Energy is a form of motion rather than a substance. The realization that no transformation is perfect, that some energy is always lost or degraded, was profound, the first sign that the clockwork universe so long visualized

Right: The 1911 international physics conference in Brussels was a most formidable gathering of scientists. From left standing: Victor Goldschmidt, Max Planck, Rubens, Somerfeld, Lindemann, Louis de Broglie, Knudsen, Hasenohrl, Hostelet, Herzen, James Hopwood Jeans, Ernest Rutherford, Heike Kamerlingh-Onnes, Albert Einstein, Paul Langevin. From left seated: Walther Nernst, Marcel Louis Brillouin, Ernest Solvay, Hendrik Lorentz, Otto Heinrich Warburg, Jean Baptiste Perrin, Wilhelm Wien, Marie Curie, Henri Poincaré.

Left: Aristotle (384–322 BCE) was one of the greatest philosophers of ancient times and his work was influential for centuries after his death. He was a seminal figure in the study of physics, metaphysics, logic, biology, music, and even theater.

by physicists might one day run down. The very potent notion of "entropy" or loss of information was born, as was the study of thermodynamics. The notion that matter was comprised of mostly tiny indestructible particles called atoms gained enhanced credibility when the model was used to explain the behavior of gases.

In the twentieth century, increasingly sophisticated experiments found that light travels at the same speed in all directions, contrary to expectations. So the tenuous medium called the "aether" required to carry light and other radiation through space could not be found. And the failure to find the "ultraviolet catastrophe," which should have caused the gentlest fire to pour out searing ultraviolet light, meant that our understanding of how energy moves around and within nature was fatally flawed.

Yet from those failures were born the extraordinary triumphs of twentieth-century physics. From the unchanging speed of light, Albert Einstein created Special Relativity. Max Planck solved the ultraviolet conundrum by making energy transferable only in tiny indivisible packets, the first step towards quantum theory. It wasn't long before Einstein redefined gravity in his famous General Theory of Relativity, and physicists were dismembering the previously indestructible atom, revealing its inner structure and in time liberating the energy of its nucleus. Long-held distinctions blurred. A wave could be a particle, a particle a wave. Matter could be transmuted to energy and energy to matter. And however predictable was the behavior of large objects, in the nano-world of atoms and molecules fundamental uncertainty held sway.

Into the twenty-first century, we still push these boundaries. Particle accelerators now probe for fleeting fragments of matter that could explain why we have matter at all. Exotic ideas such as string theory seek to make a connection between the structure of matter and the architecture of the universe.

Left: Polish-born physicist Marie Curie (1867–1934) won the 1903 Nobel Prize for Physics, which she shared with her husband Pierre Curie. She won a second Nobel Prize in 1911 for the discovery of the elements radium and polonium. She was the first woman to win a Nobel Prize.

Physics investigates the essential nature of the world …

Willard Van Orman Quine, mathematician and philosopher, 1908–2000

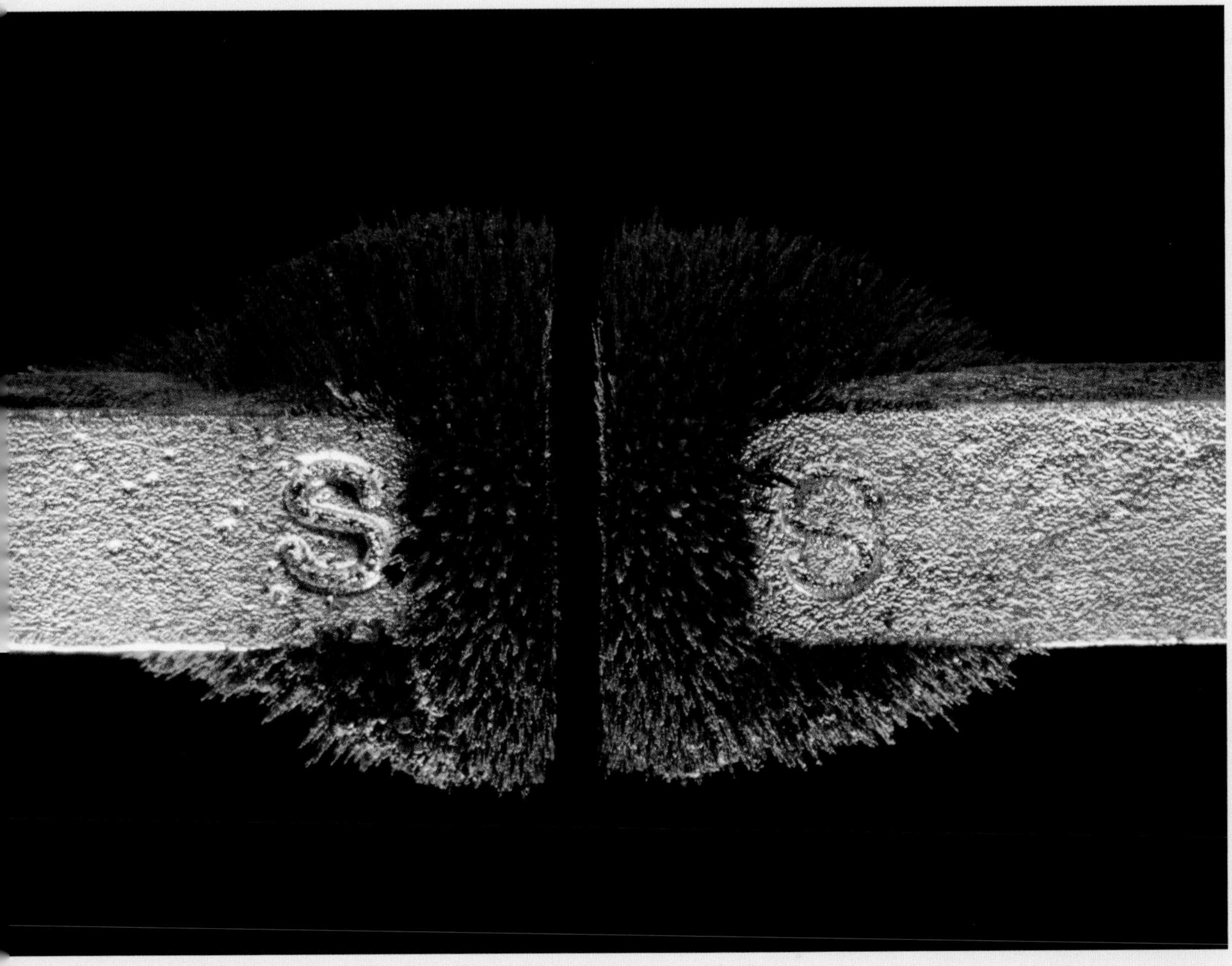

Above: Carbon comes in a number of forms, including as diamonds, the hardest natural mineral. Because of their high thermal conductivity, diamonds have many industrial applications, including being used in semiconductors.

Left: Magnets both attract and repel. Opposites attract, which means that a north pole is always attracted to a south pole. On the other hand, a magnet repels similar poles.

Matter surrounds us. The water you drank this morning, the keyboard keys you touched while at work, the air you breathe every day: They are all various forms of matter. Explained simply, matter is anything that takes up physical space in the world around us.

Understanding matter

Physical or chemical?

Matter possesses a number of properties that can identify it. These properties might be the physical characteristics of a particular metal or mineral, or they may be related to the chemical changes that occur under certain conditions.

Physical properties of matter are those properties that can be observed without changing the chemical identity of the substance. For example, boiling water causes it to change to steam, however it still retains its identity as water (H_2O). Other physical properties of matter include color, odor, melting point, boiling point, hardness, magnetism, density, and state.

Chemical properties of matter relate to chemical reactions that cause a change in the identity of matter. For example, the combination of hydrogen and oxygen to form water constitutes a chemical reaction, and alters the atomic structure of both the hydrogen and oxygen atoms, to create a new structure.

Changing states of matter

Matter has five physical states or phases: Solid, liquid, gas, plasma, and the recently discovered Bose–Einstein condensates. The state of a particular material is determined by the structure of its atoms, and their

Right: The inner globe of a plasma ball has a voltage source that makes a current flow to points with a lower voltage. The ball is filled with inert gases that glow when the electricity flows through it. If nothing is touching the ball, the current flows anywhere around the ball, but if you touch the ball with your hand, some of the current flows through the glass and into you.

> The most incomprehensible thing about the world is that it is at all comprehensible.
>
> Albert Einstein, physicist, 1879–1955

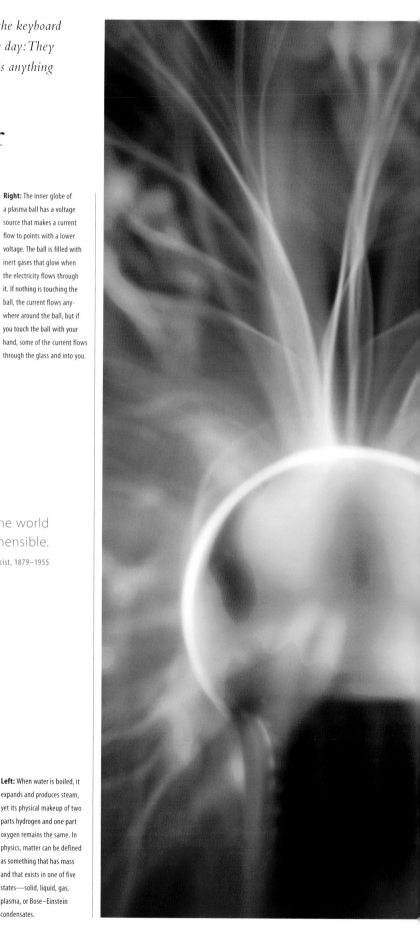

Left: When water is boiled, it expands and produces steam, yet its physical makeup of two parts hydrogen and one part oxygen remains the same. In physics, matter can be defined as something that has mass and that exists in one of five states—solid, liquid, gas, plasma, or Bose–Einstein condensates.

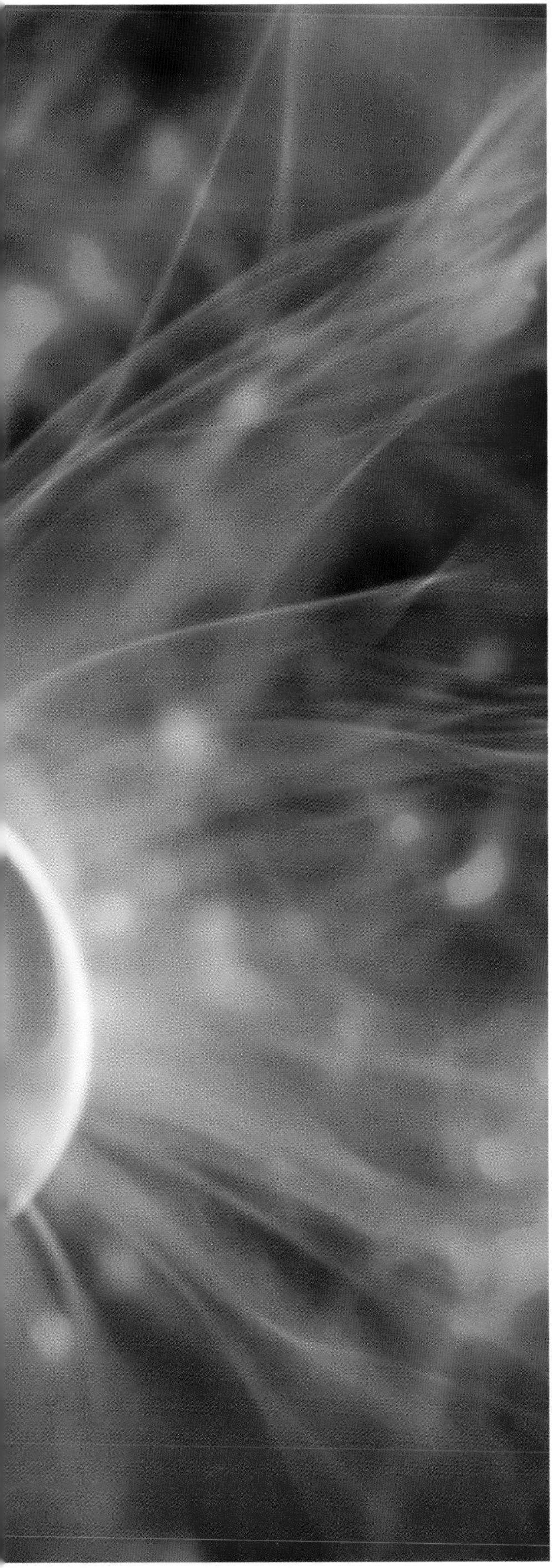

Right: Fiber optics in the black cables (left) sit alongside the insulators and cooling lines (right) in one of the converter bridge rooms at the Celilo Converter Station, Oregon, USA. The Celilo plant is a transfer station for power traveling to and from California.

Above: A large consignment of fire extinguishers await delivery. Different fire extinguishers are used for different types of fire, but most contain pressurized gas that interrupts or stops the chemical reaction that occurs when things burn (see also page 145).

ability to move about within the structure. By adding energy to a material, often in the form of heat, the atoms within the structure become excited and begin to move around. With enough energy, elements and compounds can shift from one phase to another, reflecting a physical, but not chemical, transition. As the temperature increases, matter shifts to a more active (high energy) state; as the temperature decreases, matter shifts to a lower energy state.

Solids are made up of atoms that are arranged in a rigid structure. Until the recent discovery of Bose–Einstein condensates, solids were assumed to be the most stable (low energy) state of matter. Solids can be categorized according to two structure types—crystal and amorphous—and also according to their thermal or electrical properties, with some solids being excellent conductors (such as metals), and others being excellent insulators.

When a solid is exposed to sufficient heat, its atoms may break loose from their fixed positions and turn into a liquid. One obvious property of liquids is their ability to flow. Liquids can also maintain their volume and, notably, they can take the form of the vessel in which they are contained.

Gases are easily compressed, expand to fill their container, and occupy more space than liquids or solids. However, one of gas's most utilitarian properties is that it generally behaves the same way in response to temperature and pressure by expanding and contracting in a predictable manner.

Plasmas are a gas of charged particles and are the most highly energized form of matter. They consist of atoms that have lost electrons, and they are heavily influenced by electromagnetic forces. Because plasmas require extreme temperature to strip the electrons, they will eventually revert to a neutral gas state if the energy source becomes unsustainable.

Bose–Einstein condensates are essentially the complete opposite of plasmas, as they will form only at extremely low temperatures.

A NEW STATE OF MATTER

In 1995, over 70 years after its possible existence was predicted, a new state of matter of was born. Bose–Einstein condensates (BEC) are an unusual quantum phenomenon, named after Satyendra Nath Bose and Albert Einstein, who in the 1920s joined forces to alert the scientific community of this new, low energy state of matter. Unfortunately, at the time, they were unable to prove their theory, as technology had not yet provided the means to form the condensate. However, in 1995, the BEC became an observable reality at last.

Bose–Einstein condensates are less energetic than solids, and occur at extremely cold temperatures—less than millionths of a degree above absolute zero. As the temperature drops, atoms in a BEC begin to clump. These groups of atoms behave identically, and are indistinguishable from one another.

BEC shares some qualities with laser light, in that all the atoms within the BEC are behaving the same way. For this reason, BECs are likely to be very useful. However, because their discovery is so recent, practical applications are still to be uncovered.

Understanding how energy is transferred in mechanical systems goes a long way toward building useful simple machines that make life easier. If an object is moved—such as a ball being lifted into the air—that object gains energy. This energy is known as mechanical energy, and is comprised of potential energy and kinetic energy.

Mechanical energy

Kinetic energy comes from the Greek word "kinesis" which means "movement," and is the extra energy an object possesses if it is moving. Potential energy can be thought of energy stored in the system, to be released as kinetic energy. It is called potential energy because it has the potential to change the state of the system. For instance, if you pick up a ball and hold it above the ground, it has gained potential energy. If you let go of the ball, as it starts to fall to the ground, the potential energy starts to convert to kinetic energy. The moment before it hits the ground, the ball has kinetic energy, which is equivalent to the potential energy at the start. The total mechanical energy is equal to the potential and kinetic energy in the system, and should remain constant through the whole process.

This idea was first intimated by Galileo after he studied objects under the influence of gravity. He was specifically interested in the pendulum after watching the swinging motion of a chandelier in a cathedral (see page 26). From his investigations came the use of a pendulum as a timing mechanism in mechanical clocks.

Conservation of energy

If all of the energy was transferred back to moving the ball again, then it will bounce back to the height from which it was dropped. This would happen if both the ball and the ground were completely solid and did not deform. However, in the real world this is not the case. Some of the mechanical energy is lost as it converts to other forms of energy; some of the mechanical energy goes into deforming the ball if it is squishy, or the ground if it is soft, and the energy is dissipated as heat and sound. As the mechanical energy dissipates, the ball bounces lower and lower until it comes to rest on the ground.

This example is a simple mechanical system, but even for more complex systems such as the movement of levers, or the workings of an engine, the same principle still applies. The mechanical energy diminishes as it converts into other forms, and does not increase unless more energy is introduced into the system. Energy is not create or destroyed, but instead converts from one form to another. This is the Law of Conservation of Energy, and was first stated by German physicist and physician Hermann von Helmholtz in 1847.

Give me a place to stand and I shall move the Earth.

Archimedes, mathematician, physicist, and astronomer, *c.* 287–*c.* 212 BCE

Above: A simple machine, like a pulley, is one that requires the application of a single force to work—it does not contain any internal sources of energy, and works by application of mechanical energy. Using simple machines can reduce the amount of force needed to accomplish a task.

Below: A simple machine, such as a cog wheel, is any instrument that applies mechanical energy at one point and goes on to yield a more useful form of energy at another point. This is known as a mechanical advantage.

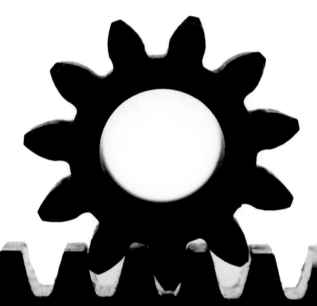

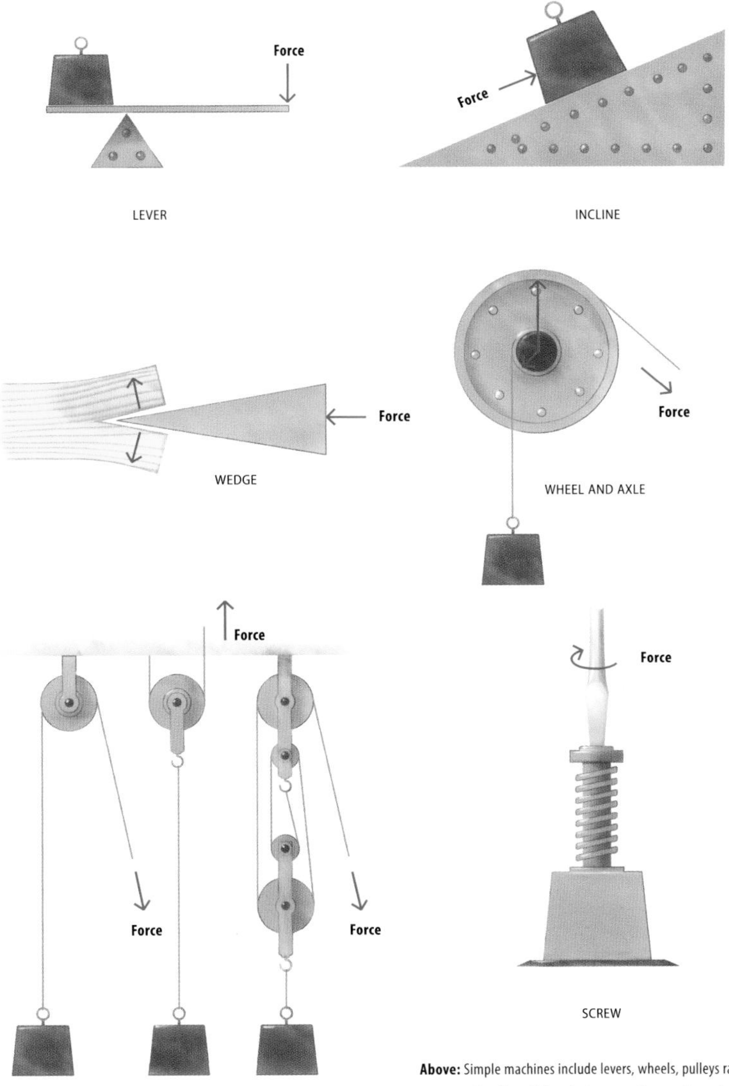

Above: Simple machines include levers, wheels, pulleys ramps, screws and wedges. Although these machines make work a great deal easier, no machine is able to output more mechanical energy than was put into it.

ARCHIMEDES OF SYRACUSE

The Greek mathematician Archimedes was born in Syracuse, Sicily, in 287 BCE. During his lifetime Archimedes made many major discoveries in mathematics, but he is equally remembered for his practical machines. He invented what is still known as the "Archimedes screw" (below) used to raise water or other things that act like fluids, such as grain. Although he did not invent the lever, he did make a rigorous examination of its operation. He is said to have been so impressed by the power of the lever that he remarked, "Give me a place to stand and I shall move the Earth." The historian Plutarch claims Archimedes demonstrated a system of pulleys that allowed him to pull a fully laden ship onto shore single-handedly.

Archimedes was a celebrated genius in his day, and so the whole of the Mediterranean world greeted with grief the news of his death at the hands of a Roman solider. The last words attributed to him are "Do not disturb my circles," a reference to the mathematical drawing in the dirt that he was studying at the time he was killed.

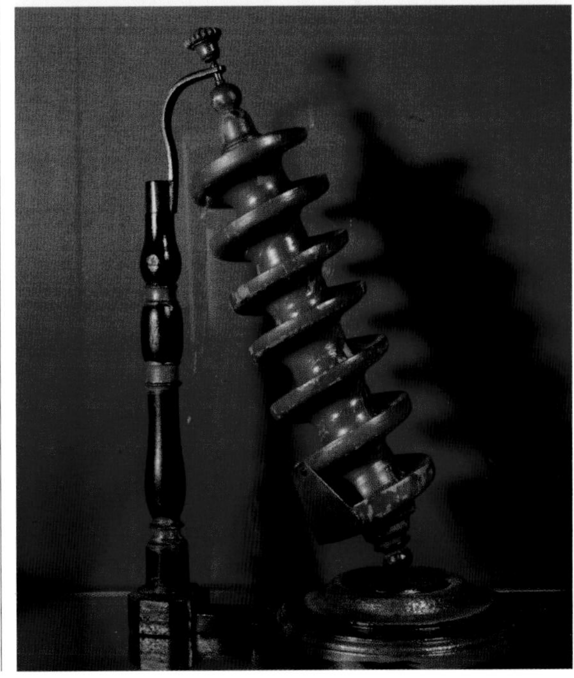

The ancients believed that the world consisted of four elements: Earth, fire, air and water. Modern science has shown that there are 117 different elements, as listed in the periodic table of the elements (see pages 138–139). So it might be thought that there exist as wide a range of forces in nature.

The fundamental forces of nature

In fact there are only four distinct forces, and scientists are working to reduce this number further. In everyday life, we constantly observe and experience a multitude of forces, pushing, pulling, twisting, and rubbing. Close investigation of these forces reveals that they all end up being due to one of four fundamental forces of nature.

Gravity

Isaac Newton's theory of gravitation brought together advances in mathematics with astronomical observations. His theory says that the attractive force between two objects is greater the heavier the two objects, and reduced as the distance between them increases.

This theory stood up for over two centuries, matching precisely with almost all observations, with the notable exception of the orbit of the planet Mercury. Einstein's general theory of relativity added the subtle corrections required to explain the motion of Mercury, once again giving us a complete theory of gravitation.

The idea behind general relativity is that any object with mass changes the shape of space–time itself, which in turn affects the motion of all other objects.

Right: The ancients believed in the four elements of earth, air, fire, and water and attributed phenomena such as thunderstorms to these natural forces. Thunderstorms are one way that the atmosphere releases energy (see also page 415).

> … all things are made of atoms—little particles that move around in perpetual motion, attracting each other when they are a little distance apart, but repelling upon being squeezed into one another. In that one sentence you will see an enormous amount of information about the world …
>
> Richard Feynman, physicist, 1918–1988

UNIFICATION THEORY

One of the loftiest goals of theoretical physics today is to come up with a single theory that encompasses all four known forces. In the 1960s, Sheldon Lee Glashow, Steven Weinberg, and Abdus Salam (pictured) came up with the Electroweak theory, and were awarded the 1979 Nobel Prize for Physics for their work.

Their theory incorporates both the weak and electromagnetic forces, showing that these forces are simply different aspects of essentially the same thing, in much the way that the electric and magnetic forces were originally combined into the electromagnetic force.

Promising work has been done in combining the strong force with the Electroweak force, however the most challenging hurdle is bringing gravity into line with the other three forces. Doing so would require a quantum theory of gravity, bringing together both quantum mechanics and Einstein's general theory of relativity. All attempts so far have been unsuccessful, leaving this Holy Grail of modern physics unsolved.

Right: A technician works on an apparatus in the HERA Tunnel in Hamburg, Germany. HERA is an underground circular particle accelerator facility with a tunnel over 3½ miles (6 km) long. HERA is mainly used to study the structure of protons and the properties of quarks.

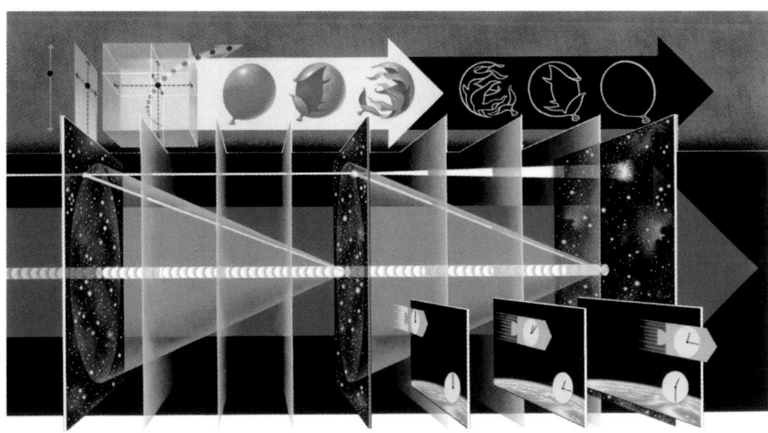

Above: An artist's impression of one way to imagine the concept of space–time. The general theory of relativity states that any object with mass changes the shape of space–time itself, and this in turn affects the motion of all other objects.

Electromagnetic force

The second fundamental force to be discovered was the electromagnetic force, which acts between electrically or magnetically charged particles, such as protons and electrons. It was thought that the electrical and magnetic forces were distinct, however Scottish scientist James Maxwell realized that the mathematical description of each of these forces could be consolidated into a single force.

In doing so, he also came to the remarkable conclusion that light must consist of a combination of both electric and magnetic waves, bringing the three concepts together under a single theory. The discovery of quantum physics allowed the theory of electromagnetism to be refined, and the theory of quantum electrodynamics (QED) was formulated in the 1950s. According to QED theory, all electromagnetic interactions are due to the exchange of photons (particles of light) between the particles involved.

Quatum electrodynamic theory explains the properties of not only charged particle interactions, but also the behaviors of light, from reflection to refraction.

Weak nuclear force

The weak nuclear force, possibly the least familiar of the four fundamental forces, is only effective over a range one-thousandth the size of an atomic nucleus. It is responsible for beta radiation, where a neutron within a nucleus spontaneously changes into a proton, emitting an electron and an anti-neutrino in the process.

Strong nuclear force

The strong nuclear force governs the interactions between quarks, the subatomic particles that make up protons and neutrons. The theory of quantum chromodynamics (QCD) explains how quarks interact by exchanging "gluons."

The strong nuclear force holds quarks together within protons and neutrons and, in turn, is able to hold these protons and neutrons together within atomic nuclei.

Under normal conditions, when two protons are brought together, the electromagnetic force would cause them to repel each other, however at very short distances, the strong force is greater than the electromagnetic force, and this allows the protons to remain together within the nucleus.

Isaac Newton (1643–1727) left an enormous legacy. His views did not just change mathematics, physics, and astronomy, but revolutionized the way we view nature and how we relate to it.

Newton's laws of motion

The formulation of his laws of motion had profound consequences. To achieve this, he did use ideas on dynamics that others before him had gradually developed, but turning these ideas and his own into a coherent, simple and elegant yet complete set of laws, powerful enough to make predictions about the motion of any material object, was a monumental achievement.

Left: Isaac Newton, as painted by Enoch Seeman in 1726. Newton was one of the most brilliant scientific minds of all time. His laws of motion changed our perception and comprehension of the universe.

The law of inertia

What does an object do if left completely undisturbed? This may sound a simple question, but answering it correctly is crucially important because it is the step upon which everything else is based. From the work of Descartes and Galileo, it was clear that Aristotle's view on the subject were inadequate. Objects do not "stop unless a force keeps pushing or pulling them forward." It is friction, a force, that causes them to slow down and stop. In fact, objects always move in the same direction at the same speed unless an applied force changes one or the other.

Right: When a rocket (such as *Apollo 11*) is launched and expels exhaust, it is pushed upward by the force from the exhaust, according to Newton's third law, but its much greater mass makes it move much more slowly (second law).

If I have seen a little further it is by standing on the shoulders of giants.

Isaac Newton, physicist and mathematician, 1643–1727

The force law

If an object always keeps moving the same way when left alone, then the effect of a force on it cannot be to make it move at a certain velocity. Newton realized that what it does instead is gradually change the velocity of the object in the direction of the force. The rate at which the velocity changes, the acceleration, varies; the greater the force, the greater the acceleration, but the larger the mass of the object, the smaller the acceleration.

The law of action and reaction

Any force that is acting on an object in nature can be traced to some other object which is exerting an influence on it, but this influence is never just a one-way action. Every time that you push or pull on something you get pushed or pulled back; every force is actually an interaction. This observation—that Newton turned into his final law—tells us that nothing can affect other objects without being itself subject to their influence.

Every motion involves a constant interplay of Newton's three laws and their implication is that, if we know what an object is doing at some time and what forces act on it, we can predict everything it will do. This realization came at around the same time as the discovery that every material object, out to the planets and stars, is involved in dynamics and must therefore follow these laws. Humanity gained confidence that it could in principle understand and control this "clockwork universe," and it would be more than 200 years before new developments started casting doubts on this ability.

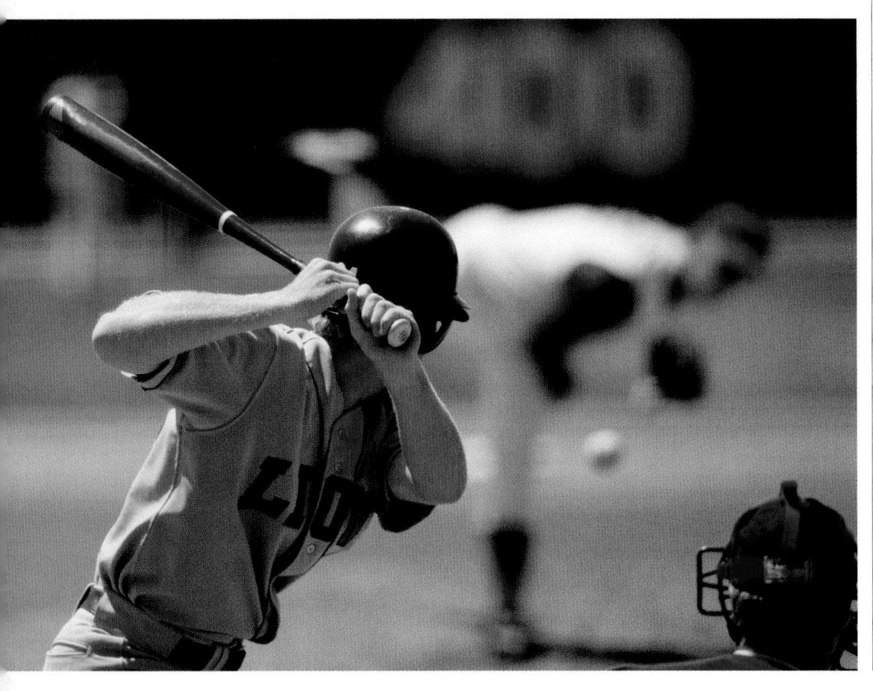

Left: A baseball game in action is a good illustration of motion—from the position and speed of the ball and the position of the bat, we can apply Newton's second law in order to predict what will happen during and after the collision.

Opposite: The inside of a particle accelerator, where the accelerated particles are moving almost at the speed of light. A modified Newton's second law takes into account the much greater force required to produce a small acceleration at these speeds.

IS THERE A LIMIT TO THE VALIDITY OF NEWTON'S LAWS?

The simplicity and relatively intuitive nature of Newton's laws can lead us to forget that things could have been very different. Natural laws are not like mathematical theorems—they all have a range of validity, outside which they must be modified or replaced by a different set of laws.

There are two situations that have been known since the beginning of the twentieth century, in which this applies. One comes from Einstein's theories of relativity. In special relativity, a material object cannot move at the speed of light or faster, and when its speed is a considerable fraction of the speed of light this fact shows up in that the same force produces a smaller acceleration; while in general relativity, space–time becomes warped, and it is more difficult to even define the concept of force. Yet one can modify Newton's second law to take these facts into account. The other kind of situation is the one in which quantum theory must be used, and here is where predictability really breaks down.

Then there are the situations we do not yet understand, for example, ones involving accelerations so small that we have not been able to set up the appropriate experiments on Earth, but for which observations of motions of stars around the centers of galaxies and of spacecraft in interplanetary space show anomalies that could be due to a different relationship between force and acceleration.

Dynamics is the study of the motion and other possible changes of any physical system, which it seeks to understand in terms of the effects of other objects it interacts with. Ultimately, its goal is to be able to make testable predictions about the behavior of the system in any given situation.

Dynamics

It might be natural to try to understand how objects around us behave, but it turns out that exactly what we mean is not so obvious. For example, understanding why a person is flying to a distant city may mean knowing the purpose of the trip. However, in physics the most useful kind of understanding is different. We will not understand the path a stone follows when flying through the air if we think it has a goal, or the motion of a planet by assuming that its location is trying to tell us something. Objects simply interact with each other, and react to forces as they go along.

Early development

It took a long time to develop this point of view. The ancient Greeks, with their emphasis on a naturalistic philosophy and their experience with mathematics, realized that to understand nature beyond observing patterns and interpreting them, they needed to make systematic use of logic and geometry. They proposed models for phenomena as diverse as the smallest parts of matter—atoms—and the motion of planets on the celestial sphere. Aristotle (384–322 BCE) even proposed a theory of dynamics. In his famous analysis of a stone moving through the air, he argued that because rest is the natural state for it, a force needs to keep pushing the stone along until it comes to rest on the ground. But such arguments based on dynamical laws did not apply to heavenly bodies, which follow the rotation of their spheres and are not subject to change.

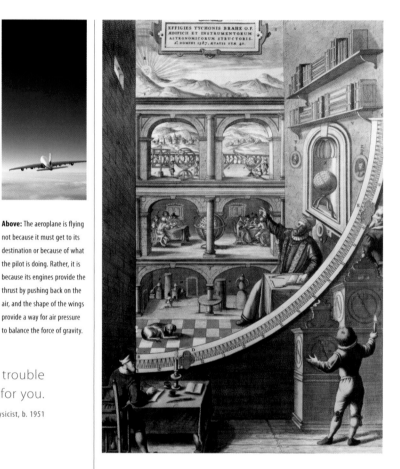

Above: The aeroplane is flying not because it must get to its destination or because of what the pilot is doing. Rather, it is because its engines provide the thrust by pushing back on the air, and the shape of the wings provide a way for air pressure to balance the force of gravity.

> In physics, you don't have to go around making trouble for yourself—nature does it for you.
>
> Frank Wilczek, physicist, b. 1951

Above: Astronomer Tycho Brahe (1546–1601) in his observatory at Uraniborg, the "Castle of the Heavens," on the island of Hven between Denmark and Sweden. He determined that the heavens were not immutable.

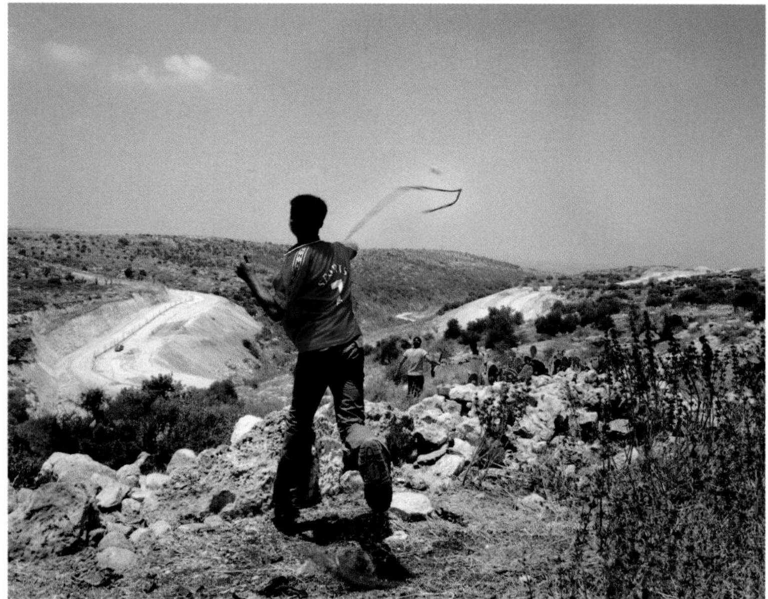

Left: In Aristotle's analysis of a stone being thrown through the air, he argued that because rest is the natural state for it, a force must keep pushing the stone along until it comes to rest on the ground some distance away.

The universal reach of dynamics

Only in the seventeenth century, after increasing evidence made it impossible to maintain the distinction, were celestial phenomena recognized as being essentially the same as earthly ones. Tycho Brahe showed that things can change among the planets because comets are at about the same distance and they come and go; Galileo saw that the Moon's surface had mountains like Earth; Newton explained the way planets move around the Sun by assuming that they are subject to the force of gravity. From then on, dynamical laws described the motion of every material object, from the smallest known ones to the farthest known ones. Over the next two centuries, dynamics was successfully extended to electricity, magnetism, and even light; thermodynamics was added to describe heat and fluids, and the overall knowledge gained powered the Industrial Revolution. Questions about the nature and reach of dynamics did not appear again until the early twentieth century.

Left: Dynamics is the branch of physics that is concerned with the effects of forces on the motions of bodies. It describes how all kinds of matter, such as air, water, and solid objects, interact, for example the interaction between the air and a person bungee jumping.

Below: This Hubble Space Telescope image shows a cluster of galaxies acting as a gigantic gravitational lens, distorting distant objects. The five bright spots are images of a single distant quasar. This cluster of galaxies is some 7 billion light years away from the constellation Leo Minor.

WHAT ARE THE BOUNDARIES OF DYNAMICS?

By the end of the nineteenth century, dynamical laws applied to all known forms of matter and to light. Then, during the first three decades of the twentieth century, two dramatic changes occurred.

The first change was that, just like stars and planets before, the universe as a whole as well as space and time themselves came to be seen as dynamical. We know that space–time both affects the motion of objects and is warped by the objects present in it from Albert Einstein's theories of relativity; this is our current view of how gravity works. But even Einstein had not anticipated American astronomer Edwin Hubble's 1929 discovery that the universe, rather than being the place where things happen, changes and is now expanding together with its contents of matter and radiation according to dynamical laws.

The second change came when quantum theory showed that the positions and velocities of particles have unavoidable uncertainties that severely limit the extent to which dynamics can make predictions. Physics is still trying to fully comprehend the implications of these developments, which both extend the boundaries of dynamics and limit the kinds of questions it can answer. Can dynamics be extended even further? Could it be, for example, that not just the shape but the type of space–time we live in is subject to dynamical laws? To answer this perplexing question, we will almost certainly need to understand how the two recent revolutions in dynamics fit together.

The first thermometer with a numerical scale is credited to Italian scientist Santorio Santorio in 1613. Since then, different temperature scales have been developed and discarded. Still in use today is the scale developed by Swedish scientist Anders Celsius in 1742. He divided the interval between the freezing and boiling points of water into 100 degrees. Interestingly, he chose 0 degrees for the boiling point and 100 degrees for the freezing point. One year later, Frenchman Jean Christin inverted the scale.

Taking temperatures

The temperature scale

While most of the world uses the Celsius scale, scientists prefer to use the Kelvin scale when dealing with very low temperatures. The scale is named after British physicist William Thomson, Lord Kelvin, who in 1848, proposed the need for an absolute temperature scale. Zero Kelvin is the equivalent of –459.67 degrees Fahrenheit (°F) or – 273.15 degrees Celsius (°C), with one Kelvin (K) being the same size as 1 degree Celsius. The temperature of 0K is referred to as absolute zero and is the point when things cannot get any colder.

The more energy a substance has, the faster its particles move. A gas therefore is a form of matter with more energy than a liquid. Heating a substance causes its particles to move faster and the reverse is true when a substance is cooled. The theory of absolute zero is that if a substance is cold enough its particles should stop moving. Theoretically, absolute zero is the temperature at which all molecular motion ceases, but according to the laws of quantum mechanics, it is impossible to actually reach this point.

Temperatures here and beyond

The coldest official temperature on Earth was –128.56°F (–89.2°C), recorded in July 1983 at the Vostok research station in Antarctica. By comparison, the hottest temperature on record is 136.4°F (58°C), recorded in Libya in 1922. Not surprisingly, the temperature range in space is more extreme than on Earth. The average temperature of the universe is a chilly –454.77°F (–270.43°C or 2.73 Kelvin). Due to the remnant heat left over from the Big Bang, the temperature in even the very depths of space is always greater than absolute zero.

The coldest place in the universe is believed to be the Boomerang nebulae, a huge cloud of gas and dust, some 5,000 light years away from Earth. It has an estimated temperature of –457.6°F (–272°C), just a little above absolute zero.

Above: When certain materials, such as zinc, are cooled to temperatures close to absolute zero, they lose electrical resistance and become superconductors; electrons can then travel freely through the material without losing energy as heat. Superfluidity, another effect of very low temperatures, is created by the cooling of liquid helium. When cooled to a critical point, a small fraction of the liquid helium will display unusual qualities, such as zero viscosity and an infinite capacity for thermal conductivity. Here, a magnet floats above a superconductor.

Right: Liquid nitrogen is very cold and is commonly used to cool things down to 77K, well below the freezing point of water. It has many industrial applications, including the freezing of food products and biological samples.

Below: Air temperatures across most of the world are measured using the Celsius scale (°C), although in the USA the Fahrenheit scale (°F) is preferred. Scientists, however, tend to work with the Kelvin (K) scale (far right).

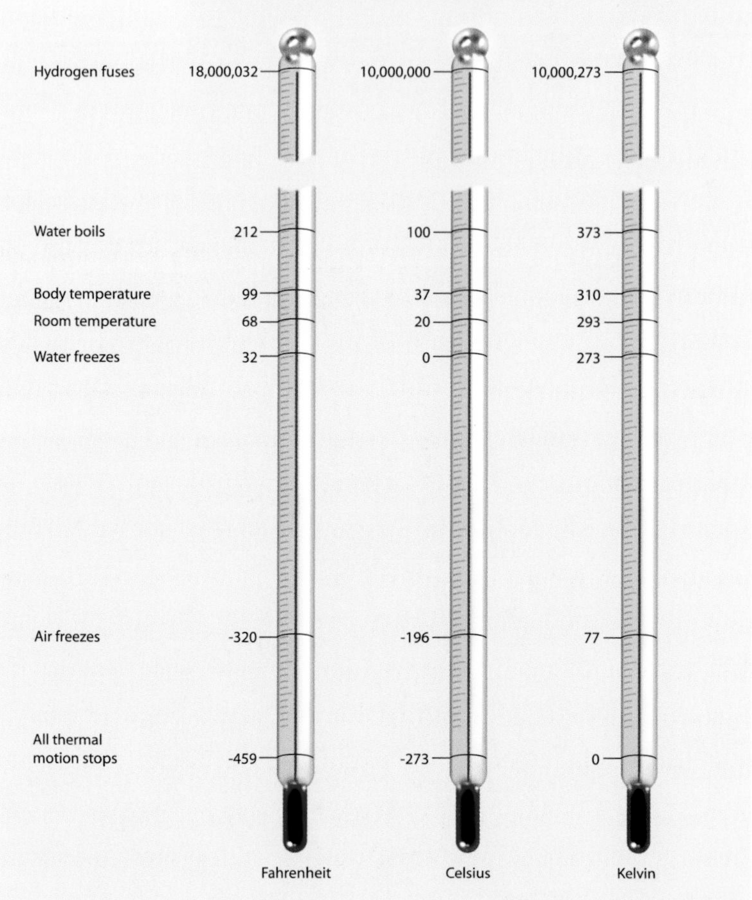

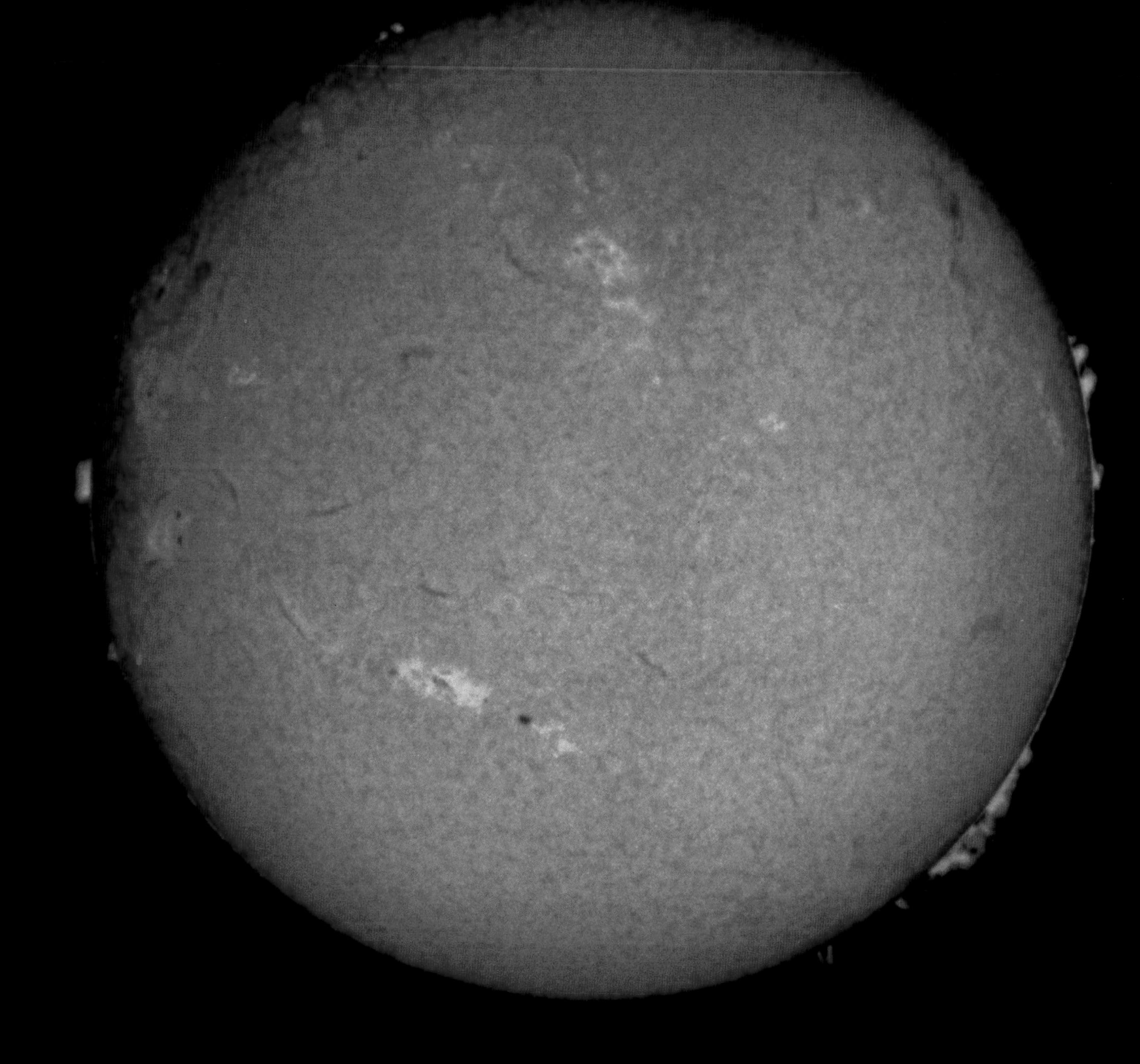

At the other extreme, scientists have estimated that the Sun's core is 56.3 million °F (13.5 million °C). Yet this is cool compared with the cores of red and blue supergiants, which can have temperatures in excess of 212 million °F (100 million °C). The hottest temperatures that have ever occurred anywhere were probably during the very first moments of the universe during the Big Bang, with astronomers and physicists suggesting temperatures of up to 1886°F (1030°C)—that's 1 million million million million millon degrees!

Above: The Sun is a great ball of hot gas, mainly hydrogen (74 percent by mass) and helium (25 percent), with small amounts of other elements. The massive core pressure causes hydrogen atoms to fuse and form helium, releasing energy.

There are no physicists in the hottest parts of Hell, because the existence of a "hottest part" implies a temperature difference, and any marginally competent physicist would immediately use this to run a heat engine and make some other part of Hell comfortably cool. This is obviously impossible.

Richard Davisson, physicist, 1922–2004

HOT STUFF

On Earth, the dominant states of matter are solids, liquids, and gases. Yet these constitute only a tiny part of the universe. A different state, plasma, makes up more than 99 percent of the visible universe. Stars, including the Sun, are predominantly made of plasma. Being the highest energy form of matter, plasma is extremely hot. It is distinct from the other states because it contains a large number of electrically charged particles which makes it behave very differently. If a normal gas is exposed to very high amounts of energy, a significant number of its atoms will lose some, or all, of their electrons. This leaves some atoms with a positive charge and the released negative electrons are able to move around freely. This process is known as ionization. Plasmas are therefore ionized gases that will behave according to their temperature, density, and amount of electrical charge. Plasmas were first identified by Sir William Crookes in 1879 and can be found on Earth in neon signs and fluorescent lights. In 2006, the American laboratory Sandia used a "Z machine" to produce a plasma for less than a second with a temperature exceeding 2 billion degrees Kelvin.

What do ice cubes and nuclear power plants have in common? While one is cooling down your drink and the other is converting uranium into electricity, they are both taking advantage of thermal energy to do their job.

A matter of heat—thermal energy

To understand how thermal energy works, we must first consider one of the simplest forms of energy—kinetic energy. Any object in motion has a certain amount of energy associated with that motion. This kinetic energy depends on the mass of the object and the speed at which it is moving.

If we take that same object while stationary and observe it under a microscope we notice that, even though the object as a whole is not moving, the molecules and atoms it is made of are all in constant motion. Each of these atoms and molecules has its own certain amount of kinetic energy. Taken together, the total kinetic energy of the atoms and molecules in the object make up its thermal energy.

When we hear the word "thermal" we usually think of temperature, and indeed thermal energy is closely related to temperature. As we increase the thermal energy of an object, its temperature increases at the same rate. This rate is not the same for all materials however, and some materials heat up faster than others.

Latent heat

To melt a sample of ice at 0°C, we need to expend some energy. The amount of energy required to transform a sample from a solid to a liquid (or a liquid to a gas) is known as its latent heat.

This latent heat is another component of the thermal energy of a sample; however, instead of being stored in the motion of molecules, the latent heat is stored as energy in the bonds between molecules.

Right: A view into the control rods unit for fuel elements while the reactor is turned off for a routine inspection at the nuclear power plant in Kruemmel, Germany. Nuclear power plants use thermal energy to generate electricity.

Below: Iceland sits atop the Mid-Atlantic Ridge where basalt lava is continously emplaced, so the country has access to abundant cheap geothermal energy. Power plants take advantage of this natural thermal energy.

> The law of conservation of energy tells us we can't get something for nothing, but we refuse to believe it.
>
> Isaac Asimov, writer and biochemist, 1920–1992

POWER GENERATION

Thermal energy is critical in providing power to people around the world. Coal and nuclear power plants both take advantage of thermal energy to generate electricity. In both cases, water is heated to create steam, which in turn dissipates its thermal energy in turning turbines.

In coal-based power plants (pictured) the coal is burnt to heat the water, while in nuclear power plants the energy of the nuclear reactions heats the water. In both cases the conversion to thermal energy is critical.

The energy that we put into a sample to melt it is used to break the bonds which keep it in its solid form.

Geothermal energy

Another source of heat, and subsequently electrical power, is Earth itself. At certain points on Earth the crust is particularly thin, with the underlying mantle being relatively close to the surface. In these locations hot springs and geysers can be found. Here Earth's heat is transferred to water that makes its way to the surface, bringing the thermal energy with it.

By harnessing the thermal energy of this water it is possible to power turbines and generate power. Two of the primary advantages of geothermal energy over conventional power sources are that it is both significantly safer and less polluting. There are at least 20 countries around the world that use geothermal energy as part of their national power supply. (See also pages 402–405.)

Thermodynamics—it's the law

The first law of thermodynamics states that the increase in thermal energy of an object is equal to the energy added to the system by heating, minus the heat lost by the system to its surroundings. Put simply, this means that if something heats up or cools down, the thermal energy has to either come from, or go to, somewhere.

You can see this in action in a refrigerator. The food and drink inside have thermal energy removed from them as they cool down. The first law of thermodynamics tells us that this energy must have gone somewhere, and if you (carefully!) reach around behind your fridge you will find that it's very warm, as the energy that used to be in your food is pumped out into the rest of your kitchen.

This constant give or take of energy is a result of a much wider principle known as the conservation of energy, which says that energy cannot be created or destroyed, only converted from one form to another. (See also pages 42–43.)

Above: When water freezes, its molecules move farther apart, making ice less dense than water, which is why ice floats in water. As ice melts, it uses thermal energy to break the bonds between the molecules.

Right: Refrigerators work to remove the thermal energy from foods and drink thus cooling them down. This thermal energy is pumped out through the refrigerator and into the surrounding air.

In the 1700s the first practical steam engines were being produced. By studying how to make them more efficient, scientists of the day discovered the laws that were to become the basis of a new branch of physics.

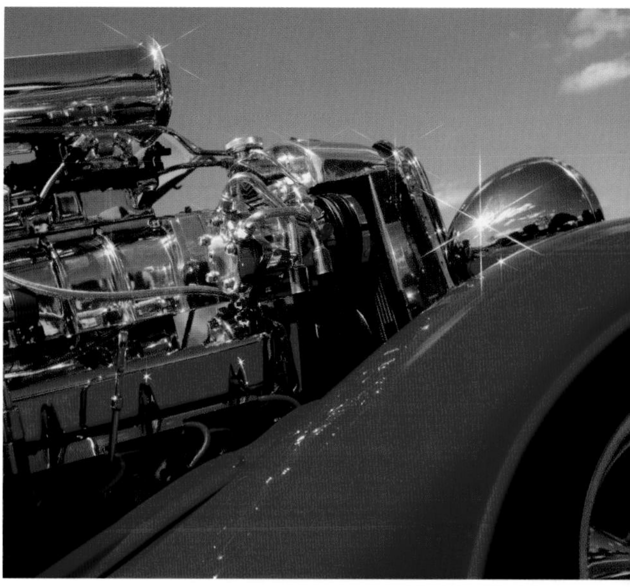

Thermodynamics

Thermodynamics is a branch of physics that came out of engineers' attempts to make steam engines more efficient. Real progress in steam engines began in the early eighteenth century, when they were able to be used in industrial applications such as pumping water from deep mines. This gave incentive to financial backers willing to pay for the elaborate machines. Financial difficulties aside, the main problem for developers at the time was the relative inefficiency of the engines. No matter how much steam was fed into the engine, the amount of work being produced by the machine was tiny compared with the amount of energy used for making the steam.

Right: Many devices operate on thermodynamic principles, including a car engine, such as this hot-rod engine. A moving car converts kinetic energy into heat energy.

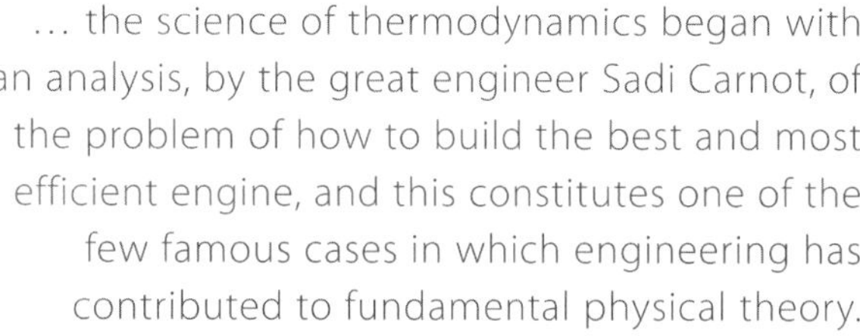

... the science of thermodynamics began with an analysis, by the great engineer Sadi Carnot, of the problem of how to build the best and most efficient engine, and this constitutes one of the few famous cases in which engineering has contributed to fundamental physical theory.

Richard Feynman, physicist, 1918–1988

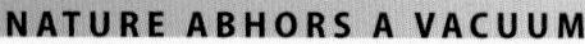

NATURE ABHORS A VACUUM

In 1650, the world's first vacuum pump was built by German inventor and politician Otto von Guericke (pictured). He was compelled to make it in order to disprove the long-held assumption postulated by Aristotle that "nature abhors a vacuum." In the experiment he joined two copper hemispheres and pumped out the air between them. A team of eight horses was harnessed to each hemisphere, yet they were unable to pull them apart until the air was let back into the enclosure.

In 1656, inspired by the experiment, Irish scientist Robert Boyle, with help from English scientist Robert Hooke, built an air pump. Together they used the pump to conduct a series of experiments and found that air allowed things to burn, and was important in the transmission of sound. Boyle's main discovery with the pump was the relationship between volume and pressure—a relationship we now know as Boyle's Law. It was from work that was based on these concepts that eventually led to the development of the steam engine.

The first and second laws

During this time, heat was thought of as a type of fluid called "caloric" that would flow from hotter bodies to cooler ones. In 1824, French military engineer Sadi Carnot theorized that it was not only the amount of heat that mattered, but also the temperature. Carnot's revolutionary step was to imagine an idealized engine that could be used to understand the fundamental principles of all steam engines. What he found was that the efficiency of that engine was directly related to the difference in temperature of the steam coming into the machine and that exiting the machine. The greater the difference in temperature between the steam coming into an engine and the steam coming out, then the greater the efficiency of the engine.

At the time Carnot's thought experiment could have been revolutionary. The steam engine was fairly advanced in its development by this time, but as yet no one had studied the problem scientifically. However, reaction to Carnot's book was small. It wasn't until 1850, 20 years after his death, that his work was used in a famous paper by German physicist Rudolph Clausius, who resurrected the ideas of Carnot in something that later became the second law of thermodynamics. He did away with the caloric view of heat and derived the notion of "entropy." Entropy is a measure of the unavailability of energy to do work.

Before Clausius's paper, British natural philosopher James Prescott Joule put another nail in the coffin for caloric theory of heat. It had been noted at the time that heat was coming from sources other than that suggested by caloric theory. For instance, beating water with a paddle could cause it to heat up. Joule set a paddle that was driven by a falling weight. He realized that lifting the weight stored energy in it, and as the weight fell this energy was transformed through the beating paddles to the water in the form of heat.

In 1849 Joule presented a paper to the Royal Society in London in which he stated what was soon to become the first law of thermodynamics, that is, that work and heat are both ways of transferring energy from one place to another. Joule is immortalized in the term "joule" as a unit of energy.

Left: If ice is placed into a hot cup of coffee, heat flows from the coffee to the ice and melts it. When the heat energy in the cup is evenly distributed, the coffee, which has cooled, has reached a state of entropy, that is, there is no more thermal energy available to do work.

Below: An early English steam locomotive, *c.* 1845, hauling a tender and three carriages. The design of this engine was adapted from horse-drawn coaches. Steam engines use thermodynamic principles to convert heat energy to work.

Overzealous spectators at sporting events often "do the wave;" they generate a spontaneous wave of enthusiasm by standing and raising their arms at precisely the right moment. Viewed from afar, this wave of up-thrust hands is seen to sweep slowly through the crowd. The speed of this wave depends on the response time (and sobriety) of the individual spectators.

Making waves

Generally speaking, the speed of a wave depends on the springiness of the connection between the individual particles forming the wave. For example, in a flaccid rope waves travel only a few tens of meters per second, but in a steel guitar string, which is much stiffer, waves travel some 50 times faster.

Transverse and longitudinal waves

Spectator waves (sometimes called "Mexican" waves) are transverse waves, where the wave energy flows sideways to the motion of the individual elements. In the stadium the wave propagates sideways through the spectators, but the individual hands move up and down. Transverse waves can be generated in countless materials, including ropes, guitar strings, and the surface of water. In water, a surface wave travels at surfing speed, but in the deep ocean a tsunami wave can race at speeds of up to 620 miles per hour (1,000 k/ph).

Sound waves traveling through the air are longitudinal waves, where the individual particles travel back and forth in the same direction as the direction of propagation of the wave. A blast from a trumpet compresses the air in the trumpet's bell, which knocks into the air next to it, and so transmits the energy of the sound wave. Sound waves travel faster in hot air than in cold, because hot air molecules naturally bump into each other more often and therefore transmit the wave more quickly.

In a solid material such as Earth, both transverse and longitudinal sound waves can propagate, although with different speeds. An earthquake generates both kinds of these sound waves (these are referred to as S and P waves, respectively), and the difference in the arrival time of these two waves at a distant monitoring station reveals the location of the earthquake.

Above: When a trumpet is blown, a sound wave is produced, which travels the length of the instrument. When it reaches the bell, there is a drop in resistance and the wave changes its direction, going back toward the mouthpiece. The player's lips control the frequency of the sound wave.

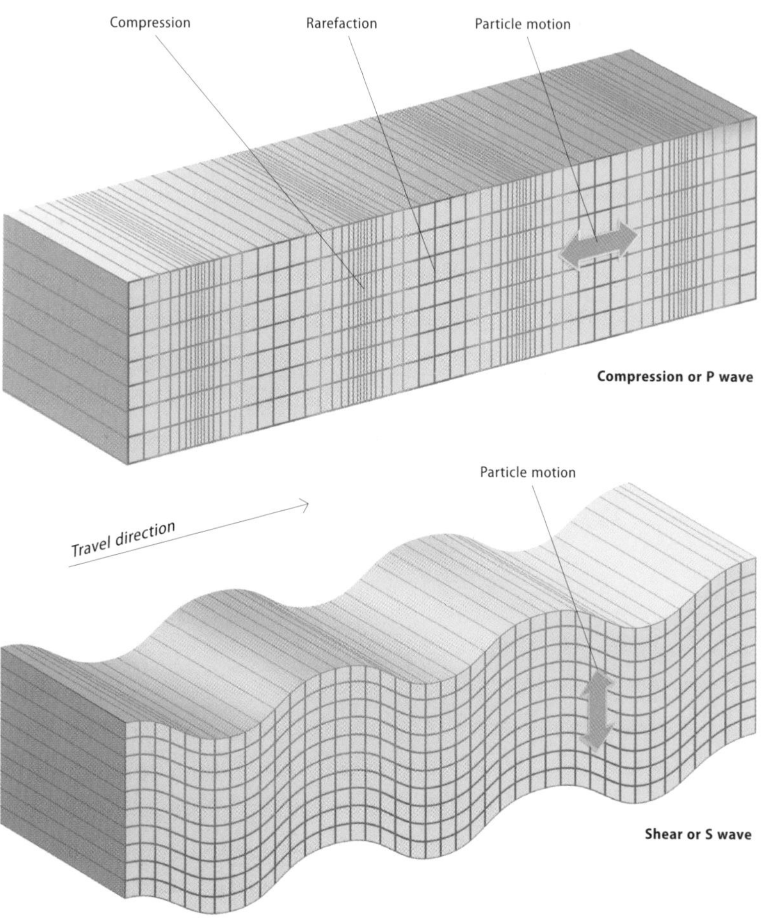

Left: Earthquakes generate two kinds of sound waves that propagate through the Earth. In "P" waves (longitudinal waves) the displacement of the soil is along the direction of wave travel. In "S" waves (transverse waves) the soil is displaced transverse to the direction of wave travel. P waves travel almost twice as fast as S waves.

Above: A spectator or "Mexican" wave is a common sight at sporting events around the world. The wave forms as spectators stand up from their seats, raise their hands up into the air, and then fold them back before they sit down again. The dramatic effect of a rolling wave is generated.

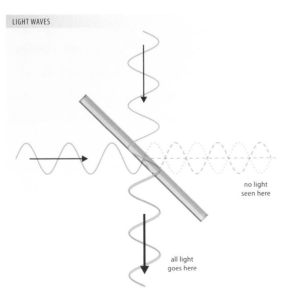

SOUND WAVES AND THE HUMAN BODY

Sound waves travel easily through the human body, and they reflect whenever they encounter a change, such as a bone, or a blood vessel, or a baby (below). By detecting these reflected sound waves, a three-dimensional image of the body's interior can be created. Much higher-energy sound waves are used to disintegrate kidney stones (usually crystals of calcium oxalate) lodged in the kidneys or bladder.

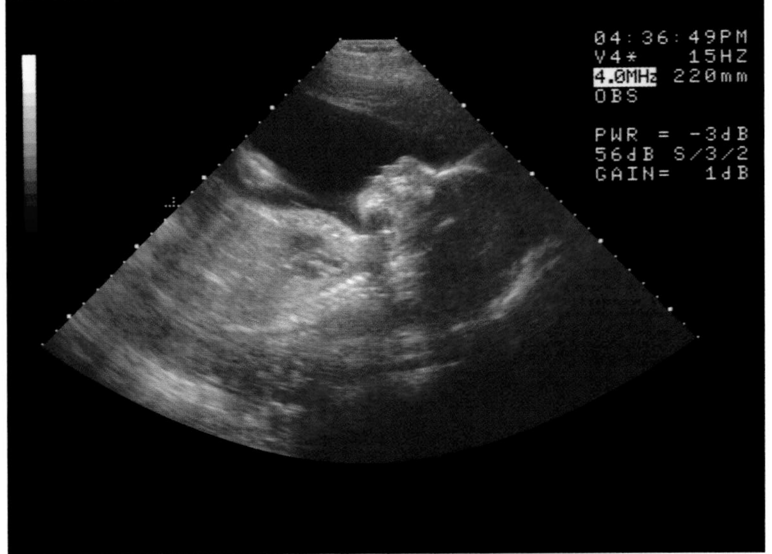

You can't stop the waves, but you can learn to surf.

Jon Kabat-Zin, scientist and writer

Wave superposition and interference

All waves obey the principle of superposition: The effect of two waves is simply the sum of the effects of each wave separately. A surfer, for example, if poised off the coast with waves coming at him from two directions, can find pockets where the water is moving neither up nor down. While one wave may be pushing him up and down, the second wave can be pushing him down and up. The two waves interfere, with the net result that the surfer feels no wave at all.

In a plucked guitar string, waves travel in both directions at the same time. These two waves interfere to create no motion of the string at its two fixed ends, but create a large transverse motion in the middle of the string. Only those wave frequencies that reinforce each other in the middle of the string will exist, which gives a stretched string of a particular length its particular musical pitch.

Light waves also obey superposition. Two light waves can interfere with each other and produce darkness! Because energy is conserved and cannot simply disappear, the energy of the light waves must always appear somewhere else, as shown in the figure on the left.

Interference is the signature of any wave; the demonstration that light beams can interfere is proof that light is a wave.

Radio waves, light waves, X-rays, and gamma rays are all waves of electromagnetic energy; they differ only in their frequency. These electromagnetic waves all travel at exactly the same speed, which is 300,000,000 meters per second. Unlike sound waves, which need a medium to travel through, electromagnetic waves can travel through empty space. In a triumph of physics over reason, electromagnetic waves travel by continuously collapsing and then regenerating themselves a short distance ahead.

LIGHT WAVES

no light seen here

all light goes here

Left: Two light waves, one incident from the top and another from the left, partially reflect at the half-silvered mirror in the center of the illustration to interfere with each other. Here the two light waves traveling to the right cancel out each other, with all of the light energy being directed downwards.

What's the connection between probing the furthest reaches of the universe and the distinctive noise of a racing car shooting by at high speed? The answer is the Doppler effect, which describes the way that waves—sound or light—from a moving source change their pitch.

The Doppler effect

All waves normally spread out equally in all directions, like ripples on a pond. The number of waves being sent out each second determines the frequency of the wave—the waves in high frequency waves are very close together; the waves in low frequency waves are further apart. What this frequency determines depends on whether we're considering sound or light waves.

Above: A Hubble Space Telescope image of the center of globular cluster M15 reveals a new population of about 15 very hot blue stars isolated at the core. The Doppler effect shows us these stars as blue in color because they are moving toward us.

Right: When struck on a surface, a tuning fork emits a "vibrating" sound which shows the Doppler shift in frequency. When the sound is moving toward you the frequency goes up; it goes down while sound is moving away. The Doppler effect is named for Austrian physicist Christian Doppler (1803–1853).

Sound waves

The frequency of each sound wave determines the pitch you hear. The waves of very low notes are a long way apart, whereas the waves of high-pitched notes are very close together. When something noisy, like a car, is standing still, the waves spread out evenly in all directions and it sounds the same no matter where you're standing.

If the car is moving, however, the waves will be different depending on whether you're in front of, or behind the car. The waves at the front will end up pushed closer together—the car is catching up with the waves it's already sent out—while the waves at the back will be stretched out. So standing in front of an oncoming car you'll hear a higher pitch sound, while standing behind it you'll hear a lower pitched sound. This shift in frequency is known as the Doppler effect.

For racing cars or police or ambulance sirens, when the car is approaching you the sound waves are being compressed and you hear a higher pitched sound. But when it passes you, it's now moving away and the stretched sound waves have a lower pitch, giving the distinctive sound we know from movies and real life alike.

An analogy for the Doppler effect is two people rolling balls to each other. Person A rolls one ball every five seconds, so there's an even distance between each ball. If Person A starts walking toward Person B, they'll be chasing the balls they've already rolled. When the time comes to roll the next ball, the space between balls will be less and Person B will receive the balls closer together—a higher frequency. Likewise, if Person A walked away from Person B, the balls would be rolled even further apart, just like the soundwaves of a receding car.

> Happy is he who gets to know the reasons for things.
>
> Virgil, poet, 70–19 BCE

The lighter side of Doppler

It's not only sound waves that are affected by the Doppler shift. Light waves can also be Doppler shifted when the source is moving. But for light waves, what changes is the color of the light. Because light travels so fast, this is hard to see in day-to-day life. But astronomers looking at the light from distant stars can detect the shift in the star color. Stars moving away from us have their light stretched out and "red shifted," the blues and greens turning into reds and yellows. Similarly, a star moving towards us becomes "blue shifted," turning reds into greens and blues. If you were traveling fast enough toward a red traffic light, it would appear green, due to the Doppler shift! By carefully measuring how much Doppler shift each star has, astronomers can map the motion and location of distant stars and even galaxies, helping to understand the overall structure of the universe.

Below: If a car with a siren is moving toward us, we hear a high pitched sound; if it is moving away from us, the sound is at a lower pitch. The different frequency is the Doppler effect.

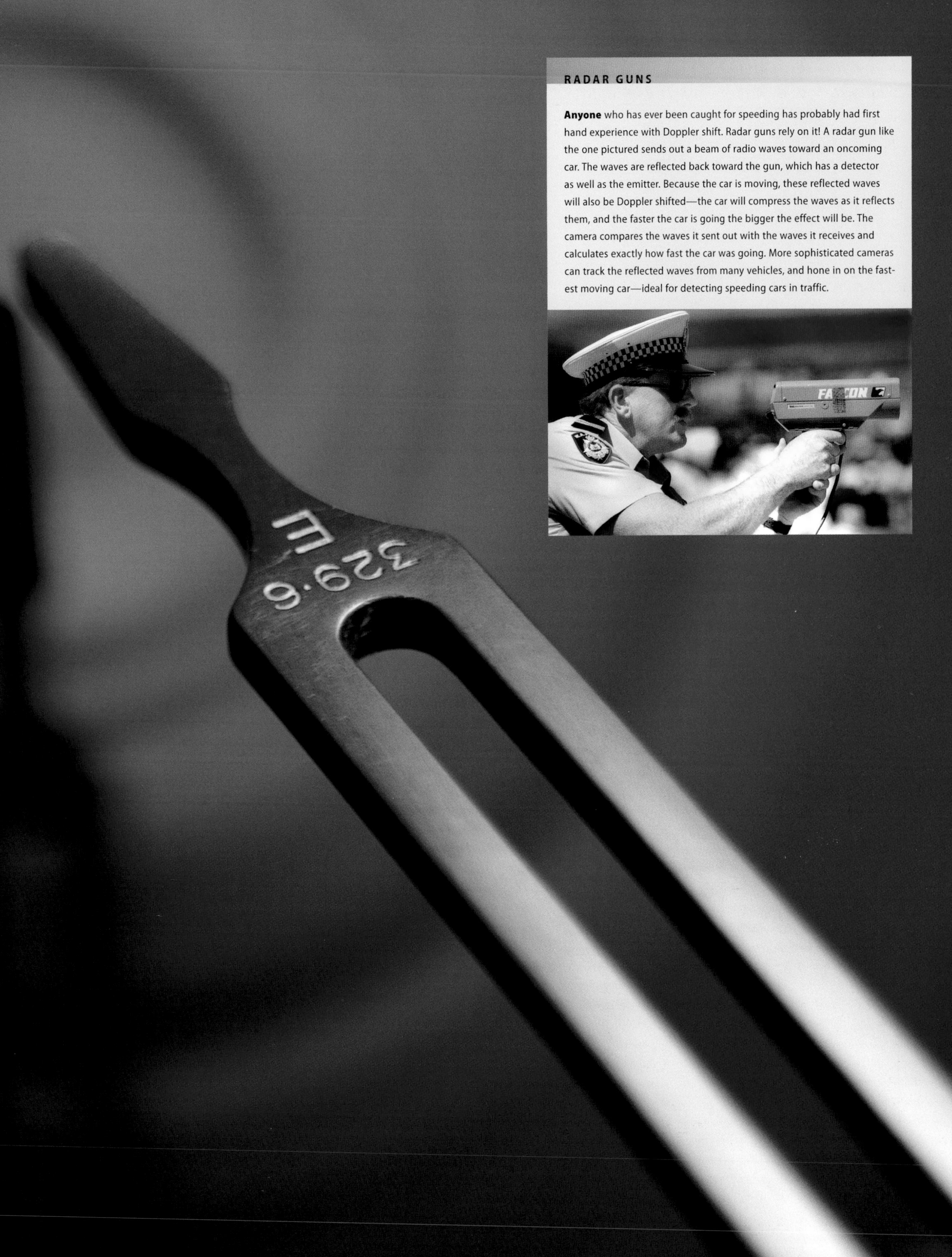

RADAR GUNS

Anyone who has ever been caught for speeding has probably had first hand experience with Doppler shift. Radar guns rely on it! A radar gun like the one pictured sends out a beam of radio waves toward an oncoming car. The waves are reflected back toward the gun, which has a detector as well as the emitter. Because the car is moving, these reflected waves will also be Doppler shifted—the car will compress the waves as it reflects them, and the faster the car is going the bigger the effect will be. The camera compares the waves it sent out with the waves it receives and calculates exactly how fast the car was going. More sophisticated cameras can track the reflected waves from many vehicles, and hone in on the fastest moving car—ideal for detecting speeding cars in traffic.

Left to itself, a ray of light will travel through a uniform material in a perfectly straight line. Optics are used to bend, deflect, or focus light. Telescopes, microscopes, and eyeglasses all use optics to form images of objects.

The science of optics

The objects can be distant, or very small, or right in front of our noses. In all of these cases, the light spreading out from each point on the object is captured by the optics and guided to a corresponding point on the image.

Guiding light with lenses

A light ray passing from one material into another will change its direction, depending on the angle of the incident light ray with the boundary between the two materials. The amount of bending (which is called refraction) also depends on the ratio of the indices of refraction of the two materials. The law of refraction was worked out by Willebrord Snell and first published by René Descartes in 1637, and all optical lenses use this law. See Figure 1.

A crude lens can be constructed by stacking a number of glass wedges (Figure 2). Such a stack forces the light rays striking near the edges of the lens to bend more than those passing through the center of the lens, and thereby sends all of the rays toward a common focus. A more sophisticated lens uses continuously curved glass surfaces in order to bend the light rays.

A perfect converging lens would make all of the light rays captured from a single point come together again at another point (see Figure 3). As simple as this might seem, it is impossible to accomplish except for a single point object with a single color of light. For imaging extended objects over a range of colors, a series of lenses must be used to create a reasonably good image. If the image needs to be incredibly sharp, for example, in photolithography, when the images are used to manufacture microscopic computer chips, then upwards of 30 different lenses must be stacked together to create the required near-perfect image. It is more difficult to focus light in the deep ultraviolet range of the spectrum, because lenses made of glass absorb too much of this light's energy. Lenses for deep ultraviolet light are instead crafted out of special crystals, such as calcium fluoride. For even higher-energy light, such as X-rays, the light can be guided only by bouncing it off of the planes of atoms in metal crystals.

Changing the color of light

It was once thought that an optical prism could create new colors of light. But Isaac Newton showed that a prism merely directs different colors of light into different directions, and thereby proved that a beam of sunlight contained all of the colors of the visible spectrum.

However, it is possible to change the color of a light beam with optics, but only if the light is sufficiently intense. Lasers produce intense light, and, within a year after the invention of the laser, optical materials were discovered that could alter the color of light from a laser. For example, an optical crystal of lithium niobate can double the light frequency of an infrared laser and thereby produce the green laser-pointer beam preferred by sophisticated seminar speakers.

> It is possible that some other science may be more useful [than optics], but no other science has so much sweetness and beauty of utility.
>
> Roger Bacon, philosopher and scientist, 1214–1294

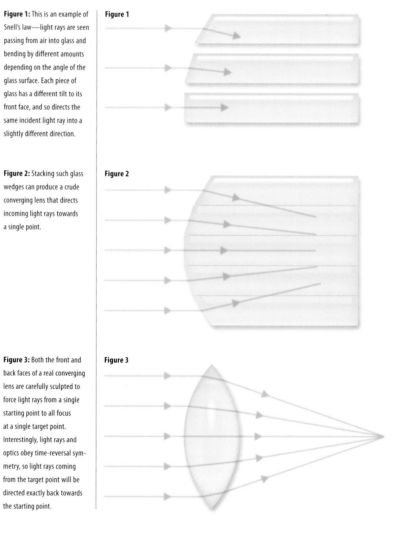

Figure 1: This is an example of Snell's law—light rays are seen passing from air into glass and bending by different amounts depending on the angle of the glass surface. Each piece of glass has a different tilt to its front face, and so directs the same incident light ray into a slightly different direction.

Figure 1

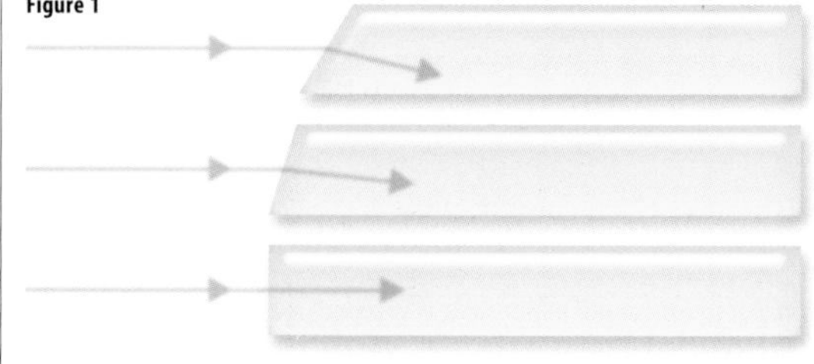

Figure 2: Stacking such glass wedges can produce a crude converging lens that directs incoming light rays towards a single point.

Figure 2

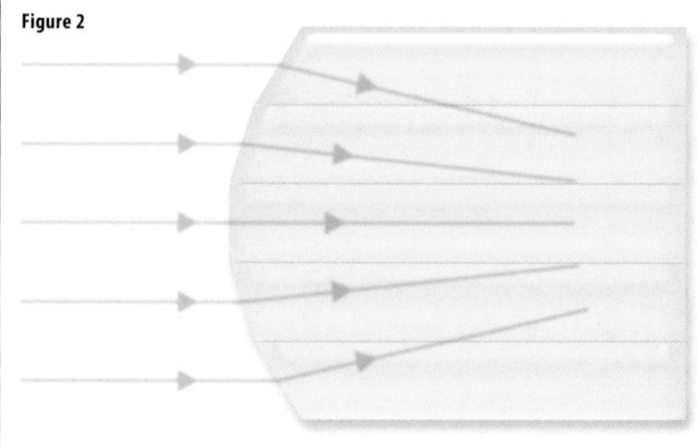

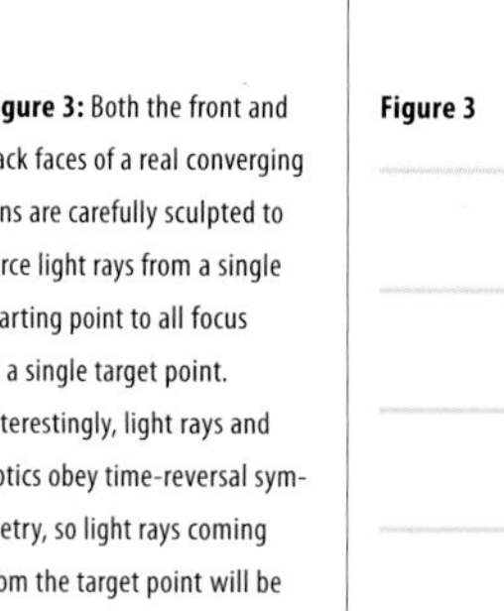

Figure 3: Both the front and back faces of a real converging lens are carefully sculpted to force light rays from a single starting point to all focus at a single target point. Interestingly, light rays and optics obey time-reversal symmetry, so light rays coming from the target point will be directed exactly back towards the starting point.

Figure 3

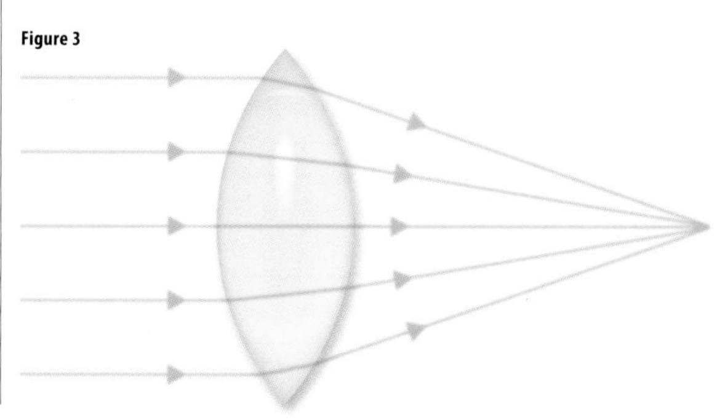

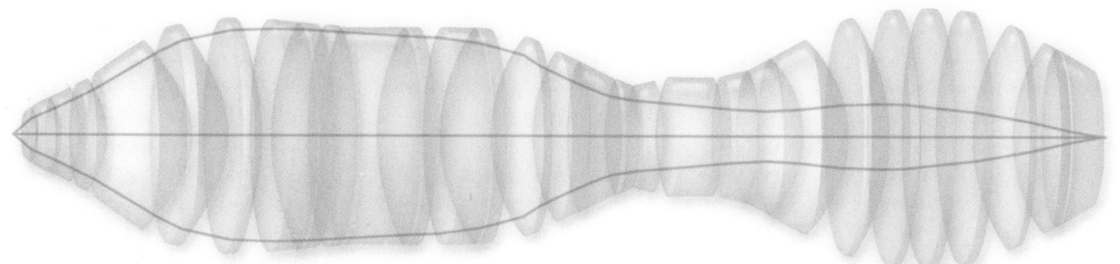

Opposite: A beam of white light is separated into its different colors by a prism.

Left: Computer chips are fabricated using complex lens systems costing millions of dollars and containing more than thirty individual lenses.

GUIDING LIGHT WITH REFLECTION

The Hubble Space Telescope uses a precisely curved optical mirror to reflect incoming light rays and form images of distant galaxies. The surface of the Hubble's mirror is specified to within 30 billionths of a meter, and, ironically, was launched into space with the wrong shape.

Light can also be guided by total internal reflection. When light traveling in one medium encounters another medium at a glancing angle, and the second medium has a smaller index of refraction than the first, the light can be totally reflected from the boundary, with no light transmitted across the boundary (below). Optical fibers no thicker than a human hair use total internal reflection between concentric layers of glass to confine light inside the fiber. Such fibers can transmit light signals for hundreds of miles with barely any loss. Optical fibers on our ocean floors now carry most of the world's transcontinental telephone traffic in the form of rapid pulses of differently colored light beams.

Light can also be guided by internal reflection inside straight channels called waveguides. Minuscule waveguides of pure silicon can guide infrared light; such channels can be sprayed or painted onto chips and molded into the modulators and detectors used for optical telecommunications. Miniature rings, only a few millionths of a meter in diameter, can shuttle light between adjacent waveguide strips, or can trap light and make it circulate for billions of bounces.

Light is weird stuff. A physicist might say that light consists of coupled electric and magnetic fields propagating through space. A non-physicist probably wouldn't be able to provide a better definition.

Into the light

All light is produced by shaking charges, usually electrons. Electrons that shake in a periodic manner at a particular frequency produce light at the same frequency. Humans have exquisite sensors (eyes) to detect light only in a narrow frequency range of approximately 4×10^{14} to 7×10^{14} oscillations per second, and not outside it. In this visible range, the different light frequencies appear to us as different colors, from red to yellow to green to blue to violet as the light's frequency gradually increases. Frequencies slightly less rapid comprise infrared light, which we cannot see—but bees can. Frequencies a bit more rapid comprise ultraviolet light, again which we cannot see but will tan our skin.

A mixture of all of the visible colors of light appears white. Magazines and books (such as this one) take advantage of the many colors that are inherent in white light. Using inks that preferentially absorb only some colors of the ambient white light, the remaining colors are scattered into our eyes. For example, red ink is composed of compounds that absorb the higher-frequency colors (blue and green) from white light, so that only the lower frequency color (red) scatters into our eyes. By combining different inks, different

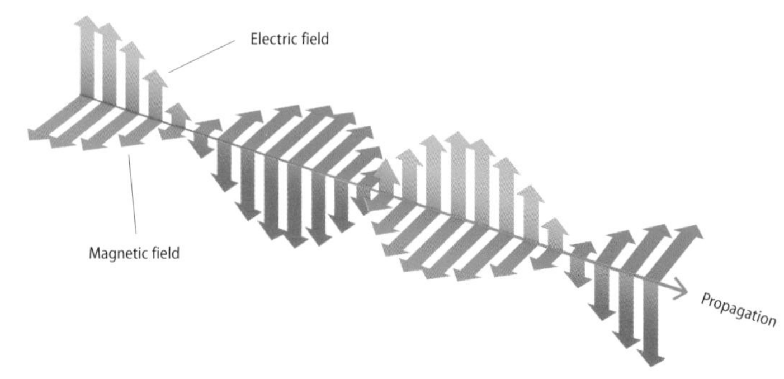

Below: The view at sunset from Blouberg Beach toward Cape Town, Western Cape Province, South Africa. Water vapor and dust in the atmosphere scatter blue light from the Sun, which is why the sky appears blue. At sunset, the unscattered red light appears orange to red.

Above: Light is an electromagnetic wave hurtling through space. Visible light is the only part of the electromagnetic spectrum that humans are able to see.

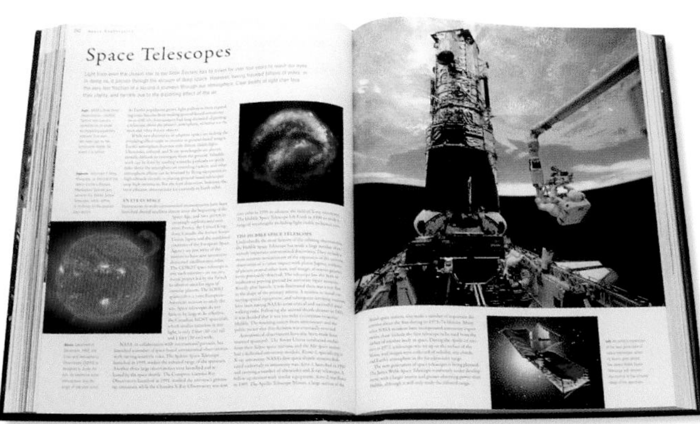
Space Telescopes

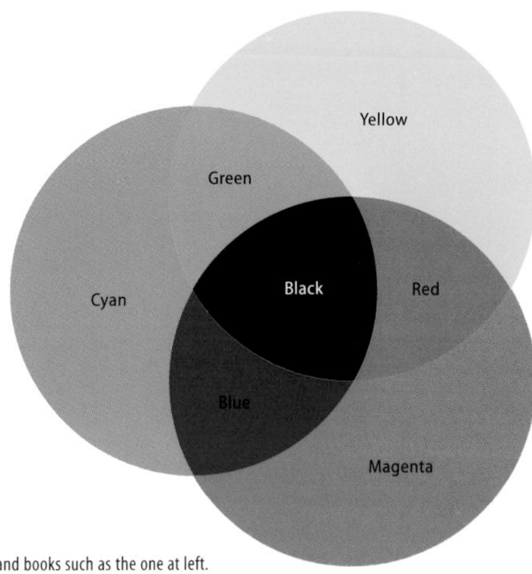

Above: Subtractive color mixing is used by all magazines and books such as the one at left. Cyan, magenta, and yellow inks, when illuminated by white light, subtract out some of its colors and scatter the remaining ones into your eye. For example, where the yellow and magenta inks overlap in the figure on the right, all of the high-frequency (blue and green) colors in the white light are absorbed. This leaves only the color red to be scattered into your eyes.

colors can be removed from the white light, with the remaining colors forming the desired colored image.

The speed of light

Light can propagate through empty space. Only a short 100 years ago it was assumed that light, like sound, needed some kind of medium to travel through. But this is not the case. In space, light travels at the fixed speed 3×10^8 meters per second (aproximately 186,000 miles [300,000 km] per second). Each day we see light from the Sun that has traveled about eight minutes through the empty vacuum of outer space to reach Earth.

Surprisingly, the speed of light is independent of any motion of the light source or of the observer of the light. Compare this with the speed of a bottle thrown from a passing car, which, to a person standing on the road, depends of the speed of the car. In contrast, the speed of the light from a flashlight held in a passing car and detected by a person standing on the road is independent of the speed of the car. It is always 3×10^8 meters per second. The unchanging speed of light is one of two fundamental postulates underlying Einstein's theory of special relativity.

Above: Transparent materials and substances such as translucent liquids, allow light to pass through, yet as it does so it disperses and reflects.

Light is a wave

Light can act as a wave. The wavelength (distance between its peaks) of the light is determined by its frequency and its speed, according to the formula: wavelength = speed /frequency. Visible light has a wavelength ranging from 0.4 to 7 microns (1 micron = 1 millionth of a meter). In contrast, radio waves are a few meters in wavelength. Antennas are usually about as long as the wavelength of the wave they are detecting, which is why car radio antennas are a meter or so long. The rods and cones in our eyes are only a few microns long, and so form ideal antennas to detect visible light waves.

> I could compare my music to white light, which contains all colors. Only a prism can divide the colors and make them appear; this prism could be the spirit of the listener.
>
> Arvo Part, composer, b. 1935

A STREAM OF PARTICLES

Light can also act as a stream of particles. The energy of each particle (a photon) depends on the frequency (that is, the color) of the light, while the number of particles arriving per second depends on the brightness of the light. For example, when light hits a metal plate, some of the light particles collide with electrons in the metal and knock the electrons loose, creating an electrical current. Bright blue light causes lots of electrons to be released, which makes sense because bright light has lots of photons arriving per second, and the light's blue color means that each photon has enough energy to liberate the electron. But bright red light liberates no electrons, since the red light particles, although numerous, are each too weak to knock an electron loose. In every experiment where light is asked to act as a stream of particles, it does.

Electricity is easy to describe, because its effects are all around us, but hard to explain, because it involves profound understanding of science built up over many years by many famous physicists.

The power of electricity

A short definition of electricity might be "the presence and effects of a charge that is moving or displaced," but this is not particularly useful to most of us. Electricity is a form of energy resulting from the behavior of electrically charged particles such as protons and electrons. Of course, we know the word more commonly as the supply of power to our homes and workplaces.

Static electricity

When amber or various plastics are rubbed, they attract small pieces of paper. This experiment has been known for thousands of years. In fact the word "electricity" comes from the ancient Greek work meaning "amber." The attractive material can also deliver a small spark. We call the stuff that gives rise to the spark and that attracts the pieces of paper "static electricity."

Benjamin Franklin's famous experiment involving flying a kite in a thunderstorm proved that lightning was the same static electricity. Static electricity is also responsible for the shock you sometimes get as you alight from a car on a dry day, and the crackling noise as you pull off certain pieces of clothing.

The spark of a static electric discharge creates sound, from the tiny crackling you hear as you pull off your favorite woollen jacket to thunder in a storm. Some charged particles are moved from one

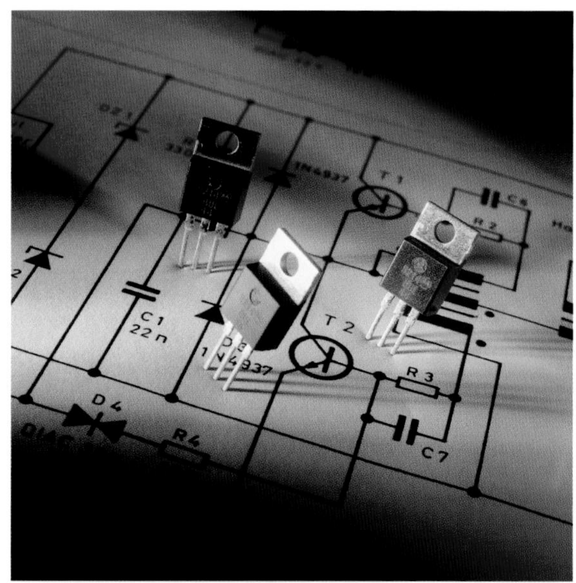

Right: Transistors on a wiring board. Transistors amplify an electrical current by using a small amount of electricity to control the flow of a larger supply of electricity. This works in a similar way to a valve that is used to control the flow of water in a hose or pipe.

> Imagination has brought mankind through the dark ages to its present state of civilization. Imagination led Columbus to discover America. Imagination led Franklin to discover electricity.
>
> L. Frank Baum, novelist, 1856–1919

DYNAMIC EFFECTS

Static electrical effects were known by the ancients, but dynamic electric effects were only discovered about 400 years ago. The great chemist Robert Boyle noted the connection between chemical reactions and electricity in the seventeenth century, and later Alessandro Volta (at left) invented the first reliable battery. Most famously, Galvani observed that electricity activated muscles. Electricity could be produced by chemical reaction in batteries, and it was known to be able to stimulate chemical reaction in return, for example, in the splitting of water into hydrogen and oxygen. By the nineteenth century, electricity was a well-investigated scientific curiosity.

Above: American inventor and scientist Benjamin Franklin (1706–1790), and his famous kite experiment where he proved that lightning is electricity. He is credited with the invention of the lightning rod.

Below: Lightning is a sudden dazzling flash of electrical current that usually accompanies a thunderstorm. It happens when vertical air movements and various interactions between cloud particles bring about the separation of positive and negative charges.

Above: Utility poles and power pylons support the overhead electricity conductors that channel heat and power into our homes, schools, and offices. There are insulators, often made of glass, between the cables to stop any possible leakage of current from the pylons.

place to another, and the spark is what you see when those charged particles suddenly return to their starting place.

Generating power

In the nineteenth century, electricity moved from being a scientific phenomenon to being a tool for the delivery of both power and information. Michael Faraday invented the motor and generator, and Gaston Plante the lead-acid battery. Power could be generated from steam, stored, and reconverted to do mechanical work. By the twentieth century, power was being sent directly from generator to application by wires, removing the need for the battery in static applications. Edison was famous for the invention of the incandescent light globe, allowing electricity to replace candles and lamps. Before long, electricity was being used to light streets and houses, pump water, move cargo, and to send messages by means of the telegraph and telephone.

Conductors and insulators

The application of electricity involved a simple observation of physics. All materials are made up of atoms, and atoms are made up of electrons, protons, and neutrons. In some materials, the electrons move from atom to atom relatively easily. These materials are called conductors, and are mostly metals. Materials that do not easily allow electrons to move around are called insulators; typical examples are wood, plastic, dry thread, and air. We can direct electricity by creating a conductor surrounded by insulator, just as a hose directs the flow of water. This is an insulated wire. We can force electrons through the wire by pushing electrons onto the end of the wire, just as a pump pushes water through a hose. To do work, the charge, usually carried by electrons, is pushed hard, and eventually it creates motion, light, or heat. This is the transfer of energy through the medium of moving charges—which is exactly like lightning, but controlled and harnessed by human ingenuity.

Magnetic energy dictates the behavior of substances from the smallest electrons to the biggest planets, and is fundamental to the nature of light. Named after the region Magnesia in Asia Minor, magnetism was known to both the ancient Greeks and Chinese, and was probably discovered by the Chinese around the fourth century BCE *as a property of the mineral lodestone.*

The mysteries of magnetism

How magnets work

At an atomic level, much about magnetism remains mysterious. Each electron within an atom, and even the nuclei of atoms, are potential miniature magnets. Electrons in an atom usually exist in pairs and seek out unpaired electron partners. In natural materials such as iron, unpaired electrons in the outer orbital area of the atom have what is known as large magnetic moments—the magnetic strength of a magnetic field at a distance from a magnet. In other words, these electrons possess a large magnetic energy.

Magnetic energy is driven by the motion of electrons, such as in an electrical current, by the inherent properties of electrons themselves; and in certain cases, by the motion of electrons around the nucleus of an atom. Even the bonds between atoms are created by magnetic energy. The strongest natural magnets are ferromagnetic materials, for example, cobalt, iron, and nickel. Manufactured magnets are often made of steel. But even materials such as carbon and water are slightly magnetic.

Metallic objects can be magnetized by exposing them to a magnet. Rub a magnet along a paper-clip or a nail, and the electrons begin to spin in such a manner that small magnetic domains (like miniature north–south magnets) line up and the object begins to act like a magnet itself, picking up a bundle of nails or paper-clips.

Poles apart

Magnetic energy is strongest at its poles. Two south or two north poles repel, and north and south poles attract each other. The force

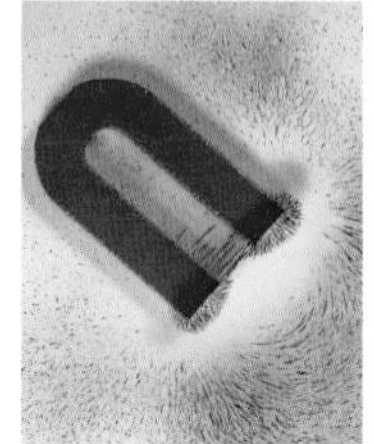

Above: Magnets produce magnetic fields and attract metals such as nickel, cobalt, and iron, like these iron filings. All magnets have a north pole and a south pole.

Right: The Maglev (Magnetic Levitation) train gliding into Longyang Lu Metro Station, Shanghai, China. The Maglev is propelled by electromagnetic force, and can reach speeds of 360 miles per hour (580 km/h).

> Magnetism, as you recall from physics class, is a powerful force that causes certain items to be attracted to refrigerators.
>
> Dave Barry, humorist, b. 1947

Right: Michael Faraday's magnet, *c.* 1830. Faraday found that electricity not only affected magnets, but that a wire moving through a magnetic field developed a current running through it, a process known as induction. Thus the link between the electricity and magnetism went both ways—a moving magnetic field produces a current and a moving electric field produces magnetism.

ELECTROMAGNETISM AND THE QUANTUM REVELATION

Nineteenth-century scientists discovered the link between electricity and magnetism and began a scientific exploration of the concept that led to the use of magnets in everything from fridges to sophisticated medical equipment. Studying electromagnetic energy and its effects also led scientists to the development of quantum theory. One of the revelations made by Albert Einstein and other early twentieth century physicists was the wave–particle duality theory of matter. English scientist Thomas Young proved that light acts as a wave in his double slit experiment, in which a beam of light shines through two slits. The pattern of interference formed shows that light interacts with itself, like a wave. However, Einstein described how light acts as a particle in his treatise on the photoelectric effect, which describes how an electrical current is produced by shining light on certain metals. As light "hits" the metal, electrons are ejected from the atomic structure, producing a flow of electrons—an electrical current. This and subsequent discoveries led to the formation of the theory of quantum electrodynamics, which also describes how electromagnetic fields and matter interact. (At left is an electromagnet.)

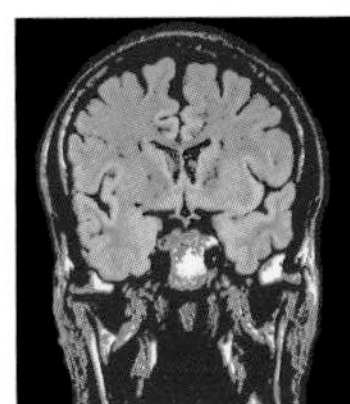

Above: Magnetic resonance imaging, or MRI, is an imaging technique that is used primarily in medical settings to produce high quality images of the inside of the human body. This scan shows the inside of a normal human head.

of the push or pull varies depending on the strength of the magnetic field, and may be powerful enough to drive a Maglev train (the world's fastest train), or subtle enough to image the human brain (as in nuclear magnetic resonance imaging).

The equation for force between two magnetic poles is given by

$$F = \frac{\mu q_{m1} q_{m2}}{4\pi r^2}$$

where F is force, q_{m1} and q_{m2} are the pole strengths, μ is the permeability of the intervening medium and r is the separation.

Pinpointing north

Earth's magnetic field is created by motion of the conductive iron metal in Earth's core. Heat-driven convection currents within Earth's outer core interact with the motion of the liquid metallic inner core to generate moving electrical charges, which in turn drive Earth's magnetic field. The field is strongest at the magnetic north and south poles of Earth, which move depending on the wobble of Earth about its axis, and may be several thousand miles from the geographic North or South Poles.

A compass utilizes a magnet's tendency to stabilize into the lowest energy possible state. The magnetic needle thus aligns itself with the stronger force of Earth's magnetic field, and the north-seeking pole of the compass needle rests aligned with the north–south dipole of Earth's magnetic field.

Have you ever wondered why we don't fly off Earth into the atmosphere every time the Earth rotates on its own axis? The answer is gravity. Gravity refers to the forces of attraction that occur between two masses. The bigger the mass of an object, the stronger the force of attraction. Hence, because Earth's mass is so much greater than our own, the attraction is great enough to keep us grounded.

A question of gravity

Newton's notion of universal gravitation

English physicist, mathematician, and natural philosopher Isaac Newton is considered one of the most important contributors to our modern understanding of gravity.

At the heart of Newton's work was his deep curiosity about the forces that held the Moon in its orbit around Earth. According to the old legend, Newton was inspired to develop his theory about gravitational forces when he observed an apple falling from a tree. He then went on to propose that the same forces that attracted the apple to fall to the ground were also responsible for the Moon orbiting Earth.

Above: English mathematician and physicist Isaac Newton (1642–1727) contemplates the force of gravity, as the famous story goes, on seeing an apple fall in his orchard. His theory of gravity was the accepted theory for centuries.

Using the example of a cannonball being fired horizontally from a very high mountain, ("Newton's Mountain"), Newton proposed that in the presence of gravity, the cannonball would eventually fall to Earth. However, if the speed at which the cannonball was fired were increased, the cannonball would travel further before falling to the ground. Newton suggested that if the cannonball could be fired so that it followed a trajectory that matched the curvature of Earth, and that if sufficiently high speeds could be achieved, the cannonball would display similar behaviour to the Moon—circling Earth, but never actually colliding with it.

So why doesn't the Moon collide with Earth?

Newton's inverse square law explains this mystery. Newton demonstrated mathematically that while the attraction between Earth and the Moon is sufficiently large to maintain the Moon's orbit, it is diluted by the distance between the two masses. In short, Newton's law concluded that the force of gravity between Earth and any given object is inversely proportional to the square of the distance that separates the object from Earth's center.

Einstein versus Newton

While Newton's theory made huge inroads into the understanding of gravity, his theory of universal gravitation had some flaws. It did not explain the orbit of Mercury, and it did not explain why gravitational acceleration is independent of the mass or composition of the object.

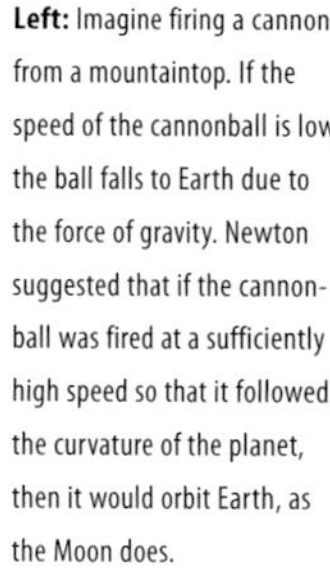

Left: Imagine firing a cannon from a mountaintop. If the speed of the cannonball is low, the ball falls to Earth due to the force of gravity. Newton suggested that if the cannonball was fired at a sufficiently high speed so that it followed the curvature of the planet, then it would orbit Earth, as the Moon does.

It's a good thing we have gravity, or else when birds died they'd just stay right up there. Hunters would be all confused.

Steven Wright, comedian, b. 1955

ARISTOTELIAN THEORY OF GRAVITY

Before Newton and Einstein, the great Greek philosopher Aristotle (384–322 BCE) proposed his own explanation for the movement of objects around Earth. Aristotle's theory states that all bodies move to their natural place—some to Earth, and others to the heavenly spheres, away from Earth.

Aristotle's theory was supported by his belief that the universe was composed of four elements—fire, earth, air, water, as well as aether, the divine substance that makes up the heavenly spheres. Each earthly element had its natural place and motion—bodies sink in water, air bubbles up, rain falls to Earth, and flames rise in the air—while heavenly elements have perpetual circular motion.

After centuries of criticism, Aristotle's theory was finally discredited by the work of Italian mathematician Galileo Galilei, who after dropping balls of different densities from the Tower of Pisa (right) concluded that they fell at the same speed, irrespective of their mass.

Left: Astronauts experience feelings of weightlessness in space because Earth's gravity has less effect in space. It is the same sort of feeling we would have inside an elevator if it fell from the 40th floor of a building. We would fall at the same rate as the elevator and would float inside.

However, perhaps most significantly, Newton's theory was inconsistent with Einstein's special theory of relativity, which proposed that distance and time were not absolute (see pages 70–71).

Einstein went on to develop the general theory of relativity, which proposes that gravity and motion can affect the intervals of time and space. The basic premise of general relativity is that gravity pulling in one direction is equivalent to acceleration in the opposite direction. An example of this is an elevator accelerating upwards. It feels just like gravity pushing you to the floor. Einstein's theory implied that if gravity is equivalent to acceleration, and if motion affects measurement of time and space, then it follows that gravity does as well. Therefore, the gravity of any mass, such as the Sun, has the effect of altering the space and time that surrounds it.

While considered by many as one of the greatest leaps of faith in scientific history, Einstein's relativity theory has been proven experimentally, and consequently has had tremendous implications on modern science (see pages 72–75).

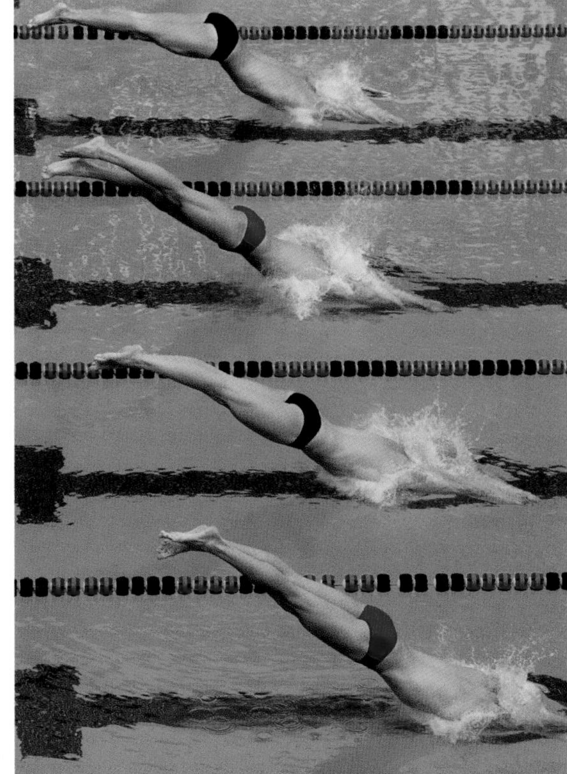

Right: Gravitation is the ability of objects made of matter to attract all other objects. The strength of gravity increases with increasing mass and as the distance between objects decreases. Swimmers are subject to two forces—gravity and buoyancy. Water's buoyancy makes us feel lighter.

Imagine life without sunshine, television, radio, X-ray machines, rainbows, and microwave ovens. This would be a life devoid of radiation. While radiation usually conjures up negative thoughts and connotations, it is actually a very important part of our everyday lives.

The role of radiation

Radiation is energy traveling through space in the form of high speed particles and electromagnetic waves. Most types of radiation cannot be seen, smelled, tasted, or even felt, unless the dose is high enough to damage your skin or body.

Everything on our planet is constantly being exposed to low doses of radiation coming from space, the air around us, and Earth itself. These days, humans can also make various types of radiation, used in medicine, research, and the supply of electricity.

Ionizing radiation is one particular type of radiation that can cause damage to matter. It has enough energy to interact with an atom by removing electrons. Ionizing radiation can damage living tissue and at high levels is dangerous.

Where does radiation come from?

Most of the radiation that we are exposed to is background radiation, which is naturally present in the environment. Levels of natural radiation vary greatly. Sources of natural radiation include materials like granite or mineral sands, radon gas (which is released from Earth), and cosmic radiation (radiation from space).

As well as natural radiation, there is also man-made radiation. This radiation comes from things like X-rays, medical radiation therapy, and radium in clocks and dials.

The average annual dose of radiation an average person is exposed consists of about 88 percent natural radiation and 12 percent man-made radiation, usually from medical procedures.

Right: Used radioactive fuel rods can be safely sheltered in water. The rods must be placed into a large, deep pool filled with water and located inside a building at the nuclear power plant site. The walls of these pools are about 6 ft (1.8 m) thick and are made of concrete with steel lining.

> I am one of those who think like Nobel, that humanity will draw more good than evil from new discoveries.
>
> Marie Curie, physicist, 1867–1934

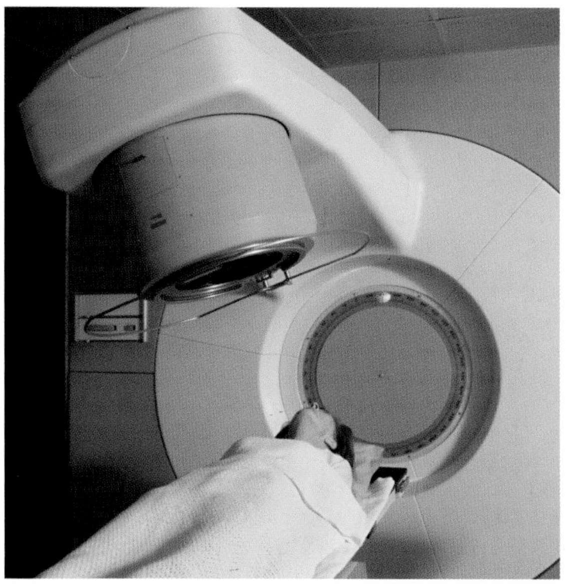

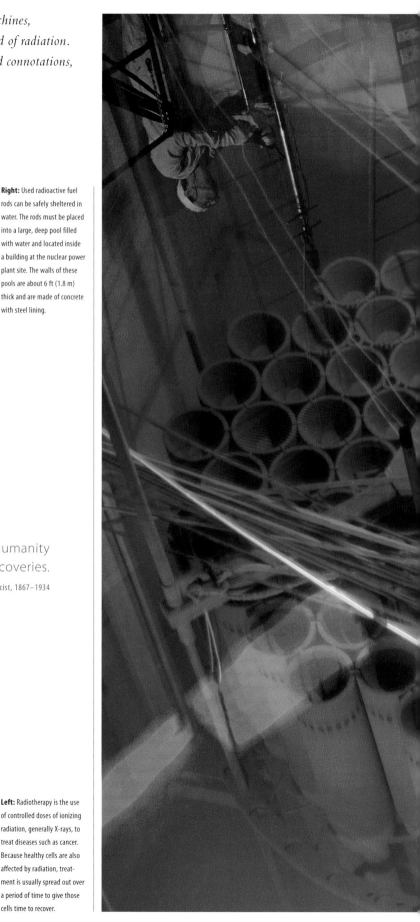

Left: Radiotherapy is the use of controlled doses of ionizing radiation, generally X-rays, to treat diseases such as cancer. Because healthy cells are also affected by radiation, treatment is usually spread out over a period of time to give those cells time to recover.

Above: The French physicist Henri Becquerel (1852–1908) shared the 1903 Nobel Prize for Physics with Marie and Pierre Curie, for his discovery of spontaneous radioactivity. His name has been immortalized—the Becquerel is a unit of measurement of radioactive material.

Measuring radiation

There are two factors that influence how radiation affects substances. The first is how long something is exposed to radiation. The second factor is the level of radiation present. Being exposed to low-level radiation over a long period usually is not as dangerous as being exposed to a short but highly radioactive dose.

Three common units are used in relation to measurements of radioactivity. Becquerel (Bq) is the unit used to measure the amount of radioactive material. It is used instead of the normal measures of mass and volume. Using Becquerels allows scientists to measure and then compare the radioactivity of materials.

Gray (Gy) is the dose of ionizing radiation that is received by a substance. It measures the amount of energy absorbed.

The third unit of measurement is Sievert (Sv). Although equal exposures to different types of radiation are expressed in Grays, they do not necessarily produce the same biological effects. One Gray of one radiation type may have a greater effect than another type. Therefore a term that is used to express the effects of radiation is called an effective dose. A Sievert is the unit for an effective dose. Regardless of the type of radiation, one Seivert of radiation causes the same biological effect.

How much is too much?

Over-exposure to radiation can have detrimental effects on the human body. These effects can be short-lived or long-term, and can even lead to death. Symptoms of radiation sickness include nausea and vomiting, skin burns, fatigue and weakness, hair loss, ulceration, and vomiting blood. High doses of radiation (around 100 Grays) can cause central nervous system damage, which results in death within hours. Doses of 10 to 50 Grays will cause death in one to two weeks through gastrointestinal tract damage.

Right: While radiation can be a useful tool, it can also cause serious damage. It can even make paint peel off walls. In 1986, an explosion at the Chernobyl nuclear power plant in the Ukraine, released a mixture of radioactive material into the atmosphere, containing at least 100 times more radiation than the atomic bombs dropped on Nagasaki and Hiroshima in 1945.

SMOKE DETECTORS

Household smoke detectors are one of the most common uses of radioactivity (in the form of radioisotopes). Most smoke detectors contain small amounts of americium-214. This radioisotope emits particles that ionize surrounding air. This allows for a current to pass between two electrodes. If smoke is present in the air, particles of smoke enter the detector and block the current. This causes an alarm to go off. The amount of radioactive material in smoke alarms is in no way dangerous to humans. The amount in the alarm is too small to cause damage, even if swallowed.

To scientists and engineers, microwaves are an incredibly useful part of the electromagnetic spectrum, perfect for mobile phone communication and for sending the radar signals used for weather forecasting. Yet most people think of microwaves as a useful kitchen appliance that can cook food in a flash, an accidental discovery that we owe to a block of chocolate!

Microwaves

Microwaves are part of the electromagnetic spectrum at frequencies between 1 and 300 GHz, placing them between regular radio waves and infrared waves. Like radio waves, microwaves can pass through walls and buildings, which means that the sender and receiver do not need to be in the line of sight of one another. The higher frequencies of microwaves, however, means there is more bandwidth available—more data can be transmitted simultaneously—in the microwave band, making it preferable for large scale communications.

Mobile telephones in many countries around the world operate on the microwave spectrum, as do other short-range communication technologies such as BlueTooth, used for hooking up devices wirelessly to laptop computers or mobile telephones. The allocation of part of the spectrum to a particular device or technology is usually done on a government level, as care must be taken that devices don't interfere with each other (we already see this interference every time a mobile telephone is placed near a radio or speakers).

Cooking with microwaves

Aside from communication, microwaves are used for cooking food. A regular oven cooks by heating up an element at the top of the oven which gives off infrared light. The food in the oven absorbs this heat, and eventually cooks through. A microwave oven, on the other hand, cooks food by bombarding it with microwaves, heating it by a process called dielectric heating.

Some types of molecules, most notably water, are dipoles, and are positively charged on one end and negatively charged on the other. When the microwave is switched on, these molecules start to spin as they try to line up their charges with the alternating microwave field. As the molecule rubs against other molecules, even ones without dipoles, friction causes the whole piece of food to heat up.

Because only molecules with dipoles absorb the microwave energy, this can lead to uneven heating of food. A common example is defrosting meat. Although water dipoles can absorb the microwaves, they can't move much while frozen and heating is slow. As soon as part of the meat defrosts, however, the freely moving water molecules absorb more than their share of the microwaves. So some parts of the meat will cook, while others remain frozen. The "defrost" setting

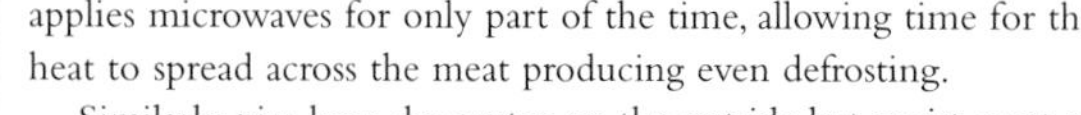

applies microwaves for only part of the time, allowing time for the heat to spread across the meat producing even defrosting.

Similarly, pies have dry pastry on the outside but moist meat and vegetables on the inside. The inside absorbs most of the microwaves heating up rapidly, while the outside may remain cool. This probably contributed to the popular misconception that microwaves heat from the inside out. In fact, the microwaves must penetrate in from the outside and unless, like a pie, the outside won't absorb them, the food heats up just like it would in a regular oven.

Above: The world's first BlueTooth-equipped wristwatch, the Citizen "i:VERT," which also operates as a mobile telephone. BlueTooth uses the microwave radio frequency.

Opposite: Microwaves are high frequency radio waves with more bandwidth available than other radio waves. Using microwave technology on radio towers allows quick and easy transmission of very high volumes of data.

Below: The electromagnetic spectrum is the complete range of electromagnetic radiation and is arranged according to wavelength or frequency. Microwaves sit between infrared and regular radio waves.

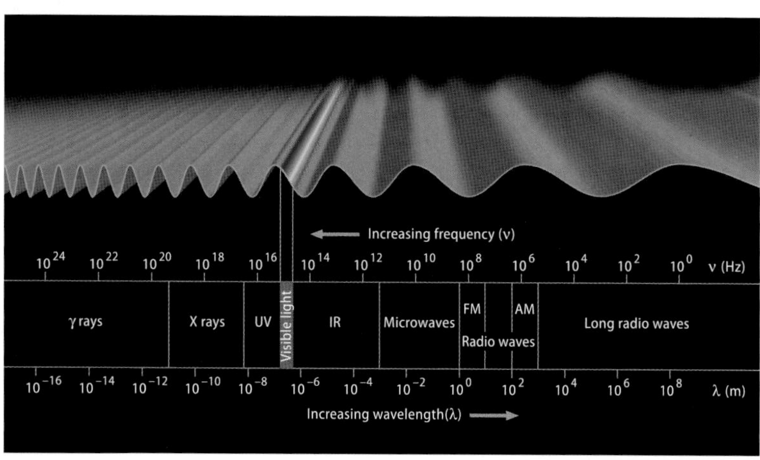

> The microwave oven is the consolation prize in our struggle to understand physics.
>
> Jason Love, humorist

THE DISCOVERY OF THE MICROWAVE OVEN

The possibility of cooking food with microwaves was discovered by scientist Percy Spencer, who worked for Raytheon, a company that built radars. Touring one of their laboratories one day, he stopped in front of an active magnetron, the part of a radar set that produces microwaves. He felt a strange sensation, and discovered that a chocolate bar in his pocket had melted. Although not the first to have noticed this, Spencer was an experienced inventor who quickly realized the possibilities of this discovery. His second experiment was with unpopped popcorn. When held in front of the magnetron, it popped right in front of his eyes. His third experiment is reported to have been an egg, which promptly exploded due to the sudden build up of pressure from being heated rapidly, an experiment that can be easily, if rather messily, repeated in today's microwaves.

The first microwave oven, the Radarange, pictured at right, was built by Raytheon in 1947. It was around 6 feet (1.8 m) tall, and weighed 750 pounds (340 kg). The first successful commercial microwave oven was not produced until two decades later, but falling costs quickly introduced microwaves as an essential kitchen appliance. Estimates put the current annual production of microwave ovens at more than 30 million units!

Gamma rays are a form of high energy emitted from the nucleus of an atom when the nucleus releases energy to move to a state of lower energy. Although damaging to living tissue, gamma rays can also be used beneficially in medicine for both treatment and diagnosis.

Gamma rays

Left: A gamma ray burst is a short-lived burst of gamma ray photons and is considered the most energetic form of light. According to NASA, gamma ray bursts shine about a million trillion times as bright as the Sun, which makes them, for a brief period, the brightest source of cosmic gamma ray photons in the universe.

Source of gamma rays

The emission of a gamma ray from a nucleus generally follows the emission of alpha or beta radiation from an unstable nucleus. The nucleus that results from the emission of alpha or beta radiation may be in a high-energy state and it moves to a lower energy state by the emission of energy in the form of gamma rays. Alpha radiation is a helium nucleus; beta radiation is either an electron or a positron and consequently in these two cases, the original nucleus is changed to another element when either of these two types of radiation is emitted. As gamma radiation is a form of energy, the original nucleus remains the same element, but in a lower energy state.

Gamma rays are also emitted when an electron and a positron interact and all of their masses are converted to energy in the form of two gamma rays moving in opposite directions. A positron is the antimatter equivalent of the electron and has the opposite electric charge to that of the electron.

Interaction with matter

Gamma ray energy is higher than the energy of X-rays, therefore a very dense material like lead or concrete is required as shielding to attenuate gamma rays. Gamma rays are known as ionizing radiation, as when they penetrate into any matter or living tissue, they interact with the atoms in the material and dislodge electrons from the atoms and molecules in the tissue. This is called ionization. If the radiation dose is high and the amount of ionization in cells is high, then death of some cells may occur; alternatively, some cells may become defective and these cells produce more defective cells when they divide.

It's a Geiger counter … Listen to it! It's going wild!

Bruce Banner, "The Incredible Hulk," after being exposed to gamma rays, Marvel Comics, 1962

Left: A Geiger counter is a portable device for detecting and measuring radiation. It works by detecting radioactive particles as they cross a glass or metal tube that has been filled with gas, resulting in ionization of the gas molecules and the prodution of an electrical discharge.

Detecting gamma rays

One of the simplest methods employed for the detection of gamma rays is to detect the ionization that the gamma rays produce in a gas. This technique is employed in a Geiger counter, which has a gas in a small tube with a wire down the center. The wire has a high positive voltage applied to it with respect to the tube. When gamma rays enter the tube, ionization occurs—the electrons are attracted by the

Right: This high voltage generator at the Sandia plant, Albuquerque, New Mexico, USA, photographed in 1969, produces 5 billion trillion watts by simulating the flash of gamma radiation. At 80 ft (24 m) long and 22 ft (6.5 m) in diameter, it is one of the largest of its kind in the world.

positive voltage and the positive ions are attracted by the negative voltage. This triggers the other atoms with which they collide to become ionized. The consequence is that a current is generated for a very short time; this can be used to trigger a counter or to generate a click on a loudspeaker.

Radioactivity

Nuclear radiation was first discovered in 1896 by Henri Becquerel when he noticed that a uranium compound that had been left in a drawer on a photographic plate darkened the developed plate. The term radioactivity was coined later by Marie Curie. Researchers determined that the radiation was of three types, called alpha, beta, and gamma radiation. The SI unit of the activity of a radioactive sample was given the name of the person who discovered radioactivity and it is the Becquerel (Bq) that represents one radioactive decay per second. In this case, the word decay represents a nucleus that is undergoing a change. This unit replaced the commonly used unit of the Curie (Ci) which is 37,000 billion decays per second and is based on the activity of a gram of pure radium.

Above: Marie Curie (1867–1934) is one of the great names in science. A Curie (Ci), named for her and her husband Pierre, is a measure of the radioactivity of a material equal to 37 billion decays per second. It has been replaced by the Becquerel (Bq).

RADIATION THERAPY

Gamma rays are used in radiation therapy to kill cancerous cells in a tumor. A very sophisticated form of radiation therapy is the gamma knife, pictured below. In this technique, very narrow beams of gamma rays are directed at a tumor from multiple directions. These beams are aimed to meet at the location of the cancer. Consequently, the cancerous growth receives a high radiation dose in order to kill the cancerous cells, while the surrounding tissue receives a much lower dose in order to minimize the damage to the healthy tissue.

X-rays—their discovery dramatically changed how we see the world. X-rays are now one of the most important tools in medical diagnostics and materials investigations. Doctors use X-rays to detect tuberculosis and heart diseases; scientists and industry use them to study how to clean up environmental pollutants or improve production processes for the next generation of cars.

Finding the invisible—X-rays

On November 8, 1895, German physicist Wilhelm Conrad Röntgen was experimenting with passing electric discharges through partially evacuated glass tubes. When he covered a tube with black paper to block out the visible light of the discharges, astonishingly, a nearby fluorescence screen lit up. Something invisible was penetrating the black paper, causing a reaction in the fluorescence screen. Röntgen had just discovered X-rays.

A boon for medicine

Although light in a general sense, X-rays are invisible to the eye. Furthermore, they can readily shine through solid matter; a doctor will examine an X-ray image of a broken bone to determine the best healing therapy. X-rays were used in medicine quite soon after their discovery. Health practitioners used X-rays to treat some forms of cancer as early as the 1920s and to diagnose tuberculosis routinely from the 1930s. The medical use of X-rays has, of course, advanced dramatically over the years.

Versatile X-rays

A striking example of X-ray technology is called "computed tomography" (CT). X-ray images of a patient or an object are taken from various angles. Each image provides a two-dimensional projection in one direction. Scientists then computer-generate a full three-dimensional representation, which can be sliced open in any way without touching the patient or the object.

X-rays have wavelengths that correspond to the distances between atoms in matter, so they are ideal to unravel the structures of crystals. A technique known as X-ray diffraction is used for structure determination. It is an important tool for geologists, materials scientists, the minerals industry, and many others.

In biotechnology, researchers use X-ray diffraction to analyze the structures of proteins, essential building blocks of life. This work is of great importance in the development of new pharmaceuticals.

Right inset: This is one of the first X-ray photographs. Taken on January 23, 1896 by the German physicist Wilhelm Conrad Röntgen (1845–1923), it shows his wife's hand. The first X-rays were given the name "röntgenograms" after their discoverer.

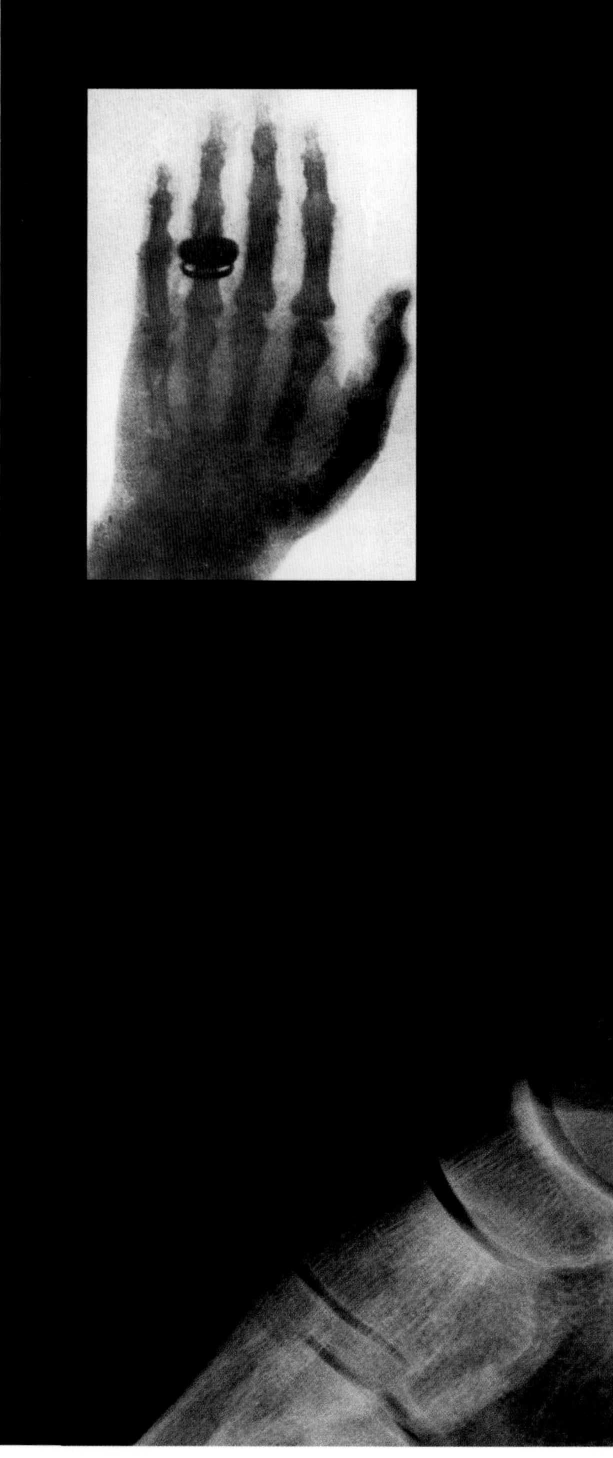

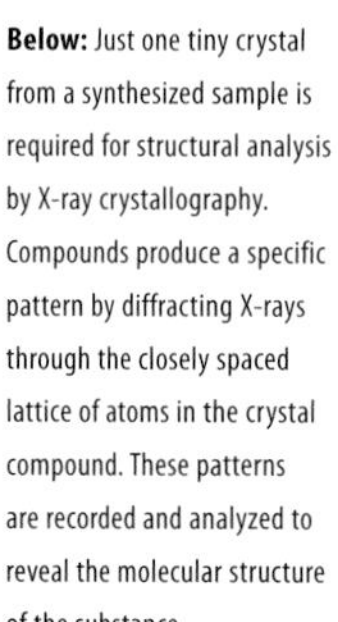

Below: Just one tiny crystal from a synthesized sample is required for structural analysis by X-ray crystallography. Compounds produce a specific pattern by diffracting X-rays through the closely spaced lattice of atoms in the crystal compound. These patterns are recorded and analyzed to reveal the molecular structure of the substance.

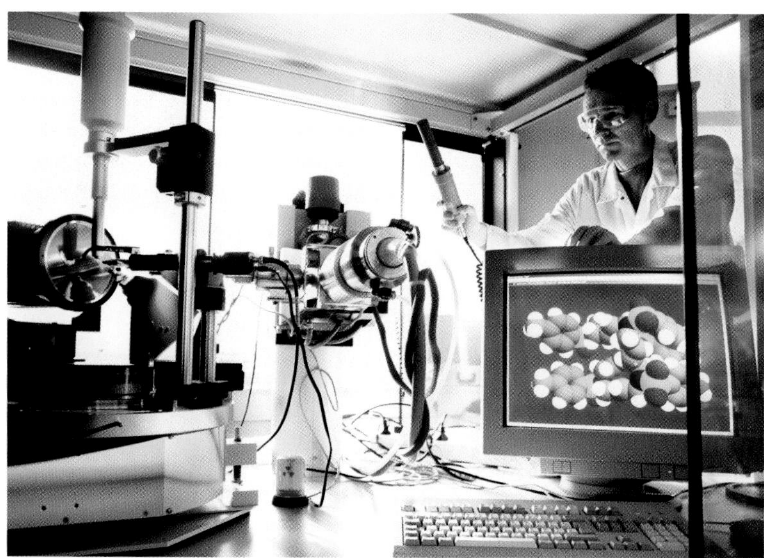

Left: X-ray technology is the most common form of baggage screening at airports and other venues where strict security is necessary. The X-rays can detect explosives, firearms, and drugs.

The uses of X-rays extend far beyond the sciences of medicine and biotechnology. X-rays screen luggage at airports to detect drugs and weapons. With X-ray imaging engineers can locate microscopic cracks in welds, making construction safer. Industry uses X-ray spectroscopies to improve the efficiency of catalysts. Galleries and museums use X-ray technology to analyze artefacts and objects to discover masterpieces hidden under other paintings.

Keeping it safe

While extremely useful in a number of areas, X-rays are ionizing radiation and can pose health hazards. In the early days most people were unaware of the hazards, although the potential harmful effects of X-rays were discussed as early as 1898. Today, radiation protection laws around the world strictly regulate the use of X-rays. Fortunately, X-rays are readily absorbed by concrete or heavy metals such as lead. Doctors and dentists use lead shielding to protect the parts of your body other than the area exposed to X-rays.

> Mozart's music is like an X-ray of your soul—
> it shows what is there, and what isn't.
>
> Isaac Stern, violinist, 1920–2001

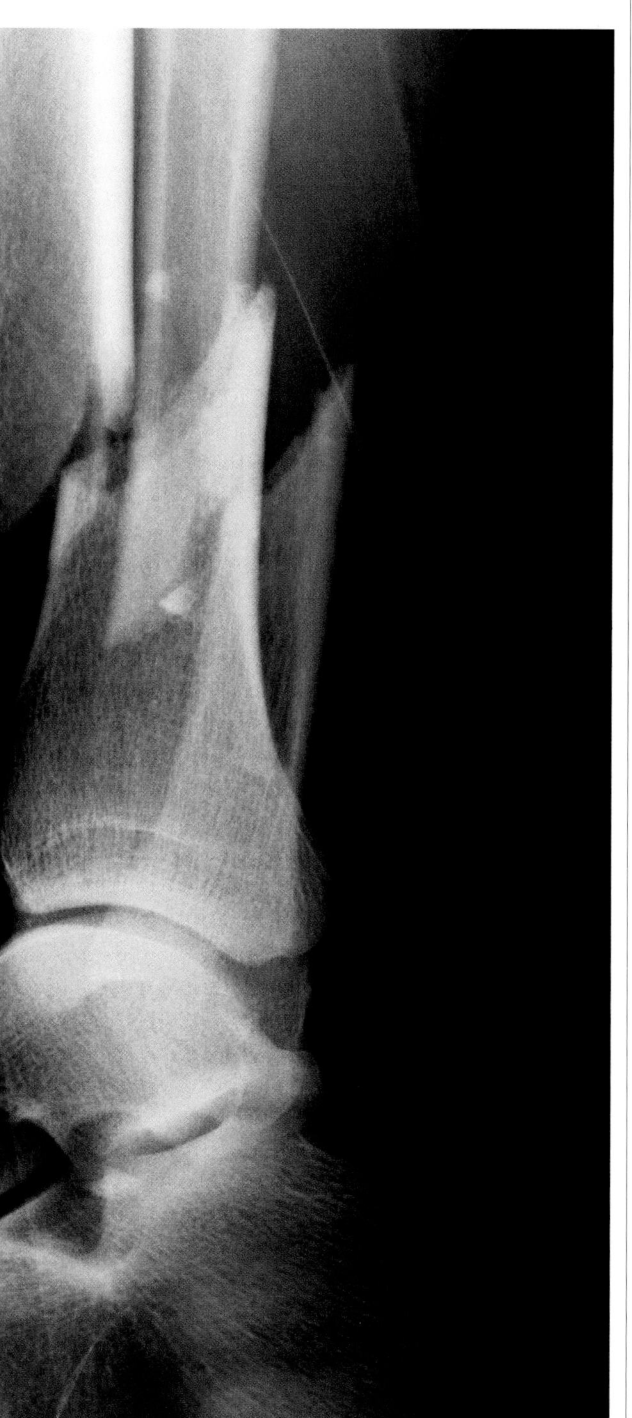

Above: The discovery of X-rays by Wilhelm Conrad Röntgen earned him the very first Nobel Prize for Physics in 1901. Röntgen also worked on the specific heats of gases and the conduction of heat in crystals.

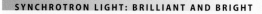

SYNCHROTRON LIGHT: BRILLIANT AND BRIGHT

X-rays millions of times brighter than a conventional X-ray tube are produced at a synchrotron facility (below). Large machines accelerate electrons almost to the speed of light, and big electromagnets keep the electrons on a circular path. Each magnet combines the strength of 10,000 fridge magnets. Synchrotron light contains X-rays so bright that experiments can be completed in minutes, rather than weeks using a normal X-ray tube. For instance, you can follow chemical reactions in a catalyst in real time, while the catalyst is working. Such process monitoring is useful for making catalysts more efficient.

Synchrotron light has much to offer. The X-rays (and the rest of the light) come in a continuous spectrum, and you can select specific wavelengths for an experiment. This allows materials to be investigated with very high accuracy. For example, proportions of toxic and non-toxic chromium compounds in contaminated soils can be analyzed. This is important to develop remediation techniques for a cleaner environment.

Synchrotron light can also be narrowed to very fine beams, allowing images of very high quality to be recorded. The intensity of the light speeds up the analyses. Synchrotron imaging is very important for many fields, including medicine.

About 40 synchrotron facilities exist worldwide, and the network is expanding. Synchrotron light is a key for modern research with X-rays, offering almost limitless possibilites.

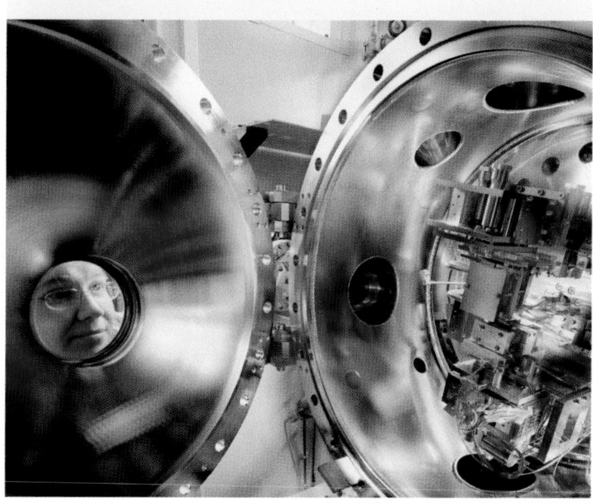

Left : An X-ray of a broken tibia and fibula in a human leg. Because tissues absorb X-rays at different rates, different parts of the body appear light or dark. Bones look white because calcium is a good absorber of X-rays. Soft tissue and fat deposits absorb less, so they appear gray.

Ultraviolet rays and infrared are two forms of electromagnetic radiation that influence our lives every day. Ultraviolet rays are shorter (and higher in frequency) than those associated with the color violet in the visible light spectrum. Infrared radiation lies between the visible light and microwave portions of the spectrum, just after red visible light.

What a spectrum—ultraviolet light and infrared

UV—origins from the Sun

While we cannot see ultraviolet light with the naked eye, many of us are familiar with the painful effects of sunburn, a result of ultraviolet radiation. Yet not all UV is alike. Ultraviolet is defined as having wavelengths of less than 400 nanometers, and is divided into three categories, according to wavelengths.

UVA radiation has wavelengths between 315–400 nanometers, and is the most common form of UV radiation to reach Earth. It is a contributing factor to city smog, and is known to cause damage to synthetic materials such as plastics, paint, and fabrics. This category of UV is associated with cancers.

UVB has wavelengths of 290–315 nanometers, and presents the greatest risks to us. Although it makes up only 1 percent of the ultraviolet radiation produced, UVB has been shown to cause significant damage to human skin.

UVC radiation is the shortest of the ultraviolet categories with wavelengths between 220–290 nanometers. This form of ultraviolet is very readily absorbed by the ozone layer surrounding Earth's atmosphere, so is not generally found on Earth's surface.

Some useful applications of UV

While UV may be associated with some negative health implications, there are many positive uses for it. One particularly useful application is as a sterilization tool. Due to its ability to alter the structure of DNA, UV is commonly used to destroy unwanted bacteria. Medical and biological laboratories often use UV-based lamps to disinfect and sterilize medical implements.

People with a tendency to develop skin-related disorders such as psoriasis or eczema have also found a use for UV. Exposure to UVB, also known as phototherapy, has been found to reduce the reliance on topical skin treatments for relief from these skin-based disorders.

What is infrared?

What does channel surfing, discovering new galaxies, and military surveillance have in common? They have all been revolutionized by the application of infrared radiation. Infrared is a type of radiation, its primary source usually being heat or thermal radiation. Radiant heat, such as the heat that can be felt from a radiator or the energy emitted from concrete after a hot day, is infrared. This radiation is produced by the motion of molecules and atoms within an object. The hotter an object, the faster its molecules and atoms will move and the greater amount of infrared radiation it will emit. Any object with a temperature of more than absolute zero will emit infrared radiation.

Infrared, as with all forms of radiation, travels at the speed of light: 186,000 miles (300,000 km) per second in a vacuum. Infrared has longer and lower frequency wavelengths than those of visible light, but shorter and higher frequency wavelengths than microwaves.

There is not an exact point on the electromagnetic scale where visible light becomes infrared, since the electromagnetic scale is a smooth continuum of changing wavelengths. Generally, infrared is found at wavelengths from 0.7 micrometers (μm) to 1 millimeter. Equipment using infrared radiation is usually designed to detect a small window of infrared wavelengths.

Above: William Herschel discovered infrared radiation in 1800. Best known for his discovery of Uranus in 1781, Herschel noticed that telescope filters of different colors seemed to transmit different amounts of heat. His experiments showed that infrared could be reflected, refracted, absorbed, and transmitted in the same way as visible light.

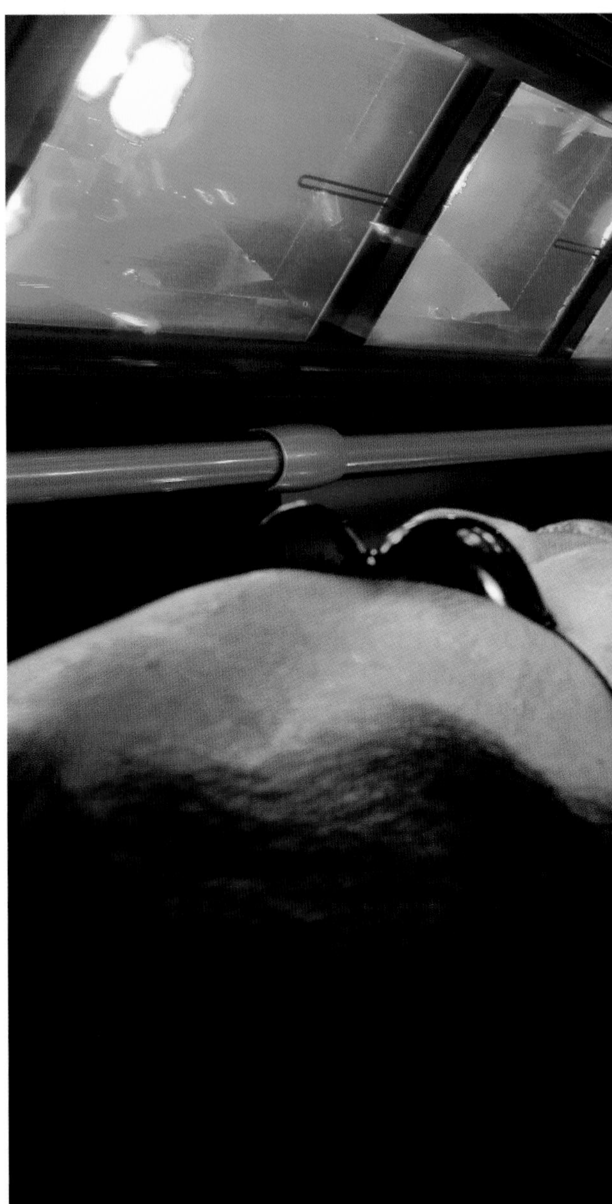

Right: In a tanning booth, UV light bulbs emit radiation that causes tanning. Tanning booths emit UVA radiation, the form of ultraviolet light that is closely associated with an increased risk of skin cancer.

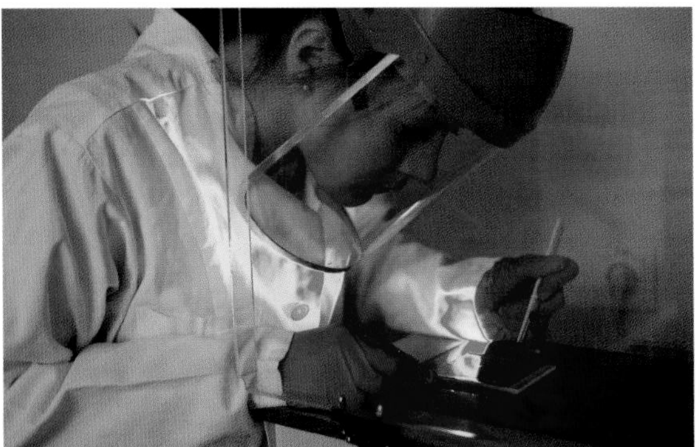

Left: A scientist uses ultraviolet light to isolate DNA in order to detect bacteria or viruses, or to diagnose a genetic disorder. The presence of DNA can be detected by electrophoresing on a fluorescent dye that reacts with the DNA. This is then examined under UV light.

APPLICATIONS OF INFRARED

Infrared imaging, often called thermal imaging, is one of the most important and most used applications of infrared. Infrared images not only reveal the existence and position of an object emitting infrared radiation, they also indicate the temperature of the radiation being emitted. Infrared images are used in meteorology (below), land and water management, ecology, in military operations, surveillance, and medical imaging to name but a few. Art historians use infrared imaging to look through the layers of a painting to reveal the artistic process used. Recent infrared analysis of the *Mona Lisa* has shown that the initial painting by Leonardo da Vinci is quite different from the finished product. Similar analysis of the Dead Sea Scrolls is revealing the text, which is otherwise illegible on the 2,000-year-old papyrus.

Right: An aviation systems warfare operator searching for and tracking surface contacts using radar and the Infrared Detection System (IRDS) of his P-3C Orion patrol aircraft. An infrared detection system works by transmitting a beam of infrared rays toward possible targets; the rays reflected from the target are then detected.

What appeared remarkable was that when I used some of them, I felt a sensation of heat, though I had but little light; while others gave me much light, with scarce any sensation of heat.

Sir William Herschel, astronomer, 1738–1822

Although Alexandre Becquerel (1820–1891) was the first person to observe the photoelectric effect, it was Heinrich Hertz in 1887 who showed that a spark could be made to pass between two electrodes at a lower voltage if the electrodes were illuminated with ultraviolet light.

The photoelectric effect

It remained to J. J. Thomson in 1898 to discover that electrons are emitted by a metal surface when it is irradiated with UV light. This is known as the photoelectric effect. Different metals begin to emit electrons at different wavelengths of incident light.

The explanation for the effect came in 1905, a momentous year for Albert Einstein in which he published a series of papers, covering such diverse subjects as the special theory of relativity and the photoelectric effect. Study of the photoelectric effect led to important steps in understanding the quantum nature of light and electrons and influenced the formation of the concept that all matter exhibits both wave-like and particle-like properties.

Above: German physicist Heinrich Hertz (1857–1894) unintentionally discovered the photoelectric effect, whereby ultraviolet radiation releases electrons from the surface of a metal. He didn't pursue this, and the effect was later explained by Einstein.

The photoelectric effect and solar cells

The technique of photoelectron spectroscopy is a powerful analytical tool for studying the surfaces of materials. The analysis must be conducted in a vacuum because electrons are scattered by air.

A variety of devices depend on the photoelectric effect of which the solar cell is very important. It has been reported that the sunlight that reaches Earth's surface delivers 10,000 times more energy than we consume. Solar cells offer the promise of tapping into this endless energy source. Solar cells depend on the photoelectric effect to produce electricity from the Sun.

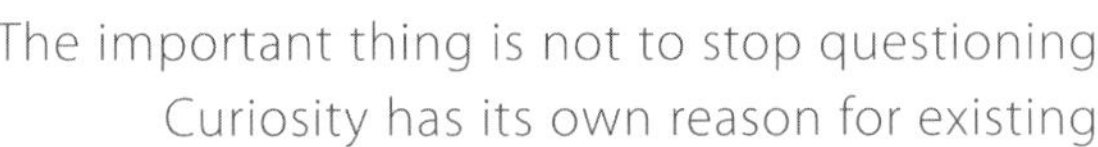

The important thing is not to stop questioning. Curiosity has its own reason for existing.

Albert Einstein, physicist, 1879–1955

Solar cells are composed of various semiconducting materials. Semiconductors are materials that become electrically conductive when supplied with light or heat, but which operate as insulators at low temperatures. Semiconductors can absorb light of relatively low energy, such as visible light, to promote the electrons into a higher energy conduction band, without ejecting them, where they can be harnessed to create electric current. Most the solar cells are composed of the semiconductor material silicon. Fortunately, silicon is the second most abundant element in Earth's crust. In order to produce a solar cell, the semiconductor is contaminated or "doped." "Doping" is the intentional introduction of chemical elements, with which one can obtain a surplus of either positive charge carriers (p-conducting semiconductor layer) or negative charge carriers (n-conducting semiconductor layer) from the semiconductor material. If two differently contaminated semiconductor layers are combined, then a so-called p–n-junction results on the boundary of the layers.

Like other processes for harnessing energy there are some limits to the efficiency of solar cells. Because the different semiconductor materials or combinations are suited only for specific spectral ranges, a specific portion of the radiant energy cannot be used. On the other hand, a certain amount of surplus photon energy is transformed into heat rather than into electrical energy. There are also some optical and mechanical losses. The theoretical maximum level of efficiency is approximately 28 percent for crystal silicon. In practice, efficiency is much lower. At present the biggest hurdle to the use of solar power is the cost compared with traditional sources of energy; but that could change in coming years.

Left: The production of pure silicon rods of metallurgical-grade silicon are mechanically broken into chunks and then melted at 2593° F (1423° C) to produce pure silicon used for computer chip production. Here the chunks are beginning to be melted into liquid form.

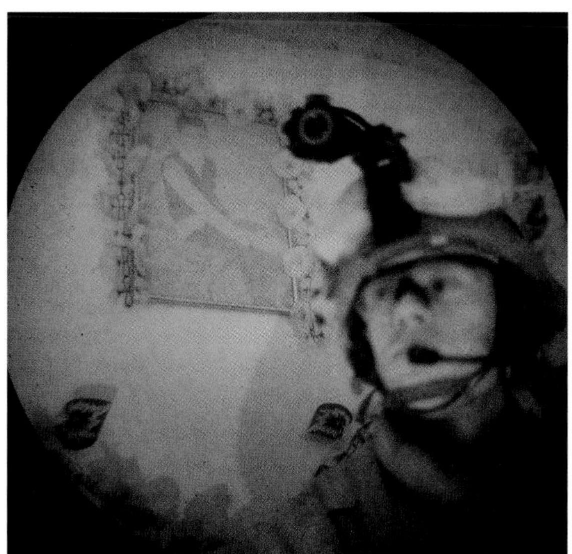

Above: Night vision devices, such as goggles, rely on the photoelectric effect, where photons collide with a metal plate. This produces electrons which are amplified so that more electrons are produced, with the result that the screen lights up.

Below: Solar power panels, such as these in California, USA, use the photoelectric effect to convert light into electricity. When light hits certain materials, electrons are released. Those electrons then create a current that flows through the material and provides power.

WAVE–PARTICLE DUALITY OF MATTER

Contrary to ideas of the day on the wave-like nature of light, Einstein assumed that the light impinging on a metal consists of light quanta or discrete energy packets called photons with energy hf, where h is a constant called Planck's constant, and f is the frequency of the incident light. When the metal absorbs the light, all of the energy of one photon is converted into energy of the photoelectron. For the electron to escape from the metal it requires a certain energy, represented by the symbol Ei. The remaining energy is kinetic energy ($\frac{1}{2} mv^2$) of the escaping photoelectron: hence the Einstein photoelectric equation:

$$hf = Ei + \tfrac{1}{2} mv^2$$

Einstein's view seemed at odds with the wave theory of light and the assumption of infinite divisibility of energy in physical systems. For some time his views were not accepted. Although his equation predicted that the energy of the ejected electrons increases linearly with the frequency and is independent of the intensity of the light, it was not until 1915 that Robert Millikan (pictured between Einstein and his wife) experimentally verified Einstein's predictions.

One day, Albert Einstein asked himself what would happen if you were to run alongside a light beam? What would it look like? The answer could not be explained by the current theories of physics, so Einstein developed his theory of special relativity, turning our understanding of the universe upside down.

Einstein's special theory of relativity

Reference frames

Reference frames are central to understanding Einstein's theory. We experience the concept many times each day—say, in a moving train or a car. A reference frame is how the world looks from where you are standing. Imagine standing on the side of a freeway, watching a car drive past at 70 miles per hour (112 km/h). In your reference frame, the road, trees, and shrubs appear stationary, and the vehicle appears to be moving at 70 miles per hour (112 km/h).

Now imagine the same scene, but you are in a truck, moving at the same speed, in the same direction as the car. Your reference frame has changed. Now, the car appears stationary, and the road, trees, and shrubs are moving at 70 miles per hour (112 km/h).

Finally, imagine you are in a truck traveling at 70 miles per hour (112 km/h), in the opposite direction to the car. The car appears to be moving at 140 miles per hour (225 km/h), and the road, trees, and shrubs are moving at 70 miles per hour (112 km/h) in the opposite direction to the previous situation.

In these scenarios, the only thing that has changed is your frame of reference. To put Einstein's question into different terms—what would a light beam look like if your reference frame were traveling at 671,080,888 miles per hour (1,080,000,000 km/h)—the speed of light? The answer requires the theory of special relativity.

Above: Albert Einstein was arguably one of the greatest scientific minds of all time. He had a deep understanding of the problems of physics and spent his life attempting to come up with solutions for them.

> When you are courting a nice girl an hour seems like a second. When you sit on a hot stove a second seems like an hour. That's relativity.
>
> Albert Einstein, physicist, 1879–1955

The basis of special relativity

Special relativity is based on two postulates. The first is that *the laws of physics hold true in all reference frames*, so no matter what speed you are moving, you will always measure a meter to be the same length. Note that we assume the reference frames are traveling at a constant velocity, and not accelerating. The second postulate is that the speed of light is *the same in all reference frames*.

The reason we consider that the speed of an object changes as our relative speeds change (that is, as we travel faster or slower compared with the object, as with the example of the car) is because we assume that distance and time are constants, no matter how we look at them. A meter and a second to an observer watching a car are the same as a meter and a second to the driver of the car, no matter how fast or slow the car is driving.

However, if we are now assuming that the speed of light remains constant in all reference frames, one of these assumptions must be wrong. Remember, speed = distance/time.

Either distance or time must change as the relative speeds of two objects change, for the speed of light to remain constant. That is, either a meter or a second must be different for an observer compared with the driver of the car. In fact, *both* distance and time vary for the observer compared with the driver.

Opposite: One of the more interesting consequences of the special theory of relativity is that space bends around objects of large masses, such as planets and stars.

Left: A reference frame is how the world looks from where you are standing. If you are sitting in a truck, watching a car traveling in the same direction, it looks as if the car is stationary. Frames of reference were used by Einstein to explain his special theory of relativity.

TIME DILATION AND LENGTH CONTRACTION

Time dilation and length contraction are strange phenomena that come with Einstein's theory of special relativity. These two fascinating concepts refer to the differences in length and time measured by, for example, a stationary observer and a person moving at speed. They are best explained with an example.

An observer sits on the sidelines of a racetrack, timing a car travel around the track. The car is 20 feet (6 m) long. When it is stationary, the observer agrees that the car is 20 feet (6 m) long. However, the faster the car travels, the shorter the observer measures it to be. The observer sees the length of the car shrink in the direction it is traveling! If it were capable of traveling at the speed of light, the driver would still measure his car to be 20 feet (6 m) long, but the observer would not see it. He would observe the length of the car to be zero feet!

What you would see if a car travelled at the speed of light

Speed = stationary

Speed = half the speed of light

Speed = the speed of light

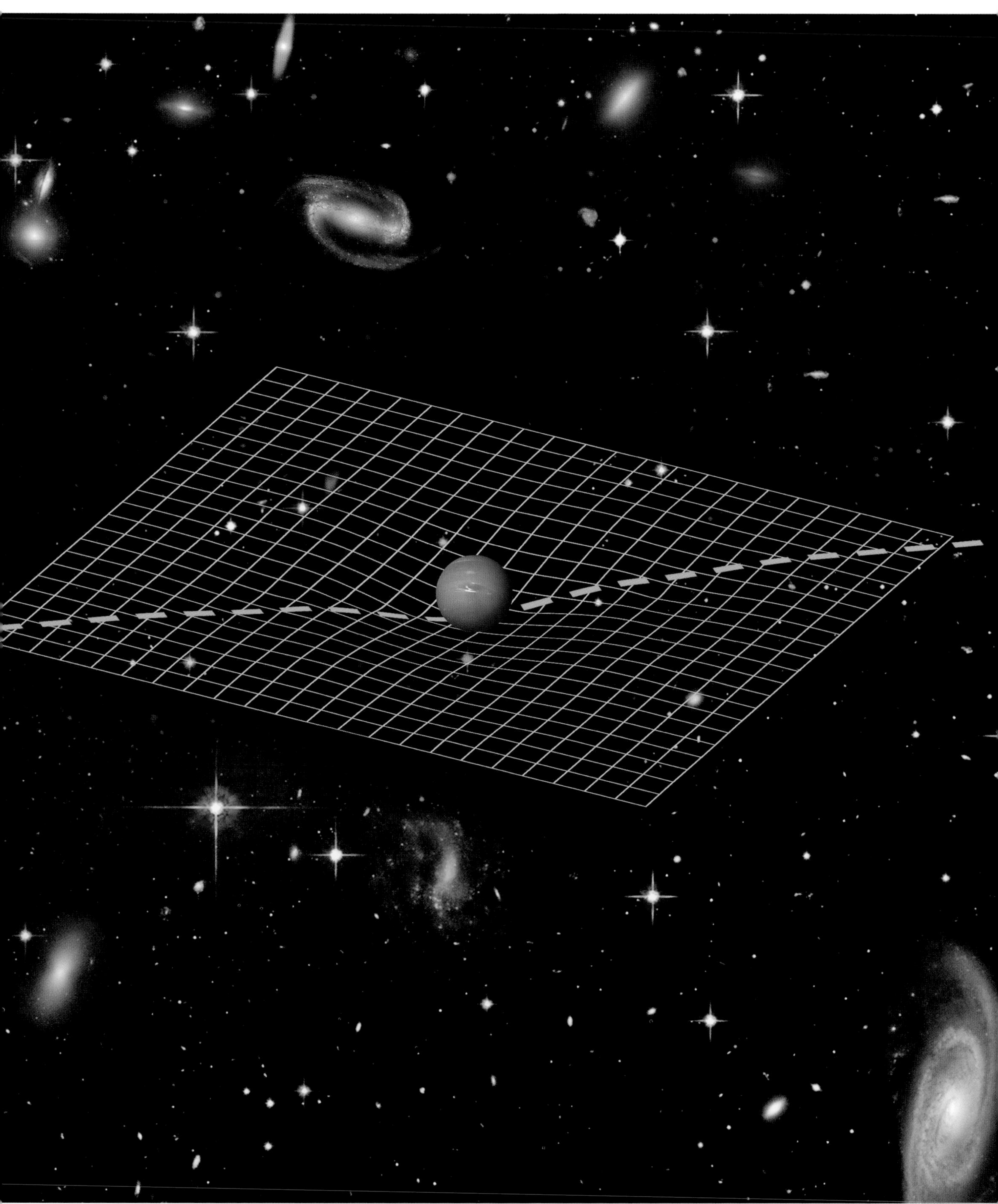

General relativity is a theory of gravity developed by Albert Einstein, in an attempt to incorporate gravity and acceleration into his theory of relativity. When published in 1915, it revolutionized scientists' understanding of space and time, and has gone on to become one of the most thoroughly tested theories of modern physics.

Einstein's general theory of relativity

Left: Albert Einstein and his wife Elsa, *c.* 1935, photographed in Egypt. In 1999, *Time* magazine named the brilliant physicist the "person of the century."

Not only is it an essential part of astrophysics, helping us to understand both the formation of our universe and the death of stars, it has also proven essential for technologies such as GPS.

Einstein published his theory of general relativity in 1915, ten years after his paper on special relativity, motivated by the need to combine gravity into the startling new picture of our world that his earlier theories had produced.

In the same way that special relativity changed our ideas about the existence of absolute time and speeds, general relativity changed our understanding of gravity and acceleration, primarily through what is known as the equivalence principal.

In this famous thought experiment, Einstein imagined a person in "free fall," such as what can now be experienced on some theme park rides where a carriage drops for a time without any brakes or restraints. In Einstein's case, he imagined a lift free-falling down an elevator shaft with a person inside it. The person would feel weightless, as both he and the objects around him would be falling and accelerating at the same rate as the floor below. If he was unaware he was in a falling elevator, he might instead believe that he was in sealed box drifting through space, away from any source of gravity.

Equivalently, passengers in a rocket ship (or indeed an elevator) out in space, away from Earth, might be tricked into thinking they were experiencing gravity by continually accelerating the rocket ship.

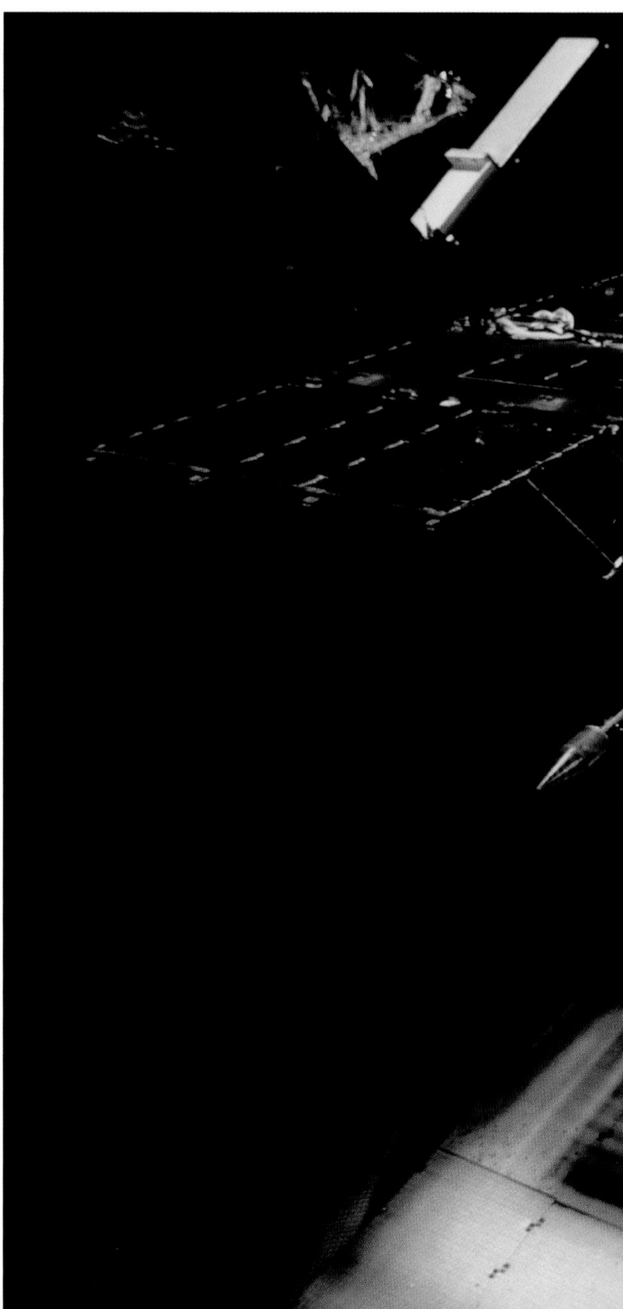

Right: A global positioning satellite is checked out at the Arnold Engineering Center in Tullahoma, Tennessee, USA. GPS systems would not work if relativity was not taken into account.

> I don't understand my husband's theory of relativity, but I know my husband and I know he can be trusted.
>
> Elsa Einstein, 1876–1936

EINSTEIN'S PAPERS

Albert Einstein, arguably the greatest physicist in history, is best known for his theories of relativity, which revolutionized our understanding of space and time. Yet this work was neither his most referenced paper, nor the work that earned him the Nobel Prize. In fact, at the same time as he published his paper on special relativity in 1905, Einstein produced three other equally significant works.

The first was a paper explaining Brownian motion—the random motion of very small objects when placed in water—as direct evidence for atoms and molecules, and it became his most cited work. Einstein also published a paper outlining his famous equivalence between energy and matter, $E = mc^2$, where c is the speed of light (below).

Finally, Einstein published the paper for which he received the Nobel Prize in Physics in 1921, on the nature of light. He showed that several puzzling experimental results could be explained if light only interacted with matter in discrete packets, called photons. This later formed the foundation of quantum mechanics, the theory explaining matter on the very small scale. And all this while working as a patent officer in Bern!

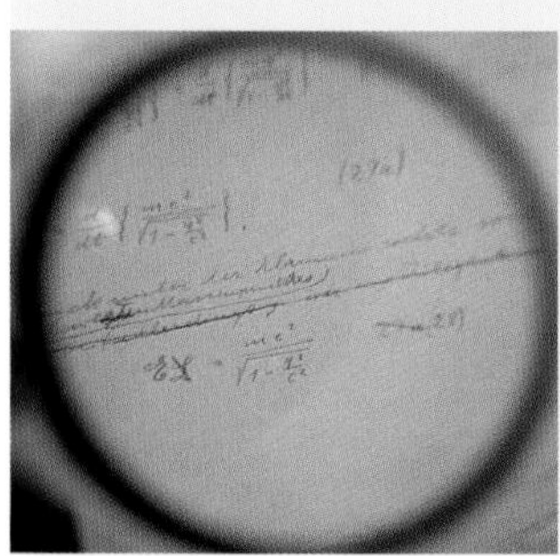

Above: During a solar eclipse in 1919, Arthur Eddington demonstrated that light from distant stars bent as it traveled through space–time. This was in agreement with Einstein's theory of relativity and brought Einstein's name and work to the public's attention.

Just as we feel ourselves pressed into our seats when accelerating a car, passengers in a floating elevator accelerating upwards would feel themselves pushed into the floor. With the right acceleration, it would be indistinguishable from Earth's gravity. This basic idea is used by science fiction writers to create giant circular space stations whose constant rotation provides a sort of artificial gravity for its occupants.

Einstein's great insight in 1907 was to hypothesize that these represented a fundamental principle of the universe—the laws of physics should appear exactly the same to a weightless experimenter as to one who was in free fall, such that no experiment could tell the two situations apart. This meant there was no fundamental difference between gravity and an acceleration, and experiments should give the same results regardless of which situation actually applied.

Newton and space–time

It took Einstein eight years to develop this idea into a fully-fledged theory. His second breakthrough was to realize that the principle of equivalence was incompatible with the traditional notion of gravity as described by Newton. The fact that acceleration and gravity could appear identical meant that gravity being just a force between two masses could not fit.

Above : British astronomer Sir Arthur Eddington (1884–1944) was one of the first scientists to recognize the significance of Einstein's theories of special and general relativity.

Instead, building on work of earlier scientists such as Hermann Minkowski, Einstein proposed that what we perceive as gravity is in fact a warping of space and time itself caused by objects in it, and introduced the concept of space–time.

The traditional picture of gravity as a force like magnetism was abandoned. Instead, Einstein imagined a universe where everything attempted to travel in a straight line. Each object's mass, however, would bend the space around it a little, so that this supposed "straight line" might in fact include bending around or even circling other objects—just like a planet in orbit around the Sun. The more massive the object, the more it would bend space and the more curvy the paths of passing objects would be, the same as would have been predicted by a stronger gravitational force.

Clearly, this was a radical idea, and was initially approached very cautiously by other scientists. Soon, however, experimental evidence began to accumulate which could not be ignored.

Tests of general relativity

In most cases, the predictions of general relativity are very close to what would be predicted by Newton's theories—an essential criterion, since Newton's theories had served well for over 200 years.

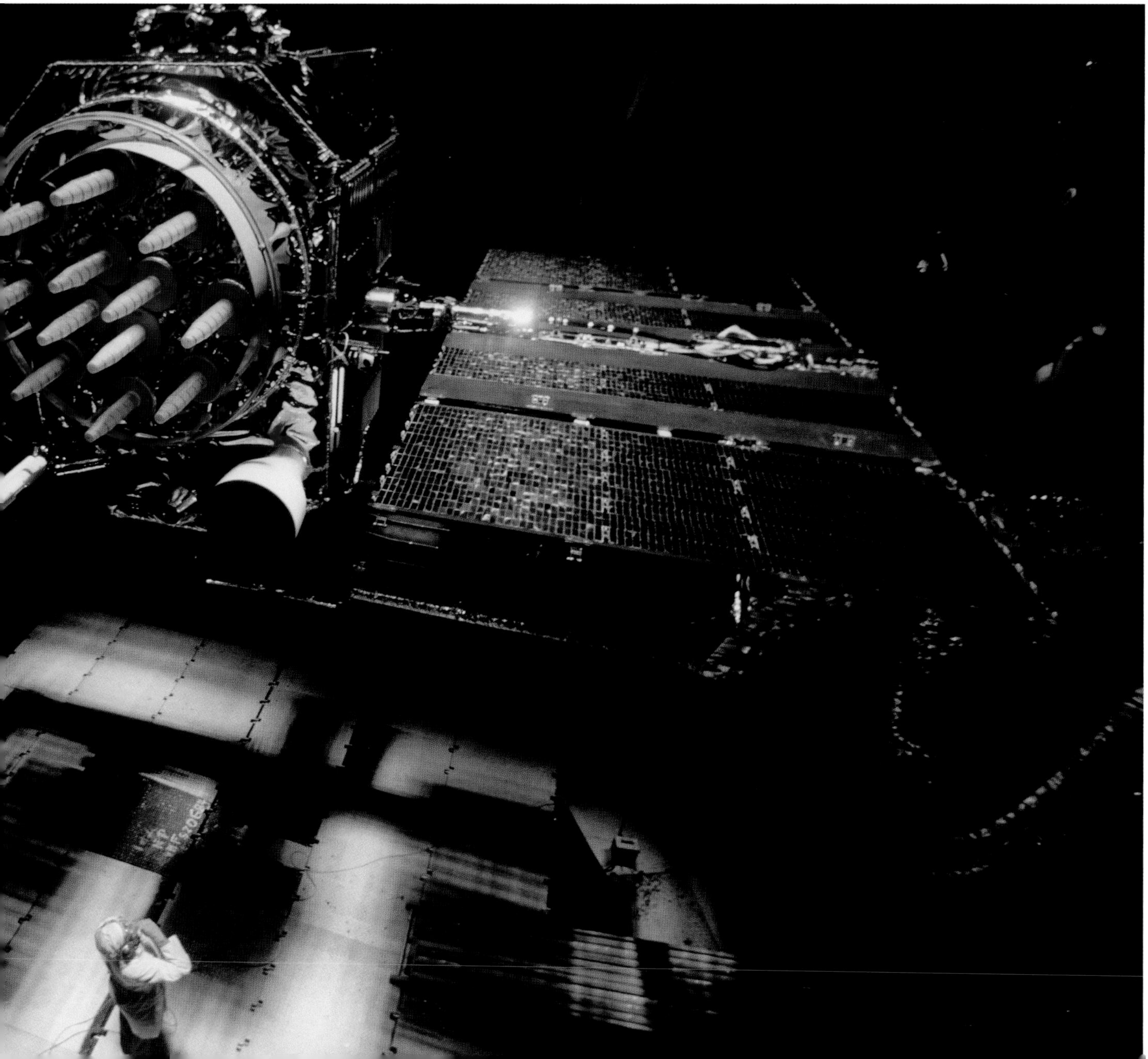

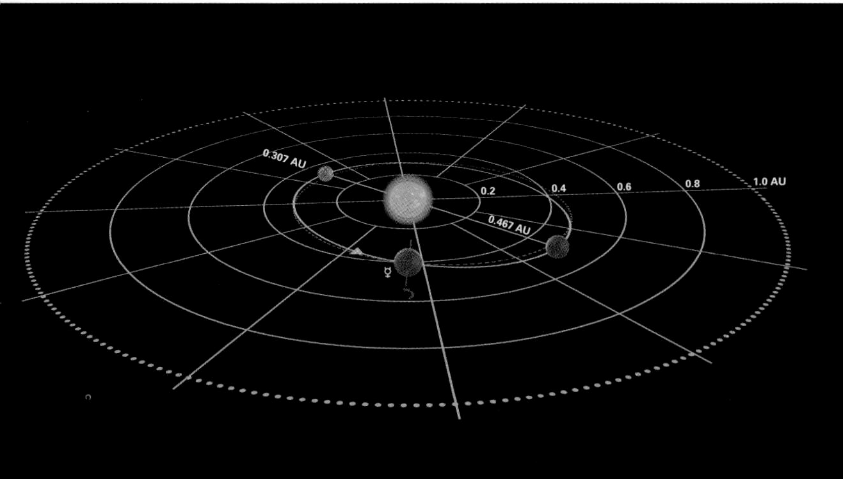

Above: The discrepancy between the motion of Mercury and Newton's theory was rectified by Einstein's general theory of relativity. Mercury takes 88 Earth days to complete an orbit of the Sun. The dash line near the pink circle indicates the degree of tilt of Mercury's orbit from the plane of the Solar System (not to scale).

However, Einstein made several key predictions where general relativity produced different results and these have been confirmed to high levels of accuracy.

The first was a well-known discrepancy between the motion of Mercury and what was predicted by Newton's theory. When general relativity was included, the predictions matched experiments exactly, and this was a powerful argument in support of general relativity.

Relativity (and Einstein's) popularity soared, however, after an experiment conducted during the solar eclipse of 1919. Using simultaneous observations, from Brazil and Africa, of stars close in the sky to the eclipsed Sun, Sir Arthur Eddington demonstrated that light from distant stars was bent as it traveled through the Sun's curved space–time. The precise agreement with general relativity made news headlines around the world, and brought general relativity to the public.

These and other predictions have made general relativity one of the best-tested theories of modern physics, finding uses ranging from GPS technology to predicting the ultimate fate of the universe.

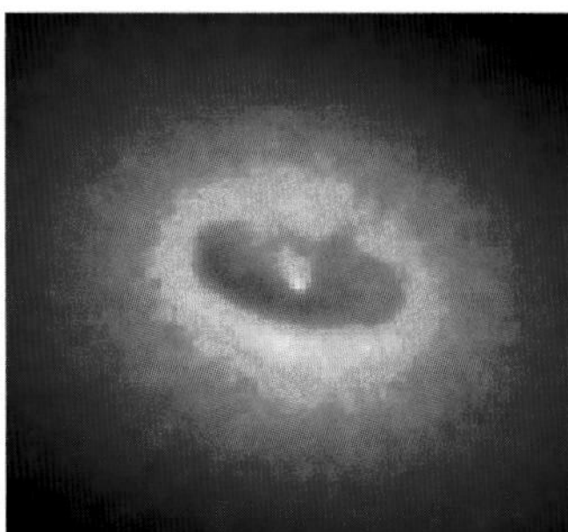

Above: A Hubble Space Telescope image of a giant disk of cold gas and dust fueling a possible black hole at the core galaxy NGC 4261. General relativity predicts that time runs slower near a black hole.

Gravity, time and GPS

In special relativity, the rate at which time passes for a person depends on their speed. A traveler in a rocket ship, or even a plane, ages more slowly compared with someone who is stationary. A similar effect is predicted by general relativity—the stronger the gravity, the slower time passes relative to someone in weaker gravity. Each person, however, would think their own time ran at the "correct" rate (they would see their own watches tick off the seconds correctly) and that the other person was ageing faster or slower.

On Earth this effect is very small, but it can still be measured by high precision atomic clocks, capable of measuring billionths of a second intervals. Clocks sent high above Earth's surface in rockets or planes experience less gravity and indeed run measurably faster compared with clocks left on the surface.

Even this small difference, however, is significant enough to affect the Global Positioning System (GPS) and must be accounted for. GPS works by different satellites sending out timing signals which are picked up by a GPS unit on Earth's surface. By comparing the delay from the different satellites, the unit is able to triangulate its position, often to within 3 feet (90 cm). If, however, the effects of general relativity were not included, the satellites would not be synchronized and this triangulation would fail.

On an astronomical scale, general relativity predicts that near an incredibly massive and dense object, such as a black hole, time would run so much slower that an explorer who spent just one hour nearby might return to find 100 or even 1 million years had passed.

Right: An artist's impression showing a black hole and its yellow companion star being sent out on a long journey through the Milky Way galaxy by the explosive kick of a supernova, one of the universe's most titanic events. Einstein's theory explained the very existence of black holes.

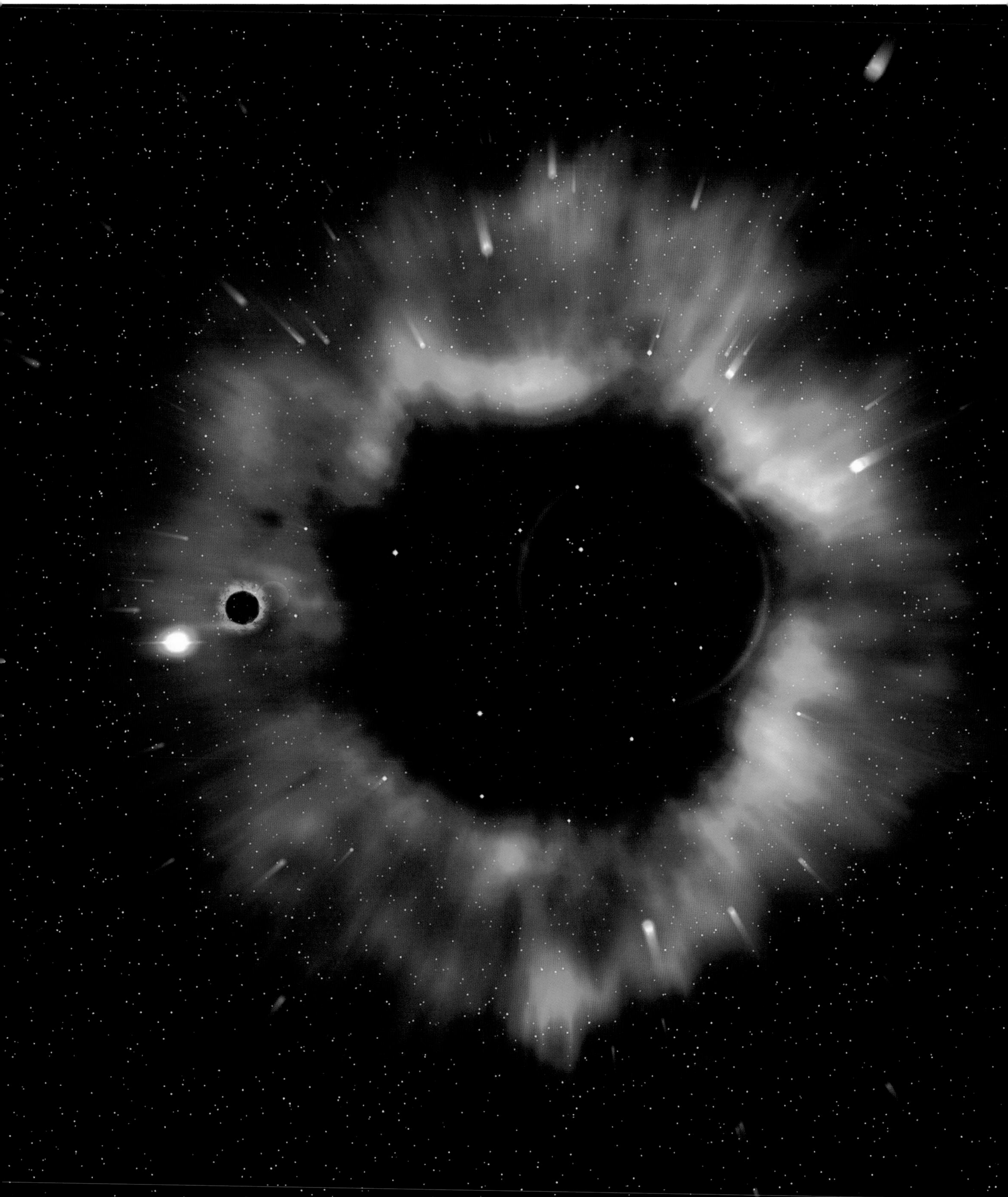

For centuries, scientists had been developing science as a tool to be able to specifically measure and quantify the world around them. Quantum mechanics turned this view of the world on its head.

Quantum mechanics

Quantum physics had it roots at the turn of the nineteenth century. German physicist Max Planck, after trying to determine why hot objects glow the colors they do, came up with a theory that energy was being emitted as light in discrete chunks, or "quanta." Five years later in 1905 Albert Einstein showed that light really did come in pieces—what we know today as photons.

Developing theories

This new "quantum" theory was hotly debated in the physics community, but it helped solve a key problem with the model of the atom. At the time, the model of the atom was electrons orbiting a nucleus in much the same way that the planets orbit the Sun. Classical mechanics, as developed by Newton, predicted that electrons would lose energy, and as their orbits decay would collide with the nucleus. This was a problem for physicists, because in the real world electrons defy classical physics and remain in an unknown orbit around the nucleus.

Right: In 1935, Japanese physicist Hideki Yukawa (1907–1981) predicted the existence of mesons, particles with masses between those of the electron and the proton. In 1949, Yukawa became the first Japanese person to be awarded a Nobel Prize.

Left: German physicist Max Planck (1858–1947) was one of the twentieth century's most influential physicists. His work on quantum theory marked a turning point in the study of that science. He won the Nobel Prize for Physics in 1918.

In 1913, looking at how atoms emit and absorb light, the Danish physicist Niels Bohr took the old atom model and combined it with the new ideas of quantum theory. In Bohr's model, the electrons would orbit the nucleus with certain fixed energies. This was a huge conceptual leap, but it also came with a paradox. Up until this time, scientists knew that light acted like a wave, but both Planck and Einstein described light as coming in particles. If electrons and other particles of matter acted in the same way as light, then they too could also act as a wave. The question was, how could something be both a wave and a particle at the same time?

> If anybody says he can think about quantum physics without getting giddy, that only shows he has not understood the first thing about them.
>
> Niels Bohr, physicist, 1885–1962

SCHRÖDINGER'S CAT

To help illustrate Heisenberg's Uncertainty Principle, physicist Erwin Schrödinger came up with a thought experiment: A cat, a vial of poison, and a lump of radioactive material are sealed inside a box, along with a small mechanism that will open the vial if it detects the decay of one atom in the radioactive material. It is impossible to predict when this decay might occur, and since there is no way of looking inside the box, there is no way of knowing if the cat is still alive or dead.

According to the theory of quantum mechanics, the cat is said to be in a "superposition" of both states, that is, alive and dead at the same time. It is not until someone opens the box and observes what is inside that the system stops being a superposition of both states and becomes either one or the other.

Strange as this thought experiment may sound, the result is borne out in real experiments in detecting the position of particles such as an electron.

Left: Nobel Prize-winning theoretical physicist Niels Bohr (1885–1962) proposed a model of atomic structure where electrons orbit the atom's nucleus. This was a great conceptual leap in science.

In the 1920s, rather than try to visualize what was going on, the German physicist Werner Heisenberg studied the mathematics to make sense of what was going on. In studying the orbits of electrons he realized that if you knew the position of the electron, you could not determine its momentum. On the other hand, if you detected the electron's moment, you would not be able to measure its position. Heisenberg's Uncertainty Principle became a major cornerstone of the new physics (see pages 78–79).

This led Heisenberg, with Austrian scientist Erwin Schrödinger and British scientist Paul Dirac, to reformulate Isaac Newton's mechanics into a new theory that was called quantum mechanics. It was based on the Uncertainty Principle and extended to all microscopic particles of matter.

Waves and particles

According to quantum mechanics, everything is both a wave and a particle at once. Its position is predicted only as a number of different possible outcomes and the probability of each of these outcomes. It is only when you observe it that the particle obtains a specific location "chosen" at random from the possibilities available.

The strange view of the world that is provided through quantum mechanical theory can be confusing and even disturbing. Einstein himself had some trouble with it, and is quoted as saying that he believed "God does not play dice with the universe." However, quantum mechanics works, and explains what happens inside atoms as well as other phenomena in our universe.

QUANTUM COMPUTING

We may soon be able use quantum mechanics to our advantage in computers. Currently, computers represent numbers and do calculations in bits, either 1 or 0. These are represented by transistors on a silicon chip that act like switches, where on is 1 and off is 0. In a quantum computer, instead of switches, the bits are represented by individual electrons, and the direction of the electron's spin is 1 and 0. The advantage is that because of quantum superposition, the bits could be 1 and 0 at the same time. In doing so these computers could perform calculations at phenomenal speeds, unprecedented compared with today's computers.

Below: Research laboratories around the world, such as the Los Alamos National Laboratory in New Mexico, USA, have already been experimenting with ways to construct such a quantum computing device, making significant advances over the last few years. The quantum computer could come sooner than expected.

A BRIEF TIMELINE OF QUANTUM MECHANICS

1896

Henri Becquerel discovers radioactivity. Pieter Zeeman shows that when light is shone through a magnetic field, the emission spectral lines change.

1897

J.J. Thomson discovers the electron.

1900

Max Planck suggests that electromagnetic energy is quantized—that it comes in discrete amounts.

1905

Albert Einstein proposes that light consists of discrete quantum particles, later called photons.

1911

Ernest Rutherford puts forward the nuclear model of the atom.

1912

Albert Einstein describes the curvature of space–time.

1913

Niels Bohr proposes a planetary model of the atom, as well as the theory of stationary energy states.

1919

Ernest Rutherford finds evidence of the proton.

1923

Arthur Compton discovers the quantum nature of X-rays and confirms that photons are particles. Louis de Broglie postulates that particles of matter have wave properties.

1924

Albert Einstein and Satyendra Nath Bose work out a new way to count quantum particles, later known as Bose–Einstein statistics. They also show that with enough energy, elements and compounds can shift from one phase to another, later known as Bose–Einstein condensates.

1925

Wolfgang Pauli formulates the exclusion principle for electrons. Werner Heisenberg, Max Born, and Pascual Jordan develop matrix mechanics, an early version of quantum mechanics. Hans Geiger and Walther Bothe show that mass and energy are conserved in atomic processes.

1926

Erwin Schrödinger develops wave mechanics. Enrico Fermi and Paul Dirac devise Fermi–Dirac statistics.

1927

Heisenberg states his Uncertainty Principle. Niels Bohr presents the Copenhagen interpretation of quantum mechanics.

1928

Dirac combines the notions of special relativity and quantum mechanics to describe the electron.

1935

Hideki Yukawa suggests the theory that nuclear forces are mediated by particles called mesons.

1946-48

The term "lepton" is first used to describe objects that do not interact strongly.

1948

The first theory of quantum electrodynamics is postulated by Richard Feynman, Julian Schwinger, and Sin-Itiro Tomonaga.

1960

Theodore Maiman builds the first practical laser.

1970s

A standard model of particle physics is proposed, in which matter is said to be made up of quarks and leptons that interact through physical forces.

When German physicist Werner Heisenberg formulated his Uncertainty Principle in 1927, its implications were discussed by scientists around the world. It soon became clear that the principle, which sets a limit on how precisely we can measure scientific results, affects the very fundamentals of classical physics, paving the way for the quantum mechanical world of the twentieth century.

Heisenberg's Uncertainty Principle

Left: Werner Heisenberg (1901–1976) won the Nobel Prize for Physics in 1932 for his work on the development of quantum mechanics. His Uncertainty Principle is at the center of modern physics.

Heisenberg's Uncertainty Principle examines our inability to precisely measure physical attributes at the quantum scale—the scale of an atom and beyond. In Heisenberg's day, technical manipulations at the molecular and atomic scale were theoretical, and at the forefront into investigations of the very nature of matter. Nowadays, particle physicists and nanotechnologists are able to manipulate molecules to examine how individual atoms interact, create novel ways to deliver drugs into the human body, and develop new materials that are capable of reacting and responding intelligently to their environment.

The early days

While the scientific community now accepts that matter behaves in a way that is constrained by Heisenberg's Uncertainty Principle, it was a very different matter back in 1927. The seeds of quantum mechanical theory were being sown by such people as Albert Einstein, Erwin Schrödinger, and the French physicist Louie de Broglie. These early twentieth century physicists were interested in the fundamental structure of light and matter (de Broglie suggested that light and matter behaved like a wave, a theory Einstein supported, while Schrödinger assigned certain quantum energies to the vibrational frequencies of the "matter waves"). The scientists used a series of equations to draw out these arguments on how matter behaved at an atomic scale. What Heisenberg discovered was a problem with the very method of measuring the physical attributes of atoms—casting doubt on our ability to know how matter can and does behave.

Simply put, Heisenberg's Uncertainty Principle states that an observer cannot know both the position and momentum of a particle at any given instant.

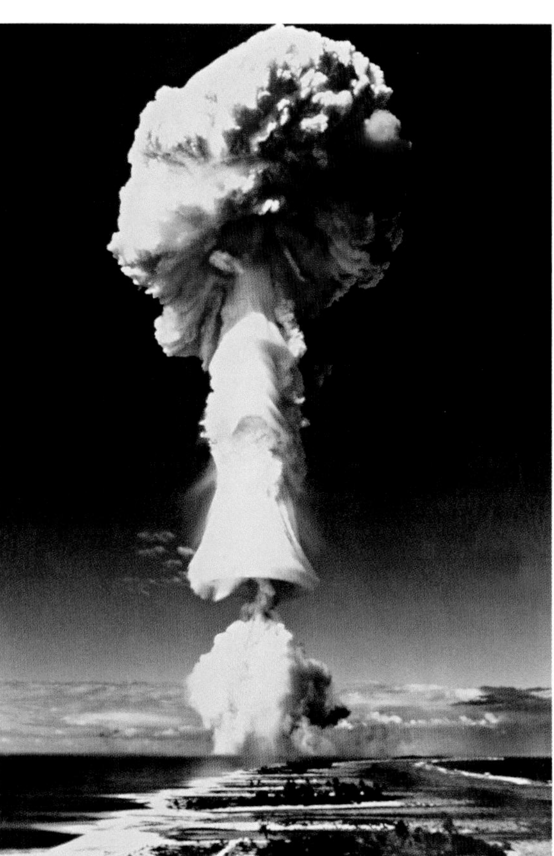

Left: This photograph taken in 1970 shows a French nuclear test at Mururoa in French Polynesia. Knowledge of the quantum physical world led to the development of the atomic bomb.

Right: The world's largest superconducting solenoid magnet (CMS) at the European Organization for Nuclear Research (CERN) in Geneva, Switzerland. The CMS is used to measure quantum particles. CERN is the world's largest particle physics laboratory.

The more precisely the position is determined, the less precisely the momentum is known in this instant, and vice versa.

Werner Heisenberg, physicist, 1901–1976

In a quantum world

It is in the quantum realm where Heisenberg's principle becomes important—for the physical world we see around us, the rules of classical physics still hold. But as early physicists came to examine the very nature of matter (incidentally discovering the knowledge that led to the formation of atom bombs), quantum physics and its strange consequences come into play.

Although initially more exciting to quantum physicists, Heisenberg's Uncertainty Principle was soon seen to have philosophical implications, in particular, the idea that the observer affects the state of a system by the act of observation. According to Heisenberg, the existence of a classical physical path (such as the orbit of an electron around the nucleus of an atom) only comes into existence when we observe it. This idea must have shocked classical physicists used to fundamental physical laws such as those described by Newton.

Above: Erwin Schrödinger (1887–1961) was one of the architects of quantum mechanical theory. His great discovery, now known as Schrödinger's wave, won him the 1933 Nobel Prize for Physics.

HEISENBERG'S THOUGHT EXPERIMENT

To explain the principle to the layman, Heisenberg offered a thought experiment where an imaginary, extremely powerful microscope focuses on a single electron. The microscope utilizes a beam of light fine-tuned to very small wavelengths to illuminate a single electron. Theoretically, at very small wavelengths of light, a photon (a wave-like particle of light described by Einstein) will "strike" the electron, causing the electron to jump into a new energy state. (Absorbing the energy of the photon, the electron jumps into a new quantum energy state.)

Heisenberg said that at the moment that the photon of light strikes the electron, an instrument set up to detect the electron can measure the electron's original position, but not its momentum (its shift into a discrete energy state), which changes as soon as the light strikes it. If the electron's momentum is measured at the moment the light hits, its position is uncertain—it can only be known as sitting within a certain range of energy states. The Uncertainty Principle gives the bounds within which the outcome (the position or the momentum of the electron) can be measured, and states that the more certainly one state is known (for example, the position), the less certain we become about other state (for example, the momentum).

By the 1930s, the inner structure of the atom was thought to be known. But further discoveries were to be made through the use of a device that smashes atoms apart.

Accelerating particles

Atom smashing

In the early twentieth century, scientists thought they knew what lay inside an atom. The theory was that there were three fundamental particles—the electron, the proton, and the neutron. However, the second half of twentieth century saw the subatomic structure of the atom as much more complex. These so-called fundamental particles were made up of even smaller particles. Helping to reveal the interior of the atom was a device called a particle accelerator.

The principle behind a particle accelerator is to take a small particle, such as an electron or proton, and get it to smash into an atom at close to the speed of light. An analogy could be made with dropping a television off the top of a building and examining the wreckage to find out what was in it.

Right: A side view of the "atom smasher" at Notre Dame University, Indiana, USA in 1945. The atom smasher was used for experimentation in the disintegration of nuclei by high speed electrons, and in the production of radioactive metals by X-ray.

Left: At Fermilab, Illinois, USA, the Cockcroft Walton pre-accelerator is where protons begin their journey through the Fermilab accelerator complex. Fermilab specializes in high energy particle physics.

In the 1920s the only way to collide high-energy particles with atoms was to use cosmic rays. Earth is being bombarded with these particles from space all the time, so experiments were set up on mountaintops where cosmic rays were more common. Although there was some success with these experiments, it became necessary for a more controlled way to conduct these experiments.

With a particle accelerator, the particle could be artificially accelerated by pushing it through using a series of electromagnets. Each push would increase the speed of the particle until it hit a target at the other end. Once the collision had occurred, what was left of the atom could be detected.

Cyclotrons and more

Such a device laid out in a straight line would be far too long and unwieldy for laboratories at the time. Thinking about how to overcome this problem, Ernest Lawrence, an American physicist, realized that he could accelerate a particle in a much more compact accelerator by making it circular. Called a "cyclotron," the circular accelerator would be set between two poles of an electromagnet that would hold the charged proton in a spiral path. After a hundred or so turns, the protons would hit the target. In 1931 he successfully constructed the device, and in doing so had invented a method for creating high-energy particles without the use of high voltages and which could fit in his university laboratory. For his efforts, Lawrence was awarded the 1939 Nobel Prize in Physics, and the chemical element "lawrencium" was named in his honor.

During the second part of the twentieth century, particle accelerators grew much larger and were used to discover the inner workings of the atom. It turned out that the proton, electron, and neutron were not the most fundamental of particles; it is now thought that they are made up of further particles to be called quarks and leptons. The use of particle accelerators helped to confirm theories of the standard model of matter, but also opened up new areas of research.

> Certainly, it may bring to light such a deeper knowledge of the structure of matter as to constitute a veritable discontinuity in the progress of science.
>
> Ernest Lawrence, physicist, 1901–1958

Left: A technician works on the cap of the world's largest superconducting solenoid magnet (CMS), one of the experiments that is preparing to take data at European Organization for Nuclear Research's Large Hadron Collider (LHC) particle accelerator in Switzerland. The magnet weighs 11,000 tons (10,000 tonnes).

Far left: A technician performs maintenance on an accelerator at Stanford University. Stanford Linear Accelerator Center research involves experimental and theoretical research in particle physics, and research in atomic and solid-state physics, chemistry, biology, and medicine using synchrotron radiation.

LARGE HADRON COLLIDER

Coming online in 2008 will be the Large Hadron Collider, and with a circumference of 16½ miles (26.7 km) it will be the world's largest particle accelerator. Built at the European Organization for Nuclear Research (CERN) in Switzerland, the collider is so big that it actually crosses the border of Switzerland with France four times. It is possible to enter the facility in one country and exit in another. Using such high energies, physicists hope that the collider will produce hitherto undiscovered particles that will confirm predictions and fill gaps in the standard model of matter. Physicists also predict a chance that the collider may produce tiny black holes. A black hole in outer space forms when enough matter from a collapsing star is squashed into a small enough space to reach a critical density. According to theory, the same critical density could be reached if two particles slam violently together. Luckily, the theory also predicts that if such a tiny black hole is produced, it will immediately evaporate. (At right is a dipole magnet, part of the LHC.)

The atomic theory of matter was first postulated by ancient Greek philosophers, in particular Democritus (born 460 BCE). He held that the atoms (from the Greek, meaning "indivisible")—of which the universe is composed—are eternal, but the bodies containing the various types of atoms decay and perish.

Nuclear know-how

In the seventeenth century, Isaac Newton reiterated Democritus's hypothesis, that "The atoms do not wear out or break in pieces, no ordinary power being able to divide what God Himself made one in the first creation." In 1802, John Dalton enshrined the law of the indestructibility of matter as the first postulate of his atomic theory: "Matter can neither be created nor destroyed." The belief that matter is indestructible suggested that heat and light have no mass. Along with the law of the conservation of mass stands the law of the conservation of energy, which implies that mass and energy are two distinct separate entities.

Albert Einstein's contribution

In the early years of the twentieth century, Albert Einstein observed that the speed of light was independent of the effects of motion, and developed the special theory of relativity (see pages 70–71). Any measurement of the velocity of light will always be 186,000 miles per second (300,000 km per second) in any inertial frame.

Einstein was also convinced that the momentum of colliding bodies is always conserved as required by Newton's laws of motion. To maintain this position he was required by his special theory of relativity to postulate that the mass of a moving body depends on its velocity. From this postulate he derived the following relationship between the mass of a body at rest, m_0; and its mass, m; moving at a velocity, v; with c the velocity of light. This relationship is known as the relativistic mass formula:

$$m=m_0/[1-(v/c)^2]^{1/2}$$

The effect is undetectable at ordinary velocities, but as the object approaches the velocity of light the mass increases without limit.

Above: German-born Swiss citizen Albert Einstein was only 26 years old when he developed the formula for which he is best known.

Einstein demonstrated that the relativistic mass is a direct measure of the total energy of a body. This is the famous mass–energy equivalence formula:

$$E=mc^2$$

where E represents the energy, m the mass, and c the velocity of light.

It follows that if a body emits a certain amount of energy, then the mass of that body must decrease by a proportionate amount.

More confirmation

Confirmation of this relationship was slow coming. In Paris in 1933, chemistry researchers Irène and Frédéric Joliot-Curie were able to photograph the conversion of energy into mass.

In Cambridge in 1932, John Cockcroft and Ernest Walton observed the reverse process—the conversion of mass into energy. They broke apart an atom and found that the fragments had slightly less mass in total than the original atom. In the process, energy was released. The fission of uranium was first discovered in 1938, by German scientist Otto Hahn (1879–1968), working with Lise Meitner and Franz Strassman.

> The release of atomic energy has not created a new problem. It has merely made more urgent the necessity of solving an existing one.
>
> Albert Einstein, physicist, 1879–1955

NUCLEAR FISSION AND FUSION

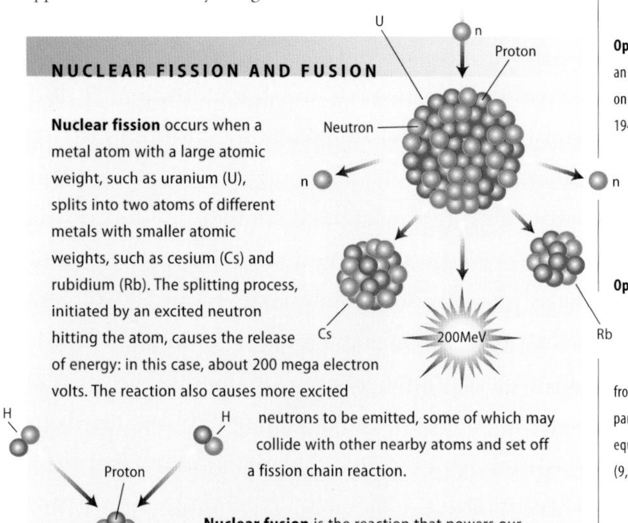

Nuclear fission occurs when a metal atom with a large atomic weight, such as uranium (U), splits into two atoms of different metals with smaller atomic weights, such as cesium (Cs) and rubidium (Rb). The splitting process, initiated by an excited neutron hitting the atom, causes the release of energy: in this case, about 200 mega electron volts. The reaction also causes more excited neutrons to be emitted, some of which may collide with other nearby atoms and set off a fission chain reaction.

Nuclear fusion is the reaction that powers our Sun and other stars. At very high temperatures, hydrogen (H) atoms, with a single proton in the nucleus, smash together to form helium (He) atoms. As in the process of nuclear fission, the collision causes the emission of a small amount of energy (about 3.2 mega electron volts) and an excited neutron.

Opposite page inset: When an atomic bomb was dropped on Hiroshima on August 6, 1945, almost all the buildings within 1 mile (1.6 km) of the impact point were flattened.

Opposite: The United States conducted more than 900 nuclear weapon tests at its Nevada Test site from 1951 to 1962. This particular bomb was the equivalent of 10,000 tons (9,071 tonnes) of TNT.

The A- and H-bombs

It is in nuclear processes that the mass changes are most evident. The enormous amounts of energy released by radioactive elements had been known for some time. Thus the atomic bombs, both uranium and plutonium (fission) and hydrogen (fusion), release vast quantities of destructive energy. The fission bomb, or A-bomb, contains the radioactive isotope of uranium, uranium-235, or plutonium-239. When a critical mass of the isotope is brought together, a violent explosion occurs as the result of a chain reaction. The fusion bomb, or H-bomb, is triggered by a fission reaction, which in turn fuses the nuclei of various hydrogen isotopes to form helium nuclei. The H-bomb is even more destructive.

Right: The nuclear fusion process taking place in the Sun's atmosphere periodically gives rise to violent eruptions of radiation and matter.

HISTORY OF THE ATOM AND NUCLEAR ENERGY

c. 400 BCE
Theory of an indivisible "atom"

1805
Discovery that atoms of each chemical element are different

1895
Discovery of X-rays

1897
Discovery of the electron

1898
Discovery of radioactive elements

1903
Theory of nuclear reactions

1905
Special theory of relativity

1911
Discovery of the nucleus

1927
Uncertainty Principle

1932
Discovery of the neutron

1932
Splitting of the atom

1933
Particle accelerators

1939
Fission of uranium atom

1942
Controlled nuclear chain reaction

1942–45
Manhattan Project

1945
Atomic bomb

1951
Electricity from nuclear energy

1952
Hydrogen bomb

1955
Nuclear-powered submarine

1957
International Atomic Energy Agency

1986
Chernobyl reactor meltdown

The first laser (an acronym for Light Amplification by the Stimulated Emission of Radiation) was built in 1960 by American Theodore Maiman, mostly to prove that it could be done. He used a ruby rod, but today lasers can use a wide range of solids, liquids, and gases as the "medium" to generate their unique radiation.

The light fantastic

To understand how a laser works, we must review how atoms deal with light and other radiation. Energy in atoms is stored a bit like you might store boxes, in a series of shelves, one above the other. If one shelf gave way, a box could drop without any help to a lower shelf, but lifting a box to a higher shelf takes some effort.

In atoms and molecules, the shelves are called energy levels and they hold electrons. When an atom or molecule absorbs energy, say from a passing light wave, electrons are hoisted up to higher levels, and when the light wave has gone, the electrons can fall back down again, giving out exactly the same amount of energy they took in.

Of course, all the atoms or molecules of the one chemical have the same arrangement of shelves, with the spacing between the shelves quite precisely determined.

Coherent light

A laser's talent depends upon making the atoms or molecules in the "medium" behave in exactly the same way and in step. When they are stimulated, say by a flash of light, all the particles in the medium (usually atoms or molecules) release energy (in the form of light) simultaneously, that is, all the electrons in billions of atoms drop to

Below: The Conservatory Building in Kiev, Ukraine, was dramatically transformed by the "Sculptures by Light" hologram show. Lasers provide the coherent light source that give holograms their three-dimensional effect.

LASERS—HOW THEY WORK

The first example of a laser was produced using a ruby rod as the lasing medium, although nowadays other media—such as carbon dioxide gas—are also used. A flash lamp provides the initial energy that excites atoms in the lasing medium, causing them to emit photons. The photons, originally emitted in random directions, bounce back and forth between two mirrors until they are all lined up in the same direction and emitted through the partially reflective mirror as a laser beam. The resulting beam can be focused with pinpoint accuracy for a broad range of applications.

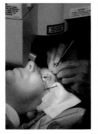

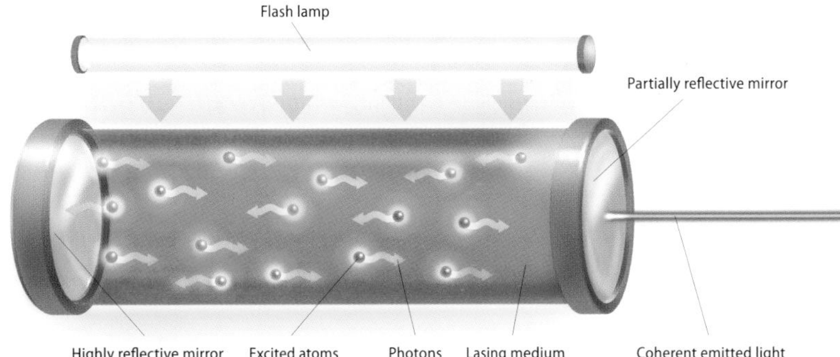

a lower energy level precisely in step. That makes the light coherent, with the peaks and troughs of all the light waves lined up. And they all release precisely the same amount of energy, so that all the light given out is the one pure color.

Coherent light has some amazing properties. It has little tendency to spread out, so can be focused into a very tight spot, less than a thousandth of a millimeter across. This is the quality that makes lasers ideal for creating and then detecting the tiny marks that code information in the grooves of a CD or DVD. Every CD or DVD player or burner has a laser in it.

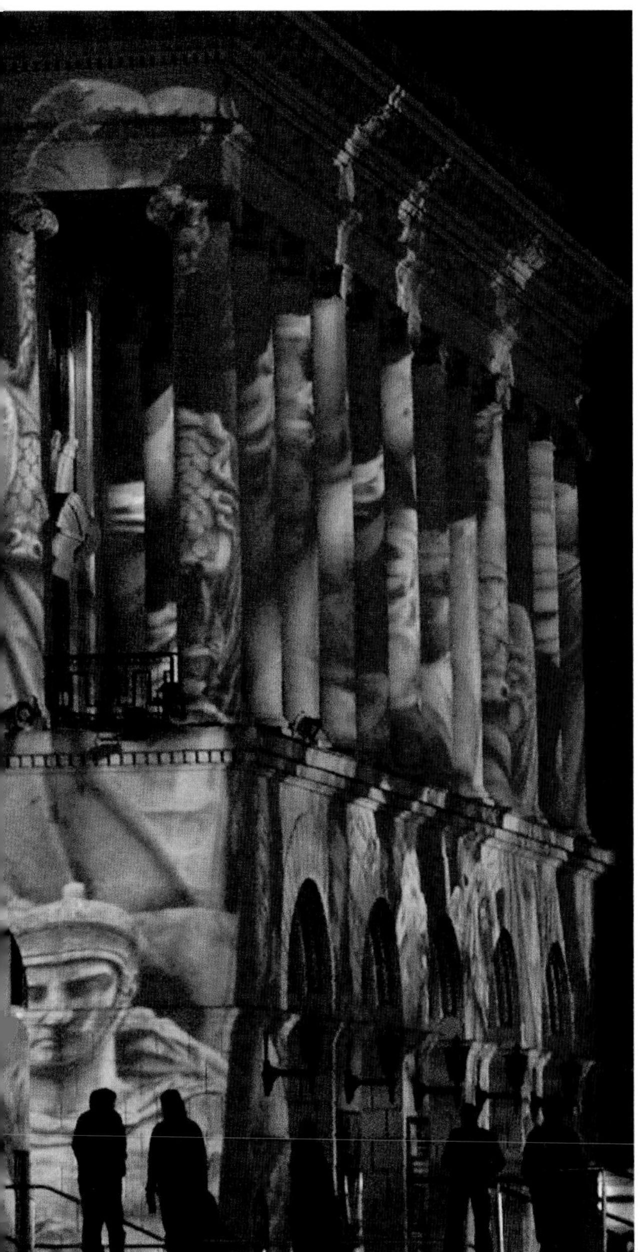

Above: More than ten million laser eye surgery procedures have been performed worldwide, and lasers are also used in cosmetic and plastic surgery, podiatry, and dentistry.

New applications

One exciting development in recent years has been finding ways to make lasers give out blue light. Blue is harder to make, because the energy levels needed to create it are much further apart. Blue laser light can be focused even more tightly than red light, so the dots in a DVD can be even smaller and the disk can hold more information.

Tight focusing allows a laser to deliver its energy into a tiny area, able, say, to burn minute and perfectly formed holes or to shape metal parts very precisely. So high-powered lasers, often using carbon dioxide to make the infrared (heat) laser radiation, have an important role in precision engineering.

The precise color of a laser beam means that many beams can be sent together down an optic fiber without confusion—significantly increasing the fiber's capacity to carry data coded as tiny lumps of radiation. The "colors" used in optic fibers are in the infrared range, a bit lower in frequency than red light, since those frequencies encounter the least resistance as they push through the glass in the fiber.

Researchers are currently investigating different ways to use laser light to carry information around inside computers rather than using electric currents. This technology is called photonics, and it is very likely to revolutionize computers yet again, making them even more powerful and useful tools.

> [Science is] a great game. It is inspiring and refreshing. The playing field is the universe itself.
>
> Isidor Isaac Rabi, physicist, 1899–1988

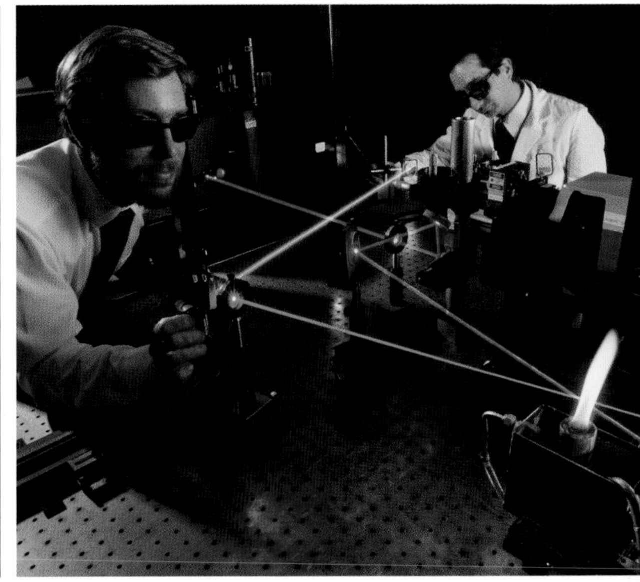

Right: Lasers are now being used in an amazing variety of research areas, ranging from air pollution to energy generation, optics to communications, and weaponry to surgery. Because lasers surgically cut without physical contact, the risk of infection is lowered.

String theory is hailed as one of the turning points for modern physics, much like Einstein's general relativity, or Maxwell's electromagnetism. Yet to understand string theory, it is important to appreciate one of the biggest problems physicists have faced over the last 50 years.

What is string theory?

General relativity and quantum mechanics

The theories of general relativity and quantum mechanics are the foundations of modern physics. Einstein's general theory of relativity (see pages 72–75) describes the dance of objects thousands of miles across and light years apart. It describes the behavior of stars, planets, galaxies, and beyond.

Quantum mechanics (see pages 76–77) is at the other end of the scale, describing the buzzing of particles so small that many million of them inhabit just one grain of sand.

Both theories describe the same thing—the movement and interaction of the matter of the universe—yet in vastly different sizes. Logically, if we extend quantum mechanics to larger objects, we should end up with the general theory of relativity, and vice versa.

Yet with our current understanding of physics, both theories cannot be right. One of the two foundations upon which our understanding of the universe is based, must be wrong. As you can imagine, this is a rather large problem.

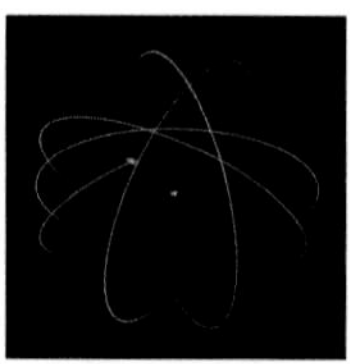

Above: Electrons orbiting around a nucleus. While electrons are fundamental particles, protons and neutrons are made of smaller particles which are known as quarks.

Left: Particle accelerators such as the Tevatron are designed to smash protons together, and it is possible that this will enable scientific tests of string theory. The Tevatron is a circular particle accelerator with a diameter of 4 miles (6 km).

Theory of everything

Einstein tried to find a single theory to unify quantum mechanics and general relativity, and thus explain the universe. He called it the "Theory of Everything." Though he spent 30 years searching, he could not find it.

That one of the greatest minds in physics was not able find this elusive theory indicates the difficulty of the task. Almost 50 years later, the combined minds of many physicists came up with string theory. It looks likely that string theory could be this remarkable "Theory of Everything." It has already solved one major problem—within the theory's framework, both quantum mechanics and general relativity are correct. With a theory to describe the universe, a whole new era of science, full of unimaginable possibilities, opens up.

> String theory is a part of twenty-first-century physics that fell by chance into the twentieth century.
>
> Edward Witten, theoretical physicist, b. 1951

So, what exactly is string theory? We know that there must be fundamental particles, the smallest units of matter in existence. These particles are the tiny building blocks of the universe. Everything, from DNA, to water, to stars and planets, is formed from different combinations of these particles.

String theory is a beautifully simple and elegant way of describing these fundamental particles.

Fundamental units of matter: The smallest particle

Since ancient times, the existence of a fundamental particle, a smallest unit of matter, has fascinated scientists. Our understanding of this particle, however, has come from just the last two centuries.

Just a century ago, an atom was thought to be was the smallest unit of matter, a solid particle, indivisible. Different combinations of atoms were thought to make up all the different substances, such as air, water, and so on.

As technology advanced, a closer look revealed that atoms were in fact made up of three smaller particles—protons, neutrons, and

Below: This "scale" shows where quantum mechanics works, and where general relativity works, giving us some idea of just how huge the differences are.

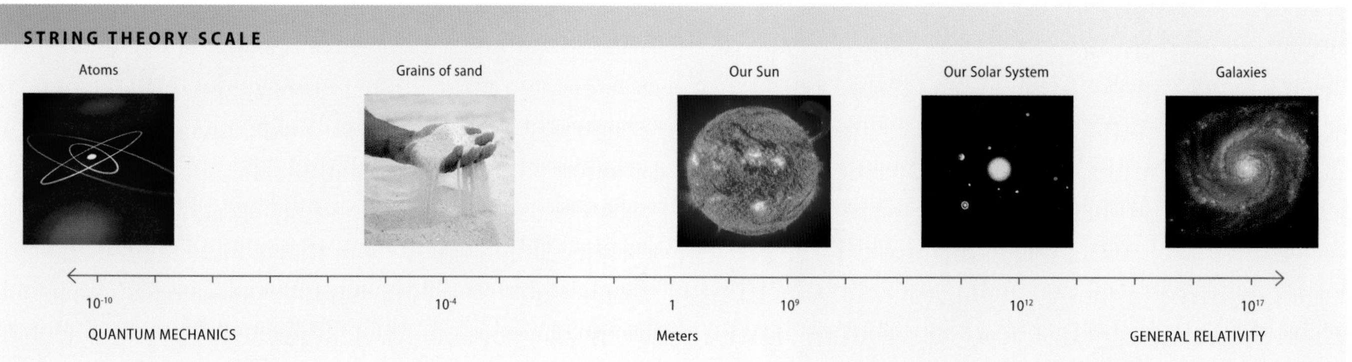

electrons. These fit together in a neat pattern—the proton had a positive charge, the electron had a negative charge, and the neutron had no charge. Interestingly, the proton and the neutron were about 2,000 times heavier than the electron, which seemed to fit no pattern at all.

A much closer look at protons and neutrons revealed that they are made up of even smaller particles, which come in several different varieties—quarks, leptons, and bosons.

These families of particles replaced protons and neutrons as our smallest unit of matter. However, they were a collection of particles with seemingly random masses, charges and groupings, and with no particular relation to each other. There also seemed to be an endless stream of them being discovered. Indeed, there are still many elementary particles that have been predicted to exist, but their detection is beyond the limits of current technology.

Below: String theory says that if we could look closely at fundamental particles, we would find that they were not point particles, but infinitessimal vibrating loops of string. In the scale of things, strings are almost unimaginably small. Each box in this chain represents a reduction of 100 times the size of the previous box.

AN ELEGANT THEORY

String theory describes, in simple, elegant terms, how these apparently random particles are related to one another. Until the formulation of string theory, all particles were considered point particles—tiny solid spheres. However, string theory says that if we were able to look even closer at the fundamental particles, we would find that they were not point particles, but tiny vibrating loops of string.

The difference between particles, such as mass and charge, are caused by the strings vibrating at different rates. The beauty of string theory is that with one simple idea, two of this century's biggest physical problems have been solved.

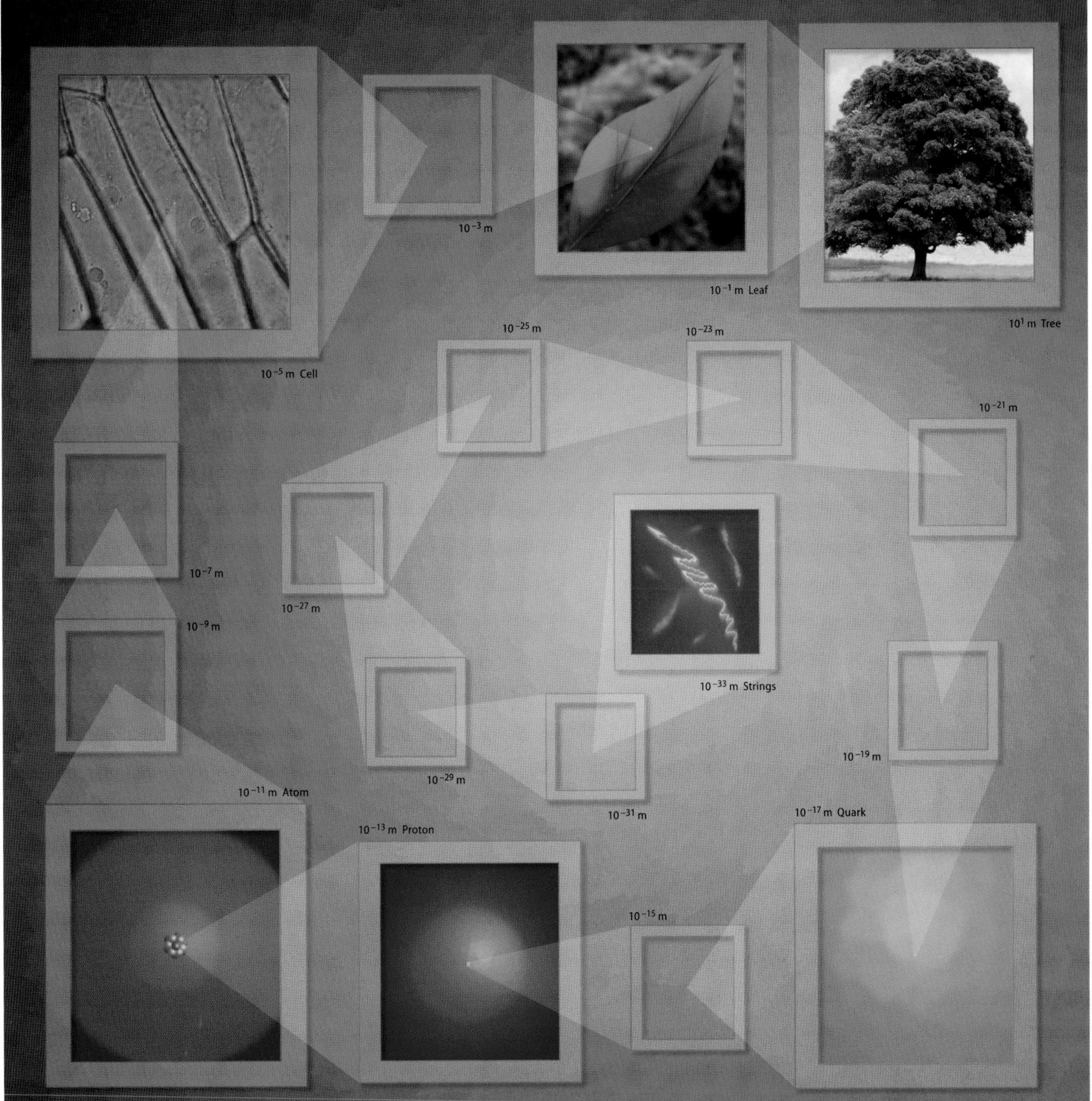

From the oldest telephone to the very latest in wireless technology, the goal of telecommunication is the same—to transport information from one point to another. The way in which we communicate is constantly being revolutionized by the continual development of new and improved telecommunication technologies.

Communicating with the world

The birth of telecommunication

All of today's communication technologies have evolved from the telegraph, first patented in 1837. Samuel Morse (1791–1872), capitalizing on the work of earlier inventors, demonstrated in 1835 that electromagnetic currents could be transmitted by wire. He created the operator key, that when pressed, sent an electrical signal. Upon reaching its destination, the receiving operator key embossed paper with the dashes and dots of Morse code.

Below: A statue of the Madonna sits incongruously amid communication towers. When you make a call on a mobile cell phone, it is wirelessly linked to the telephone network via such towers. The call is then connected.

Right: While estimates vary as to the number of people around the world who regularly use the internet, it seems that in early 2008, more than 1,400,000,000 people logged on. English is the most widely used internet language, followed by Chinese, Spanish, and Japanese. By continent, Asia is the biggest user of the internet, followed by Europe, then North America. Internet usage is expected to increase over the next decade and beyond.

The first commercial electric telegraph line commenced in London in 1839. Attempts to improve the telegraph led to the invention of the telephone, with credit going to Alexander Graham Bell following the patenting of his machine in 1876. Bell demonstrated that different tones varied the strength of an electric current in a wire. Within a year, he had created a functioning telephone that transmitted speech electrically. Sound vibrations were converted to electrical currents in the mouthpiece, which traveled along a wire to the receiving phone, which converted the currents back to sound in tones that replicated the original voice. The fundamental design of Bell's invention remains much the same today.

From the telegraph to the mobile phone

In 1886, German physicist Heinrich Hertz demonstrated that fast variations of electric currents could be sent into space in the form of radio waves. By the beginning of the twentieth century, "spark gap" transmitters were used to send wireless telegraphs. In 1906, Lee de Forest invented amplitude-modulated (AM) radio by amplifying radio signals, which eventually led to radio broadcasting in the 1920s. Frequency-modulated (FM) radio was invented by Edwin Howard Armstrong in 1933. This improved the quality of sound by varying the frequency of the signal, thereby controlling the noise static caused by other sources of electromagnetic radiation.

Mobile phone technology is in many ways an extension of radio technology. The phones themselves act like two-way radios by transmitting and receiving radio waves. The transmitter inside the phone converts the voice into radio waves. These waves travel through the air until they reach a receiver at a base station. This is then sent through the network until it reaches the recipient phone, where the signals are converted back into data or voice. Base stations are strategically placed to create areas of coverage known as cells. The current generation of mobile technology, 3G, uses higher frequency radio waves than those used in previous 2G networks. Consequently, the radio waves travel shorter distances resulting in more base stations and smaller cell sizes. By using this broader, higher frequency band, larger amounts of data can be delivered which means the system has the capacity for greater voice and data transfer. 3G networks are complex and utilize packet based (internet) networks more than the 2G generations.

Right: Todays' mobile telephones have the capacity to send and receive picture and video files, and can also connect to the internet. Text messaging remains the most popular form of mobile phone communication.

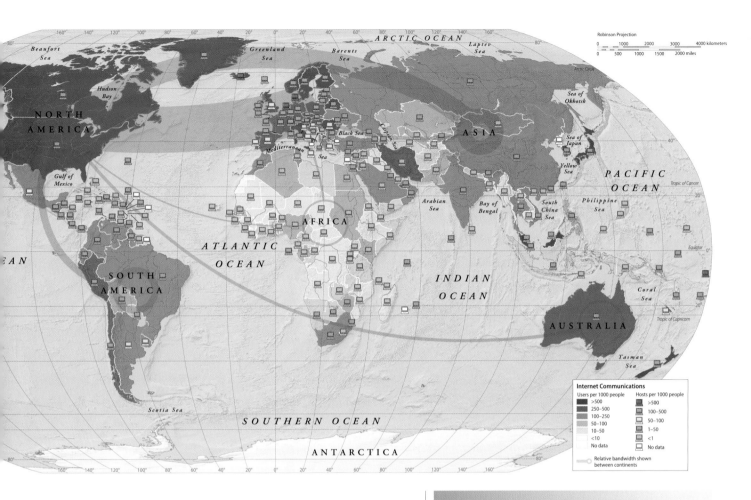

Robinson Projection

0 1000 2000 3000 4000 kilometers

0 500 1000 1500 2000 miles

ARCTIC OCEAN · NORTH AMERICA · SOUTH AMERICA · AFRICA · ASIA · AUSTRALIA · ANTARCTICA · ATLANTIC OCEAN · PACIFIC OCEAN · INDIAN OCEAN · SOUTHERN OCEAN

Internet Communications

Users per 1000 people	Hosts per 1000 people
>500	>500
250–500	100–500
100–250	50–100
50–100	1–50
10–50	<1
<10	No data
No data	

Relative bandwidth shown between continents

Wireless [telegraphy] is all very well but I'd rather send a message by a boy on a pony!

William Thomson, 1st Baron Kelvin, mathematician and physicist, 1824–1907

Above: Inventor Samuel Morse recognized that the pulses of electromagnetic currents were capable of transmitting information over wires. His operator key heralded the beginning of the telecommunications age.

Left: Alexander Graham Bell (1847–1922), who invented the telephone in 1876, photographed here in 1892, making the historic first telephone call from New York to Chicago, USA.

THE INTERNET—A GLOBAL REVOLUTION

The internet is a global network of interconnected computer networks. The first interconnected network, ARPANET, was established in 1969 and consisted of four networked computers. Today, the internet is a network of millions of computers. The internet has numerous applications (below) including email, file sharing, and voice telephony.

The ubiquitous World Wide Web was invented by British scientist Tim Berners-Lee in 1989. The words "web" and "internet" are often used interchangeably, yet they are not the same thing. The web is a collection of resources and documents connected by hypertext links, whereas the internet is the physical connection of computer networks. The development of the internet became possible from the 1960s because of the implementation of packet-switching networks, the means through which the internet delivers information from one computer to another. At the source computer, information is divided into small blocks of data called packets, which are defined by a language which is known as the standard Internet Protocol (IP). Packets can be compressed and encrypted and are transmitted individually, often using different routes to make the transfer of information more efficient. The packets are transported using physical connections made of copper wire or fiber optics. At the destination computer, the packets are recompiled.

MATHEMATICS

The number system that we use in everyday life, based on tens, was invented some 1,500 years ago in India. This reached Europe via the Middle East, along with Arabic words such as "algebra." That route also provided us with translations of the great mathematical writings of the ancient Greeks, such as the famous geometry collated by Euclid and Pythagoras's Theorem. That knowledge had been lost in western Europe following the fall of the Roman Empire.

Introduction

Below: The exact origin of the abacus, a wooden frame strung with colored beads, is unknown. Historical records show that it was used in Babylonian times, in ancient and medieval China, and also in Greece. The abacus is still used for simple calculations.

Mathematics in symbols

To manipulate numbers, we also needed a number of other symbols, and these came steadily into use through the sixteenth and seventeenth centuries: Symbols for "square root," "equals," "multiply by," "divide by," "greater than," "less than," "infinity." Through those same centuries we started to use the decimal point and to use letters to represent qualities in algebra (such as "x" for the unknown). These are familiar to most of us now but someone had to invent them.

Mathematics can be a matter of technology as well as ideas. In counting we began with fingers and toes and then moved on to devices like the abacus. The first mechanical calculating machines appeared in the seventeenth century as a way of easing the labor of computation, but they were unreliable. The same motivation led Scotsman John Napier to devise logarithms, later engraved onto wooden sticks to become the first slide rules. That technology had an impressive longevity. A slide rule was to be found in the breast pocket of every engineer until the arrival of the pocket calculator less than 50 years ago.

A number of famous mathematicians were caught in the development of what today we call "computers." With modern software, computers seem able to do an extraordinary range of things, from word processing to manipulating pictures, from playing computer games to helping you compose music. These multifarious activities perhaps might better be called "information processing," yet deep down, every computer does what it does by manipulating numbers, simply by adding and subtracting numbers and by comparing them.

Right: The Greek astronomer and mathematician Hipparchus (*c.* 190–*c.* 120 BCE) calculated distances between the Sun, the Moon and Earth, and in the process is said to have sown the seeds for what later became trigonometry.

Mathematical minds

The most famous mathematical names include nineteenth-century Englishman Charles Babbage, who devised and started to build (though he never completed) the first programmable mechanical computer. His intention was that it would be able to calculate mathematical tables, then done by hand, automatically and without any errors. He called it a "difference engine." In homage to Babbage, we still talk about search "engines."

A few decades later, his countryman George Boole devised a way of manipulating symbols, which later became the starting point for computer programs as they developed 100 years later. The French philosopher René Descartes, best known for saying "I think, therefore I am," combined arithmetic and geometry to create "coordinate geometry," in which groups of two or three numbers are used to describe a location in two- or three-dimensional space. This has proved a powerful tool with many practical applications.

A little later, German mathematician Gottfried Leibniz and his English contemporary Isaac Newton independently devised a system called "calculus" by the former and the "theory of fluxions" by the latter. With the new technology (known today as calculus) we can very easily deal with changing physical quantities, such as speed or position, and write down various groups of symbols ("equations") to describe the motion of objects in time and space. Calculus also allows us to simply describe intricate curved lines and surfaces.

In the sixteenth century, another German mathematician, Georg Rheticus, worked out the relationships between the three sides and three angles of a triangle, so that if only some of these are known the others can be calculated. This became the basis of trigonometry, another highly useful branch of mathematics. Surveyors use trigonometry all the time in making charts and maps, and to calculate distances, say between two mountains, without needing to roll out a long measuring tape.

We cannot pretend that all mathematics has a practical application. Much of the more advanced mathematics is abstract, with no real use to which the knowledge can be put. Trying to figure the distribution of "prime numbers" (a number not divisible by any number other than itself or one) is an example. Yet such exercises are immensely challenging. It has been said that mathematicians are often creators of patterns, much like artists or composers.

Below: Before the late 1980s, computers were found mainly in workplaces and government offices, such as the UK Board of Trade. This programmer, working in 1965, uses English Electric's supercomputer to calculate trade figures.

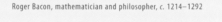

Mathematics is the door and the key to the sciences.

Roger Bacon, mathematician and philosopher, *c.* 1214–1292

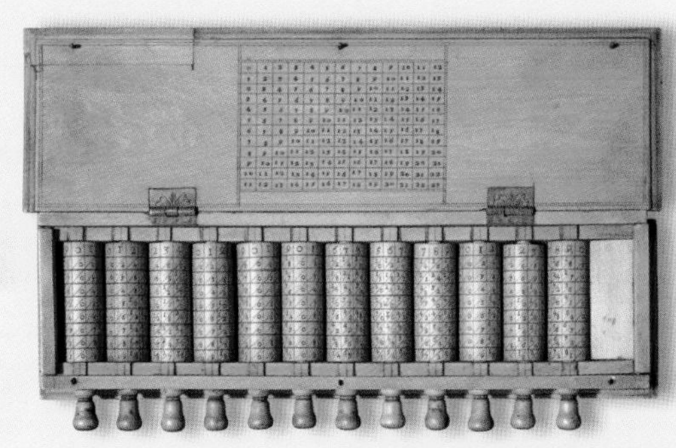

Above: Napier's rods, also known as Napier's bones, devised by mathematician John Napier around 1617, were rods—made of wood or bone— divided into nine spaces that contained the numbers of a column of a multiplication table.

Left: At a primary school in England in 1963, a teacher shows three students how to use their mechanical adding machines, after a survey at the time found that children learn arithmetic more quickly and with deeper understanding when they have access to mechanical aids.

The disciplines within mathematics have a fundamental role in the development of cultural, social, and technological advances and in empowering individuals to be significant members of society. In everyday life, in personal, work, and study situations, people experience or use elements of mathematics through the use of number, space, measurement, chance, and data. Regardless of nationality, culture, or religion, mathematics is shared and utilized by all human beings.

Making mathematics meaningful

Mathematics is the science that gives us the resources to explore our bodies, the universe, and our planet. It also helps us with simpler activities, such as cooking and buying daily necessities on a budget. It is integral to some of the world's most beneficial advances, such as computer technology. With the use of mathematics and the many technologies it supports, we can financially prepare for retirement, protect our lives and possessions with insurance, and access our bank accounts from anywhere in the world.

Cryptography

Cryptography is the science of hiding information. Many early cryptographic practices, such as those used by ancient civilizations to conceal military information, were based on linguistics. But in the modern era, cryptography is very much based on mathematical processes. Mathematical cryptography is now used for many situations in which information is to be hidden or secure, or messages are to be kept confidential, or the identities of senders and receivers need to be authenticated. There are many processes that we use in everyday life that are reliant on these practices. Without cryptography we would not be able to undertake the numerous daily tasks that depend on the security of our identity and information. We would not be able to safely use credit cards over the internet, protect our computer through passwords, withdraw money from an cash machine, or purchase and pay for items from around the world via the internet.

Right: Whenever we shop, we use basic arithmetic to make informed purchasing decisions. Weight, packaging, and price of items are compared in order to get the most value for money.

Opposite: Credit cards were introduced in the late 1950s. Since then, credit card use has soared, with recent figures indicating that in 75 percent of US households at least one person owns a credit card.

> Cryptography began in mathematics. Codes were developed, even from Caesar's time, based on number theory and mathematical principles.
>
> James Sanborn, sculptor, b. 1945

WEIGHING UP MATHEMATICS

Mathematics is involved in most aspects of our lives. Even one's health is assessed using mathematical principles. Body mass index, or BMI, is a measurement used by physicians and researchers use to assess a person's weight and height. A mathematical formula is used to calculate a person's body mass index by taking their weight and dividing it by their height in meters squared: (BMI = weight (kg)/height^2 (m^2)). This calculation is used to assess how much an individual's body weight departs from what is normal or desirable for their height based on average body composition. The World Health Organization (WHO) has assigned cut-offs that reflect the relationship of BMI to the risk of diseases and mortality. The WHO advises that a BMI of less than 18.5 indicates the person is underweight; 18.5 to 25 indicates optimal weight; above 25 suggests the person is overweight; and a BMI above 30 is suggestive of obesity. This should only be used as a guide, however, since professional athletes can have an overweight BMI value if they have large amounts of muscle, which is heavier than fat.

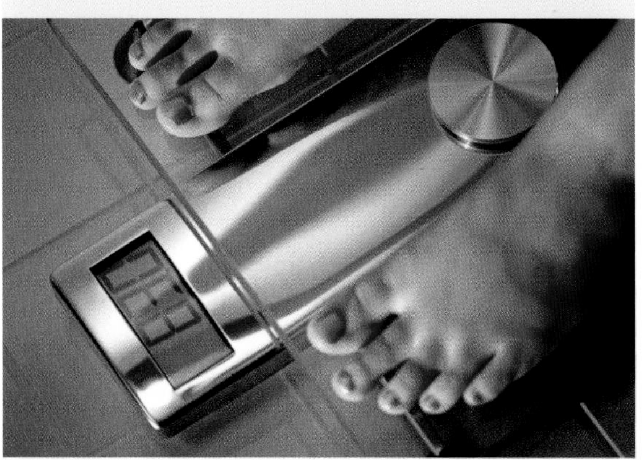

Above: Mathematical cryptography has helped to make online banking a safe process. Financial transactions can be conducted on a secure website maintained by the bank or financial institution.

Opposite inset: Automatic cash machines have become an integral part of our lives. Accessed by a computer-coded card and a personal password, bank customers can withdraw cash any time of day or night.

Bank interest

Most interactions with a bank involve interest, whether you are paying it or receiving it. By placing your money into a savings account, the bank is essentially borrowing your money and paying you interest on the amount you have deposited. The reverse is true if you borrow money from the bank or use a credit card. Understanding the mathematical principles behind bank interest can help you to make sound financial decisions that can pay off handsomely over time.

The principle of compound interest can either help to make you wealthy or greatly magnify your debt to a bank. Most loans from banks, including housing loans and credit card loans, are based on compound interest. If the interest on a $10,000 credit card loan is charged at, say, 20 percent annually, and calculated and compounded annually, after one year the amount owed would be $12,000. If the interest is calculated monthly, the amount owed after one year would be $12,194, and if calculated daily it would be $12,207. Although the amount of interest shown in these examples may not seem too large, with a larger debt and over a longer time period, the amount of interest owed or paid can actually be greater than the original loan. Compound interest is what makes loans so much harder to pay off.

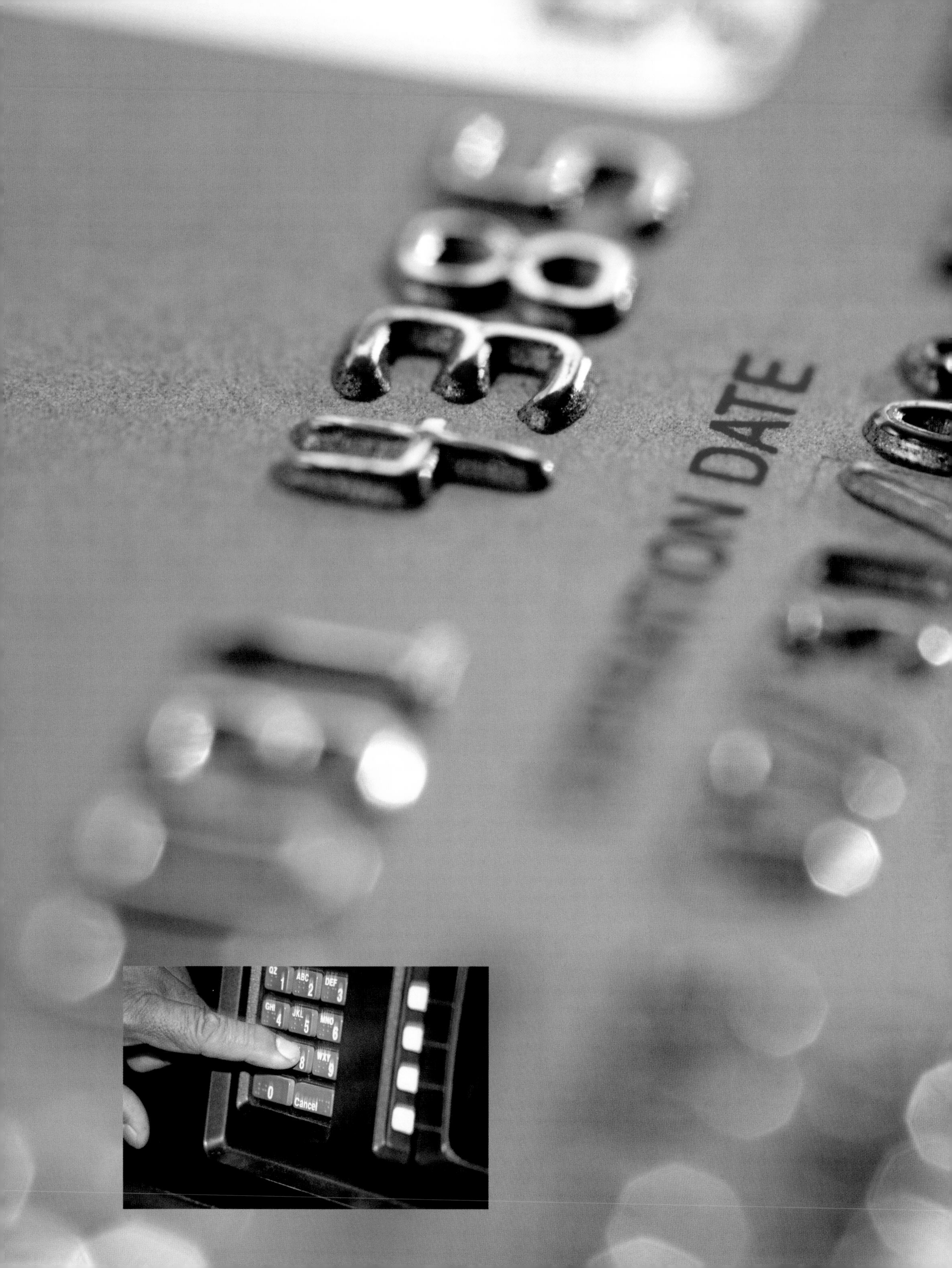
ON DATE
QZ 1
ABC 2
DEF 3
GHI 4
JKL 5
MNO 6
WXY 9
0
Cancel

If mathematics is the science of patterns, famous mathematicians are often those who bring the most famous of these patterns into the light. The mathematicians we remember somehow create, recognize, or publish patterns that others have not.

A fraction of mathematicians

There have been many important mathematicians through history, from Euclid, the ancient Greek geometrist, to the Persian algebrist Omar Khayyam, to the Italian Fibonacci, who in the early 1200s, introduced the use of Hindu–Arabic numbers to mathematical equations.

Left: In 1642, French mathematician Blaise Pascal (1623– 1662) invented his first mechanical calculator, which he hoped would reduce his father's workload as a taxation commissioner. The machine performed basic addition and subtraction tasks.

John Napier

Not surprisingly, the most famous mathematical patterns are those that are thought to be the most important. These important patterns are often those that make sense of recognized problems. For example, sailors used to find navigating the seven seas difficult, so if someone recognized the mathematical patterns to make navigation easier, he would become famous for his work. As history tells us, John Napier (1550–1617) was that man. His work with logarithms, published in 1614, led to safer and more predictable sea travel. Logarithmic methods replaced the old methods of prosthaphaeresis and "dead reckoning," which had been the major forms of ocean navigation for centuries. Napier's mathematics drastically increased a ship's chance of reaching a destination safely and also on time. He is famous not only because of the brilliance of his mathematics, but also because it had such an impact on the world.

Above: Charles Babbage was the first to come up with the concept of the programmable computer. In the 1830s, he was professor of mathematics at Cambridge University.

Charles Babbage

English mathematician Charles Babbage (1791–1871) achieved much in his life, including deciphering the supposedly undecipherable Auto Key code. Yet he is most famous as the inventor of the programmable computer. Although not constructed in his lifetime, Babbage invented what he called a "difference engine" that would weigh about 15 tons (13½ tonnes), designed to mechanically calculate long, tedious mathematical problems. Years later a working prototype was completed which proved the effectiveness of his concept. He continued his design work and soon developed what he called "analytical engines." Although these engines would have been run by steam power, they included most of the essential logical processes of modern computers. Babbage introduced the world to the possibilities of computers.

ALAN TURING

Alan Turing (1912–1954) was a mathematician credited as one of the world's first computer scientists. He developed the concept of a primitive but powerful type of computer that became known as a Turing Machine. He is also famous for his work in cryptography. His research and work in deciphering German war communications led to the production of the Bombe, a machine specifically designed to decode messages sent using German Enigma machines, used to encrypt messages. The Bombe was a logic machine that tested intercepted coded messages using various combinations of circuits to detect logically correct German language sentences. When a likely message was discovered, a bell sounded and that combination was further investigated by skilled cryptographers. While Turing is remembered as a cryptographer, a computer scientist, and a logician, he was essentially a mathematician who found effective ways to turn mathematical theory into practical science.

> Mathematical reasoning may be regarded rather schematically as the exercise of a combination of two facilities, which we may call intuition and ingenuity.
>
> Alan Turing, mathematician, 1912–1954

Right: The Enigma coding machine used by the Germans during World War II resembled an old-fashioned typewriter. By scrambling the letters, important messages could be encrypted. Literally millions of permutations were possible.

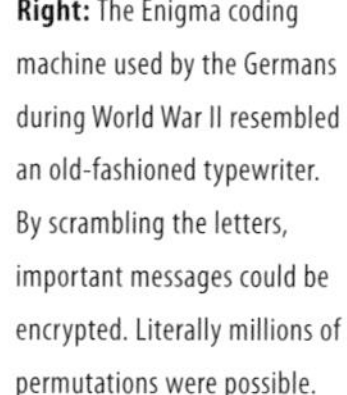

Left: A polished gold model of Charles Babbage's "difference engine," a mechanical device designed to calculate values of polynomial functions. Construction was never completed in Babbage's lifetime, but in 1989–1991 in London, the difference engine was built using Babbage's plans. It performed better than today's pocket calculators.

Isaac Newton

A poll of the famous Royal Society of London for the Improvement of Natural Knowledge conducted in 2005, found that Isaac Newton (1642–1727) was considered to have had the greatest effect on the history of science of any person. In 1687 he published his theory of gravity and the three laws of motion. These three laws of inertia, acceleration, and reciprocal actions are not laws that we must obey, but rather, laws that occur around us whether we are aware of them or not. Every action and every movement in our world occurs in the context of these laws, but it was not until Newton formalized them that mathematicians and physicists became fully aware of the forces in action. In addition to these achievements, Isaac Newton has also been credited with pioneering the field of calculus, in tandem with the German mathematician Gottfried Leibniz (1646–1716) (see page 108.) Newton developed a generalized binomial theorem and several other major advances in mathematics. However he is often remembered as the man who suddenly discovered gravity on the day he was hit by an apple falling from a tree. The story may be apocryphal, but gravity is solid science.

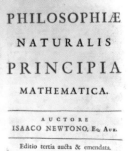

ISAACUS NEWTON EQ. AUR. ÆT. 83.
PHILOSOPHIÆ
NATURALIS
PRINCIPIA
MATHEMATICA.
AUCTORE
ISAACO NEWTONO, Eq. Aur.
Editio tertia aucta & emendata.
LONDINI:
Apud Guil. & Joh. Innys, Regiæ Societatis typographos.
MDCCXXVI.

Right: Isaac Newton's *Philosophiae Naturalis Principia Mathematica* (*Mathematical Principles of Natural Philosophy*) was published in 1687. It is still considered one of the seminal scientific works of all time.

From New York to New Delhi, Cape Town to Cape Horn, high school students learn and apply the Theorem of Pythagoras—that "for any right-angled triangle, the square of the hypotenuse (the longest side) is equal to the sum of the squares of the two shorter sides." But who was Pythagoras, and was the theorem really his discovery?

The Theorem of Pythagoras

Born on the Aegean island of Samos around 572 BCE, the Greek mathematician Pythagoras established the Pythagorean School, which quickly developed into a powerful and politically influential brotherhood. Although many mathematical discoveries were made there, it was the theorem that bears his name that is most closely associated with Pythagoras. However, calling it Pythagoras's Theorem may be a misnomer. Earlier civilizations had knowledge of this result—there is evidence that it was used over 1,000 years before Pythagoras's time by the Babylonians. Yet it is generally accepted that the Pythagoreans probably established the first formal proof of the theorem.

A simple demonstration of the theorem can be made by considering a wire running from a point 12 ft (3.5 m) from the base of a flagpole, to a point on the ground 5 ft (1.5 m) from its base. The wire acts as the hypotenuse of a right-angled triangle with the pole and the ground as the two shorter sides. To find the length, *l*, of the wire we apply the theorem thus:

$$l^2 = 12^2 + 5^2$$
$$= 144 + 25$$
$$= 169.$$

The square root of 169 is now taken to give the length of the wire as $l = 13$ ft (4 m).

Above: Acknowledged as one of the ancient world's greatest mathematicians, Pythagoras the man has remained something of a mystery. His interests ranged beyond mathematics into philosophy and astronomy.

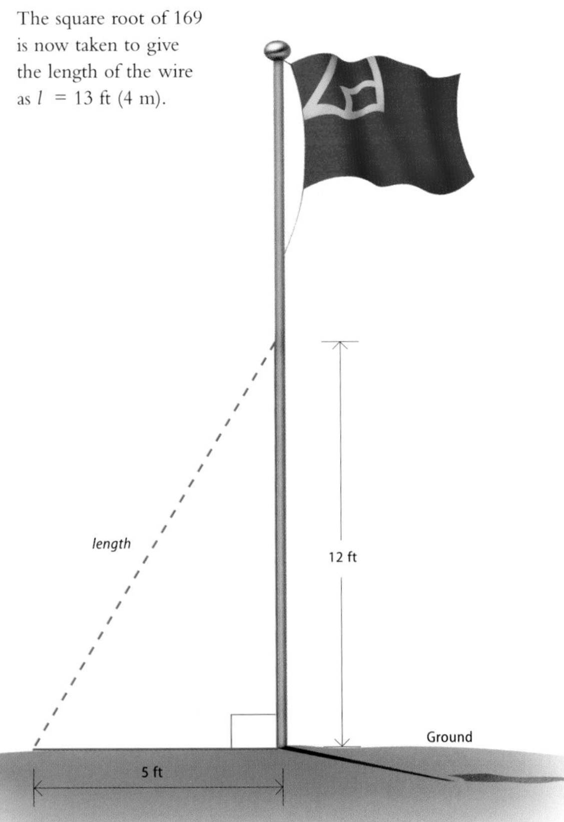

Opposite : Using the formula provided on page 99 to work out how far away a yacht is from rescue, we can determine that the distance to the horizon for a person standing on the ocean shore 5 ft (1.5 m) above sea level is 2.8 miles (4.5 km).

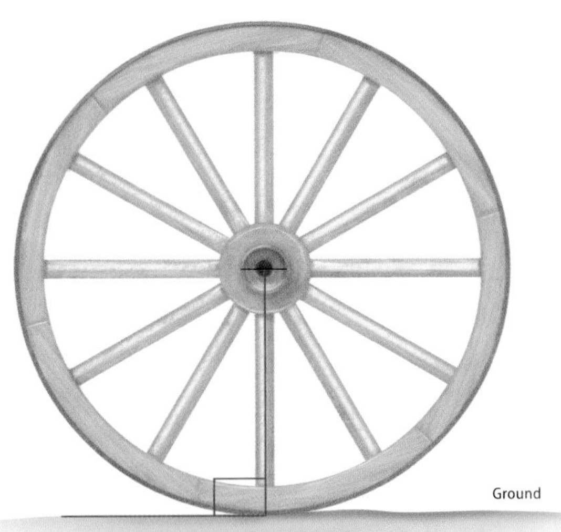

How far to the horizon?

A sailor sits in the crow's nest of a stranded yacht, 53 ft (16 m) above the ocean surface, scouring the horizon for signs of a rescue craft. Looking through his telescope, he sees another yacht. How far does this yacht need to sail to reach him? By using Pythagoras's Theorem, a general equation can be established which can then be applied to specific cases in order to find such unknown distances. Logic dictates that distance to the horizon depends on height above sea level of the observer. If we consider this height as miles, a right-angled triangle may be formed. One basic concept of circle geometry states that a tangent to a circle makes a right angle with the radius at the point of contact, much like the spokes of a wheel form a right angle with the ground (see above). The angle made by joining the center of the Earth to the horizon and then to the observation point is 90 degrees.

THE CONVERSE THEOREM

A converse theorem, or statement, assumes the result of the original theorem to qualify its assumption. Not all converses are true. For example, it would be reasonable to state that all people called Emily are female. The converse of this would be that all females are called Emily! This is clearly not true. The converse of Pythagoras's Theorem (demonstrated by this teacher) states that if the square of the hypotenuse is equal to the sum of the squares of the two shorter sides, then the triangle is right-angled. This can be shown to be true.

The story of the knotted rope used by the Egyptians to form right angles is a good illustration of the converse theorem in practice. The tale is told, that by taking a rope of 12 equivalent intervals separated by knots, Egyptian workers would form a triangle with sides of three, four, and five intervals. According to the converse theorem, the angle between the two shorter sides would be 90 degrees, which could then be employed in building and land measurement.

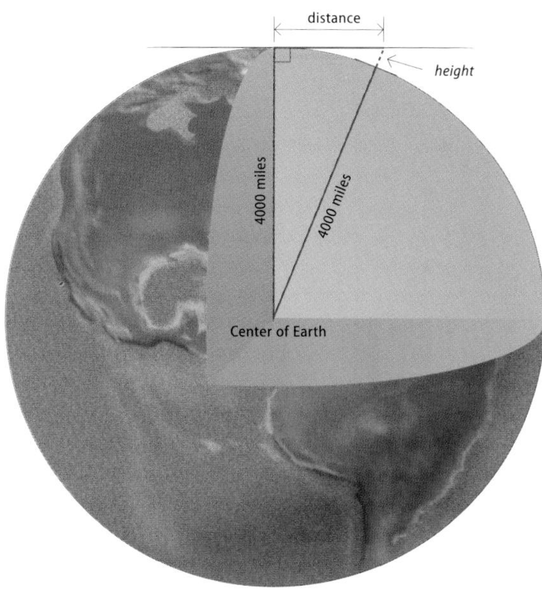

Bulging slightly at the Equator, Earth is not a perfect sphere, although we can consider its radius is approximately 4,000 miles (6,400 km). So the radius joining the horizon (H) and the center of Earth (C), is 4,000 miles (6,400 km), thus making the first side of the triangle. The second side is the line of sight (OH), which joins the point of observation to the visible horizon. This line is a tangent to Earth's surface with the horizon as the point of contact and is the unknown distance *d*. Our point of observation (O) is $(4000 + h)$ miles from the center of Earth and the line OC is the hypotenuse of the triangle. We now have a right-angled triangle as shown in the diagram on the left.

Applying Pythagoras's Theorem:
$(4000 + h)^2 = 4000^2 + d^2$,
squaring the left hand side,
$4000^2 + 8000h + h^2 = 4000^2 + d^2$,
subtracting 4000^2 from both sides
$8000h + h^2 = d^2$.
To find we take the square root of both sides:
$d = \sqrt{8000h + h^2}$.
This is the equation for the visible distance to the horizon from a point h miles above sea level. For our stranded sailor, as he is 53 ft, or 0.01 miles above the surface of the ocean the distance to the yacht upon the horizon is:
$d = \sqrt{8000 \times 0.01 + 0.01^2}$
$= 8.9$ miles (14.3 km).

There is geometry in the humming of the strings, there is music in the spacing of the spheres.

Pythagoras, mathematician and philosopher, *c.* 572–*c.* 490 BCE

Numbers are tools to help us solve problems and to understand the world better. Even more fundamental to our survival, we use numbers every day in countless ways. We love to see them increasing in our bank accounts and hate to see them increasing on a bill. But where did the concept of numbers and the symbols we use to represent them originate?

Numbers: the language of science

Very early civilizations, such as those found in Africa and Australia, had words for "one," "two," and "many," but appeared to have no need for further numerical detail. Their primary concern would have been food and shelter, so any numerical concepts that developed would have been linked to this. The first instances of "counting" were probably more a matter of matching objects than a sense of ascending or descending order. To make sure that the entire herd returned, a stone representing each animal was moved from one pile to another while the animals passed by. Stones left over meant that an animal was missing. These first "numbers" indicated quantity, but because there was no concept of order, it cannot really be considered as true counting. Numbers that only indicate quantity are called cardinal numbers. Numbers that reflect order (first, second, third, etc.) are called ordinal numbers and these came into use later in our history.

Above: This terracotta tablet dating from *c.* 2400 BCE is a record of accounts from Mesopotamia. Created in the fourth millennium BCE, cuneiform is recognized as the first known written language.

Finger counting

As human relationships developed, so did opportunities for business and trade. This required a jump from cardinal numbers to ordinal numbers. The use of ordinal numbers implies an ability to count.

We all carry devices that enable us to count—our hands and feet. Evidence of finger counting exists in many primitive languages where the word for "five" was "hand." Up until a few hundred years ago, mathematics textbooks from western Europe had pages of instruction on the art of finger counting. Even though our children still learn to count using their fingers today, this art has largely-been replaced by written figures and words.

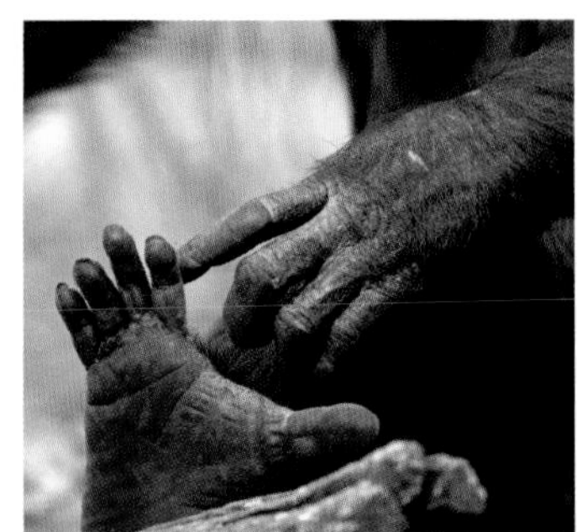

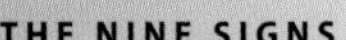

THE NINE SIGNS

Our current Hindu–Arabic numerical system based on nine symbols and zero had its origins somewhere between the fourth and seventh centuries CE. It gets its name from the Hindu Indians who developed the numerical system and the Arabs who passed it on to the rest of the world. Examples of written numbers thought to date from before 800 BCE look like those we use today and were given special names. The way the system developed in India is largely unknown, but the system appeared in Bagdad around 800 BCE. The Arabs carried it with them as they voyaged across northern Africa and into Europe. Here it was transformed into the current system by scribes (such as this one, *c.* 1460), mathematicians, and printing houses.

> ... when you can measure what you are speaking about, and express it in numbers, you know something about it; but when you cannot measure it, when you cannot express it in numbers, your knowledge is of a meager and unsatisfactory kind.
>
> William Thomson, 1st Baron Kelvin, mathematician and physicist, 1824–1907

Above right: Are we the only ones who can count? There is evidence that some animals have some comprehension of quantity—there are examples of chimpanzees that are able to add numbers and that can distinguish the answer from a group of numbers.

Opposite page: Binary numbers are represented here as hieroglyphics carved into a rock. A binary numerical system has 2 as its base, rather than 10. Only two numeric symbols, 0 and 1, are used.

Keeping count

When the primary concern of food and shelter was replaced by the desire to make a profit, record keeping developed. Some of the oldest artefacts found indicate that numbers were recorded as stroke markings between 20,000 and 30,000 years ago. In the 1930s, a wolf's leg bone was found in Czechoslovakia (present day Czech Republic) with notches cut into the bone, some 55 in all. The notches are in two groups—one of 25 and one of thirty. These are then assembled into subgroups of five. Other bones of similar age have been found in Africa with similar markings. While the meaning behind the markings is not known, the markings themselves indicate that the use of numbers and a basic understanding of some mathematical concepts were vital for the survival of early humans.

In ancient Babylon, Egypt, and Greece, numbers were written as a single stroke to represent each object counted. As societies and businesses multiplied, the number of objects that needed counting increased too. Single stroke counting became inaccurate. New symbols to represent 10 developed.

Around 3400 BCE, Egyptians used a hieroglyphic system based on 10 but lacked a symbol for zero, using a space instead. This meant that many numbers could only be understood by their context. The earliest recorded Roman numerals from around 500 BCE were also based on a system of ten.

Formulas (or formulae) are rules, facts, or principles that are written using symbols, including letters. When we use a formula, we replace the symbols with numbers to calculate the desired answer. The beauty of formulas is that they are portable and flexible; and they can be used and re-used in all sorts of different situations.

Phenomenal formulas

Using formulas to solve mathematical problems is a little like pilots using flight simulators as part of their training. When pilots use a simulator, they can try all kinds of different situations, such as storms or equipment failure, without worrying about crashing. Formulas give us the same power. We can try all kinds of different combinations of numbers on paper or at our desk to give different answers before we use the formula in the real world. For example, if we want to paint a wall and we need to calculate how much paint we need to buy, we can use the formula for the area of a rectangle—length × height. If we know the capacity of the paint tin, we can calculate how many tins of paint we need altogether, rather than having to buy tins one at a time. Looking at it in reverse, if we have a tin of paint already, we can work out how large a wall we can paint with it.

The famous

Formulas show us the relationship between different quantities. Geometry is the mathematics of shapes and their position and size. It uses formulas to calculate area, width, and volume. But formulas are not restricted to geometry—they extend far beyond the boundary of mathematics and can be found in all areas of science. Some of the more famous scientific discoveries are recognized by their formulas. Probably the most famous equation of all is $E = mc^2$. This formula was developed by Albert Einstein in the early twentieth century and illustrates the relationship between matter (physical substance) and energy (the ability to do work, to make physical changes).

The fantastic

Formulas reach even beyond the world of science and into the world of art, music, history, and architecture. The "golden ratio" is a formula, that when applied to an artwork or a musical composition, produces a work generally thought to be pleasing to the senses. First appearing in the fifth century CE, this formula is thought to play a role in the appearance of the Parthenon, the music of Debussy, and the art of Leonardo da Vinci. So what makes a ratio golden? Let's take two numbers called *a* and *b*. If we add *a* and *b* together, then the ratio of this number to *a*, is the same as the ratio of *a* to *b*. We can write this as a formula:

$$\frac{a + b}{a} = \frac{a}{b}$$

This formula is given the Greek letter ϕ or phi, reflecting in a small way the esthetic pleasure that this formula is responsible for.

The fabulous

Vedas mathematics is a system based on a number of formulas called sutras. These can be used in some complex mathematical problems to

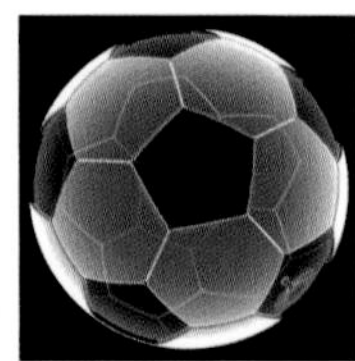

Above: Formulas are used for countless reasons and in many disciplines. For example, to find the area of a sphere, such as a football, the formula $4\pi r^2$ is used, where r is the radius.

Below: This spiral arrangement of circles resembling a nautilus shell can be described as a logarithmic pattern. The pleasing proportions of the pattern can be equated with the golden ratio .

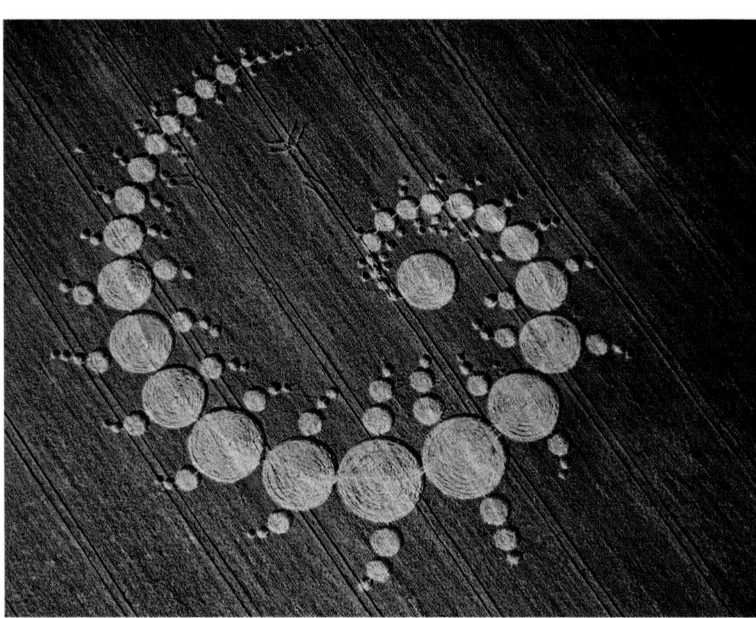

Left: An essential part of pilot training includes using a flight simulator, so certain conditions can be experienced without risk. In the same way, formulas give us the opportunity to test our hypotheses before applying them in the real world.

Below: The Greek sculptor Phidias was one of the men who built the Parthenon in the fifth century BCE using his concept of the golden ratio. The letter phi φ is named for him.

THE FAULTY

Formulas are only as good as the information that is put into them. There have been some spectacular examples where formulas have appeared to fail but where the disaster can actually be traced back to the data used. In 1998, NASA launched its Mars Climate Orbiter, designed to gather information about the climate of the red planet. Shortly before it was due to begin its orbit around Mars, it disappeared and was never found. The reason for the failure? While the formulas were found to be correct and appropriate, it was discovered that some scientists had entered inches and pounds into their formulas while others had used the metric meters and kilograms.

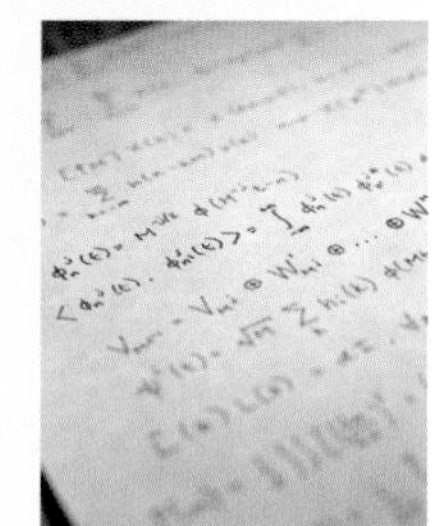

solve them without paper and pen. One of the formulas reads "by one more than the previous one." This can be applied to calculating the result of multiplying large numbers ending in 5 by themselves. One example is 65 × 65. In this case, the formula tells us to multiply the first digit, 6, by 6 + 1 = 7 to get 42. The second digit is 5 and is multiplied by 5 to give 25. The answer is reached by writing these two numbers together to give 4,225.

No human investigation can be called real science if it cannot be demonstrated mathematically.

Leonardo da Vinci, scientist and artist, 1452–1519

If subtraction is the opposite of addition and division can reverse multiplication, then logarithms are the opposite of exponentials. In other words, if you take any number and find its logarithm, then take this answer and find its exponential, then you will be back where you started.

Using logarithms

The number 2, multiplied by itself seven times can be written as 2^8. In this case the 2 is the base and the 8 is the exponent and the result is 256. If we were working backwards from 256 and we wanted to know how many times our base 2 had to be multiplied by itself to get 256, we simply take the logarithm of 256 to the base 2 and we come up with the exponent 8.

Logarithms were initially developed as a way of turning complex multiplication and division into more simple addition and subtraction. This ability, combined with pre-printed tables of logarithmic values, enabled complex calculations to be performed quickly and relatively accurately. Since their development, logarithms have proved to be useful tools in all forms of calculation, ranging from navigation to bank interest, and from chemistry to astronomy. It is claimed that logarithms are the most useful arithmetic concept in science and it all began with a simple problem.

Right: The slide rule—one of the most useful and portable calculating devices ever invented. A slide rule consists of a ruler with a central sliding piece. Both parts are marked with logarithmic scales, allowing for speedy calculations.

For the things of this world cannot be made known without a knowledge of mathematics.

Roger Bacon, mathematician and philosopher, *c.* 1214–1292

The slide rule

Shortly after the Scottish mathematician John Napier published his theory of logarithms, a mechanical device for calculating logarithms was invented. For over 300 years the slide rule was the primary tool for logarithmic manipulation. With a small, accurate slide rule and knowledge of how to use it, it was possible to find solutions that would otherwise have to be carried in a large book of log tables. These tables sometimes contained over 100,000 figures, were clumsy, and subject to deterioration. A slide rule was compact and durable and provided a means to calculate logarithms quickly and simply, much like electronic, scientific calculators of today.

Above: Logarithms can be used to describe sound intensity and frequency. When musical notes are played together, they can be pleasing to the ear or not, depending on the relationship between the frequencies.

Left: Interest rates are the amounts charged or paid for the use of money. Using logarithms can help determine how long it will take to pay off a loan to a financial institution.

Who uses logarithms?

Many different scientific measuring systems use logarithms. For example, the Richter scale of earthquake intensity is calculated by taking the base 10 logarithm of the horizontal part of an earthquake. The negative base 10 logarithm of the concentration of hydronium ions in a chemical is a measure that we call pH. Even the decibel that we use to measure the pressure of sound waves is expressed logarithmically. Scientists use logarithms for these purposes because they find it more convenient to deal with logarithms than numbers that could be either very small or very large.

Perhaps the most obvious use of logarithms for people today would be in calculating the interest on an investment. If you know the interest rate, the size of your bank balance, and that the bank compounds your interest continually, logarithms can help you work out how long it will take to double your money. Better still, ask your banker, because banks have computers specifically programmed with these logarithmic functions to do the work.

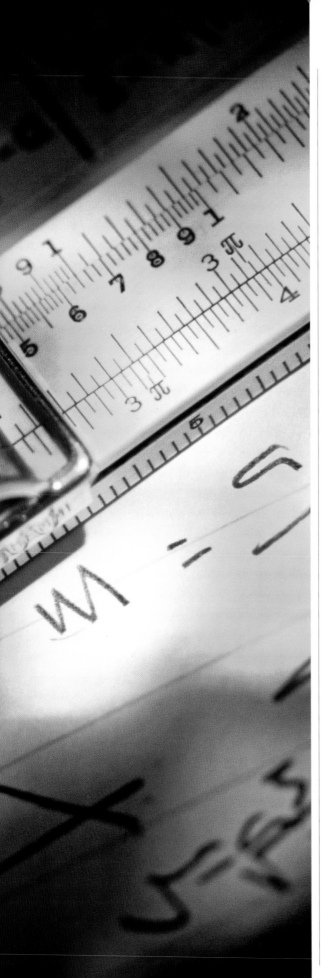

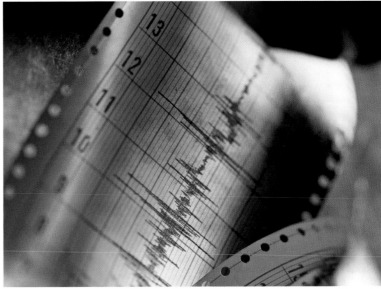

Below: There are various scientific measuring systems that regularly make use of logarithmic principles. The Richter scale uses a seismograph to determine the scale of earthquake intensity.

CELESTIAL NAVIGATION AND LOGARITHMS

The logarithm principle was first published in 1614 by John Napier. Napier's aim was to develop a system to help astronomers with their massive calculations. He thought that by reducing the amount of work that they had to perform, he would increase their productive lifespan. At the time, logarithms replaced the much more complex route to long multiplication and division called prosthaphaeresis. Prosthaphaeresis involved addition and subtraction of trigonometric formulas, resulting in approximations that would otherwise have been too tedious. It was still time-consuming, but it enabled charts to be developed that described the locations and movements of the stars and planets. These charts were used by navigators. However, after a relatively short period of popularity, prosthaphaeresis fell into disuse as logarithms became more convenient. The primary advantage of logarithms was that they enabled the navigator to calculate the large numbers required assisted only by a table of log values. This both increased the speed and the accuracy of the calculations, enabling a ship to stay closer to a chosen course. This in turn greatly reduced travel times and increased safety, both massive and important gains for maritime enterprise.

When Muhammed ibn Musa al-Khwarizmi wrote a handbook on how to distribute an inheritance according to complicated rules under Islamic law in Persia in about 820 CE, he created the earliest known algebra textbook. The Book of al-jabr and al-muqabala *literally translates as "restoring" and "balancing," two of the fundamental principals of algebra.*

Algebra

Algebra explained

Algebra is a system of mathematics for solving problems, in which variables, those numbers that can vary or are unknown, are given symbols or letters of the alphabet, such as x and y. These symbols are called pro-numerals because they take the place of numerals. A mathematical statement uses numbers and variables to make an algebraic expression, e.g. $2x + y + 2$. A statement with expressions on either side of an equals sign is called an equation. When a problem is solved, it may be expressed as an algebraic equation or rule called a formula.

From the formula for the area of a rectangle $A = l \times b$ we can calculate the area of any rectangle given its length l and its breadth b by substituting their known values. For example, if $l = 6$ and $b = 2$ we can calculate the unknown value $A = 6 \times 2 = 12$. Formulas have been derived for numerous geometry, applied mathematics, and accounting functions.

Solving equations

To solve equations we need to simplify expressions by collecting "like terms." Like terms are terms that are the same, such as a or a^2 or a^3. To collect the same terms we add the numbers in front of these terms: $3a + 2a = 5a$. This is the "restoring" referred to in the title of al-Khwarizmi's book.

In more complex equations it is possible to substitute all the known values and manipulate them to find the value of an unknown. At every step in the process the equation has to balance about the equals sign like a balance-beam, so whatever we do to one side of the equation we have to do to the other. For instance, in a rectangle with an area of 48 square ft (4.5 sq m) and length of 4 ft (1.2 m) we want to find the value of b the breadth. Substituting $A = 48$ and $l = 4$ in the equation $A = lb$ gives $48 = 4b$. First, we have to get rid of the 4 by using the opposite function to multiplication, which is division. We divide $4b$ by 4. But if we divide by 4 on one side we have to divide by 4 on the other. 48 divided by 4 = 12. Thus the breadth is 12 ft (3.5 m). This process of transferring quantities from side to side is the "balancing" in the title of al-Khwarizmi's book.

Left: Algebra uses mathematical statements to describe the relationships between things over a period of time. Letters or symbols known as variables replace numbers.

Above: Swiss mathematician Leonhard Euler (1707–1783) contributed enormously to the understanding of mathematics in the eighteenth century and beyond. As well as algebra, Euler wrote works on calculus, trigonometry, and geometry.

A brief history of algebra

Al-Khwarizmi was not the first to grapple with algebraic problems. The ancient Babylonians, Egyptians, Greeks, and Chinese also used algebraic methods. The Rhind Papyrus found in Egypt, dating back thousands of years states: "A quantity and its half are added together and become 16. What is the quantity?"

Since al-Khwarizmi, many mathematicians have contributed to the development of elementary algebra as it is taught in high school today: Del Ferro in Italy, Harriot in England, Viete in France, Leibniz in Germany, to mention just a few. It was Leonhard Euler in 1765 who formalized much of our algebra in a textbook called *Elements of Algebra*. More complex forms of algebra studied at higher levels of pure mathematics have been developed including fields, groups, rings, matrices, vectors, and abstract algebra.

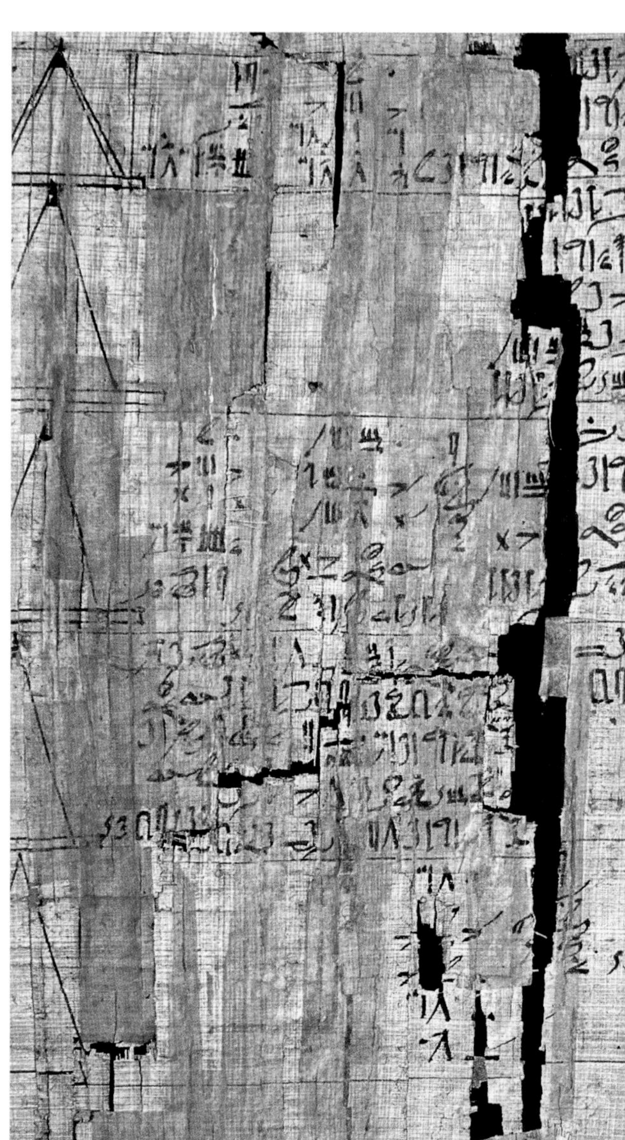

Left: Leonardo Fibonacci (1170–1250) helped spread knowledge of algebra and the use of Hindu–Arabic numbers throughout the west. He is also associated with the concept of the golden ratio (see page 102), seen here in the regular pattern of a nautilus shell.

Below: A section of the Rhind Papyrus, dating from at least *c.* 1650 BCE, but possibly hundreds of years older. The Papyrus is one of our major sources of information about Egyptian mathematics.

Algebra, the means of expression which is the simplest, most exact and best adapted to its object, is both a language and an analytical method.

Antoine Lavoisier, chemist, 1743–1794

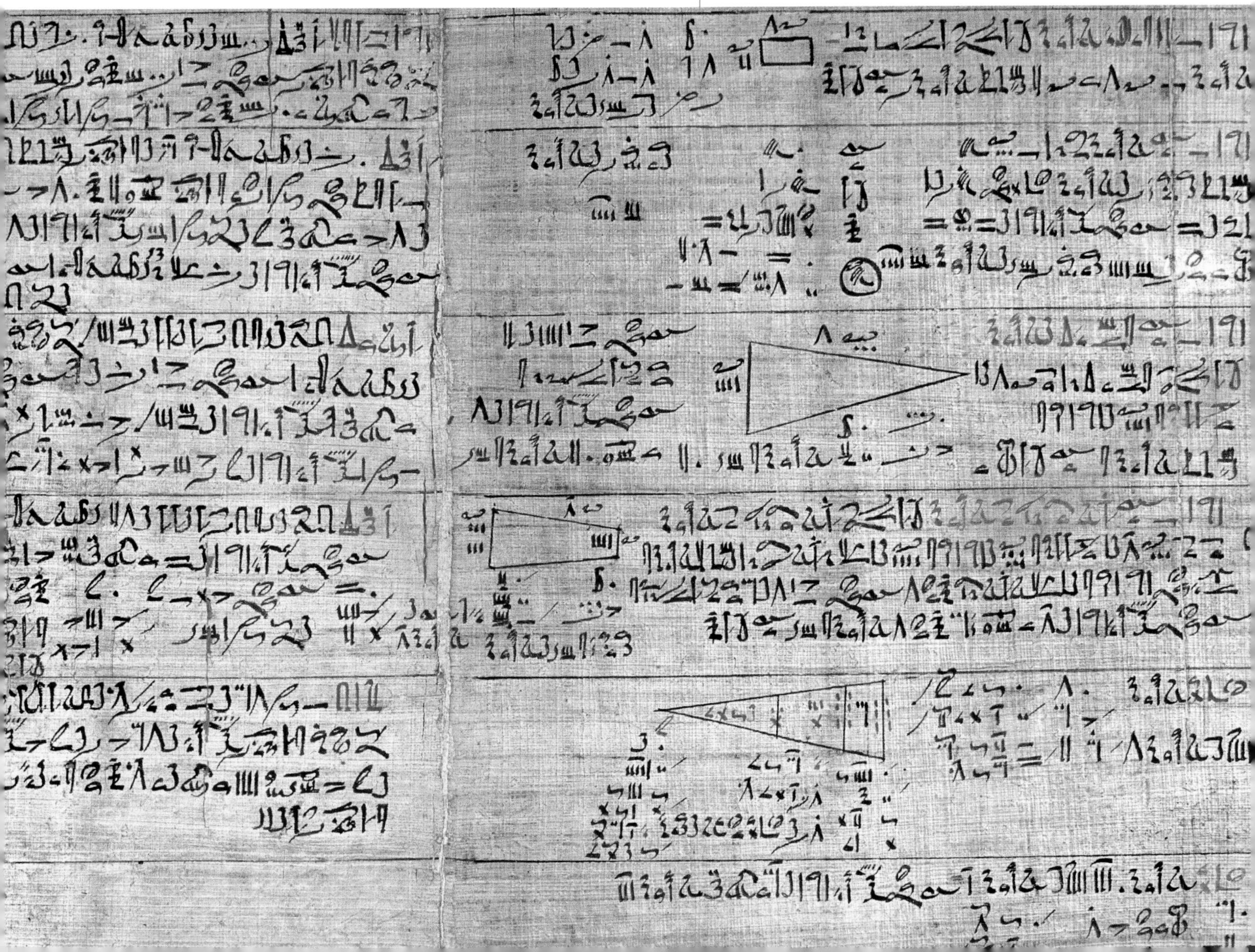

APPLICATIONS OF ALGEBRA

Algebra is integral to modern life, underpinning everyday situations, and is used in scientific fields ranging from medicine to physics. In fields like engineering, structural formulas and the principles of electronic circuits (such as the one below) depend on algebra. Banking and financial matters such as accounting and superannuation use formulas to calculate such things as compound interest, mortgages, and payouts when people retire. In addition, the techniques of algebra can be applied to probability in areas such as health statistics, genetics, and games of chance. Cooks, accountants, builders, and engineers all use algebra.

Not only is algebra fundamental to our world, the mental exercise of solving problems helps to develop logical thinking.

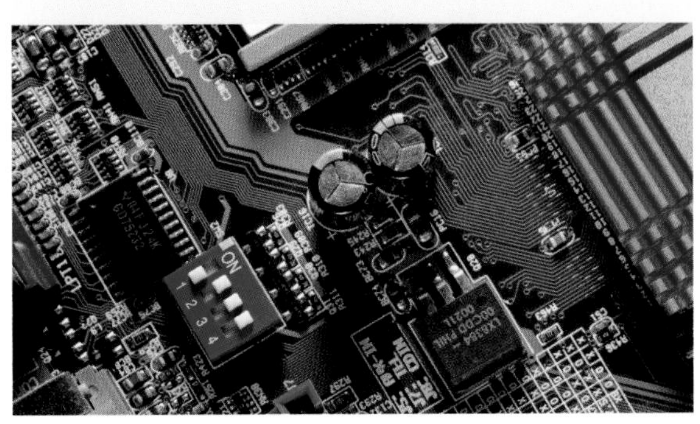

Since the Stone Age, humans have invented tools. From finding better ways to keep warm in a cave, to creating instruments enabling communication with people in faraway places, when a need arises, great thinkers have taken the seed of an idea to the germination of a solution. The development of calculus is no exception.

Calculating calculus

Although the word alone strikes fear into many hearts, without calculus, our world would be a lot more difficult to comprehend. Calculus is the branch of mathematics that we use to understand change and motion and how alterations in one quantity can affect another, related quantity. From space exploration to maximizing rental car bookings during busy holiday periods, calculus is the mathematics for solving complex problems.

From stones to stars

The word "calculus" is thought to have originated from the Latin word for pebble. This may to refer to the ancient practice of using stones to count objects. By moving a stone from one pile to another, shepherds in times gone by could be sure that all their animals had returned. Modern calculus however, is better known for its ability to make sense of changing quantities and to measure rates of change. It is used widely in all fields of science, as well as engineering, medicine, computers, and business. Calculus uses mathematical functions or rules that can predict or anticipate changes in a variety of situations. Mathematicians can then draw graphs that represent these functions and can examine the function over a wide range of numbers. These situations include times when the changes are very small or very large. One major use of calculus is in the field of space exploration. Scientists and mathematicians need to be able to understand how the relationships between quantities such as speed and temperature will change when the time scale changes too.

The development of calculus

Although Isaac Newton and Gottfried Leibniz are credited with the invention of calculus in the seventeenth century, the Egyptians are known to have found the volume of a pyramid over 3,000 years earlier. Mathematicians such as the Greeks Eudoxus (*c.* 400–355 BCE) and Archimedes (287–212 BCE) and Italy's Bonaventura Cavalieri (1598–1647) continued to touch on various elements of the early foundations of integral calculus. It wasn't until Pierre de Fermat in the early seventeenth century that concepts involving differential calculus began to be explored on a deeper level.

Above: Archimedes was one of the great scientific minds of antiquity. Almost 2,000 years before Gottfried Leibniz and Isaac Newton announced their mathematical findings, Archimedes had anticipated the advent of integral calculus.

In 1665 the bubonic plague rampaged through Britain and many public institutions closed their doors. Cambridge University was no exception and for a few years Isaac Newton left Trinity College to work at home. During this time he pursued the development of his "methods of fluxions" known today as differential calculus. This was a rigorous dissection of the foundations of calculus including verifiable proofs. Isaac Newton's career was marked by a reluctance to publicize results due to fear of possible controversy. His delay in the release of his latest findings proved disastrous.

German mathematician Gottfried Leibniz also released works on calculus and the debate as to who was the originator of the work raged for years. Much of Leibniz's notation is still used today, such as the sign for the integral: an elongated 's' ($\int$) used to represent the sum of Cavalieri's indivisibles. Many of the rules Leibniz that derived for differentiation are still taught to school students today.

Calculus for all

Calculus can be divided into two branches. The first one is integral calculus, which primarily deals with areas and volumes. The second is differential calculus, which is applied to functions and phenomena in order to determine rates of change and maximum and minimum values. These two seemingly diverse applications are actually strongly

Below: Calculus is applied practically in many fields including engineering and bridge-building. Calculus can predict possible stresses and other changes to a bridge over a long period of time.

linked. When we square a number, then take the square root of the answer, we return to the original number. Squaring and taking square roots are thus known as inverse operations. In the same manner, when the integral of a function of x (used here as the independent variable) is taken (written as

$$\int f(x)dx)$$

the result is known as the primitive of $f(x)$. By differentiating this primitive with respect to x,

$$(\frac{d}{dx}\int f(x)dx)$$

we are led back to the original function and thus differentiation and integration may also be considered as inverse operations.

Integral calculus

Whereas areas bounded by straight lines or volumes of solids encased in flat sides have a logical simplicity to their solution, once the edges and sides become curves, a less obvious method of evaluation is required. Integral calculus takes a curve defined by a function and calculates the area bounded by the curve, an axis of the Cartesian plane and two straight lines. See Figure 1.

Above: It might be hard to believe, but without calculus, computer games wouldn't be nearly as exciting and realistic as they are. Programmers use the principles of calculus to build and maintain action-packed games.

FIGURE 1

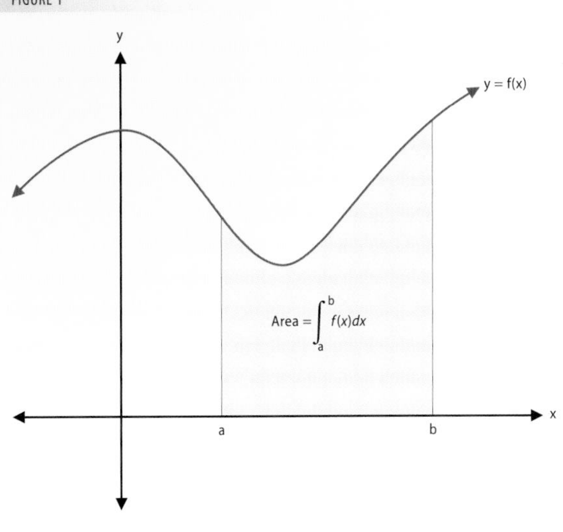

Top left: Mathematician and philosopher Gottfried Leibniz (1646–1716), one of the founders of modern calculus, introduced several notations to mathematics, including the integral sign (∫). Leibniz also invented a calculating machine.

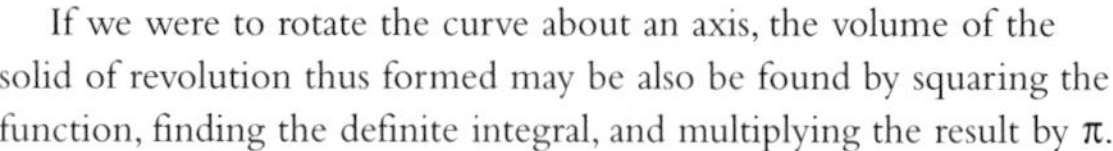

If we were to rotate the curve about an axis, the volume of the solid of revolution thus formed may be also be found by squaring the function, finding the definite integral, and multiplying the result by π.

Differential calculus

Differential calculus is a method by which this rate of change for a curve can be measured. The process of differentiation allows us to find the gradient of a tangent to a curve at a point of contact. See line 1 on Figure 2.

As a consequence, we can also investigate local maximum and minimum values of a function by looking at points at which the gradient of the tangent is equal to zero (line 2 on Figure 2), as these represent turning points on the curve.

Zeno's paradox

A fundamental concept underlying calculus is that of infinitesimals. The areas required for an integral are based upon the sum of strips with infinitely small widths and the increments in the change in x used to find the derivative are also infinitely small. The philosopher Zeno challenged seemingly logical notions involving infinite sums in the fifth century BCE. Prior to the tale of the tortoise and the hare, one of Zeno's paradoxes could be well illustrated by the story of the tortoise racing Achilles, the great athlete. Achilles, being fair minded, offers the tortoise a head start of 100 ft (30 m). The tortoise then claims that he will win the race offering this logical explanation: "After running 100 ft, you will reach my starting point by which time I shall have moved on to a second point closer to the finish line. You will arrive at this position after I, once again, will have moved on. You will continue to narrow the gap in such a fashion; however, I will always retain that ever decreasing advantage resulting in my crossing of the finishing line first!" This story is based upon the concept of time being made up of infinitely divisible parts.

FIGURE 2

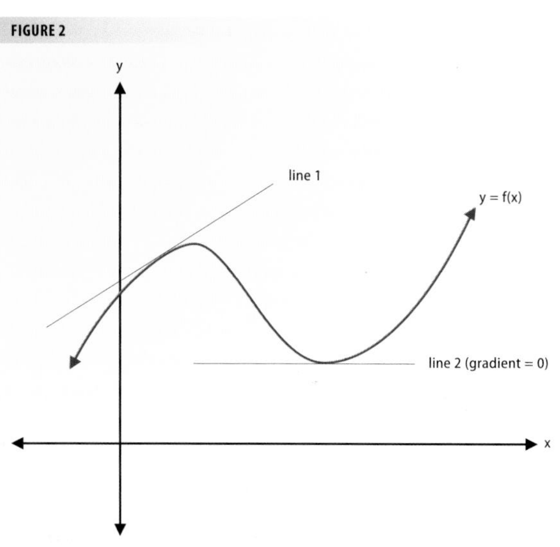

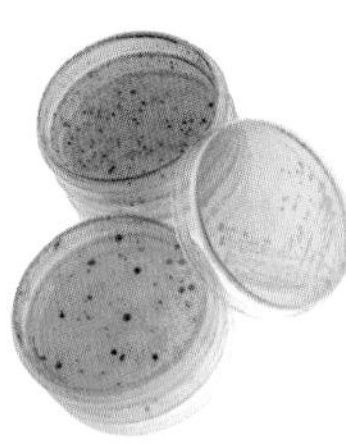

Above: When microbiologists and medical researchers study bacteria, one of the tools they use is calculus, which allows them to come up with valid formulas to determine the rate of bacterial growth.

Calculus has its limits.

Keith Devlin, mathematician and writer

The mathematics behind the games

Calculus plays an important role in making computer games. In the artificial world of gaming, the action needs to be fast and realistic, and the graphics spectacular and cutting edge. In games involving outer space, the visual effects of gravity need to be considered and included. Those games involving water need to recreate the right amount of viscosity to appear realistic. A graphics programmer needs to use calculus and its derivatives as the tools to build, maintain, and advance the world of the computer game.

Medical calculus

Calculus is also useful for predicting how populations will behave over time. This is particularly useful when looking at bacteria because the population numbers are so high and the rate of growth so rapid. Calculus can help create a formula that can be used to estimate the population size at any given time. This means that the effectiveness of antibiotics on the bacterial population can be examined, which can help in testing and designing new antibiotics for the future.

MAKING SENSE OF IT ALL

Differential calculus studies the effects of one small change, while integral calculus looks at the accumulation of many small changes. One way to look at these two categories is to use an everyday example—a car dashboard. The dashboard has an odometer which measures how far the car has gone in miles (or kilometers). Close by is a speedometer, which measures how fast the car is going in miles per hour (k/ph). Calculus is about the relationship between these two quantities.

If we only know how far the car has gone, then we can use calculus to work out how fast we went by using differential calculus or differentiation. By drawing a graph of the distance traveled and the time taken, the velocity can be calculated by working out the slope of the graph. Integral calculus or integration uses the speed we travelled at, or velocity, to calculate the distance we traveled. Again, by using a graph of the velocity and time taken, we can use the area under the graph to calculate the distance travelled.

Opposite: Calculus is used to predict the size and growth of bacterial populations such as this one—here, *Micrococcus* bacteria (on an agar plate) magnified 21,000 times. Knowing how bacteria behave helps medical science to design and test effective antibiotics.

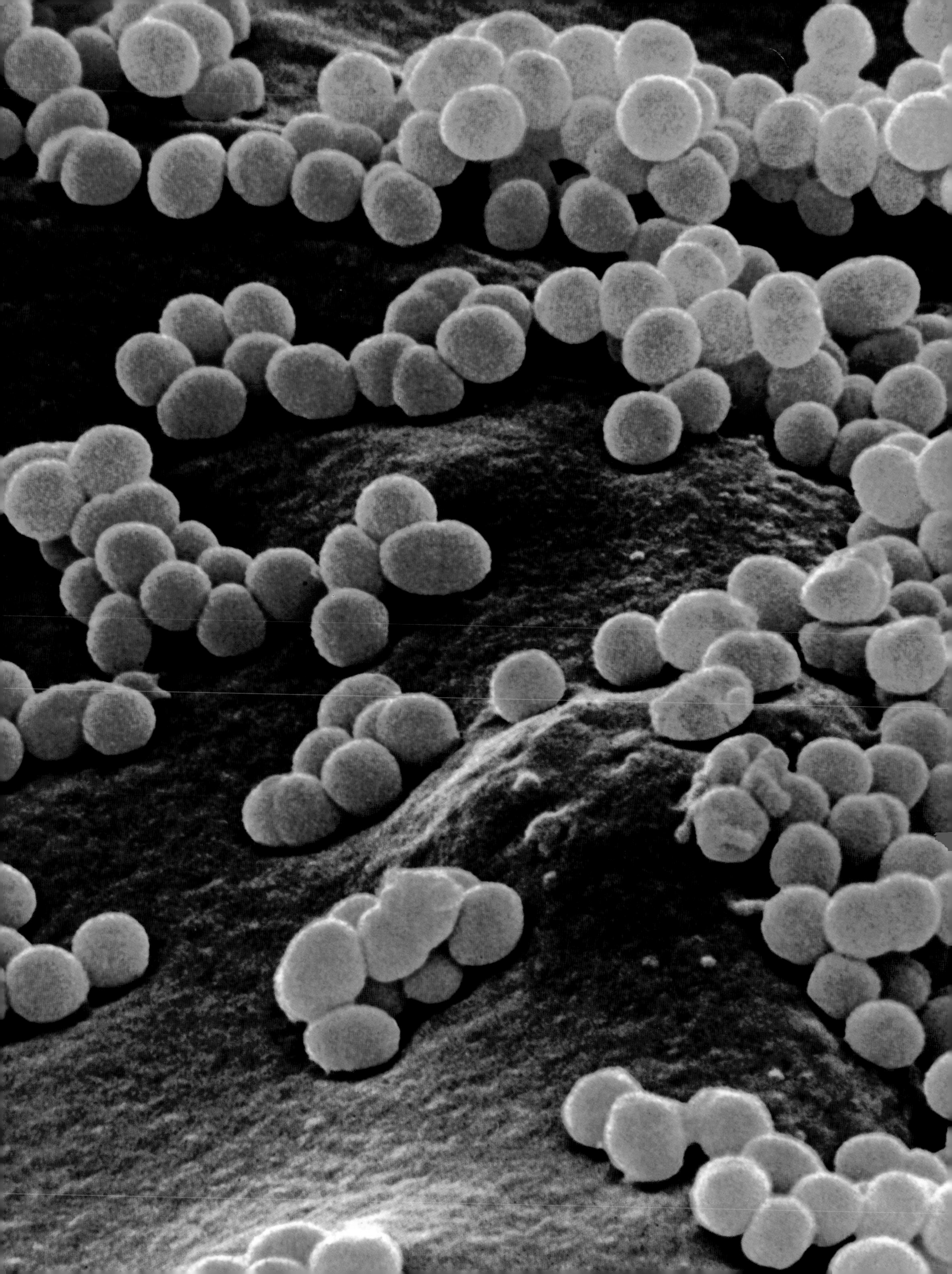

Differential equations are a specific type of equation that relate to change. They describe how changes in one thing are related to changes in another. One differential equation might describe the relationship between daily temperature and the number of ice creams sold by a street vendor; another might describe the rate of decay of a radioactive element over time.

What are differential equations?

Many people who buy a house are confronted with borrowing money from a bank. They borrow what they need and pay it off in equal installments over the life of the loan. However, the bank will also want to charge interest as a percentage of the amount left owing. This becomes more complicated when we realize that interest may be calculated based on the amount remaining, as well as any interest that has accumulated since the previous payment. Thankfully, banks use computer software that can model the rate of change of what we owe as they input various combinations of interest rates, repayment size and frequency, and any other variables. These models are powered by differential equations.

Right: Most people need to borrow money to buy a home and this is probably the biggest financial commitment they'll ever make. The installment amounts we repay the bank at regular intervals are based on models determined by differential equations. Mathematics plays a huge role in our lives.

Modeling

Differential equations are often used in scientific modeling. Whether scientists are studying the rate of growth of a population, the rate of decay of an element, or the rate of popular acceptance of a new technology, all can potentially be described using differential equations. To do this, scientists try to recreate the relationships that exist in the real world to use in their model. Researchers study the entire system in close detail to discover the relationships in play. Then they limit the complexity of the relationships by reducing the number of factors involved so that they can be modeled. These relationships are then used to construct a set of differential equations. The equations are tested by using them on known data to see if they produce accurate results. If not, the differential equations are adjusted and refined, then tested again. This process can go on many times until the researchers are satisfied that their model sufficiently represents the real phenomenon. If a set of differential equations has been demonstrated to describe a current situation, then scientists have a very reasonable expectation that the equations can be used to predict future events.

Modeling with differential equations is often used to describe the growth and decline of populations worldwide. It can easily be seen that the change in the size of a population is dependent on the current size, age, and death rate. All these factors must be included in the model to predict future population size. However, it could also be argued that the income level of the population, added to the standard of living that the income affords, could also impact on population change. Other factors include population optimism, social grouping, education, and the standard of health care. The more of these factors that are included in the model, the more likely that it will accurately represent the real world. However, the more that is included, the more complex the differential equations become and the more data will be needed. This is the compromise that must be made in using differential equations. To develop a set of equations that is both manageable in complexity and data volume but also acceptably accurate in representing the real world is a balancing act that scientists and mathematicians must constantly perform. There will usually be many more variables that could be included in an equation, but which of these should be chosen?

Right: The Biblical tale of the Tower of Babel tells of the time when languages suddenly diversified. Since ancient times, there has been growth and decay in languages, and even today, many languages are still being lost. This growth and decay can be modeled using differential equations.

Below: Differential equations can be used to determine the rate of growth of mushrooms, or indeed any plant. Factors considered include soil type, climate, rainfall, sunlight, and pests. There are advantages for farmers in such knowledge.

> Among all of the mathematical disciplines, the theory of differential equations is the most important ... It furnishes the explanation of all those elementary manifestations of nature which involve time.
>
> Marius Sophus Lie, mathematician, 1842–1899

Left: Differential equations are used to find the shortest path between two points on a curve. This is known as a geodesic. A geodesic dome uses differential equations to make structures like domed buildings. They look round but are actually made up of straight lines.

NEWTON'S SECOND LAW

Differential equations describe many situations that involve change. Probably the most famous differential equation was formulated by Isaac Newton to describe the rate of change in momentum of a body. His second law of dynamics, often abbreviated to $F = ma$, describes the relationship between the acceleration of an object (a), its mass (m) and the force applied to it (F). This relationship is explained simply by stating that the acceleration (the change in velocity) of the object is proportional to the force acting on it but inversely proportional to its mass. In other words, as the force increases, the acceleration increases, but as mass increases, acceleration decreases.

When we think of the word "chaos," we think of something more likely to be found in a science fiction movie than a laboratory. But chaos is the new mathematics, and research into the science behind chaos is providing new ways to understand such diverse and complicated systems as the weather, financial markets, and how populations grow and decline.

The order of chaos

Early chaos

Until the late 1800s, scientists believed that long-term predictions about anything could be made with accuracy, as long as the initial information was correct. This way of thinking is called determinism, an ancient way of looking at the world that became part of scientific thinking during the 1500s. In the seventeenth century, Isaac Newton used determinism to develop his laws of motion, which describe movement in the physical world. But it soon became obvious that the idea of cause and effect was not always so clear-cut.

In the late nineteenth century, French scientist Henri Poincaré discovered that the orbits of some of the planets did not follow the deterministic idea. Poincaré saw that even when the information concerning the planet's orbit was recorded more accurately, the long-term predictions about how the planet would move became more uncertain and actually appeared almost random. Research into this phenomenon continued, but it was not until computers became available for mathematical research that the full extent of chaos theory could be understood.

The butterfly effect

Because chaos theory depends on being able to look at long-term predictions while making small changes in the initial information, computers that could repeat calculations quickly were needed for the progression of research into chaotic systems, such as the weather. Very small changes in atmospheric pressure in one place can have big consequences days later, many miles away.

In the 1960s, while American meteorologist Edward Lorenz was taking a shortcut through a computer program that modeled how air currents changed in the Sun's heat, he discovered that by slightly changing the initial information, the long-term prediction changed dramatically. Describing his results, he used the analogy of the flap of a seagull's wing being responsible for changing the path of the weather. By 1972, the seagull had been replaced by the more picturesque butterfly and a flap of its wings in Brazil was held responsible for a tornado in Texas, USA.

Right: A frond of the maidenhair fern (*Adiantum pedatum*). Ferns are a wonderful example of the characteristic of self-similarity—each fern leaf consists of even smaller leaves that resemble the larger leaf. The closer you look, the more detail becomes apparent.

Below: Many plants, such as the Teddy Bear cholla cactus (*Opuntia bigelovii*), show something of the infinite detail of fractals. The closer you look at the parts of the plant, the more you can see the repetitive pattern of its growth.

Right: The work of French mathematician Henri Poincaré (1854–1912) on the so-called three-body problem dealing with planetary orbits became the basis of modern chaos theory.

Below: There are many objects occurring naturally, apart from plants, that display the self-similar characteristics of fractals, for example, snowflakes, lightning bolts, and these quartz crystals.

The fractal fantasy

For a system such as the weather to be called chaotic, it needs to be iterative, that is, made up of a set of repeating processes. There are geometric shapes that also have this quality and can help us understand the nature of chaos. These shapes cannot be described using traditional geometry. They are called fractals, are chaos in geometric form, and are made by repeating a shape over and over again; there is no end to their complexity, beauty, and usefulness.

> It is far better to foresee even without certainty than not to foresee at all.
>
> Henri Poincaré, mathematician and physicist, 1854–1912

WHAT MAKES A FRACTAL A FRACTAL?

Fractals are shapes with three important features. First, they have self-similarity. A shape that is self-similar is made up of smaller copies of itself. As you zoom in or out, the image looks the same. Look at a snowflake. Compare a rugged rocky mountain, a patch of rocky soil, and an enlarged image of the surface of a rock. Even though their size is very different, their appearance is much the same.

Second, fractals have a special dimension all of their own. The traditional geometric shapes are classified as one-, two-, or three-dimensional. Fractals have dimensions that are not whole numbers, so compared with two- or three-dimensional shapes, they are much more complicated.

The third feature is repetition, or iteration. This repetition means that fractals have a very unusual length. They have a definite area but are made up of infinitely long lines. This can be visualized by thinking about measuring a mountain range or a coastline. Using a ruler that is one foot long will give you a smaller measurement than if you use a ruler that is one only inch long. This is because the smaller ruler is able to measure the really small curves and dips of the coastline that the bigger ruler cannot.

In fact, the smaller the ruler that is used, the longer the length becomes. This is called the coastline paradox and is part of what makes fractals such useful shapes.

Two mirrors placed face-to-face will show ever-decreasing images of the objects reflected by each other. When viewing these images, we are looking into infinity. The Mandelbrot set is an image utilizing these conceptual realms to mathematically create objects of incredible beauty and genuinely infinite intricacy.

The Mandelbrot set

The magnitude of a real number may be demonstrated by placing it in an appropriate position on a number line. When two variables are required, for example, when finding a location using a street directory, a set of perpendicular axes may be used. Here, two components, a horizontal reference and a vertical reference (sometimes referred to as an ordered pair), are combined on the grid to locate the required site. Generating sets of ordered pairs according to a given rule is a concept familiar to most high school students. Such sets may be plotted on axes to form straight lines, parabolas, exponentials, and other conventional curves. It is in the same manner by which a complex number (see box) may be represented. The real part is represented on the horizontal axis and the imaginary part on the vertical axis. The Mandelbrot set, named for its creator, Benoit Mandelbrot, is a collection of such points on this plane. It has become the icon of chaos theory and much of the allure of the Mandelbrot set stems from the fact that an object of great beauty and complexity is generated by a charmingly simple formula.

The creation of each point in the Mandelbrot Set is a process of iteration (continuous repetition of the same procedure). To generate points, a complex number, C, is "tested" as to how it behaves when

Above: Using the power of computers while investigating the properties of Julia sets, Benoit Mandelbrot was able to bring to life the esthetic exquisiteness, created by his simple formula, of the set that now bears his name.

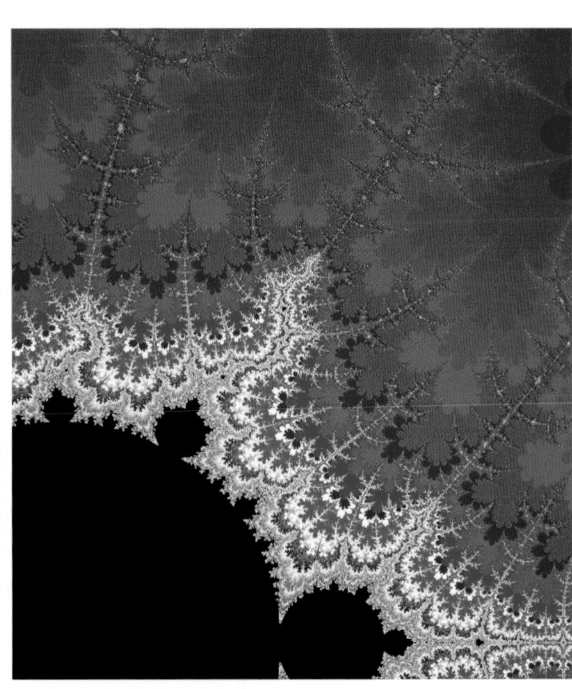

> A cloud is made of billows upon billows upon billows that look like clouds. As you come closer to a cloud you don't get something smooth, but irregularities at a smaller scale.
>
> Benoit Mandelbrot, mathematician, b. 1924

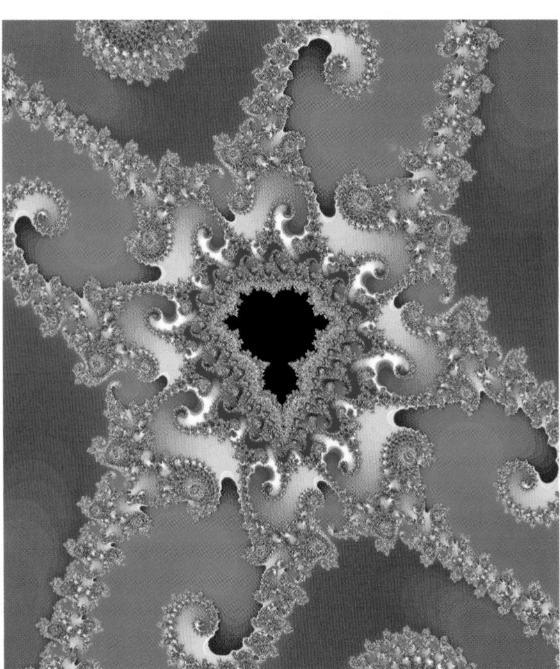

Top right and left: These magnified fractal images from the Mandelbrot set show something of the fine detail and intrinsic beauty of the entire set. Mandelbrot studied the phenomenon of fractals, observing them in objects ranging from blood vessels and lungs, to plants, mountains and up to that of galaxies. He also found that fractals exist in made-made phenomena such as music, architecture, and even the stock market.

COMPLEX NUMBERS

A complex number consists of the sum of two parts—a real part and an imaginary part. The real part is any real number. It can be rational (that is, it is possible to write it as a fraction) or irrational such as π or ÷ 5. The second element is more esoteric. It is the product of the value known as *i*, and to understand the concept of *i*, we need to forget that our high school math teacher told us that we could not take the square root of a negative number. This is because *i* is defined as the square root of minus 1. The value was specifically created so that such a square root of a negative could be utilized. For example:

$$\sqrt{-25} = 5i$$

$$\text{as: } 5i \times 5i = 5 \times 5 \times i \times i = 25\times(-1) = -25$$

Regular mathematical operations such as addition can be performed on complex numbers. For example $(4 + 5i) + (7 + 2i)$ can be found by adding the real parts $(4 + 7)$ followed by the imaginary parts $(5i + 2i)$, giving the result $11 + 7i$. Complex numbers may also be squared (as required for the creation of the Mandelbrot set) in the same manner as any binomial.

For example:

$$\begin{aligned}(2 + 6i)^2 &= (2 + 6i)(2 + 6i)\\ &= 4 + 24i + 36i^2\\ &= 4 + 24i - 36 \text{ (as } i^2 = -1)\\ &= -32 + 24i.\end{aligned}$$

Hence the square of a complex number is another complex number.

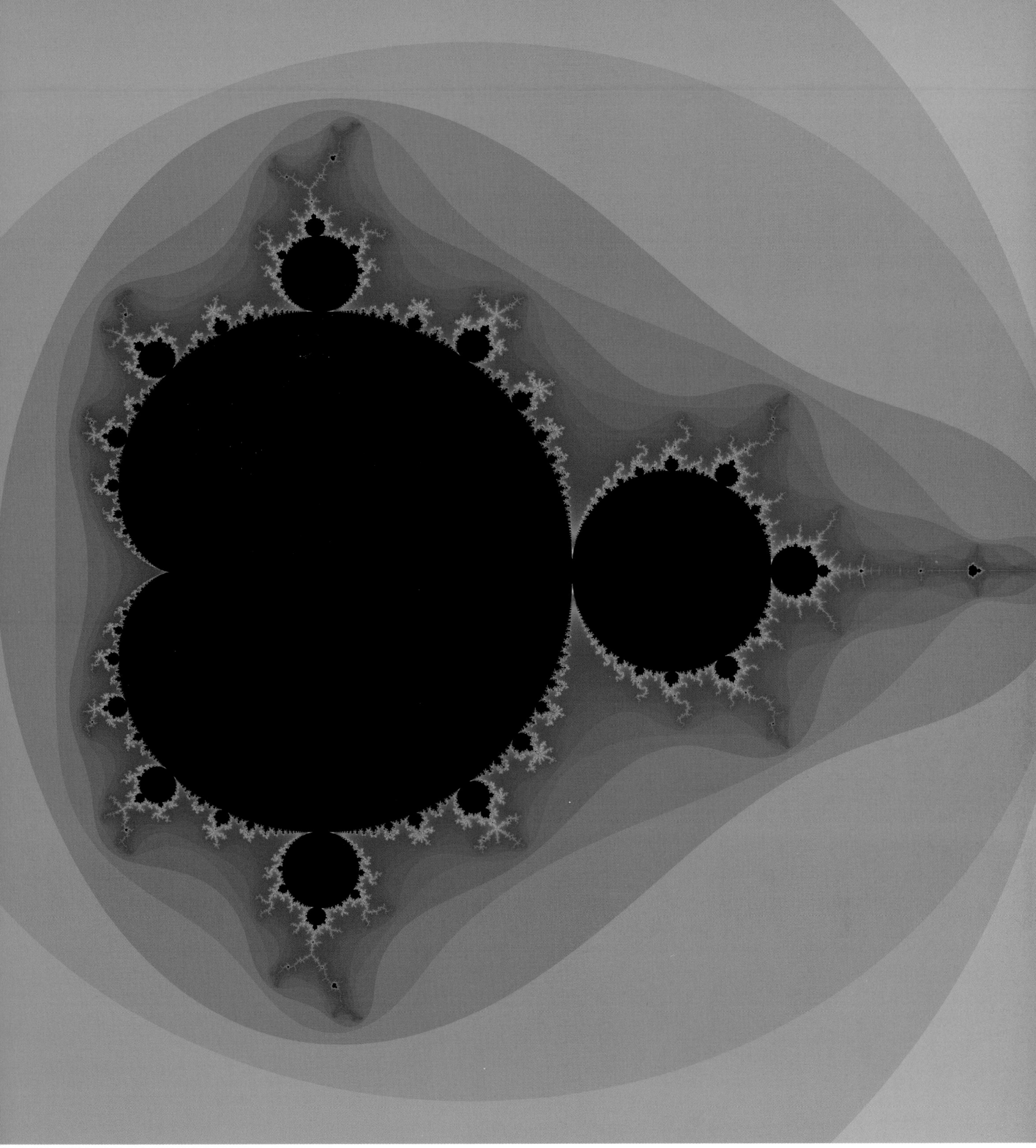

placed in the formula $Z = Z^2 + C$. The initial value of Z is zero, which is then squared and the result added to C. Hence a new value for Z is created, which in turn is squared and added to C to generate the next value of Z. With each subsequent iteration, Z changes while C remains constant. As the number of iterations increases there are two possibilities for Z. The first is that the magnitude of Z—its distance from the origin (0, 0)—oscillates about a fixed value. (Consider a ball bouncing between two planks of wood, one of which is moving downward, continuously reducing the distance of each bounce.)

Above: The Mandelbrot set is a set of all points in a complex plane computed by iterating a certain mathematical function. The set is also popular with non-mathematicians for its essential beauty and its complex structure that eternally repeats itself.

Values of C satisfying this criterion become the border, or "shoreline" of the Mandelbrot set and are generally colored black. By constantly zooming in on this shoreline, we see a structured series of variations and mutations consistently returning to a main theme. The second possibility is that the value of Z rapidly becomes increasingly larger. For each of these points a color is allocated dependent upon the number of iterations required for the value of Z to shoot off towards infinity. These points make up the stunning colored "ocean" outside the border of the set.

"There are three kinds of lies: lies, damn lies and statistics." Attributed to Benjamin Disraeli, this statement in itself is not quite true. It is the interpretation and display of statistics where the true lies begin! One of the best ways to display numeric information is by graph and fundamental to the interpretation of the graph is our understanding of the scale.

Interpreting scales and graphs

Central to any discussion of scales is the creation of graphs. The representation of information using a basic graph generally involves two elements known as the independent and dependent variables (values that change relative to each other). If we consider graphing the distance traveled by a free falling object against time, distance traveled would be the dependent variable, as its value is determined by the number of seconds (time—the independent variable) for which the object has fallen. The vertical axis could show distance while time would be represented on the horizontal axis.

Paramount to a graph's accuracy is the selection and appropriate display of the scale. The scale of both axes is determined by the range of time required. If the graph were to display the first 10 seconds of the fall, then the horizontal scale would be best represented using increments of individual seconds. These would be spaced evenly along the axis. During this period the object would travel 1608 ft (490 m) (ignoring air resistance) and as such it would be appropriate to display distances from 0 to around 1,700 ft (0–520 m) on the vertical axis. A suitable scale would utilize increments of 200 ft (60 m). Using the table of values below, the graph can be drawn:

Time (seconds)	0	1	2	3	4	5	6	7	8	9	10
Velocity (feet/second)	0	16	64	145	257	402	579	788	1029	1302	1608

Distance (feet)
Time (seconds)

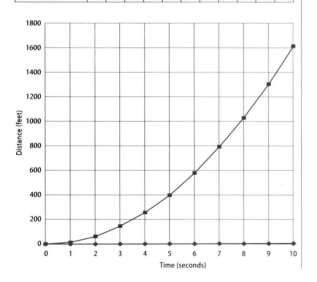

I see it, but I don't believe it.

George Cantor, mathematician, 1845–1914

What should be noted is the even grading of the vertical axis. A common error is to use the values for the dependent variable (0, 16, 64, … 1,302, 1,608) as the evenly spaced incremental values. This would result in a straight line graph (see above) and a hence a very inaccurate representation of the information.

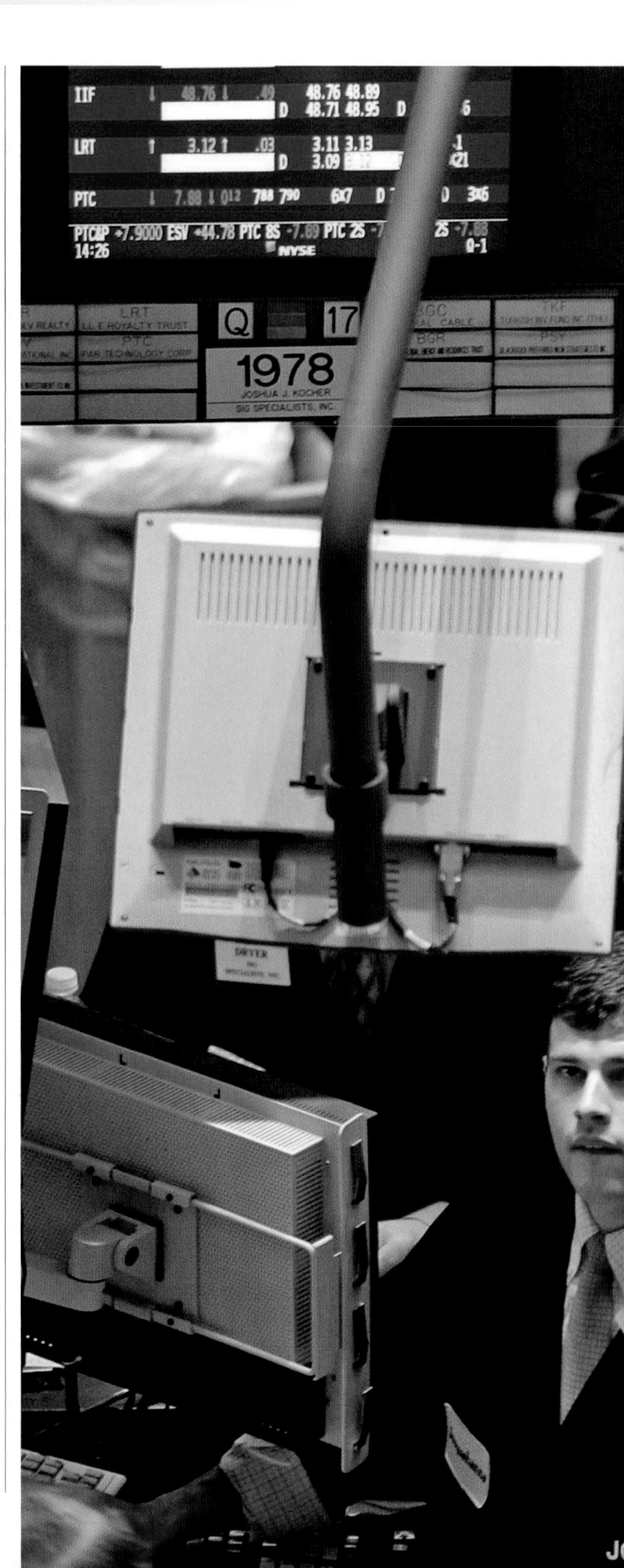

Right: Traders on the floor of the New York Stock Exchange constantly monitor what is happening in the share market. One of the most effective ways of gauging the performance of the overall stock market is to watch the Dow Jones Index, which shows how a selected number of large companies are performing.

OTHER SCALES

Scales are not only used for graphing. One familiar scale is a musical scale. The human ear is sensitive to variations in pitch. As a scale is played, the notes ascend or descend in an ordered, even, and melodic fashion allowing us to recognize when a wrong note is played. It is this even progression that defines the set of notes as a scale. There even exists a scale calibrated to measure a chilli's heat. Their intensity is measured in Scoville units, based upon the number of parts per million of capsaicinoid—what gives chilli its bite. On this scale the humble bell pepper scores a maximum of 100 Scoville units, while some habeñero chillies score a mouth-scorching 300,000!

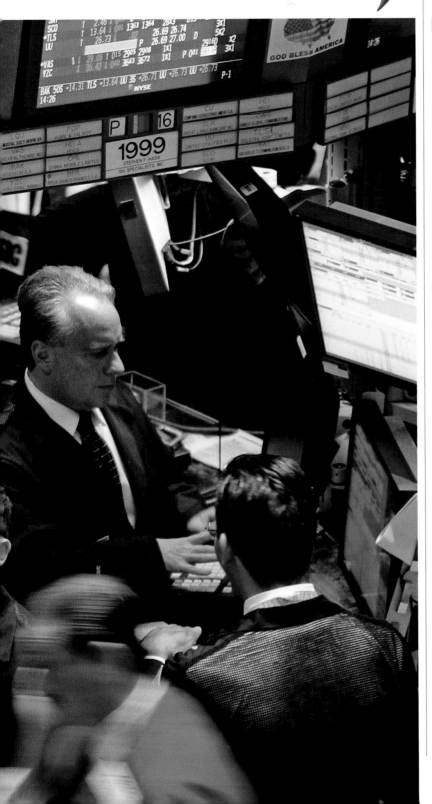

Misleading graphs

In cases of misleading graphs, it is often the representation of the scale that plays the villain. A simple demonstration can be made of the persuasive power of a graphical display by looking at a column graph of the rise and fall of Dow Jones index (a scale in itself) over the period of the year 2006, a good year for investors. The Dow rose by a healthy 16 percent from a December 2005 close of 10,717, to a December 2006 close of 12,463 points. Representing this on a graph shows the December 2006 column as 16 percent larger than that of the corresponding period of the previous year. See Figure 1.

FIGURE 1

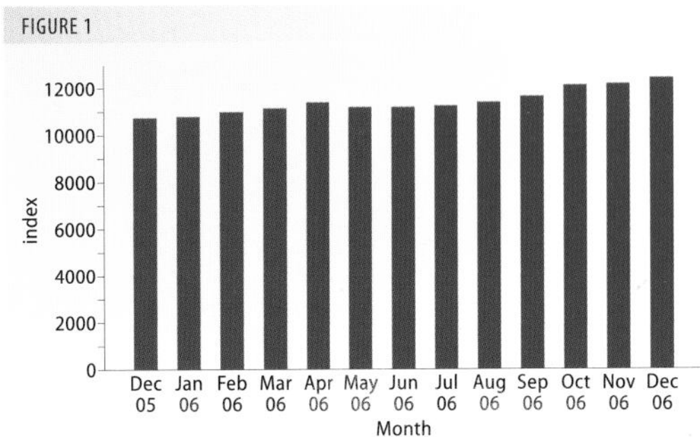

If wishing to convince a nervous investor of the benefits of a share portfolio, a little graphical creativity can transform a good year into a seemingly brilliant one. By virtue of a truncation of the vertical axis at 10,500 points, the December 2006 column has become over 11 times the height of that for December 2005. See Figure 2.

FIGURE 2

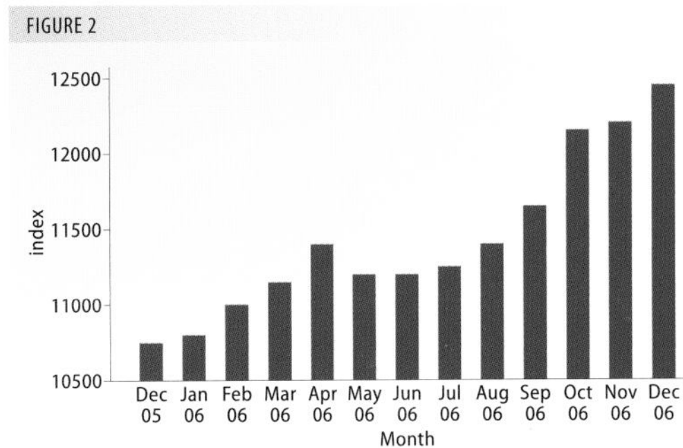

Such a graph such might drive even the most skeptical of investors to convert their hard earned cash into shares!

Finally by bumping the display into three-dimensions, the desired effect is even further enhanced. See Figure 3.

FIGURE 3

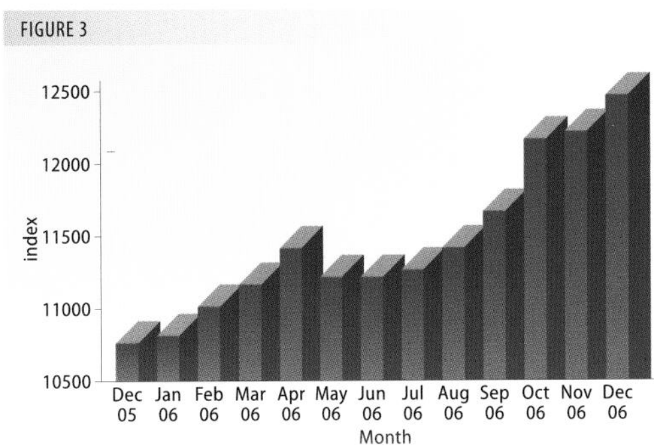

Such manipulations of scale provide support for even the most feeble of positions. Interpretations of the display of such data should always be treated with caution.

While you sit patiently at traffic lights, a computer controls how long you wait. The computer in a washing machine controls the cycles so that you can do something else while it washes. When you are in a car accident it is a computer that triggers off the airbag that saves your life. Computers are everywhere—not just where we expect.

Computers and mathematics

The history of computers

A computer is a machine designed to compute mathematical and other problems and generate accurate results. Today, computers are very elaborate, programmable, electronic devices that carry out the mathematics without the operator knowing it, but it wasn't always this way.

The first computers were simply counting devices like the abacus. Merchants, government officials, and indeed anyone who had an interest in calculating quantities, needed a device that could calculate beyond the 10 fingers and toes that most of them had. While an abacus was a massive step forward in computational efficiency, it wasn't until about 100 BCE that computers took a leap forward in sophistication. The Greeks of that time were fascinated by astronomy and developed an analog computer called the "Antikythera mechanism" to do their calculations. It is the first recorded machine of its type and the last to appear for about 1,000 years.

Left: A reproduction of the Antikythera mechanism which was invented during the first century CE. It consisted of at least 29 gears that were moved by turning a handle. An ancient yet effective calculating device, the scales of the Antikythera mechanism performed precise astronomical calculations.

> I think there is a world market for maybe five computers.
>
> Thomas J. Watson, IBM president, 1874–1956

The next massive jump forward in practicality, but backwards in complexity, was the invention of the slide rule in the 1620s (see pages 104–105). This device took the new mathematics of logarithms and made it usable in an affordable and reliable way. A slide rule could do the mathematical computations that would have otherwise taken a long time or required reference to massive tables of logarithmic data.

Who is the computer?

By the seventeenth century, it was still uncommon for any of these devices to be called a "computer." Since the person who used them had to be involved intimately with the mathematics, he or she was usually called the Computer (that is, the person who computes), and the machine was a known as a computational device. Consider the difference with today, when an ordinary person, not mathematically inclined, sits down at a computer and unwittingly performs elaborate mathematical operations in order to listen to a Podcast, email a friend or colleague, or simply write a letter. The term "computer" has been removed from the human and given to the machine. This tells us a little about who does most of the work today.

Computer programming

Computers would not be the valuable tools they are today if it were not for sensible programming. Programming is the set of rules that has been given to a computer, that describe what is to be done to any information entered into it. Early computer programming was achieved by physically changing the hardware of the computer. In the late 1950s, computer scientists began to develop coded languages that more closely resemble the programming that is used today. These days computer programming is carried out in one of 2,000 languages that dictate what a computer will do with the information that it is given. Most of these languages are distant relatives of three early languages named Fortran, Cobol, and Algol.

Right: The advances in computer technology have been phenomenal. Only a few decades ago, personal computers were few and far between. Nowadays, many people can't get by without a personal digital assistant—a small hand-held computer that has all the features and capabilities of a regular computer.

Opposite: The world's first successful digital computer was built in 1945 in Pennsylvania, USA. ENIAC (Electronic and Numerical Integrator and Computer) weighed more than 60,000 lbs (27,000 kg) and contained some 18,000 vacuum tubes, of which about 2000 per month were changed by ENIAC technicians.

HOW COMPUTERS WORK

Mathematics is an integral part of computers. With modern computers, everything is either a "1" or a "0." It can be nothing else. Everything that a computer is presented with—paragraphs typed into a word processor, data received from a camera, or temperatures read by a sophisticated thermometer in the engine of your car—is converted by the computer into a simple binary code made up of a string of either 1s or 0s. Once coded, every operation that a computer is asked to carry out is simple mathematics. If the letter "B" is typed into the keyboard of a computer, the machine will immediately code the keystroke in binary form and will store the data as a series that looks like this: "01100010." This explains why a human (even an accomplished mathematician) can provide a computer with a difficult logical problem and receive an accurate answer before he has finished sharpening his pencil. A computer can do the calculations faster and more accurately than a human because it breaks every operation down into this simple code.

The world is full of statistics. Newspapers, television, advertising, product packaging, and in particular the internet, seem to be overflowing with statistics of all sorts. Who, but the statistician, can ever expect to really understand all of these numbers?

Sampling statistics

Statistics is the science of data. A statistician collects data, analyzes it, interprets it, and presents it in a way that is supposed to be clear to those who read it. It is easy to forget that when we see a statistic, it usually relates to something physical in the real world. For example, unemployment statistics relate directly to people who are out of work and road accident statistics describe the collision of vehicles containing real people. In this way, we can understand that statistics are simply a way of bringing numbers together in a form that is meant to be more useful than raw data.

Statistical literacy

Statistical literacy is the ability to understand what statistics mean. Just as it may be possible for you to read words in a foreign language, without understanding what they mean or how to use them, statistics need a level of understanding. Statistical literacy requires that we have a basic understanding of what a statistic tells us about the reality it represents. For example, if we ask every person alive how tall they are, after many individual measurements, we would have a massive set of data that would provide us with accurate statistics. We could calculate the average or mean height of the world's population by adding all of the heights together and dividing this total by the number of people.

> There are two kinds of statistics: the kind you look up and the kind you make up.
>
> Rex Stout, novelist, 1886–1975

We could calculate the median height by finding the height that has half of the world's people taller and half shorter. We could calculate the modal height, which is the most commonly occurring height, and this would all just be the beginning. We could reasonably guess that the heights of the world's population, when put in graph form, would take the shape known as a bell curve. This bell shape is derived from the fact that most of the heights would be centered around the average. However as the heights moved further from this average, either toward the taller or shorter end, we would see fewer and fewer people. It would eventually get to the point where the tallest people and the shortest people ended the curve to the right and the left of the bell respectively.

Suppose, however, we wanted to find the average height of the world's people but didn't want to go to a lot of trouble. We could measure members of our family or the members of our office and work out some averages and other statistics. Is it likely that these statistics would be representative of the entire world's population? Hardly. For data taken from a sample to be truly representative of an entire population, each member of the larger population must have an equal chance of appearing in the sample. A sample made up of people with whom you work excludes all the people in the world with whom you don't work. A sample that only included Japanese males (perhaps the members of a local sports team) naturally excludes all people who are not Japanese and are not male. This is an error of sampling that good scientists go to great lengths to avoid. Statistics that are not properly sampled are often useless and sometimes they can even be outright dangerous.

Right and opposite: Statistics are used for myriad reasons, including determining who makes the team and how the stock market is performing over a given period. Careful analysis of results assists us in making important decisions.

Below right: A Hollerith machine used to process the results of the 1931 census in Great Britain. A census not only collects the basic statistics of a population, it gathers other demographic information such as home ownership, religion, and special needs.

Above: We hear a great deal about road accident statistics. These can be categorized by driving experience, location, age, and so on. By making the statistics meaningful, we have the opportunity to change our behavior and drive more safely.

Left: Tourist operators and governments conduct statistical studies in order to attract as many visitors as possible, thus increasing benefits to the local and national economies. The Eiffel Tower has been a popular tourist destination since it was built in 1889.

STATISTICAL TYPES

There are two main types of statistics. The first are descriptive statistics—they describe a situation as it is or as it was. A common example of a descriptive statistic would be the number of times that a particular football game has been won by a particular margin, or the average stock market price change over a trading day.

The second type is inferential statistics. Here a smaller subset of the population being studied is recorded and it is inferred that this statistic also applies to the entire population. For example, a sample of television viewers is surveyed regarding what they watch and when they watch it. If this sample is seen as representative of the whole population, then it is inferred that the whole population has similar viewing patterns.

Geometry may not make the world go round, but ancient Greek mathematician Eratosthenes used it to determine how round the world actually is. By watching the Sun and measuring shadows, he noticed that the Sun appeared at different angles in different places at any given time.

Another angle on geometry

Using geometry, Eratosthenes was able to demonstrate that this meant Earth was actually about 25,000 miles (40,000 km) in circumference, which is surprisingly accurate. He was able to calculate this massive distance at his desk without needing a very long tape measure. It was this ability to use geometry to go from what people knew to what they needed to know that excited ancient mathematicians so much.

All at sea

In Eratosthenes' day, maritime navigation was in definite need of an upgrade and geometry was used to supply many of the solutions. Before geometry came into use, sailors used "dead reckoning" to cross the seas. This involved logging time and distance traveled from a known fixed point and estimating the effects of winds and currents. Out at sea, this form of navigation can gradually take a ship well off course in between known fixed points. Celestial navigation, on the other hand, relies on geometric relationships between the positions of the stars and planets. Whenever there is a clear sky, a sailor can use these relationships to determine a latitude and longitude and to plot a course that the ship must sail.

Above: The basic tools of geometry include protractors, which measure angles, compasses, which are used to draw circles at different radii, and set squares, which are used to draw parallel and perpendicular lines.

Modern laser surveying techniques still rely on the ancient principles of geometry. The more complex the design, the more crucial the geometry. Consider the geometry of the sails of the Sydney Opera House or the Guggenheim Museum, Bilbao. Despite the intricacies of these curves and angles, each of these striking buildings is based on geometry both in their design and in their construction. Although observers might not be able to see a geometric form or pattern to a design, it is usually there.

> It is the glory of geometry that from so few principles, fetched from without, it is able to accomplish so much.
>
> Isaac Newton, physicist and mathematician, 1642–1727

Essential geometry

Today many people use geometry in their everyday lives. Geometry, for example, is used in all forms of construction and surveying. A house, road, bridge, or skyscraper is built in exactly the right place to exactly the right design because of a series of geometric measurements and calculations occurring throughout the building process.

In this way, a project such as a massive domed ceiling or a spiral ramp onto a freeway can be accurately designed on a computer and then constructed on site. Precise geometric calculations ensure that a bridge or tunnel can be built from both sides at the same time and meet in the middle. The 80-mile (128 km) Channel Tunnel running beneath the sea bed of the English Channel was drilled from both ends in this way. Continual surveying ensured that when the two ends met after travelling some 40 miles (64 km) each, they were just 14 inches (35 cm) out of alignment.

Opposite right: Measuring the angles between the horizon and the stars and planets was known as celestial navigation. Sailors had recourse to star charts and atlases, such as this one published in 1822 and showing the Hydra.

Right: The Guggenheim Museum in Bilbao, Spain is one of Frank Gehry's architectural masterpieces. Mathematical software helped to define its impressive geometric facade.

Left: The famous triangular shells of the Sydney Opera House, designed by architect Jorn Utzon and opened in 1973. Spherical geometry informed the design of the spectacular roof vaults.

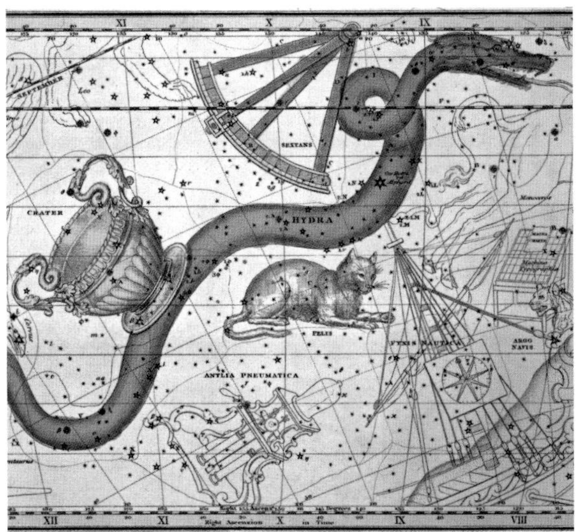

NAVIGATION

In days gone by, celestial navigation and astronomy went hand in hand as scientists came to understand the relationships between the orbits of various planets. They were able to calculate the size and distance of these stars and planets from their relative positions and were then able to plot their paths across the night sky. In a strange reversal of this situation, geometry has now helped us to develop a new way to navigate on the surface of our Earth and sea. Instead of relying on observations of natural stars, we have now installed our own set of satellites that transmit microwaves down to us on Earth. These microwaves can be picked up by a Global Positioning System (GPS) receiver anywhere on the globe and once transmissions from four satellites have been received, the GPS receiver will calculate, using geometry and trigonometry, the position of that receiver. Using a GPS receiver (right), navigators can accurately determine where they are, without doing any mathematical calculations—because all of the trigonometry and geometry takes place inside the electronic box they hold.

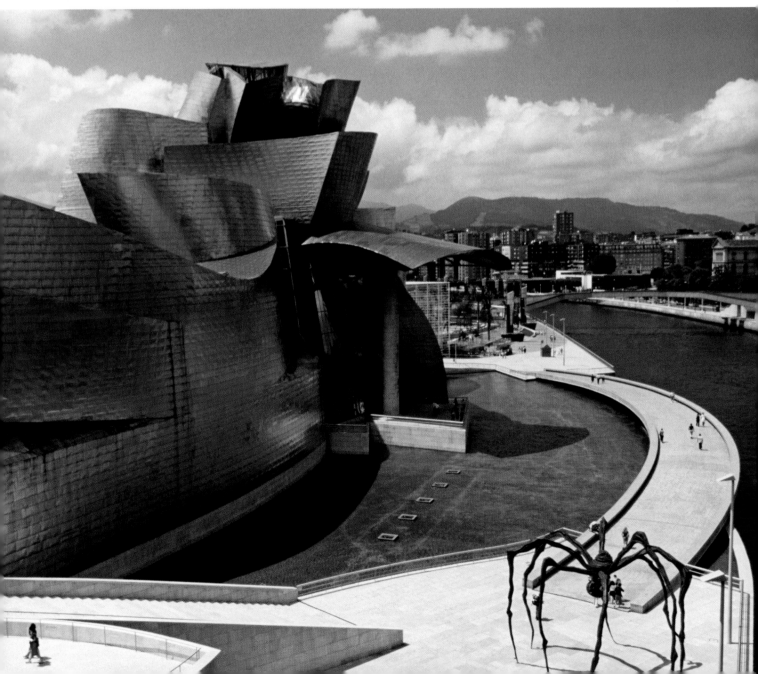

In our daily lives, plans and actions are often dictated by subconscious estimations of chance. Even a young child understands the statement, "there is a good chance it might rain today." But what is "a good chance"? Probability is the application of values to such likelihoods.

Chance and probability

Prior to 1700, probability was mainly used for games of chance. During the eighteenth century the insurance business rapidly expanded, creating a need for mathematicians to make complex calculations to reckon premiums. These mathematicians, known as actuaries, used probabilities to provide the data for much of their work. In the language of probability the outcomes of experiments or phenomena are called events. The application of a value, be it percentage, decimal, or fractional, to the likelihood of events, allows the user to make comparisons between their probabilities. The range of values starts at 0, describing an impossibility, and progresses to 1 (100%) which is the given probability of a definite occurrence.

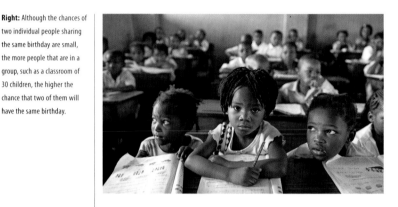

Right: Although the chances of two individual people sharing the same birthday are small, the more people that are in a group, such as a classroom of 30 children, the higher the chance that two of them will have the same birthday.

Combined events

Combining events requires a variation in the arithmetic that is used to evaluate a probability. Let's assume that one in 14 Californians is under the age of five. What is the probability of two randomly selected Californians both being under five? An arithmetic method of evaluation is to consider this problem using a statement describing the outcome, such as "the first person is under five and the second is also under five." The word "and" directs us to *multiply* the individual probabilities thus:

$$\frac{1}{14} \times \frac{1}{14} = \frac{1}{228} \text{ (0.44\%)}$$

Now assume the largest group of Californians by age is the 45–54-year-olds, representing approximately 1 in 6 residents. The chance of the two selected individuals being in this age group is:

$$\frac{1}{6} \times \frac{1}{6} = \frac{1}{36} \text{ (2.78\%)}$$

What if we considered the likelihood that the two selected are both under five or both 45–54? The key word here is "or." When this is employed we add the individual probabilities. (This requires the sets to be mutually exclusive.) Thus the probability would be:

$$\frac{1}{228} + \frac{1}{36} = \frac{1}{342} \text{ (3.22\%)}$$

Above: One of the methods that weather forecasters use to predict rain is to divide the number of days in their database that it has rained by the total number of days in the database. The answer gives the probability of rain.

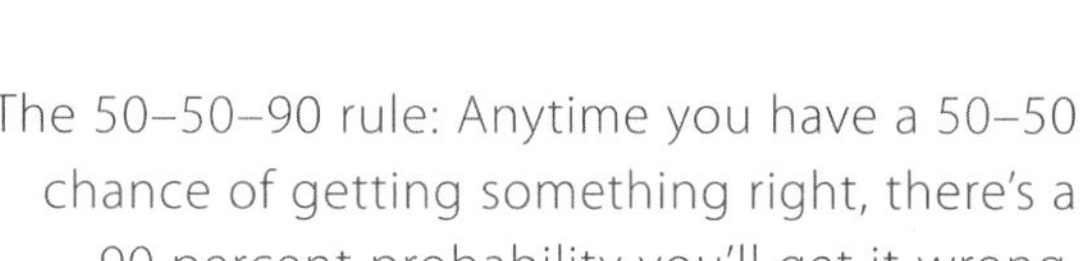

The 50–50–90 rule: Anytime you have a 50–50 chance of getting something right, there's a 90 percent probability you'll get it wrong.

Andy Rooney, journalist, b. 1919

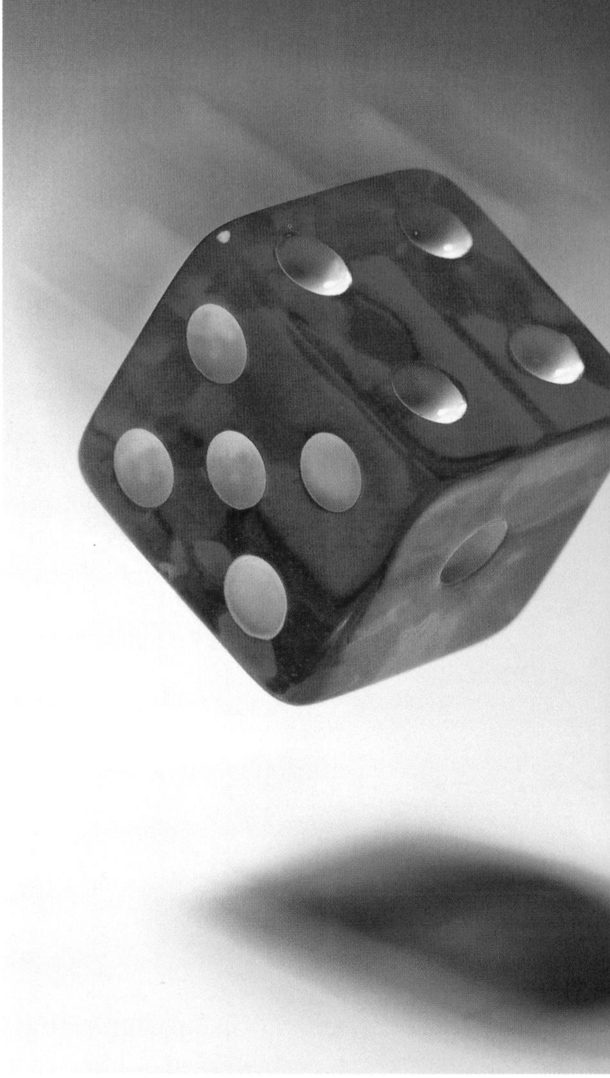

What is the chance that in a classroom of 30 students, at least two of them share a birthday? Many will assume 30/365 as there are 30 birthdays out of 365 possible days in the year, but this is not correct. The value is actually a little over 70 percent. To calculate this, we use the concept of the "complementary event." The complement of an event is that same event not occurring—the complement of throwing an even number on a die is to throw an odd number. As the event and its complement encompass all possible outcomes, when we add their individual probabilities, the result is 1. The complementary situation of at least two of the group sharing a birthday is that that none share a birthday. The arithmetic required to evaluate this condition can be extrapolated by looking at the case of just four students.

Right: Games of chance often rely on the roll of one or more die. If one die is thrown, all outcomes are equally probable, yet if two dice are thrown, the probabilities change as there are more ways to get some numbers than others.

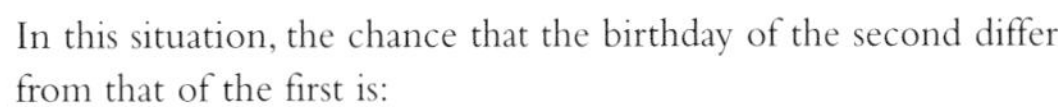

The possible combinations of birthdays of four students— Kate, Sophie, Tom, and Nick—can be shown diagrammatically.

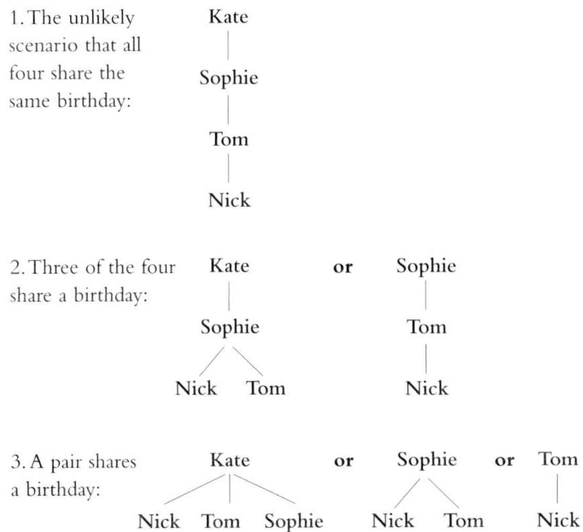

(We would also consider two pairs sharing two different dates.)
4. The final situation is that is no birthday is shared, making it the complement of at least one shared birthday.

Above: Car license plates can also be used to demonstrate the principles of probability. Amazingly, in a group of only 25 plates, the chances of two cars sharing the last two digits is 96 percent.

In this situation, the chance that the birthday of the second differs from that of the first is:

$$\frac{364}{365}$$

(This exercise ignores leap years.) For the third student not to have the same birthdate as either of the first two is:

$$\frac{363}{365}$$

Finally, the probability that the fourth pupil does not share a birthday with any of the other three is:

$$\frac{362}{365}$$

All permutations of no one sharing a birthday have now been considered. To calculate the combined chance we now multiply the individual probabilities, that is:

$$\frac{364}{365} \times \frac{363}{365} \times \frac{362}{365}$$

As each new individual is considered, the numerator decreases by 1, while the denominator remains at 365. So the calculation for 30 people is:

$$\frac{364 \times 363 \times \ldots \times 337 \times 336}{365^{29}} = 0.2937\ (29.37\%)$$

As this is the chance of the complementary event, this value is subtracted from 1 to give 0.7063 (70.63 percent).

A simple way to test this principle is by performing a similar experiment using car registration plates. Next time you are on a long car journey, ask your passenger to note the last two digits of the first 20 cars to pass in the opposite direction.

The expectation is that it is unlikely that there will be a matching pair. Yet the probability of at least one matching pair can be evaluated as follows:

$$1 - \frac{99 \times 98 \times 97 \times \ldots\ldots \times 82 \times 81}{100^{19}} = 0.8696$$

That is, there is an 87 percent chance that you will have a match. Take it up to 25 plates and the likelihood of a match increases to over 96 percent!

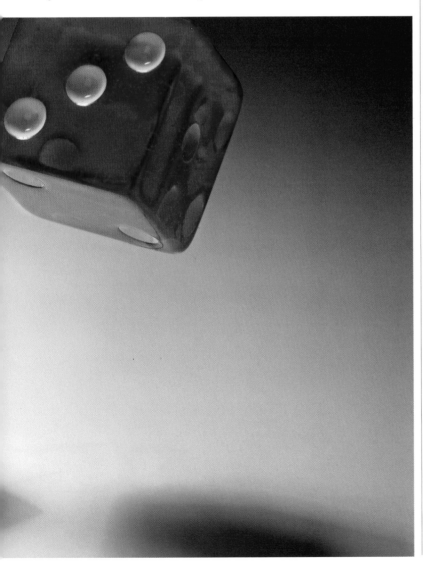

EVALUATION OF A PROBABILITY

A probability is allocated a value by creating a fraction with the number of desired outcomes as the numerator, and the number of possible outcomes as the denominator. This represents the theoretical probability of the event, e.g., the probability of randomly selecting an ace from a regular deck of 52 cards is 4/52, as there are four desired outcomes (aces) out of 52 possible outcomes.

Some probabilities can only be calculated based upon the results of trials and documented data. The accuracy of such an estimation is dependent upon the number of trials. In 1856 *New York Times* journalist E. Meriam researched the July 4th New York City weather records of the preceding 67 years. He found that it rained on 13 occasions and so predicted a one in five (20 percent) chance of rain on Independence Day that year. With over 150 more years of data, the probability of a rainy July 4th is now more like 12 in 31 or 39 percent!

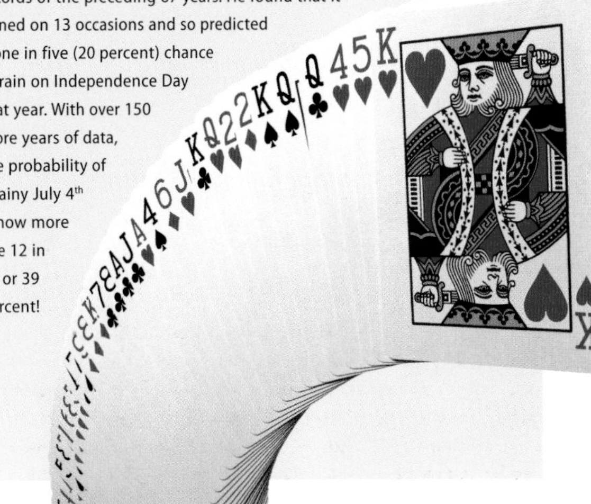

Take a large collection of essentially abstract mathematical concepts and combine them in a highly structured way, and what do you get? Not something even more abstract, but rather an enormously powerful branch of mathematics known as functional analysis. Functional analysis is applied to a wide range of real-world problems in physics and engineering.

Functional analysis

There are many problems in the real world of physics and engineering requiring complicated differential equations that involve many different quantities (such as time, distance, mass, temperature, etc.). Such equations are very difficult to solve exactly, but the nature of their solutions can be studied using functional analysis. This has provided breakthroughs in such diverse areas as fluid dynamics, control theory and optimization, and quantum mechanics.

Right: Equations are central to functional analysis, which is, as its name suggests, the study of identifying and describing the functions of systems. Apart from its use in mathematics, functional analysis is important in the applied sciences.

Classifying things into spaces

One of the basic principles of functional analysis is to group things according to how they "behave" mathematically. These groups, mathematically known as "spaces," can contain various things: Sometimes numbers, sometimes spatial coordinates, sometimes even mathematical functions and curves. For example, one possible space is simply the collection of all whole numbers $\{0, 1, 2, 3, \ldots\}$, and another space may be all simple functions with one variable $\{x, x^2, 1 + x, \ldots\}$.

Although there are many types of these "spaces," there are two particular ones that can be applied to a wide variety of important problems. These are called Hilbert spaces and Banach spaces, named after the mathematicians who studied them. Each of these has specific mathematical rules to define how to add and multiply the members of the space. This is important, since although adding and multiplying numbers is straightforward, it is somewhat more complicated to add and multiply curves or functions.

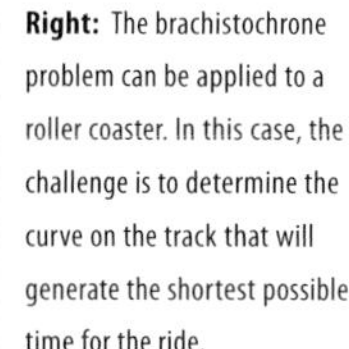

Right: The brachistochrone problem can be applied to a roller coaster. In this case, the challenge is to determine the curve on the track that will generate the shortest possible time for the ride.

THE HILBERT SPACE

German mathematician David Hilbert (1862–1943), one of the most influential mathematicians of his time, is frequently referred to as the father of functional analysis. His genius lay in being able to think beyond mathematical conventions and penetrate to the depths of a problem drawing out new insights and interpretations.

The study of differential and integral equations became his passion in 1909, 14 years after he became Chairman of Mathematics at Göttingen University, Germany. He devised the concept of the Hilbert space as a new approach to these equations. This work became the basis of functional analysis and was vital in the development of quantum mechanics.

Hilbert's headstone aptly reads, "We must know. We will know." All his life he applied himself to the understanding and explanation of the most difficult problems in mathematics and physics. In his famous 1900 address at the Second International Congress of Mathematicians, he issued a list of 23 fundamental mathematical problems that he considered of prime importance. Hilbert charged the mathematicians of the world with the task solving them. These problems have influenced the direction of mathematics and some still remain unsolved.

David Hilbert's mathematical genius, incredible insights, and elegant approach touched almost every area of practical and pure mathematics from the abstract world of quantum mechanics, relativity and algebra to geometry, and a practical attempt to formalize all mathematical enquiry.

Below: Farmers usually use machinery to sow seeds for crops such as corn. How precise the machine is, in terms of the distance between the seeds, is often determined by functional analysis methods.

> Mathematics, rightly viewed, possesses not only truth, but supreme beauty.
>
> Bertrand Russell, mathematician and philosopher, 1872–1970

Turning one problem into another

A very famous mathematical problem, known as the brachistochrone problem, was to calculate the shape of the curve that allows a ball to roll from one location to another in the shortest possible time. This type of problem is called optimization, and is very similar to many engineering and economic situations where materials or costs must be minimized. Instead of testing every possible curve in the brachistochrone problem, functional analysis uses a mathematical space (a Banach space) of all curves. By studying the overall properties of the Banach space, the desired curve, that is, the one with the shortest roll time, can be readily found. The power of this technique is that totally different problems can have Banach spaces with identical properties, so the same type of solution can be used for many situations.

Right: To reduce or compress a CD-quality, 100-megabyte song without reducing the music's quality, a functional analysis technique called truncated spectral analysis is used. The human brain ignores some frequencies when they are "masked" by others. Truncated spectral analysis allows masked frequencies in music files (such as MP3 music) to be deleted.

Another, similar situation applies to the physical theory of quantum mechanics, which describes the universe in terms of waves. The equations governing the behavior of these waves can be difficult to solve, but they can instead be considered as a Hilbert space. Many of the important results of quantum mechanics, for example, those leading to the development of lasers, are only possible using functional analysis with a Hilbert space.

Imagine that you are given a task to complete, and instead of doing it properly you only make a rough attempt. Then when your boss is upset, you reassure her that although the task is not really finished, you know precisely how much you haven't done. Believe it or not, this way of working is at the heart of the mathematical field of numerical analysis.

Numerical analysis

In many scientific and engineering problems it is inconvenient (or sometimes even impossible) to make exact calculations. It may be that the problem is so complex we do not actually know how to solve it, or it could be impossible to collect all the required information. For example, to examine how the strength of a bridge truss changes over the course of 100 years, we do not want to spend a century taking measurements. Instead, we need an approximate solution and, just as importantly, a way to in which to determine the accuracy of that approximation. Numerical analysis is the branch of mathematics that is concerned with algorithms—systematic mathematical procedures for generating approximations.

Above: To make a starting point for numerical analysis, such as the problem of modeling the structure and behavior of molecules, sometimes a random set of numbers is used to simulate possible starting values. This technique employs random numbers, a little like rolling a ball around a roulette wheel. It was used to model nuclear reactions in the development of the atomic bomb at Los Alamos, where physicists called it a Monte Carlo calculation, after the Monaco casino.

Some simple approximations

To study the growth of a child you might measure their height every Christmas. If you want to use these measurements to find the child's height halfway through the year, then you could use the numerical analysis tool of interpolation, which finds values in between known quantities. A similar technique, called extrapolation, can be used to predict how tall the child will be in several years. If the child's growth is reasonably constant, interpolation and extrapolation will be quite simple. This is called linear numerical analysis. In more complicated situations, such as turbulent fluid flow or financial forecasting, the variations between known values will not be so straightforward. This requires more difficult non-linear algorithms, which can involve the complex effects of chaos theory (see pages 114–115).

Keeping track of error

When designing an irrigation system for a farm, the calculations of the fluid flow can probably be off by an inch or two without any dramatic consequences. On the other hand, when studying the flow of blood through the human heart, the calculations must be much more accurate, since the heart is only several inches across. Different situations require drastically different levels of accuracy. The difference between the exact answer and the approximation is called the error, and an important aim of numerical analysis is to understand the error. There is often a compromise between obtaining the smallest possible error and performing the calculation in the shortest time. This is especially important when the cost of computing time may be a major expense in solving a problem.

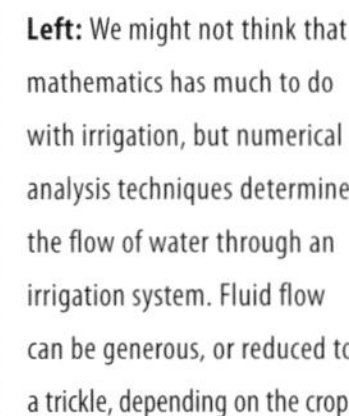

Left: We might not think that mathematics has much to do with irrigation, but numerical analysis techniques determine the flow of water through an irrigation system. Fluid flow can be generous, or reduced to a trickle, depending on the crop.

Mathematics is a more powerful instrument of knowledge than any other that has been bequeathed to us by human agency.

René Descartes, mathematician and philosopher, 1596–1650

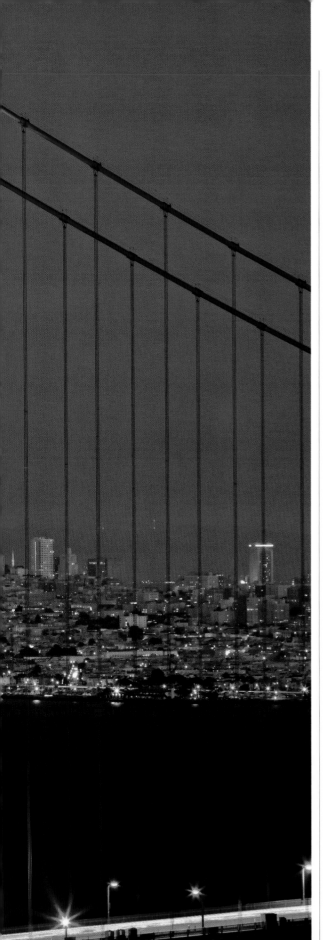

Above: A supercomputer at the Center for Computational Visualization at the University of Texas, USA has created a simulation of ocean circulation temperatures, information that will help in the fight against global warming.

Left: When will San Francisco's famous Golden Gate Bridge require major maintenance? Will it be as strong 100 years from now as it is today? By using numerical analysis, engineers can predict changes in the strength of the bridge's trusses over time.

Super computers to the rescue

Numerical analysis becomes most useful when a large number of measurements, or a large number of guesses, can be used in the algorithms. Enormously repetitive tasks are ideally suited to computers, and it is no surprise that numerical analysis has become widely used as better computer technology has been developed. Some of the most powerful supercomputers in the world are used to run numerical analysis programs for such things as the weather.

PREDICTING THE WEATHER

Meteorologists and atmospheric physicists predict the weather based on the collection of known atmospheric data and the manipulation of this "real weather" information using numerical analysis. Groups of equations extrapolate a set of weather conditions (weather prediction); others are used to look at the likely error or accuracy of a forecast. This computer modeling is known as numerical weather prediction (NWP).

Initially, meteorologists construct a "virtual" three-dimensional grid of the atmosphere over the area where the weather is to be predicted. Thousands of pieces of data from "real" sources—weather stations and radio telescopes (below), satellite images, and weather balloons—are plotted into this model. Then the information is manipulated and inserted into a set of equations which are solved using numerical analysis methods and supercomputing. The answer will be a close approximation of future weather conditions for that particular area.

NWP results are affected significantly by miniscule changes in the initial data placed into the model. Even though NWP has become more accurate as computers have become faster, there is a limit to how far into the future meteorologists will be able to accurately predict the weather. This is because of the chaotic and constantly changing nature of the atmosphere, as well as the "sensitive dependence on initial conditions" of numerical analysis.

CHEMISTRY

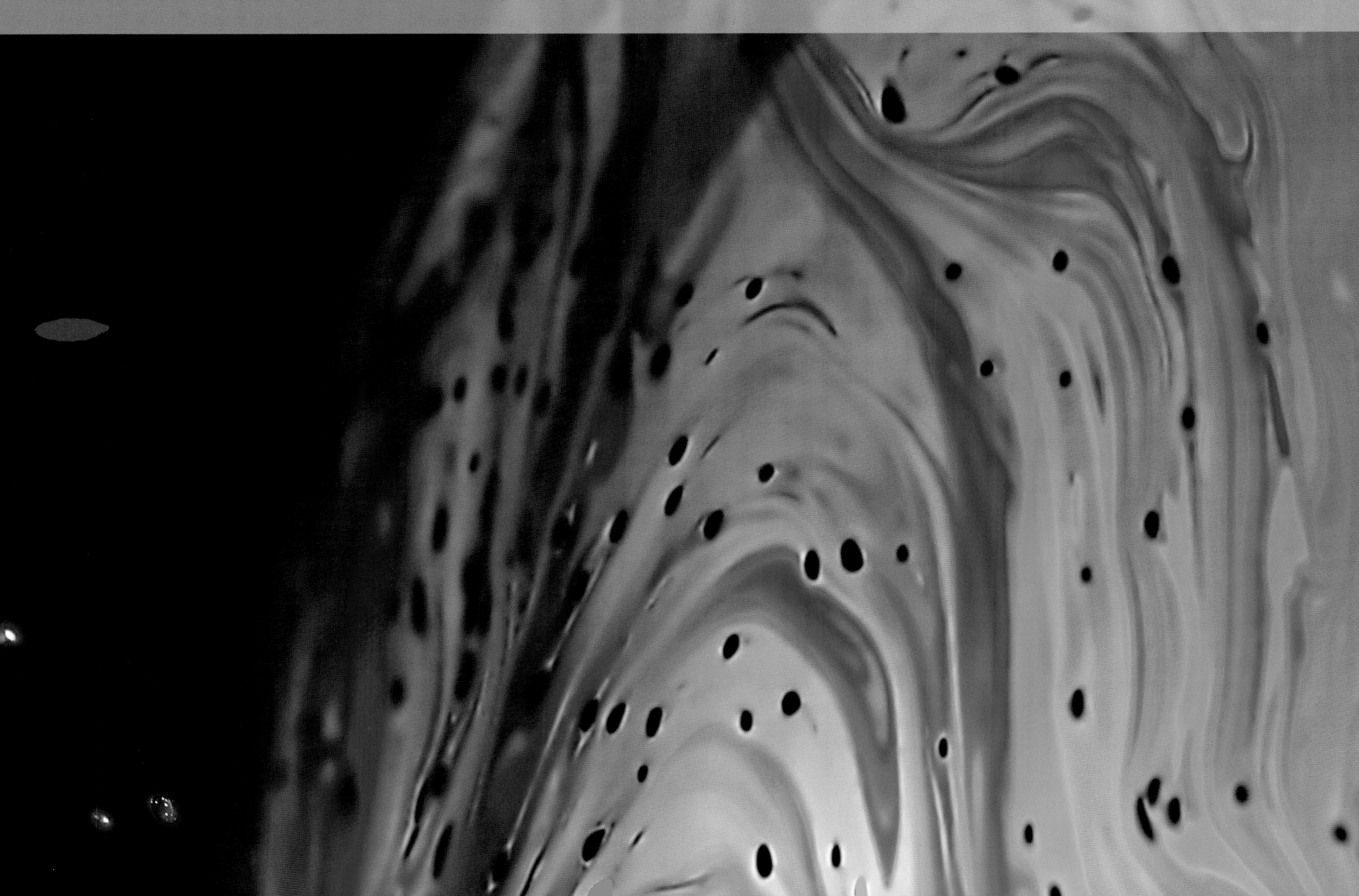

The science we now call chemistry has ancient origins. It probably began when we first observed what the heat of a fire could do to naturally occurring materials, baking clay into pottery, extracting dyes from plants and metals such as iron and copper from their ores, changing the nature of foods through cooking. It was all essentially practical.

Introduction

Below: Quinine was introduced into Europe during the seventeenth century. A bitter alkaloid, it is derived from the bark of the cinchona tree found in South America. Once used to treat malaria, quinine has largely been replaced by other chemical compounds.

Yet any solid understanding of the properties of matter and energy that make chemistry possible was a long time coming. We learnt much from Arab scientists during the Middle Ages. Much important chemical technology came from them, such as distillation, along with words like "alcohol." In later years they began to apply such understanding as they had to protecting human health.

But such studies were entangled with alchemy, with attempts to turn base metals into gold or to find the "elixir of life." They believed that matter was composed of three "elements," mystical entities displaying the properties of sulfur, salt, and mercury, much as the ancient Greeks thought all things were compounded of earth, air, water, and fire in varied amounts.

Left: An early domestic radio in a Bakelite case. Named for its inventor, the Belgian–American chemist Leo Baekeland in the early years of the twentieth century, Bakelite is a combination of phenol and formaldehyde; it was the first totally synthetic plastic.

Below: Alchemy was the forerunner of modern chemistry. Alchemists were concerned with the transformation of matter, in particular the conversion of base metals into gold and silver. They also sought the so-called "elixir of life."

The beginnings of modern chemistry

Probably the first real evidence of modern chemistry came in seventeenth-century Europe, when Irish scientist Robert Boyle defined an element much as we do today (although he was not sure how many there were) and who first emphasized the importance of experimental evidence combined with accurate measurement.

This path was extended early in the next century by men like Stephen Hales, who devised technology for collecting and describing "gases" (itself quite a new term). By the end of the eighteenth century many previously unrecognized gases had been discovered and characterized, including oxygen, nitrogen, carbon dioxide, hydrogen, and chlorine.

Many of the early chemists believed in the erroneous notion of "phlogiston," a fiery substance released by materials as they burn. It was not until French chemist Antoine Lavoisier banished phlogiston,

Somewhere, something incredible is waiting to be known.

Carl Sagan, astronomer, 1934–1996

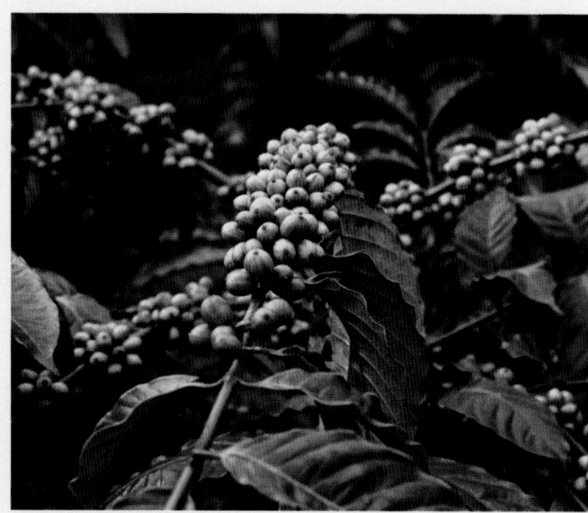

Right: Caffeine is naturally produced in nature—occurring in cacao beans, the source of chocolate, and in coffee beans (*Coffea* species), shown here in immature form. Consisting of carbon, hydrogen, oxygen, and nitrogen, caffeine is classified as an alkaloid.

replacing its role in combustion with that of oxygen, that rational chemistry became possible. Lavoisier devised a methodical way to describe the increasing number of chemical compounds known, a nomenclature little changed today.

Synthetic materials

Chemistry made immense strides through the nineteenth century, becoming the basis of major industries and beginning to penetrate all aspects of life through the application of many newly discovered materials. By 1900 we had the first synthetic materials, starting with celluloid and later Bakelite, gun cotton and dynamite replacing the venerable gunpowder, the first artificial dye stuffs beginning with "Perkins mauve," and synthetic drugs pioneered by aspirin. We had learnt to extract from traditional plant remedies and mixtures their active ingredients, such as quinine and caffeine.

To meet the rising demand for "soda ash" (sodium carbonate), vital in many industries including paper and soap making, many factories were built to extract it from common salt. But these were environmentally disastrous, fouling air, water, and soil with noxious wastes such as hydrochloric acid.

The chemists of the day were beginning to fill in the theoretical background to these developments. Through John Dalton's "atomic theory," the pioneering researchers of Justus von Liebig and Jöns Jacob Berzelius, and the work of many other inquiring minds, we began to understand why chemical elements and compounds behave as they do and how they are put together.

A notable scientific triumph was the classification by Dmitri Mendeleev of the chemical elements known as the periodic table. So powerful was this understanding that we were able to predict the existence of elements still to be discovered. The nineteenth century also saw the first subdivisions of chemistry into the multiplicity of disciplines we see today.

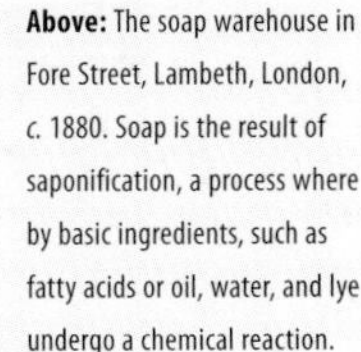

Above: The soap warehouse in Fore Street, Lambeth, London, *c.* 1880. Soap is the result of saponification, a process where by basic ingredients, such as fatty acids or oil, water, and lye undergo a chemical reaction.

The twentieth century saw the beginnings of polymer chemistry, with chemical compounds built out of many identical simple units. Some chemicals became known as "plastics," though they quickly penetrated every part of our lives, from packaging to paint, from clothing to building materials to synthetic rubber. We also saw the first composites, with some chemical compounds embedded in others. Fiberglass was an early example.

The search was also on for new chemical sources of energy, for example, converting plant fats into the equivalent of diesel fuel, or more efficient means to convert carbohydrates from plants into ethanol. These were merely some examples of the concern by chemists for the health and quality of the environment—a startling contrast to the environmental destructiveness of the nineteenth century.

Right: German chemist Justus von Liebig (1803–1873) was a prominent scientist of the nineteenth century. A highly respected educator, he published highly influential works on organic chemistry during the 1840s, paving the way for radical changes in food production .

Immediately following the Big Bang, the simplest of the elements, hydrogen and helium, were created, but it was another billion years at least before more complex elements were formed in any quantity. As stars were born and cycled through to their deaths, atoms were progressively smashed together and split apart and new elements were formed. The elements of which we are made have the stars as their ancestors.

The chemical elements

At the time of writing, we recognize 117 known elements. Amazingly, just these few elements make up all of the matter around us. Consider, for example, something as complex as the human body: Only four elements—hydrogen, carbon, oxygen, and nitrogen—make up 99 percent of our bodies (the other elements account for the remaining one percent).

Of these 117 elements, at least 94 occur naturally on Earth. However, their abundance varies widely. Oxygen, for example, is the most abundant element in Earth's crust. It makes up 49.5 percent of the crust by weight. By contrast, the element astatine has been dubbed the rarest naturally-occurring element—there is estimated to be less than one ounce (25 grams) of astatine in the entire Earth's crust at any time.

The remaining 23 elements have not been discovered as occurring naturally on Earth but have been prepared artificially. These artificial elements are typically radioactive and they can be amazingly short-lived. Samples of some of them last only milliseconds before half of the atoms have decayed.

Above: Fuel pellets of uranium oxide are used primarily to power nuclear reactors. The pellets are formed by cold-pressing uranium oxide and plutonium oxide.

What distinguishes one element from another?

All matter is made up of very tiny particles called atoms and each element is made of one type of atom only. So, saying that there are 117 elements is exactly the same thing as saying that there are 117 different types of atom.

Atoms are made up of subatomic particles such as protons, neutrons, and electrons. Protons and neutrons are found in the nucleus of an atom. It is the number of protons that is largely responsible for determining the properties of an atom. This gives rise to the precise definition of an element: "All atoms of an element contain the same number of protons." Carbon atoms, for example, are defined as atoms that contain six protons. The number of neutrons and electrons can change but as long as there are six protons in an atom it is carbon and will behave as such.

Uranium, with 92 protons, has the heaviest atoms of the naturally occurring elements. Because of the large number of protons in its nuclei, uranium can be unstable and will break apart over time to form smaller elements. However, within and beneath Earth's crust radioactive uranium breaks down at a more sedate pace. With a half-life measuring in the millions to billions of years, the breakdown of uranium is believed to be one source of the energy that keeps Earth's mantle molten. While helium is the second most common element in the universe, on Earth the only natural supply is as a result of the radioactive breakdown of elements such as uranium.

Right: Gold is arguably the most sought-after and visually appealing of all the elements. Each element is assigned an atomic number based on the number of protons in its nuclei; gold has 79 protons, so its atomic number is 79.

> To say that a man is made up of certain chemical elements is a satisfactory description only for those who intend to use him as a fertilizer.
>
> Hermann Joseph Muller, geneticist, 1890–1967

Left: Mercury is one of only two elements, the other being bromine, that is in liquid form both at room temperature and 1 atmosphere pressure.

Opposite: There is a great deal of science behind the flash and excitement of fireworks. Certain chemical compounds give fireworks their vibrant colors. For example, copper compounds are added for blue and sodium salts are added for yellow.

SPECTRAL FINGERPRINTS

Astronomers can confidently pronounce that hydrogen is the most abundant element in the universe. Chemists can analyze a water sample and prove it is contaminated with arsenic or some other toxin. Pyrotechnicians know that if they want a lilac color for a firework they need only add a potassium salt. If they'd like a brick-red color they can add a calcium salt.

Each of these feats is possible because of one remarkable fact: Each element has its own spectral fingerprint, as unique to it as our fingerprints are to each of us. If the atoms of an element are given enough energy (for example, via heating or by using light) they will momentarily absorb the energy only to then give it out again as light. This released light can be in the visible range, ultraviolet light, or even X-rays. The atoms of the same element reliably give out the same colored light and each element can be identified by the colors it emits.

From the time of the ancient Greeks to the emergence of the science of chemistry, all matter was considered to consist of four elements—earth, air, fire, and water. Greek philosopher Empedocles (fifth century BCE*) accepted these elements not only as physical manifestations or material substances, but also as spiritual essences.*

The periodic table of the elements

Aristotle (384–322 BCE), perhaps the greatest Greek philosopher of all time, further developed the doctrine of the four elements.

The father of modern chemistry

One of the most important developments in science was the division of matter into two classes—elements and compounds—by the father of modern chemistry, French chemist Antoine Lavoisier (1743–1749). He provided us with the first extensive list of elements containing 33 elements (which included heat and light and some substances later shown to be compounds), and distinguished between metals and non-metals. Lavoisier defined a compound as a substance that can be decomposed into two or more substances, and an element as a substance that cannot be decomposed. Since the discovery of the electron and the atomic nucleus, a substance consisting of atoms that have nuclei with the same electric charge is called an element.

Mendeleev's periodic table

Scientists gradually developed a table of elements that illustrated periodic similarities between the elements. But in 1869, Russian scientist Dmitri Mendeleev produced what is recognizable as the forerunner of the modern periodic table.

Mendeleev's table was based on atomic weights—the relative masses of the elements—but arranged "periodically" with elements with similar properties under each other. Gaps were left for elements unknown at that time and their properties were predicted (such as gallium, scandium, and germanium). The order of elements was re-arranged if their properties dictated it, for example, tellurium is heavier than iodine but comes before it in the periodic table.

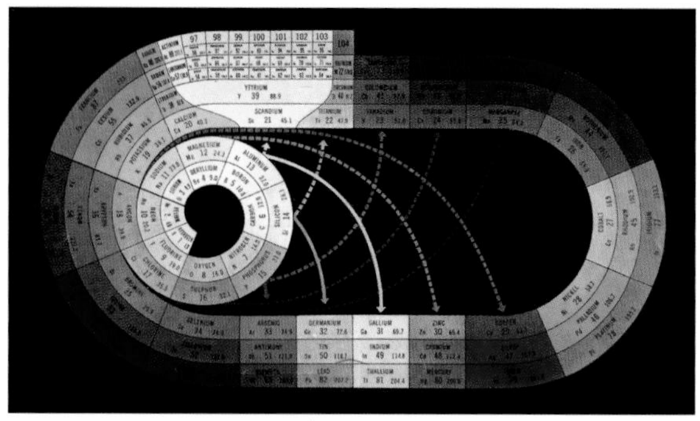

Right: The periodic table of the elements can be represented in various forms other than the traditional chart we all know (see below). Elements are grouped together according to shared chemical properties.

Above: When he categorized the elements according to their atomic weights, Russian-born scientist and teacher Dmitri Mendeleev (1834–1907) changed the face of science.

> Certain characteristic properties of elements can be foretold from their atomic weights.
>
> Dmitri Mendeleev, chemist, 1834–1907

Below: The periodic table of the elements is a central reference in chemistry. The elements are organized from left to right and top to bottom in order of increasing atomic number.

1 H Hydrogen																	2 He Helium
3 Li Lithium	4 Be Beryllium											5 B Boron	6 C Carbon	7 N Nitrogen	8 O Oxygen	9 F Fluorine	10 Ne Neon
11 Na Sodium	12 Mg Magnesium											13 Al Aluminum	14 Si Silicon	15 P Phosphorus	16 S Sulfur	17 Cl Chlorine	18 Ar Argon
19 K Potassium	20 Ca Calcium	21 Sc Scandium	22 Ti Titanium	23 V Vanadium	24 Cr Chromium	25 Mn Manganese	26 Fe Iron	27 Co Cobalt	28 Ni Nickel	29 Cu Copper	30 Zn Zinc	31 Ga Gallium	32 Ge Germanium	33 As Arsenic	34 Se Selenium	35 Br Bromine	36 Kr Krypton
37 Rb Rubidium	38 Sr Strontium	39 Y Yttrium	40 Zr Zirconium	41 Nb Niobium	42 Mo Molybdenum	43 Tc Technetium	44 Ru Ruthenium	45 Rh Rhodium	46 Pd Palladium	47 Ag Silver	48 Cd Cadmium	49 In Indium	50 Sn Tin	51 Sb Antimony	52 Te Tellurium	53 I Iodine	54 Xe Xenon
55 Cs Cesium	56 Ba Barium	57-71 Lanthanides	72 Hf Hafnium	73 Ta Tantalum	74 W Tungsten	75 Re Rhenium	76 Os Osmium	77 Ir Iridium	78 Pt Platinum	79 Au Gold	80 Hg Mercury	81 Tl Thallium	82 Pb Lead	83 Bi Bismuth	84 Po Polonium	85 At Astatine	86 Rn Radon
87 Fr Francium	88 Ra Radium	89-103 Actinides	104 Rf Rutherfordium	105 Db Dubnium	106 Sg Seaborgium	107 Bh Bohrium	108 Hs Hassium	109 Mt Meitnerium	110 Ds Darmstadtium	111 Rg Roentgenium	112 Uub Ununbium	113 Uut Ununtrium	114 Uuq Ununquadium	115 Uup Ununpentium	116 Uuh Ununhexium	117 Uus Ununseptium	118 Uuo Ununoctium
		57 La Lanthanum	58 Ce Cerium	59 Pr Praseodymium	60 Nd Neodymium	61 Pm Promethium	62 Sm Samarium	63 Eu Europium	64 Gd Gadolinium	65 Tb Terbium	66 Dy Dysprosium	67 Ho Holmium	68 Er Erbium	69 Tm Thulium	70 Yb Ytterbium	71 Lu Lutetium	
		89 Ac Actinium	90 Th Thorium	91 Pa Protactinium	92 U Uranium	93 Np Neptunium	94 Pu Plutonium	95 Am Americium	96 Cm Curium	97 Bk Berkelium	98 Cf Californium	99 Es Einsteinium	100 Fm Fermium	101 Md Mendelevium	102 No Nobelium	103 Lr Lawrencium	

Alkali metals	Alkaline earth metals	Transition metals	Rare earth metals	Other metals	Semi-metals	Other non-metals	Halogens	Noble gases

The periodic table and atomic structure

Why is there periodic resemblance in the chemical properties of families of elements? The explanation finally came with the discovery of the structure of the atom with its nucleus and its outer rings of electrons. J. J. Thomson (1856–1940) discovered the electron in 1897, and in 1906 Ernest Rutherford (1871–1937) showed that the mass of the atom is concentrated in a very small particle, which he named the atomic nucleus. Atoms resembled miniature solar systems with a nucleus and concentric rings of electrons. The nucleus consisted of protons, which are positively charged, and neutrons, with the same mass as the proton but carrying no charge. Further discoveries into the architecture of the atom linked the chemical behavior of the elements primarily to outer electronic shells of the atom.

The reason for some of the anomalies observed by Mendeleev in his periodic table—for example tellurium (atomic weight = 127.8) is heavier than iodine (atomic weight = 129.9) but comes before it in the periodic table—came to light with the discovery of isotopes. Isotopes are atoms with the same number of protons in the nucleus of the atom but with a different number of neutrons. The periodic table is based on the nuclear charge, not the atomic mass of the elements. Mendeleev had the courage to place iodine where the chemical properties indicated it should go.

Well over 100 elements make up the terrestrial periodic table. The chemical elements could not have always existed. Their nuclei cannot have a gargantuan age, because if they did, the radioactive elements such as radium and uranium, which decompose into other elements, would have already ceased to exist. Such elements could be twice as old as Earth, for instance, but not ten times as old. Each radioactive element is characterized by a property called "half-life"—the unique time for half the element to decay to another element. Whatever their age, half will decay to another element within the next half-life, leaving half remaining.

Below: A technician works with 7,000-watt xenon lamps in Las Vegas, USA. The lights are said to be the brightest in the world, powerful enough to read a book 10 miles (17 km) out in space. With an atomic number of 54, the element xenon is one of the noble gases. Its name comes from the Greek word for stranger.

WILLIAM RAMSAY'S DISCOVERY

In 1894, Scottish chemist William Ramsay (1852–1916) began isolating the noble gases—so-called because of their lack of chemical reactivity—beginning with argon. In 1895 he discovered helium as a decay product of uranium and matched it to the emission spectrum of an unknown element in the Sun that was discovered in 1868. He went on to discover neon, krypton, and xenon, and realized these represented a new group in the periodic table. Ramsay won a Nobel Prize in 1904 for his discoveries.

Good conductors of electricity and heat, metals are lustrous, malleable, and ductile. Each metallic element also has its own unique properties. In truth, the metals are an enormously diverse group of elements. Nanotechnology highlights some of this diversity.

Metals and alloys

Most elements are metals or have some metallic properties. Only about 18 of the 117 known elements are non-metals. Grouping elements according to whether they are metals or non-metals has enabled chemists to make profound discoveries as to the structure of matter.

Rather than focus on the similarities between metals, here we highlight some of the differences, by looking at silver, gold, and iron—metals often overlooked because of their familiarity. Each of these metals has unique properties that are yielding exciting applications in nanotechnology.

Above: Metals can be ranked according to their reactivity or how readily they combine with other elements. Gold and silver are at the bottom of the reactivity scale and are found in their native form in nature. Copper (at left) is slightly more reactive and as well as being found native, is also found combined with other elements in minerals such as malachite (shown at right).

Nano-silver

Metallic silver has germicidal properties—a characteristic not often associated with metals. Yet silver is one of the oldest known anti-microbial agents: It kills bacteria, fungi, and yeasts. Silver has an advantage over many other germicidal agents in that it affects the physiology of microbes in several deleterious ways at once. So the likelihood of microbes developing a resistance to silver is very slim as it would require multiple advantageous mutations to occur in a microorganism all at once.

Because of its ability to kill microorganisms, the manufacture of silver nanoparticles has become one of the fastest areas of growth in nanotechnology. Such nanoparticles are made of very small clusters of silver atoms. Washing machines, stuffed toys, food containers, some fabrics, and even brooms are now being manufactured with silver nanoparticles in them to limit the growth of mold and kill bacteria.

Nano-gold

Did you know that it's possible to make gold that is red or blue? This is because the color of gold is entirely dependent upon how large the particles of gold are. At the scale on which we live, solid lumps of gold appear yellow, because large lumps of gold reflect yellow light.

Left: Its malleability and ductility make gold a highly serviceable metal. It is used in a variety of applications, ranging from jewelry to dentistry. If gold is shrunk to a nanoparticle, its color not only changes, but the gold becomes a semi-conductor.

> In thinking about nanotechnology today, what's most important is understanding where it leads, what nanotechnology will look like after we reach the assembler breakthrough.
>
> K. Eric Drexler, scientist, b. 1955

FERROFLUIDS

Rethinking the ability of iron to be attracted to a magnet using nanotechnology leads to some surprising areas for investigation. Ask someone how to turn a liquid into a solid; usually the answer will be to "cool it!" Yet there are other ways to turn liquids into solids. Ferrofluids, for example, are a remarkable type of liquid that can take on solid-like properties just by placing them near a strong magnetic field.

Away from a magnetic field, ferrofluids behave like any other liquid, but in the presence of a magnetic field ferrofluids take on structure: Becoming denser and forming gravity-defying spikes. In effect, they become more solid-like.

Ferrofluids are comprised of a suspension of very small iron-containing particles, often in the form of hematite (above), in a fluid called a surfactant. The surfactant allows the particles to move freely over each other while also remaining suspended in a liquid state. The smaller the particles the better. Nano-sized particles are ideal.

Ferrofluids received some attention by NASA as part of investigations to control the flow of liquids in space. There are probably ferrofluids in your computer where they are used to seal the hard-drive from dust. The fluid is held in place by the magnets of the hard-drive. Ferrofluids have potential to be used in everything from smart shock absorbers for vehicles through to medical uses such as the treatment of tumors and cancer.

However, if we take small clusters of gold atoms, we see something very different indeed. Nano-sized clusters of around 1000 gold atoms emit red light, while nano-sized clusters made of about 10,000 gold atoms appear blue.

Other elements have also been found to change color at the nanoscale. This is because nanoscale objects are on the same scale as the wavelength of visible light, so they interact with light very strongly. Small changes in size at the nanoscale make a big difference to the way particles interact with light.

Molecular biologists use nano-gold to tag DNA molecules in test tubes. Simple methods are needed to test when different pieces of DNA have joined together. Unjoined, gold-tagged DNA molecules look red. However, once the DNA molecules have joined together they turn blue, showing that the reaction is complete.

Amazing alloys

Alloys are mixtures of either two or more metals, or of metals and some other material. Pure elemental metals are often either too soft, too brittle, too chemically reactive, or have other disadvantages to be useful by themselves. These disadvantages can often be overcome when metals are combined into alloys. Alloys can have surprisingly different properties from their elemental constituents.

Left: Steel is an alloy composed mainly of iron and a small amount of carbon, and is used widely in building and construction. The Sydney Harbour Bridge is a steel-arched bridge that connects the two sides of the biggest city in Australia.

Alloys are tailored to suit specific needs. Low melting point alloys are not uncommon. NaK, an alloy of sodium and potassium metals, provides an interesting example of this.

In cars, water is usually used as a coolant, transferring heat away from the engine, but in nuclear reactors water is not so effective and molten metals and alloys are often used as coolants. In research-focused reactors, NaK is one such coolant that is commonly used. NaK is a liquid from -12°C to 785°C. It is able to rapidly transfer heat away from the reactor without expanding or boiling away.

One rather significant drawback of using NaK is that sodium and potassium metals are highly reactive. It is essential that such a mixture of reactive metals be kept away from water and air to avoid the risk of explosion. However, in a reactor the advantages of using an alloy such as NaK outweigh this disadvantage.

At the other end of the reactivity and melting point spectrum to NaK are found the superalloys. Designed for high temperature (up to 1100°C), highly corrosive environments, their development has been pushed by aerospace and military industries that need such materials for use in extreme environments such as gas turbines in rocket engines. Superalloys are based on metals such as nickel, cobalt, and iron to which a rich suite of other common and exotic elements are also added. Interestingly, not all of these elements are unreactive. Reactive metals such as aluminum are deliberately added to superalloys to react with oxygen in the air and so form an unreactive, ceramic-like, protective oxide surface on the alloy.

Opposite: Where would we be without metals and alloys? They are applied to countless aspects of our lives. Here, technicians carry out maintenance work on a large gas turbine generator, which is made of a number of metals and alloys including steel and nickel.

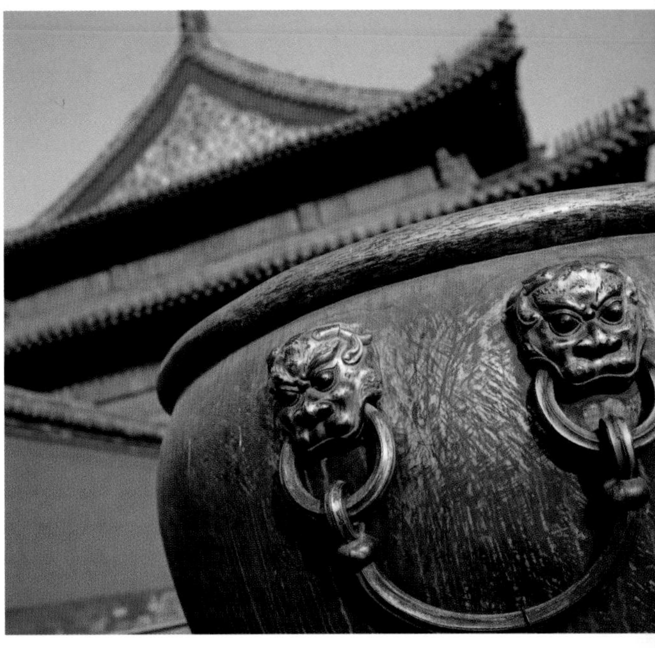

Right: This bronze urn is from the Forbidden City in Beijing, China. Humans have made good use of bronze, an alloy of copper, since the Bronze Age, some 4,000 years ago.

Left: Brass, an alloy of copper and zinc, is very malleable making it relatively easy to cast. It is commonly used to make musical instruments such as French horns (shown), tubas, trumpets, and trombones.

Richt: Spectacle frames can be made of various metals or alloys including nickel, chrome, and steel. Memory metal alloys, such as an alloy composed of nickel and titanium, are also used to make frames.

Memory metal alloys

Wet a curly piece of animal hair and it straightens. Now let the hair dry and it goes back to its original, curly shape. The same process can be repeated over again and the hair always "remembers" its curly shape. This property of hair is unusual and unique in nature. Now, however, solid metal alloys exist that share this ability.

Bend a piece of memory metal alloy and it will easily flex into whatever shape it is given. Now gently warm the same piece of metal (by using an electric current, a naked flame, or even a warm glass of water) and it will spring back into its original shape. Memory metal alloys alternate between two different solid "phases." In the low temperature phase, the alloy is soft and pliable. In the high temperature phase it is rigid and springy.

Complex mechanical gears and mechanisms can now be replaced with a single piece of memory wire. Pass an electric current through the wire and it heats, thereby deforming. Turn off the current and the wire will relax. Hence, memory metal alloys are finding applications in robotics where they can simulate the flexing of muscles. Memory alloys are used in a range of applications, such as in electric kettles, spectacle frames, dental braces, and surgical stents.

METALLIC GLASSES

New materials find an extraordinary range of applications. Metallic glasses are used in armor-piercing ammunition, in surgeons' knives, and even in modern golf clubs (right) to help hit the ball further. Metallic glasses are suitable for these diverse uses because they are "bouncier" than normal metals. This has to do with the way the atoms are arranged within the substance.

Solid metals are usually crystalline—their atoms are arranged in a regular pattern. A disadvantage of this arrangement is that the atoms can slip and slide past each other when the metal is put under stress. As the atoms slide, a kind of atomic friction is produced and energy is lost. For a golfer, some of the energy put into the swing is lost as friction, so the ball doesn't go as far.

Metallic glasses are not crystalline. Instead, the metal atoms are arranged in a jumble, causing the atoms to be locked into place so that they cannot slide past each other. When the golfer hits the ball we get what is called an elastic collision. More of the energy from the swing is passed to the ball. Hence, the ball travels further.

Metallic glasses are made by cooling a liquid metal quickly (this is called quenching). This process locks the atoms into place before they have time to arrange themselves in a regular way.

Gases are all around us, and although they are generally invisible we are definitely aware of their properties. Breathing, pumping air into tires, gas stoves, blowing up balloons, checking atmospheric pressure to predict the weather—all involve gases.

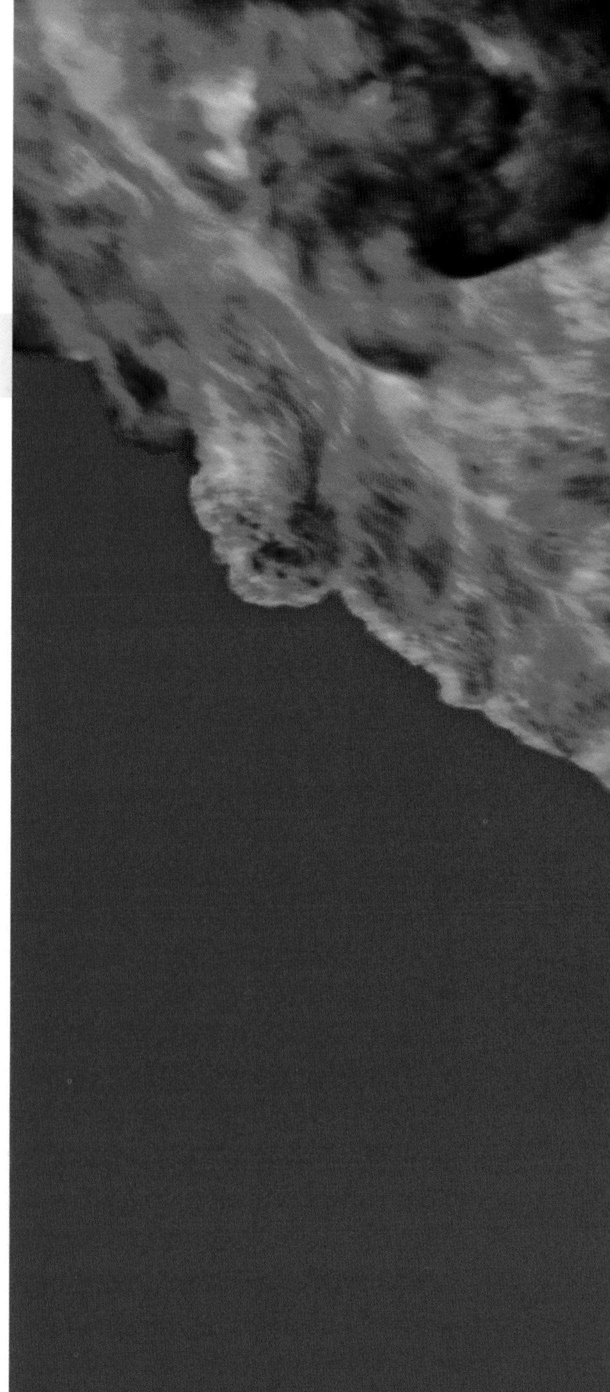

Gases

Matter can be classified according to its physical state—gas, liquid, or solid. A gas (or vapor) has no fixed volume or shape, rather it expands spontaneously to fit the volume and shape of its container (in a vacuum environment), it flows freely, and is easily compressible.

Kinetic–molecular theory of gases

A basic understanding of gases requires a quick introduction to the kinetic–molecular theory of gases, which talks about the properties of gases (e.g. pressure, temperature, and volume) by considering their molecular composition and motion. Briefly, the theory supposes that gases are composed of tiny particles, either atoms or molecules. Since they are so small, they are considered to take up a negligible volume in an empty space (gas molecules in the atmosphere occupy about 0.1 percent volume of air where the remainder is empty space); this is why gases are easily compressed compared with the denser liquid and solid states of matter. The theory also supposes that the particles in gases are energetic and are constantly whizzing around in straight lines after bouncing off the inside walls of the container or rebounding off other gas particles. This explains why different gases mix easily when combined. The force of many gas particles hitting the inside wall of its container creates pressure.

A final consequence of the theory is that the hotter the gas, the higher the average speed and energy of the gas particles. This property is exploited in hot air balloons. When the air inside the balloon is heated, the gas particles become more energetic and the gas expands. Since the same number of gas particles in the balloon now occupy a larger space, the density of the gas is decreased, causing it to be

Above: Hot air balloons are popular with people wanting a bird's-eye view of some of Earth's natural wonders. As the air inside the balloon is heated, the gas expands and the balloon rises. The hotter the gas the faster the balloon. To come down, heat is reduced, causing the balloon to descend.

NOBLE GASES

At the far right of the periodic table sit the noble gases—helium, argon, neon, krypton, xenon, and radon. This group of unreactive, colorless, and odorless elements is distinguished by their lack of activity, but their unique properties make them suitable for a range of applications.

Helium is less dense than air and is often used in weather balloons and blimps (below). Helium is also used in the air tanks of deep-sea divers to avoid the health problems associated with breathing nitrogen under pressure. A poor conductor of heat, argon is used as an insulator in double glazing, to fill incandescent light bulbs, in welding and steel manufacture, medical lasers, and ultraviolet lamps.

Neon is best known for its use in neon lights and advertising signs, but it is also used in cryogenics and in the lasers in barcode readers. Krypton is used in fluorescent light bulbs and krypton lasers are used in eye surgery.

Xenon is used in lasers and a number of specialty lighting applications. It has also been successfully used as an anesthetic. Radon has been used in radiotherapy but is generally considered hazardous to human health, having been linked to lung cancer.

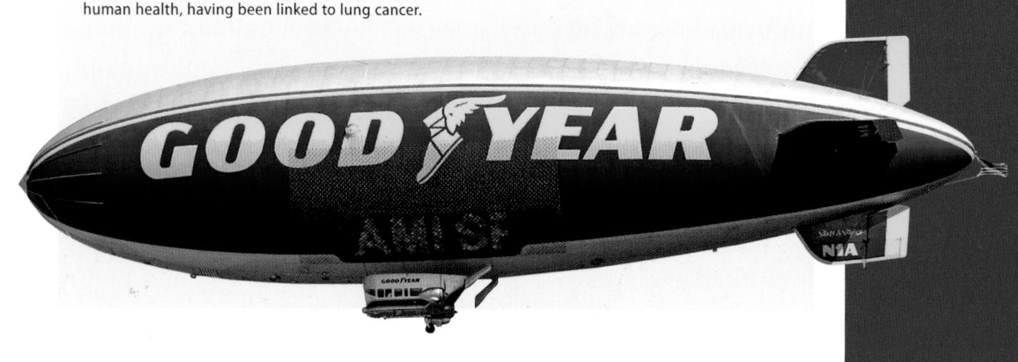

Right: Fires thrive on oxygen. Fire extinguishers are often filled with pressurized carbon dioxide, a non-flammable gas. Because carbon dioxide is heavier than oxygen, it displaces oxygen from the immediate area and stops the fire.

What in the world isn't chemistry?

Anonymous

Left: Raw natural gas consists largely of methane along with varying amounts of propane, ethane, and other gases. Gas refineries take this raw natural gas, and purify and convert it into fuel gas that is suitable for residential, commercial, and industrial purposes.

buoyed up by the cooler air outside the balloon. The hotter, less dense air inside the balloon provides the lift.

Gases in the atmosphere

Earth's atmosphere is made up of a mixture of nearly 20 different elements and compounds. Nitrogen makes up 78 percent of the dry atmosphere at sea level, while oxygen constitutes 21 percent; the remaining one percent is made up of a mixture of argon, carbon dioxide, and other elements in trace amounts. Due to Earth's gravitational field, lighter atoms and molecules rise to the higher parts of the atmosphere, which partly explains why the composition of the atmosphere is not uniform.

Versatile gases

Gas that comes out of our stoves can be natural gas or petroleum gas compressed in large metal cylinders. These gases are flammable and readily undergo the combustion reaction where they combine with oxygen to form heat and light. Carbon dioxide gas is inflammable and can be used in fire extinguishers because it displaces atmospheric oxygen and prevents it from feeding the fire. In dentistry, nitrous oxide, or "laughing gas," is used as an anesthetic and analgesic. Gases have been used in chemical warfare, especially during World War I where mustard gas was used.

Below: Lasers fan out above a crowd at the Las Vegas Sports Center, Nevada, USA. Gas lasers work when an electric current is passed through a gas thus producing light. One of the most common gas lasers in use is the helium–neon (He–Ne) laser.

Everything you see or touch is made of atoms, but you cannot see or touch an individual atom. You are made of atoms. The tens of millions of known chemical compounds are made up from about 90 different types of atom that occur in nature.

Atomic and molecular structure of matter

The idea that matter, the stuff you can see and touch, is made up of atoms goes back to ancient Greece; however, the modern atomic theory of matter is credited to a colorblind English school teacher named John Dalton who, in the early 1800s, published his theory linking the atomic nature of matter with observations of the masses of substances consumed and formed in chemical reactions. It was known well before Dalton's time that some substances could never be broken down into simpler substances in chemical reactions—these were the chemical elements (see pages 136–137). In addition to proposing that all matter was made of atoms, Dalton's theory stated that each element was composed of identical atoms, but those atoms were different from the atoms of another element. So, for example, all the atoms of iron were identical, but different from any atom of sodium. The atoms of various elements could combine, always in the same number ratio, to form a compound. For example, every molecule of water contains one oxygen and two hydrogen atoms, whereas every molecule of hydrogen peroxide contains two oxygen atoms and two hydrogen atoms.

Dalton used icons to represent atoms of different elements, but a notation using a few letters to represent an atom of a particular element, very similar to the modern notation, was developed by a Swedish chemist, Jons Berzelius, and was in common use within a few decades after Dalton published his theory. This notation is what chemists use today, giving the familiar H_2O formula for water, and H_2O_2 for hydrogen peroxide.

Inside atoms

People wondered if atoms were truly indivisible, since they must change in some way when they combine to form compounds.

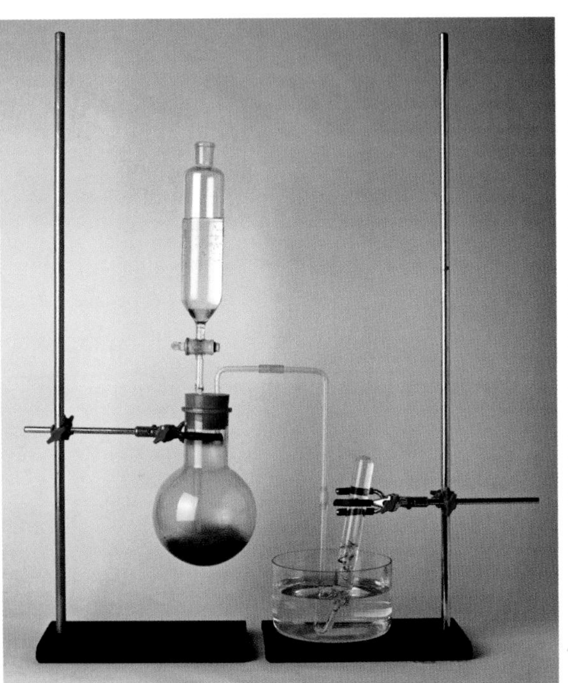

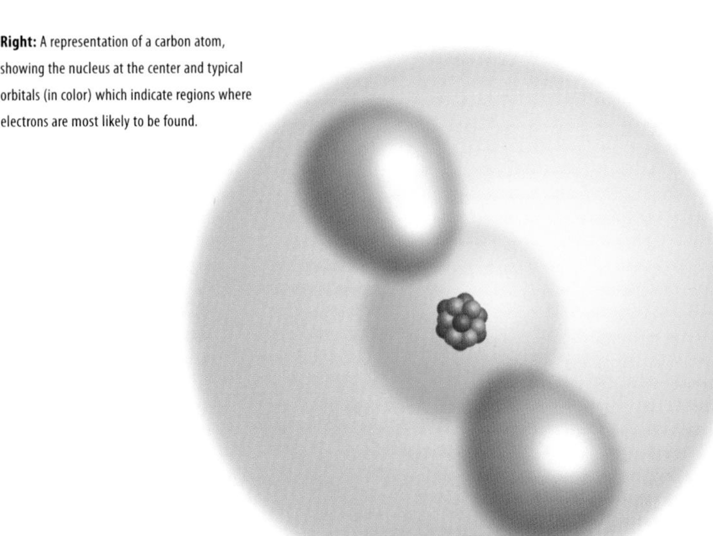

Right: A representation of a carbon atom, showing the nucleus at the center and typical orbitals (in color) which indicate regions where electrons are most likely to be found.

> When it comes to atoms, language can be used only as in poetry. The poet too is not nearly so concerned with describing facts as with creating images.
>
> Niels Bohr, physicist, 1885–1962

Opposite: Carbon atoms in HOPG (Highly Oriented Pyrolytic Graphite). Although this is an artistic drawing, it uses 3-D topographic data taken with an atomic force microscope, a 3-D surface imaging tool that is able to image atoms.

Left: Every molecule of hydrogen peroxide consists of two oxygen atoms and two hydrogen atoms. This laboratory apparatus is for the production of oxygen from catalytic decomposition of hydrogen peroxide.

HOW TO SEE AN ATOM

You cannot see an atom with your eyes, but there are experiments that can create images showing individual atoms lying on the surface of a solid or bonded into molecules. Scanning probe microscopes use a sharp tip which is moved back and forth over a surface while the force of attraction of the tip to the surface (in atomic force microscopy, or AFM) or the size of an electric current flowing between the tip and the surface (in scanning tunnelling microscopy, or STM) are measured. This allows a map of the surface to be drawn, like a contour map, but with such fine detail that individual atoms and molecules can indeed be recognized. Atomic force microscopes can even move individual atoms on a surface to form nano-structures which could not be created in any other way.

STM and AFM have revolutionized the way chemists and biologists work with molecules. These methods have been applied to measuring the roughness of surfaces (important in the development of "self-cleaning" surface coatings), studying the surfaces of semiconductors used in micro-electronics, and even detecting the differences between the surfaces of normal and cancerous cells. Unlike many other methods for imaging molecule-sized structures, AFM can be applied to samples in water, giving a huge range of applications in biology and biochemistry.

Experiments in the late 1800s involving high voltages applied to metals gave a clear sign that under the correct conditions atoms could be separated into smaller particles. These experiments showed that particles with less mass than an atom could be pulled out of atoms in a large electric field. The particles (which were first referred to as "cathode rays" and discovered by J. J. Thomson) were negatively charged and were later named electrons.

Experiments by Ernest Rutherford during the early 1900s showed that atoms were mostly empty space, with a positively charged central nucleus which held most of the mass of the atom but had a diameter of only about a thousandth that of the atom.

The composition of the nucleus was resolved when Rutherford discovered the proton and in 1932 James Chadwick discovered the neutron. Dalton's original ideas about atoms were proven to be true—with one minor correction. Many elements have slight variations in the number of neutrons in their nuclei. This variation gives rise to isotopes where the differing number of neutrons (but not protons or electrons) changes the mass of the atom but has little effect on how the atom behaves in chemical reactions.

Above: A model of an atom, close up view. Atoms are the fundamental units and smallest components of the chemical elements. Atoms consist of a central nucleus surrounded by an arrangement of electrons.

The electrons posed a problem for scientists in the 1900s because an atom where the electrons orbited the nucleus like planets orbit the Sun was realized to be impossible. Because the electrons continuously changed direction as they moved around the nucleus they would give off electromagnetic radiation and spiral inwards. The development of quantum mechanics around this time provided a picture of how the electrons could exist in the space around the nucleus, but that picture was at odds with the way the world worked for larger objects.

The quantum mechanical atom

According to quantum mechanics (see pages 76–77), objects like electrons behaved not like particles, but more like something spread out in space, like a wave, where the wavelength of that wave depended

on the electron's energy. In some experiments the wave nature of the electron was evident; in others the wave died away quickly with increasing distance so that the electron behaved like a particle. For an electron in an atom to remain stable, quantum mechanics imposed restrictions on its wavelength and thus on its energy. This resolved all the unanswered questions about the nature of atoms but required that the picture of each electron having an orbit around the nucleus had to go, so the area where each electron is spread over became known as an "orbital" (meaning like an orbit, but not exactly the same).

The quantum mechanical model of how the electrons fit into orbitals does more than just predict the energy changes an atom can undergo; it predicts how the atom will react with other atoms. This is the basis of all chemistry. With a powerful computer, and the methods developed using quantum mechanics over the last 80 years, it is possible to predict the properties of most simple molecules and at least get usable values for the properties of large biological molecules.

Above: James Chadwick (second from right), pictured here in 1945, established the liaison between British and American scientists in the development of the atomic bomb. Chadwick was awarded the Nobel Prize for Physics in 1935 for his discovery of the neutron.

LIGHT FROM ATOMS

The energy levels allowed for electrons in an atom are different for atoms of different elements. This means that the light emitted by an atom of an element when an electron drops from one energy level to a lower level has a wavelength characteristic of the element. Sodium atoms give off yellow light (below) when they lose energy (and some other wavelengths with less intensity). This is why sodium street lamps have their characteristic color. Those same streetlamps glow red when they are first switched on because other elements carry the electric current before the sodium metal vaporizes as the lamp heats up. Sodium atoms high in the atmosphere can be excited by a laser on the ground to produce their intense yellow light and form an artificial star, which astronomers can use to unscramble the optical distortion of distant objects, such as planets, galaxies, and nebulae, caused by the atmosphere.

The beautiful colors we see in fireworks are caused by chemical compounds added to the gunpowder which contain elements that emit light with different wavelengths. Compounds of strontium or lithium emit red light, barium compounds emit green light, and copper compounds give off blue light. Because the yellow light from sodium is so intense and sodium is such a common element, care must be taken not to contaminate colored firework mixtures with sodium compounds.

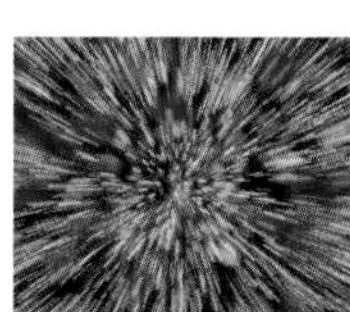

Above: The rainbow of colors we see in fireworks comes from the addition of particular chemical compounds to the gunpowder (see page 137). Different chemicals yield different colors.

Chemical bonds hold the world together. Without them there would be nothing but atoms and none of the substances that make up our world and us would exist. When a chemical bond forms between two atoms the electrons in those atoms rearrange slightly so that the energy of the atoms decreases. This means that energy is needed to separate the atoms and this is the origin of the chemical bond.

Holding the world together

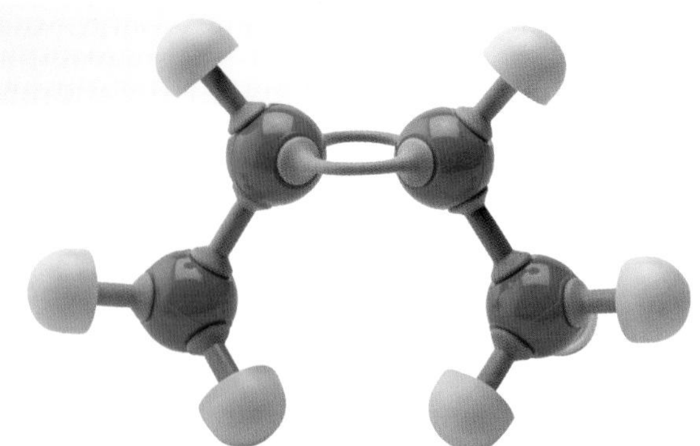

Right: A molecular model made of balls and rods, where the balls represent atoms and the rods represent the chemical bond between them. Different colors are used to show different atoms.

Quantum mechanics tells us that electrons do not circle the nucleus like planets orbit the Sun. Electrons are spread or smeared out, with some electrons more likely to be found closer to the nucleus than others. These inner (or core) electrons are largely unaffected when an atom forms a chemical bond. The outer (or valence) electrons are the ones which can be redistributed as a bond forms or breaks. Atoms of different elements have different numbers of electrons and so have different numbers of valence electrons and this is why different elements undergo different reactions (and why atoms of a single element all react in the same way).

What is a chemical bond?

If an atom of potassium comes close to an atom of bromine in a bromine molecule, an electron jumps from the potassium to the bromine, leaving the potassium atom with an overall positive charge (it has become a positive ion) and the bromine molecule negatively charged, with an ionic bond forming between the two. Ionic bonds hold together substances like sodium chloride, common salt. The exchange of electrons dramatically changes the properties of the atoms, so, for example, sodium chloride is nothing like the elements (sodium and chlorine) from which it is made.

Not all pairs of atoms will form chemical bonds. If two atoms of argon come close together, because of the number and way the electrons are arranged, there is no change in the electron distribution that can lower the energy of the atoms and form a chemical bond.

Chemical bonds do not have to form as a result of the complete transfer of electrons from one atom to another. Often the rearrangement of electrons in the two atoms is more subtle, with one or more pairs of electrons shifting into the space between the nuclei of the two atoms, resulting in the energy of the two atoms becoming lower. This means that some energy will be required to pull the two atoms apart, and this energy is called the bond energy.

Bonds that are formed by sharing electrons are called covalent bonds and usually one, two, or three pairs of electrons are shared between the bonded atoms in what are called single, double, or triple bonds. As the number of shared electrons goes up, the bond energy

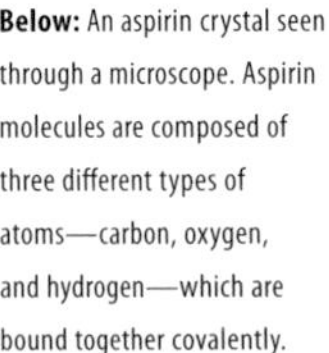

Below: An aspirin crystal seen through a microscope. Aspirin molecules are composed of three different types of atoms—carbon, oxygen, and hydrogen—which are bound together covalently.

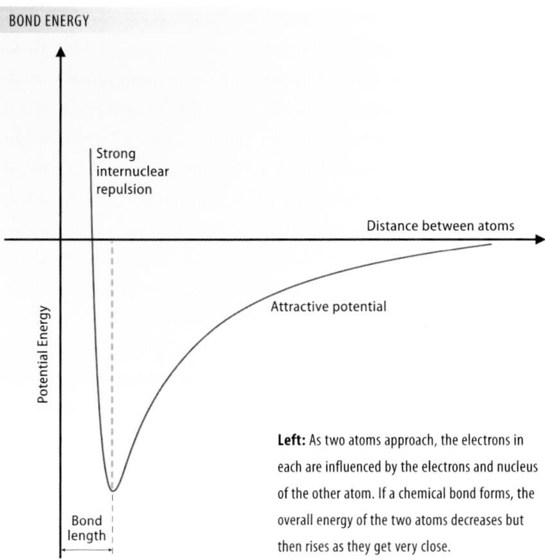

Left: As two atoms approach, the electrons in each are influenced by the electrons and nucleus of the other atom. If a chemical bond forms, the overall energy of the two atoms decreases but then rises as they get very close.

gets larger and the distance between the bonded atoms gets smaller. Large biological molecules such as proteins and DNA are held together by covalent bonds.

Metallic bonds

Sharing of electrons also holds together the atoms in metals, but in metals the shared electrons are not restricted to the region between the two atoms they bond. The shared electrons in metals are free to move through the whole piece of metal and they do just that when the metal is carrying an electric current. The special properties of metals, their strength, electrical conductivity, ability to conduct heat, malleability, and ductility, all come from these mobile electrons which bond the metal atoms together.

Above: Rock salt (NaCl) is composed of sodium and chloride that are held together by an ionic bond. The outer sodium electron enters the chlorine atom to form salt.

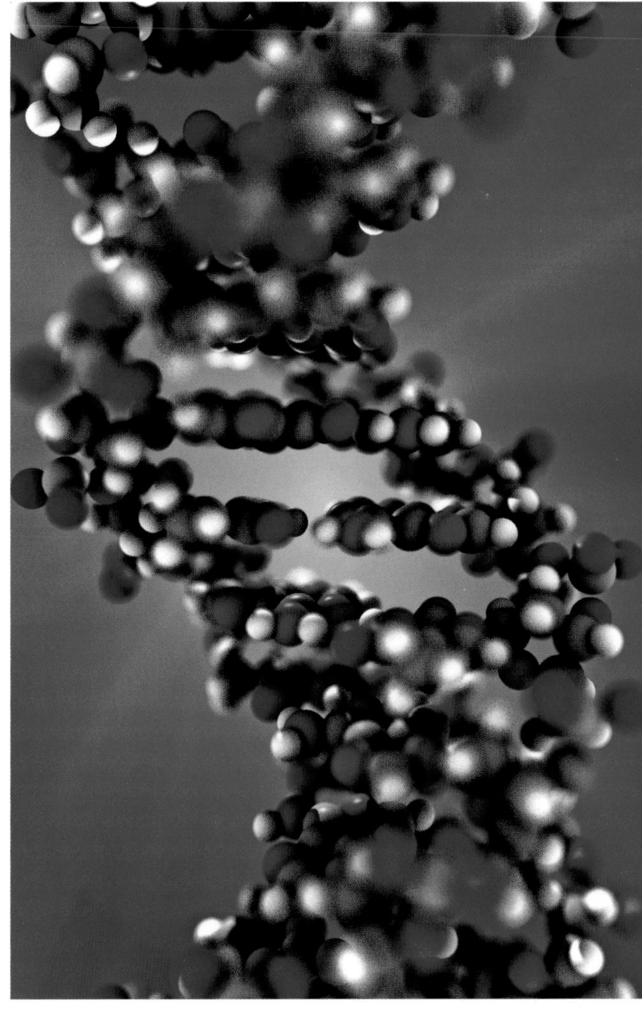

Right: A covalent bond is a chemical link between two atoms that share one or more electrons between them. Large molecules such as DNA, which carries our genetic information, are held together by covalent bonds.

> The important thing in science is not so much to obtain new facts as to discover new ways of thinking about them.
>
> William Bragg, physicist, 1862–1942

DIAMOND—THE HARDEST SUBSTANCE

Diamonds are sought after as gems because of their luster and sparkle, but industry values diamonds because of their hardness. Hardness measures the resistance of a substance to scratching. When you damage a crystal of diamond you are breaking the chemical bonds that hold atoms in the crystal together. Diamond is made up of carbon atoms held together by a three-dimensional network of covalent bonds which extends through the whole crystal. A perfect diamond crystal is one giant molecule and the only way to break that crystal is to break a huge number of covalent bonds.

Very recently, another form of carbon called aggregated diamond nanorods has been produced with a density and hardness greater than that of diamond; this may replace diamonds in some industrial applications.

Kinetics is concerned with measuring the rates of chemical reactions and studying the factors that affect those rates. Some reactions proceed spontaneously but very slowly, taking days, weeks, or even years to achieve conversion, an example being the natural conversion of carbon to diamond.

Kinetics

Other reactions can be much faster, taking minutes or even seconds to undergo conversion, while others can proceed so fast that conversion appears almost instantaneous. An understanding of the kinetics (rate) of reaction provides opportunities to manipulate the chemistry to our advantage.

Combine a fast reaction with one that involves a major release of energy, along with a large evolution of gaseous products, and the outcomes can be spectacular, such as nitroglycerin exploding.

$$2C_3H_5(NO_3)_5(l) \rightarrow 3N_2(g) + \tfrac{1}{2}O_2(g) + 6CO_2(g) + 5H_2O(g)$$

Decomposition of nitroglycerin is highly exothermic, instantly releasing 1.5 calories (6,260 J) of thermal energy per gram.

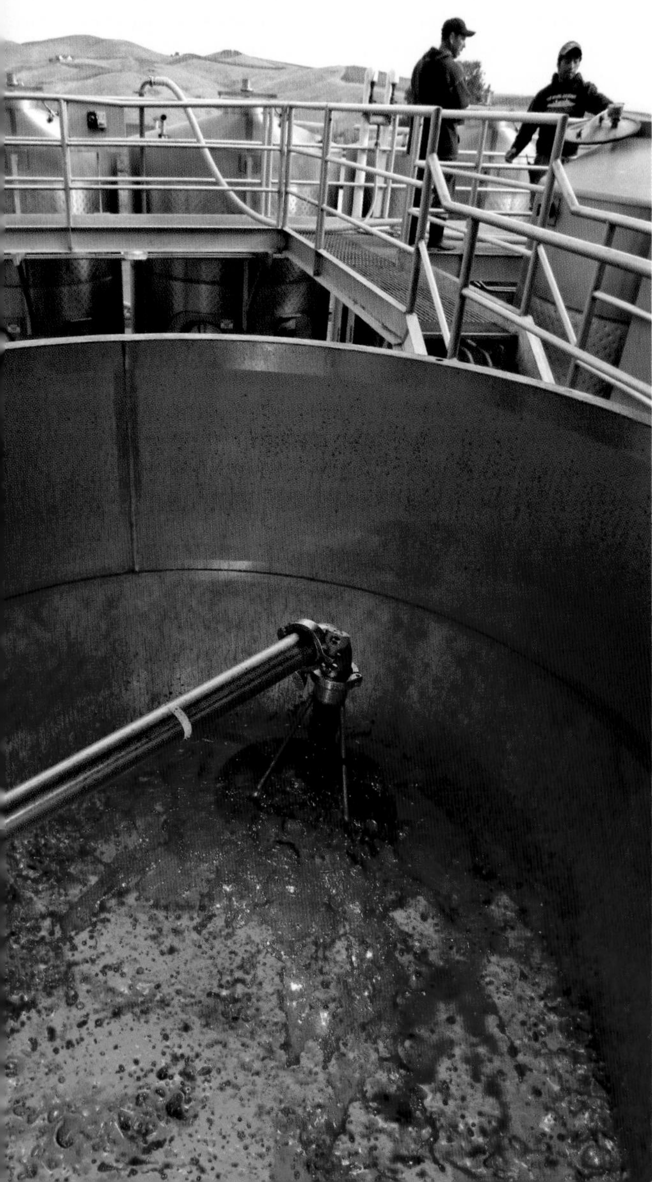

Bottom left: A fermenting tank is filled with pinot noir at a winery in California, USA. The kinetics of fermentation involve, among other factors, the time taken for sugar to be converted into alcohol.

Order of rate reactions and half-life

The reaction of nitrous oxide with chlorine to form nitrosyl chloride (nitrogen oxychloride) is:

$$2NO_2(g) + Cl_2 \rightarrow 2NOCl(g)$$

The rate = $k\,[NO]^2\,[Cl_2]$, which is second order with respect to NO concentration and first order with respect to Cl_2 concentration.

For actions with a single reactant:

$$A \rightarrow \text{Products}$$

graphical techniques can be used to relate the concentration A at a time t ($[A]_t$) to the rate constant (k) and the initial concentration at initial (zero) time $[A]_0$ and thus determine the order.

The half-life of a reaction ($t_{1/2}$) is the time which concentration of reactant (A) takes to reduce to half its initial concentration $[A]_0$ and this is representative of the rate.

Zero Order	rate	=	k
	where $[A]_t$	=	$-kt + [A]_0$
	and $t_{1/2}$	=	$0.693/k$
First Order	rate	=	$k[A]$
	where $\ln[A]_t$	=	$-kt + \ln[A]_0$
	and $t_{1/2}$	=	$[A]_0/2k$
Second Order	rate	=	$k[A]2$
	where $1/[A]_t$	=	$-kt + 1/[A]_0$
	and $t_{1/2}$	=	$1/k[A]0$

RATE LAWS AND THE SPEED OF A CHEMICAL REACTION

The rate of reaction is measured by the rate of loss of molar concentration (mole per liter) of reactant divided by its stoichiometric coefficient, or alternatively the rate of gain of molar concentration of product divided by its stoichiometric coefficient. (The stoichiometric coefficient is the number that comes before a substance in a chemical equation; these values are positive for products of the reaction and negative for the reactants.)

In the following reaction, three moles of hydrogen react with one mole of nitrogen to produce two moles of ammonia gas:

$$3H_2(g) + N_2(g) \rightarrow 2NH_3(g)$$

$$\text{The rate of reaction} = -\frac{1}{3}\Sigma\frac{d[H_2]}{dt} = \frac{d[N_2]}{dt} = \frac{1}{2}\Sigma\frac{d[NH_3]}{dt}$$

The rate is usually measured during the initial stages of a reaction where loss of reactant is minimal compared with the initial concentration. The formation of product is also low, so does not interfere in the reaction.

An overall rate law (rate equation) is found experimentally in terms of reactant concentrations. The stoichiometric coefficients do not influence the overall rate law; however, they are relevant to rate equations of elementary reactions that combine to make up the overall reaction. Elementary reactions occur exactly as written in the chemical equation, without any intermediate steps. Usually the slowest elementary reaction influences the overall rate of reaction.

$$aA + bB \rightarrow yY$$

The rate law for the general reaction shown above consists of the rate constant (k) and reactant concentrations (mole per liter):

$$\text{Rate} = k[A]^m[B]^n$$

The terms m and n give the order of reaction.

The influence of temperature

The rate constant dictates the rate of reaction, which is linked to a minimum energy requirement—Activation Energy (E_a). This minimum energy is required to break chemical bonds thus allowing the reaction to proceed. This applies to both endothermic and exothermic reactions as shown in the two diagrams below. The activation energy maximum corresponds to the formation of a highly unstable activated transition complex.

ENERGY USED IN AN ENDOTHERMIC REACTION

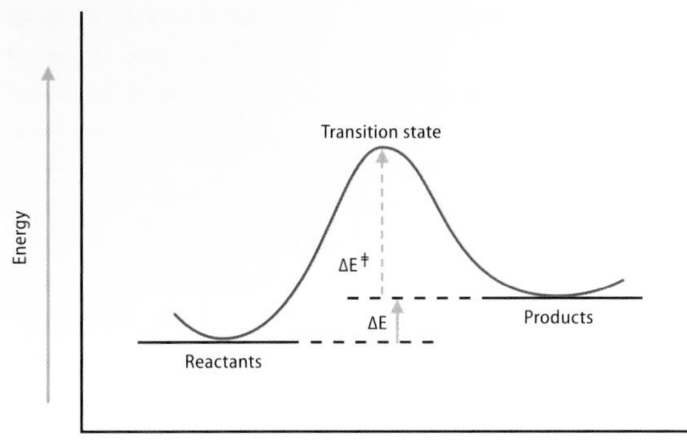

ENERGY USED IN AN EXOTHERMIC REACTION

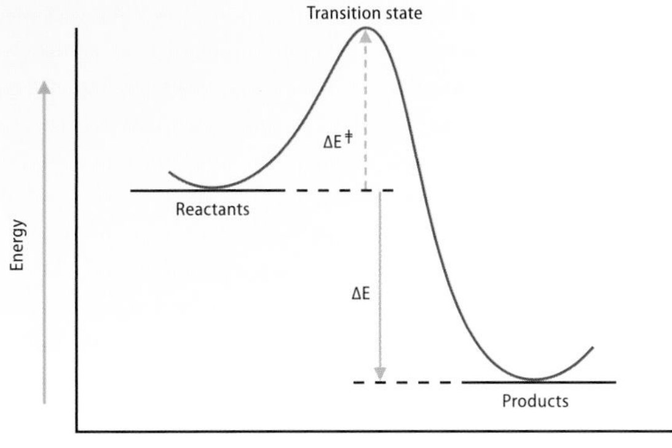

ENERGY DIAGRAM SHOWING AN EXOTHERMIC REACTION

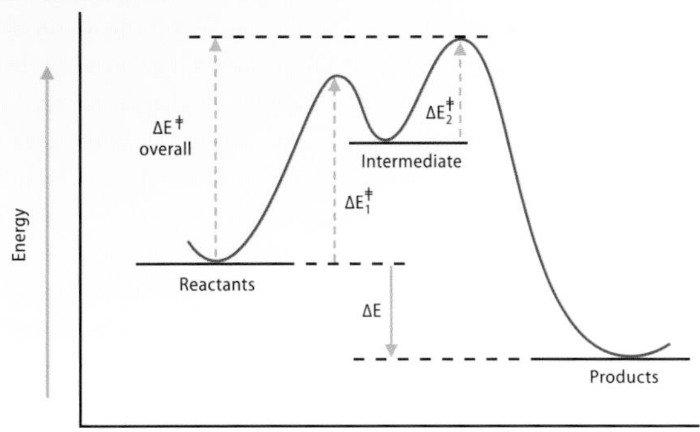

Above: This graph depicts a reaction, involving two elementary reactions, in which the second reaction must cross a higher activation energy barrier, and becomes the rate determining reaction.

Above: Ahmed Zewail won the 1999 Nobel Prize for Chemistry for his studies of the transition states of chemical reactions using femtosecond spectroscopy, that is, the rates of very fast reactions.

Below: Bystanders watch as a building is demolished using nitroglycerin-based dynamite. When nitroglycerin explodes, a huge amount of energy is released. The decomposition is highly exothermic.

The range of kinetic energies that molecules have follows a Maxwell–Boltzmann distribution and thus only a fraction of the molecules have an energy greater than the minimum activation energy required for the reaction to proceed.

However, as the temperature increases, a larger fraction of molecules have energies exceeding this minimum activation energy, and so the reaction proceeds faster. See the diagram below that demonstrates this situation.

THE MAXWELL–BOLTZMANN DISTRIBUTION

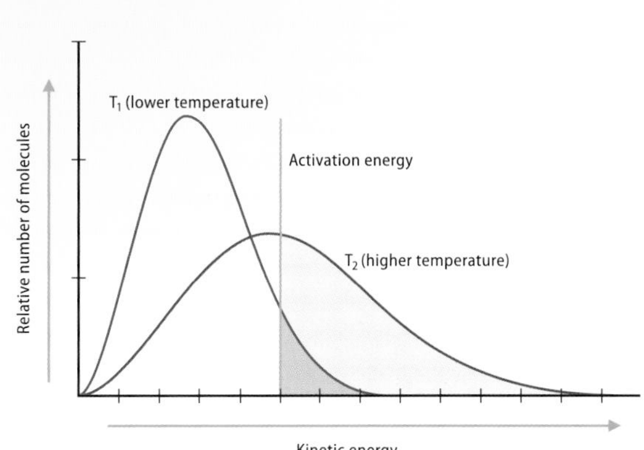

Above: The Maxwell–Boltzmann distribution of molecule kinetic energies at two different temperatures.

For me, chemistry represents an indefinite cloud of future potentialities ...

Primo Levi, chemist and novelist, 1919–1987

Catalysis is the acceleration of a chemical reaction by means of a substance called a catalyst, which is added to the reaction but remains unchanged after the reaction has occurred. A catalyst speeds up a reaction by lowering the amount of energy required to initiate the reaction.

Catalysis—a need for speed

Why use a catalyst? There are some chemical reactions that can easily be accelerated, for example, by increasing the temperature of the reaction. This approach is not suitable, however, in living organisms where an increase in temperature may result in the destruction of important proteins. Catalysts provide an alternative means of achieving the same result.

Right: An oil refinery processes and refines crude oil into more useful petroleum products such as gasoline, diesel fuel, liquid petroleum gas, and kerosene. Catalysts are used to speed up cracking, which is when large hydrocarbons are broken down into smaller ones, an important part of the process.

Lowering the activation energy

For a chemical reaction to occur there must be sufficient energy to initiate it. This energy threshold is called the "activation energy." The purpose of a catalyst is to lower the activation energy required to initiate a chemical reaction, thereby speeding up the transition from reactant (material consumed during a chemical reaction) to product. Lowering the activation energy by adding a catalyst can happen in a number of ways. For example, the catalyst may alter the structure of a reactant, allowing it to react more readily, or the catalyst might hold the reactant molecules in a particular orientation, which reduces the amount of energy that is required for the reaction. Either way, it is important to remember that catalysts remain unchanged after the chemical reaction occurs.

Yet this is not to say that the catalyst does not participate in the chemical process—in fact, there are many catalysts that work by breaking down the reaction process by creating intermediate compounds. Typically, these intermediate compounds are unstable, and a second reaction occurs, producing the desired end products, and returning the catalyst to its original form. For this reason, catalysts are recyclable, and can be used many times, making them incredibly useful for industrial applications.

Enzymes: biological catalysts

Living cells also share a need for catalysts. These biological catalysts are called "enzymes" and are usually proteins produced within the cell. Some animal cells are known to contain up to 4,000 different types of enzymes, each responsible for catalyzing a specific chemical reaction. While some enzymes are common to all cells, many are specific to the function of the cell in which they are produced. In an enzyme reaction, the reactants are called "substrates," and each enzyme acts selectively on one substrate or set of related substrates to reduce the activation energy of the desired chemical reaction. Enzymes are an important feature of normal cell activity, controlling the metabolic pathways that underlie cell structure and function.

As proteins, enzymes are sensitive to the effects of temperature, so it is vital that an enzyme is able to reduce the activation energy of a chemical reaction to enable it to occur at physiological temperatures.

Above: A close up of iron (III) chloride granulate, which is also known as ferric chloride. This chemical compound is used widely as a catalyst in various industrial applications.

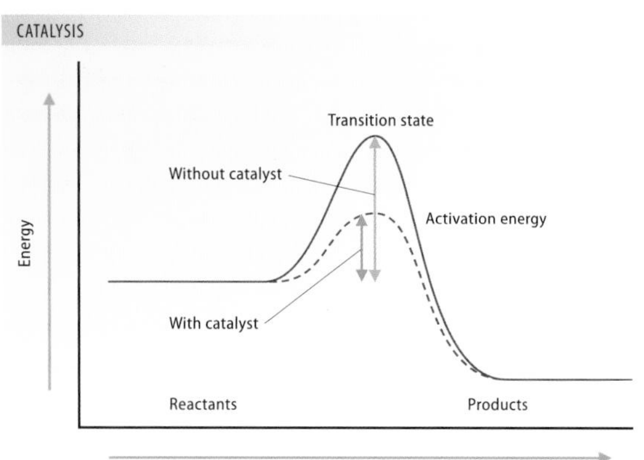

Above: A catalyst lowers the activation energy needed for a chemical reaction to occur, thus speeding up the transition from reactant to product. Catalysts themselves do not change during a chemical reaction.

> The development of a rational view of the nature of catalysis was thus absolutely dependent on the creation of the concept of the rate of chemical reaction.
>
> Wilhelm Ostwald, chemist, 1853–1932

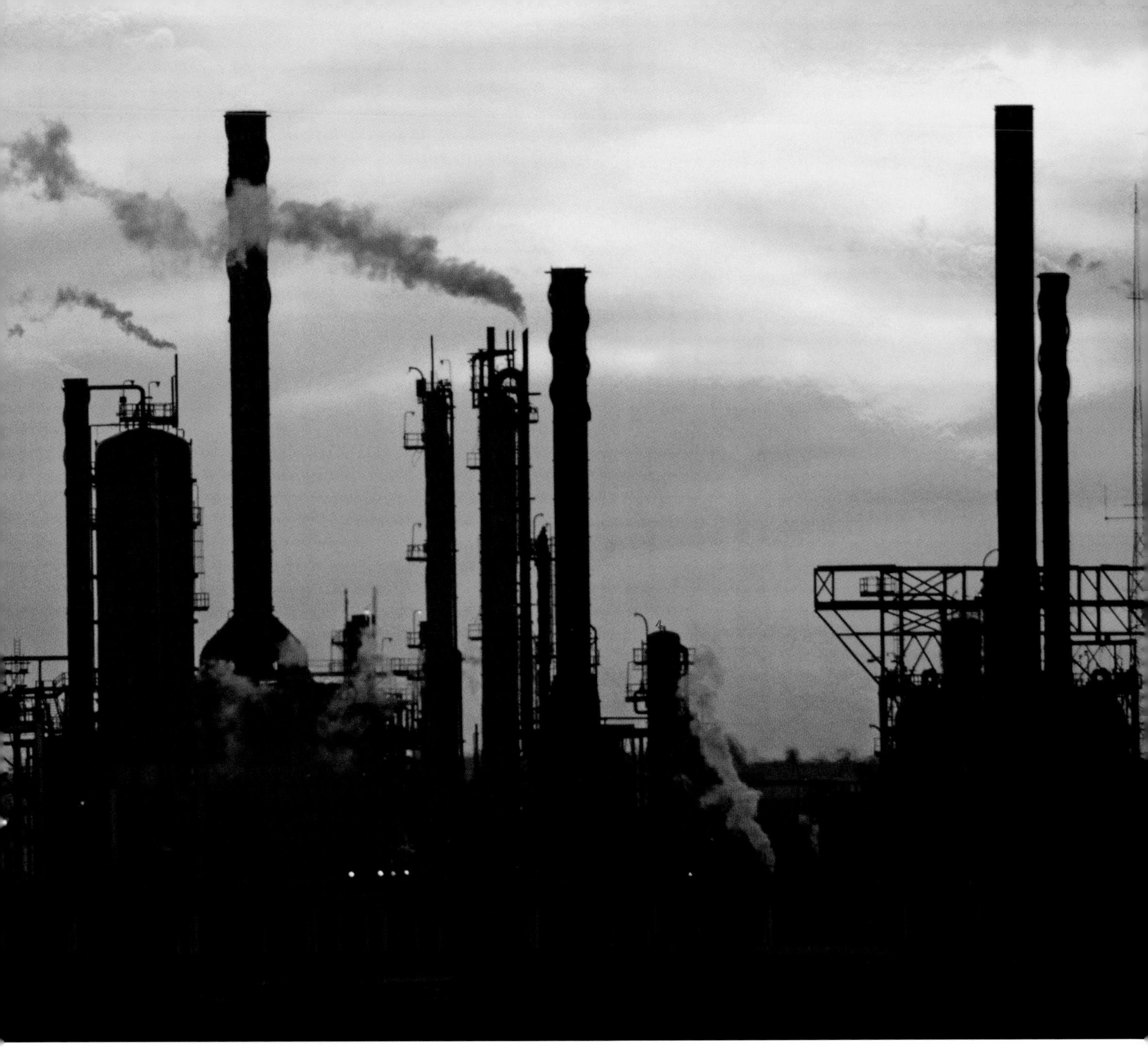

CATALYTIC CONVERTERS

First introduced to the automotive industry during the 1970s, the catalytic converter uses the principle of catalysis to help reduce the toxic emissions that are produced by car engines.

There are two types of catalytic converters—two-way and three-way catalytic converters. Each of these converters performs a similar task, and their function can be explained chemically according to a few simple reduction and oxidations (redox) reactions.

A two-way catalytic converter results in two outcomes. First, it performs the oxidation of carbon monoxide (a highly toxic substance) to carbon dioxide.

$$2CO + O_2 = 2CO_2$$

It also results in the oxidation of any unburnt hydrocarbons (left over from the internal combustion of fuel), resulting in the production of carbon dioxide and water.

$$2C_xH_y + (2x + y/2)O_2 = 2xCO_2 + yH_2O$$

A three-way catalytic converter goes one extra step to reduce nitrogen oxides to nitrogen and oxygen.

$$2NOx = xO_2 + N_2$$

The catalyst that is used in automotive catalytic converters is most often a precious metal. In most cases platinum is used (as in the catalytic converters shown on the left), because of its high levels of activity. However, other precious metals can also be used, such as rhodium, palladium, iron, copper, nickel, and manganese.

Thermodynamics was developed in the 1800s by engineers wanting to improve the efficiency of steam engines and later diesel/internal combustion engines. Chemists quickly realized the importance of understanding heat energy and how it related to understanding the spontaneity of chemical reactions.

Thermodynamics and chemical equilibrium

Energy change drives processes, and there are many forms of energy. Energy of motion (kinetic energy) is the energy of moving objects, including the random thermal motion of molecules, and the oscillating motion of gas particles in sound. Stored (potential) energies include gravitational energy, chemical potential energy (for example, a battery not in use), or electrostatic energy.

There are three laws of thermodynamics, the first being that energy and mass cannot be created or destroyed. (See also pages 42–43.) Obviously in Einstein's physics, which describes creation of energy from mass (E = mc2), this does not hold, but in the conventional world of chemistry, the first law of thermodynamics applies.

Above: When a battery-operated device is switched on, its stored potential chemical energy is converted into other forms of energy, such as sound.

> Nothing in life is certain except death, taxes, and the second law of thermodynamics.
>
> Seth Lloyd, mechanical engineer, b. 1960

The flow of heat energy

Heat energy and temperature are not the same. A thermometer can measure temperature (in degrees Fahrenheit (°F), degrees Celsius (°C), or in Kelvin (K)). As the temperature of the system increases, atoms or molecules move faster and collide with each other more frequently, thus raising its internal thermal energy. The heat capacity equation is used to calculate the heat (q) (in joules or calories) accompanying a temperature change:

$$q = C \times m \times \Delta T$$

where C is the heat capacity of the substance, m is the mass and ΔT is the temperature change.

The system consists of all object(s) of interest, for example the substances undergoing a chemical reaction, with the surroundings being everything else, and the universe naturally being the system plus the surroundings. Heat can transfer either into or out of the system, and out of or into the surroundings, but energy must still be conserved (first law). Any process in which heat flows out of the system into the surroundings is called exothermic. Any process in which heat flows into the system from the surroundings is described an endothermic. Burning a match is an exothermic process because heat is lost from the system (match) to the surroundings (air). Boiling water is an endothermic process because heat is supplied from the surroundings (flame) to the pot of water (system).

Opposite: McBride Glacier, Alaska. An endothermic reaction is one that absorbs heat from the environment, for example when ice melts. Although this is a spontaneous reaction, it is still classsed as endothermic.

Spontaneity of a chemical reaction

Not all spontaneous reactions are exothermic. Ice melts spontaneously but in doing so absorbs heat from the surroundings. This is a spontaneous, endothermic process. A factor contributing to whether a process is spontaneous is the change in randomness of the system during the process. Liquid water is much more random than its highly structured, crystalline state of ice. Even though energy is required to melt ice, the higher probability of a more random state (water) allows the melting process to happen spontaneously. This factor is referred to as entropy (S).

This results in all substances having positive entropy, with gases having more entropy (randomness) than liquids, which in turn have more entropy than solids. Large molecules generally have more entropy than small molecules. The second law of thermodynamics states that for any spontaneous process, the overall change in the entropy of the universe must be a positive one. The third law of thermodynamics states that any perfect crystal at absolute zero (K) has zero entropy.

Right: Water boiling on top of a stove is an example of an endothermic process because the chemical reaction requires the absorption of heat from the surroundings.

ENTHALPY—THE HEAT ENERGY OF A CHEMICAL REACTION

The heat lost or gained from a reaction under conditions of constant pressure (reactants and products) is known as enthalpy change (ΔH, in calories (joules) per mole). A reaction that has a negative enthalpy of reaction ($-\Delta H$) is an exothermic reaction because heat is lost by the reactants (system). Burning materials such as wood, coal (at left), liquid petroleum fuel, or hydrogen gas in air or in oxygen are usually exothermic reactions because heat is spontaneously given out. For example, the ΔH for burning propane in oxygen is –9,297 calories (–2220 kJ).

Reactions that require an input of heat energy to force the reaction are endothermic reactions. A considerable amount of energy is required to force water to split into hydrogen and oxygen: $\Delta H = +1{,}013$ calories (+241.8 kJ).

Usually, but not always, a negative enthalpy of reaction means that a reaction is likely to proceed spontaneously.

Nanotechnology refers to science and technology, nominally on a scale of 1 nm to 100 nm (nanometers) where a nanometer is one-billionth of a meter (10^{-9} meters). Although nanotechnology is considered to be a new science, aspects of nanotechnology have been around for many years. It is the unique, multidisciplinary linking of chemistry, physics, and biology that characterized the growth of nanotechnology from the mid-1980s onward.

Nanotechnology—a new science?

The nanotechnology concept was first raised by Richard Feynman at an American Physics Society meeting at Caltech in 1959. Feynman did not actually coin the term nanotechnology, but suggested that it would be possible to manipulate atoms and molecules at the nano level. He also suggested that scaling issues would occur and phenomena such as gravity would become less important, and that surface tension, Van der Waals attractions, and quantum phenomena would become more important. At this scale, the nano-scale, there is a unique combination of macro and quantum behavior, providing new materials with unique properties.

The term "nanotechnology" was defined in 1974 by Tokyo Science University Professor Norio Taniguchi at a speech given to the International Conference on Production Engineering in Tokyo. K. Eric Drexler subsequently promoted the concept of nanoscale phenomena and the potential to develop nano-devices, probably most famously in the book *Engines of Creation: The Coming Era of Nanotechnology* published in 1986.

The first nanotechnology conference was introduced at Stanford University in 1989, while the first nanotechnology academic course for graduate students started at the University of Southern California in 1994. The University of Washington is credited with introducing the first post-doctoral degree in nanotechnology and Flinders University (Adelaide, Australia) the first undergraduate (honors) degree in nanotechnology.

> Nanotechnology will let us build computers that are incredibly powerful. We'll have more power in the volume of a sugar cube than exists in the entire world today.
>
> Ralph Merkle, nanotechnologist, b. 1952

Above: Richard Feynman (1918–1988), who introduced the concept of nanotechnology to the scientific community in the late 1950s, won the Nobel Prize for Physics in 1965.

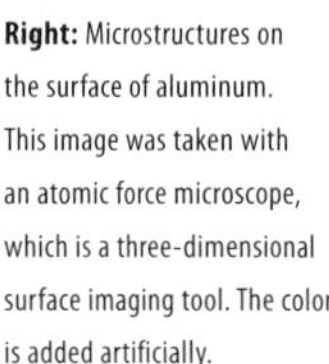

Right: Microstructures on the surface of aluminum. This image was taken with an atomic force microscope, which is a three-dimensional surface imaging tool. The color is added artificially.

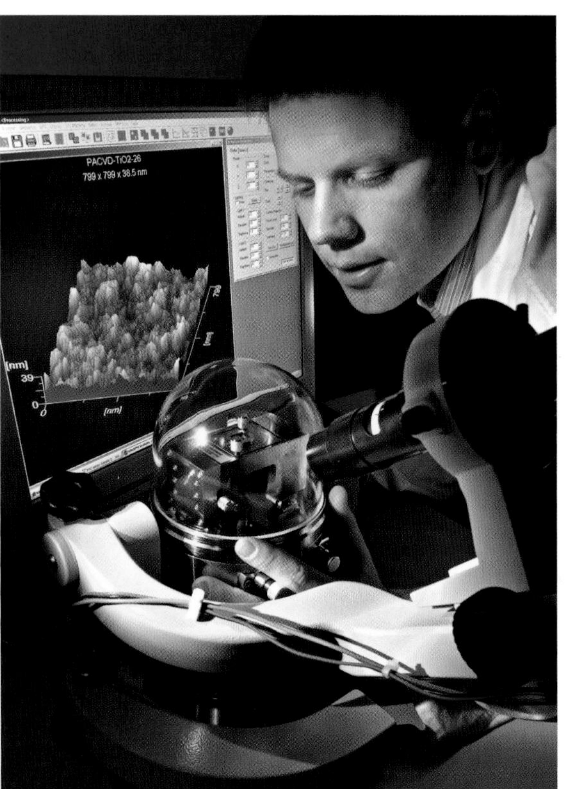

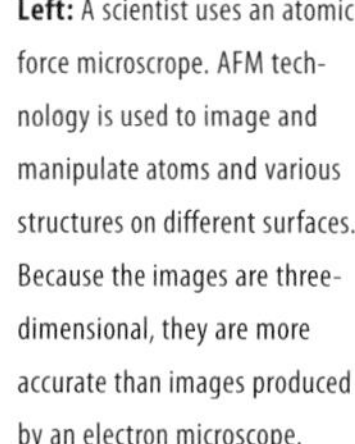

Left: A scientist uses an atomic force microscrope. AFM technology is used to image and manipulate atoms and various structures on different surfaces. Because the images are three-dimensional, they are more accurate than images produced by an electron microscope.

Scanning tunneling microscopy

STM imaging was developed by Gerd Binnig and Heinrich Rohrer (at IBM) in 1981 (for which they won the Nobel Prize for Physics in 1986). STM employs tunneling current from a conducting tip, through a sample on a conducting substrate. The electrical current change is used to visualize samples with a lateral resolution of 0.1 nm and a vertical resolution to 0.01 nm.

Atomic force microscopy

Gerd Binnig, Calvin Quate, and Christoph Gerber invented AFM in 1986. The major tool for imaging, measuring, and manipulating matter in the nano region, AFM uses a tip with a radius of curvature (nm wide) attached to a silicon or silicon nitride cantilever. As the tip is brought close to a surface, attracting or repelling forces cause the cantilever to bend, with nanoscale motion. A laser beam reflects off the back into a quadrant of photodetectors, which is used to simultaneously measure deflection and forces using Hooke's Law (strain in an elastic body is proportional to the force acting on it).

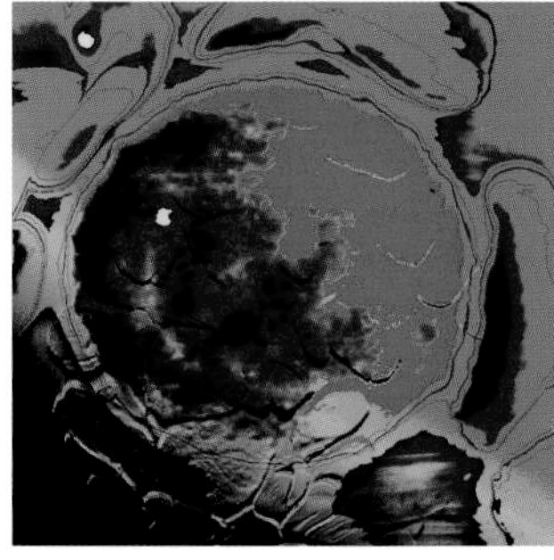

Below: An AFM image of a lymphocyte (a white blood cell). All branches of science are interrelated. The developments in nanotechnology are also being utilized in biological and medical research.

VISUALIZING NANOMATERIALS

Until relatively recently, scientists could only speculate about nano-sized materials, as visualization was limited to optical microscopes, which operated in the micro region (millionths of a meter). The advent of scanning electron microscopy (SEM) saw the first visualization of the nano region with imaging possible of very large polymeric molecules, and biological species such as bacteria and some viruses. SEM was eventually commercialized in 1965 by Cambridge Instrument Company, but SEM resolution was limited to about 5 nm. High-resolution SEM improved resolution to 2 nm and field emission SEM (FESEM) to 0.5–1 nm. Modern transmission electron microscopy (TEM) now approaches the theoretical electron wavelength resolution limit of 0.22 nm.

Scanning tunneling microscopy (STM) and atomic force microscopy (AFM) ultimately galvanized nanotechnology interest in the early 1980s. These techniques allowed the visualization of molecules and even atoms (at least larger atoms) in the nano region. These techniques also provided topographical data with AFM being able to provide valuable direct force-interaction data between particles.

Nanomaterials are new materials, developed from the science of nanotechnology, which have properties that rely on the fact that they exist on the nano-scale, nominally from 1 nm to 100 nm (nanometers). A nanometer is one-billionth of a meter. The development of nanomaterials requires a linking of chemistry, physics, and biology science disciplines to engineer these new materials.

Nanomaterials—technologies for the future

Nanotechnology is providing many new nanomaterials that may help to supply clean water, and provide engineering solutions for food and crops with improved low-cost agricultural productivity along with the provision of smart foods. Energy generation, efficient manufacturing, improved drugs and diagnostics, improved information storage and communication, smart appliances, and increased human performance are all possible outcomes.

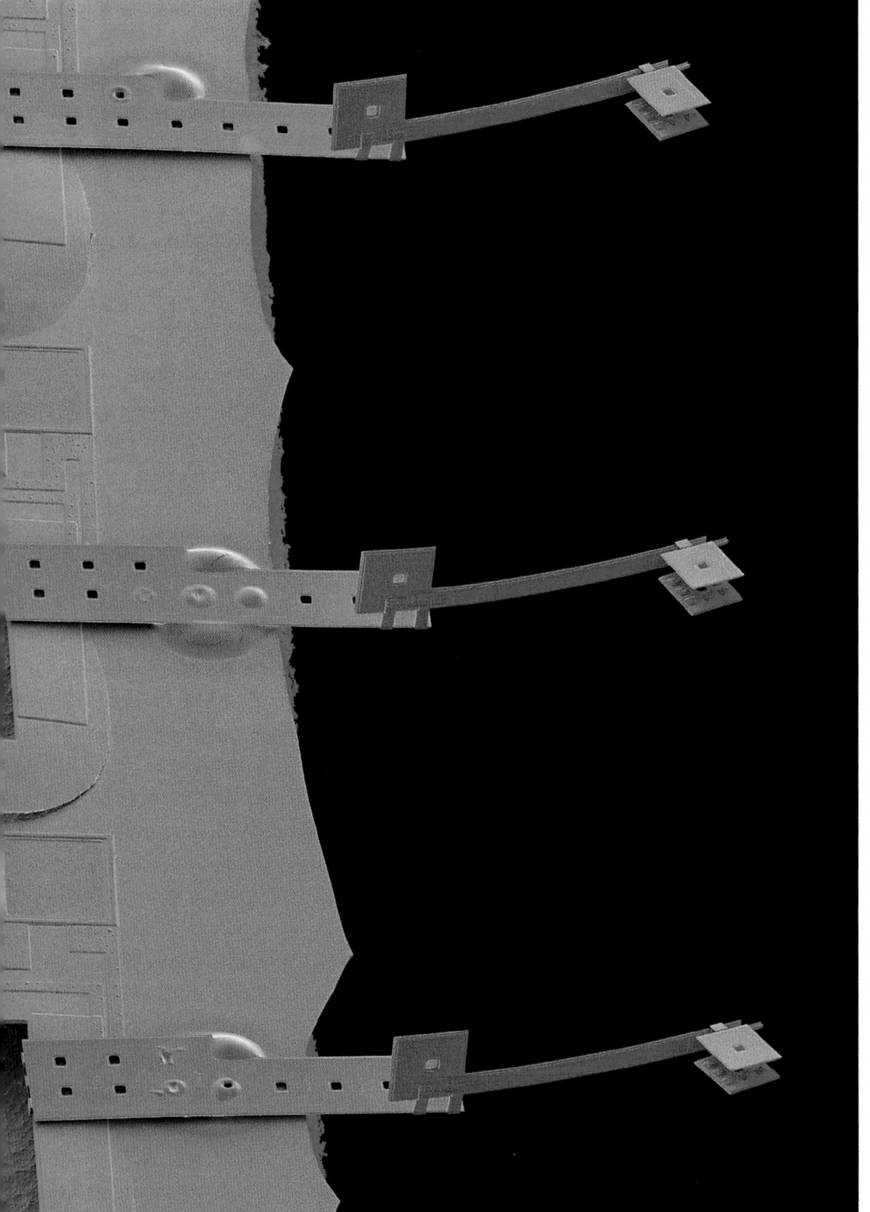

Below: MEMS (Micro-Electrical-Mechanical Systems) micro force gage beams. These useful devices were created by nano-fabrication and micro-machining. MEMS is the technology of the exceedingly small.

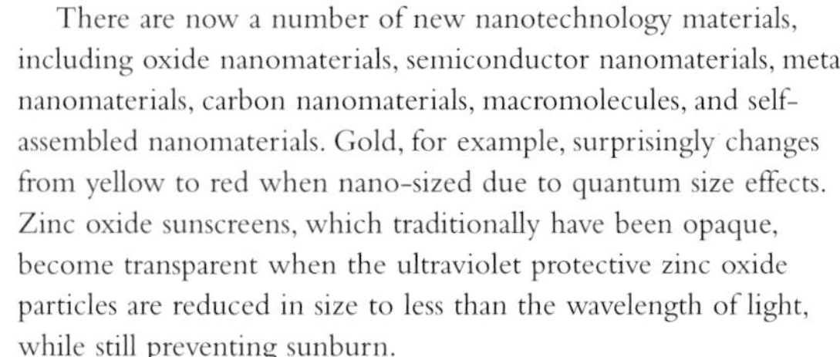

There are now a number of new nanotechnology materials, including oxide nanomaterials, semiconductor nanomaterials, metal nanomaterials, carbon nanomaterials, macromolecules, and self-assembled nanomaterials. Gold, for example, surprisingly changes from yellow to red when nano-sized due to quantum size effects. Zinc oxide sunscreens, which traditionally have been opaque, become transparent when the ultraviolet protective zinc oxide particles are reduced in size to less than the wavelength of light, while still preventing sunburn.

Nanocomposites are producing new plastics with fire-resistant properties, improved structural applications, and improved conducting and biodegradable properties. Nanotechnology is resulting in new membrane applications for gas separation, desalination, and the hydrogen fuel cell development. Specialist coatings can also result in improved anti-corrosion properties.

THE DANGERS OF NANOMATERIALS

This technology has the potential for unseen hazards. Skeptics believe that barriers of life might be broken by nanobiotechnology and redefine human existence. K. Eric Drexler postulated potential problems with molecular nanotechnology and described nano-chemicals or nano-robots (such as this one in a human artery) that might uncontrollably self-replicate with disastrous outcomes—the so-called nanotechnology "gray goo" scenario. Parallels to "gray goo" are real-life bacteria and viruses, which provide an example of biological nanotechnology, or "living goo," leading to natural human disasters such as HIV and ebola. While "gray goo" seems to be science fiction, there have been real concerns about nanomaterials, which exist on dimensions equal to that of living cells. Materials previously considered harmless, when reduced to the nano-scale, might be able to enter the human environment and cause unforeseen damage. When working in the nano-realm scientists need to be mindful of such dangers.

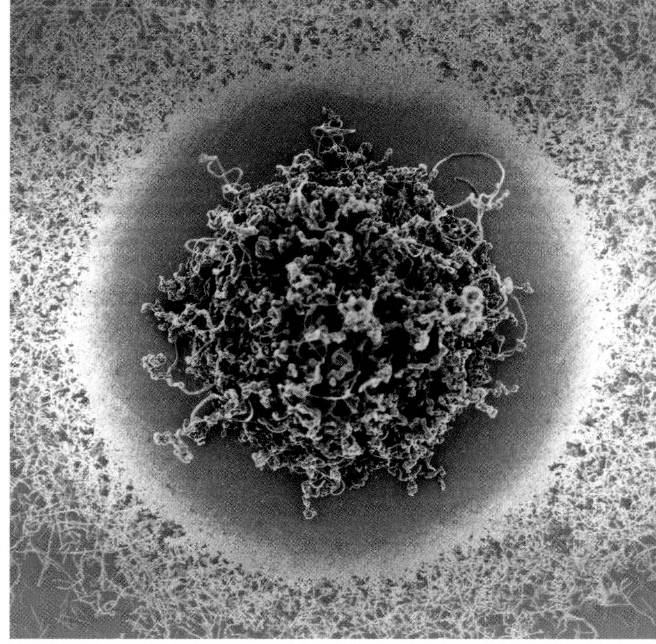

Nanotechnology is an idea that most people simply didn't believe.

Ralph Merkle, nanotechnologist, b. 1952

How nanomaterials are produced

There are two approaches to the science of nanotechnology—a "top-down" approach and a "bottom-up" approach. The "top-down" engineering approach has been to reduce materials down to the required size or shape. For example, a laser can focus light in a tight beam to burn away material using a process known as micro machining. This process can be difficult or expensive. The "bottom-up" approach manipulates atoms or molecules on the nano-scale. One fascinating example of "bottom-down" nano-fabrication was the use of scanning tunneling microscopy (STM) by IBM scientist Don Eigler to position 35 xenon atoms on a smooth nickel substrate to spell out the name IBM. The laws of chemistry or physics can also be employed to have atoms or molecules come together in a "bottom-up" predefined manner to bond and form a new nanomaterial. Molecular self-assembly and supramolecular chemistry now allows molecules to arrange themselves through a "bottom-up" approach. The Watson–Crick base-pairing rules in DNA are a result of molecular self-assembly.

Complex nanocomposites

Although less exotic, but certainly more economically significant, is the production of complex nanocomposites. One classic example is the development of clay/nylon-6 polymer nanocomposites by Toyota scientists in 1990 in order to produce timing belt covers. The bonding between the polymer and silicate layer results in excellent mechanical properties compared with the unmodified polymer, along with improvements in mechanical tensile strengths and heat resistance and chemical resistance. Such nanocomposites are now in wide use by the automotive and aeronautical industries.

DNA nanotechnology uses Watson–Crick base pairing to construct well-defined DNA structures, rotaxane chemical molecular switches and motors, and quantum dots that fluoresce visible light depending on their size.

In the future molecular electronics will endeavour to develop molecules with useful electronic properties, synthetic molecular motors, and many more interesting nanomaterials.

Above: These fragrance microcapsules are made of plastic and are packaged into an ultra thin nanofilm. When subjected to pressure, they burst like a balloon, releasing the incorporated fragrance.

Above right: Carbon nanotubes on the surface of a silicon chip. This image has been taken with a scanning electron microscope so it is a true image, although the color has been added artificially.

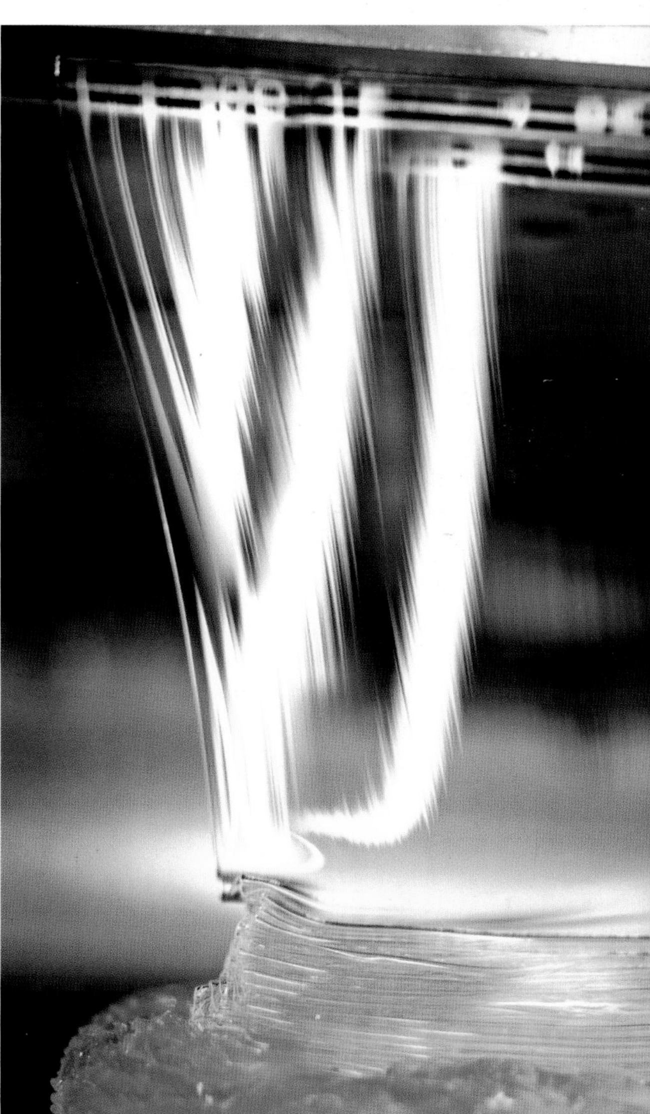

Right: In the production of high tech plastics, melted plastic extrudes into very fine layers. Nanoparticles within the polymer of the plastic must be evenly distributed, in order to make the plastic airtight as well as glossy in appearance.

When you take your car to the carwash you've probably noticed that the water forms bead-like droplets on the paintwork's surface. This phenomenon can be explained by surface science—the study of physical and chemical phenomena that occur at the interface (common boundary) between two phases of matter.

The science of surfaces

Matter has different states based on its physical properties. These states include solids, liquids, gases, and plasmas, and each state has specific qualities that define them.

There are some substances where specifying the state of matter does not completely specify the properties of the substance. For example, solid carbon can exist as graphite or diamond and these two forms have very different properties. Although both are solid carbon, they are referred to as different phases of carbon. When these phases come into contact with each other, their interactions become the stage for a number of important processes and behaviors, which form the basis of surface science.

Surface tension

An important concept of surface science is surface tension. This property helps explain the behavior of many liquids, and is responsible for a number of phenomena involving liquids in our environment.

Surface tension is created by intermolecular forces of molecules at the surface of a liquid. These forces differ from those experienced by molecules beneath the surface, which are surrounded by other molecules providing evenly distributed intermolecular forces in all directions. By contrast, the surface molecules are only attracted by the molecules below or beside them, forcing the liquid surface to contract, and adopt skin-like qualities.

Above: Early morning dew drops on a blade of grass. The beading of the water drops is caused by surface tension.

The strength of this liquid surface is called surface tension. The stronger the attraction between molecules, the greater the surface tension of the liquid, meaning that the amount of energy required to stretch the liquid surface is higher. Water, for example, has a very high surface tension compared with most other liquids due to the extensive hydrogen bonds holding water molecules together. This is why small insects can walk across a pond's surface.

Adsorption

Adsorption occurs when a gas or liquid adheres to the surface of a solid or liquid, forming a molecular or atomic film. Just like surface tension, adsorption is the result of surface energy. When atoms at the surface of a material have a bond deficiency, they can fulfill these bonding requirements by adhering to whatever is available.

This is useful in industrial applications, as it allows the selective capture of certain particles during industrial processes, such as during filtration. The material used to capture the particles is the "adsorbent," and is usually selected according to its specific bonding properties. Examples of adsorbents are silica gel and activated carbon.

In all science, error precedes the truth, and it is better it should go first than last.

Hugh Walpole, novelist, 1884–1931

Left: Certain solutes can have different effects on surface tension. For example, inorganic salts increase surface tension, while alcohols and surfactants, such as detergents (pictured) decrease it.

Right: Various forms of carbon, including a piece of coal, charcoal, graphite, and diamonds. Even though these are all forms of carbon, each possesses different properties, so they are said to be different phases of carbon.

Opposite: Surface tension is that property of water that causes it to act like an elastic coating. When a heavy object is thrown into the water, bubbles form because the surface tension is broken.

Surface area to volume ratio

In the science of chemistry, the surface area to volume ratio can be a critical factor for a material's chemical reactivity, affecting the rate at which a chemical reaction takes place.

Solids that have a small diameter or are porous typically have a high surface area to volume ratio, and react more quickly due to their bigger surface area. Conversely, a bulk solid with a low surface area reacts more slowly.

By virtue of the principles of surface tension, liquids tend to reduce their surface area spontaneously by forming spherical droplets when they come in contact with hydrophobic (water-repellent) solid surfaces—just like beads of water on a newly washed car. By devising superhydrophobic surfaces research chemists are working toward the development of self-cleaning surfaces.

The surface area to volume ratio can be readily calculated by dividing the surface area of the object by its volume.

WALKING ON WATER

Water striders (*Gerris remigis*) have perfected the art of walking on water and for a long time these small insects baffled scientists with their unique abilities. However, scientists have now concluded that the water strider's capabilities can be explained by surface tension. The hairs on the water strider's legs trap air, which gives them water-resistant (hydrophobic) qualities, and enables the insect to form indentations on the water's surface without breaking through.

Being lightweight insects, the downward gravitational force on a water strider is not large enough to stretch the water surface enough to allow its legs to sink into the water. By having a high surface area to volume ratio, they are more responsive to surface forces and are supported by the surface tension of the water.

Water is one of the most common compounds on Earth, but it is also one of the most unusual. Water is the only compound that naturally exists in three states of matter on Earth's surface and the only compound to become less dense when it changes from liquid to solid. We depend on the unique chemical and physical properties of water for all life as we know it.

The fascinating properties of water

A molecule of water consists of two hydrogen atoms covalently bonded to an oxygen atom. Water is a tasteless, colorless, odorless liquid at room temperature and pressure. Water is found on Earth as solid ice, liquid water, and water vapor. It will readily change state according to temperature and pressure changes. For example, the ice in a glass will melt on a warm day and water droplets from your breath will quickly evaporate. Water can also change directly from a gas to a solid when frost is formed and from a solid to a gas.

The temperature and pressure at which all three states of water can coexist is known as the triple point and occurs almost exactly at 32°F (0°C) at sea level (1 atmosphere). Under these conditions it is possible for water to change from ice, liquid water, or vapor as a result of infinitely small changes in pressure and temperature. The balance of this change will differ when you climb higher in altitude, where water will boil at a lower temperature. On the other hand, if you increase the pressure, as in a pressure cooker, water will boil at a higher temperature. If the temperature of Earth were not as it is, for example if Earth were closer to or further from the Sun, then we would not have the vast quantities of solid, liquid, and gaseous water that we have on our planet.

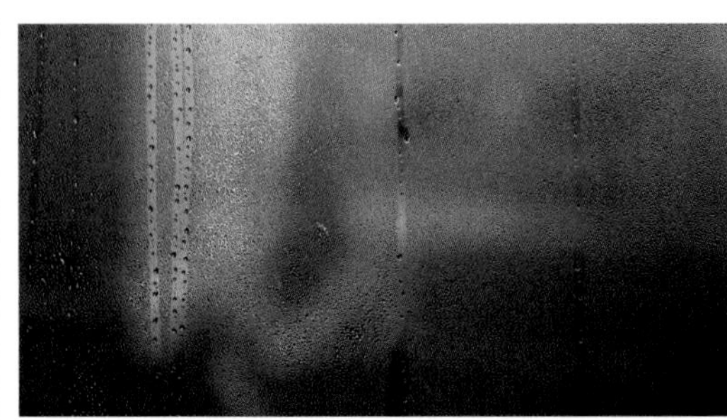

Above: Condensation is the process whereby water in vapor form is converted into liquid form. A condensation reaction occurs when two molecules combine to form a larger molecule which in turn produces water as a consequence.

Right: When water freezes, the molecules arrange themselves differently so that frozen water becomes less dense and also expands by up to 9 percent volume. This is one of the reasons why ice floats on top of water. Water freezes at 32ºF (0ºC), even large bodies of water such as this waterfall.

> Every human should have the idea of taking care of the environment, of nature, of water.
>
> Tenzin Gyatso, 14th Dalai Lama, b. 1935

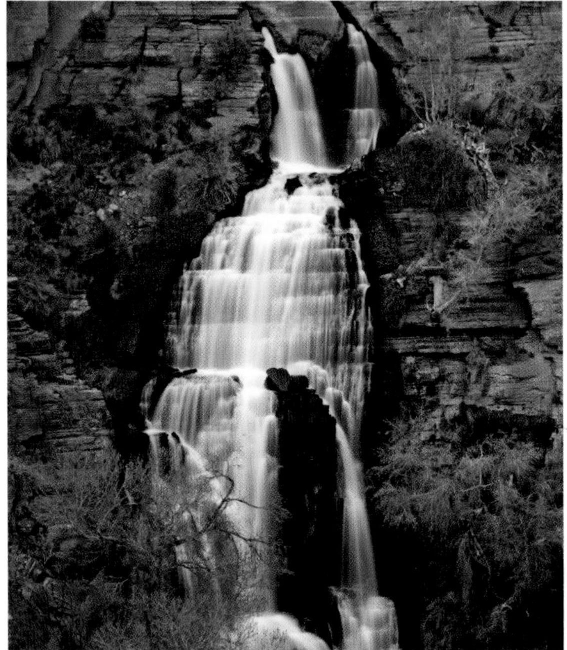

Left: Water is present on Earth in three states—as a solid, as gas, or as liquid, such as this waterfall undulating down a canyon wall. A hydrogen bond exists between water molecules. In liquid water more molecules are present.

The bonds within a water molecule

The chemical bonds that hold a water molecule (H_2O) together are important in giving water the shape and structure that makes it so useful. Oxygen has higher electronegativity than hydrogen, which means that it will have a stronger attraction for electrons in the O–H bonds than hydrogen. This difference in electronegativity results in an intramolecular charge difference, with the oxygen part of the molecule being negatively charged and the hydrogen ends positively charged. The small, positive hydrogen nuclei will be exposed, which allows for the formation of a unique intermolecular force—the hydrogen bond.

Hydrogen bonds, which give water many of its special properties, form between the negatively charged oxygen atom in one molecule and the small, positive hydrogen atoms in a neighboring molecule. Hydrogen bonds are the cause of the high boiling point, melting point, surface tension, and viscosity of water, and the lower density of solid water compared with liquid water.

Aqueous solutions

An aqueous solution is one where a substance is dissolved in water. A substance will be able to dissolve in water if it can break the strong hydrogen bonds between water molecules. Salts, sugars, acids, and bases as well as some gases mix easily with water to form a homogeneous liquid. Water is actually a strong solvent, often referred to as the universal solvent, as it can dissolve many different types of substances. Most components of our cells are dissolved in water and we rely on water for the chemical reactions that occur within our bodies.

WATER AND LIFE ON EARTH

Liquid water is very unusual as, unlike most substances, it expands when it becomes solid. Solid water is less dense than liquid water and this allows ice to float. Subsequently life can exist under the relatively stable conditions *below* solid ice. The transparency of water is also important for all aquatic plants and animals as sunlight can penetrate underwater to provide life-giving energy. If water was like other substances, all ice would sink below the surface and we would have very different conditions on Earth. The ability of water to insulate against large changes in temperature also accounts for the livable conditions on Earth, as without water vapor in our atmosphere fluctuations between night and day temperatures would be severe.

Acids and bases are related to the concentration of hydrogen ions present—pH. The pH scale goes from 0–14. Acids have a pH below 7; bases have a pH above 7. Pure distilled water has a pH of 7.0, which is neutral.

Acids and bases

Acid strength, in aqueous solution, is a measure of the concentration in moles per liter of the positive hydronium ion $[H_3O^+]$, or more simply, but less realistically, the concentration of the positive proton $[H^+]$—mathematically $pH = -\log_{10} [H^+]$.

A number of highly colored compounds act as pH indicators, changing from one color to another at a specific pH. Indicator paper, prepared from these compounds, can be dropped into a solution to identify its pH. Litmus paper, which is red in acid and blue in a base, is one well-known indicator paper. One of the most fascinating pH indicators is red cabbage. Juice from red cabbage is bright red, reflecting the natural acidic nature of cabbage, but the color changes to purple, then green, and finally to yellow as it is made increasingly basic. Of course technology has stepped in and pH can be accurately measured using a portable pH meter.

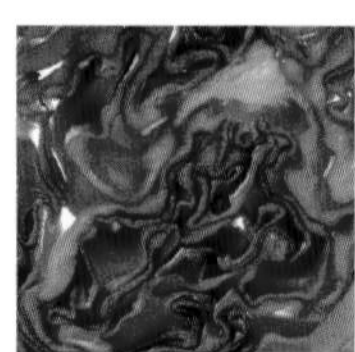

Above: The juice of a red cabbage turns red when it is mixed with something acidic such as vinegar, but green if it is mixed with a base, such as detergent. Red cabbage juice itself is a great pH indicator.

Uses of acids and bases

The most common industrial acid is sulfuric acid, which is highly corrosive. Hydrochloric acid, another strong acid, is used as "pool acid" and in the construction industry (spirits of salts) for dissolving and cleaning mortar from brickwork.

A base is also referred to as an alkali, a term that comes from an Arabic word describing ashes—a source of compounds of some of the Group 1 and 2 elements in the periodic table, now known as the alkali metals and alkaline earths respectively. These metals react with water, sometimes violently, to form basic (or alkaline) metal hydroxides and hydrogen gas. Similarly Group 1 and 2 metal oxides react with water to form alkaline (basic) hydroxides. Baking powder is a common base used as a raising agent in cooking. It releases CO_2 when neutralized with a weak organic acid in the dough. Cloudy ammonia (NH_3), a dilute solution of ammonia gas dissolved in water, is commonly used around the home for cleaning.

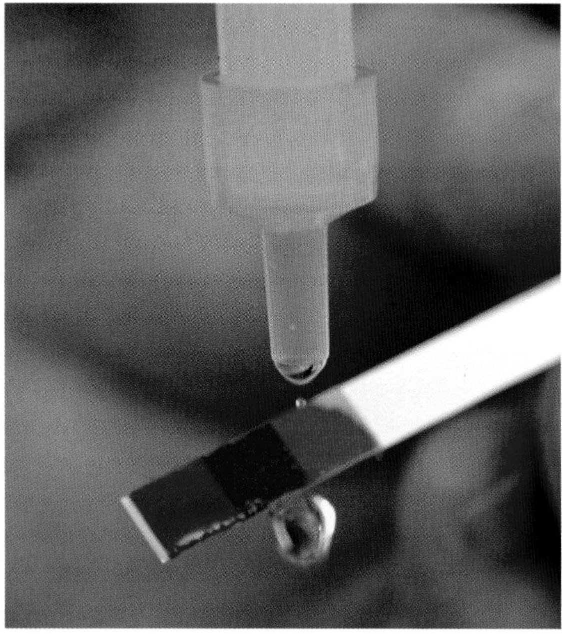

Left: A solution is dropped onto a litmus test paper. The fact that is has turned red indicates that it is an acid. Litmus paper is filter paper treated with a water-soluble dye that is obtained from lichens, in particular from *Rocella* species.

Right: Soap is made by chemically combining fats and oils (or their fatty acids) with sodium hydroxide and an alkali. The alkali acts to neutralize the acid.

Classifications of acids and bases

A Brønsted-Lowry acid is defined as a compound that donates a proton. For example, hydrochloric acid donates a proton (H^+) into solution as it dissociates, so is classified as an acid:

$$HCl \rightarrow H^+ + Cl^-$$

A Brønsted-Lowry base is defined as a compound that accepts a proton. When sodium hydroxide (NaOH) dissolves in solution it dissociates into sodium (Na^+) and hydroxide (OH^-) ions. The negative hydroxide ion can accept a proton (H^+) and is therefore a Brønsted-Lowry base, as shown below:

$$NaOH \rightarrow Na^+ + OH^- + H^+$$

$$OH^- + H^+ \rightarrow H_2O + Na^+$$

Another technical classification of acids and bases is the Lewis acids and bases. A Lewis acid is defined as a compound that accepts a pair of electrons, for example aluminum chloride ($AlCl_3$). Ammonia gas ($:NH_3$), not dissolved in water, is an example of a strong Lewis base. Lewis bases can also act as Brønsted-Lowry bases. For example, when ammonia is dissolved in water the ammonia abstracts a proton from the water to form the NH_4^+ ion.

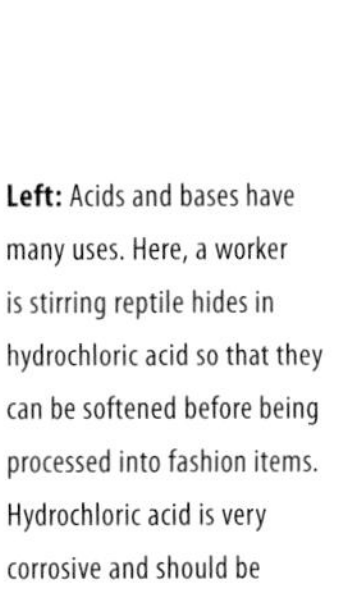

Left: Acids and bases have many uses. Here, a worker is stirring reptile hides in hydrochloric acid so that they can be softened before being processed into fashion items. Hydrochloric acid is very corrosive and should be handled with great care.

> Science is part of the reality of living; it is the way, the how, and the why for everything in our experience.
>
> Rachel Carson, marine biologist and writer, 1907–1964

COMMON OCCURRENCES OF ACIDS AND BASES

We all use acids and bases. For example, vinegar has a pH of 2.8, being an aqueous solution of acetic acid. Acids are characterized by a sharp, biting taste, for example, citric acid gives citrus fruits their tart, bitter flavor. Another common acid is carbon dioxide (CO_2)—the gas we breathe out. Carbon dioxide is also used to carbonate fizzy (soda) drinks. Carbon dioxide dissolves in water to form carbonic acid and in carbonated (soda) drinks has a pH of 2.9. The ocean is a CO_2 sink, having a pH of 8.3, and it absorbs excess CO_2 from the atmosphere. Since pre-industrial times the pH of seawater has dropped by around 0.1 pH units, and is ultimately expected to be reduced by around 0.35 pH units.

Can you imagine life without cars, mobile phones, MP3 players, or any other portable devices? If it wasn't for electrochemistry we wouldn't have any of these objects! While it was Italian physician Luigi Galvani who first discovered a new form of electricity from a biochemical reaction using a dead frog, it was Alessandro Volta who built the first known electric battery that produced electricity from a chemical reaction.

Electrochemistry

In the common type of battery, the "dry-cell" battery, also called the primary cell battery, chemical energy is converted into electrical energy. To understand how batteries work we first need to understand oxidation–reduction reactions, or redox reactions. Oxidation is a reaction where electrons are being generated during the course of the reaction and reduction is a reaction where electrons are being consumed during the course of the reaction. These reactions occur simultaneously, for example corrosion of iron is a redox process, where electrons are being transferred from the oxidation site to the reduction site through the metal. If we physically separate these two reactions then we can generate electricity.

Below: The corrosion of iron is a simultaneous process of oxidation of iron and reduction of oxygen, shown at left. At right, the iron ions then react with hydroxide ions and produce insoluble iron hydroxide complexes, or rust.

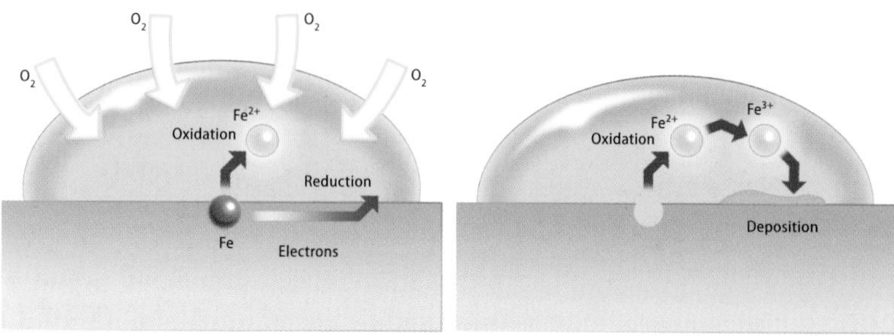

Use of electrochemistry in analytical chemistry

With the significant improvements in electronics in the last 50 years, electrochemical processes are now becoming more important in the area of analytical chemistry, in the detection and identification of substances in a matrix. Electrochemical techniques, such as "anodic stripping voltammetry," are now being used to detect heavy metals in water. In this process the heavy metal ions in water are reduced onto a surface and then stripped off or oxidized back into the matrix, and the electrons that are produced during the reduction process are measured as current and can be calibrated to reflect the concentration of those metal ions in water.

Right: In 1783, Luigi Galvani (1737–1798) dissected a frog, exposing its sciatic nerve with a scalpel, which caused a spark of electricity. The dead frog's leg kicked as if it were alive. This was solid evidence that electricity could come from a biochemical reaction.

Structure of a battery

A battery is an "electrochemical cell," where the oxidation and reduction processes occur at separate sites, called "half cells," connected by a conductive electrolyte, and electricity is produced until the chemicals are used up. The oxidation half-cell is called the "anode" and reduction half cell is called the "cathode." The most common types of battery are the "dry cell" battery used in toys, etc and the "lead-acid accumulator" battery used in cars.

Importance of electrochemistry

Being able to produce electricity from chemicals and vice versa, that is, being able to induce chemical change from electricity, has greatly affected the metal industry and the way we live. For example, we would not be able to drive a car if it wasn't for electrochemistry; in fact, we wouldn't even have a car because to build a car you need metal, which needs to be extracted from ore via an electrochemical process.

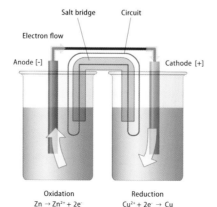

Salt bridge
Circuit
Electron flow
Anode [-]
Cathode [+]
Oxidation
$Zn \rightarrow Zn^{2+} + 2e^-$
Reduction
$Cu^{2+} + 2e^- \rightarrow Cu$

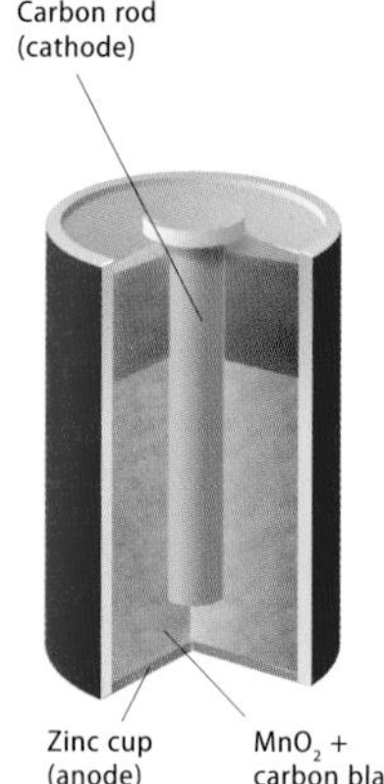

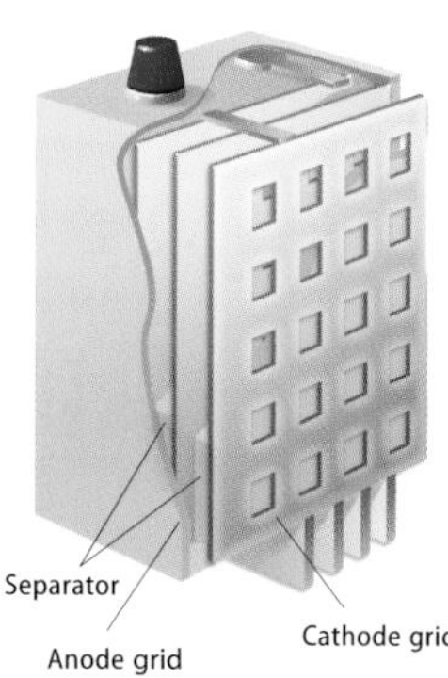

Right: Electrochemical cells. The first diagram shows an electrochemical cell where the oxidation of zinc reaction is separated from the reduction of copper reaction, so the electrons are forced to flow through a wire, producing electricity. The diagram in the middle is an example of a dry cell where the anode is zinc metal and the cathode is a carbon rod. The third diagram shows the inside of a car battery.

REDUCING ARSENIC POISONING IN BANGLADESH

Heavy metal contamination of groundwater is a very serious environmental problem, particularly in Bangladesh. For over 35 years people have been drinking groundwater that has been contaminated with arsenic. Arsenic in its elemental state is non-toxic, but most arsenic compounds are highly poisonous. For example ingestion of only 1.05 grains (70 mg) of arsenic trioxide, As2O3, will kill an adult.

Millions of people are now suffering from health problems such as cancer of the skin as a direct result of drinking arsenic-contaminated water from tube-wells. It is crucial to stop people from using tube-wells that have arsenic concentrations above 0.75 grains (50 mg) per liter. With over six million tube-wells and many more formed each day, the current techniques used for detecting arsenic are expensive, complicated and not field-based, or are highly inaccurate.

The ideal solution would be to remove the arsenic from the groundwater, but currently all that can be done is to identify tube-wells that have unsafe arsenic concentrations and prevent people from using them by painting them red.

Currently researchers are involved in developing a cheap and accurate infield method to detect arsenic in groundwater. The method uses anodic stripping voltammetry, and a Portable Digital Voltameter, the PDV6000, which uses a solid gold electrode to detect arsenic ions in water samples. To ensure that reliable and accurate results researchers must investigate the processes occurring on the surface of the gold electrode.

Above: The dark spots on this young Bangladeshi man's body are indicative of arsenic poisoning. Much of the groundwater in Bangladesh is contaminated with arsenic.

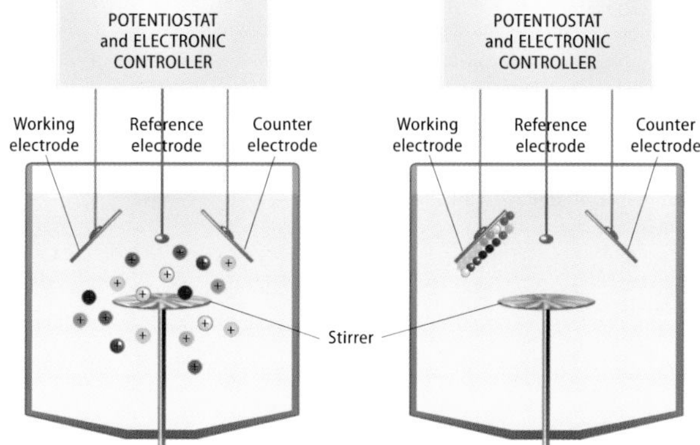

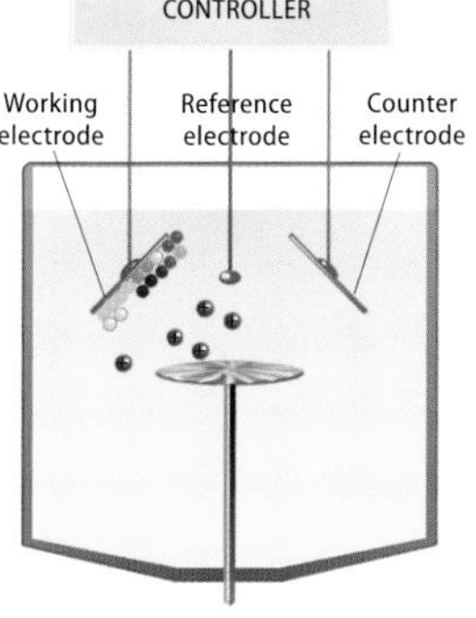

Left: A sample of water (far left) contains a mixture of different heavy metal ions. The middle diagram shows the depostion step or reduction of metal ions onto an electrode surface. The diagram on the right shows the stripping step or oxidation of metals into solution with the production of a very small electrical current.

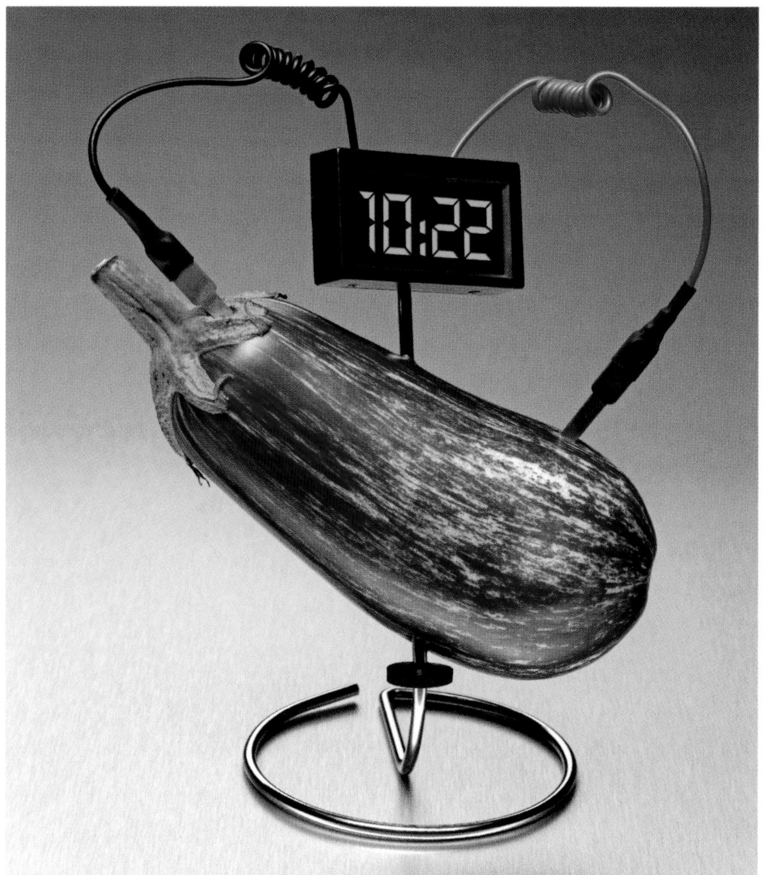

THE REDUCTION PROCESS

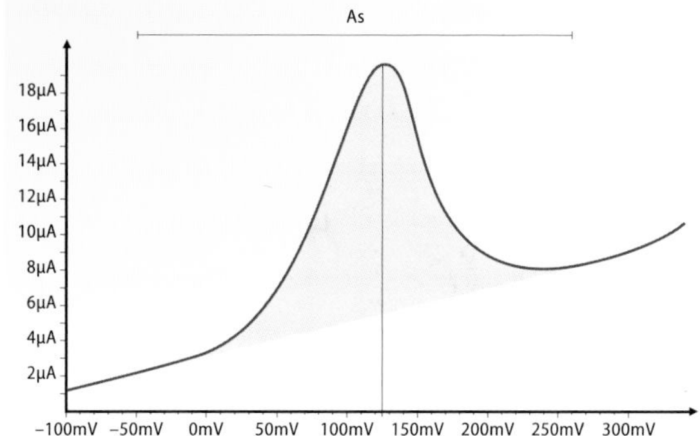

Left: An electrochemical cell requires two partial reactions, one producing electrons, the other consuming electrons, and a solution joining these two reactions. In this case, the reactions involve two different metals that are embedded in an eggplant.

Above: The graph shows that the measured current in the diagrams above corresponds to a particular concentration of metal ions.

> It is the tension between creativity and skepticism that has produced the stunning and unexpected findings of science.
>
> Carl Sagan, astronomer, 1934–1996

Inorganic chemistry is concerned with the preparation and properties of compounds that are not organic, that is, that don't contain carbon to hydrogen bonds. As such it encompasses the chemistry of the majority of the elements in the periodic table. Inorganic compounds are used in diverse applications such as aircraft construction, anti-cancer drugs, semiconductors, and breast implants.

Inorganic chemistry

Industrial inorganic chemistry

The smelting of ores (most often metal oxides, sulfides, or carbonates) to extract metals such as iron, lead, and tin is an example of inorganic chemistry that goes back to ancient times. In modern times, inorganic compounds such as ammonia (NH_3) and sulfuric acid (H_2SO_4) top the list of industrial chemicals in terms of annual production. Most people know that sodium chloride is in many of the foods we eat, but another simple inorganic compound, titanium dioxide (TiO_2) is also added to some candies to provide an opaque white background to make surface colorings more vivid.

Metals and their compounds figure prominently in inorganic chemistry. A broad class of inorganic compounds feature one or more metal atoms or ions surrounded by other molecules or ions somewhat loosely bound to the central metal. These clusters containing the central metal and surrounding molecules or ions ("ligands") are called complex ions. The simple act of dissolving a metal compound in water forms complex ions as water molecules surround each metal ion. Complex ions are crucial in many industrial and biological processes. Because the ligands are bonded a little less tightly than in other forms of chemical bond they can be released or exchanged with other ligands. Complex ions are at the heart of transport mechanisms in biological systems, for example the transport of oxygen molecules by hemoglobin involves a complex ion with an iron ion at the center which can bond to oxygen molecules or water molecules. Ligands that bind strongly to metals can be used to remove metallic pollutants or serve as antidotes for metal-containing poisons.

Above: Inorganic chemistry is the study of inorganic compounds, which have very diverse applications, including being used for breast implants.

When the ligands in a complex ion are organic molecules, the compound is called an organometallic compound. Many catalysts used to speed up industrial chemical reactions are organometallic compounds, such as Grubbs's catalyst, which earned the inventor Robert Grubbs a share in the Nobel Prize for Chemistry in 2005.

BIO-INORGANIC CHEMISTRY

Bio-inorganic chemistry is the study of chemicals containing metals in biological systems. Hemoglobin (shown below) is a metalloprotein—a large molecule that is made up of over 500 amino acids, containing thousands of carbon, hydrogen, nitrogen, and oxygen atoms but also containing four iron atoms crucial to its role in transporting oxygen molecules. The binding of oxygen molecules to hemoglobin is a complex process involving subtle changes to the shape of the molecule after one oxygen molecule binds so that another oxygen molecule can bind more readily. Other metalloproteins contain small clusters of iron and sulfur atoms including some with a cubic arrangement with iron and sulfur atoms at diagonally opposing corners of the cube.

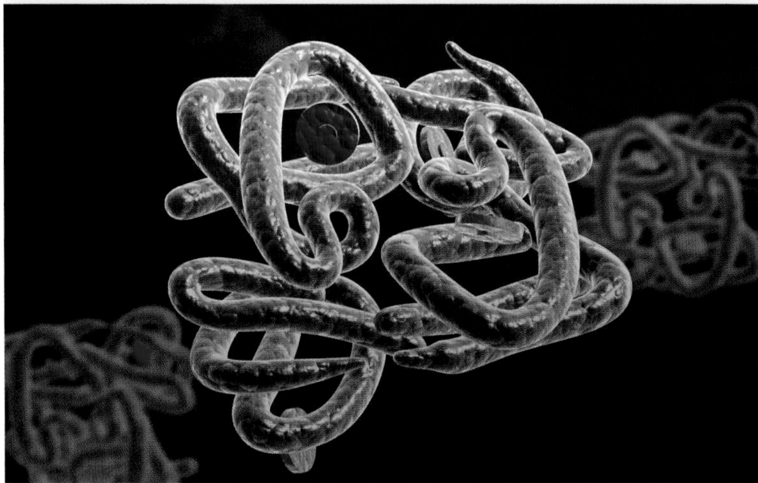

Inorganic compounds in biology

In 1828, German chemist Friedrich Wöhler (1800–1882) converted an inorganic compound (ammonium cyanate) into an organic compound (urea), which led to the realization that organic compounds did not possess any special connection with living organisms. Since then, the complementary notion that inorganic compounds were restricted to inanimate matter has given way to a realization that

Science is the desire to know causes.

William Hazlitt, essayist, 1778–1830

these substances are essential to the chemistry of living organisms and can function as powerful pharmaceuticals.

In the early 1900s arsenic compounds were found to be effective against syphilis, replacing mercury compounds, but were replaced in a few decades by penicillin. Very simple inorganic compounds may also act as potent pharmaceuticals. When lithium salts are dissolved, they release the lithium ion, which acts as a mood stabilizer. A complex containing platinum surrounded by two ammonia molecules and two chloride ions in a square shape with the ammonias on one side of the square and the chlorides on the other (with the platinum in the middle) is used as a potent anti-cancer drug (cisplatin).

Above: Inorganic compounds are used in the construction of aircraft such as this Boeing 737-800, to produce, for example, high temperature-resistant (oxidation-resistant) fibers.

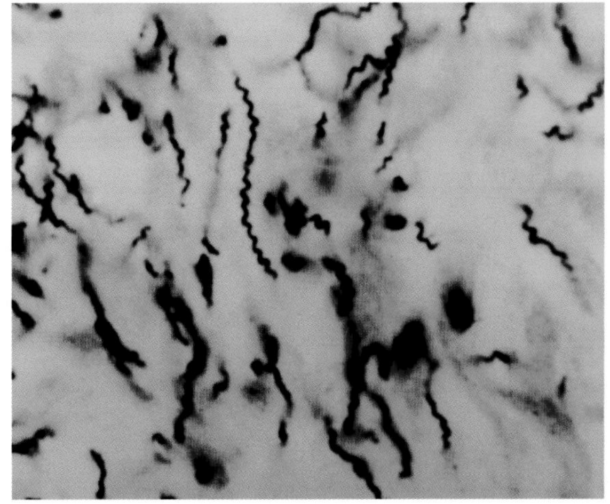

Right: The bacteria that causes syphilis, *Treponema pallidum*, magnified 600 times. Arsenic, an inorganic compound, was once used to treat this disease. Nowadays penicillin is the preferred treatment.

We live in an organic world. Anything living owes it all to a little element called carbon and a huge range of reactions known as organic chemistry. It warms our homes, holds our clothes together, cures us when we're sick, and keeps pests at bay. Even today, chemists continue to make exciting breakthroughs using its many possibilities.

Organic chemistry

Carbon is an element that forms the basis for all living things—and many non-living things as well. Carbon is also the essential element in the science of organic chemistry.

An organic compound is any chemical that contains carbon. Because there are also many chemicals considered non-organic that also contain carbon, the definition of organic compounds now includes chemicals that are held together with covalent bonds. These are strong bonds in which the atoms share electrons.

What's in a name?

Naming might seem an arbitrary thing in some areas, but in organic chemistry it can mean the difference between picking up a bottle of medicine or a bottle of poison. Some chemicals that look extremely similar (even on the molecular level) can have very different properties.

Left: Fume cabinets, such as this one photographed in 1968, are used when chemists are doing experiments that may involve or produce dangerous gases. A fan or vacuum inside the cabinet sucks away the dangerous gas and makes the experiments safer.

This is because it's not just *what* atoms an organic chemical contains, but *how* these atoms are arranged in each molecule. Even with very large, complex molecules, moving one atom can completely change how the chemical behaves.

To get around this, organic chemists came up with an internationally understood way of naming compounds. They can describe the location of atoms in a molecule, how many atoms there are and of what type, the types of bonds between the atoms and where they are, and even the shape of parts of the molecule (such as rings or chains).

Take, for example, the chemical 1,4-butanediol. "Diol" explains that the molecule has two alcohol groups. "Buta" indictates that the molecule has exactly four carbons, connected in a chain. The 1 and 4 explain that one alcohol group is bound to the first carbon in the chain, and the second group is bound to the second carbon.

Right: A model of the chemical 1,4 butanediol. Carbon is shown as dark blue, hydrogen is white, and oxygen is red.

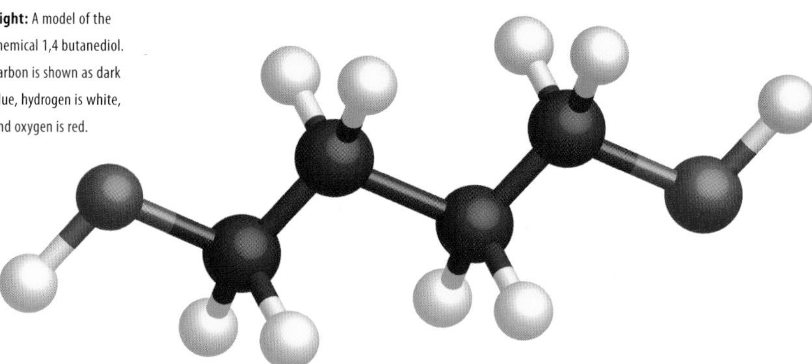

Organic chemistry is the chemistry of carbon compounds. Biochemistry is the study of carbon compounds that crawl.

Mike Adams

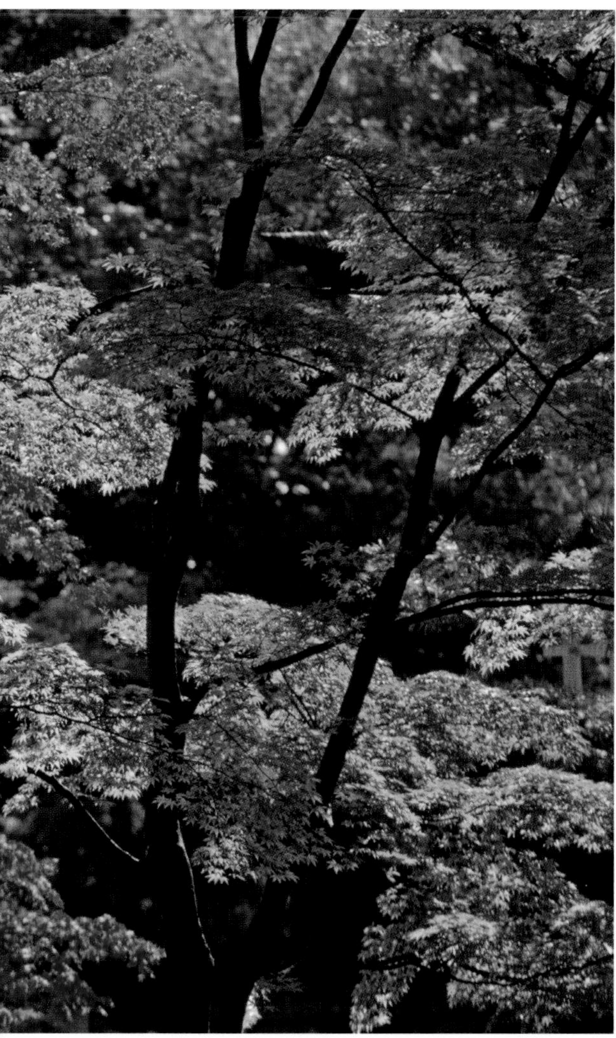

THE DIAGNOSIS

Organic chemistry plays an important role in the pharmaceutical industry. The human body is an organic environment, so understanding how chemicals will behave in the body helps chemists to design drugs that do what they want them to.

One subtle feature of organic chemicals which can have a huge effect on their behavior as a drug is a concept known as "isomerism." This describes chemicals that have the same chemical structure but a different three-dimensional arrangement of the atoms. In many drugs, only one of the isomers is effective as a drug and the other can have adverse effects. An infamous example is the drug thalidomide, marketed in the 1950s to relieve morning sickness during pregnancy. However, only one of the isomers was effective. The other isomer caused major birth defects in the unborn child (see page 175). Organic chemists must therefore take special care not only in determining a chemical's composition, structure and bonding, but also the effects of its overall three-dimensional shapes.

Left: Every living thing, from the tiniest cell to a sprawling forest, owes it existence to carbon and the huge range of reactions studied under the umbrella of organic chemistry.

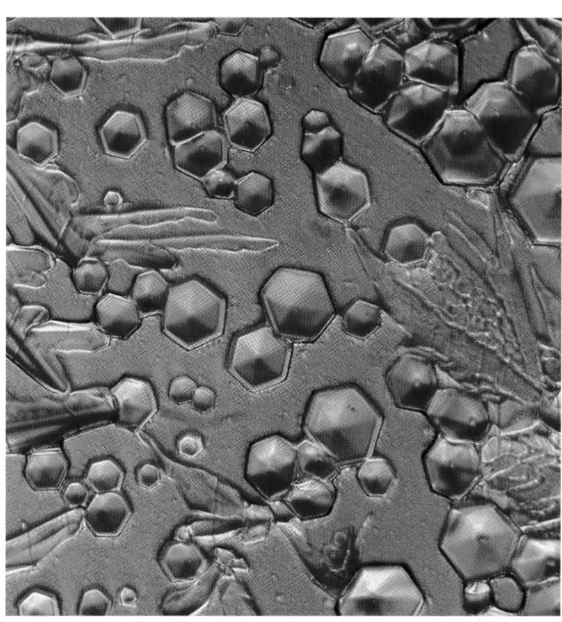

Right: One of the biggest industrial applications of organic chemicals is through the manufacture of polymers, such as this polyester fiber. Polymers are large chains made up of smaller molecules called monomers. Each monomer is usually identical and has a relatively simple structure. Held together in long chains through covalent bonding, these rather "simple" chemicals can have extraordinary properties.

Above: Dichloro-diphenyl-trichlorethane or DDT was one of the first synthetic insecticides. A highly efficient organic compound pesticide, its use was banned when evidence of its negative effects on the environment became known. Here, unhatched ibis eggs have been damaged by the pesticide.

Right: A model showing the structure of the carboxylic acid group. The carboxylic group is weakly acidic. A common example of a member of the carboxylic group is vinegar (acetic acid).

Keeping it together

Carbon is able to form up to four strong covalent bonds with neighboring atoms, allowing for the formation of large, stable molecules. Sometimes two single bonds can be sacrificed to form one double bond. Double bonds are extremely strong and can help a molecule keep its shape, and they react easily and give scientists the opportunity to perform a greater variety of more sophisticated reactions.

Mix it up

By understanding the structure and composition of an organic molecule, chemists are able to predict what will happen when two organic chemicals come into contact.

To try and simplify things, scientists have named some of the common arrangements of atoms with predictable behaviors. These arrangements are known as "functional groups." Common examples are the alcohol (or "hydroxyl") group and the carboxylic acid group. Each group behaves dramatically differently in the same environments.

Understanding how functional groups work helps organic chemists work in two directions. They can predict the outcome of mixing certain chemicals together, and they can work out which chemicals they need to mix to produce the particular material they want. This makes designing new chemicals (such as drugs and products for industry) more of a science and less of an art.

Stereochemistry is an area of science interested in the shapes and structures of molecules. Two molecules may have the same components but still be different compounds due to the relative spatial arrangement of atoms in the molecule. This is important as the different forms of the molecules may have very different effects in physical or biological situations.

Three-dimensional arrangements—stereochemistry

Many objects in nature appear as two versions of the same thing. If you look at a pair of shoes there is a distinct difference between the left and the right shoe. You cannot swap the left and the right shoe and still walk properly. We describe objects like these as "chiral," which comes from the Greek word for hand, *cheir*, as our hands are one of the most familiar chiral objects.

Another simple example of two-dimensional chiral objects can be seen in the alphabet. Letters like A, B, and C are achiral (the opposite of chiral) because a mirror image of the letter can be superimposed on top of the original. However, letters like F, G, and J are chiral as no matter how we rotate them we cannot make their mirror image match the original letter.

Three-dimensional objects like screws, snail shells, and gloves are also all examples of chiral objects.

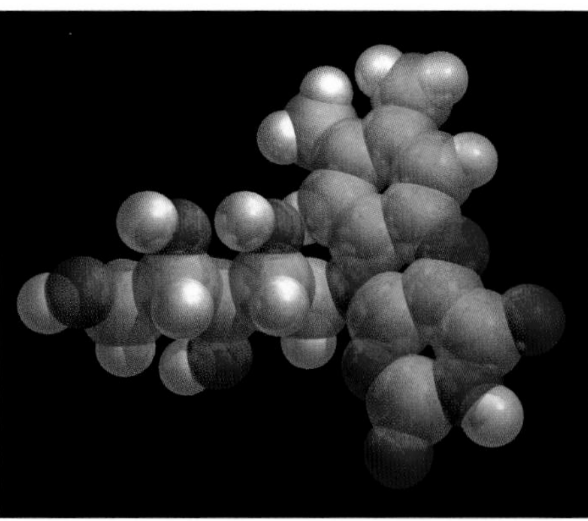

Below: The arrangement of atoms in a riboflavin (vitamin B2) molecule (in particular along the horizontal carbon chain shown here) makes this molecule non-superimposable on its mirror image. Despite this very subtle difference between these two stereoisomers, only one form is found in nature.

What do stereochemists do?

Stereochemistry is concerned with the different spatial arrangements of atoms between molecules, which can influence their large-scale shape, which in turn affects their function. It is the role of stereochemists to investigate the spatial arrangement of molecules and how this affects physical and biological systems.

It is difficult to explain the difference in molecular structure to another person, so stereochemists had to devise systems, first to work out the arrangement of atoms in the molecule, and then to describe and classify those arrangements. This is done by modeling and experimental methods such as X-ray crystal structure analysis.

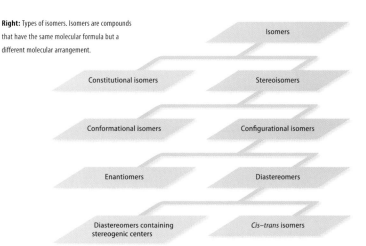

The first thing a stereochemist needs to do is establish the molecular formula of the compound and then determine the bonds between the atoms, or the constitution. Compounds with the same molecular formula but different constitution are called isomers. There are many different types of isomers but all fit into two important categories: Constitutional isomers and stereoisomers.

Constitutional isomers have the same number and types of atoms, but they are connected differently. For example, ethanol and dimethyl ether both have the molecular formulae C_2H_6O, but the pairs of atoms joined by chemical bonds is different, namely CH_3CH_2-OH and CH_3-O-CH_3.

Stereoisomers are different in that the formulae and the sequence are identical but the spatial arrangements of the molecules are different. Stereoisomers can either be conformational or configurational. There are many types of configurational stereoisomers.

Right: The shell of a common garden snail (*Helix aspera*) is chiral, that is, its structure and mirror image are not able to be superimposed. In chemistry, chirality refers to molecules. Enantiomers are molecules that cannot be superimposed on one another.

Right: Types of isomers. Isomers are compounds that have the same molecular formula but a different molecular arrangement.

Isomers
Constitutional isomers
Stereoisomers
Conformational isomers
Configurational isomers
Enantiomers
Diastereomers
Diastereomers containing stereogenic centers
Cis–trans isomers

THALIDOMIDE

The significance of stereochemistry is perhaps best known through the effects of thalidomide. Thalidomide is a drug that was prescribed in the late 1950s as a morning sickness preventative. Due to the positive effect of preventing morning sickness, thalidomide quickly gained popularity and was sold in many countries under many different names. Thalidomide was successful as a sedative but an isomer of thalidomide was eventually found to result in fetal abnormalities.

At the time there were few requirements for pharmaceutical companies to carry out safety tests and the drug's developers promoted the product with little knowledge of its tragic impact (see right).

Approximately 10,000 babies were born with malformations and many more babies would have been lost due to miscarriage. A connection between using the drug and a high incidence of deformations in babies was established after a few years and the product was gradually removed from circulation.

Thalidomide is still used today to provide pain relief for leprosy sufferers, however it has to be used in conjunction with contraceptives in order to prevent damage to a fetus.

The discovery of the unexpected side effects of thalidomide led to strict regulations that require thorough testing of new drugs and checks for safety during pregnancy before public release of a product on the market.

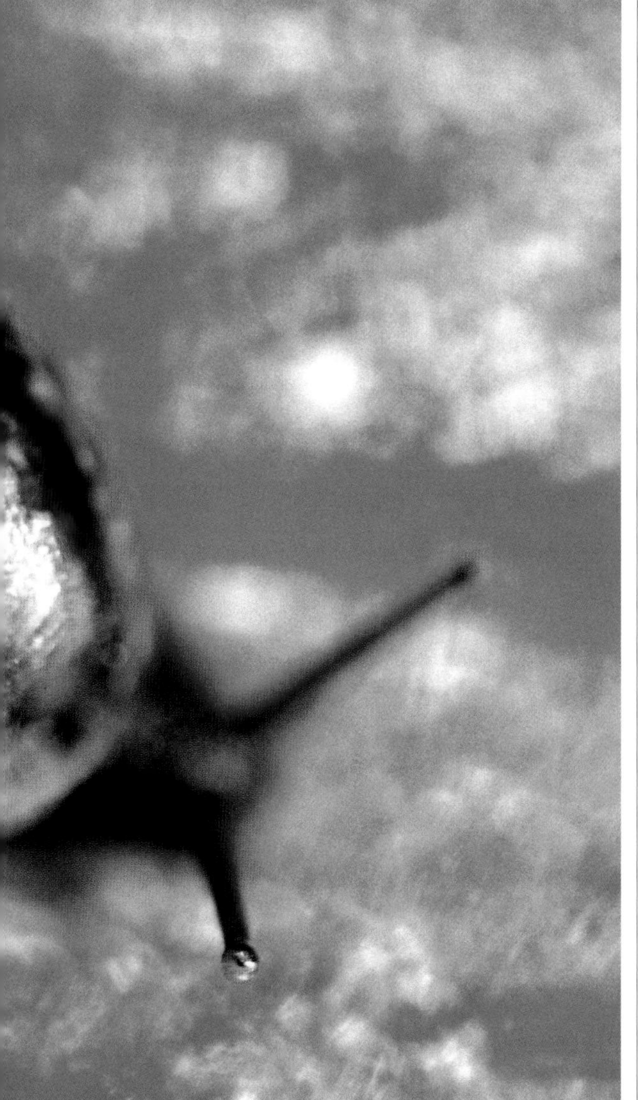

Below: Curious as to the reason why tartaric and para-tartaric acids showed different properties in solution even though they had the same chemical makeup, scientist Louis Pasteur (1822–1895) conducted a number of experiments that laid the foundation for the science of stereochemistry.

Life is a relationship between molecules.

Linus Pauling, chemist, 1901–1994

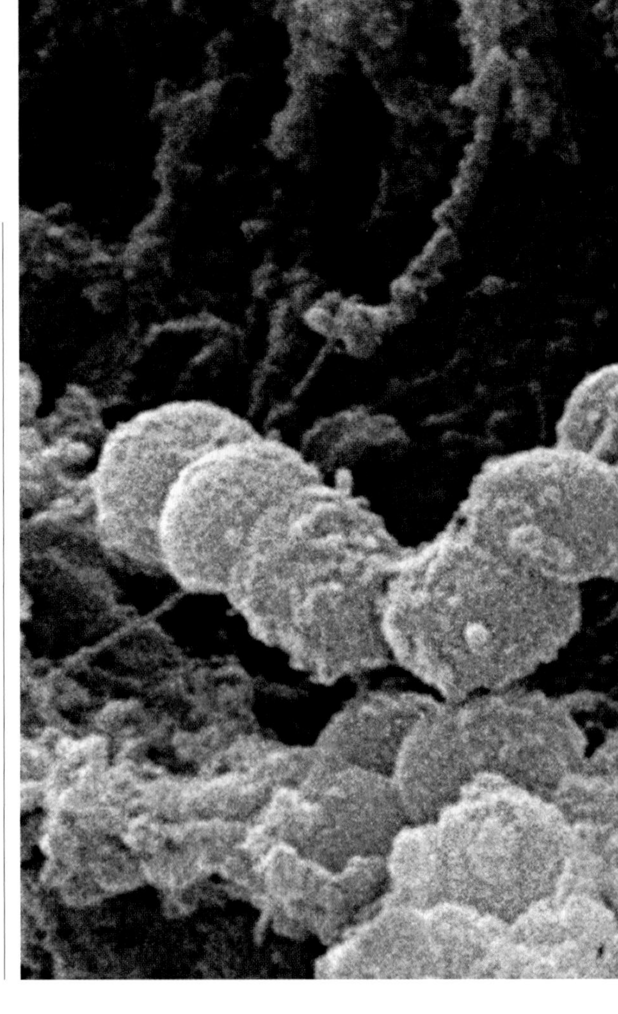

Within the cells of every living organism a large number of chemical reactions occur, which assist in the growth and survival of all life forms. Biochemistry deals with these chemical reactions. Biochemistry is a very broad and complicated field of study, but this is hardly surprising considering the complexity of living organisms!

The chemistry of life

Metabolism

Anyone who has been on a diet will be familiar with the term metabolism. Every day our bodies require energy to survive and grow. Metabolism refers to the set of physical and chemical processes that create the energy and molecular structures required for life.

Metabolic processes can be split into two categories—catabolism and anabolism. Catabolism involves breaking down large molecules into smaller units, which are used for energy or nutrition, and potentially recycled by the body for other purposes. An important catabolic process in living organisms is the process of cellular respiration. This is a set of biochemical reactions, which lead to the breakdown of fuel molecules in order to generate energy. Common fuel molecules include glucose, amino acids, and fatty acids, which all undergo oxidation, with the energy released stored by the cell until required.

Anabolic processes result in the building of molecules by the cell. These reactions require energy, and the larger molecules are built using the smaller molecular units left over during catabolism. Products produced by anabolic processes include proteins, most of which are used up by the body to build up and repair organs and tissues.

Below: A molecule of vitamin C, which is essential for humans and many animals. A deficiency in vitamin C causes the disease scurvy.

The building blocks of life

The human body requires the synthesis of certain molecules to ensure healthy organ and tissue structure and function. These molecules, also known as biochemicals, are the "building blocks" of life, forming the basis for metabolic processes.

Every protein created by the body is made of a string of units called amino acids, linked together by a bond called a peptide bond. Proteins created by the body serve many functions. Some have a structural function, assisting in maintaining cell shape. Other proteins are involved in specific cellular functions such as cell signaling, immune responses, cell adhesion, and active transport.

Many other proteins created by the body are enzymes—biological catalysts that help activate chemical reactions in the cell. The structure and function of a protein is determined by the amino acids that make it up, which is determined by the cell's DNA. Lipids are another

> It is one of the more striking generalizations of biochemistry … that the 20 amino acids and the four bases are, with minor reservations, the same throughout Nature.
>
> Francis Crick, biologist and physicist, 1916–2004

HOW AMINO ACIDS DETERMINE PROTEIN FUNCTION

As the basic blocks of proteins, amino acids form polymer chains, linked together by peptide bonds. There are 20 amino acids involved in protein synthesis, and the sequence in which they join together is determined by the genetic code detailed in the cell's DNA. These amino acids are called proteinogenic or standard amino acids. However, equally important to the linear composition of the protein, called the primary structure, is the three-dimensional structure of the final protein. This structure is created by post-translational modifications (that is, changes after the protein has been formed from the RNA template), which are facilitated by another group of amino acids. These post-translational changes in protein structure are important, as they determine the function of the protein, and are influenced by the chemical bonds and attractive forces between different parts of the protein chain. (The picture shows a mass spectrometry technique for detecting biomolecules such as proteins and peptides.)

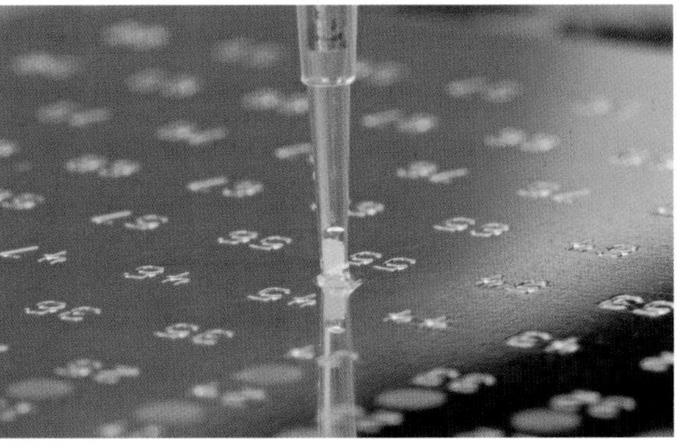

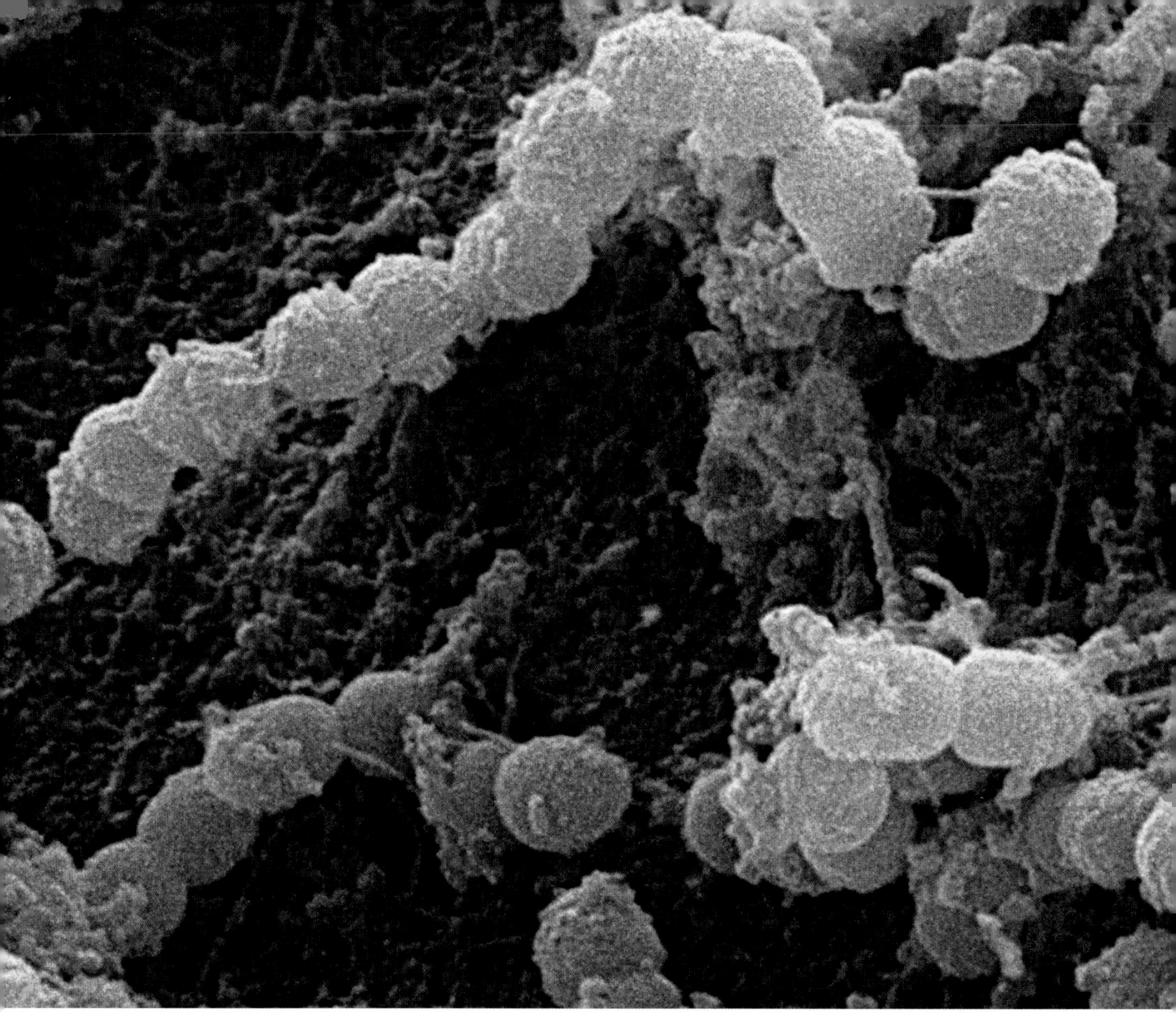

Above: *Streptococcus mutans* growth on dental plaque on a human tooth. This bacterium metabolizes sucrose so that it becomes lactic acid. This chemically reacts with tooth enamel leading to tooth decay.

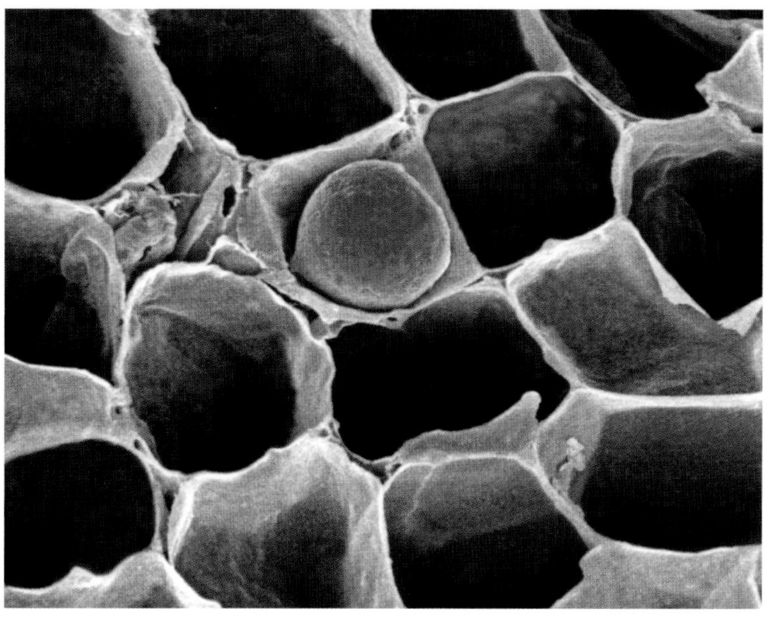

Right: Magnified 500 times, adipose tissue showing adiposites, one with a lipid droplet consisting primarily of triglycerides. Triglycerides are broken down by enzymes called lipases which convert fatty acids to glycerols.

important group of biochemicals. They play a structural role in the membranes in the cell, and are used for energy storage. Lipids include naturally-occurring fat molecules, for example oils, waxes, cholesterol, sterols, as well as fatty acids such as triglycerides, diglycerides, monoglycerides, and phospholipids.

Carbohydrates are the largest group of molecules in the cell, and serve a variety of functions, from energy transport and storage, to structural components. Carbohydrates are made up of many basic units called monosaccharides, also known as simple sugars. One of the most significant carbohydrates used by the body is glucose.

Nucleotides are the building blocks of the genetic structures DNA and RNA. These structures store the cell's genetic information, and are critical to the processes of transcription, the translation of DNA genetic code to RNA, and protein biosynthesis, the translation and creation of protein structures from the RNA code.

The 2007 Intergovernmental Panel on Climate Change (IPCC) report documents that atmospheric greenhouse gases now stand at 430 parts per million (ppm) and are continuing to rise at a staggering 3 ppm per year. When greenhouse gas levels reach 450 ppm, air temperatures will ultimately rise by around 3.6 °F (2 °C), considered a point of no return.

Biofuels chemistry

Right: Beakers of biofuel on a laboratory table. In order to be considered a biofuel, the fuel must contain more than 80 percent of renewable materials. Biofuels are CO^2-neutral, which means that they do not add to the carbon dioxide level in the atmosphere.

The use of fossil fuels by humans has resulted in carbon dioxide (CO_2) emissions from several sources—coal, petroleum fuels, natural gas, flaring gas, cement production, and other sources. Coal and oil are two major sources of CO_2 with coal being primarily used for power generation. Alternative and renewable technologies have already been developed for generating electrical power, for example hydro, wind-energy, wave-power, geothermal, nuclear, solar panels, and solar hot water.

Alternatives to fossil fuels for cars include the so-called third-generation hydrogen- and battery-powered vehicles. But these are a long way from commercial reality and the carbon neutrality of such technologies is questionable. The biofuels biodiesel (replacing petroleum

> If you think mitigated climate change is expensive, try unmitigated climate change.
>
> Richard Gammon, chemist and oceanographer

RISING SEA LEVELS

Since 1950, air temperatures have increased by 1.35ºF (0.75ºC), sea levels have risen around 4 inches (100 mm) and Northern hemisphere snow cover has been reduced by 7.9 percent. So far, minor sea-level increases have been attributed to Arctic ice-melt and thermal expansion of water; however, when the Greenland glaciers melt, and this has already started, sea levels will increase by 23 feet (7 m). Many coastal environments will disappear under a flood of rising water. If the Antarctic ice melts, then sea levels will rise an additional 33 feet (10 m). Scientists claim that we have five to ten years to start stabilizing greenhouse gases if we want to maintain our current climatic conditions.

diesel) and bioethanol (replacing gasoline/petrol) are the only CO_2-neutral alternative fuel technologies currently available.

Biodiesel and bioethanol, as pure B100 and E100 (not blended with petroleum diesel or gasoline), are carbon neutral. The feedstock used to make these biofuels removes CO_2 from the atmosphere by the natural plant and animal life cycle and, when these biofuels are burnt, an equivalent amount of CO_2 is returned to the atmosphere. B100 biodiesel can be produced from triglycerides (naturally-occurring fat or vegetable oils) such as tallow, vegetable oils, rapeseed, waste cooking oil, and fish oil. E100 bioethanol can be produced from first-generation feedstock like sugarcane, sugar beet, sorghum, maize (corn), wheat, barley, and rye.

The chemistry of biodiesel production

Biodiesel triglyceride lipids react with alcohol (commonly methanol) using the base-catalyzed, transesterification chemical reaction (shown below) to produce FAME (fatty acid methyl ester) biodiesel plus glycerol. Each 2.6 gallons (10L) of triglyceride used produces 2.2 gallons (10L) of FAME biodiesel and 0.26 gallons (1L) of glycerol.

$H_2C-O-C(=O)-CH_2\sim R$ / $HC-O-C(=O)-CH_2\sim R$ / $H_2C-O-C(=O)-CH_2\sim R$ $\xrightarrow[\text{base catalyst}]{CH_3OH}$ $3\ H_3C-O-C(=O)-CH_2\sim R$ + H_2C-OH / $HC-OH$ / H_2C-OH

Triglyceride — Fatty acid methyl ester (FAME) — Biodiesel — Glycerol

The chemistry of bioethanol production

Bioethanol fuel is produced by the biological fermentation of aqueous sugar. This two-step reaction involves yeast, which contains an enzyme (invertase) that converts sucrose into two simple sugars, fructose and glucose. Zymase, a second enzyme in yeast, converts these sugars to bioethanol and carbon dioxide.

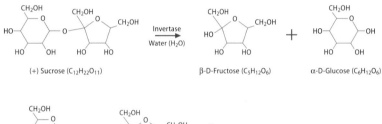

(+) Sucrose ($C_{12}H_{22}O_{11}$) $\xrightarrow[\text{Water } (H_2O)]{\text{Invertase}}$ β-D-Fructose ($C_5H_{12}O_6$) + α-D-Glucose ($C_6H_{12}O_6$)

Glucose + Fructose $\xrightarrow[\text{Water } (H_2O)]{\text{Zymase}}$ $4\ CH_3CH_2OH + 4\ CO_2$

Left: Cooling towers at the Jaenschwalde lignite coal-fired power station in Germany. Built in the 1980s, this power plant emits 25 million tons (23 million tonnes) of CO[2] annually and is among the biggest single producers of CO[2] emissions in Europe.

Bioethanol can also be made by fermenting hexoses (six-carbon sugars) that have been produced by the hydrolysis (chemical breakdown reaction with water) of starch from corn and wheat using a process known as saccharification:

$$(C_6H_{10}O_5)n + nH_2O \rightarrow nC_6H_{12}O_6$$

Fuel versus food debate

The controversial "fuel for food" debate argues that biofuel feedstock should be more productively, and more humanely, used to grow food rather than fuel. This has accelerated scientific research toward the development of second-generation lipids from microalgae, which can be grown in environments such as saline water. If production costs can be driven down then microalgae have the potential to supply the world's B100 biodiesel requirements with no impact on food growth and supply. Value-added co-products might also be extracted from microalgae, with the remaining protein meal being used for seafood and animal feed.

For the production of bioethanol there is considerable scientific interest in the hydrolysis of lignocelluloses to sugars. Lignocellulose is available from willow, bagasse, cereal straw, and grain-processing byproducts, and energy crops including miscanthus, switchgrass, and reed canary grass. Unfortunately, at the moment it is difficult to extract the sugars from lignocellulose. Research is ongoing to improve knowledge of this issue.

Left: Maiden grass (*Miscanthus sinensis*) is one of a number of species of perennial grasses that have been trialed as a biofuel in Europe and elsewhere. Miscanthus species are considered more productive than other grasses because they grow quickly and have a high biomass yield.

Of all the billions of cubic miles of water on Earth, 97.5 percent is undrinkable seawater. The opportunity to produce cheap water by desalination could allow us to green our deserts and even produce vast quantities of low-cost feedstock to manufacture renewable biofuels. This will lower atmospheric carbon dioxide, reduce global warming, and even grow more crops.

The science of desalination

The water cycle uses energy from the Sun to naturally remove salt from seawater by evaporation, constantly feeding fresh water back into our rivers and lakes. Each year, the human race removes thousands of cubic miles of fresh water from the environment, but we only use about half of that.

As desalination prices drop, water can be obtained by distillation or by seawater reverse osmosis (SWRO). At the moment only a small amount of desalinated water is produced each year, of which Arab countries produce 65 percent.

SWRO systems

Commercial reverse osmosis (RO) systems, developed in the early 1960s, were based on Loeb-Sourirajan, anisotopic (asymmetric) cellulose acetate membranes. These were obtained by chemically modifying cellulose, which is a rigid, linear, polymer derived from natural plant materials such as cotton.

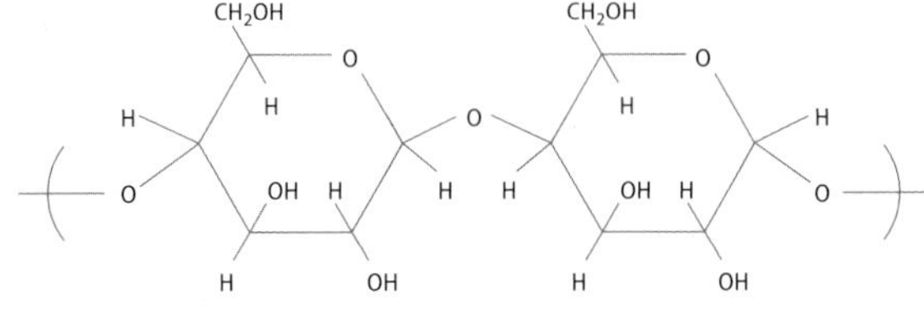

Above: The chemical structure of modified cellulose, a polymer that is obtained from natural plant materials, for example cotton.

If we could ever competitively, at a cheap rate, get fresh water from salt water, then it would be in the long-range interests of humanity, which would really dwarf any other scientific accomplishments.

John F. Kennedy, 35th US President, 1917–1963

Top: Pipes in a desalination plant bring seawater into the processing area. Desalination of water is increasingly being considered around the world, especially in places where demand exceeds supply.

Left: An engineer inspects banks of reverse osmosis cells in a desalination plant. In reverse osmosis, seawater is pumped at high pressure through a filter. This removes the salt from the water, making it suitable for drinking.

SEAWATER REVERSE OSMOSIS

Conventional seawater reverse osmosis (SWRO) is essentially a filtration process on the molecular scale, and pore sizes in reverse osmosis (RO) play a significant part in the separation process. Various membrane types having different surface porosities are used. RO membranes have porosities that range from 0.1 nm to 1 nm (nanometers).

These membranes appear almost impervious to water, and pump pressures (energy) increase with decreasing pore size. The nominal concentration of seawater is 2.5 percent sodium chloride (salt) and to meet World Health Organization (WHO) standards it needs to be reduced to 0.005 percent for human consumption. A difficult task indeed!

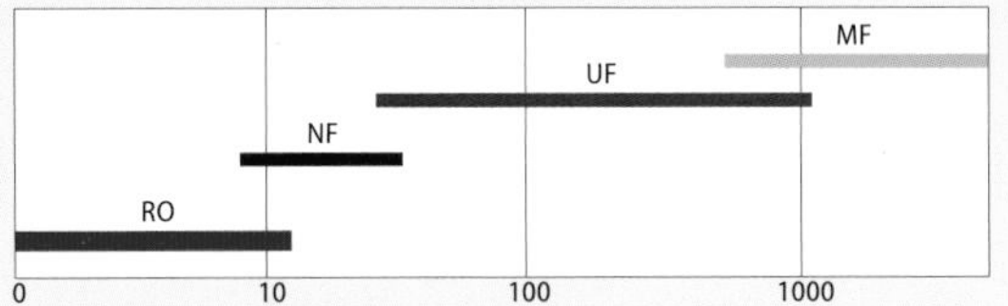

Above: Four types of water treatment membranes: microfiltration (MF), ultrafiltration (UF), nanofiltration (NF) and reverse osmosis (RO) membranes.

Hydroxy (OH) groups on each cellulose ring can be acylated (that is, reacted with acetic acid or a similar substance) using the chemistry shown below. The greater the degree of acylation, the better the salt rejection, but the rate of water transfer through the membrane will decrease.

Original, asymmetric, cellulose acetate or polyamide film membrane technologies have good chlorine resistance and are still in use for this very reason.

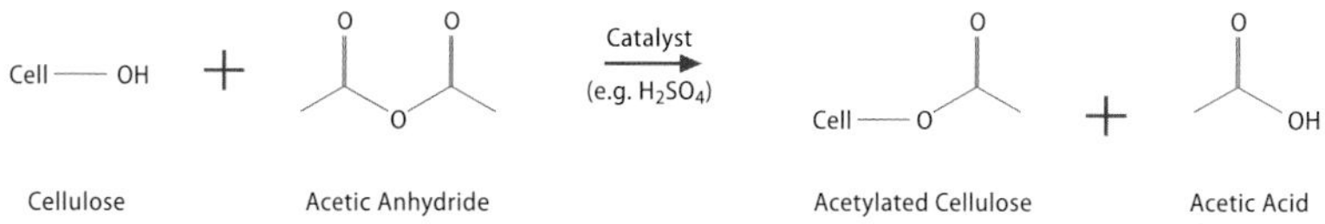

Cell—OH + (Acetic Anhydride) → Catalyst (e.g. H_2SO_4) → Cell—O— (Acetylated Cellulose) + OH (Acetic Acid)

Cellulose | Acetic Anhydride | Acetylated Cellulose | Acetic Acid

However, modern SWRO systems are commonly based on thin-film composite membranes, consisting of three layers: A polyester reinforcing sail-cloth substrate, a 50μ (micron) thick porous polysulfone intermediate layer, and an active 100 nm aromatic polyamide surface skin deposited on the polysulfone substrate by interfacial polymerization. This layer is the active salt rejection component.

The mechanics of salt separation

The mechanics of salt separation by these membranes is poorly understood and there is a need to reduce their operational energy costs. This is particularly important given consumer sentiment about excessive energy use and the associated greenhouse gas emissions. SWRO is fundamentally governed by osmotic pressure across the porous membrane, from natural equilibrium forces attempting to push pure water through the membrane and dilute saline water on the other side. This osmotic pressure pushes against the water being driven through the RO membrane by large pumps—hence the terminology reverse osmosis.

Using a reworking of the ideal gas equation (PV = nRT, where P is pressure, V is volume, n is the amount of gas (moles), T is temperature, and R is the ideal gas constant), the osmotic pressure for seawater with a nominal salt concentration of 2.5 percent can be calculated as 21.5 atmospheres, and is equal to 313 psi (pounds per square inch) (22 kg per square cm). This is considerable and must be overcome to push sea water through an RO membrane. Typically pump pressures required to operate SWRO systems range from 800 to 1,180 psi, which is two and a half to four times the fundamental osmotic back-pressure from seawater osmosis.

There is no doubt that, as the influence of climate change effects grows, there will be greater research focus on reducing the operating costs of SWRO technologies.

Left: More cost-effective and environmentally friendly desalination methods are currently being developed, with the aim of making potable drinking water readily available in even the most remote places.

Top: Large-scale desalination requires large amounts of energy and critics say that there will be an increase in greenhouse gas emissions from power plants supplying electricity to desalination operations.

Spectroscopic experiments rely upon the absorption or emission of electromagnetic radiation (such as visible light and radio waves) from atoms and molecules. Spectroscopic methods are used to obtain information about the structure of molecules, and to determine the amounts of substances present in mixtures from the environment, industrial processes, and living organisms. Spectroscopy is the core technique in most molecular sciences.

Spectroscopy in chemistry

The word "spectroscopy" comes from the word "spectrum" and we are taught in school that a rainbow shows us the spectrum of visible light. When we look at a rainbow, what we are seeing is the range of wavelengths in white light from the Sun spread out by reflections in water droplets. Just as a wave moving across the surface of water has peaks and troughs, so in many ways light behaves like a wave with peaks and troughs. The wavelength of light is the distance between adjacent peaks in the wave. What we perceive as different colors are just waves with different wavelengths. When we see the colors in a rainbow we see the waves spread out from those with longer wavelengths (red) to shorter wavelengths (violet).

Right: A rainbow is a spectacular arc of light. Caused by the refraction of the Sun's rays with rain, a rainbow is made up of all the colors of the visible spectrum.

Electromagnetic waves

Waves of electromagnetic radiation, just like any other wave, also have a frequency that is defined by the number of waves passing by a location per second. Frequency and wavelength are related because of the constant speed at which all electromagnetic waves travel. Imagine watching an electromagnetic wave pass you by: If the wave has a long wavelength then fewer complete waves will go by you per second compared with radiation with a shorter wavelength. So waves of red light with their relatively longer wavelengths have lower frequencies than violet light with its shorter wavelength.

The range of wavelengths does not stop at the red and violet ends of the visible spectrum. Beyond visible light the spectrum continues to shorter wavelengths and higher frequencies (ultraviolet, X-rays, gamma rays) and to longer wavelengths and lower frequencies (infrared, microwaves, radio waves).

Atoms and molecules can absorb an electromagnetic wave, but only if the energy of the wave, which is related to its frequency, matches the gap between two allowed energy levels of the atom or molecule. The energies of atoms and molecules are very accurately predicted by quantum mechanics and are very specific to particular types of atoms and molecules. For example, the allowed energies of a potassium atom are different from those of a copper atom, but the allowed energies of all potassium atoms are identical.

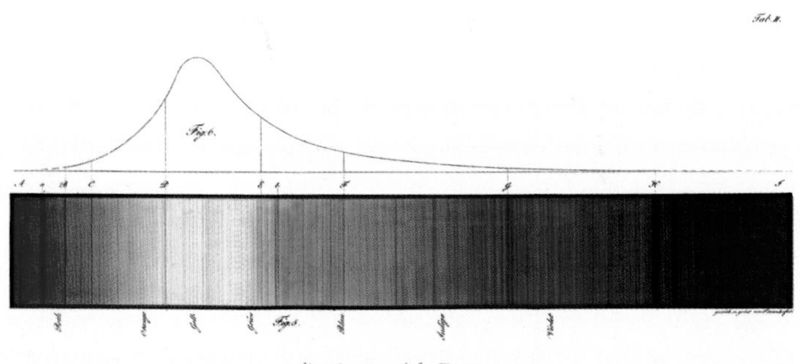

Below: This sketch of the emission spectrum of the Sun, as originally drawn by Josef Fraunhofer in 1814, shows a background of varying color emitted from the hotter, deeper regions of the Sun, with dark lines superimposed where atoms in the cooler regions have absorbed very specific frequencies.

MOLECULES IN SPACE

One application of spectroscopy is the use of rotational spectroscopy to detect molecules deep in space. As molecules change their rotational energy they emit or absorb radio waves. Radio telescopes can detect these signals and comparison of the frequencies with the rotational spectra of known molecules allows the molecule to be identified. About 130 molecules have been detected in the space between stars, ranging from simple two-atom molecules like carbon monoxide (CO) to molecules containing more than ten atoms. Collisions between molecules in space are rare because the molecules are spread over much greater distances. This allows very reactive or unstable molecules to exist for much longer than they would on Earth, so some molecules found in space have been unknown in the laboratory, and chemists have had to produce them in order to confirm their structure.

Atoms and molecules can hold energy in various ways; for example, the electrons in atoms and molecules can exist in different energy levels and this is expressed by saying that atoms and molecules can have varying amounts of electronic energy. Molecules have two or more atoms bonded together and molecules can rotate or spin and vibrate (chemical bonds act like springs, allowing the atoms to move back and forth). The gaps between the allowed electronic energies are usually much larger than the gaps between the allowed vibrational energies, which in turn are larger than the gaps between the allowed rotational energies. This means that the frequencies of electromagnetic waves which can cause an atom or molecule to increase its electronic energy (visible light or ultraviolet) are higher than those which cause an increase in vibrational (infrared) or rotational energy (microwaves and radio waves).

What is a spectrum?

A spectrum is a graph showing the amount of electromagnetic energy absorbed (or emitted, depending on the method used) as the frequency of the energy is varied. If a molecule absorbs energy at a particular frequency, this will be shown as a peak in the spectrum. In

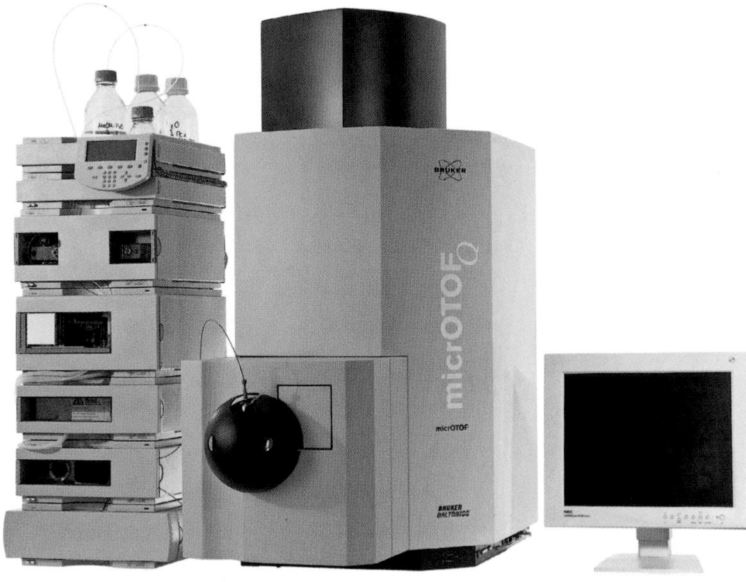

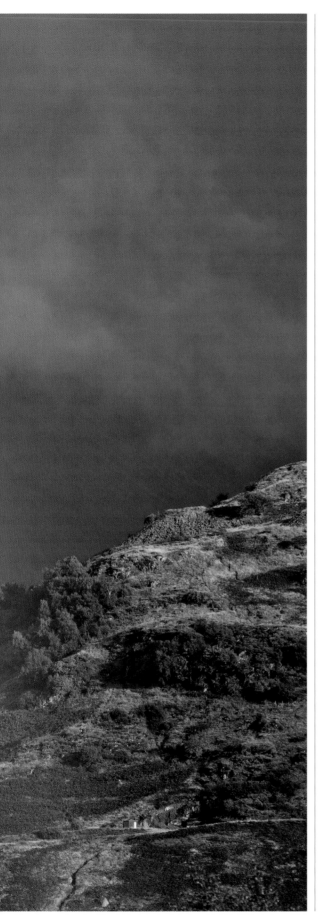

Below: A high-performance spectrometer is used to detect and measure wavelengths of electromagnetic radiation. Wavelengths are indicated on a calibrated scale.

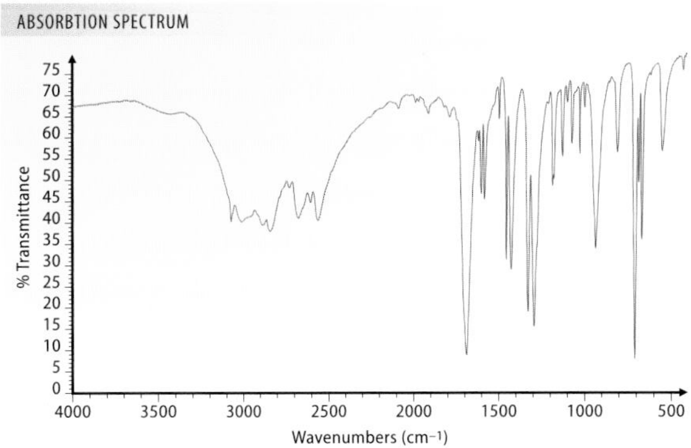

Left: An absorption spectrum is a graph showing the amount of light absorbed at different frequencies. In this spectrum of benzoic acid (a common food preservative), the wavelengths are in the infrared portion of the electromagnetic spectrum and the valleys in the graph indicate wavelengths that are absorbed. Absorption of infrared radiation occurs when the radiation frequency matches a frequency at which the molecule vibrates, so a spectrum such as this can provide information about the structure of a molecule.

a microwave spectrum the frequency range covered matches changes in the rotational energies of molecules, which depend on the shape of the molecule and the masses of the atoms in the molecule. Rotational spectra are therefore very specific to particular molecules and are used by radio astronomers to identify molecules in the gas and dust clouds in outer space.

Infrared spectra show the frequencies at which molecules vibrate, and these frequencies are determined by the masses of the atoms in the molecule and the strengths of the chemical bonds between them. Light atoms (such as hydrogen) or atoms joined by strong bonds (such as double or triple covalent bonds) usually vibrate at higher frequencies and produce absorption peaks at the high frequency end of an infrared spectrum. Combinations of atoms in a molecule often absorb infrared waves at a characteristic frequency (called a group frequency) which means that an infrared spectrum can be a quick method of determining if a particular group of bonded atoms is present in a molecule.

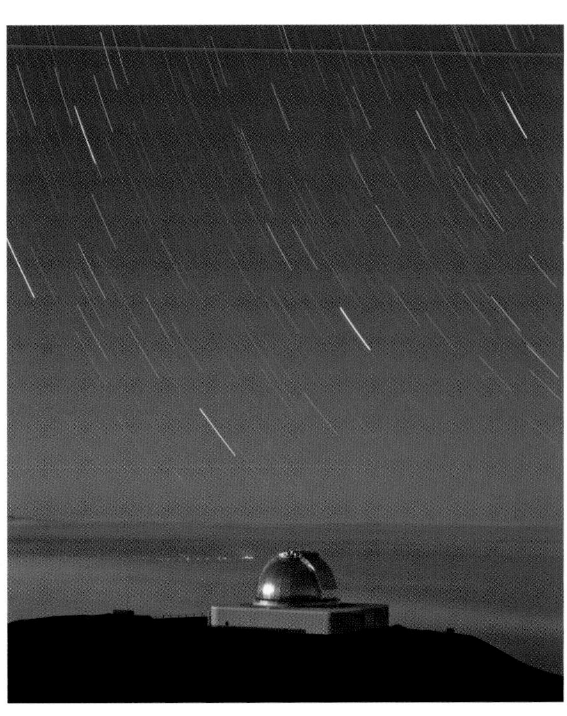

Right: Initially developed for support of the *Voyager* mission, the NASA Infrared Telescope Facility, located on the summit complex of Hawaii's Mauna Kea, utilizes a 10 ft (3 m) diameter telescope to explore the infrared part of the spectrum of space.

Nuclear magnetic resonance

Electronic, vibrational, and rotational spectroscopy all use the allowed energy levels possessed naturally by molecules. But there is one type of spectroscopy where the separation of energy levels is produced by a magnetic field being applied to the molecules. This is nuclear magnetic resonance spectroscopy, or NMR. NMR relies upon the magnetic properties possessed by many types of nuclei—they act as tiny magnets which can be aligned in a magnetic field. Without a magnetic field the energy of a magnetic nucleus doesn't depend on the orientation of the nucleus, but in a magnetic field the nucleus can be aligned with or against the field and these orientations have different energies. The energy gap is tiny so electromagnetic waves with low frequencies (radio waves) can cause the nucleus to flip its orientation and so absorb the wave.

Opposite: In nuclear magnetic resonance (NMR) spectroscopy, electromagnetic radiation is absorbed by an atomic nucleus placed in a strong magnetic field. Used to study molecular structure, NMR and other magnetic resonance technologies have applications in chemistry, physics, biology, and medicine.

> Single molecule absorption spectroscopy is an extremely sensitive technique for analytical chemistry …
>
> Martin Gruebele, chemist and biophysicist

ANALYSIS VIA SPECTROSCOPY

Most spectroscopic methods can measure the quantities of substances present in mixtures. For example, the amount of a colored substance like a food dye dissolved in a drink could be determined by measuring the amount of light absorbed by a sample of drink at the wavelength most strongly absorbed by the dye. By comparing the amount of light absorbed by the drink with the light absorption of solutions containing known amounts of the food dye, the amount of dye in the drink can be determined. The instrument used to measure the strength of light absorption is called a spectrophotometer, usually prefaced by the region of the spectrum it works in, for example an IR (infrared) spectrophotometer. Samples for a spectrophotometer are often solutions, sometimes solids, and, less commonly, gases. Because spectrophotometry is a rapid analytical method it can be used to track changing concentrations, for example the concentrations of pollutant gases in a car exhaust as the car starts up.

Left: The different colors of the solutions in these dishes come from the dissolved dyes, which absorb different frequencies of visible light.

In most molecules each atom of a particular element will produce an NMR signal at a slightly different frequency due to the magnetic fields of the other atoms around them. For example, if the NMR spectrometer is tuned to detect signals from carbon atoms, and a molecule contains 20 carbons atoms, it is likely (although not always the case) that the NMR spectrum will contain 20 separate peaks. These slight changes in frequency are what make NMR a highly important method for deducing molecular structure. For large molecules (e.g. a protein molecule containing thousands of atoms), it is likely that the NMR signals from different atoms will overlap, making it difficult to use the NMR spectrum to learn about the structure of the molecule. Because of the way NMR works, NMR signals are more spread out when a more intense magnetic field is used and so modern NMR spectrometers use very strong magnets, often using electromagnet coils cooled to close to absolute zero in liquid helium to produce magnetic fields 100,000 times as strong as Earth's field.

Many materials, particularly polymers (plastics), pharmaceuticals, food products, metals, and even some inorganic materials, can be better understood by analyzing changes in their properties while increasing or decreasing their temperature.

Changing the temperature—thermal analysis

Polymers are unique because they exist in the nano-realm and display thermal properties that might not be obtained from conventional materials. Global warming, being directly related to energy, has also resulted in an increase in thermal analysis for the development of improved biofuel products.

Some common thermal analysis techniques include:

- DSC—differential scanning calorimetry
- DTA—differential thermal analysis, which is a simplified version of DSC
- TGA—thermogravimetric analysis, which measures weight loss with temperature
- TMA—thermomechanical analysis, which measures dimensional change with temperature
- Rheology—which measures viscosity (thickness/resistance to flow of molten polymer liquid) change with temperature
- DMA—dynamic mechanical analysis, which measures varying dynamic (oscillating) dimensional change with temperature
- DEA—dielectric analysis, which measures electric properties with temperature
- µTA—microthermal analysis, which measures thermal properties on a micro scale.

> Research is the process of going up alleys to see if they are blind.
>
> Marston Bates, zoologist, 1906–1974

Above: The properties of polyethylene terephthalate (PET) make it ideal as a drink container. When melted then quickly cooled, PET produces a clear plastic material that is lightweight, shatterproof, and easily recyclable.

Below: Thermal analysis techniques such as modulated temperature differential scanning calorimetry (MTDSC) measure the heat released during the freezing process, for instance when fresh food is frozen.

Differential scanning calorimetry

DSC measures the temperature of a sample, compared with an inert reference (usually an empty DSC pan), when the sample and reference are both heated or cooled at identical rates. "Power-compensation" DSC was the first DSC technique introduced and this technique directly measured heat flow in and out of a sample. The second type of DSC introduced was the so-called "heat flux" DSC instrument, where the sample and reference are thermally connected and enclosed in a single furnace. Small temperature differences between the sample and the reference are indirectly converted to heat flow of the sample. Both instruments provide virtually identical thermal heat energy information.

UNDERSTANDING YOUR DRINK BOTTLE

Polyethylene terephthalate (PET) is a polymer that is widely used to make modern, clear, plastic, drink bottles, as well as 60 percent of artificial synthetic fibers. PET has important thermal behavior, as shown in the DSC trace below, when the polymer is melted then rapidly cooled (quenched). Quench cooling prevents growth of crystals in the polymer and therefore produces a clear, amorphous (random) plastic material. Because of this, a strong reversing glass transition at 167 ºF (75 ºC) is seen, as the material changes from hard material to a soft plastic material.

In contrast, if molten PET is cooled very slowly, it becomes highly crystalline. When crystalline, PET is opaque and white colored, due to reflection of light from the large polymer crystals in the material, and therefore is completely useless for use as a drink bottle.

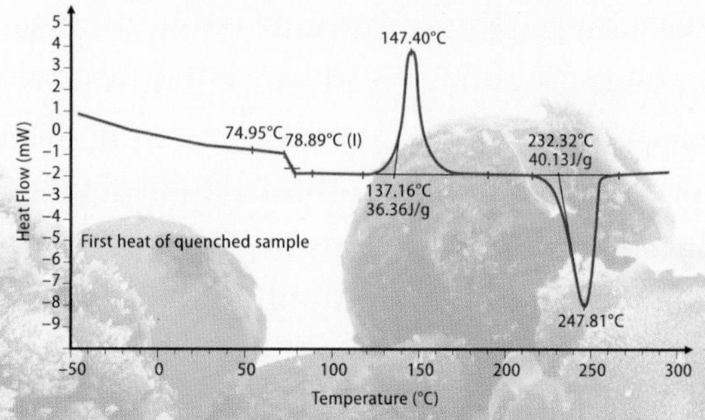

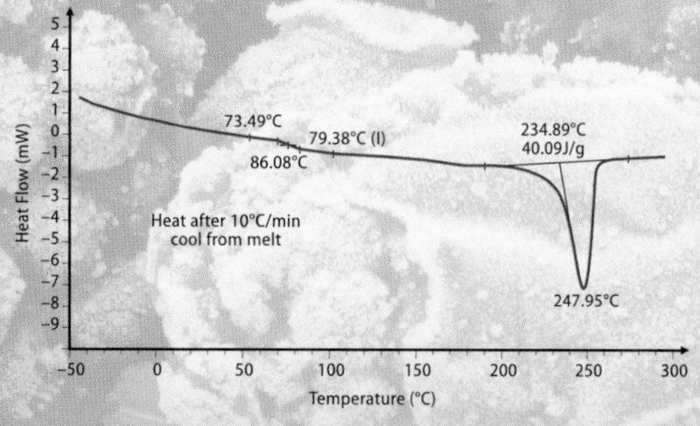

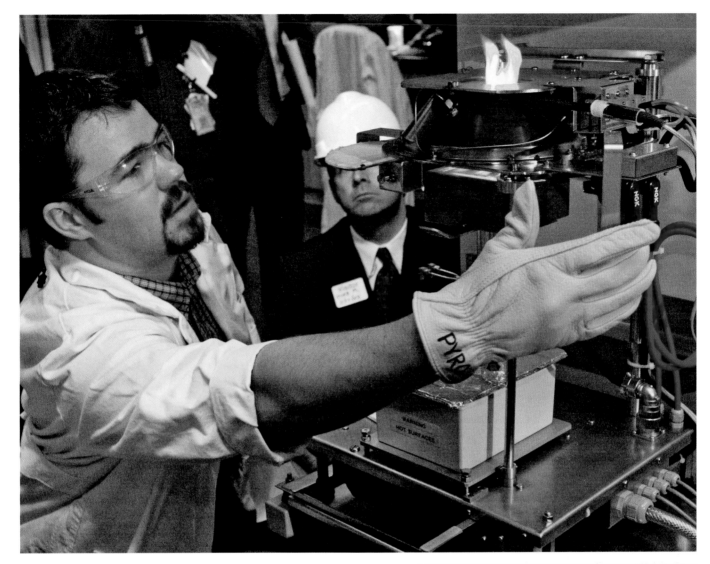

Heat capacity and kinetic heat flow

Conventional DSC measures total heat flow in and out of a sample as the sample is heated or cooled. Total heat flow (dQ/dt) is a mix of two different types of heat flow, which, in conventional DSC, is seen only as a single signal. The two heat flows mentioned are heat-capacity-related heat flow (so-called reversible heat flow) and the kinetic heat flow (the so-called non-reversible heat flow).

Interestingly, reversible heat flow is influenced by the heat capacity of the material and also how fast the material is heated. Heat capacity is the ability of the material to absorb heat energy. The non-reversible heat flow is not affected by such factors, and independently releases or absorbs heat energy, at a specific temperature and time.

Modulated temperature DSC

A novel DSC modification, which was introduced by Mike Reading in 1992, and is currently referred to as modulated temperature DSC (MTDSC), was for the first time able to separate reversing (heat capacity) heat flow and the non-reversing (kinetic) heat flow, revolutionizing DSC analysis. Kinetic and heat capacity events can overlap, and MTDSC allows separation of these thermal events into reversing (heat capacity) and non-reversing (kinetic) heat flow signals, which is not possible using conventional DSC. However, there is considerable controversy about this technique. In spite of sound mathematical validation from Bernhard Wunderlich in 1994, and follow-up work by Stephen Clarke et al in 2000, there are still some pockets of resistance to the MTDSC technique.

Above: A research engineer at the Bureau of Alcohol, Tobacco, Firearms and Explosives (ATF) Fire Research Laboratory in Maryland, USA, demonstrates a heat release rate test using a cone calorimeter.

Right: The food industry makes use of thermal analysis techniques. Much of the food we eat is subject to variations in temperature, which happen during production, storage, preparation, and so on. One such temperature variation happens during pasteurization which destroys viruses such as avian influenza, and harmful bacteria such as salmonella. Here, eggs are ready for the pasteurization process.

Science gives a simplified picture of whatever it studies. This is unavoidable. The best science can give is an approximation to how things work. Scientific studies conducted in a laboratory where confusing details are eliminated can give brilliant results. However, for the environmental chemist who must make sense of their laboratory results in a real world context this presents a stunning challenge. Complexity can soon overwhelm.

Environmental chemistry

The task environmental chemists have set themselves is huge. Natural systems are multi-layered and the answers to investigations are often inconclusive. This uncertainty is one reason why historically the environmental sciences have not received the levels of funding or esteem enjoyed by the more traditional pursuits of physics, chemistry, and engineering. In this science is like any other human pursuit—money and effort are invested where they can be most guaranteed of a return. It is easier and more productive to fund making a better "anything" than it is to look for environmental effects that may or may not eventually prove to exist. Individuals and companies are also more inclined to fund research if they can own the results and gainfully profit from them. In the case of the environment, results are generally useless unless they are shared.

Above: Aerosol cans often contain chlorofluorocarbons, which have been recognized as major contributors to the depletion of the ozone layer. Chemists have been developing alternative compounds that will have less impact on the environment.

Helping out

It is one of the extraordinary successes of environmental chemistry that, despite the complexities of the subject, chemists around the world have been able to reach consensus on the damaging effects of chlorofluorocarbons to atmospheric ozone; and on the broad impacts of human greenhouse gas emissions. On a local level, environmental chemists help guide policy makers in establishing and following appropriate water, air, and soil quality standards or in advising on conservation and biodiversity issues.

It must be acknowledged: Science is invaluable for understanding nature and for providing guidance on options for management. It is a diagnostic tool that lets us understand where we are and lets us ask informed questions about where we might be going. However, the decisions as to how we choose to act to protect or enhance the environment are ultimately political and ethical choices.

Below: Recycling paper is one way to reduce water and air pollution. Paper is recycled by using a series of filtration steps to extract the cellulose fiber and to disperse the inks.

Green chemistry

Green chemistry aims to make products and use chemical processes that are not hazardous to human health or the environment. Green chemistry uses environmentally benign chemicals that result in less waste, safer processes and products, and reduced energy and resource use. The principles of green chemistry can partly be summarized as "reduce, reuse, recycle." To this could be added "reform" or "rehabilitate" in that green chemistry aims to change both the processes that are used to manufacture goods as well as the types of products made.

Green chemistry should not be confused with environmental chemistry. They are complementary, but different, disciplines. While the focus in environmental chemistry is on the natural environment,

green chemistry focuses on industrial applications. Every day we are surrounded by useful, but potentially harmful chemicals. The usual approach to managing these chemicals is to try and limit exposure to them. Regulations are made to control their use. These regulations then need policing. Frustratingly, this is often both expensive and ineffective. The approach of green chemistry is to question the wisdom of making harmful chemicals at all if they can be replaced with those that are more benign.

Globally, waste management costs industry and governments billions of dollars each year. Industries that adopt green chemistry practices gain an edge over their competition—they can make their products cheaper by using fewer materials and by making less waste.

WHAT IS ENVIRONMENTAL CHEMISTRY?

Environmental chemistry studies the chemistry of the natural environment as well as the human impacts upon it. It examines the sources, reactions and fates of chemicals in the air, soil, and water—and then sets out to explain what is going on. Environmental chemistry is a multidisciplinary subject using knowledge from almost every other discipline imaginable, from the physical sciences, to the humanities and social sciences. Here, soil samples are being taken for analysis.

Science cannot resolve moral conflicts, but it can help to more accurately frame the debates about those conflicts.

Heinz Pagels, physicist, 1939–1988

Right: A HFCEV, or hydrogen fuel cell electric vehicle, is essentially an electric car. Under the floor pan is a hydrogen fuel cell and auxiliary energy system that supplies electricity to the powertrain, while in the trunk is a hydrogen tank that carries the car's renewable hydrogen fuel.

Left: Many industrial processes leave behind an environmental trail of destruction. When high levels of certain chemicals such as phosphates are discharged into a river system, it can result in some long-term detrimental effects on the health of local flora and fauna.

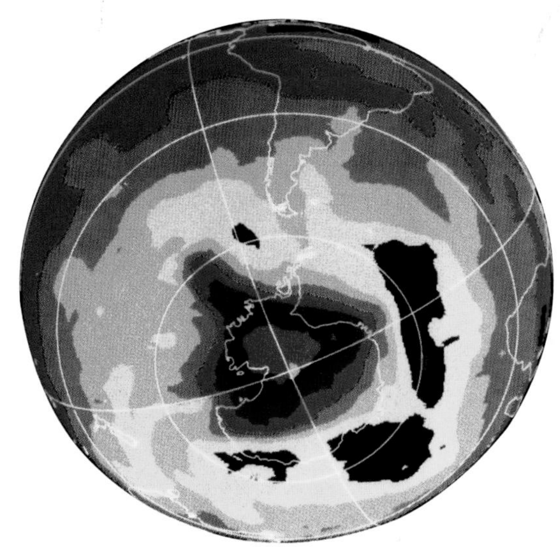

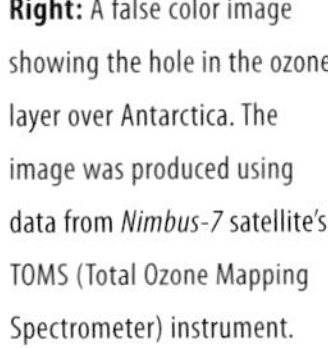

Right: A false color image showing the hole in the ozone layer over Antarctica. The image was produced using data from *Nimbus-7* satellite's TOMS (Total Ozone Mapping Spectrometer) instrument.

The advance of computer technologies in recent times has been more than phenomenal. The relatively newfound ability of computers to store huge amounts of data efficiently and rapidly perform calculations has provided scientists with a new tool for investigating problems. Computational chemistry has revolutionized the way scientists approach research as they can now solve large and complex problems that, until recently, were quite simply beyond their reach.

Computational chemistry

Two factors contribute to the complexity of computational chemistry. The first is the large number of particles (nuclei and electrons) in all but the simplest molecules. This gives rise to a huge number of repulsions and attractions to take into account. The second is the need to use quantum mechanics to calculate molecular properties. Having to take into account the wave-like nature of electrons increases the complexity of the calculations enormously.

Right: The BioServer is a powerful new computer developed by Fujitsu, that will be able to handle the large amount of calculations currently needed in many fields of development including chemical and medical research.

What computational scientists do

Traditionally, science has been conducted through three primary approaches—observation, experimentation, and the development of theories. Advances in computing power and calculation and vastly improved hardware and software have opened up a fourth area of scientific research, computational science. Scientists can now search for the solutions to complex problems using a range of skills from science, computer science, and mathematics. Computational scientists use digital technology to solve equations that help define a particular theory or model and this helps them to study various scenarios.

Computational chemists develop models that help make predictions about what might happen in laboratory or real-life situations.

> ... without a supercomputer, no leading chemist could do significant computational chemistry in a reasonable amount of time.
>
> Robert Panoff, computational scientist

Left: Computational investigations are now a huge part of science and research chemists are finding computational modeling an incredibly useful tool. Vast amounts of information are stored on computer memory chips.

SUPERCOMPUTERS HELPING HUMANKIND

Advanced computational methods allow researchers to predict the structure and dynamics of molecules. These predications can be tested by researchers using state of the art supercomputers (such as the one below) as they design molecules that they expect to be effective compounds in various systems. In medicine, for example, scientists have been able to synthesize molecules that they predict would act as anti-breast-cancer compounds. Modeling of the compounds can test to see if they inhibit cancer production and, if promising, the models can then be verified experimentally.

Predictions on the structure and dynamics of molecules are also important in biological systems. Experimental measurements can also be combined with computational modeling to help scientists understand complicated events better. Processes such as the decay of uranium, when investigated with computational modeling, help us understand the parameters effecting radioactive contamination. Uranium is one of the heavier elements, so has many more electrons than simpler atoms like carbon or nitrogen. The large number of electrons makes the behavior of the element harder to study. With advanced calculating power through supercomputers, chemists have recently been able to understand and interpret data relating to the chemical and physical properties of uranium complexes. Obtaining results from supercomputer calculations minimizes the need for hazardous experiments and can have a significant impact on the way we address problems of radioactive waste and nuclear cleanup.

Above: Physicist Walter Kohn (pictured) and chemist and mathematician John Pople shared the 1998 Nobel Prize for Chemistry for their work in computational methods in quantum chemistry.

These models help scientists to make better observations and understand their results more clearly. It also allows experiments that would otherwise be too expensive, difficult or dangerous to conduct to be investigated and analyzed. For example, many computational studies are conducted in order to predict how a newly developed drug will react in a system, which helps to reduce the number of clinical trials or animal tests that are carried out. While these computational models cannot replace the results that come from laboratory investigation, they have become an integral part of the process in which scientists gain knowledge and solve problems.

Interests of computational chemists

The kinds of predictions made by computational chemists usually relate to the structure of a molecule. Computational chemists try to determine the relative positions of all the atoms in space. The most favorable connectivity between atoms, or bonding, is usually determined as the structure that will require the lowest possible energy.

The potential energy surfaces of a molecule are also of great interest to computational chemists. Localized levels of high or low energy often affect how enzymes or drugs interact with specific molecules. The arrangement of atoms in three-dimensional space and their relationship to each other defines the bond length and angles between atoms. These features are important when predicting how molecules will interact with each other.

ASTRONOMY

We have had a fascination with the night sky for thousands of years. Early on, we distinguished between the "fixed stars," those that held their places relative to one another, and the "wandering stars" or "planets." The old observers measured the positions of the stars to construct maps, and their observations were accurate enough to detect slow long-term changes in the pattern of the heavens such as the 22,000-year long "precession of the equinoxes."

Introduction

Expanding our horizons

The invention of the telescope four centuries ago transformed astronomy. By gathering much more light than our eyes can, even the most primitive telescopes revealed totally unexpected wonders, and soon brought into question the millennium-old view that the Earth stood unmoving at the center of all things. Polish cleric Nicolaus Copernicus was bold enough to set down on paper his conviction (heretical in the eyes of the Church) that the Earth went round the Sun, rather than the reverse.

Since then, astronomy has had ever-expanding horizons. Even for Copernicus and his supporters, like Galileo, the Sun and its retinue of planets (some with their own attendant satellites) was still the center of things. The telescope had revealed increasing numbers of fixed stars, but their distances were unknown, and their importance was little considered.

By the nineteenth century, all that was changing. Telescopes had improved since Galileo's day, and new and powerful methods had come into use. The analysis of starlight through spectroscopy confirmed that stars are objects similar to our Sun, reduced to pinpoints only by their vast distances from us.

The first direct measurement of the distance to a star was achieved by detecting the almost indiscernible movement of the star back and forth as our viewing point on Earth changed throughout the year. The distances were vast, with light travel times of several years to even the nearest stars, compared with the mere hours taken for light to travel between the Sun and its planets.

Astronomers began to conceive all the visible stars as members of a great wheel-shaped congregation known as the Milky Way galaxy. It was soon apparent that our Sun was nowhere near the center of this system, but rather more towards the outskirts, merely one star, and a fairly typical one at that, among many millions. Earth's position in the scheme of the cosmos was further downgraded.

Below: A person observes the inner workings of Palomar Observatory in San Diego County, California. Palomar came online in 1948 with the 200-inch Hale Telescope. Since then, the Palomar telescopes have been modernized to detect fainter and fainter signals from more distant sources.

Right: This artist's conception shows the spacecraft *Cassini* in orbit around Saturn. The tiny orbiter has spent a decade performing amazing feats such as diving through the planet's rings and sampling some of the water-ice geysers on the moon Enceladus. The images and data transmitted back to Earth have deepened our knowledge of the Solar System.

Independent star systems

Debate soon began as to whether our galaxy was all the universe contained. Philosophers visualized a multiplicity of "island universes" spread through space, but hard evidence took a century to collect. By the early twentieth century, individual stars could be seen within nebulae, bright wisps of gas among the nearer stars, and their distances were being measured by new techniques. The stunning conclusion, reached first by American astronomer Edwin Hubble, identified many of the nebulae as independent star systems, rivaling in size our own Milky Way. The distances to even the nearest were measured in millions of light years. The known universe continues to expand before our eyes.

We now know that galaxies in almost uncountable numbers extend through space to distances of more than 10 billion light years. Most of these galaxies are moving away from each other, and our universe is expanding. The available evidence points to a cataclysmic beginning to our universe some 13 billion years ago—the Big Bang. It is sobering to ponder what Copernicus or Galileo would make of it all.

> For everyone, as I think, must see that astronomy compels the soul to look upwards and leads us from this world to another.
>
> Plato, philosopher, *c.* 423–*c.* 348 BCE

Astronomy has gained many new tools. Telescopes can gather radio waves, infrared and ultraviolet images, X-rays, gamma rays, and even gravitational waves, adding to our understanding, since these different forms of radiation commonly arise from different processes.

Much nearer at hand, we are exploring our Solar System, with spacecraft launched from Earth to fly by, to orbit, or even to land on all the planets. The *Cassini* mission to the Saturn system, which has been in progress for a decade, is typical; likewise the current intense interest in Mars, with several simultaneous missions. Early telescopes showed fuzzy disks with barely discernible markings. These are now known worlds, photographed in detail, mapped and monitored, and increasingly better understood.

Further out, we are now finding planets orbiting not our own Sun but other stars. Fifteen years ago we did not know of "exoplanets." How much more will we learn in the coming years?

Left: An ancient astronomical map is covered with fantastic figures representing the constellations of stars in the Southern hemisphere sky. Twentieth century measurements have shown that the stars in many constellations are not associated with each other at all.

It's hard to imagine, but 200 years ago, astronomers believed that the universe consisted of our Solar System and the stars, an indeterminate distance away. Less than 100 years ago, the Milky Way was thought to be the entire universe.

Cosmology and the Big Bang Theory

In 1923, Edwin Hubble, using the 100-inch (250 cm) Mt Wilson telescope, discovered that the fuzzy Andromeda "nebula" was actually a huge city of stars about one million light-years away—a distance more than three times the known diameter of the Milky Way. Astronomers quickly found that other fuzzy patches were also distant galaxies. By 1925, astronomers had realized that the universe consisted of hundreds, maybe thousands, of galaxies, many of which were rushing away from us at enormous speeds—that is, the universe seemed to be expanding in all directions.

WHAT IS COSMOLOGY?

Cosmology is the study of the origin, structure, and evolution of the universe. It is a science that is always under construction as new discoveries cause revisions to old theories. Cosmology tries to answer the big-picture questions: how and when was the universe born, what is its overall shape, what is it made of, and how will it ultimately end?

Different theories

The first scientist to take these findings and extrapolate backwards was Abbé Georges Eduard Lemaître (1894–1966). In 1927, he theorized that the universe had a definite beginning, a time in which all its matter and energy were concentrated into a single point. When that point exploded like a burst of fireworks, it marked the beginning of time and space and caused the universe to expand.

Twenty years later, an alternative idea arose. Called the Steady State theory, it assumed that the cosmos was the same at all times, in all places, and in all directions. The universe had no beginning and would have no end, and the expansion that had been observed was simply an artefact of the creation of new matter that pushed old matter away. British astronomer Fred Hoyle, a vociferous proponent of the Steady State model, coined the term "Big Bang" as a sarcastic comment on the opposing theory.

Around the same time, physicist George Gamow proposed that some of the chemical elements observed today were created within the first few minutes of the birth of the universe. Further, he argued that the early universe was very hot, and cooled as it expanded. If the Big Bang theory was correct, Gamow predicted that today's cosmos should be filled with just a little of the heat left over from its birth. The Steady State theory made no such prediction.

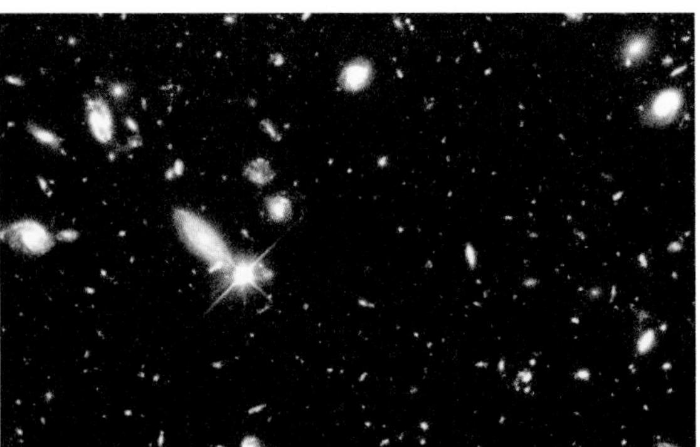

Above: George Gamow predicted that heat from the Big Bang would still be detectable, and the 1965 discovery of CMB radiation proved him correct. Arno Penzias and Robert Wilson were later awarded the Nobel Prize for their work, and Gamow lived just long enough to see his theory proved.

Right: The Andromeda galaxy, nearest neighbor to the Milky Way, was once thought to be a fuzzy nebula. Edwin Hubble first recognized it as a city of stars to rival our home galaxy in size. Andromeda and the Milky Way are rushing toward each other, and will collide in about three billion years.

Left: Several hundred galaxies never seen before are visible in this "deepest-ever" view of the universe—called the Hubble Deep Field (HDF)—made with NASA's Hubble Space Telescope. Some of the galaxies in this image may have formed less that one billion years after the Big Bang.

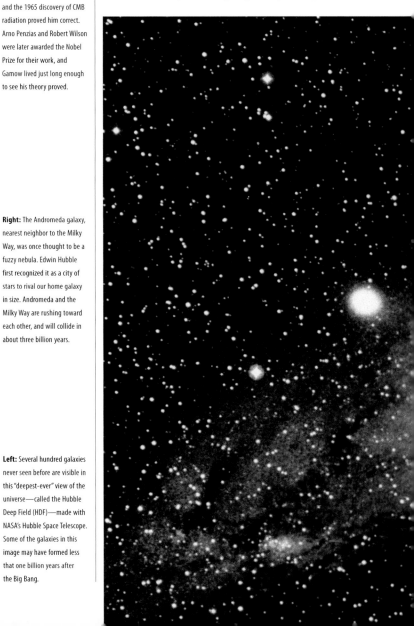

Astronomers called this predicted remnant heat the cosmic microwave background, or CMB radiation. When it was accidentally discovered in 1965, CMB radiation hammered the final nail into the coffin of the Steady State theory.

The Big Bang conundrum

Even though it became the generally accepted hypothesis, the Big Bang theory raised many questions. Cosmologists have come up with some answers, but many mysteries remain.

In the early 1980s, Alan Guth proposed that in the minuscule fraction of a second immediately after the birth of the universe, the size of the cosmos expanded by a factor of at least 10^{30}. Space itself inflated dramatically. His theory helps to explain why the universe is

> One thing we've learnt from astronomy is that the future lying ahead is more prolonged than the past. Even our Sun is less than halfway through its life.
>
> Sir Martin Rees, astronomer, b. 1942

so large, even though it is only 13.7 billion years old. At the same time, astronomers were concluding that most of the mass of the universe is invisible, consisting of so-called dark matter. What this is composed of nobody knows, but if there is enough dark matter, the expansion of the universe will stop and reverse direction, leading to a "Big Crunch" in the far-distant future.

To further muddy the situation, in 1998 it was discovered that the rate of expansion of the universe is actually speeding up. It seems the universe is full of an unknown force called dark energy that is pushing everything apart. The stuff we can actually see—from atoms and molecules to planets, stars, and galaxies—comprises a mere four percent of the universe. No wonder it is hard to understand our cosmos. We cannot see most of it.

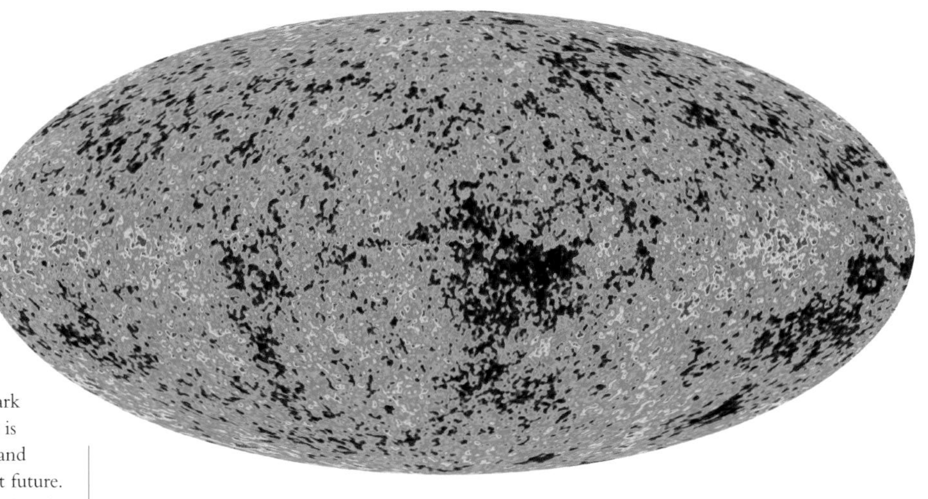

Above: This image, showing the first all-sky microwave image of the universe soon after the Big Bang—the Cosmic Microwave Background radiation—uses data from the Wilkinson Microwave Anisotropy Probe (WMAP) and shows minute temperature variations, indicated by color-coding. The measurements made by WMAP continue to reveal more about the age, composition, and geometry of the universe, such as the estimate that 72 percent of the universe is made up of dark energy.

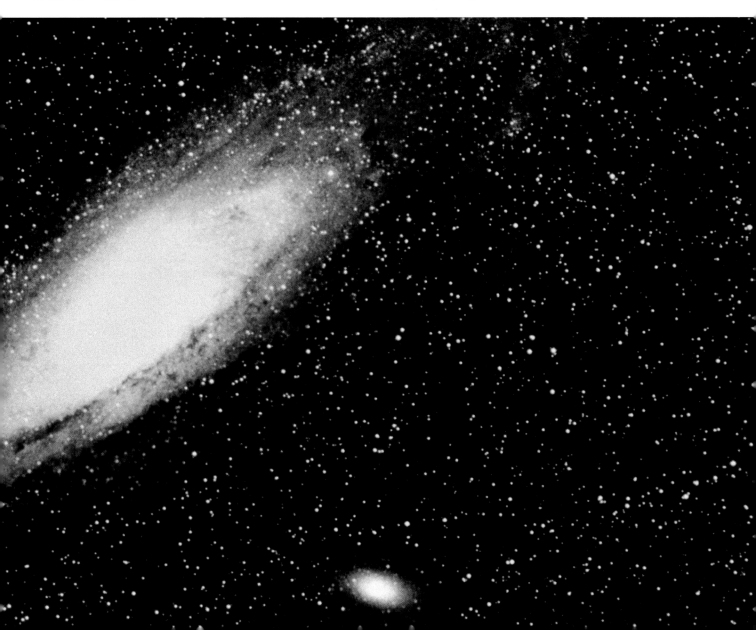

Much about the science of cosmology is counterintuitive, and completely alien to our everyday ways of thinking. Astronomers have now realized that the Earth and our Solar System constitute a minute fragment of an immense long-lived universe, the nature of which is turning out to be almost completely different from all that we find familiar.

How the universe began

The universe: early days

The Big Bang theory explains how the universe came to be, but it does not describe the precise moment of cosmic birth. Cosmologists do not know what caused our universe to come into existence. So let's assume that the universe just is—a tiny speck, perhaps a millionth of a foot across—and go on to the next fraction of a millisecond of its life.

If the universe didn't "explode" into existence, what happened? Think of the birth of the cosmos as an unfolding of matter, energy, and time that took place everywhere simultaneously. One problem that faced cosmologists late in the twentieth century was explaining how the universe expanded rapidly enough to end up as big as we see it today. In the 1980s, Alan Guth and others proposed that a period of extremely rapid expansion—inflation—occurred in the first 10^{-34} seconds of its life. During this inflationary burst, the universe doubled in size at least 85 or 90 times, creating a super-hot, super-dense mixture of matter and energy.

The driving force behind this rapid expansion is believed to be vacuum energy. Although it defies common sense, physicists know that empty space has energy. In a universe as small as ours at its birth, a minuscule amount of vacuum energy would be powerful enough to inflate the cosmos many times. But as the universe grew, the influence of that energy waned, inflation weakened, and the expansion slowed.

Below: *An Allegory of the Creation of the Cosmos*, by Domenicus van Wijnen, shows a seventeenth century Christian explanation of the birth and early history of the universe. Twentieth century science tells the complex tale differently, but explanations of the "first cause" of the universe are still elusive.

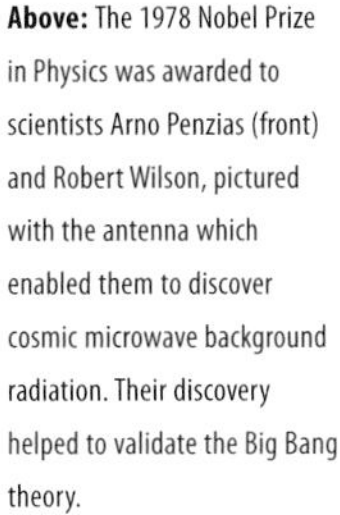

Above: The 1978 Nobel Prize in Physics was awarded to scientists Arno Penzias (front) and Robert Wilson, pictured with the antenna which enabled them to discover cosmic microwave background radiation. Their discovery helped to validate the Big Bang theory.

As the universe expanded, it cooled and became less dense. One second after its birth, the cosmic temperature was roughly 10 billion degrees. Two seconds later, the universe was a scorching plasma of photons (radiation) and sub-atomic particles smashing together to form protons and neutrons.

The next several hundred thousand years of cosmic history is called the Radiation Era. During this time the universe was dominated by the remnant energy from its birth. The cosmos was a searing sea of energetic electrons and protons, too hot to allow matter to form. Little is known about this high-energy era because the plasma that filled the universe prevented radiation (in the form of light) from escaping, leaving the newly born cosmos completely opaque to our prying eyes.

It all becomes clear

Some 380,000 years after the Big Bang, the universe had cooled to less than 4,000 degrees and hydrogen and helium nuclei began capturing electrons to form electrically neutral atoms. This stable form of gas is not as good as plasma at blocking radiation, and so, as the amount of hydrogen and helium increased, the cosmos slowly became transparent, which meant that astronomers could finally "see" what was going on.

As the universe expanded, the radiation from its early days was stretched (red shifted) and cooled. Today it can be observed only as a cold (2.7° above absolute zero or –459°F, –263°C), universe-filling, faint glow of microwaves: the Cosmic Microwave Background (CMB) radiation. The presence of CMB radiation was predicted by the Big Bang theory, but its discovery was accidental. Arno Penzias and Robert Wilson, at the Bell Telephone Laboratories in Murray Hill, New Jersey, found it in 1965. They were building a specialized antenna, and an annoying hiss they could not eliminate turned out to be CMB radiation. (They shared the 1978 Nobel Prize in Physics for their discovery.)

In the early 1990s, the COsmic Background Explorer satellite (COBE) measured the distribution and temperature of the microwave radiation, and detected faint irregularities in both. So sensitive were the satellite's instruments that they recorded temperature differences of a mere one/hundred-thousandth of a degree between the "hot" and "cold" regions. (For their work on Cosmic Microwave Background radiation, COBE team leaders, John Mather and George Smoot, won the 2006 Nobel Prize in Physics.)

Smoot and others argued that these differences in temperature were fluctuations in the density of matter in the early universe that eventually led to the formation of galaxies and galaxy clusters. But what caused those variations? Better data was needed, so in 2001 a follow-up mission was launched: WMAP, or the Wilkinson Microwave Anisotropy Probe.

> The universe contains vastly more order than Earth-life could ever demand. All those distant galaxies, irrelevant for our existence, seem as equally well ordered as our own.
>
> Paul Davies, physicist, cosmologist, and astrobiologist, b. 1946

Top right: This is a false-color image of the star AE Aurigae (near the center of the image) embedded in a region of space containing filaments of carbon-rich dust. The dust might be hiding deuterium that was formed in the early moments of the universe and foiling scientists' attempts to trace the evolution of stars and galaxies.

Right: Detailed analyses of the Hubble Ultra Deep Field (HUDF) have identified what may turn out to be some of the earliest star-forming galaxies in the visible universe. These faint sources illustrate how astronomers can begin to explore when the first galaxies formed.

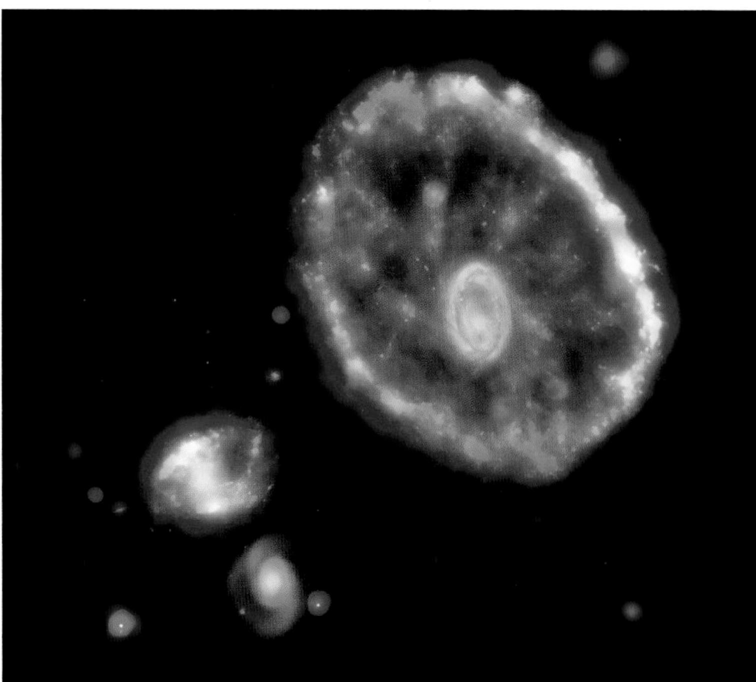

Above: The Cartwheel Galaxy's ring structure, pictured in this composite image from the Galaxy Evolution Explorer, the Hubble Space Telescope, the Spitzer Space Telescope, and the Chandra X-Ray Observatory, shows structures that are invisible to the human eye, including some ripples in an ultraviolet-bright blue outer ring of star formation and an orange, infrared-bright center.

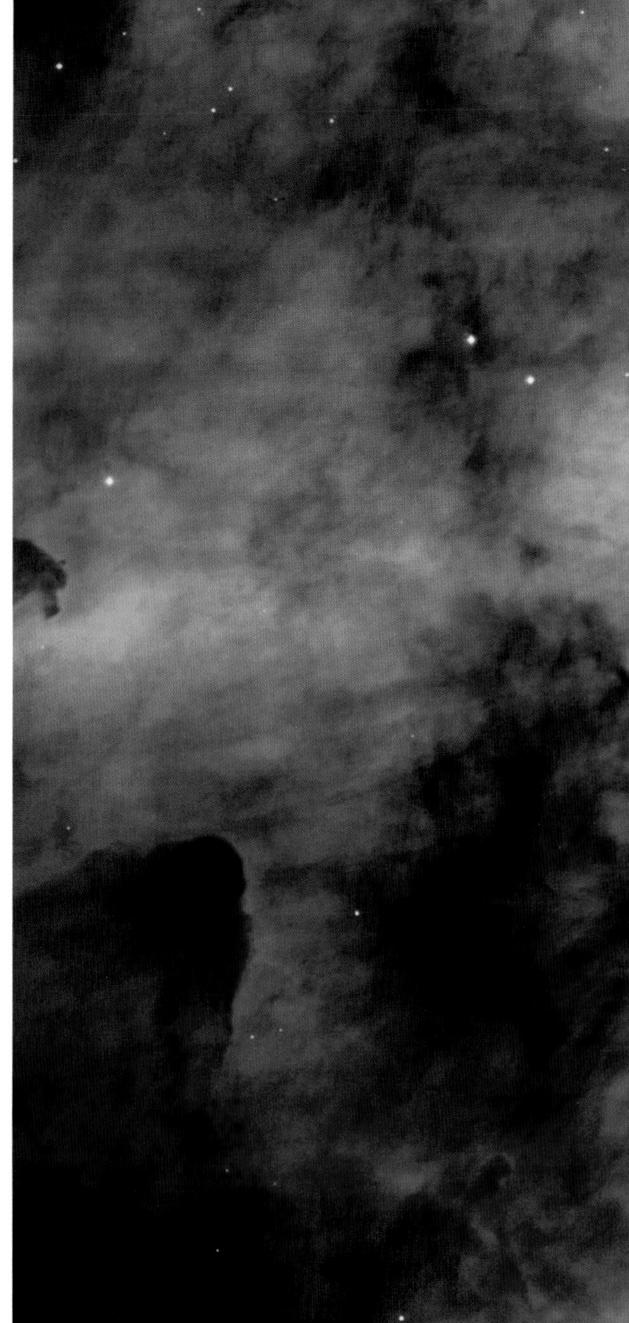

Dark matter enters the equation

While Alan Guth was proposing his inflation theory to explain how the universe got to be so big so quickly, other astronomers were asking: Why can't we see all of the universe?

In 1933, Swiss astrophysicist Fritz Zwicky observed the Coma cluster of galaxies and estimated the cluster's total mass in two ways: By the motions of galaxies near its edge, and by the number of galaxies in, and the total brightness of, the cluster. He discovered that it possessed about 400 times more mass than expected. The gravity of all the visible galaxies was too small to keep the cluster together, and yet there it was—a stable cohesive group. Zwicky concluded that there must be a significant amount of "invisible" matter with enough mass (and therefore gravity) to prevent the cluster from flying apart.

For 40 years nobody paid much attention to Zwicky's findings. Then, in 1975, astronomer Vera Rubin announced that stars living outside the core of many spiral galaxies orbit at roughly the same speed. Rubin calculated that for this to happen, upward of 50 percent of the mass of these galaxies must be contained in a dark halo enveloping each galaxy. By the turn of the century, astronomers had concluded that 85 percent of the mass of the universe consists of invisible dark matter, with the remaining 15 percent being visible baryonic matter—the protons, electrons, and neutrons that comprise all the "stuff" we see in the universe.

Dark matter and the birth of the universe

While a small percentage of this invisible material could be MACHOs, it is generally accepted that dark matter was present in the early universe, and probably played a key role in the birth of the first galaxies.

The Big Bang distributed matter (once it began to form) evenly in all directions. But the theory doesn't explain how this matter might have started clumping together. There certainly wasn't sufficient baryonic material to exert enough gravitational influence to start the clumping process. Yet something supplied enough gravity to create some minor cosmic lumps, which eventually turned into galaxies. That something was probably dark matter.

Cosmologists had theorized that buried in the Cosmic Microwave Background radiation might be signs of this initial clumping. The COBE results saw it on a large angular scale, but astronomers hoped that WMAP's better angular resolution would detect much finer features in the CMB, allowing them to fill in the details.

WHAT IS DARK MATTER?

Some dark matter could be baryonic matter that is literally too dim to see: Multitudes of brown dwarf stars, dead stars, supermassive black holes, and even massive gas clouds that astronomers simply haven't detected yet. These objects have been nicknamed MACHOs (MAssive Compact Halo Objects).

The non-baryonic possibilities are known as WIMPs: Weakly Interacting Massive Particles. (Who says physicists have no sense of humor?) Potential WIMP candidates include exotic subatomic particles that are as yet undiscovered, and neutrinos with significant mass (their mass is presently thought to be very small). If they exist, high-mass neutrinos might account for most of the dark matter in the universe, because enormous numbers of neutrinos were created during the Big Bang.

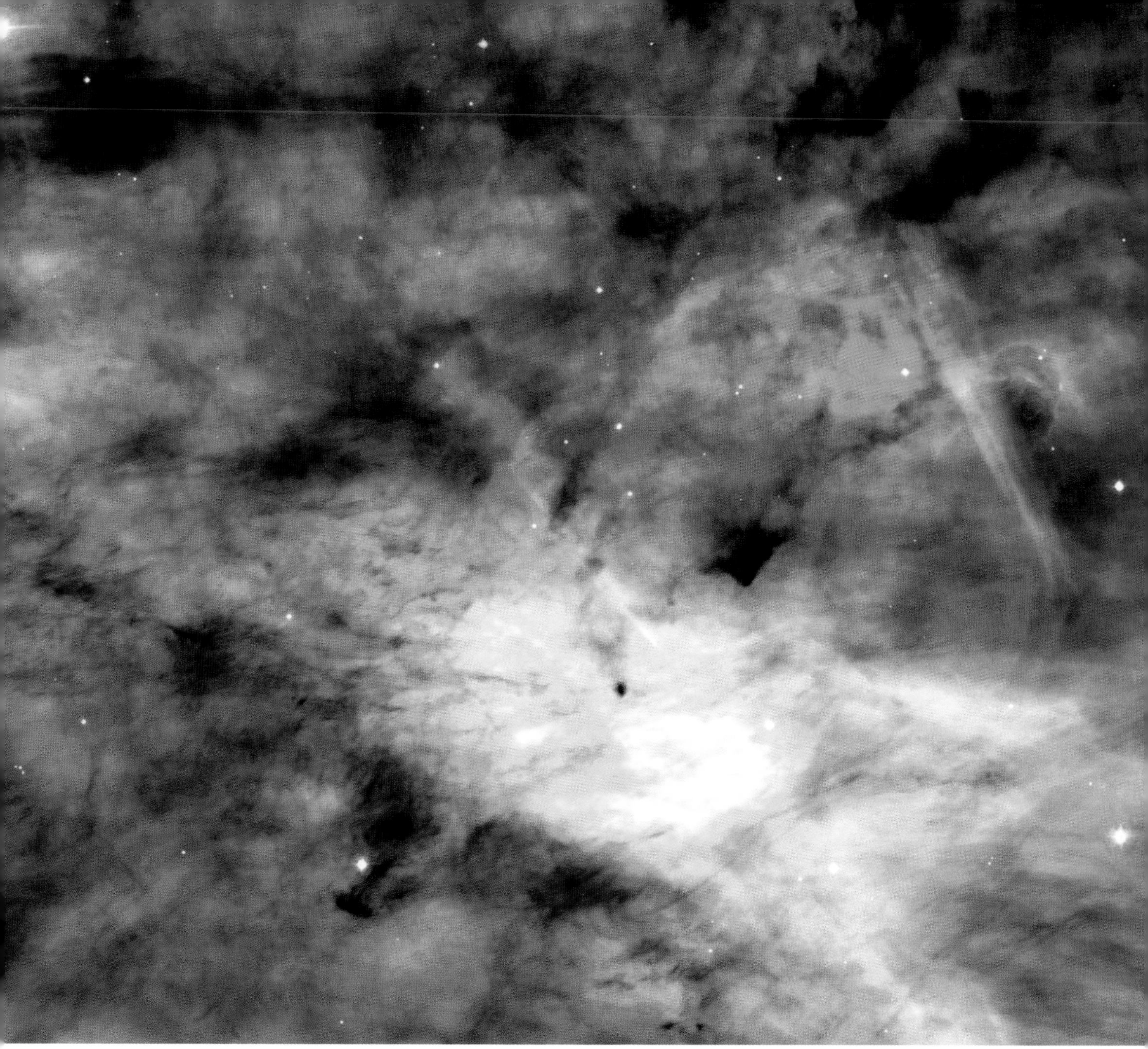

Confirmation from WMAP

Among its many important findings, WMAP gave the most precise estimate ever of the age of the universe (13.7 ± 0.2 billion years); confirmed that a mere 16 percent of the matter in the cosmos is normal baryonic material; and determined that dark matter, which makes up the remaining 84 percent of the mass of the universe, is cold. So neutrinos, which move at high speeds and prefer warm temperatures, are not the main constituent of dark matter.

WMAP also found that space everywhere is flat—confirming a prediction made by inflation theory. In addition, astronomers can use data obtained by WMAP to begin to distinguish between different versions of inflation. Some variations have been ruled out, but much work remains to be done.

One prediction of inflation theory is that there were minuscule fluctuations—at the subatomic (quantum) level—in the plasma present immediately after the birth of the universe. When inflation occurred, those tiny ripples were magnified enormously. These fluctuations ultimately became slightly denser regions of the new cosmos, which eventually developed into the galaxies and the galactic clusters that we see today.

Above: In the center of the Omega Nebula, a hotbed of newborn stars are wrapped in colorful blankets of glowing gas. Star formation continues today in many regions of the expanding universe.

Right: Extreme ultraviolet wavelengths reveal the internal structure of an active spiral galaxy. New stars blaze brightly along the spiral arms, while the central disk shows older stars packed densely together. Yet it is probable that more than half of the total mass of this galaxy is not visible in this image.

It's difficult predicting the future of the universe. Trying to forecast tomorrow's weather is a challenge, so pity cosmologists who try to peer billions of years into the future to foresee the ultimate fate of the universe. Until recently, there were a few distinct possibilities, but now everything has been thrown into disarray due to something we can't even see—dark energy.

The universe: Where to now?

In the mid-1990s, two teams of astronomers were measuring the brightness of very distant Type 1a supernovae. These stellar explosions are among the brightest events in the universe, which makes them an ideal tool for determining the distances to their remote galactic hosts.

Lights in the distance

The light from a Type 1a supernova follows a predictable path, always peaking at the same level of brightness. This lets astronomers compare the supernova's real brightness (its intrinsic luminosity) against how bright it actually appears (its apparent luminosity) in the image of the explosion. From this, the distance to the supernova can be calculated. As a side benefit, cosmologists can use Type 1a supernovae to estimate the expansion rate of the universe at different times in the past.

Amazingly, astronomers discovered that very remote Type 1a supernovae were significantly fainter than they should have been, based on their distances estimated using other techniques. This strongly suggests that these supernovae are farther away than it was originally thought, and this implies that the cosmos has expanded much more than the standard models of the universe's expansion say it should have. What that all boils down to is that the rate of expansion of the universe must be accelerating.

Above: Hubble Space Telescope image of a giant disk of cold gas and dust fueling a possible black hole at the core of galaxy NGC 4261. Estimated to be 300 light years across, the disk is tipped enough (about 60º) to provide astronomers with a clear view of its bright hub, which presumably harbors the black hole.

Dark energy makes an impression

What could be causing the universe to expand at an accelerating rate? Astronomers call this dark energy because, whatever it is, it is a repulsive energy or force and is not made of matter. WMAP data has revealed that 74 percent of the mass–energy composition of the universe is dark energy—22 percent is dark matter, and 4 percent is baryonic matter.

Recent studies have shown that matter and gravity dominated the early cosmos, causing its expansion to slow. But about nine billion years ago, dark energy began to make its presence felt, and five to six billion years ago its repulsive force overcame the force of gravity. At this point, the expansion of the universe began to accelerate.

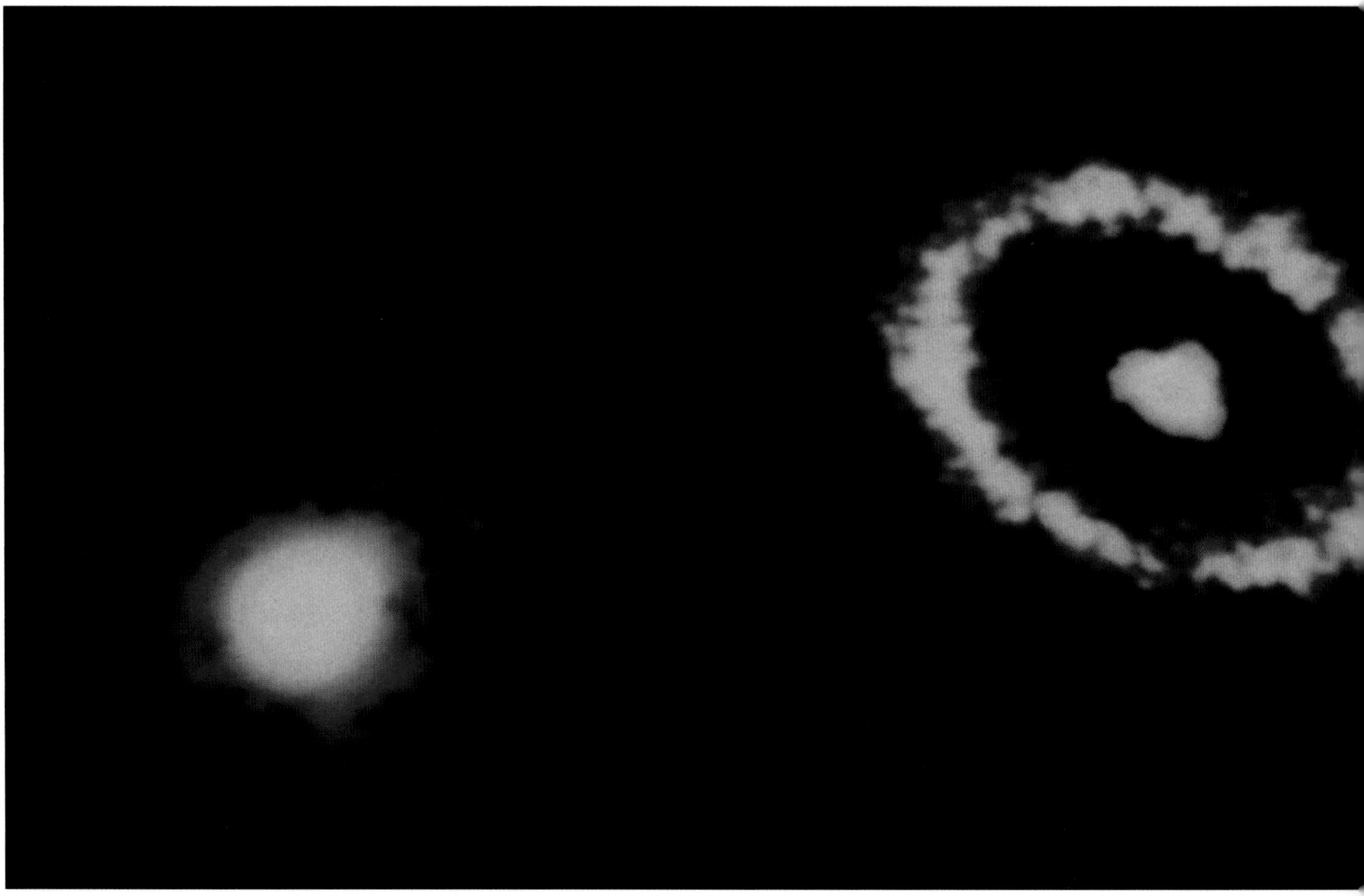

It is possible that dark energy is simply a property of space that is spread uniformly throughout the universe and it is unchanging over time. Because it is part of the fabric of space–time, cosmologists call it vacuum energy. Another possibility is that dark energy is associated with an energy field whose density can vary across time and space. In this theory it is called quintessence, and if this idea is correct there are profound implications for the future of the universe.

An uncertain future

As the universe expands, more dark energy is generated, but it is not yet clear what this means. According to quintessence, if the density of dark energy increases over time, the universe will expand more rapidly and everything—the galaxies, stars, even atoms—will eventually be torn apart in a destructive frenzy nicknamed the "Big Rip." If the dark-energy density decreases, gravity might ultimately win the battle and everything be drawn back together into a "Big Crunch."

But WMAP data suggests that the density of dark energy in any particular volume of space does not appear to be changing with time, which implies that it is an inherent property of space–time. This suggests that cosmic expansion will continue to gradually accelerate, the universe will expand forever, and our cosmos will go quietly into the night. Even as the universe inflates, its constituents will gradually age. Stars will die, and star formation will decline as interstellar gas is consumed and star-forming nebulae disappear. Galaxies will dim as their stellar lights fade. Or perhaps not. The science of cosmology is always "under construction." So in a few years we may have a much better idea of what will become of our universe.

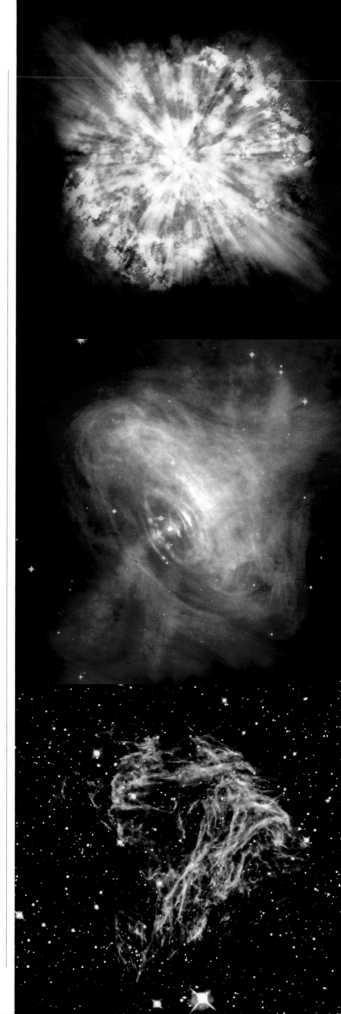

Right: An artist's interpretation illustrates the explosion of SN 2006gy, a massive star, in what scientists think is the brightest supernova ever recorded. Supernovas usually occur when massive stars exhaust their fuel and collapse under their own gravity. This star may have been 150 times larger than our own Sun.

> The progress of the human race in understanding the universe has established a small corner of order in an increasingly disordered universe.
>
> Stephen Hawking, physicist, b. 1942

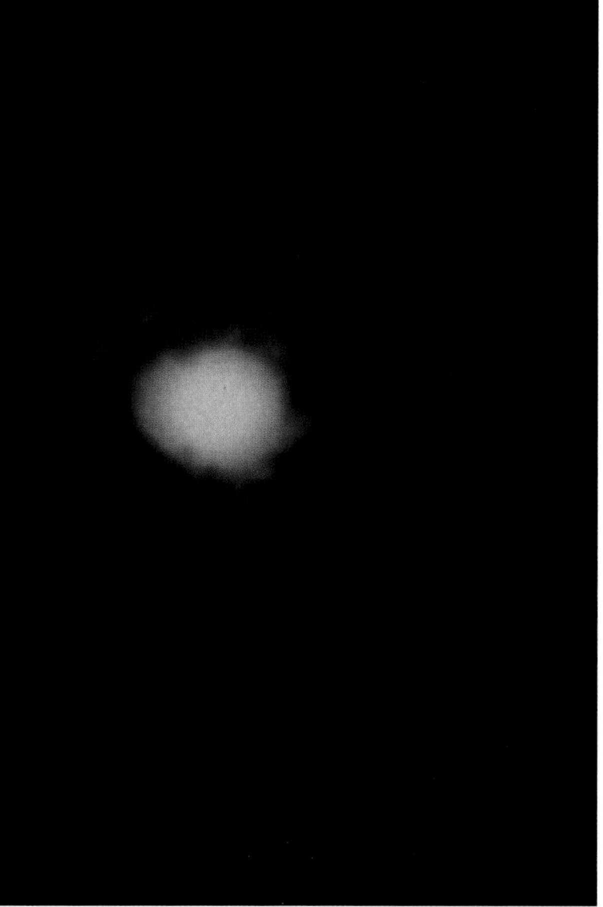

Right: A composite image of the Crab Nebula showing the X-ray (blue), and optical (red) images superimposed. The size of the X-ray image is smaller because the higher energy X-ray emitting electrons radiate away their energy more quickly than the lower energy, optically emitting electrons as they move.

Right: These tangled wisps of swirling dust are a remnant of a supernova blast. The leftover material will eventually be incorporated to form new stars in the Large Magellanic Cloud.

Left: A gaseous ring around Supernova 1987A, an exploding star 160,000 light-years from Earth. This image was taken by the Hubble Space Telescope.

Casual skywatching with the naked eye can be completely spontaneous, unlike telescopic observing, which usually requires a block of time set aside and frequently requires advance planning. Once the constellations have been learned, they become old friends reappearing at the same season every year.

Looking at the night sky

On an evening stroll your eyes turn skyward, glancing briefly at each familiar constellation. "What's that star doing there near Regulus?" you may ask yourself when you see Leo rising in the east after some months behind the Sun, but then you remember that Saturn or Mars has moved into Leo since you last saw it. Over the following months you will note the planet's slow progress through the constellation.

Telescope basics

Telescopes come in two basic types: Refractors collect light with an objective lens at the forward end of a tube, while reflectors collect light with a mirror. Telescopes are mounted on either an altazimuth mount—one that rotates about two different axes, one for altitude and one for azimuth—or an equatorial mount which has an axis parallel to Earth's axis, so that only motion around that axis is needed to follow the stars as Earth rotates beneath them.

Refractors

Simple lenses do not focus all colors at the same distance. This leads to a smearing of the image, called "chromatic aberration," and loss of contrast. The solution is to make refractor objectives from two lenses made from types of glass having different refractive indices. Standard refractors are called achromatic refractors, meaning "color-corrected." They need a long focal ratio to perform well—typically f/12 to f/15. This results in a long heavy tube requiring a sturdy mount, so achromatic refractors are not very portable beyond a 4-inch (10 cm) aperture. Although a color fringe remains—normally violet—achromats perform well on planets, double stars, and clusters.

Reflectors

Newtonian reflectors focus light with a parabolic mirror located at the lower end of the tube. A flat secondary mirror near the upper end, angled at 45°, diverts the converging cone of light to a focuser mounted outside the tube, where the image is magnified with an eyepiece. The mirrors occasionally have to be optically aligned, a procedure called collimation. Usually only a few tweaks are required to achieve optimum performance.

Newtonians are the least expensive telescope design, yet their performance is comparable with that of far more expensive telescopes.

Above: This early 20X refractor telescope was built by Galileo and is said to be the telescope with which he discovered moons orbiting Jupiter on the night of July 7, 1610.

What to look for

Galaxies, star clusters, nebulae—there's a lot to see. In the entire sky there are about 145 deep-sky objects that are true splendors. The thousands of other deep-sky objects that amateurs observe are just fuzzy glows in the eyepiece, with perhaps a few details such as a brighter core. With such objects, it is understanding the significance of what we are seeing that makes the observation worthwhile.

Through the eyepiece, you can see galaxies, which may be spirals, lenticulars, ellipticals, or irregulars, as well as open-star and globular star clusters. Most double stars are white, but the celebrated ones often have a striking color contrast. Sometimes an orange primary will make the secondary look like its complementary color, green. The brightest emission nebulae are very dramatic. Many of the swirls of nebulosity seen on photographs of the Tarantula, Orion, and Veil nebulae can be seen at the eyepiece.

> [The night sky] is the handwriting of God.
>
> Joel Primack, physicist and astronomer, 1992

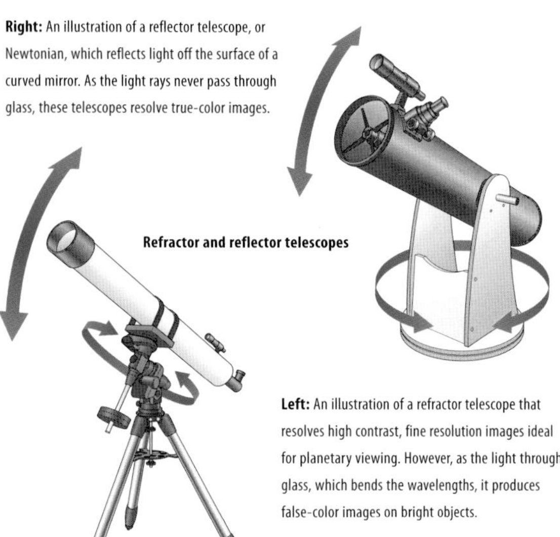

Refractor and reflector telescopes

Right: An illustration of a reflector telescope, or Newtonian, which reflects light off the surface of a curved mirror. As the light rays never pass through glass, these telescopes resolve true-color images.

Left: An illustration of a refractor telescope that resolves high contrast, fine resolution images ideal for planetary viewing. However, as the light through glass, which bends the wavelengths, it produces false-color images on bright objects.

WHICH TELESCOPE TO BUY?

All telescopes have their strong points, and all are compromises, so most hobbyists own several telescopes of different apertures. Attend a star-party or join an astronomy club and go observing with their members—either will allow you to try many different types and brands of telescope. Take the advice of knowledgeable amateur astronomers. In the meantime, most clubs rent telescopes and many have a club observatory, typically with a 14- to 24-inch (35 to 60 cm) telescope that anyone can use after some training.

New telescopes commonly include two Plossl eyepieces, which are a perfectly serviceable design. Try different eyepieces at a star-party to ensure that they work with your telescope's particular optical configuration. Premium low-power eyepieces that offer exceptionally wide fields of view, sharp to the edge, are so heavy that they cause balance problems with many telescopes, as well as being expensive.

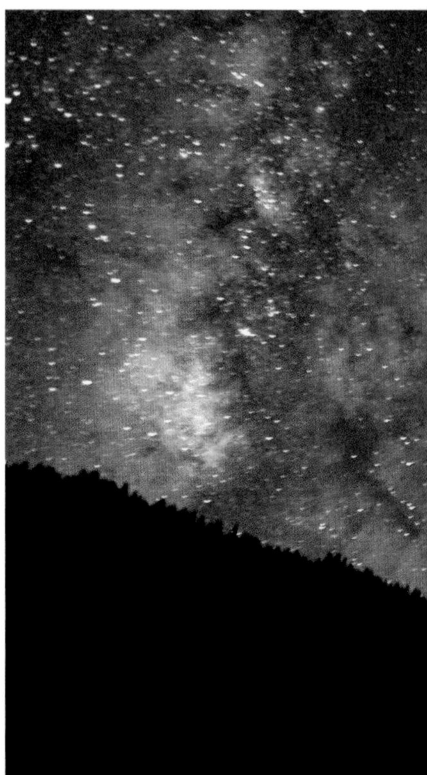

Above: Auroral colors, such as these seen above Talkeetna Mountains, Hatcher Pass, Alaska, are produced when charged solar particles strike single atoms of oxygen at altitudes of between 60 and 600 miles (96 and 960 km). This display of Northern Lights occurred during a geomagnetic storm in the early evening of November 5, 2001.

Left: Browsing the night skies for stars and constellations can bring much enjoyment to the amateur and professional alike. Only the naked eye is needed to find bright constellations such as Leo, but the binoculars will probably be uncapped as soon as you look for nearby Sextans or Crater.

The Solar System is comprised of the Sun (our nearest star), the planets, and many other smaller bodies, all held in the Sun's gravitational thrall. There are eight known major planets, three dwarf planets, and countless Small Solar System Bodies (SSSBs), such as asteroids and comets.

The Solar System

The Sun contains over 99 percent of all the mass in the Solar System, with Jupiter and Saturn making up most of the rest; the other planets and bodies comprise the tiny remainder.

Building the Solar System

The currently accepted theory for the origin and formation of the Solar System is called the "nebular hypothesis." Proposed independently by Immanuel Kant and Pierre-Simon Laplace, it says that everything in the Solar System originated in a huge interstellar cloud or nebula of gas, known as a molecular cloud.

This cloud was comprised of material ejected into space during the death throes of earlier generations of stars. It contained not only large amounts of hydrogen and helium, but also heavier elements produced by that earlier stellar generation. Slowly, local regions within this cloud began to collapse in on themselves—perhaps triggered by a shockwave from a nearby supernova (exploding star)—and began to spin. One such region became our Solar System, and is referred to as the pre-solar nebula. As it contracted and began to spin faster—held together by its own overall gravity and magnetic fields—it began to flatten into a huge spinning disk of gas, thousands of times wider than our Solar System is now. As it continued to contract, most of the material gravitated (literally!) into the center of the nebula, becoming hotter and under increasing pressure—this central portion became the proto-Sun.

As the rest of the material whirled around in the disk, gas molecules combined to form larger particles called "dust grains," and those grains stuck together to form larger and larger grains.

Over millions of years, the spinning disk of material became "clumpy." Small clumps stuck together to form larger clumps, until eventually some grew large enough to become planets.

Rocks and gas

The inner parts of the disk became too hot for ices to form or gases to liquefy, so the only materials able to remain solid were those with high melting points, i.e. rocky substances and metals. Consequently, the inner planets Mercury, Venus, Earth, and Mars, are rocky worlds. Further out temperatures were cooler, so ice could form. The gas planets, Jupiter and Saturn, became the dominant bodies in the middle reaches of the Solar System, with Uranus and Neptune further out still. Jupiter's massive bulk gave it a large gravitational influence, preventing any other rocky planet from coalescing any further in toward the Sun in the region known today as the asteroid belt.

Each of these planets has systems of rings—spectacularly so in the case of Saturn—and many moons. Some of those moons are bigger than the planet Mercury.

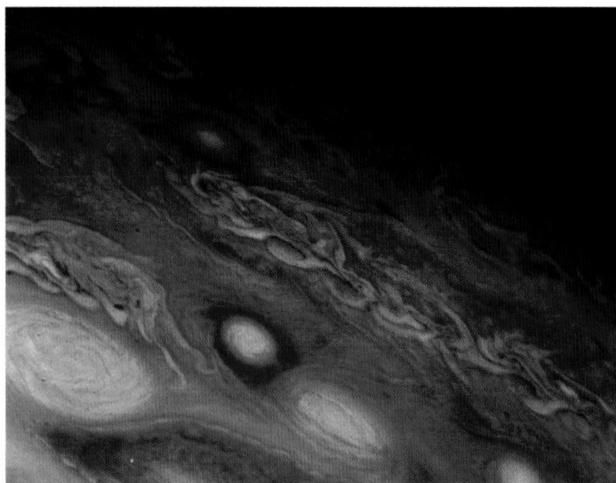

Right: One of Jupiter's best known features is the Great Red Spot. This awe-inspiring element, easily viewed through an 8-inch (20 cm) telescope, is a region of turbulence and storm activity that extends from the equator to the southern polar latitudes.

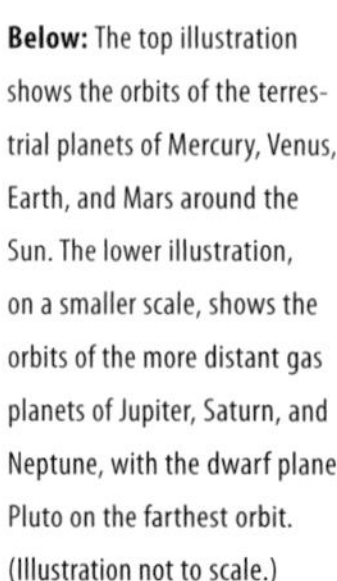

Below: The top illustration shows the orbits of the terrestrial planets of Mercury, Venus, Earth, and Mars around the Sun. The lower illustration, on a smaller scale, shows the orbits of the more distant gas planets of Jupiter, Saturn, and Neptune, with the dwarf planet Pluto on the farthest orbit. (Illustration not to scale.)

At the edge

Much further out are several different populations of smaller bodies. First is the Kuiper Belt, beyond Neptune. The leading example is Pluto, and there are dozens more Pluto-like bodies in this region. Next is what is called the "scattered disk," a region of similar icy bodies, such as Eris, that are believed to have formed closer in toward Neptune but flung into more distant orbits through gravitational encounters with that planet. Much further out is a theoretical swarm of cometary bodies that surrounds the entire Solar System, known as the Oort Cloud. It is too far away to have been detected directly, and its presence can only be inferred indirectly from the trajectories of some long-period comets.

ORBITING THE SUN

The planets orbit the Sun more or less within a plane called the ecliptic plane, an imaginary plane that joins the Sun and Earth's orbit. All the orbits are slightly elliptical. Some, such as Earth and Venus, are almost circular; Mercury, however, has a very elliptical orbit—and so does the dwarf planet Pluto.

Many SSSBs, such as asteroids and comets, have highly elliptical orbits that are also inclined at significant angles to the ecliptic plane. All the planets, and most of the other bodies, orbit the Sun in the same direction as the Sun's rotation—counterclockwise to an observer looking "down" on the Solar System from above the Sun's north pole.

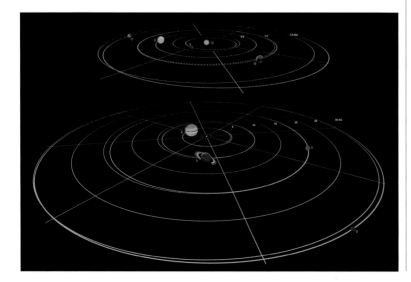

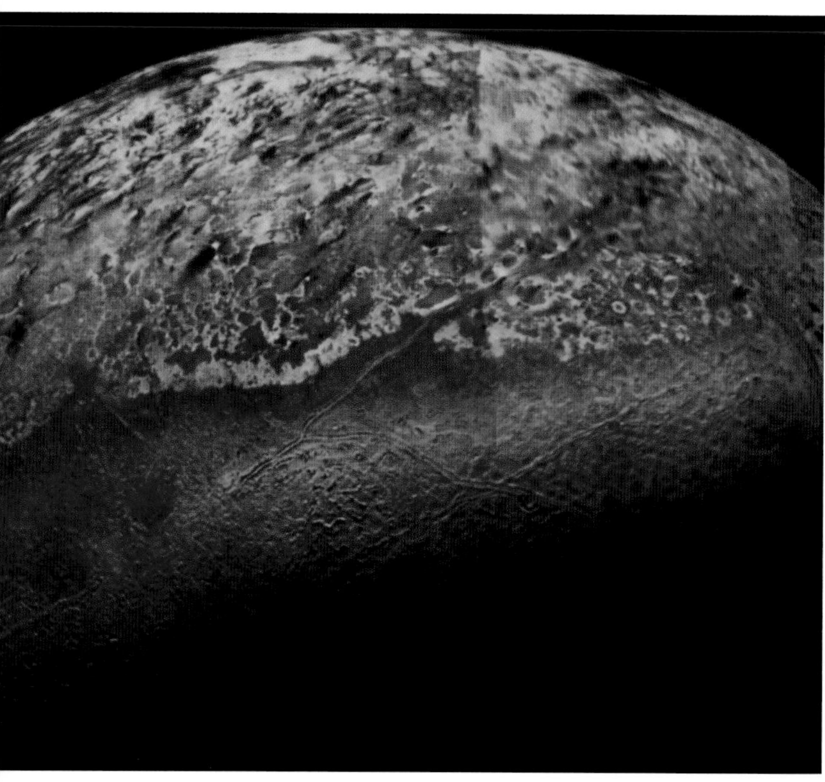

Left: This image of Neptune's moon Triton was taken by the *Voyager 2* spacecraft. Triton is the largest of Neptune's 13 moons and is notable for its icy cold surface and its retrograde orbit—it goes around Neptune the "wrong" way.

Below: The Cone Nebula taken by the Hubble Space Telescope, shows the upper 2.5 light years of the nebula, a height that equals 23 million round trips to the Moon. The entire nebula is seven light years long.

I am much occupied with the investigation of physical causes. My aim in this is to show that the celestial machine is to be likened not to a divine organism, but rather a clockwork.

Johannes Kepler, mathematician and astronomer, 1571–1630

Below: The Solar System as it is today with eight planets recognized by the International Astronomical Union: Mercury, Venus, Earth, Mars, Jupiter, Saturn, Uranus, and Neptune. Pluto has been demoted to a dwarf planet, joining Eris and Ceres.

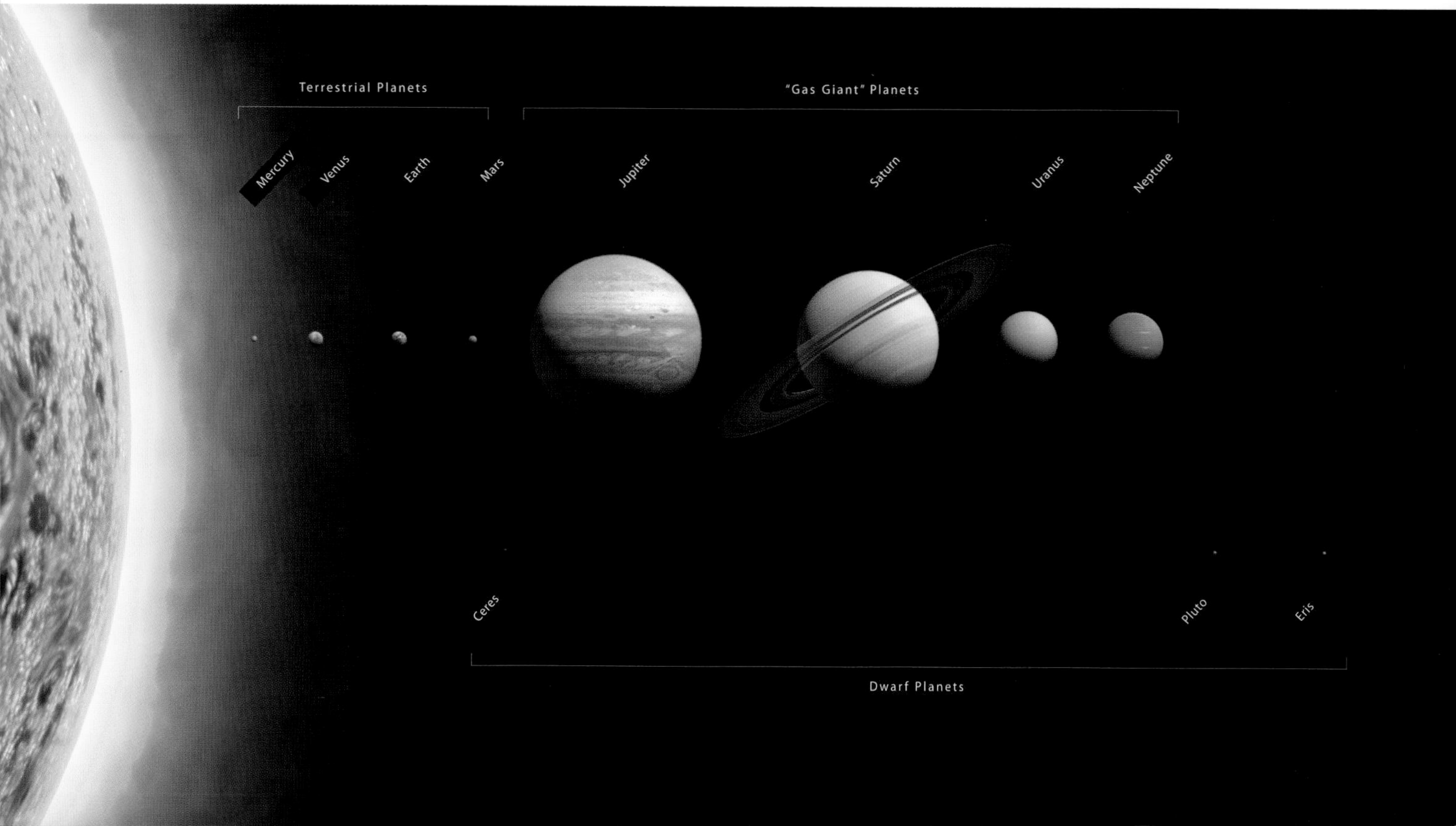

Terrestrial Planets

"Gas Giant" Planets

Mercury
Venus
Earth
Mars
Jupiter
Saturn
Uranus
Neptune

Ceres
Pluto
Eris

Dwarf Planets

The Sun is located at the center of the Solar System. All the planets, minor planets (asteroids), Kuiper Belt Objects, and most comets are gravitationally bound to it in orbits. The Sun contains 99.8 percent of all the mass in the Solar System, with the eight planets, the dwarf planets, asteroids, comets, and meteoroids making up the remaining two-tenths of a percent.

The Sun

The Sun is a massive 865 million miles (1.4 million km) in diameter; by comparison, Earth measures only 7,920 miles (12,746 km) wide. In fact, the Sun is so voluminous that you could fit Earth 1.3 million times inside, and it has about 330,000 times more mass than Earth!

Because the Sun is not a solid body, like a rocky planet, but rather is a huge globe of intensely hot gas, it does not rotate as a solid mass. It rotates faster at the equator—around 25 days for a full rotation—than it does at its polar regions where it takes 35 days. The Sun, along with its retinue of planets and other bodies, orbits the center of the galaxy, taking around 220–260 million years to complete one circuit at a speed through space of around 135 miles (217 km) per second. It, and everything else in the Solar System, is made from the material ejected during the explosive destruction of earlier generations of stars. Scientists think our star probably belongs to the third generation of stars born after the Big Bang. Although it looks big and bright in our daytime skies, the Sun is, in fact, a fairly ordinary star as stars go. Its surface temperature is a white-hot 9,932°F (5,500°C), though to our eyes it looks yellow, as part of its blue light is scattered in the atmosphere of Earth.

The Sun is a main sequence star, and this means that it is in its middle age—roughly 4.6 billion years old. Scientists estimate it has

Above: Not surprisingly, the Sun played a central role in ancient religions. Here, Nut, the Egyptian goddess of the sky and all heavenly bodies, raises the Sun. This engraving is dated at *c.* 378–341 BCE.

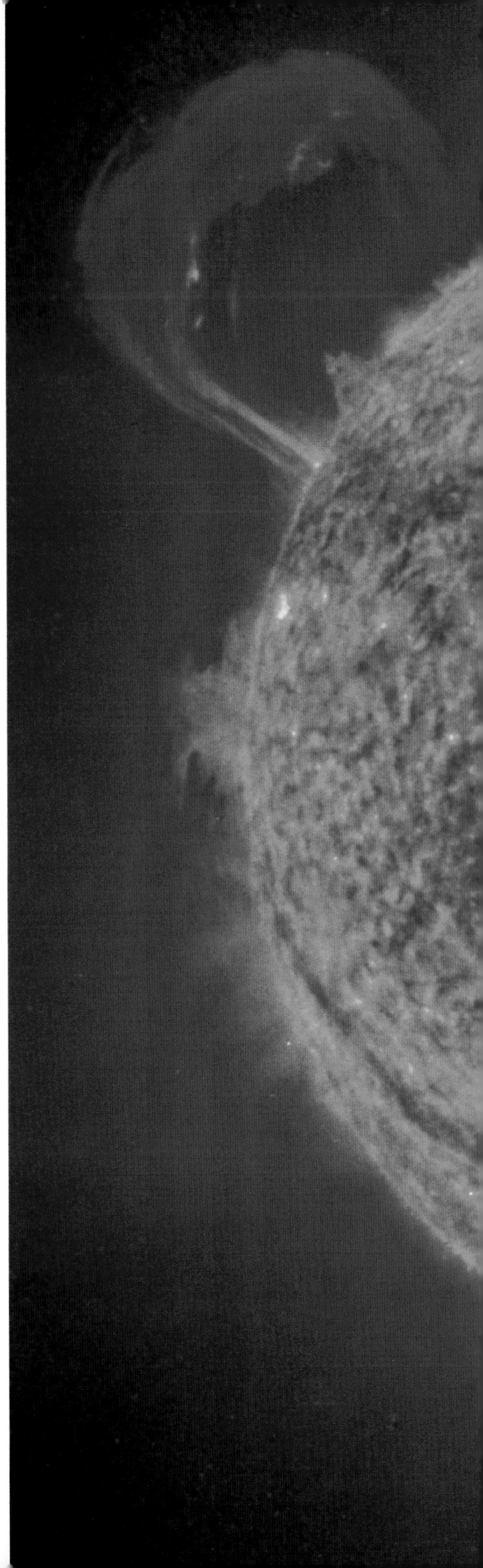

Right: A satellite view of a handle-shaped prominence on the Sun. A prominence is an eruption of gas above the Sun's visible surface (known as the photosphere) seen as bright arcs of gas at the edge of the Sun's disk.

Left: Another Egyptian deity associated with the Sun and the planets is Hathor, often shown wearing a sun disk and horns on her head. This relief depicting the Sun shining down on Hathor is from the Hypostyle Hall in Dendarah, *c.* 125BCE-60CE.

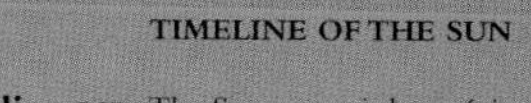

TIMELINE OF THE SUN

c. 4567 billion BCE The Sun, a star, is born (give or take a few millennia).

5000–3500 BCE The first device for indicating the time of day is built. It consists of a vertical stick casting a shadow by the Sun. The length of the shadow gives an indication of the time of day (the first sundial).

3000 BCE Known as "the cave of the sun," Newgrange, Ireland, is built. On winter solstice, the sunlight perfectly aligns with an opening in the structure to illuminate the inner chamber.

c. 2700 BCE Stonehenge, in England, is built. It is a giant circle of huge stones that are aligned to the position of the rising Sun at summer solstice.

1223 BCE The oldest eclipse is recorded on a clay tablet uncovered in the ancient city of Ugarit (in present day Syria).

c. 200 BCE Aristarchus of Samos, a Greek mathematician and astronomer, announces a theory of a Sun-centered universe. He also attempts to mathematically calculate the sizes and distances of the Sun and Moon.

965–1039 CE Muslim scholar, Abu Ali Al-Hasan, invents the camera obscura and becomes the first known person to use a device to observe the Sun.

1543 Copernicus publishes his theory that Earth travels around the Sun. This contradicts Church teachings.

1610 Galileo Galilei describes spots on the Sun that he views with his early telescope.

c. 1660 Isaac Newton shows that sunlight can be divided into separate chromatic components via refraction through a glass prism.

1687 Isaac Newton publishes *Principia Mathematica*, establishing the theory of gravitation and laws of motion. This allows astronomers to understand the interacting forces among the Sun, the planets, and their moons.

1800 William Herschel extends Newton's experiment by demonstrating that invisible "rays" exist beyond the red end of the solar spectrum.

1814 Joseph von Fraunhofer builds the first accurate spectrometer, and uses it to study the spectrum of the Sun's light.

1843 German amateur astronomer, Heinrich Schwabe, who has studied the Sun for 17 years, announces his discovery that the number and positions of sunspots vary over an 11-year period.

1845 First solar photograph is made on April 2.

1860 The total solar eclipse of July 18, 1860, is probably the most thoroughly observed eclipse up to this time.

1868 During an eclipse, astronomers observe a new bright emission line in the spectrum of the Sun's atmosphere. As a result of observations, British astronomer, Norman Lockyer, identifies and names helium.

1908 American astronomer, George Ellery Hale, shows that sunspots contain magnetic fields that are thousands of times stronger than Earth's magnetic field.

1938 German physicist Hans Bethe and American Charles Critchfield show how a sequence of nuclear reactions called the proton-proton chain make the Sun shine.

1982 *Helios 1*, a joint German and US deep space mission, sends back the last of its data indicating the presence of 15 times more micrometeorites close to the Sun than near Earth.

1990 *Ulysses*, an interplanetary spacecraft, is launched with the mission to measure the solar wind and magnetic field over the Sun's poles during periods of both high and low solar activity.

1991 Launch of the YOHKOH spacecraft, to photograph the Sun in X-ray emission over a full solar cycle (11 years).

1995 The Solar and Heliospheric Observatory (SOHO), reaches a point where the Sun's gravitational pull balances Earth's. The satellite orbits the Sun with Earth studying the Sun from its core to the outer corona, and the solar wind.

2006 NASA's two Solar TErrestrial RElations Observatory (STEREO) satellites take the first three-dimensional images of the Sun.

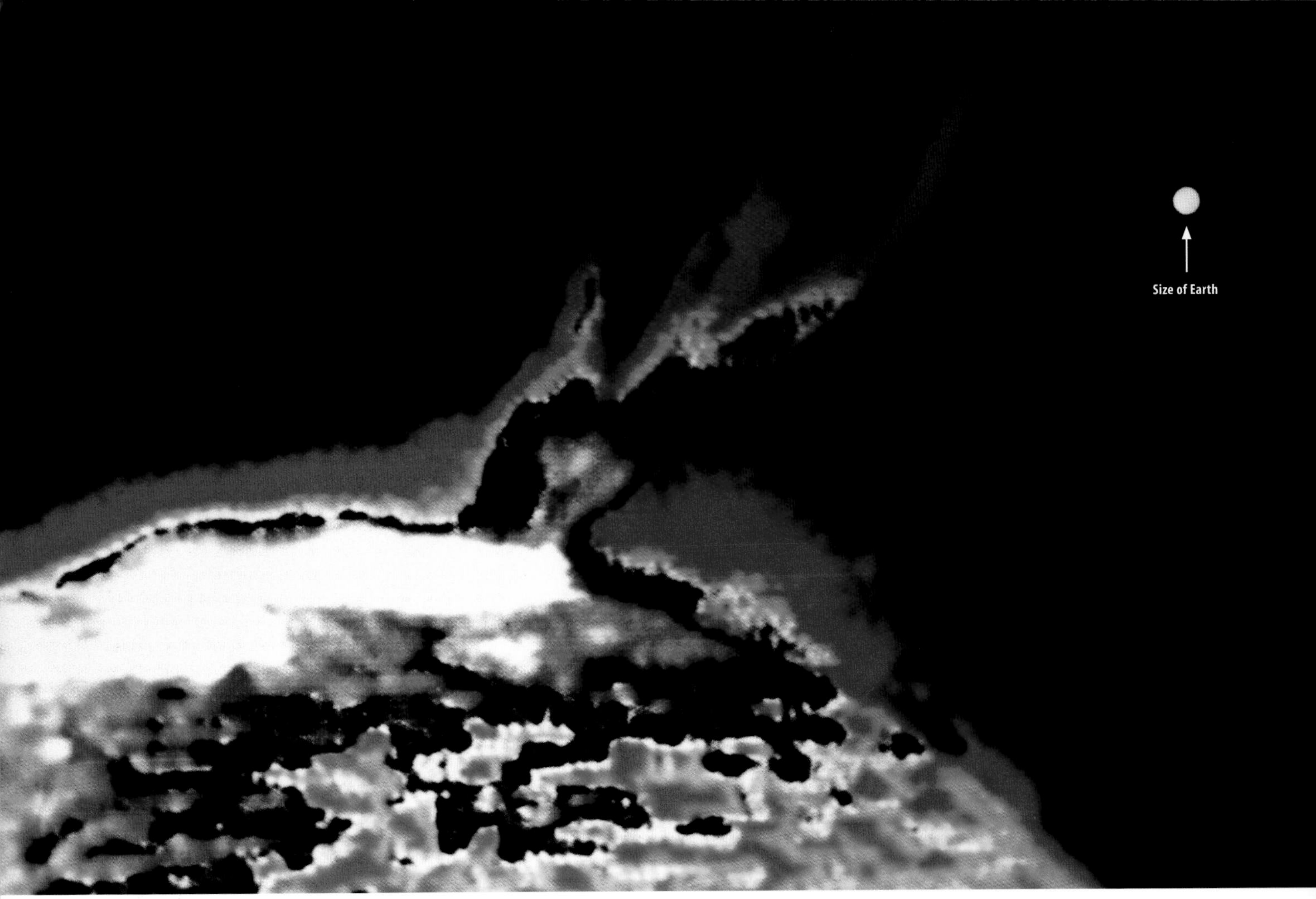

Size of Earth

Above: Occasionally there are violent explosions on the Sun, for example, solar flares and coronal mass ejections (CMEs). If Earth happens to get in the way, the effects can range from disruption to radio communications to the formation of the Northern and Southern Lights.

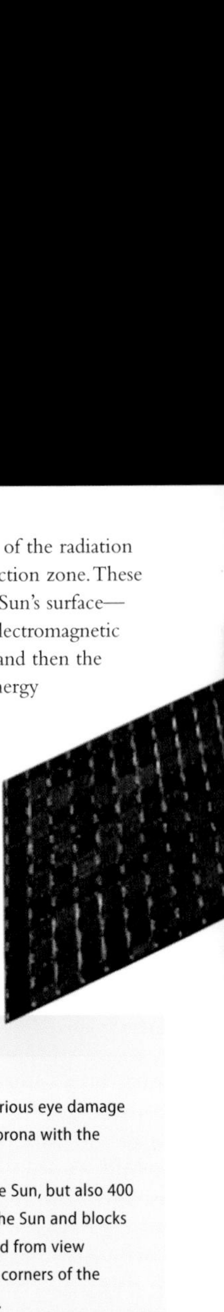

probably another five billion years of life before it peters out and becomes a white dwarf star—a dying stellar ember.

A fiery mass

The Sun is composed mainly of hydrogen (74 percent by mass) and helium (25 percent), with small amounts of some other elements. Its internal structure is broken into three zones: The core, the radiation zone, and the convection zone. In the core, the pressure is so great that hydrogen atoms get squashed together or fused to form helium, releasing prodigious amounts of energy in the process. The temperature in the core is a staggering 56.3 million°F (13.5 million°C), and every single second 4.4 million tons (4 million tonnes) of matter is converted into energy.

The radiation zone extends about two-thirds of the way out from the core. The energy released in the core's fusion reactions fights its way through the thick radiation zone, being continually absorbed and re-emitted. Once the energy has reached the edge of the radiation zone, it heats the gases at the bottom of the convection zone. These gases form into huge currents that flow up to the Sun's surface—where the energy is finally released into space as electromagnetic radiation, for example, light, infrared, ultraviolet—and then the currents sink back down again to pick up more energy from the top of the radiation zone, and start the whole process over again. The journey of energy from fusion in the core to emission at the surface can take up to millions of years.

> We need not hesitate to admit that the Sun is richly stored with inhabitants.
>
> William Herschel, astronomer, 1738–1822

ECLIPSES

While it is extremely unwise for people to look directly at the Sun, as serious eye damage or blindness can occur, there is one time when it is possible to see the corona with the naked eye—during a total solar eclipse (at left).

It just so happens that the Moon is about 400 times smaller than the Sun, but also 400 times closer. As the Moon orbits Earth, it sometimes gets in the way of the Sun and blocks its light, causing an eclipse. When this occurs, the photosphere is blocked from view and the beautiful, wispy corona stands out. Enthusiasts travel to remote corners of the world just to witness a few precious seconds of this incredible spectacle.

Total solar eclipses are of special interest to astronomers because it is the only time the Sun's corona can be seen from Earth's surface. Properties of the Sun's outer atmosphere, such as temperature, density, and chemical composition, can be observed and measured when the light of the disk is completely blocked by the Moon.

The visible surface of the Sun is the photosphere. The surface has a speckled appearance, known as granulation. These "granules" are actually the tops of the convection currents.

Extending out from the photosphere is the Sun's atmosphere, and it, too, is divided into different regions. The first region is the chromosphere. It is about 1,250 miles (2,000 km) high, and temperatures in the gas here can reach up to 180,000°F (100,000°C). Above this is a transition region, where temperatures steadily climb to around 1.8 million°F (1 million°C). Finally, the outer region or corona is reached, where temperatures peak at up to several million degrees more. The corona extends a long way out from the Sun, and can be seen from Earth during total solar eclipses. Why the temperature increases so markedly from the Sun's surface out to the corona is not fully understood, but it is tied in with the strong solar magnetic field. As well as electromagnetic radiation, particles are also emitted from the Sun's surface. The bulk of this is known as the solar wind, which "blows" through the Solar System way out to beyond Pluto. In fact, the Sun blows a huge "bubble" in space, known as the heliosphere.

Sunspots

The Sun has an extremely strong magnetic field that reaches far out beyond Pluto, gradually weakening the further out it goes (although the sparse gas in the almost-vacuum between the planets actually helps to maintain the solar magnetic field at higher strengths than it otherwise would).

Being largely made of plasma, it creates its own magnetic field. Its rotation causes the field to become twisted and distorted. This leads to phenomena like sunspots and prominences. Sunspots are small regions of cooler gas on the Sun's surface. They reach temperatures around 7,000–8,000°F (4,000–4,500°C)—it's only the comparison with the hotter adjacent gas that makes them look dark. Sunspots are tied in with complexities in the Sun's magnetic field and it is thought that the field reduces the amount of convection from the convection zone, which lowers the energy reaching the surface in the local region of the sunspot.

Sunspot numbers rise and fall in an 11-year cycle that has a definite minimum and maximum. At the time of minimum, few sunspots are seen, and those that are present are usually found at high solar latitudes (i.e. toward the poles). As the sunspot cycle moves toward maximum, the number of spots increases, and they appear closer and closer to the Sun's equator. During the period of each sunspot cycle, the Sun's magnetic field completely reverses, so after 22 years the magnetic field returns to its original orientation.

From time to time the Sun lets off violent explosions, such as solar flares and coronal mass ejections. Both types of explosion involve huge amounts of energy and can send streams of particle heading out into the Solar System.

Scanning the Sun

The human eye is too sensitive to study the Sun directly. Telescopes and film or electronic detectors can be used, but ground-based facilities have the disadvantage of having to endure the vagaries of the weather, not to mention night time. But observatories positioned in space can be set to watch the Sun essentially uninterrupted for years at a time. And indeed, a whole flotilla of scientific spacecraft is doing just that.Perhaps the most successful of these is the joint European-USA craft known as SOHO (Solar and Heliospheric Observatory). Launched in 1995, it was intended to operate for only a few years. However, it has proved far more resilient than its makers had hoped for, and is still going strong in 2008. From its vantage point 1 million miles (1.5 million km) sunward of Earth—where the Sun and Earth's gravity balance out—it has kept its electronic eyes trained on the solar disk almost continuously during this time.

Other spacecraft include the unique *Ulysses* probe, which was put on a trajectory that has taken it out of the ecliptic and into an orbit that takes it over the poles of the Sun. And launched in 2006 was STEREO, twin spacecraft that circle the Sun inside and outside Earth's orbit, providing views which, when combined, give a unique stereo perspective on the Sun's activity.

Left: An illustration of the Solar and Heliospheric Observatory satellite, SOHO, co-sponsored by the European Space Agency and NASA. Since 1995 it has been observing the Sun almost continuously.

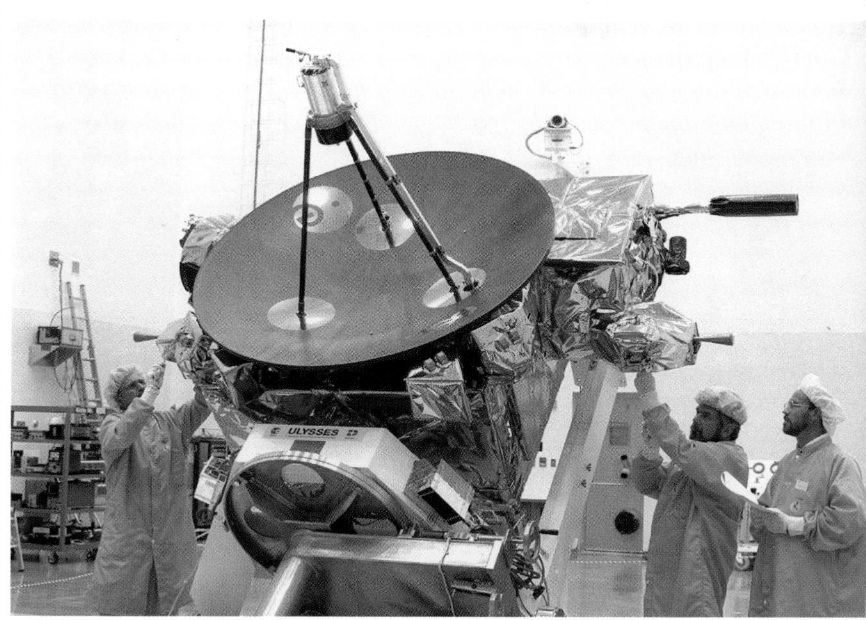

Below: Technicians in Hangar AO at Cape Canaveral Air Force Station conduct preflight checkout and testing of the *Ulysses* spacecraft. *Ulysses* is a joint NASA/European Space Agency project.

It has been known since antiquity that a handful of stars slowly move. These are the planets; the name is from a Greek word meaning "wanderers." Now we know them for what they really are—other worlds orbiting the Sun, each with its own unique characteristics and mysteries.

The planets

What is a planet?

A planet is any body that is in an independent orbit around the Sun; is big enough to have formed itself into a spherical shape; and which has cleaned up, collected in, or swept away, any competing bodies in its orbital zone. Under this new official definition the Solar System has only eight planets, from Mercury to Neptune.

Pluto, discovered in 1930, has now been placed into the newly created category of dwarf planet. This category has basically the same definition as a full planet, but with two important exceptions: That the object has not cleared its orbital zone of competing bodies, and it is not in itself a satellite or moon of another body. By this definition the Solar System has three dwarf planets: Pluto, Ceres (the largest body in the asteroid belt), and Eris.

Mercury

Although Mercury is the closest planet to the Sun, it is not the hottest. Some parts of Mercury get very hot, with a maximum in the equatorial regions of around 554°F (290°C). By contrast, in the deep shadows of craters near the poles, the temperature is a frigid −292°F (−180°C), giving the planet the greatest temperature range in the Solar System. This range is mainly due to the lack of an atmosphere, as atmospheres work to smooth out heat distribution.

Mercury's orbit is quite elliptical. The closest it comes to the Sun is around 28.5 million miles (46 million km), while the farthest point of its orbit is 43.3 million miles (69.8 million km). Mercury rotates three times for every two orbits around the Sun. Its slow rotation on its axis —one Mercury "day"—takes 58.7 Earth days, while it takes 88 Earth days to complete one orbit—one Mercury year.

Mercury is one of the four "terrestrial planets," a category that includes Venus, Earth, and Mars. Mercury is thought to be composed of approximately 30 percent rocky material (in its crust and mantle), and approximately 70 percent metallic material (in the form of a very large iron core).

Bottom: Mercury does not possess an atmosphere, but a thin exosphere made up of atoms blasted off its surface by solar winds and striking micrometeoroids. Its thin exosphere means that there is no wind erosion of the planet's surface, and meteoroids do not burn up due to friction.

Right: A photograph taken by the *Mariner 10* spacecraft showing a close-up of the cratered surface of the planet Mercury. Mercury's surface resembles that of Earth's Moon, scarred by impact craters created by collisions with comets and meteoroids.

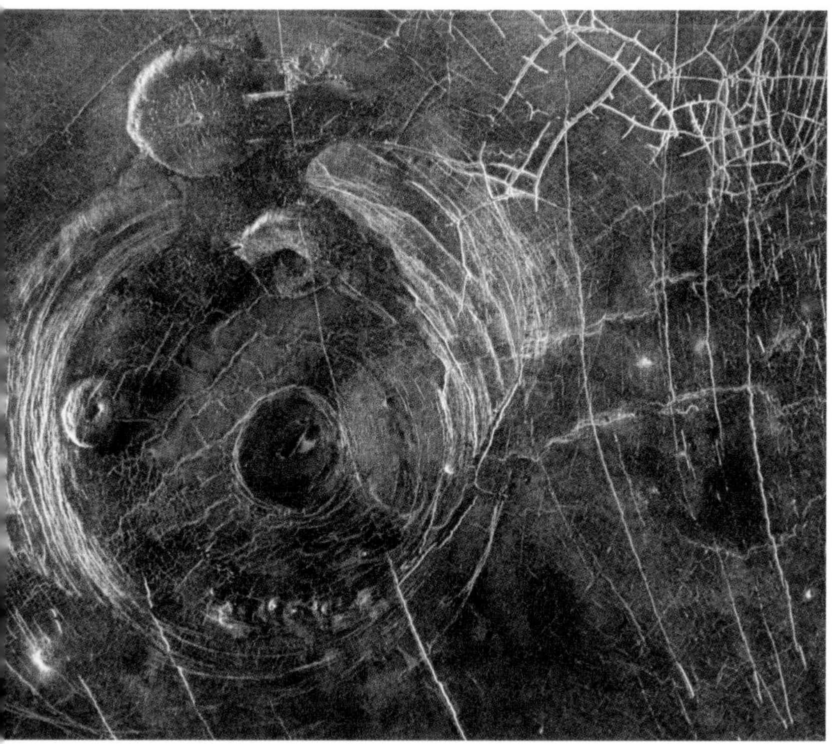

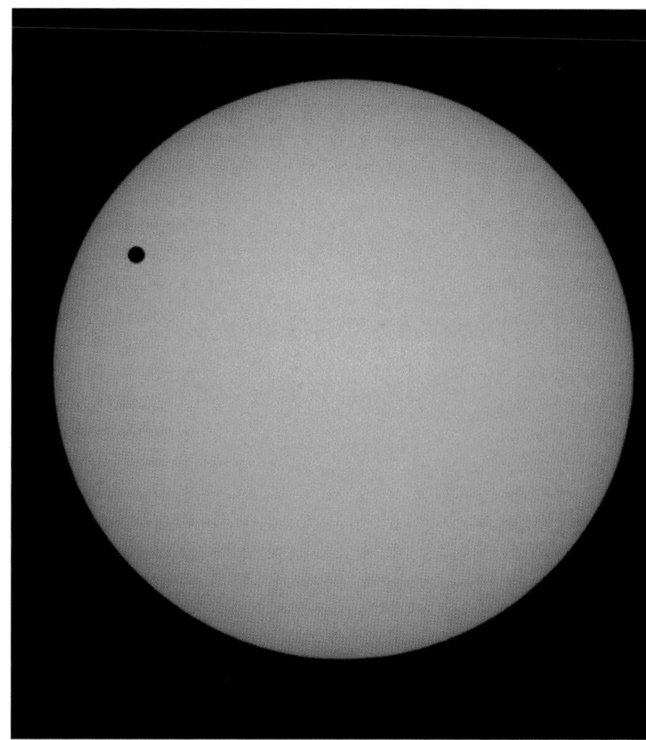

Venus

Venus is a fascinating planet. For millennia it was known only as a bright light in the sky—third brightest, in fact, after the Sun and the Moon. Even with the advent of the telescope little more was learned about the planet, for it quickly became apparent that Venus is completely covered by clouds that prevent the surface from being seen. It wasn't until space probes journeyed to Venus in the last decades of the twentieth century that we began to learn more about this planet.

Venus, the brightest planet in the sky, has been known for as long as people have admired the night sky. Being an inner planet (like Mercury), however, set it apart from the known outer planets, Jupiter and Saturn, as Venus never gets very far from the Sun in the sky—it is visible for up to a few hours before dawn in the east or for up to a few hours after sunset in the west. This led to it being called the "morning star" and "evening star," since at first it was not realized that these two "stars" were the same object. It was Pythagoras who first suggested they were one and the same. Venus is so bright that, if you know exactly where to look, it can even be seen with the naked eye in the daylight sky.

Sometimes Venus comes between Earth and the Sun, in an event known as a transit. These come in pairs, eight years apart, every 243 years. The last transit of Venus was in June 2004; the next one will occur in June 2012.

Venus has the most circular orbit of all the planets, having an eccentricity of just slightly less than one percent. (For comparison, Earth's orbit has an eccentricity of 1.7 percent, Jupiter has 4.8 percent, and Mercury a whopping 20.5 percent.) Venus orbits the Sun at an average distance of around 67 million miles (108 million km), which means it is the planet that comes closest to Earth, around 25 million miles (40 million km).

Venus has an incredibly thick atmosphere, composed mainly of carbon dioxide with traces of other gases. The pressure at the surface is a crushing 90 times the surface pressure on Earth—early Soviet spacecraft were squashed by the atmosphere even before they reached the ground. Carbon dioxide is a potent "greenhouse" gas; Venus has so much of it that the planet has a runaway greenhouse effect, leading to surface temperatures over 750°F (400°C)—hot enough to melt lead. Venus's surface is hotter than that of Mercury, even though Mercury is much closer to the Sun.

> So I saw many planets, and they looked just a little bit brighter than they do from Earth.
>
> Sally Ride, astronaut, b. 1951

Top left: A radar image of Venus, taken from the *Magellan* spacecraft, showing evidence of volcanic activity. The *Magellan* probe was launched in 1989 and for the next four years, it scanned long strips of Venus's terrain. It also mapped the planet's gravitational field.

Top right: The planet Venus appears as a black spot against the Sun in a rare astronomical event that was visible in Hong Kong in June 2004. Venus last passed between Earth and the Sun in December 1882.

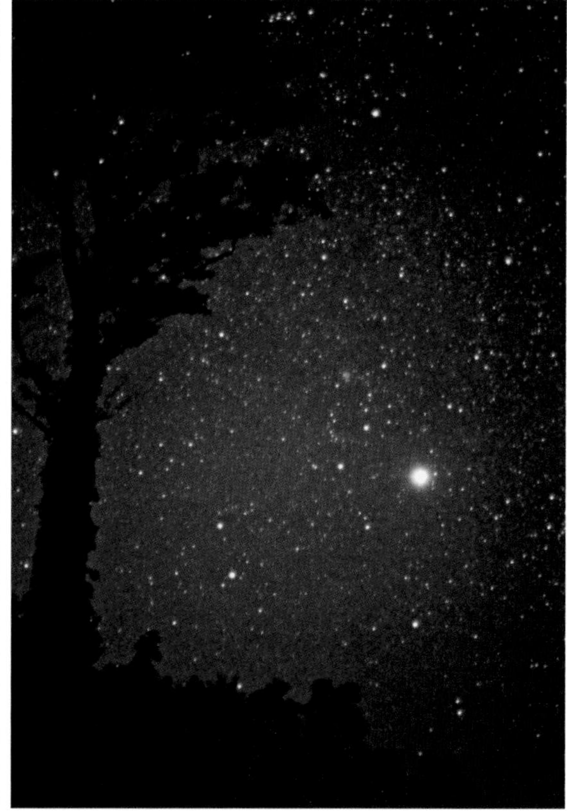

Right: The planet Venus is the brightest star in the night sky. Venus is believed to have a similar internal structure to Earth, with a core, mantle, and crust. The mantle is about 1,860 miles (3,000 km) thick, above which the crust extends around 6–18 miles (10–30 km) to the surface.

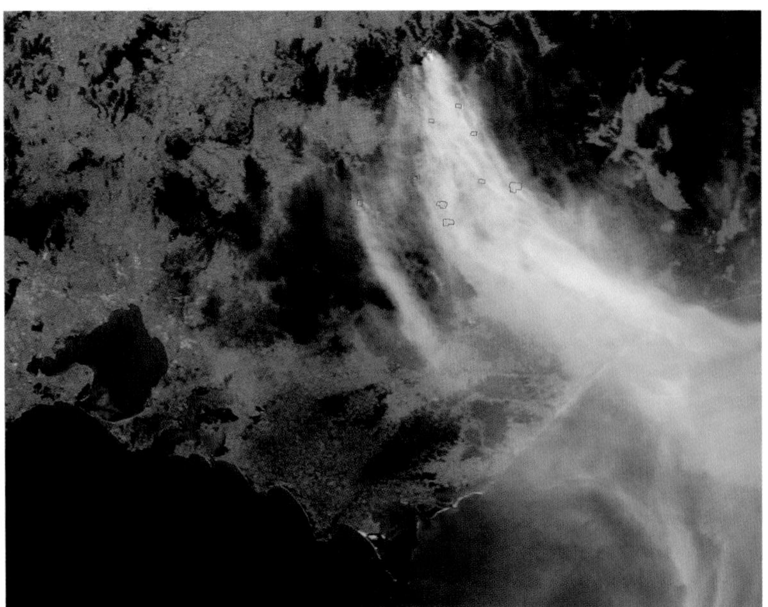

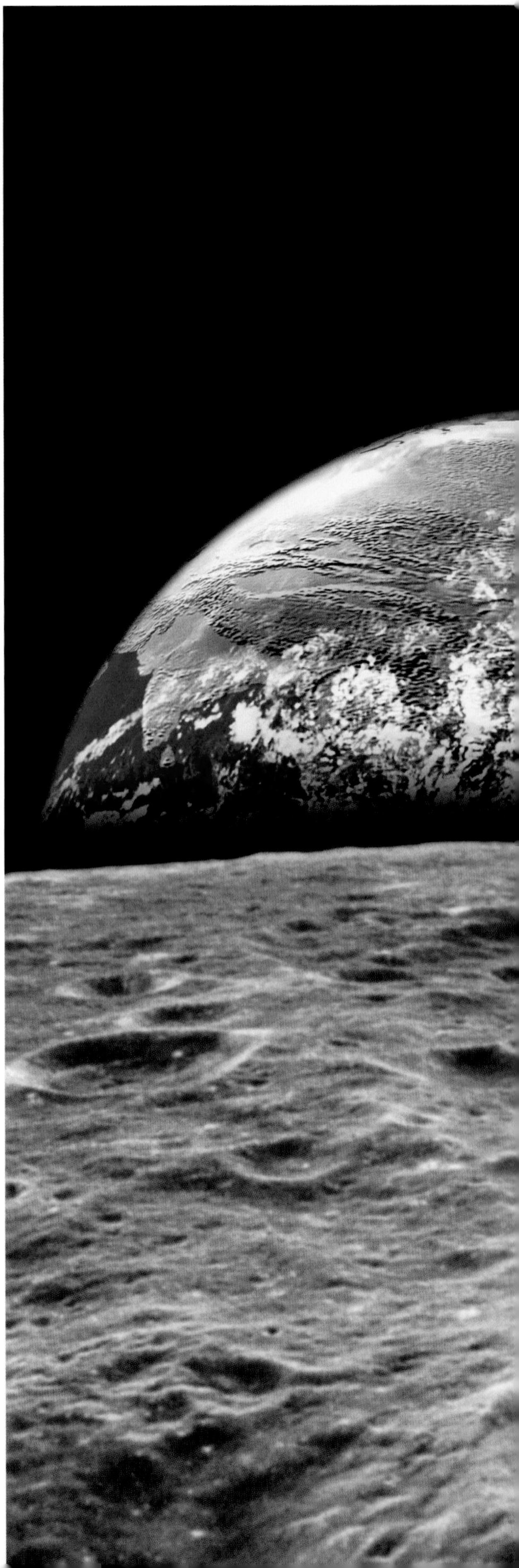

Earth

Despite being just one planet of many, Earth is exceptional. It is the only one known to support life…life that has evolved to a point where it can examine itself and the cosmos around it.

The Earth is about 4.57 billion years old. It goes around the Sun in an almost circular orbit at an average distance of 93 million miles (150 million km). One complete orbit is one year. One rotation of Earth so that the Sun returns to the meridian—an imaginary line in the sky joining North and South—is 24 hours. (This is almost four minutes more than it actually takes Earth to spin once on its axis. The difference comes because Earth is moving through space, and it takes a smidge more than one full rotation in order for the Sun to once again be on the meridian.) One full orbit around the Sun takes 365.26 days, so we need to add an extra day to the calendar every four years to keep it balanced.

Earth's axis of rotation is tilted with respect to the ecliptic by an angle of 23.4°, and it is this tilt that gives us our seasons. The spinning mass of Earth acts as a giant gyroscope, keeping its axis fixed in space. So when Earth is on one side of the Sun, its Northern hemisphere will be facing the Sun, while the Southern hemisphere will not—this is summer in the North and winter in the South. Six

Above: A satellite image from NASA taken on December 5, 2006 shows smoke from bushfires in southeastern Australia. There are many observation satellites currently in orbit, monitoring everything from weather to ocean currents to land-use patterns, to name just a few.

Right: A spectacular image of Earth taken from the Moon. Earth is the densest planet in the Solar System. Its core is mostly made up of nickel and iron, which is covered with a mantle that consists of rock containing silicon, iron, magnesium, aluminum, oxygen, and other minerals.

LUNAR ECLIPSES

The Moon is 400 times smaller than the Sun, but coincidentally 400 times closer, so at those times when the Moon's orbit makes it cross the line joining the Sun and Earth, it blocks out the Sun and we get a total solar eclipse. If the alignment is not exact, we get a partial solar eclipse. If the alignment occurs when the Moon is at the furthest point in its orbit, it won't cover the Sun completely; it will leave a thin ring of sunlight around it, which is called an annular solar eclipse.

If the point of crossing is in line with the Sun and Earth, but on the side of Earth opposite to the Sun, the Moon will pass through Earth's shadow and we will see a lunar eclipse (left). And like partial solar eclipses, if the alignment is not exact, the Moon will only partly pass through the darkest part of the shadow and we will see a partial lunar eclipse.

Right: A full moon is a wondrous sight. The Moon has become tidally locked with Earth, giving it a synchronous rotation—it rotates on its own axis in the same time it takes to orbit Earth, which means that it keeps the same face, the lunar "nearside" at all times.

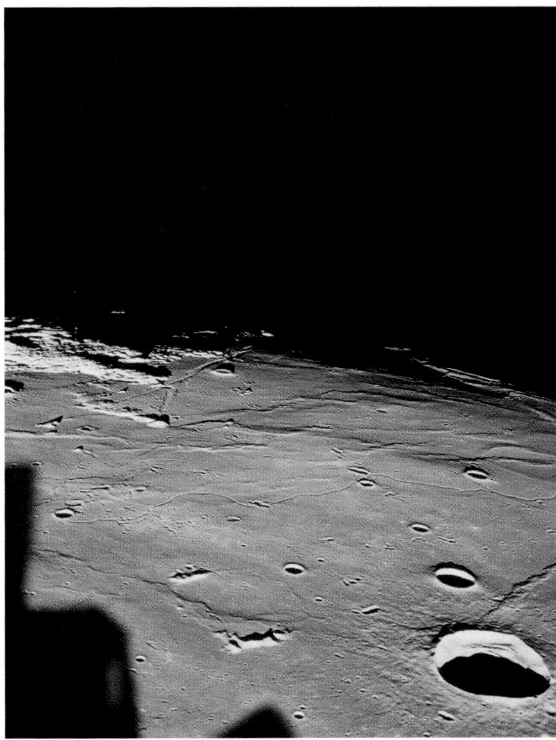

Right: The dominant features of the Moon's surface are, of course, the craters. Almost all of them are the result of asteroid, meteoroid, or comet impacts; only a few have a volcanic origin.

months later, the Earth's Northern hemisphere will be facing away from the Sun, while the Southern hemisphere now faces toward it—winter in the North and summer in the South. The autumn and spring seasons are mid-way between these two extremes. If Earth didn't have an axial tilt—if the rotation axis were straight up and down—we would not have seasons.

The Moon

The Moon is Earth's only permanent natural satellite. At 2,158 miles (3,474 km) in diameter, it is a little over one quarter the width of Earth. It's only one tenth the mass of Earth, and its surface gravity is about one sixth that of Earth's. Although these numbers might make it appear a great deal smaller than Earth, it is in fact rather large as natural satellites go. The moons of the other planets are much smaller than their parent planets—with the notable exception of Pluto, and its moon Charon.

The Moon circles Earth in an elliptical orbit that brings it as close as 225,622 miles (363,104 km) and takes it as far away as 252,087 miles (405,696 km). One lunar sidereal orbit (i.e. one based on the background stars) takes 27.3 Earth days. But because Earth is moving along in its own orbit during this time, the cycle of lunar phases (which depends on the changing Sun–Earth–Moon angle) actually takes 29.5 days.

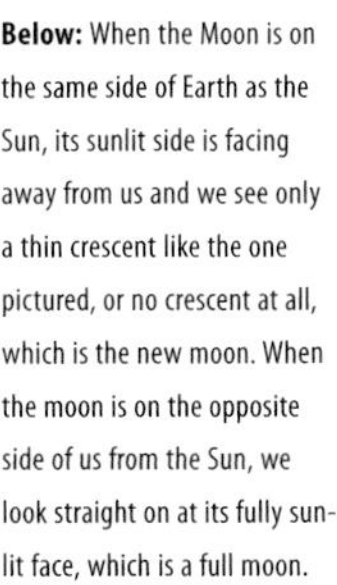

Below: When the Moon is on the same side of Earth as the Sun, its sunlit side is facing away from us and we see only a thin crescent like the one pictured, or no crescent at all, which is the new moon. When the moon is on the opposite side of us from the Sun, we look straight on at its fully sun-lit face, which is a full moon.

Mars

Mars, the small red planet half the width of Earth, takes 687 days to complete an orbit of the Sun. It has a desert-like surface, craters, and the largest volcano in the Solar System.

Mars is normally so far away that even through a large telescope it appears just a tiny featureless red dot. Fortunately, every two years or so, there is an opportunity to observe the planet close up when it is relatively close to Earth at opposition. This occurs when Earth, in its yearly circuit around the Sun, catches up with slower-moving Mars, placing it exactly on the opposite side of Earth to the Sun. As Mars has a somewhat elongated path around the Sun, on some of those occasions it is closer to Earth and brighter, and hence the opposition is regarded as more favorable. A most favorable opposition took place in August 2003, while the next similar one will be in July 2018.

Mars has a surface that is generally desert-like. Because of past impacts, numerous craters of various sizes dot the planet's surface. The Southern hemisphere is much more heavily cratered than the Northern. Olympus Mons is the largest volcano in the Solar System, rising 15 miles (24 km) above the surrounding plain. The atmosphere of Mars is so thin that average surface pressure is less than one-hundredth of that on Earth. The atmospheric composition is mainly carbon dioxide with small amounts of oxygen. Strong winds can cause major dust storms covering the whole planet. The planet's path around the Sun is somewhat elongated, so surface temperatures can vary from –207°F (–133°C) at the poles during winter, to a pleasant 81°F (27°C) on a summer's day.

Above: This image of Mars was taken by the Hubble Space Telescope. The planet's surface is generally desert-like with everything covered by red dust, which gives Mars its alternative name, the red planet.

Left: This image, taken by the High Resolution Stereo Camera (HRSC) onboard the *Mars Express* probe in January 2004, shows a channel, Reull Vallis, that was formed by flowing water when Mars was capable of sustaining liquid water on its surface.

Right: All features in this image taken by the Hubble Space Telescope are cloud formations in the atmosphere of Jupiter. They contain small crystals of frozen ammonia and traces of colorful chemical compounds of carbon, sulfur, and phosphorus.

Left: The complex layered nature of these Martian rocks record a detailed history of the physical properties that formed them. The pattern of layering indicates a history of alternating erosion and deposition that was occurring when these layers were being deposited.

Jupiter

With an equatorial diameter of 88,846 miles (142,984 km)—eleven times wider than Earth—Jupiter is, after the Sun, the most dominant body in the Solar System.

It's also one of the best-known sights in the sky for stargazers and amateur astronomers, being bright and prominent. Only the Sun, the Moon, Venus, (and sometimes Mars), are brighter. Through a telescope, at least four of its moons can be seen as tiny pinpricks of light, and the banding of its cloud patterns is easily visible. Its prominence in the night sky means that it has been known to mankind since antiquity. It is named for the chief god of Roman mythology.

Jupiter is a "gas giant," composed of roughly 75 percent hydrogen (by mass) and about 25 percent helium, with trace amounts of other gases. It is suspected that the planet has a rocky core. Jupiter rotates faster than any other planet in the Solar System, completing a rotation in just over 9 hours 55 minutes. This rapid rotation has made the planet "flatten out" into an oblate spheroid, with a sideways bulge at the equator. Jupiter orbits the Sun in a roughly circular orbit, which takes 11.9 Earth years to complete. Through even a small telescope, it can be seen that Jupiter's visible surface is divided into distinct dark colored bands known as belts, and lighter colored bands called zones. There are also two polar regions, and several other belts and zones that come and go. The different colorations of the zones and belts reflect the gases that make up the clouds. The Great Red Spot is a big, long-lasting storm. An intriguing fact is that Jupiter puts out about as much heat as it receives from the Sun.

GALILEO'S MOONS

Jupiter has the largest system of moons in the Solar System, with 63 now known. Some were discovered 400 years ago; others after spacecraft made the long journey to the planet. The Jovian satellite system comprises four large moons—the Galilean moons—plus numerous smaller bodies. The first four moons to be found were discovered by Galileo Galilei in 1610. Using his new telescope—one of the first ever built—he spotted four tiny dots near Jupiter and soon realized they were circling the planet. The four are known as Io (at left below), Europa (at right), Ganymede, and Callisto. Io, the innermost Galilean moon, is the Solar System's most volcanically active body. The volcanoes do not eject lava as we know it here on Earth; rather they project sulfur compounds high above the surface. Europa is believed by many to have an ocean of liquid water perhaps 60 miles (100 km) deep, caught between its rocky mantle and icy crust. Jupiter's largest moon, Ganymede, is the largest moon in the Solar System. Like the fourth moon, Callisto, its surface is a mixture of rock and ice covered by craters.

Saturn

With its spectacular system of rings, Saturn is the jewel of the Solar System. Orbiting the Sun in a slight ellipse that ranges from nine to ten times further out than Earth, the planet takes a little under 30 Earth years to complete one orbit. Saturn's day is thought to be about 10 hours and 45 minutes of Earth time, which is rapid as planetary rotations go. This is one reason why Saturn bulges out at its equator and is slightly flat at the poles. It's an oblate spheroid—the equatorial diameter is 74,893 miles (120,530 km), while the distance between the poles is 67,561 miles (108,730 km).

Saturn's rings are broad but flat, and are comprised of countless millions of chunks of ice of varying sizes. The orientation of the rings appears to change from year to year as the alignment of Earth's orbit, Saturn's orbit, and Saturn's axial tilt all conspire to line up in different ways. Occasionally the alignment will be such that the rings appear to be edge-on from Earth.

Uranus

Unlike the five planets Mercury, Venus, Mars, Jupiter, and Saturn, all of which are easily bright and prominent enough to be visible to the naked eye—and which therefore have been known since antiquity—Uranus was the first planet to be discovered in the modern era.

The famous English astronomer William Herschel was the first to recognize that the body was not a star, after studying it in 1781. Uranus had, in fact, been spotted previously by many astronomers from the late 1600s through to the late 1700s, but all took it to be just another star. Being so far away and therefore appearing small, it took Herschel's keen eye to see that it was not a tiny point of light, like a star, but that it had a small and discernable disk.

Noting its round shape, he initially thought that he had found a new comet, and reported it as such on April 26, 1781. But studies by Herschel and the French mathematician and astronomer Pierre-Simon Laplace soon showed that its orbit around the Sun was circular, not elliptical or parabolic—so it had to be a new planet, not a comet. Herschel named it "Georgium Sidus" (the Georgian Planet) in honor of King George III of England. Many other names were proposed however, including "Herschel" after its discoverer, and even Neptune—the name eventually given to the planet that was next to be discovered. The name Uranus began to be used in some quarters in the late 1700s, and was in wide use by the early 1800s. It did not become completely official, however, until the name was included in England's prestigious *Nautical Almanac* in 1850.

Uranus is the third largest planet (measured by diameter) in the Solar System. Like Jupiter, Saturn, and Neptune, Uranus is composed of gas—mostly hydrogen (83 percent) and helium (15 percent), with around 2 percent methane plus traces of ammonia, ethane, acetylene, carbon monoxide, and hydrogen sulfide. It is the methane in the planet's atmosphere that gives Uranus its light blue color—the methane absorbs the red wavelengths of sunlight, leaving only the bluer wavelengths to be reflected.

Above: After his discovery of Uranus, William Herschel (1738–1822) and his sister Caroline (1750–1848) went on to discover two moons of Uranus as well as two new moons of Saturn.

Being located so far from the Sun, the planet's atmosphere is very cold—the temperature of the cloud tops is a chilly −360°F (−220°C).

Internally, Uranus has far less heat than Jupiter and Saturn, both of which radiate more heat from their interiors than they receive in the way of sunlight. Uranus's core temperature is approximately 12,100°F (6,700°C)—compared with Jupiter's 70,250°F (39,000°C) and Saturn's 32,450°F (18,000°C)—which means the convection currents generated in the atmosphere are not as strong as its two larger sister planets. Because of this, Uranus does not have the same sort of strong distinct cloud bands and patterns that Saturn and Jupiter have, although if one looks carefully at the right wavelength—as the *Voyager 2* probe did, and as the Hubble Space Telescope continues to do—some cloud bands can be seen.

Uranus is unique in that the tilt of its rotation axis is so large (98°) that it is essentially lying on its side, rolling along on its orbit—most other planets have axial tilts close to a right angle to the ecliptic plane. The effect of this is that Uranus spends much of its 84-year-long orbit with one or the other of its poles facing the Sun, during which periods one polar region at a time gets far more solar heating than the planet's equatorial region.

> It would be very singular that all nature, all the planets, should obey eternal laws, and that there should be a little animal five feet high who, in contempt of these laws, could act as he pleased, solely according to his caprice.
>
> Voltaire, philosopher, 1694–1778

Right: Saturn, in an image taken by the Hubble Space Telescope in January 2004. Saturn's rings are broad but flat and are comprised of countless millions of chunks of ice, ranging from small particles up to car-sized lumps. There's also a component of rocky rubble.

TITAN: SATURN'S LARGEST MOON

Titan was discovered by Dutch astronomer Christiaan Huygens in 1655, after whom the space probe that landed on its surface in January 2005 was named. It is a moon like no other. Titan is bigger than Earth's Moon, bigger than Mercury, and it is the only moon in our entire Solar System to have a thick atmosphere (at its surface the pressure is about 1.5 times that on Earth's surface).

The atmosphere is mostly nitrogen gas (98.4 percent) with the remainder being methane plus a few trace gases. Its surface has hydrocarbon lakes of methane and/or ethane, making it the only body other than Earth to have large bodies of stable surface liquids. Some scientists consider Titan's atmosphere to be very similar to that of early Earth, hence the interest in studying it.

Above: All nine of the known rings of the planet Uranus, as seen by the *Voyager 2* spacecraft, in a false color image taken in January 1986.

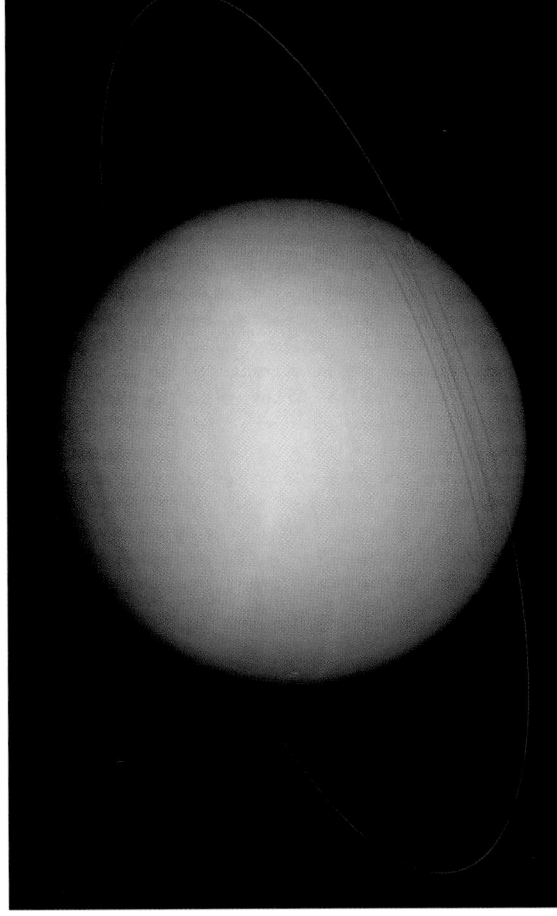

Right: A digital illustration of the planet Uranus showing its faint rings. The rings only became visible at the *Voyager* flyby and later by Hubble. The rings appear to wobble as the planet rotates, believed to be caused by Uranus's uneven gravitational pull.

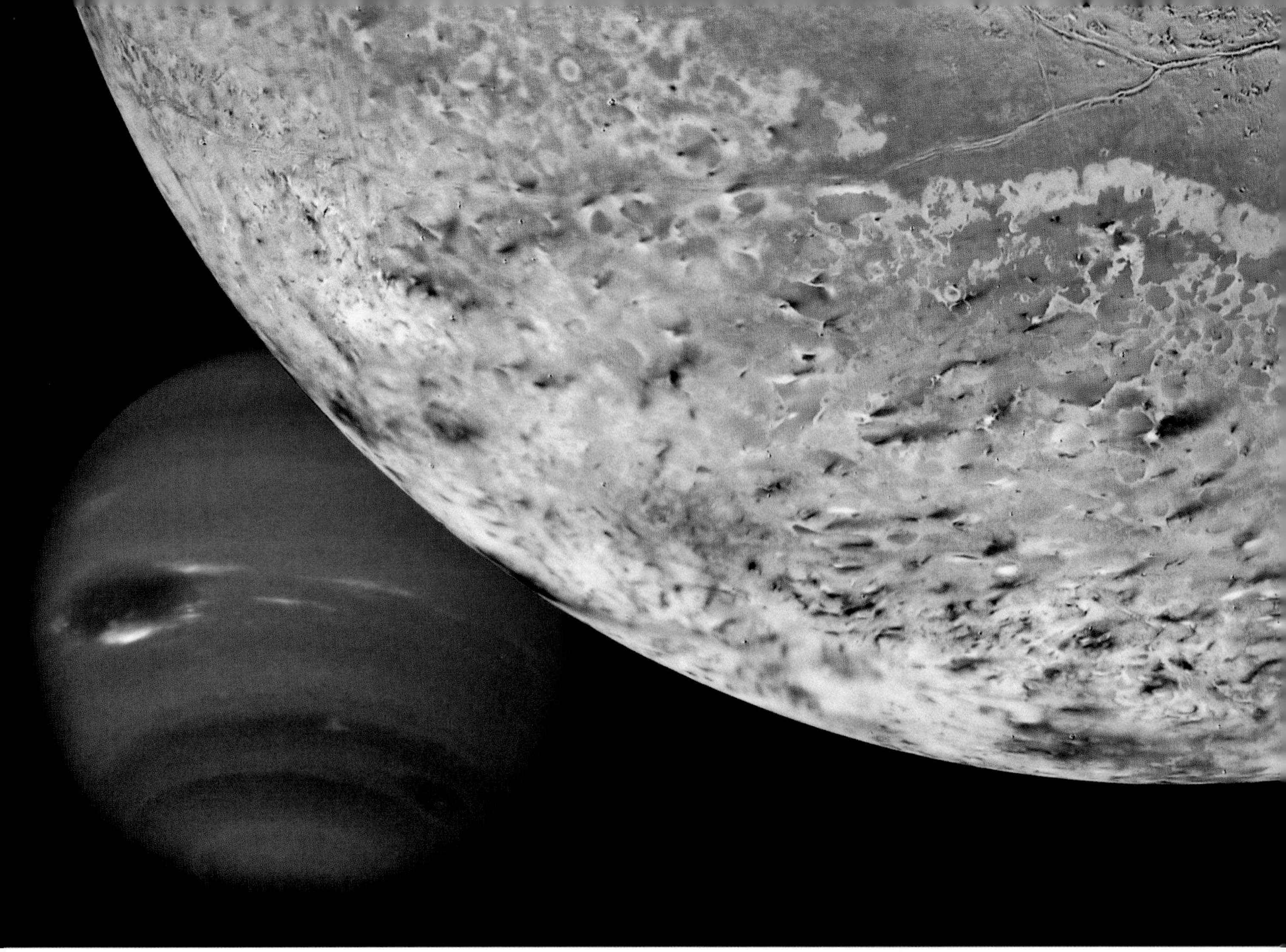

Above: This digital composite image shows Neptune as it would appear from a spacecraft approaching Triton, Neptune's largest moon. Triton is an icy world colder than Neptune and is the only large satellite in the Solar System to circle a planet in a retrograde direction, that is, opposite to the rotation of the planet.

Neptune

As NASA's *Voyager 2* space probe began to close in on Neptune in 1989, mission scientists were not expecting to find anything particularly startling. They were wrong. The story of Neptune's discovery actually begins with Uranus, when observations of that planet's motion showed that it was not keeping to its predicted orbit. There was immediate suspicion that the gravity of another, more distant planet, was pulling on Uranus and thereby affecting its orbit—and eventually that planet was found. Neptune became the first planet to be predicted and discovered mathematically.

Being so far from the Sun—30 times further than Earth—has made Neptune difficult to study. It wasn't until *Voyager 2* reached Neptune that scientists began to unravel some of its mysteries. It has a very dynamic climate system including the fastest winds in the Solar System—up to 1,300 mph (2,100 km/h).

As with the other gas giants, the surface is not solid—rather, we see the tops of continuous cloud cover. The temperature of those cloud tops is a chilly – 346°F (–210°C). Scattered across the upper reaches of Neptune's atmosphere, *Voyager 2* found bands and tufts of white clouds. Also seen was a fast-moving white cloud band that was given the nickname "Scooter." Most spectacular and unexpected, though, was the presence in the planet's Southern hemisphere of what was dubbed the Great Dark Spot, a huge cyclone similar to Jupiter's Great Red Spot.

Fifteen years later, however, Hubble Space Telescope observations showed that the Great Dark Spot had vanished. Taking over the role was a new, smaller, dark spot in the Northern hemisphere. It's uncertain whether the Great Dark Spot has actually dissipated, or whether it is being masked by some temporary affect of the atmosphere.

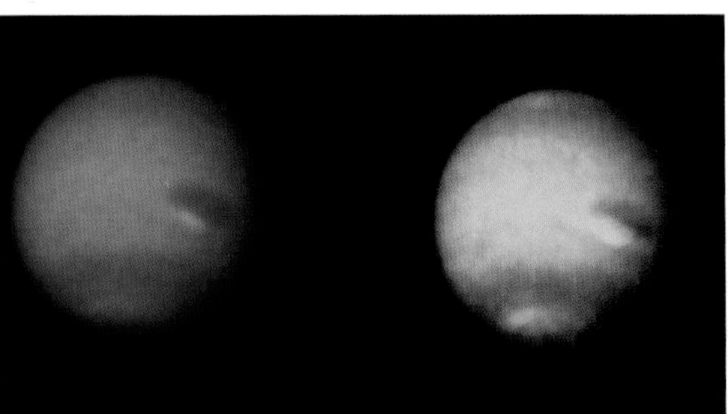

Left: Images of Neptune taken by the *Voyager 2* spacecraft. Neptune is around 57 times wider than Earth, and has more than 17 times Earth's mass. Scientists believe that it probably has a core of rock and ice, of around one Earth mass.

Dwarf planets

For many years, Pluto was our ninth planet, but there was always something mysterious about Pluto. It was much smaller than the other planets, made of ice instead of rock or gas, and with an orbit that was both highly elliptical, and highly inclined to the orbits of all the other planets. If it formed along with the other planets, why did it have such a different orbit? If it didn't form with the others, where did it come from? Was it a wandering body or had it been captured into a strange orbit around the Sun? And if so, could there be others?

Things became clearer in recent years with the discovery of more largish icy bodies orbiting the Sun beyond Neptune. One was found to be larger than Pluto itself. So, if Pluto was considered a planet, why not some of these others too? Our Solar System might actually have 10, or 15, or 40 planets, instead of just nine. Or was it time to change Pluto's status to something other than a planet?

In 2006, at a major meeting of the International Astronomical Union, a resolution was put which its sponsors hoped would solve the question once and for all. After much discussion, and unexpected worldwide popular interest, a final decision was reached. And, after all the dust was settled, most agreed that it was a reasonable compromise. It went like this. The Solar System now has eight "planets"—from Mercury out to Neptune—plus an unknown number of "dwarf planets," which was the new part of the definition into which Pluto and some other bodies would fall. A dwarf planet is defined as a body that orbits the Sun, has pulled itself into a round shape, is not a satellite of another body, but which has not cleared the rest of its neighborhood of contenting bodies. Pluto meets this definition, as does the largest of the asteroids, Ceres, and the recently discovered Pluto-like body, Eris. They were thus reclassified.

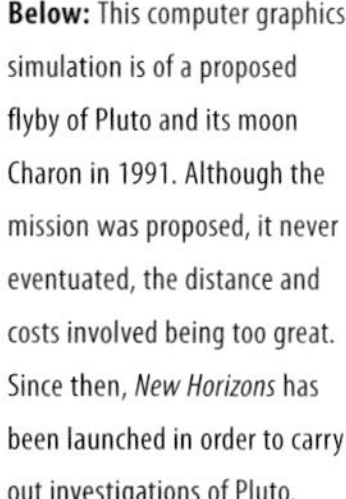

Other bodies in the solar system

Far beyond Neptune live a host of small, mysterious, icy worlds, virtually unmapped and undiscovered. We know our Solar System goes well beyond Pluto. These distant bodies have been split into several categories depending on their distance.

First is the Kuiper Belt, a population of icy bodies that circle the Sun not far beyond Neptune's orbit. Pluto and Eris, both now dwarf planets, are considered Trans-Neptunian Objects (TNOs) because of their location in space. The currently accepted cut-off point for the Kuiper Belt is around 50 astronomical units (AU—the distance between Earth and Sun) from the Sun. Then there is the "scattered disk", a region that extends from the Kuiper Belt much farther out into space. The bodies in this region typically have been "flung" onto highly elliptical and highly inclined orbits through gravitational interactions with the outer planets, mostly Neptune. Finally, much further out is the Oort Cloud, from where most comets are believed to originate. The Oort Cloud is thought to completely enclose the entire Solar System in every direction and contains billions of comets.

Right: This digital illustration shows the Pluto system from the surface of a possible moon. The other members of the Pluto system are just above the moon's surface. Pluto is the large disk at center right; Charon is to the right of Pluto.

Below: This computer graphics simulation is of a proposed flyby of Pluto and its moon Charon in 1991. Although the mission was proposed, it never eventuated, the distance and costs involved being too great. Since then, *New Horizons* has been launched in order to carry out investigations of Pluto.

Asteroids are leftover bits from the Solar System's younger days. Comets may have been responsible for bringing water to our planet and kick-starting life and may pose a serious threat to our long-term future. There's much to be learned from these space travelers.

Asteroids and comets

Where do asteroids come from?

Asteroids are bodies made of rocky and metallic substances, or either, ranging from tens of feet in diameter up to hundreds of miles. Most orbit the Sun between Mars and Jupiter in a region known as the Asteroid Belt, although some have orbits that take them further out or close in toward the Sun. Certain asteroids have orbits that cross that of Earth, making them potential Earth-colliders.

Scientists now believe that the bodies in the Asteroid Belt are fragments that didn't get the chance to combine into a full planet. As the Solar System formed from a rotating and contracting cloud of gas, small bodies collided and stuck together to form progressively larger bodies, eventuating in the planets. In some parts of the Solar System, however, the bodies were moving with too much relative velocity to stick together, and tended to smash into one another and fragment further. Asteroids can be thought of as the leftover rubble from the formation of the Solar System.

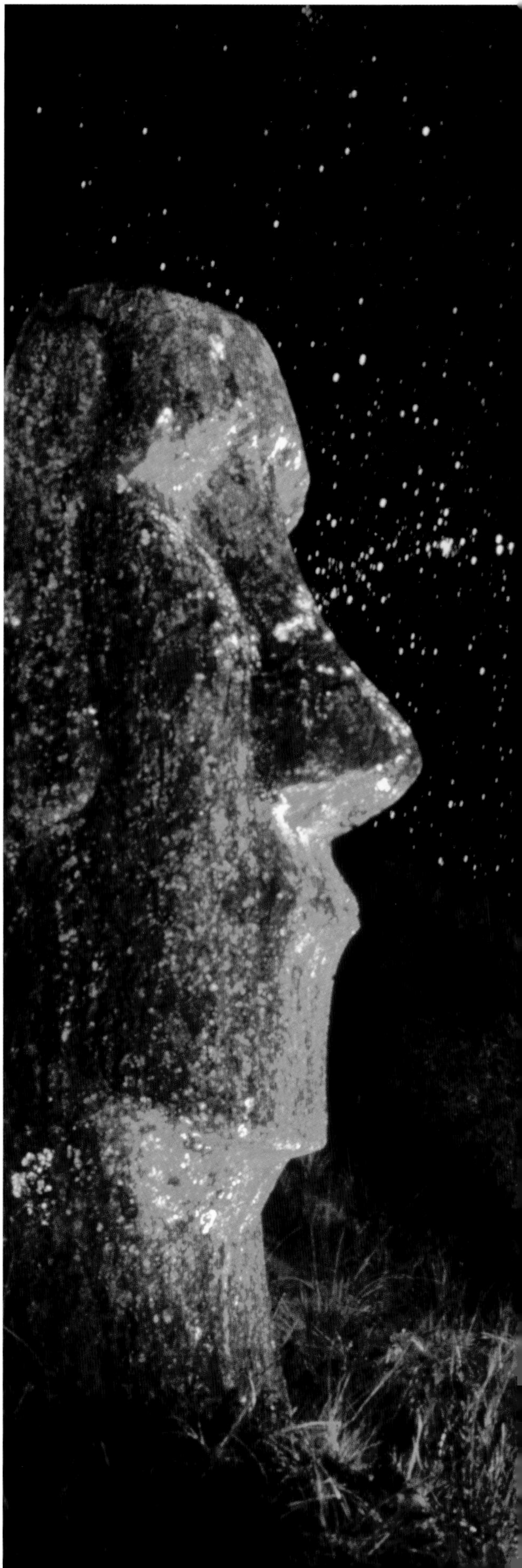

Right: Comet Halley streaks across the sky over the impassive stone statues of Easter Island in 1986. Named for Edmond Halley, the man who first realized that the comet was periodical, Comet Halley appears every 75–76 years. Its next appearance will be in 2061.

> What happens if a big asteroid hits Earth? Judging from realistic simulations involving a sledgehammer and a common laboratory frog, we can assume it will be pretty bad.
>
> Dave Barry, humorist, b. 1947

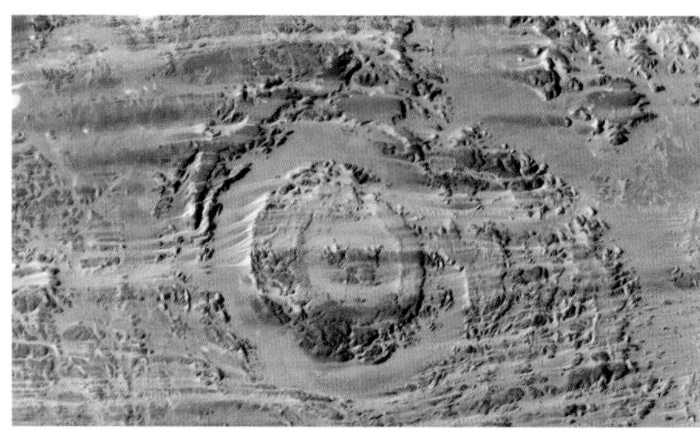

Left: Scars in the landscape in the Sahara Desert of northern Chad from the impact of an asteroid or comet hundreds of millions of years ago. The concentric ring structure is the Aorounga impact crater. The original crater was buried by sediments, which were then partially eroded to reveal the current ring-like appearance.

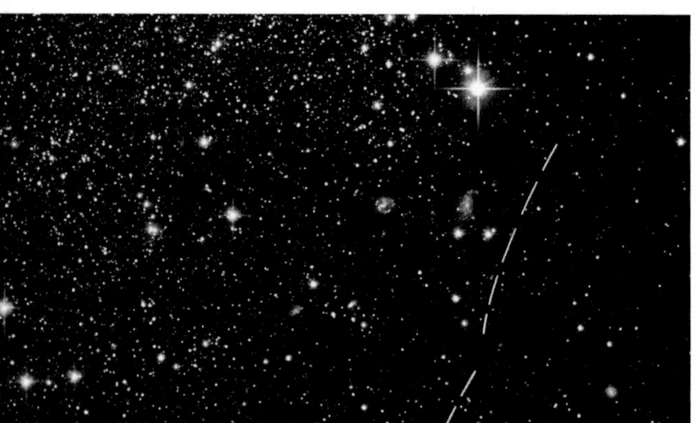

Left: An asteroid trail consisting of a series of 13 reddish arcs on the right in this August 2003 Advanced Camera for Surveys image from the Hubble Space Telescope. The brightest stars in the picture (easily distinguished by the spikes radiating from their images, produced by optical effects within the telescope), are foreground stars lying within our own Milky Way galaxy.

Above: This image shows the bright central core (the bright gas cloud surrounding the comet itself) and the coma and inner tail of Comet Hyakutake. The green colors come from ionized gas molecules, while the yellow tinge in the tail is created by dust particles.

Asteroid types

Asteroids are classified into different families according either to the types of orbits they share, or to their chemical compositions. Most asteroids are of the C, or carbonaceous type, being heavily endowed with carbon. They have a reddish appearance. A second kind, the S or silicate type, are believed to have undergone heating and melting. They have a higher measure of reflectivity. The third category is the M or metallic type, believed to be composed largely of an iron-nickel combination. It is thought that these metallic asteroids are actually the central cores of earlier larger bodies that have had their rocky exteriors smashed away during collisions. Families are comprised of bodies that follow the same orbital path around the Sun.

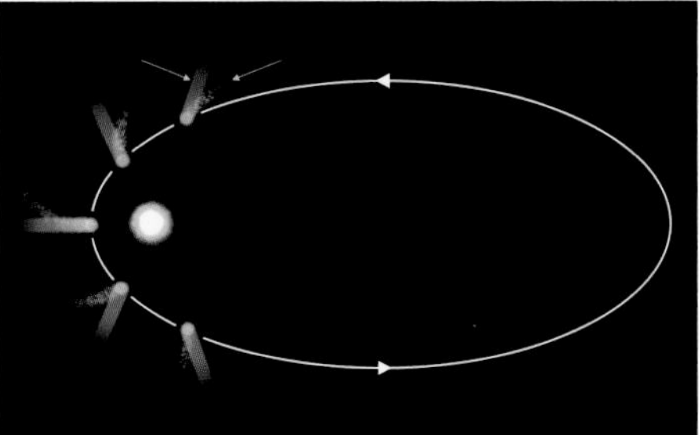

Above: This illustration shows the portion of the gas and dust tails of comets. Due to the solar wind, the tails always face away from the Sun, no matter what direction the comet takes. Not all comets develop tails but when they do, they are dazzling.

What are comets?

Comets are composed of a mixture of ices, rock, and dust. It is thought they formed far out in the distant reaches of the Solar System as the other planets were forming closer in. Because they're made of ice, and have not been significantly heated and changed during their long dark years, it is believed they may preserve the ingredients from which the Solar System was made.

The solid portion of a comet is the nucleus, and ranges in size from tens of yards to tens of miles. Sometimes when its orbit brings a comet closer to the Sun, sunlight causes some of its surface ices to sublimate—that is, convert from a solid straight to a gas without going through a liquid phase in between—forming a huge cloud, the coma, that spreads out to surround the nucleus. Some of the released gases and particles are "pushed" outward by the pressure of sunlight as well as the solar wind, into tails, sometimes spectacularly.

COMETS AS PORTENTS

Comets have been known to skywatchers since ancient times. There are records of observations going back thousands of years, when mysticism and superstition often ruled, and comets were commonly seen as harbingers of catastrophe. No doubt this was due to their sudden appearance without warning. Comet Halley, for example, is depicted on the Bayeux Tapestry, which commemorates the Battle of Hastings in 1066. The invading Normans saw the comet as a sign of their forthcoming victory; the English saw it as a sign of their defeat.

Most people have had the experience of seeing a "shooting star," a brief intense flash of light that zooms across the night sky. They can be a wondrous sight, particularly when they arrive unexpectedly—as most do. But what are they and where do they come from?

Meteors

The proper name is meteor, from the Greek work "meteoros," meaning "high up in the atmosphere." Indeed, it used to be thought that they were an atmospheric phenomenon of some kind, such as a fast moving, thin cloud. We now know that meteors are actually the result of a meteoroid entering Earth's atmosphere at high velocity.

Shooting through the skies

A meteoroid is a tiny piece of matter in interplanetary space, anywhere from the size of a grain of sand to a small boulder. The Solar System is littered with meteoroids—many are small chips of rock or iron, broken off asteroids; others are tiny particles that come from the tails of comets. Each is on a long lonely orbit around the Sun.

Sometimes the paths of these meteoroids cross Earth's orbit. And if Earth just happens to be there at that time, the meteoroid collides with our planet's upper atmosphere. Its high velocity causes a brief glow in the air it passes through, and this flash of light is called a meteor. That is, the physical object is a meteoroid; the flash of light that betrays its presence is a meteor.

Some meteoroids are large enough and strong enough to survive—or partially survive—atmospheric entry, and then fall to the ground, often in pieces. Any such fragments that make it to the surface are called meteorites.

Right: The green streak of a meteor across the southern sky of New England, USA, early November 18, 2001. This was one of thousands of meteors that entered Earth's atmosphere during a major meteor shower called the Leonids because it appeared to come from the constellation Leo.

Below: The famous Willamette meteorite, which was found in Oregon, USA, in 1902. It is the largest meteorite found in the US and the sixth largest in the world (see box at right).

> I would rather be a superb meteor, every atom of me in magnificent glow, than a sleepy and permanent planet.
>
> Jack London, novelist, 1876–1916

Meteor showers

Some meteors come from comets. As a comet enters that part of its orbit that takes it close enough to the Sun for its ices to sublimate, tiny particles of comet dust are released. These particles follow the comet, and begin to spread out along its orbit. Eventually, a whole stream of particles can be going around in the comet's orbit, with the densest, "fresher" parts closer to the comet.

When Earth hits one of these streams, it can be like someone turning a "meteor hose" onto the upper atmosphere. Thousands of tiny particles enter the atmosphere, most of them too small to produce any noticeable visual effect. But the few larger ones can put on a display of meteors all seeming to originate from roughly the same spot in the sky (the radiant)—a meteor shower.

Observing meteors

As most meteors are faint, it's best to try and spot them well away from city lights, and on nights when there is no Moon. Artificial lights and moonlight make the sky glow, drowning out all but the brightest meteors (and stars too).

Meteor showers are generally active in the early hours of the morning, which means a late night or early rise for eager meteor observers. You need to find somewhere dark and away from lights. The easiest way to observe is to find somewhere to lie on your back and look up. Make sure you allow time for your eyes to adapt to the dark—around 20–30 minutes.

AMAZING METEORITES

Meteorites come in many different shapes, sizes, and compositions. There are three main kinds: Stony meteorites (made mostly of rocky material), iron meteorites (composed mostly of an iron-nickel mixture), and stony-irons (a mixture of the two). Scientists further classify them according to chemical composition. Some meteorites have been matched with spectral analyses of asteroids—strong evidence that they are chips that have broken away from those asteroids in the past.

Meteorites weather and erode if not found and preserved. For this reason, deserts are ideal locations in which to find them, places such as the Saharan region of Africa, the vast Nullarbor region of Australia, and parts of Antarctica, which is actually a desert in terms of measured rainfall. While most meteorites are quite small and undistinguished, some are very impressive specimens indeed. The largest known meteorite, the Hoba Meteorite, was found in Namibia and weighs around 66 tons (60 tonnes). Another example is the Willamette Meteorite, found in Oregon, USA, in 1902. It weighs over 16 tons (14 tonnes).

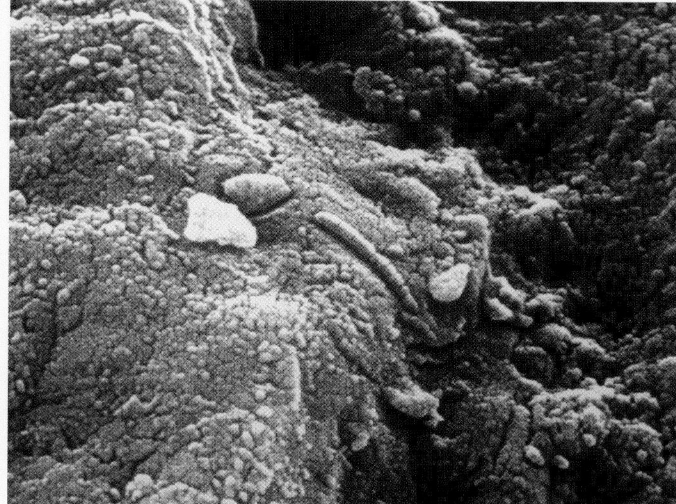

Above: Scattered across the face of our planet are large craters that bear testimony to the force with which extremely large bodies can collide with Earth. Perhaps the most famous of these is Barringer Crater, Arizona, USA, seen here from the air. More commonly known as Meteor Crater, it is thought to have been caused by a massive meteorite that hit about 50,000 years ago.

Stars are rotating spheres of hot ionized gas. After a newborn star's core is heated by gravitational contraction to 10 million°K, nuclear fusion of hydrogen into helium increases significantly. The thermal energy produced eventually builds up enough pressure to counteract the crushing force of gravity, and a normal star settles into the longest stable period of its existence.

Stars

Late in a star's life, after the hydrogen fuel in its core is depleted, and if it has sufficient mass, the fusion of heavier elements occurs.

Plotting stellar relationships

The Hertzsprung-Russell diagram is one of the most useful tools in astrophysics. It plots stellar luminosity on the vertical axis—in solar units—versus both stellar surface temperature and spectral classification on the horizontal axis. Both can be plotted on this axis, since spectral type is directly related to surface temperature.

On the diagram red dwarf stars are of spectral type M; they have a low surface temperature and low luminosity, and thus occupy the lower right corner of the belt of stars labeled "Main Sequence"—those stars whose energy comes from the fusion of hydrogen into helium in their cores.

The core of a one solar-mass protostar will be heated more rapidly by gravitational contraction than a red dwarf's will, and the core temperature will stabilize at 15 million°K. Fusion occurs more readily at this temperature, so even though solar-mass stars have much more hydrogen fuel than red dwarfs, solar-mass stars will run out of nuclear fuel sooner, in approximately 10 billion years. The Sun's position on the main sequence is at the intersection of spectral class G2 and luminosity 1—the vertical axis is labeled in units of solar luminosity.

The most luminous (and massive) main sequence stars are plotted in the upper left corner of the Hertzsprung-Russell diagram. A supergiant such as Deneb has an extreme core temperature, and fusion proceeds at such a furious pace that its lifespan will be only a few million years. The most massive stars are quite rare. This is because relatively few are created, and they live fast and die young.

The shrinking cores of red giants may eventually get hot enough—100 million°K—to fuse helium into carbon. After they lose the battle with gravity, their cores are still considered stars, the strange white dwarfs and neutron stars.

Powering the Sun

The plasma in a star's core consists of uncoupled positively charged atomic nuclei and negatively charged electrons, all moving at great speeds. The more massive the star, the greater its core temperature, and thus the greater the speed of its atomic particles. When two positively charged nuclei meet, the repulsive force of their charges normally deflects them.

However, if they have a head-on collision with enough energy behind it, the two nuclei may get close enough to bond. Within the Sun's core, the success rate for the fusion of two colliding protons—each the nucleus of the common isotope of hydrogen—is one collision in every 10 trillion trillion.

The gamma-ray photons produced during nuclear fusion power a star, although in a solar-mass star it takes approximately a million years for that energy to reach the star's photosphere and be emitted, mainly at visible, ultraviolet, and infrared wavelengths, rather than as

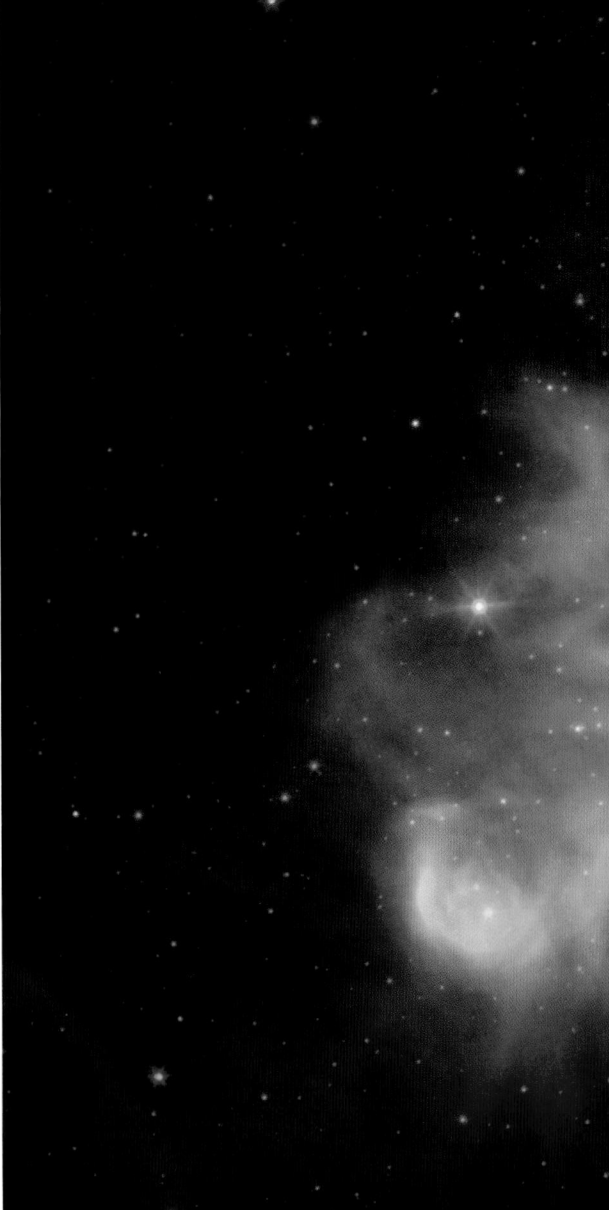

Right: Baby stars are forming near the eastern rim of the cosmic cloud Perseus, in this infrared image from NASA's Spitzer Space Telescope. The baby stars are approximately 3 million years old and are shown as reddish-pink dots to the right of the image. The pinkish color indicates that these infant stars are still shrouded by the cosmic dust and gas that collapsed to form them. These stars are part of the IC348 star cluster.

Below: Stars are the building blocks of the cosmos and have captivated us for millennia. Here the constellations of the Southern hemisphere are depicted in an engraving by Pieter Schenk (1660–1719) and Gerard Valk (1651–1726) in a book called *The Celestial Atlas*.

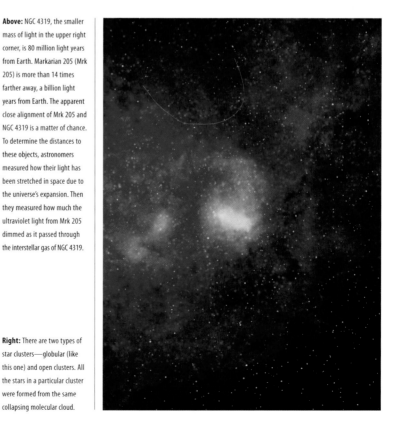

Nobody really understands how star formation proceeds. It's really remarkable.

Rogier A. Windhorst, astronomer, 1992

Above: NGC 4319, the smaller mass of light in the upper right corner, is 80 million light years from Earth. Markarian 205 (Mrk 205) is more than 14 times farther away, a billion light years from Earth. The apparent close alignment of Mrk 205 and NGC 4319 is a matter of chance. To determine the distances to these objects, astronomers measured how their light has been stretched in space due to the universe's expansion. Then they measured how much the ultraviolet light from Mrk 205 dimmed as it passed through the interstellar gas of NGC 4319.

the powerful gamma rays that fusion produced in the core. Every second, our star, the Sun, consumes 661 million tons (600 million tonnes) of hydrogen, producing 656 million tons (596 million tonnes) of helium. The remaining 4.4 million tons (4 million tonnes) are turned into energy in accordance with Einstein's equation $E=mc^2$.

In 1938, German physicist Hans Bethe and his colleagues worked out the first explanation of stellar fusion, the "CNO cycle," in which carbon, nitrogen, and oxygen nuclei act as catalysts for the fusion of hydrogen into helium. This is a more efficient process than the proton–proton chain, but requires the hotter cores of much more massive stars than the Sun. For this Bethe won the 1967 Nobel Prize for Physics.

RED DWARFS

Mass is a star's most important property. The more massive the star, the greater the pressure that its upper layers exert on its core—the greater the pressure on the core, the higher its temperature. The core of a star of at least 0.08 solar masses will slowly be heated by gravitational contraction to the threshold for efficient nuclear fusion, 10 million°K. But the rate of fusion will be so slow in such a marginal star that its hydrogen fuel will last for hundreds of billions of years. Such faint stars are called red dwarfs.

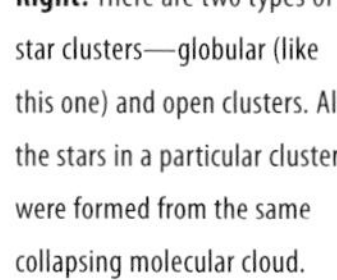

Right: There are two types of star clusters—globular (like this one) and open clusters. All the stars in a particular cluster were formed from the same collapsing molecular cloud.

The story of stars begins not long after the Big Bang. After the intense heat of the Big Bang, the matter created in it began to cool and to collect into huge clouds of gas.

Life and death of stars

Parts of those gas clouds became thicker, and in those regions, the gentle gravitational pull of the gas on itself made the thick regions collapse into huge, dense, hot balls—many times bigger than our Solar System. Eventually, enough gas collapsed into such highly condensed agglomerations that nuclear fusion reactions began. It had taken around 100 million years since the Big Bang occurred, but finally the first stars were born.

Those first stars were probably very big, very hot, and with very short life spans. The bigger and hotter a star is, the longer and faster it burns its fuel, and the shorter it lives.

Below: This image, entitled *The Pillars of Creation*, was taken by the Hubble Space Telescope in 1995. It shows gaseous pillars in M16, the Eagle Nebula. These dramatic columns of hydrogen and dust act as incubators for new stars.

Birth of a star

Scientists believe that stars such as our Sun formed from a giant molecular cloud—an enormous region of dust and gas in interstellar space. The material inside the cloud is unevenly spread; some parts are denser than others. These dense regions begin to collapse inward under their own gravity, collecting more and more material into themselves. Getting hotter all the time, eventually a huge ball of plasma—an electrically charged gas—is created. When internal pressures climb high enough, lighter elements begin fusing into heavier ones, and the star is born.

BRINGING UP BABY

The Great Nebula in Orion is perhaps the best-known example of a nearby star-forming region, being only 1,500 light years from Earth. Just visible to the unaided eye from a dark site as a tiny smudge of light, a telescope reveals it to be a stunningly beautiful cloud of gas and dust. Deep inside, in regions where the density has increased, new star systems are being born. Astronomers have caught glimpses of this process, in the form of "proplyds"—protoplanetary disks. Studying the Orion Nebula is giving science tremendous insights into the evolution of stars.

It is likely that at least several stars will form out of a single molecular cloud, so it's possible that our Sun has siblings somewhere. But so much time has passed since then—around 5 billion years, more than a third of the total age of the universe—that we are unable to identify them.

Death of a star

There are several ways that stars finish their lives, and it primarily depends on how massive they were to start with.

Our Sun, for example, is destined to become a white dwarf star—a very small dying ember roughly about the size of Earth, but with about 60 percent of its original mass. It will be about 5 billion years before this process begins, but when it does, Earth and the other inner planets will be engulfed as the Sun first swells to become a red giant star. The Sun will throw off its outer gas layers, and that gas will spread outward eventually forming a beautiful planetary nebula.

The white dwarf left in the middle of the nebula will no longer produce energy through nuclear fusion. All it will have left is its residual heat, which it will lose over billions of years, cooling down and eventually becoming a "black dwarf." This process takes so long that, so far, there has not been enough time since the universe began for any such black dwarfs to form.

The most interesting and violent forms of stellar death happen to stars that are many times the mass of our Sun. Such large stars will end their days in massive explosions, called supernovae.

Top right: This image taken with NASA's Hubble Space Telescope depicts bright, blue, newly formed stars that are blowing a cavity in the center of a star-forming region in the Small Magellanic Cloud. At the heart of the star-forming region lies star cluster NGC 602.

Center: A spiral star formation. Astronomers have identified many star-forming regions in our galaxy (and also other galaxies), including the Orion Nebula. By studying these regions, they have been able to get a handle on the different stages of star birth.

> When I follow at my pleasure the serried multitude of the stars in their circular course, my feet no longer touch the earth.
>
> Ptolemy, astronomer and mathematician, *c.* 83–161

Black holes

The ultimate fate of any star more than about eight times the mass of the Sun is a black hole. Just like the formation of a neutron star, the star's core collapses to form a dense object. The collapse is so swift and strong that the matter is squashed effectively down to infinity, becoming an object known as a singularity that is the size of a pinhead—a black hole.

The black hole's immense gravity means that it will capture anything that comes too close. The point of no return is called the event horizon. Nothing is fast enough to escape a black hole, not even light, once it has crossed the event horizon.

Right: This image shows Kepler's supernova remnant produced by combining data from NASA's three Great Observatories—the Hubble Space Telescope, the Spitzer Space Telescope, and the Chandra X-ray Observatory. It shows a bubble-shaped shroud of gas and dust that is 14 light years wide.

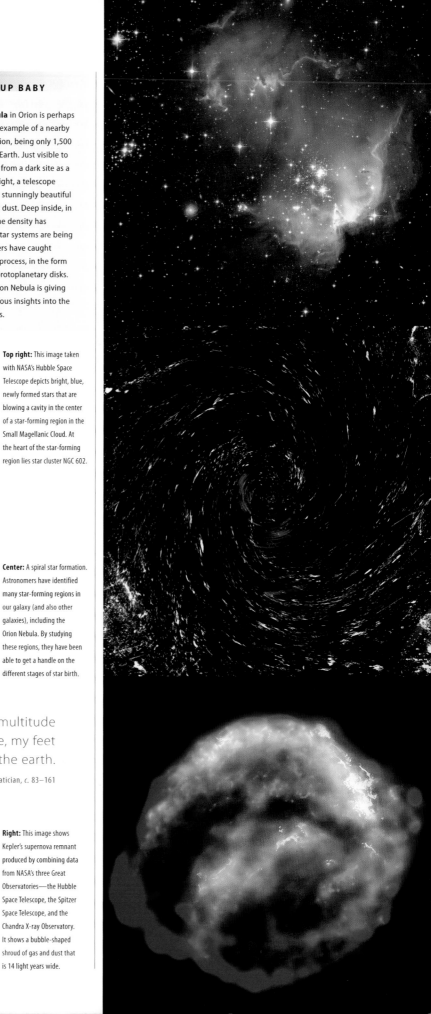

Nebulae and stars are closely intertwined. Stars are born from collapsing molecular clouds. Then, in their old age as red giants, stars enrich their surroundings with carbon created from helium fusion. Heavier elements are produced by supergiant stars when they explode as supernovae, spewing those elements into space, perhaps to form planets.

Nebulae

Our Galaxy's great molecular clouds are usually detectable only as masses called dark nebulae or absorption nebulae. While dust grains make their presence known by their absorption and scattering of starlight, they typically make up less than one percent of the mass of molecular clouds, which are primarily molecular hydrogen and helium. The microscopic dust grains are carbon soot, or may be oxides of silicon, titanium, and calcium, and provide a site for molecules to form upon.

Dark nebulae

Many dark nebulae are telescopic objects, but some, such as the Coalsack beside the Southern Cross, can be seen by the unaided eye. The Coalsack forms the head of a larger object, the dark Emu of the Australian Aborigines. The southern end of the Great Rift, from Alpha (α) Centauri through to Scorpius, forms the bird's body. The Funnel Cloud Nebula, Le Gentil 2, the great dark nebula descending from Cepheus into Cygnus, is the second-best after the Coalsack. The Funnel Cloud Nebula is still a naked-eye object with a 10-day-old gibbous moon visible in the sky.

Telescopes and binoculars reveal many smaller dark vacancies, especially in Cygnus and Sagittarius. Two of the finest are oval Barnard 92 and elongated Barnard 93, on the northwestern side of the Small Sagittarius Starcloud, M24.

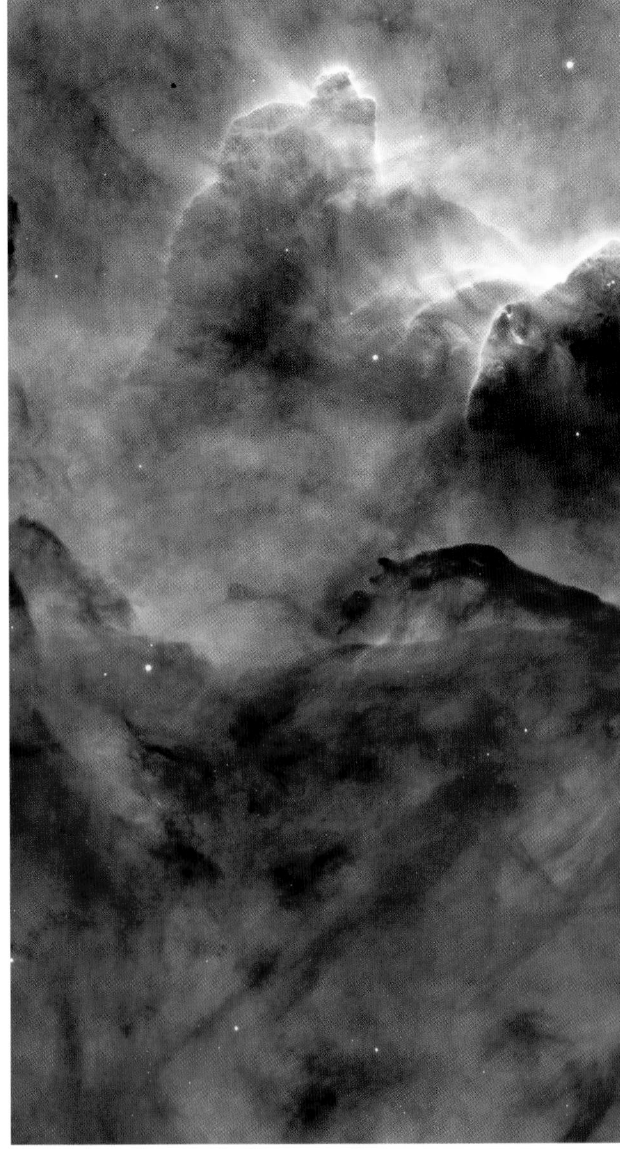

Right: These great clouds of cold hydrogen in the Carina Nebula resemble summer afternoon thunderheads. They tower above the surface of a molecular cloud on the edge of the nebula. So-called "elephant trunk" pillars resist being heated and eaten away by blistering ultraviolet radiation from the nebula's brightest stars.

Every cubic inch of space is a miracle.

Walt Whitman, poet, 1819–1892

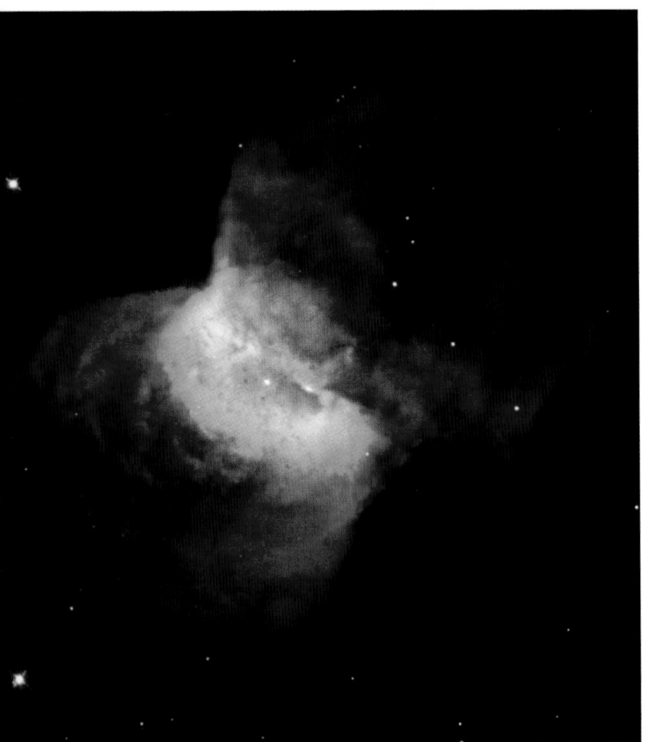

Left: NGC 2346, shown here, is a so-called "planetary nebula," which is ejected from Sun-like stars which are near the ends of their lives. It is remarkable because its central star is known to be actually a very close pair of stars, orbiting each other every 16 days. NGC 2346 lies about 2,000 light years away from us, and is about one-third of a light year in size.

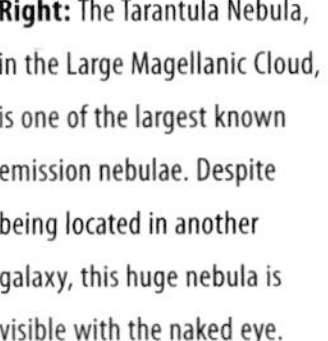

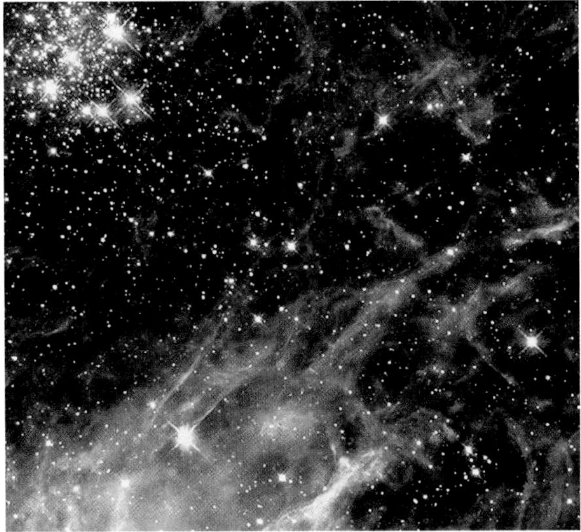

Right: The Tarantula Nebula, in the Large Magellanic Cloud, is one of the largest known emission nebulae. Despite being located in another galaxy, this huge nebula is visible with the naked eye.

Emission nebulae

When star formation begins, ultraviolet radiation from the most massive young spectral type O stars will ionize the surrounding hydrogen gas, exciting it to fluorescence, and destroying nearby volatile dust. We see this as an emission nebula, also called an HII region. The massive young stars' energetic stellar winds and ultraviolet radiation form cavities in the molecular cloud. If a cavity is on our side of the cloud, we see a showpiece object like the Orion Nebula, in which the highly luminous Trapezium stars cause the walls of the cavity to shine.

Planetary nebulae

Planetary nebulae are shells of gas that have been ejected from the outer layers of red giant stars at the end of their lives. The very hot exposed core of the star radiates strongly in the ultraviolet and causes this shell of gas to fluoresce for a relatively brief period—tens of thousands to no more than a million years—before it dissipates.

Planetaries come in many shapes, but luminous disks or rings are common among the brighter ones. An example is NGC 3918, the Blue Planetary in Centaurus. As its name implies, this small planetary has a high enough surface brightness to show color. Other small planetaries are blue-green or greenish. IC 418 in Lepus is unusual. It is called the Pink Planetary because at low power in a large telescope it has a coppery outer rim surrounding the more usual bluish disk.

Larger planetaries have lower surface brightnesses, and are colorless at the eyepiece. An example is M57, the famous Ring Nebula in Lyra. While many planetaries appear ring-shaped, their actual shape is believed to be more like a short section of tubing with flared ends. Two-lobed structures are also common.

REFLECTION NEBULAE

Sometimes dust is made visible by reflecting starlight. If stars are embedded in dust, the resulting reflection nebula (below) is usually just an amorphous glow around a star. One exception is the 15 arcseconds long, bright orange Homunculus Nebula, a double-lobed reflection nebula. The Homunculus surrounds and obscures—to varying degrees at different times—the massive star Eta (η) Carinae, at the heart of NGC 3372, a splendid complex of emission nebulosity and dark nebulae.

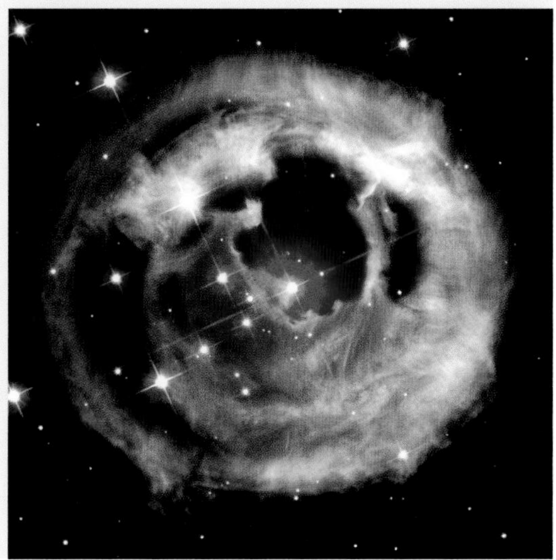

A discrete dust cloud physically near to, but not enveloping a star can form a dramatic reflection nebula. An example is IC 2118, the Witch Head Nebula, in easternmost Eridanus. It reflects the light of the supergiant star Rigel. Since dust scatters short wavelengths more efficiently than longer wavelengths, the Witch Head Nebula is bluer than its illuminating star, Rigel. This selective scattering is akin to the reason that Earth's skies are blue.

Given that our Sun, a very ordinary star with a system of planets, is just one of around a hundred thousand million stars in the Galaxy, it is possible that other similar stars have planets of the same type.

Exoplanets—planets of other stars

In 1983, it was found that the brilliant star Vega (Alpha [α] Lyrae) had "a huge infra-red excess"—because it was associated with a huge cloud of cool, possibly planet-forming material. Other stars were found to have similar surrounds, and before long one huge dust-disk, around the star Beta (β) Pictoris, was photographed. In some cases, there were indications that there might be massive orbiting bodies, which were probably planets. On the other hand, there was no definite proof, and there is always the chance of confusing a planet with a brown dwarf star.

Above: An artist's impression showing a close-up of the extrasolar planet XO-1b passing in front of a Sun-like star 600 light years from Earth. The Jupiter-sized planet is in a tight four-day orbit around the star.

Looking for extra-solar planets

To make a direct observation of an extra-solar planet is obviously very difficult. A planet is much smaller than a normal star, it shines only by reflected light, and it is so close to its parent star that it is drowned in the glare, just as a pocket torch would be overpowered if it lay beside a searchlight. However, various indirect methods can be utilized. The first of these is astrometric. An orbiting planet pulls upon its parent star, and the star describes a small circle or ellipse around the system's center of gravity. This affects the radial velocity as seen from Earth—that is to say the toward-or-away movement—and spectroscopic observations can make use of the Doppler effect in order to determine the dimensions of the system and the mass of the perturbing planet.

The first results from this so-called "wobble" technique were unexpected. In 1992, two astronomers, A. Wolszczan and D. Frail, announced that they had detected a planet moving around a pulsar. Since a pulsar is the result of a colossal outburst involving the collapse of a very massive star, the presence of a planet would indeed be hard to explain. Then, in 1995, two Swiss astronomers Michel Mayor and Didier Queloz, working at the Haute-Provence

Observatory, detected a planet orbiting 51 Pegasi, a solar-type star 54 light years away; it is slightly more massive and more luminous than the Sun. The nature of the planet was very surprising; the mass was about half that of Jupiter, but the distance from the star was a mere 4,350,000 miles (7,000,000 km), which is about one-eighth the distance between the Sun and Mercury. The orbital period was given as 4.3 days, and the surface temperature had to be around 2,372°F (1,300°C). The planet had to be a gas giant, more like Jupiter than like Earth, and when similar planets were found attending other stars they became known as "hot Jupiters."

Detection in transit

The next important method of detection involves the transit of an orbiting planet in front of its star. The result will be a slight temporary drop in the star's brightness as part of its light is blocked out, but again astronomers need to be wary of confusing a transiting planet with a transiting brown dwarf, and obviously the planet must be large enough to make its presence known. The first success here came in 1999, when keen observers D. Charbonneaux and M. Brown tracked down a planet moving around the star HD 20645 in Pegasus, some 150 light years away; the magnitude fell by 1.7 percent, which is not very much at all. Here, too, we have a "hot Jupiter," at a separation of 4,350,000 miles (7,000,000 km) and a period of 3.5 days. The temperature is of the order of 1,382°F (750°C).

In 2001, observations with the Hubble Space Telescope made it possible to analyze the atmosphere of the planet; spectra were taken first during transit and then with the planet out of view, so that the spectrum of the star was simply subtracted. No signs of water vapor were found, though there were indications of a cloud of silicate dust; however, not too much must be read into this, because investigations are still at a very early stage. By 2007, over 200 extra-solar planets had been located, mainly by the astrometric technique.

Below left: A digital composition showing the planetary system around the red dwarf Gliese 581, in the foreground. Gliese is the first planet other than Earth to be found existing in the "Goldilocks Zone," where the temperature is just right for liquid water to exist.

> To confine our attention to terrestrial matters would be to limit the human spirit.
>
> Stephen Hawking, physicist, b. 1942

Above: An artist's impression of the pulsar planet system discovered by Aleksander Wolszczan and D. Frail in 1992. Pulsars are rapidly rotating neutron stars, which are the collapsed cores of exploded massive stars. They spin and pulse with radiation, much like a lighthouse beacon.

Below: A digital photograph of the star Beta (ß) Pictoris, 63 light years from Earth. It appears to be surrounded by a halo of dust and gas. Beta (ß) Pictoris is one of the brightest stars in the constellation Pictor. Recent findings suggest it may be another solar system in the making.

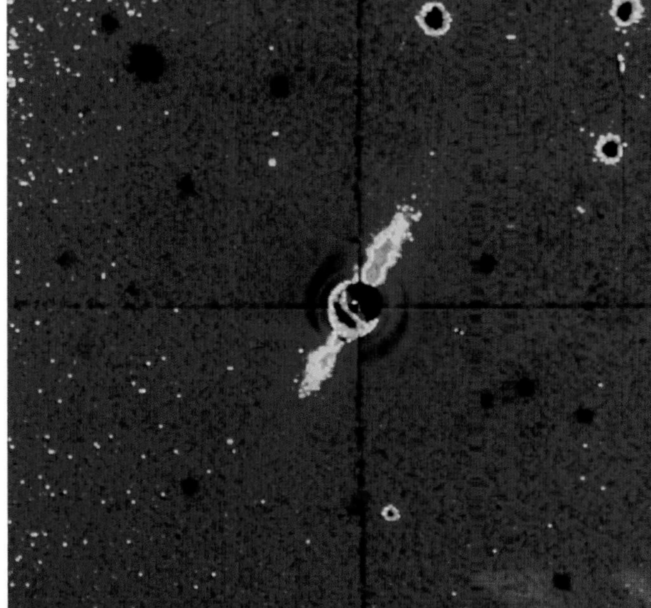

In 1845, amateur astronomer William Parsons, the third Earl of Rosse, built a giant telescope with a 72-inch (183 cm) diameter mirror, called the Leviathan of Parsonstown, in Ireland. This was a turning point in modern astronomy. Since the eighteenth century, astronomers had called the patches of light resembling clouds "nebulae," Latin for clouds, and all nebulae were thought to reside in the Milky Way.

Galaxies

William Parsons's telescope helped astronomers realize that they had been mistaken because it enabled them to see that some nebulae had spiral shapes. The "spiral nebulae" were especially interesting, since they were thought to be perhaps other solar systems forming. The introduction of the spectroscope to astronomy in the 1860s showed that some nebulae, such as the Great Orion Nebula, were composed of gas, and such objects are still called nebulae today. But the spectroscope showed that the spiral nebulae had similar spectra to those of many stars. This led some astronomers to the rather bold idea that "the spiral nebulae" were distant island universes, each comparable to the Milky Way.

In the 1920s, Edwin Hubble, after whom the Hubble Space Telescope is named, took very long exposures of the "Great Nebula in Andromeda" using the new 100-inch (254 cm) reflector on Mt Wilson in California, USA. His images resolved it into stars, and by the 1960s the word "nebula" ceased to be used to describe what we now call galaxies. The "Great Nebula in Andromeda" then became the "Andromeda Galaxy."

Hubble also devised the first classification system for galaxies, based on their appearance. The system in use today had its origins with Hubble's, but it was later modified and made more complex by Gérard de Vaucouleurs. The classification system described below is the de Vaucouleurs' system.

Every one of the billions of galaxies is different from every other one. A classification system assists in the discussion of galaxies by grouping them into similar types, but a convenient system only describes most galaxies. Some active galaxies such as Centaurus A, NGC 5128, resist being slotted into models.

Elliptical galaxies

Elliptical galaxies look bland, but they include both the most massive galaxies, the giants found at the heart of large galaxy clusters, and the least massive, the barely detectable dwarf spheroidals. Elliptical galaxies are mostly composed of old stars, and have little cool gas and dust available to create new stars.

Elliptical galaxies are classified by their degree of flattening. An E0 galaxy appears nearly round, while an E7 is extremely elongated. A galaxy's classification depends partly on its orientation to us, since a cigar-shaped galaxy seen end-on would appear round. Most elliptical galaxies have a star-like nucleus, and their surface brightness falls off steadily toward the outer fringes. When two large spiral galaxies merge, as the Andromeda Galaxy and the Milky Way are predicted to do in three billion years or so, the outcome is believed to be a giant elliptical galaxy.

Examples of giant elliptical galaxies are M87 and M49 in the Virgo Galaxy Cluster. NGC 147 in Cassiopeia, a dE5 satellite of the Andromeda Galaxy, is an example of a dwarf elliptical galaxy. Dwarf spheroidals are even smaller.

Spiral galaxies

Spiral galaxies have the most attractive form. Normal spirals have an elliptical bulge of old stars at their core, surrounded by star-forming spiral arms. Some, like the Whirlpool Galaxy, M51, have only two major arms; these are called "grand design spirals." Others, like the Sunflower Galaxy, M63, have many thin arms, sometimes with

Below: At 2.5 million light years from Earth, the Andromeda Galaxy (M31) is the most distant object that can be seen easily with the unaided eye, and was one of the "little clouds" cataloged by ancient astronomers.

Right, top: Complex loops and blobs of cosmic dust lie hidden in the giant elliptical galaxy NGC 1316. This image made from data obtained with the NASA Hubble Space Telescope shows evidence that this giant galaxy was formed from a past merger of two gas-rich galaxies.

Right, center: The aptly-named Whirlpool Galaxy (also known as M51 or NGC 5194) is viewed from the top down, showing its core and spiral arms. This image was taken by the Hubble Telescope Wide Field Planetary Camera 2.

Right, bottom: A Hubble Space Telescope (using an ESA faint object camera) image of the monster elliptical galaxy M87, which has an active galactic nucleus with a bright optical jet. M87 is located 50 million light years away in the constellation Virgo.

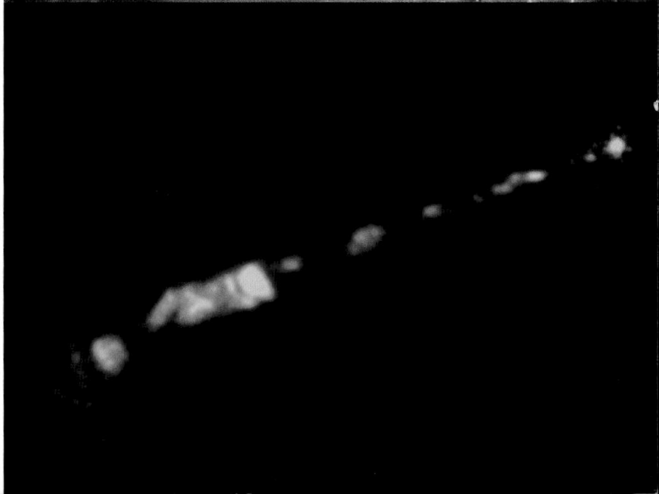

Above: This large group of radio telescopes with moveable antennas is known as the Very Large Array (VLA). Situated near Socorro, New Mexico, they detect and take in radio signals, some extremely faint, from throughout the cosmos.

branches. Normal spirals are indicated by "SA," followed by a lower case "a," "b," "c," or "d," indicating how tightly wound the arms are, and the size of the central bulge relative to the disk. Intermediate cases are indicated by a combination, for instance, "cd." Thus, M51 is an SAbc. Barred spiral galaxiess, indicated by "SB," are an even more beautiful variation.

Lenticular and irregular galaxies

Lenticular galaxies, the prototype being Sextans's NGC 3115, the Spindle Galaxy, are an intermediate type between ellipticals and spirals. They are indicated by "SA0," or "SB0" if they display a bar. Lenticular galaxies resemble spirals, having a central bulge and a disk, but they lack spiral arms.

Irregular galaxies, labeled "IA," or "IB" if barred, do not fall into any other classification. Most are relatively small, the prototype being the Small Magellanic Cloud. Many have active star-forming regions.

Left: Centaurus A, which is also known as NGC 5128, is often considered a lenticular galaxy, although it is difficult to slot this galaxy into currently accepted models. Located in the constellation Centaurus, it is the closest active galaxy to Earth, being a radio galaxy.

The Milky Way

The Milky Way—once thought to be the entire universe—is only one of many billions of galaxies. The Milky Way is a spiral, the most beautiful kind of galaxy, and also one of the two dominant members of the Local Group of Galaxies.

Within a diameter of about 100,000 light years, the galaxy has over 100 billion stars. The stars in the small central bar, in the huge spherical halo, and in the globular clusters, are mostly ancient. Most star formation now occurs within clouds of gas and dust in the spiral arms, which form a thin disk, only about 3,000 light years thick.

The surrounding disk of atomic hydrogen gas has a diameter of about 165,000 light years, while the dark matter, detectable only by its gravitational influence, extends to at least the two major satellite galaxies—the Magellanic Clouds—and probably well beyond them.

Edge-on spirals

Learning the nature of the Milky Way has been a slow process of studying other spiral galaxies. Those facing us display a brighter central region that usually contains a prominent, almost starlike, nucleus.

The larger, but fainter disk of some of the nearby galaxies looks like mottled spiral arms, with supergiant stars and little knots of light that are star clusters and nebulae like those in the Milky Way. The sky offers a continuum of spiral galaxies oriented at angles increasingly farther away from face-on, all the way to some that are almost edge-on. NGC 4565 in Coma Berenices is an edge-on spiral—its very thin disk has dark dust clouds along its length and there is a round and brighter central bulge.

Mapping the galaxy

The same dust clouds that provide these visual clues also hide our galaxy's core. While an 8-inch (200 mm) telescope easily reveals the

> ... alone with the stars: the misty river of the Milky Way flowing across the sky, the patterns of the constellations standing out bright and clear, a blazing planet low on the horizon.
>
> Rachel Carson, marine biologist and author, 1907–1964

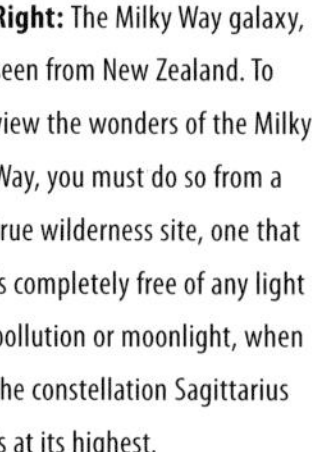

Right: The Milky Way galaxy, seen from New Zealand. To view the wonders of the Milky Way, you must do so from a true wilderness site, one that is completely free of any light pollution or moonlight, when the constellation Sagittarius is at its highest.

nuclei of hundreds of galaxies, the largest optical telescopes on the planet cannot see the Milky Way's nucleus at visual wavelengths.

Radio telescopes made the first breakthroughs in mapping the galaxy. The distribution of atomic hydrogen was mapped; then the clouds of molecular hydrogen were mapped, using the knowledge that molecular hydrogen is always accompanied by trace amounts of carbon monoxide.

Although molecular hydrogen cannot be directly detected at the temperatures of molecular clouds, which are typically around 10°K/–441°F/–263°C, carbon monoxide radiates at millimeter wavelengths. Radio telescopes eventually charted the Milky Way, revealing a barred spiral with many arms.

Below: A schematic drawing of the Local Group of Galaxies that includes our own Milky Way. The galaxies cover a 10 million light year diameter and have a binary, or dumbbell, shape.

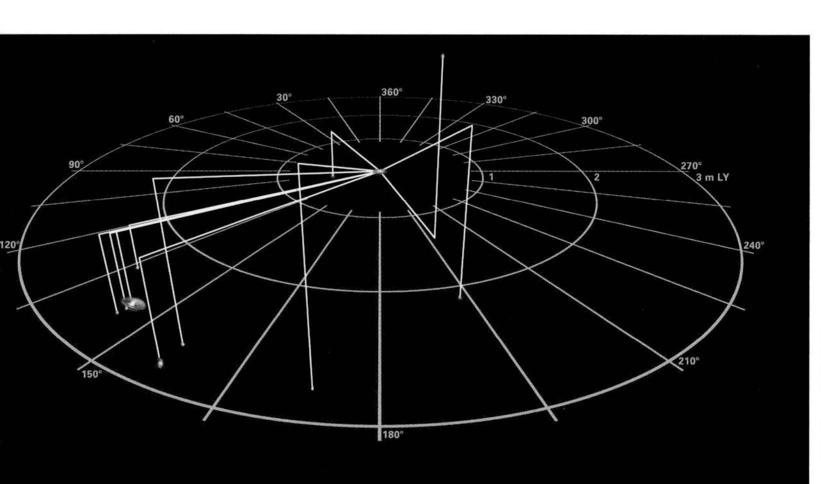

At the Sun's distance from the galaxy's center—28,000 light years—each orbit around the core at a velocity of approximately 135 miles (220 km) per second takes 230 million years. The Sun has no connection with any spiral arm—it just passes through them while orbiting the galaxy.

The Milky Way's spiral arms

The galaxy's spiral arms have been given names. The Sun is on the inner edge of the Local Arm, frequently called the Orion Arm. Three bright star clusters, the Beehive, the Pleiades, and M7, all lie near us. Toward the outer end of the Local Arm is the supergiant Rigel, the three stars of Orion's Belt, the Orion Nebula, and Monoceros's Rosette Nebula. Looking inward along the arm is the supergiant star Deneb, the Veil Nebula supernova remnant, and the North America Nebula. Supergiant stars are always associated with a spiral arm.

The next arm outward is the Perseus Arm, including the Double Cluster. Beyond is the Outer Arm. Looking toward the galactic center is the Sagittarius Arm. Inward along this arm are the Trifid Nebula, the Swan Nebula, and the Eagle Nebula. Looking outward is the Eta Carinae Nebula.

The molecular clouds on the inner edge of the Sagittarius Arm hide most of the galaxy's central regions. But we can glimpse the next two inner arms, the Scutum–Crux Arm and the Norma Arm.

Current research indicates that the Norma Arm contains the most massive molecular clouds, and exhibits the greatest rate of massive star formation within the disk.

Although six spiral arms are named, the Milky Way is currently thought to have only four major spiral arms. The Local Arm is suspected to be just a spur, attached to either the Perseus or the Sagittarius arms. Secondly, many galaxies display nearly-encircling spiral arms, and that is suspected to be the case with the innermost arm, the Norma Arm, and the Outer Arm—they are thought to be the root and the tip of the same arm.

GALAXY CLUSTERS

Almost all galaxies are found within clusters, which, typically, are bound by hot X-ray emitting gas and large amounts of "dark matter." Galaxy clusters range from small groups of a few galaxies like the Local Group, the Sculptor Group, and the M81 Group, on up to large irregular clusters like the Virgo Cluster, and even larger spherical clusters like the Coma Cluster.

The nearest large cluster, the Virgo Cluster, contains approximately 2,000 galaxies and is about 65 million light years away. Determining this distance, and thus the scale of the universe, was one of the primary science goals of the Hubble Space Telescope (HST), but the first images from orbit showed that its mirror had not been ground correctly. Corrective optics were not installed on the HST until the Space Shuttle completed the first Hubble servicing mission. This delay gave astronomers, using the excellent mirror of the Canada–France–Hawaii Telescope on Mauna Kea, Hawaii, USA, the opportunity to image Cepheid variable stars in Virgo Cluster galaxies before the Hubble Space Telescope optics were corrected.

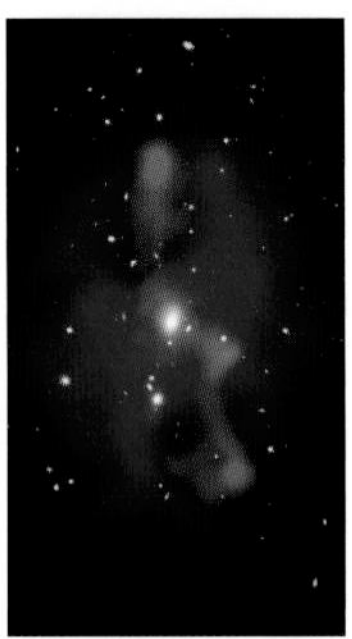

Above: A composite image of galaxy cluster MS0735.6+7421, located about 2.6 billion light years away in the constellation Camelopardalis. The three views of the region were taken with NASA's Hubble Space Telescope in February 2006, NASA's Chandra X-ray Observatory in November 2003, and NRAO's Very Large Array in October 2004. The Hubble image shows dozens of galaxies bound together by gravity.

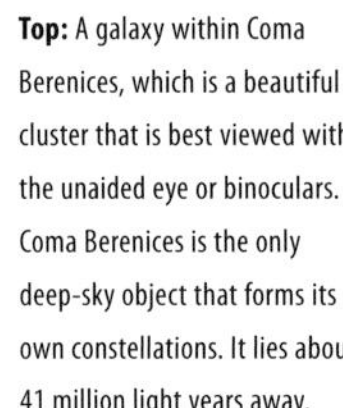

Top: A galaxy within Coma Berenices, which is a beautiful cluster that is best viewed with the unaided eye or binoculars. Coma Berenices is the only deep-sky object that forms its own constellations. It lies about 41 million light years away.

THE ANDROMEDA GALAXY

The Andromeda Galaxy (at right) is classified as an SAb galaxy, where "SA" means it is a normal spiral, and "b" means that its central bulge is average-sized. Its only unusual characteristic is that it sports two nuclei, compelling evidence that it has ingested a significant galaxy "recently." (The astronomical version of "recently" might mean many millions of years ago.) All large galaxies are thought to have grown to their present size by assimilating their smaller neighbors. Resistance is futile.

Color images show a golden elliptical bulge of elderly stars, surrounded by thick spiral arms, with stellar associations of young blue supergiant stars dotting the arms of the galaxy.

The galaxy is quite inclined to our line of sight, only 12.5° from edge-on, and this limits the detail that can be detected visually in a telescope. However, an 8-inch (20 cm) telescope easily shows a nucleus, the bright elliptical central bulge, and two long dust lanes on the side of the galaxy nearest to us.

Seeing two dust lanes means, by definition, that the brighter band between the dust lanes is one of the galaxy's thick spiral arms. With an appropriate chart, an 8-inch (20 cm) telescope can track down two bright globular clusters. One of them, known as G1, is the most massive globular cluster anywhere within the Local Group. A 16-inch (40 cm) telescope reveals the 18 brightest members of the galaxy's 300 to 400 globular clusters, as well as several open clusters.

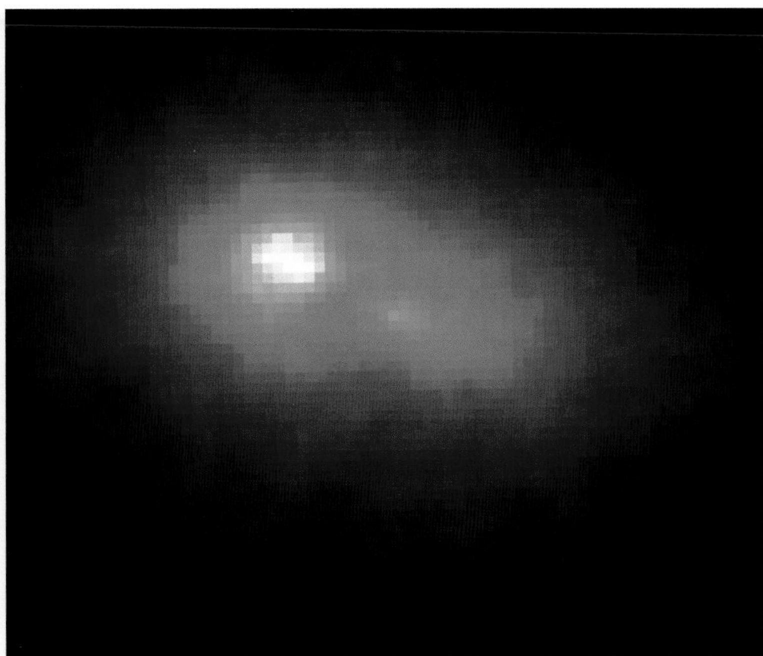

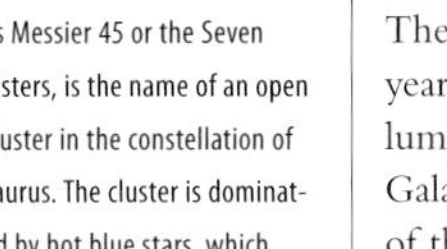

Left: The Pleiades, also known as Messier 45 or the Seven Sisters, is the name of an open cluster in the constellation of Taurus. The cluster is dominated by hot blue stars, which have formed within the last 100 million years.

The Local Group of Galaxies

The Local Group of Galaxies stretches across about ten million light years, with all its galaxies gravitationally bound together, forming a luminous neighborhood within our vast universe. Our Milky Way Galaxy and the Andromeda Galaxy, M31, are the dominant members of the Local Group of Galaxies. Both are large spirals. The Andromeda Galaxy is the queen of the Local Group. It has about 120 percent of the Milky Way's mass, so their sizes are of the same order of magnitude. The Triangulum Galaxy, M33, is a much smaller spiral.

The Large and Small Magellanic Clouds, both satellites of the Milky Way, are the fourth and fifth largest galaxies in the group. Twenty-one other galaxies are also members, but all range from small to inconsequential, and the majority are in orbit around one of the two dominant spiral galaxies.

Despite the expansion of the universe, the fact that the Local Group is gravitationally bound together means that the galaxies within the group are not moving away from each other. Nevertheless, the distance between our group of galaxies and every other clump of galaxies is increasing.

There have long been questions about whether various galaxies on the outskirts are gravitationally bound bona fide members of the Local Group, and some sources will list more galaxies than the, perhaps conservative, number of 26 given here. Those nearby galaxies that most sources no longer consider to be members are certainly gravitationally tugged by the Local Group, but do not seem to be in orbit around its center of mass.

Below: Young stars in the Magellanic Cloud. The Large and Small Magellanic Clouds, both satellites of the Milky Way, are the fourth and fifth largest galaxies in the Local Group of Galaxies.

Einstein's General Theory of Relativity predicts that if mass is squeezed into a small enough volume of space, the normal dimensions of space and time become so warped that nothing—not even light—can escape. It's there, but invisible. It's a "black hole."

Energetic galaxies

Active galactic nuclei

There is very good evidence that black holes with a few times the mass of our Sun—a few "solar masses"—exist as the final evolutionary phase of massive stars. There is also very good evidence that supermassive black holes—with millions or more solar masses—exist in the centers of galaxies.

Our galaxy apparently harbors a black hole of some four million solar masses. Although we cannot see it directly, the Hubble Space Telescope has been able to trace the orbits of stars circling it, and from their speeds we can estimate its mass. Similarly, by measuring general stellar speeds in the nuclei of other galaxies we have been able to indirectly detect many other black holes, some ranging up to around a billion solar masses.

Generally, the larger the bulge of stars in the center of a galaxy, the more massive its black hole. But whether the bulge or the black hole came first is a matter of debate. Also, not all black holes in the centers of galaxies exist as peacefully as ours. In a few percent of giant galaxies, interactions or disturbances are allowing material to be sucked toward the black hole, like bath water swirling toward the plug hole. While a black hole itself does not emit any light or radiation, it creates fierce gravitational forces in its immediate vicinity—enough to power a cosmic fireworks show. Hence, an "active galactic nucleus." There are two kinds of active galactic nuclei. The first squirts out material, but the active nuclei responsible are relatively inconspicuous—these are the radio galaxies. The other kind has luminous nuclei, ranging from those barely discernable, to the most continuously luminous objects in the universe—these are the quasars, Seyfert galaxies, BL Lac objects, and Liners.

Right: A faint blue arc of young stars can be seen in the right-upper portion of this image of the galaxy Centaurus A. Astronomers accidentally discovered this streak of young, blue stars, which were created when Centaurus A merged with a smaller neighbor galaxy over 200 million years ago.

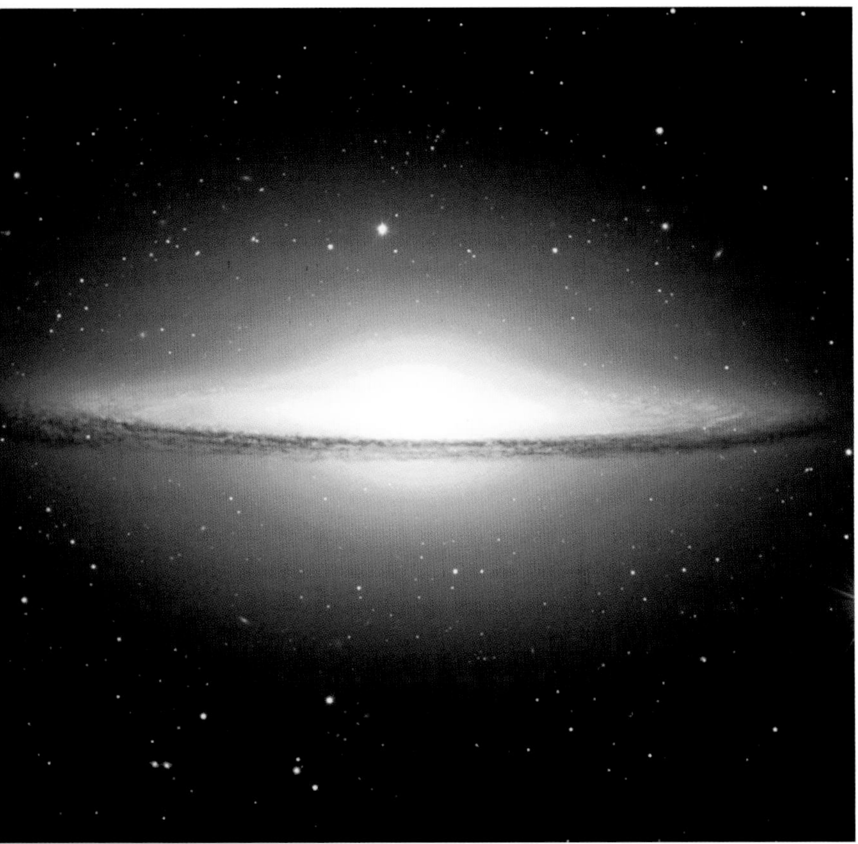

Below: Billions of old stars cause the diffuse glow of the extended central bulge in the Sombrero Galaxy. The very center of the Sombrero glows across the electromagnetic spectrum, and is thought to house a large black hole.

Radio galaxies

Most giant elliptical galaxies—the heavyweights of the cosmos—look benign. In visible light they are seen as gigantic spheroidal conglomerations of old stars, clearly dominant in mass, but seemingly mild in character. This is not so. Radio observations suggest a rather different picture. Astonishingly, two oppositely directed jets have been shot from the tiny central nucleus, with enough power to carry their contents far clear of the parent galaxy, before losing their stability and finally splattering into gigantic radio lobes.

Double radio lobes are therefore the signatures of radio galaxies. Their radio structure, sometimes spreading over millions of light years, dwarfs the physical size of the parent galaxy. Yet the size of the parent galaxy—ten- or a hundred-thousand light years across—dwarfs the size of the active galactic nucleus responsible.

Above: NASA's Hubble Space Telescope shows a group of galaxies, Seyfert's Sextet, engaging in a slow dance of destruction that will last for billions of years. The galaxies are so tightly packed together that gravitational forces are beginning to rip stars from them and distort their shapes. Those same gravitational forces eventually could bring the galaxies together to form one large galaxy.

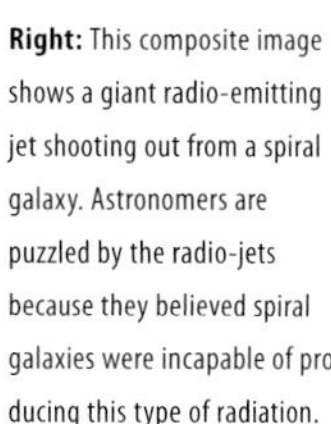

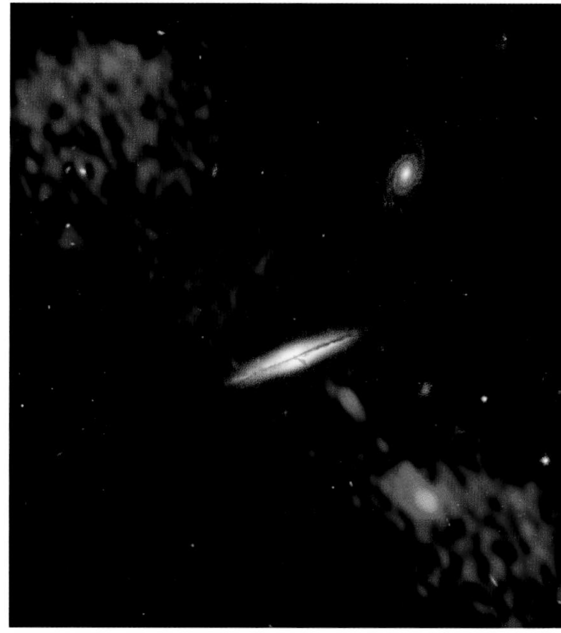

Right: This composite image shows a giant radio-emitting jet shooting out from a spiral galaxy. Astronomers are puzzled by the radio-jets because they believed spiral galaxies were incapable of producing this type of radiation.

Quasars

Quasars are active galaxies with highly luminous nuclei. Their phenomenal luminosity somehow originates in the immediate vicinity of the supermassive black holes that they harbor. They are considered the beacons of the cosmos.

The energy output is not only visible light; it probably originates as high energy X-rays, and cascades right down the electromagnetic spectrum, making the nuclei apparent in X-ray, ultraviolet, visible, infrared, and occasionally radio wavelengths. While they often exhibit jets, like radio galaxies, clearly most of the energy released has been diverted into luminosity.

The radiation output often shows variations on time scales as short as a week. To show such coherent variation, the region responsible cannot be larger than light-weeks in size, further indication that something as compact as a black hole must be involved. In physical size, the active nucleus must be tiny compared with the galaxy that hosts it. Yet the light from the nucleus may outshine all the stars in the galaxy many times over, so much so that the host galaxy is often lost in its glare. Suffice to say that when quasars were first discovered in the 1960s, they were at first mistaken for foreground stars within our own Milky Way galaxy.

Even sleepers are workers and collaborators in what goes on in the Universe.

Heraclitus, philosopher, 544–483 BCE

CENTAURUS A

The nearest radio galaxy is NGC 5128, or "Centaurus A." In visible light, it is seen as a giant elliptical galaxy, but with dust lanes and star-forming regions wrapped around its waist, it is almost certainly the remains of a smaller spiral galaxy that has been shredded by its strong gravity. Radio observations show a pair of inner radio lobes and an enormously distended pair of outer radio lobes as well. If "radio" eyesight were possible, it would appear as a conspicuous double object in the southern sky. It is obvious that the activity in Centaurus A is somehow related to its "cannibalism" of a neighboring galaxy.

The dream of space exploration is as old as human imagination. Ancient myths found in cultures around the world tell tales of heroes attempting to conquer the skies. In more recent times, the scientific romances of such writers as Jules Verne and H. G. Wells inspired dreams of traveling into space.

Space exploration

At the beginning of the twentieth century, the technological and scientific knowledge required to make the dream of spaceflight a reality was developing. By the end of the century, the simple black-powder rocket had been transformed into massive space launch vehicles, capable of taking satellites into orbit, people to the Moon, or probes to the planets. Scientists and engineers from around the world have played a part in this transformation.

Early concepts

Russian schoolteacher Konstantin Tsiolkovski was the first to realize that rockets were the ideal vehicles to make space travel a reality. Tsiolkovski's theoretical work provided the scientific foundation for astronautics by establishing all the basic mathematical laws of spaceflight and demonstrating that only a liquid-fuel rocket would have the thrust to put a rocket into Earth's orbit or to journey to another planet. In 1903, he published the first major work on astronautics, *The Exploration of Space with Reactive Devices*.

In 1919, American physicist Robert Goddard published *A Method of Reaching Extreme Altitudes*, describing the first practical sounding rocket for the investigation of the upper atmosphere. In 1926, he designed and flew the world's first successful liquid-fuel rocket, powered by liquid oxygen and petrol. The results of Goddard's research were fundamental to the development of modern rocketry. NASA's Space Flight Center in Maryland, USA, is named after him.

During World War II, German-built missiles for military use demonstrated that it was possible to develop rockets that were capable of long-distance travel while carrying strategic weapons. After the war, both the USA and the USSR sought German expertise to develop long-range missiles. As the Cold War continued, missile technology became so advanced that soon the largest missiles had achieved the capability to act as space-launch vehicles.

Launching the first satellites

In October 1957, the USSR stunned the world by launching the first Earth-orbiting artificial satellite, *Sputnik 1*. Weighing a little more than 186 lb (83 kg), *Sputnik* ("traveling companion") was a polished metal sphere with four long antennae. It contained a radio transmitter and atmospheric measuring instruments, and orbited Earth every 90 minutes. *Sputnik 1* remained in orbit for 96 days before disintegrating on re-entering Earth's atmosphere.

Sputnik 1 was massive compared with the tiny *Vanguard* satellite proposed by the USA, and a shockwave shook the west at its launch. The USA's response was to get a satellite into orbit as quickly as possible. *Explorer 1* was launched from Cape Canaveral, Florida, on January 31, 1958. It carried 18 lb (8 kg) of instruments designed to gather data on cosmic rays, meteorites, and orbital temperatures. Among other results, these instruments revealed the existence of the Van Allen radiation belts around Earth.

Vanguard 1 finally reached orbit on March 17, 1958, and made significant scientific discoveries. Weighing barely 3⅓ lb (1.5 kg)—50 times less than *Sputnik 1*—*Vanguard 1* was packed with early miniaturized electronics and used the first spacecraft solar cells, which enabled it to continue transmitting until 1964.

Achieving manned spaceflight

On April 12, 1961, 27-year-old cosmonaut Yuri Gagarin became the first person in space, riding the *Vostok 1* spacecraft into history for a

Above: American physicist Robert Goddard (1882–1945), in a photograph dated September 29, 1928, poses in a field behind a rocket set in a frame, with the combustion chamber and nozzle at the top.

Left: Army Ballistic Missile Agency scientists in Redstone Arsenal, Alabama, USA, examine a prototype of the *Explorer I* satellite, 1958. The model is actual size.

Right: Astronauts John Glenn (left) and Alan Shepard putting on their space suits. These men were part of Project Mercury, the first manned US space program, along with astronauts Walter Schirra, Donald Slayton, Scott Carpenter, Virgil Grissom, and Gordon Cooper.

single-orbit flight of 108 minutes. The first woman in space was Russian Valentina Tereshkova, on the final Vostok mission (*Vostok 6*) in June 1963.

On May 5, 1961, Alan Shepard made the first US spaceflight—a 15-minute sub-orbital lob in *Freedom 7*. The first US orbital flight occurred on February 20, 1962, when John Glenn orbited Earth for just over five hours in *Friendship 7*. The final Mercury flight took place in May 1963, when Gordon Cooper, in *Faith 7*, spent a full day in space. Before long, the ultimate goal in the space race between the two countries was to get a man all the way to the Moon.

Above: Valentina Tereshkova practices feeding in simulated flight conditions in preparation for the *Vostok 6* mission, when she became the first woman in space. Tereshkova was a textile worker before she was recruited to become a cosmonaut.

The space race

The USA concentrated its efforts on ensuring that astronauts would physically and psychologically survive long space missions. The USSR aimed to beat the US to the Moon, and took more risks. In March 1965 Soviet cosmonaut Alexei Leonov became the first person to

ANIMALS IN SPACE

Before any person could venture into space, scientists had to determine whether the human body could withstand the crushing forces of acceleration, exposure to deadly cosmic radiation, and physiological effects associated with weightless spaceflight—all of which could cause life-endangering structural and functional changes.

During the late 1940s, a number of animals were launched into the skies with varying results. Most of them died. In 1951, the Soviets sent two dogs 62 miles (99 km) into space. They returned unharmed. Although she died within hours of takeoff, in 1957, another dog Laika became the first living creature to orbit our planet.

In 1961, a chimpanzee named Ham (pictured at right) was launched on a sub-orbital mission ahead of America's first manned spaceflight by Alan Shepard. The following year, a chimpanzee named Enos flew into orbit aboard another Mercury spacecraft, this time paving the way for a successful three-orbit flight by astronaut John Glenn.

Later Apollo, Skylab, Mir, and shuttle missions carried vast numbers of biological subjects into space, including pocket mice, tortoises, frogs, fish, spiders, and a variety of non-human primates.

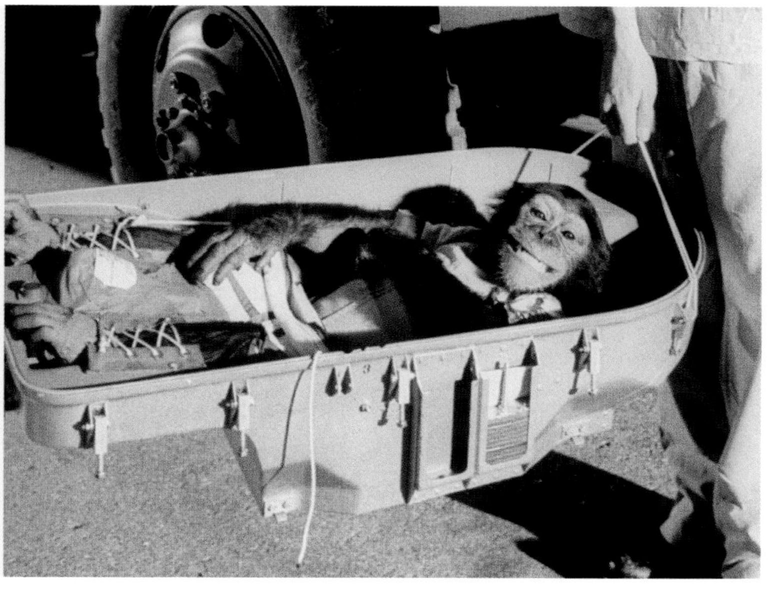

Opposite: This oblique view of the Moon's surface was photographed by *Apollo 10* astronauts in May 1969. Center point coordinates are located at 13 degrees, 3 minutes east longitude and 7 degrees, 1 minute north latitude.

Left: *Apollo 7*, launched in October 1968, spent 11 days in orbit, during which time everything on board was tested rigorously. The mission was commanded by Walter Schirra, the only astronaut to fly all three US spacecraft—Mercury, Gemini, and Apollo.

> The whole procedure [of shooting rockets into space]… presents difficulties of so fundamental a nature, that we are forced to dismiss the notion as essentially impracticable…
>
> Richard van der Riet Wooley, astronomer, 1906–1986

walk in space, but he nearly died because his space suit ballooned and stiffened, making it difficult for him to re-enter the airlock.

The first American astronaut to walk in space was Edward White II in June 1965, as part of the Gemini program that extended NASA's expertise in space operations.

NASA's Apollo program was not immediately successful—the astronauts in *Apollo 1* died during a pre-flight test in 1967, the same year that the Soviet *Soyuz 1* mission also ended in disaster when the descent module crashed killing cosmonaut Vladimir Komarov.

Aiming for the Moon

By *Apollo 7*, reaching the Moon was achievable. After 11 days in orbit in October 1968, the craft showed no significant problems. The Service Module's main propulsion engine was successfully tested and the maneuver of extracting the Lunar Module from its launch cradle was also practiced. Launched on December 21, 1968, *Apollo 8* made the first crewed flight to the Moon, arriving in lunar orbit three days later. The crew of *Apollo 8* (Frank Borman, James Lovell, and William Anders) made 10 orbits of the Moon, and on Christmas Eve (US time) made a telecast from lunar orbit. The breathtaking images taken during this six-day mission, showing Earth rising over the Moon and our world floating in the blackness of space, are some of the most important space images ever taken; they would help to encourage environmental awareness and give rise to the concept of Spaceship Earth.

Although the Lunar Module—the landing vehicle that would carry two astronauts to the surface of the Moon—was not ready to fly the *Apollo 8* mission, it was tested in Earth orbit in March 1969, during the *Apollo 9* mission. This 10-day flight was the first to test the complete Apollo spacecraft that would journey to the Moon, putting all three Apollo vehicles (Saturn V, Command Module, and the Lunar Module) through their paces.

The mission's crew, James McDivitt, Russell Schweickart, and David Scott, practiced the un-docking and docking maneuvers that would occur in lunar orbit, and also fired the Lunar Module's descent engine, and later its ascent engine in a simulated liftoff.

The dress rehearsal

With the Lunar Module successfully tested in Earth orbit by *Apollo 9*, the eight-day *Apollo 10* flight became the full dress rehearsal for the planned *Apollo 11* landing mission. If all went smoothly with the *Apollo 10* mission, the historic first Moon landing would proceed. The experienced crew of Thomas Stafford, John Young, and Eugene Cernan, all of whom had flown during the Gemini program, chose light-hearted names for their Command Module (Charlie Brown) and Lunar Module (Snoopy), which would attract criticism from some quarters as being "too flippant."

Launched on May 18, 1969, *Apollo 10* closely followed the *Apollo 11* flight plan, orbiting the Moon 31 times, and twice took its Lunar Module down to within 9$\frac{9}{10}$ miles (14.5 km) of the lunar surface. The mission was not a mere technical success; it also provided spectacular coverage of spacecraft operations and views of the Earth and Moon with 19 color television transmissions. The stage was now set for the first landing on the Moon.

Right: The docked *Apollo 9* command service module beyond and the lunar module *Spider* on the left with Earth in the background. This preparatory mission for the Moon landing was the first to feature the Saturn rocket in its full lunar configuration.

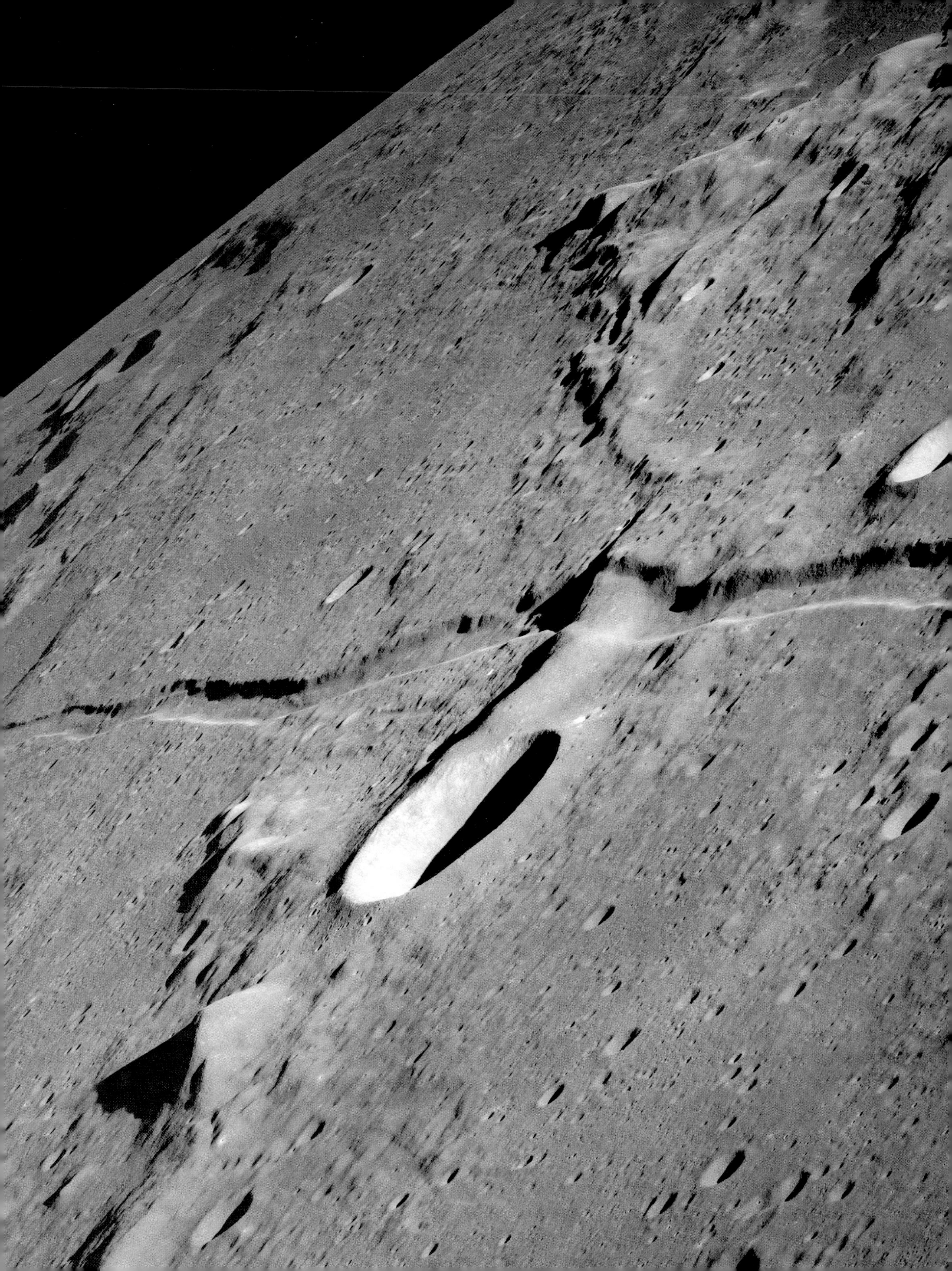

Apollo 11 was launched on July 16, 1969. Over a million people crowded the vicinity of Kennedy Space Center in Florida to watch the launch live, and more than 600 times that number witnessed it on television at home and abroad, as Apollo 11—carrying commander Neil Armstrong, Command Module pilot Michael Collins, and Lunar Module pilot Edwin "Buzz" Aldrin—thundered into space.

Man on the Moon

Apollo 11, with its Command Module *Columbia* and Lunar Module *Eagle*, passed behind the Moon on July 19, and entered lunar orbit. After several orbits, the Command and Lunar modules separated, and Michael Collins was left alone in orbit in *Columbia* as Neil Armstrong and Buzz Aldrin descended to the lunar surface in *Eagle*. The landing site was going to be the Sea of Tranquillity, but as the landing began, it was discovered that the Lunar Module was further along its descent trajectory than planned and would have to land some distance west of the intended site.

Armstrong took manual control of the Lunar Module and, with Aldrin's assistance, guided the spacecraft to a landing at 20:17 UTC on July 20, 1969, with only seconds' worth of fuel left. Armstrong's first words after landing were: "Houston, Tranquillity Base here. The Eagle has landed."

Above: Neil Armstrong stepped into history on July 20, 1969 when he left the first human footprint on the surface of the Moon. He described the surface dust as "fine ... like a powder."

Small step; giant leap

Six and a half hours after landing, at 2:56 UTC on July 21 (still July 20 in the USA), Armstrong left the Lunar Module, climbed down the ladder, deployed the Modular Equipment Stowage Assembly (MESA) that was folded against Eagle's side, and activated the TV camera. The signal was picked up at both the Goldstone tracking station in the USA and the Honeysuckle Creek Tracking Station in Australia. Pictures from Honeysuckle were broadcast to an audience of more than 600 million people worldwide.

Armstrong stepped off *Eagle*'s footpad and became the first person in history to set foot on another world. As he did so, he uttered his famous declaration: "That's one small step for (a) man: One giant leap for mankind." He took some photographic panoramas and collected a contingency sample of lunar soil.

When Buzz Aldrin joined him 18 minutes later, the astronauts set up the television camera on the surface to broadcast their activities. They collected rocks and geological core samples. After two and a half hours on the lunar surface the astronauts returned to *Eagle*. They loaded two sample boxes containing 48 lbs (21.8 kg) of lunar surface material into their craft, but lightened the ascent stage for return to lunar orbit by leaving behind their life support backpacks, lunar overshoes, and other equipment.

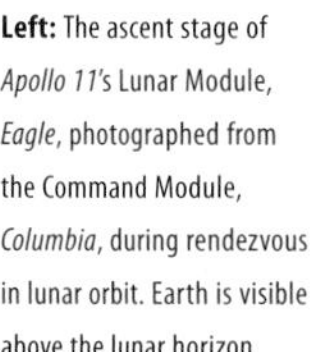

Left: The ascent stage of *Apollo 11*'s Lunar Module, *Eagle*, photographed from the Command Module, *Columbia*, during rendezvous in lunar orbit. Earth is visible above the lunar horizon.

Going home

After lifting off from the Moon and rendezvousing in orbit with *Columbia*, precautions were taken to prevent any contamination from the Moon being brought back to Earth. Before transferring to the

I think a future flight should include a poet, a priest and a philosopher … we might get a much better idea of what we saw.

Michael Collins, astronaut, b. 1930

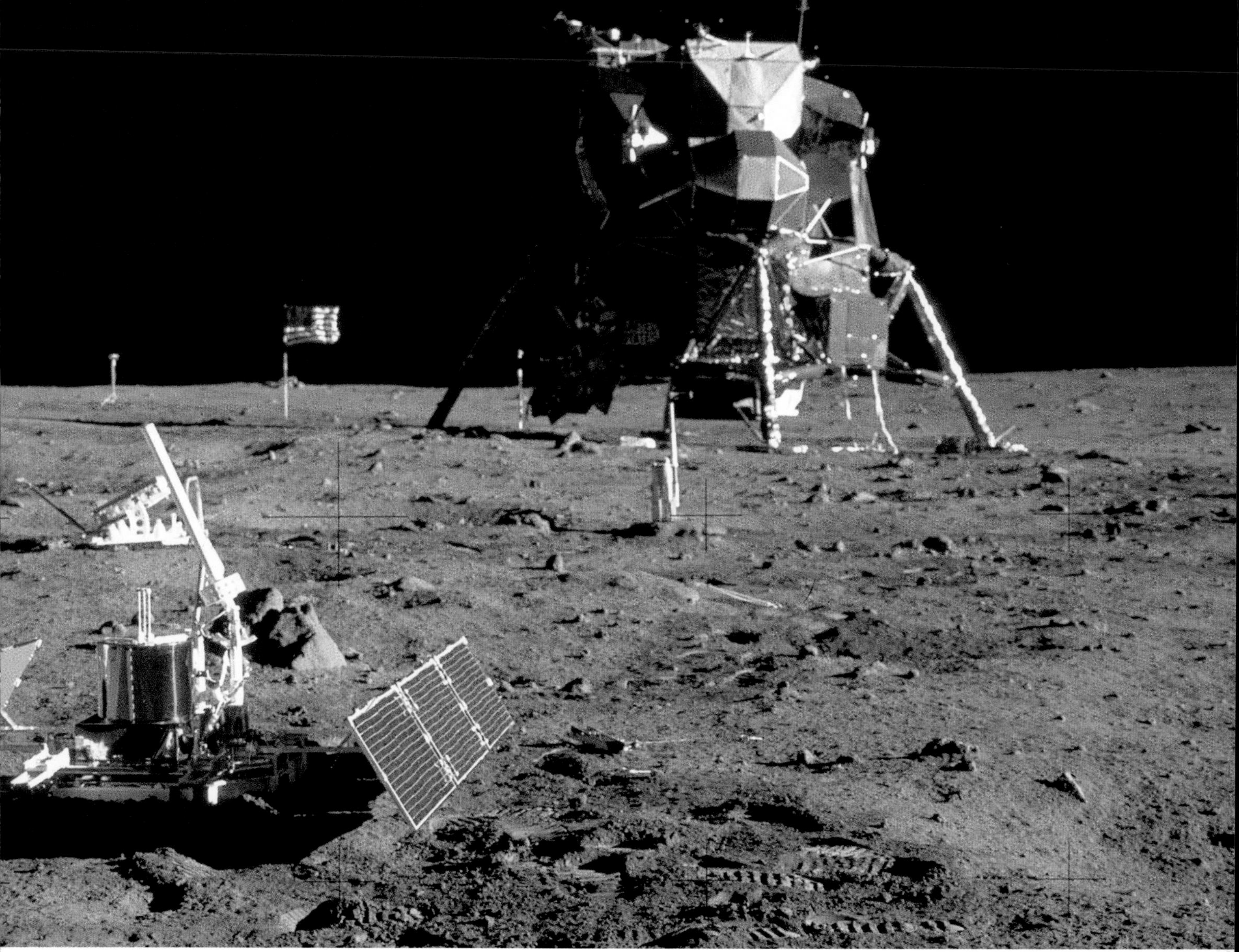

Command Module, Armstrong and Aldrin vacuumed their clothes and equipment thoroughly to remove any traces of lunar dust. *Eagle* was jettisoned and left in lunar orbit, to crash back some time later onto the Moon's surface.

On July 24, 1969, *Apollo 11* returned to Earth. The crew went into quarantine for three weeks in case the Moon contained any unknown pathogens that could affect humans. On August 13, Armstrong, Aldrin, and Collins were released from isolation to begin a long series of celebrations marking their historic achievement.

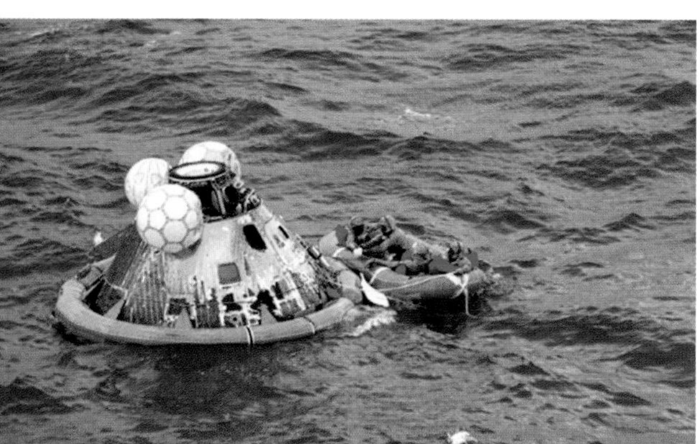

Above: Buzz Aldrin deploys the Early Apollo Scientific Experiments Package (EASEP). In the foreground is the Passive Seismic Experiment Package (PSEP); beyond is the Laser Ranging Retro-Reflector (LR-3); in the left background is the lunar surface television camera; in the right background is the Lunar Module *Eagle*.

Left: *Apollo 11* splashed down about 812 nautical miles southwest of Hawaii. The astronauts and their recovery team wore biological isolation outfits to prevent possible contamination from lunar bacteria.

THE *APOLLO 13* MISSION

Two days after its launch on April 11, 1970, *Apollo 13* was crippled by an explosion in the Service Module, which led to a rapid loss of power and oxygen; plans for the lunar landing were abandoned and the systems in the Command Module, *Odyssey*, were turned off to preserve its ability to re-enter Earth's atmosphere. The Lunar Module, *Aquarius*, with its independent power, water, and oxygen supplies, became the crew's "lifeboat" while *Apollo 13* made a looping trajectory around the Moon and returned to Earth. Under trying conditions, with little power, drinking water, or heat, the crew conducted the necessary maneuvers critical to their safe return. On the ground, flight controllers and their support teams, under extreme pressure, devised solutions to sustain the spacecraft and ensure the crew's safe return. On April 17, *Apollo 13*, having jettisoned *Aquarius* shortly before re-entry, splashed down safely; it was a "successful failure" that demonstrated the amazing resilience of the spacecraft and the program as a whole.

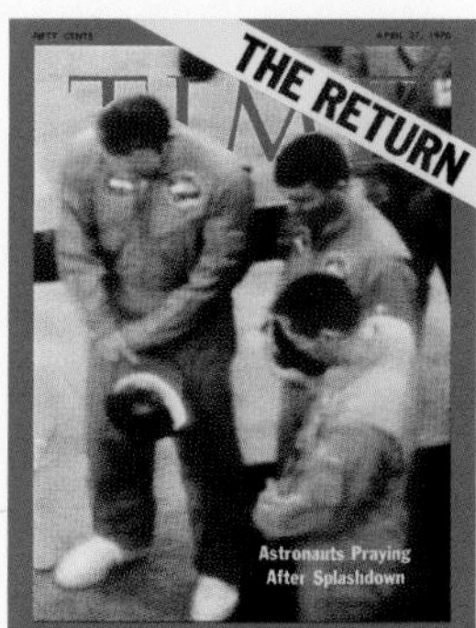

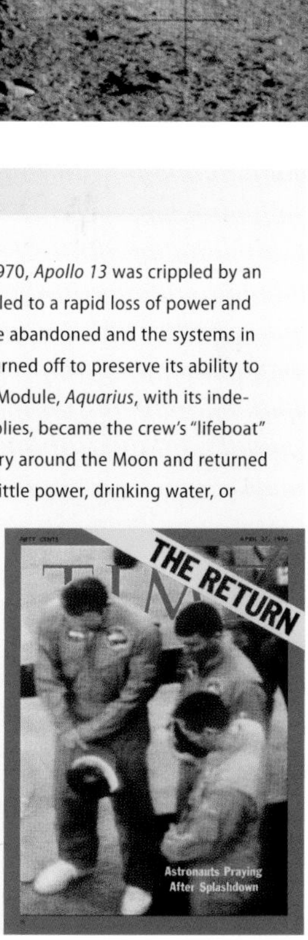

Exploring space is expensive and is also filled with danger for humans, which has made it difficult to extend our exploration in person beyond the Moon. Consequently, scientists have developed automated probes so that we can still explore the vastness of our Solar System. These probes act as our eyes and ears in traveling to distant worlds.

Robots in space

These robotic explorers have revolutionized our knowledge of our nearest neighbors in space, revealing a Solar System profoundly different from what we thought we knew just 50 years ago.

During the 1990s, the US probes *Clementine* (1994) and *Lunar Prospector* (1998) explored the Moon with new multispectral imaging technology to identify potential mineral resources and to search for evidence of frozen water. Since 2004, with the renewed US emphasis on manned lunar exploration, the Moon has once again become a target for automated spacecraft.

Right: A view of the fully deployed *Magellan* spacecraft orbiting over Earth. *Magellan* was launched in 1989 from the space shuttle *Atlantis* from Earth orbit—it was the first interplanetary probe to be launched from the shuttle. *Magellan* was designed to penetrate the Venusian clouds with its powerful radar beams and map the planet in detail.

Venus unveiled

Venus's thick cloud layer prevented direct telescopic observation of its surface and it became an early target of planetary exploration. America's *Mariner 2* (1962) was the first successful Venus probe. Its data disproved theories about a verdant or oceanic Venus, and valuable information from successive Mariner craft and Soviet Venera probes revealed that Venus is a hellish world of high temperatures and pressure, created by a runaway greenhouse effect.

Exploring Mars

The first successful Mars probe, the US *Mariner 4* (1965), revealed a planet with a cratered surface and an atmosphere that was mostly made up of carbon dioxide. Later Mariner and Soviet "Mars" spacecraft confirmed Mars to be a much harsher environment than was expected, but the first successful landers, the US *Viking* craft (1976), took some samples of Martian soil and analyzed them in onboard laboratories for signs of life, although the results were inconclusive. After the announcement in 1996 of possible evidence for ancient microbial life on Mars, it again became a focus for exploration, beginning with the first Mars rover, *Sojourner*, during the 1997 Mars Pathfinder mission. *Sojourner* operated for 90 Martian days, covering about 2,691 square feet (250 m^2).

The journey of *Ulysses*

Since it was launched in 1990, the *Ulysses* spacecraft, a joint NASA–ESA project, has orbited the Sun's poles to provide a completely new view of our closest star. In more recent times, the *Genesis* spacecraft (2001–2005) has, amazingly, trapped and returned samples of the solar wind in special collectors for study.

Until recently Mercury was a mystery, the details of its surface almost impossible to distinguish from Earth-based telescopes. But in 1974–75, *Mariner 10*, the first two-planet space probe—it also explored Venus—gave us a close-up view of this harsh and Moon-like world.

Going the distance

The exploration of the outer Solar System needed spacecraft able to navigate the hazardous Asteroid Belt and function autonomously for the incredibly long periods required to reach the outer planets. It was not until 1972 that the US *Pioneer 10* spacecraft was launched toward Jupiter, followed by *Pioneer 11* in 1973. *Pioneer 11* visited both Jupiter and Saturn, allowing astronomers to see Saturn's glorious rings up close for the very first time.

By the end of the twentieth century, robotic probes had explored every planet in the outer Solar System. Space exploration continues. In 2006 the *New Horizons* spacecraft was launched to finally explore the dwarf planet Pluto and its moons, and possibly other trans-Neptunian objects in the Kuiper Belt.

SURVEYOR AND LUNAR ORBITER

Launched between 1966 and 1968, the American Surveyor program was designed to demonstrate the feasibility of actually landing a spacecraft on the Moon. The Surveyors carried a suite of instruments to evaluate the suitability of the sites for the manned *Apollo* landings. Several Surveyor landers carried small scoops designed to test the mechanical properties of lunar soil. Some also had alpha scattering instruments that helped determine the chemical composition of the soil.

Complementing the Surveyor program were the Lunar Orbiter probes, launched in 1966 and 1967 with the express objective of mapping the surface of the Moon before the proposed *Apollo* landings. The five missions in the series were all successful, and 99 percent of the Moon was photographed with a resolution of 200 ft (60 m) or better. This is the first photograph of Earth from *Lunar Orbiter 1*.

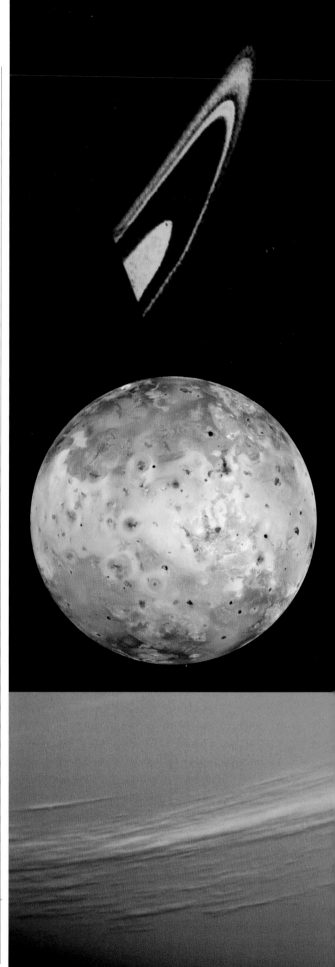

Right: An amazing image of planet Saturn's rings as seen by spacecraft *Pioneer 11*. Four spacecraft have visited Saturn, *Pioneer 11* being the first; it flew past it in 1979 and took the first close-up pictures of the planet.

Right: *Galileo* acquired its highest resolution images of Jupiter's moon Io on July 3, 1999 during its closest pass to Io since orbit insertion in late 1995. This color mosaic uses the near-infrared, green, and violet filters (slightly more than the visible range) of the spacecraft's camera.

God has no intention of setting a limit to the efforts of man to conquer space.

Pope Pius XII, 1876–1958

Left: *Viking 1* took this color picture of the Martian surface and sky on July 24, 1976. The camera, facing southeast, captured part of the gray structured spacecraft in the foreground. A bright orange cable leads to one of the descent rocket engines.

Right: A high-resolution color image showing evidence of Neptune's bright cloud streaks, taken from the *Voyager 2* spacecraft. The linear cloud forms stretch approximately along lines of constant latitude, and the sides of the clouds facing the Sun are brighter.

The first Earth-orbiting space stations were a crowded labyrinth of research modules, apparatus, and experiments, and greatly advanced our knowledge of science, medicine, weather patterns, and the Universe. That research continues to this day.

Space stations

Early space stations

On April 19, 1971, the Soviet Union launched the world's first space station, *Salyut 1*, into orbit atop a Proton rocket. The *Soyuz 10* mission to dock with the station failed. But the three-man *Soyuz 11* crew entered and occupied the station. Like the previous crew, they were not wearing cumbersome space suits in order to save precious space aboard the cramped Soyuz.

Cosmonauts Georgi Dobrovolsky, Vladislav Volkov, and Viktor Patsayev lived and worked aboard the station for three weeks, successfully completing a full program of scientific, astronomical, and biomedical studies. On June 29, 1971, they strapped themselves into their *Soyuz 10* craft to return to Earth, and separated from the *Salyut 1* station. An automated re-entry and parachute landing took place, but when recovery crews opened the spacecraft they found to their horror that all three cosmonauts had perished within seconds when the air escaped from their cabin due to a jammed valve. Subsequent crews were once again required to wear the lifesaving pressure suits. *Salyut 1* was soon abandoned, and would eventually burn up on re-entry in October 1971.

Right: Backdropped against a blanket of heavy cloud cover, the Russian-built *Zarya* control module, nears the space shuttle *Endeavour* and the US-built *Node 1*, also called *Unity* (seen in the foreground).

The Mir space station

On February 20, 1986, the core module of a mighty Soviet space station called *Mir* was launched into orbit to begin an epic 15-year journey. It was occupied by numerous Soviet and Russian expedition crews, as well as astronauts from several countries and space agencies, including the United States, Europe, and Japan. America's space shuttles docked with *Mir*, and seven NASA astronauts carried out long-duration, internationally cooperative missions working well alongside their Russian counterparts.

Before it was de-orbited, *Mir* had completed 89,067 orbits, covering 2,260,840,632 miles (3,638,470,307 km), roughly three times the distance from the Earth to Saturn.

Below: A view of space shuttle *Atlantis* still connected to the *Mir* space station was photographed by the *Mir-19* crew on July 4, 1995. Cosmonauts temporarily undocked the *Soyuz* spacecraft from the cluster of *Mir* elements to perform a brief fly-around.

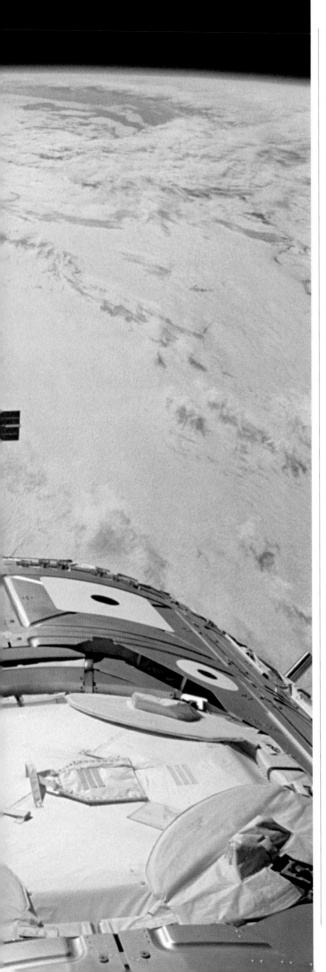

SKYLAB—LABORATORY IN SPACE

In its nine months of operation during 1973, *Skylab,* America's first space station, gained renown as an orbiting laboratory of immeasurable scientific value. During *Skylab's* 3,896 revolutions of the planet, its three visiting crews confirmed that the resources of space could offer new and unique approaches to science and technology, and the way in which we look at Earth and our universe. Among experiments carried out by manned crews was tracing the development of spots on the Sun that can cause radio interference and have other influences on life on Earth. (Here, Charles Conrad showers aboard *Skylab 1*.)

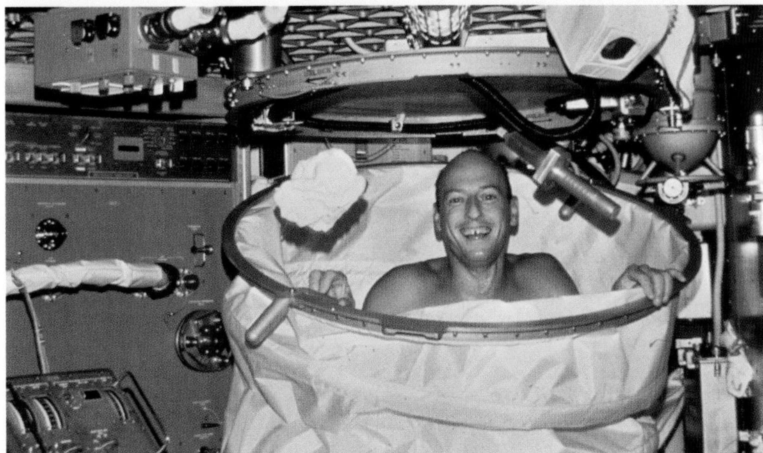

It is difficult to say what is impossible, for the dream of yesterday is the hope of today and the reality of tomorrow.

Robert Goddard, physicist, 1882–1945

The International Space Station (ISS)

In a vast and complex project, 16 nations—Belgium, Brazil, Canada, Denmark, France, Germany, Italy, Japan, the Netherlands, Norway, Russia, Spain, Sweden, Switzerland, the United Kingdom, and the United States—were involved in the creation of the International Space Station (ISS), a massive craft that became our newest star in the night sky; growing increasingly brighter as additional modules, solar arrays, and other components were delivered and attached over several years and several missions.

With assembly planned for completion in 2010, the ISS is a long-term laboratory in space, established in a realm where gravity, temperature, and pressure can be manipulated to assist in performing long-duration research in the medical, materials, and life sciences areas, while allowing vital studies and experiments that could never be achieved in ground based-laboratories. It will also serve as the first step to future space exploration.

Above: A fish-eye lens allows a unique view of the space shuttle *Atlantis* docking with *Mir*. The space shuttle fleet supplied and serviced *Mir* over 11 flights until the space station was de-orbited on March 23, 2001.

Aiming for 2010

The completed ISS—around four times the size of *Mir*—will measure 356 feet (108.5 m) across by 192 feet (58.5 m) long, and 100 feet (30.5 m) high. It will maintain an internal pressurized volume of 34,700 cubic feet (983 m^3). The 310-foot (95 m) long integrated Truss Structure and solar arrays, with an active area of 32,528 square feet (3,022 m^2), will generate more than 80 kilowatts of electrical power for the station.

A fully assembled ISS could house rotating international crews of six or seven in an area roughly equivalent to the interior of a 747 jet liner. More than 100 station elements comprising structures, equipment, and supplies will have been delivered on 45 space shuttle missions, *Soyuz* and *Progress* spacecraft, and atop Russian Zenit and Proton rockets, while multiple assembly and maintenance spacewalks will have been carried out by successive crew members.

It is difficult to imagine a time when images of news events from the other side of the world would take days to reach us. Now, the world has grown closer because of satellite applications, which allow us to send information and ideas across the world as if we were all sitting in the same room.

Satellites and space telescopes

While it is possible to transmit information around the globe using undersea cables, in many ways it is much easier to send and receive the information using a satellite orbiting the planet. This also allows the sender and receiver to be mobile, such as on a ship or aircraft, and they can also be on the remotest parts of our planet's surface and still stay in easy contact.

Some early satellites had a reflective surface, allowing signals to be bounced off their coatings, but more sophisticated satellites receive and retransmit messages electronically. Television signals, telephone calls, radio communications, and even internet services utilize the capabilities provided by satellites.

Satellites also allow us to have an overview of activities on Earth. This is important for weather reporting and forecasting. Rather than attempting to combine a large number of scattered reports from the ground, entire weather systems can be seen from orbit, and their direction and changes recorded. Large storms forming in remote parts of the ocean, that may otherwise have been unknown, can be constantly monitored. In this way, natural disaster alerts can be issued with much earlier warning, and loss of life averted. Hurricanes, forest fires, dust storms, and volcanic activity can be observed for short-term effects. In addition, longer-term changes such as ocean current direction, ice and snow cover, pollution, ozone depletion, and global warming can be accurately monitored over time.

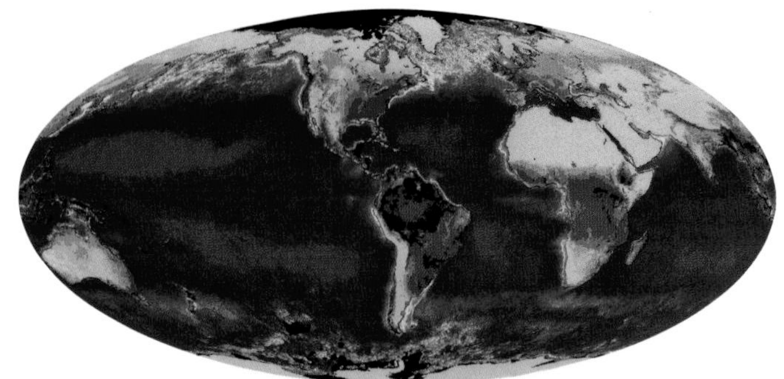

Above: An image from the National Oceanic and Atmospheric Administration's Polar Orbiting Environmental Satellite showing global vegetation cover. Dense vegetation is represented by shades of purple and green; sparse vegetation by shades of brown.

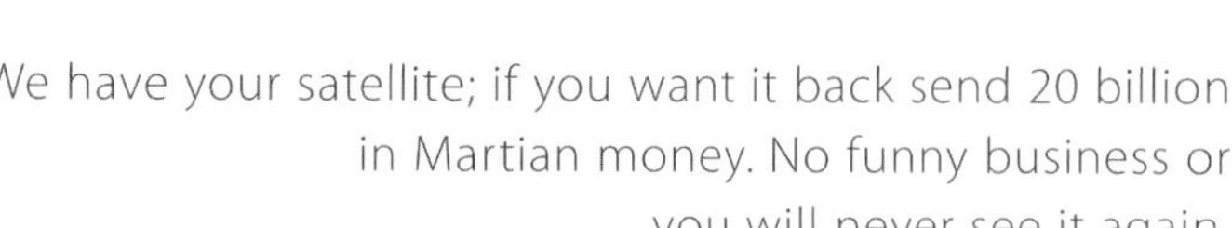

We have your satellite; if you want it back send 20 billion in Martian money. No funny business or you will never see it again.

Graffiti at NASA's Jet Propulsion Labs

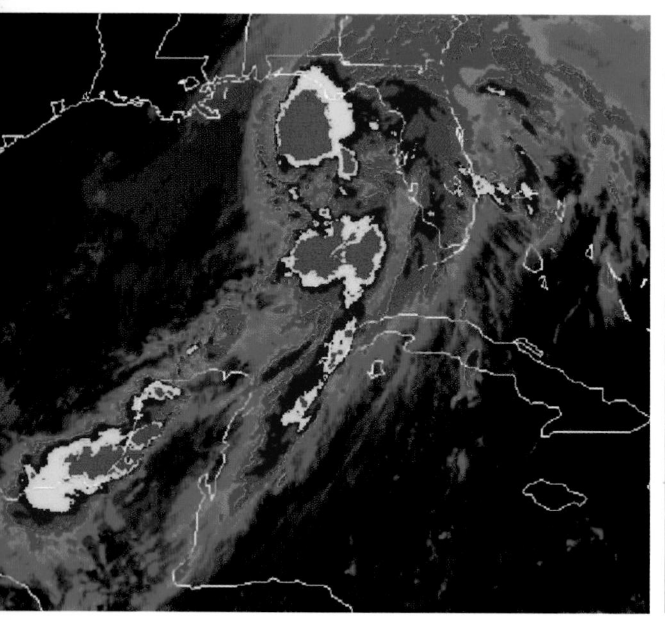

Left: Tropical storm "Josephine" raging in the Gulf of Mexico on October 7, 1996 in an enhanced infrared satellite image from the Spaceflight Meteorology Group in Houston, Texas, USA. Such satellite images allow early warning of disasters.

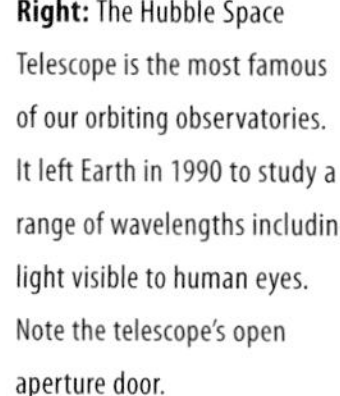

Right: The Hubble Space Telescope is the most famous of our orbiting observatories. It left Earth in 1990 to study a range of wavelengths including light visible to human eyes. Note the telescope's open aperture door.

Accurate maps can be made by using imagery from space, and long-term observations can also show human and natural changes to our planet, such as the growth of cities, major changes in agricultural land use, increase in the amount and intensity of city lights, and changes in natural vegetation.

Satellites have also been sent into orbit around other planets in our solar system, such as Mars, Saturn, and Jupiter, to study cloud and surface features, and how they change over time.

The Hubble Space Telescope

The most famous of the orbiting observatories, the Hubble Space Telescope has made a large number of extremely important astronomical discoveries, including a more accurate measurement of the expansion of the universe, observation of a comet impact with planet Jupiter, evidence of planets around other stars, and images of remote galaxies never previously observed. The telescope has also been an inadvertent proving ground for astronaut repair missions. Shortly after launch, it was discovered there was a tiny flaw in the shape of the primary mirror. Missions to install correcting optical equipment, and subsequent servicing missions, have been among NASA's most critical and successful spacewalking tasks. After the second shuttle disaster in 2003, it was decided that it was too risky to continue to service Hubble. The resulting outcry from astronomers and the public meant that this decision was eventually reversed.

EDWIN HUBBLE

In 1925, American astronomer Edwin Hubble (1889–1953), at right, proved that the Milky Way is not the entire universe when he discovered that Cepheid variables in spiral nebulae proved that they were far beyond the boundaries of our galaxy. Hubble's discovery meant that the universe was at least ten times larger than had previously been supposed. He used the new 100-inch (254 cm) telescope at Mt Wilson Observatory, California, to identify the "red shift," or Doppler effect, which causes objects moving away from us to appear red, while those approaching us appear blue. A similar effect occurs when the siren on a passing emergency vehicle seems to lower in pitch as it passes (see pages 46–47). Color and sound both travel in waves, which are compressed as they approach, thus decreasing the wavelength.

In 1929, Hubble observed 46 galaxies whose distance from Earth had been confirmed by his observations at Mt Wilson and discovered that the red shift was indeed greater at greater distances. Hubble's most significant discovery, it proved that the universe is not only expanding but also accelerating. This relationship could be applied to entire galaxies whose stars were too faint to be observed individually. The revelation increased the size of the observable universe by yet another factor of ten.

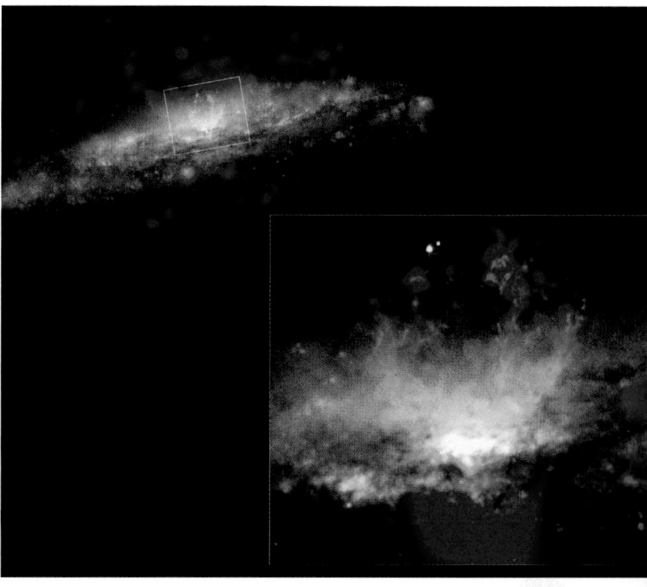

Right: This NASA composite image of galaxy NGC 3079 was created from images from the Hubble Space Telescope and the Chandra X-Ray Observatory. Large filaments of gas blown out from the galaxy's center by winds from a black hole or exploding star, are visible in the Chandra image (blue) superimposed over the Hubble image (red and green).

Below: A full scale model of the James Webb Space Telescope, a planned space infrared observatory, intended to be a vast improvement on the ageing Hubble Space Telescope. It will be built and operated by NASA with help from ESA and CSA.

BIOLOGY

Curiosity about living things is ancient. Early observers, including the ancient Greeks, were content to observe and catalogue. "Why" and "how" questions came later, stimulated at least in part by major discoveries made with the microscope, which came into widespread use in the seventeenth century.

Introduction

Above: English naturalist Charles Darwin (1809–1882) challenged the scientific community with the publication in 1859 of his most influential work, *The Origin of Species by Means of Natural Selection*.

These showed new structures in plants and animals never before seen, as well as new organisms, previously unknown or even unsuspected, such as bacteria and yeasts, so requiring the classification of the living world to be expanded beyond just plants and animals.

Rocking the boat

Many early biologists saw the natural order as divine handiwork. Young Charles Darwin probably felt this way too, but driven by the accumulation of evidence, he changed his view. The process of natural selection which he saw as powering biological evolution seemed to many to be purely random, a matter of trial and error, leaving no place for any divine plan or even any sense of purpose beyond the struggle to survive. For biologists in general, Darwin's views are now mainstream, but other people remain disquieted.

In every science a vital role is played by "standard models," ways of viewing and summarizing available data coherently so it encourages further experimentation. In biology, a key concept developed early in the nineteenth century was "cell theory." This proposed that all living things—plants, animals, bacteria, yeasts—are composed of fundamental building blocks called cells which are mostly identical within any one organism.

It therefore follows that much insight into the way the whole organism behaves can be gained by understanding the workings of the individual cells. Furthermore, it was soon apparent that all new cells come from the subdivision of existing cells. Nothing comes from nothing. The spontaneous generation of life, a popular notion 200 years ago, simply does not occur.

The quest for understanding

One great thread of enquiry running through biology over time was the quest to understand inheritance, the mechanism whereby the traits of the parents of any organism are preferentially displayed in their offspring. Unraveling the mystery of inheritance took several centuries, beginning with the realization that both the male and the female of the species contribute to the offspring's characteristics.

Below: Magnified 18,000 times through an electron microscope, bacteria are shown contaminating chicken meat. The fibers between the cells are an extracellular polymer called the glycocalyx. The glycocalyx forms the "slime layer."

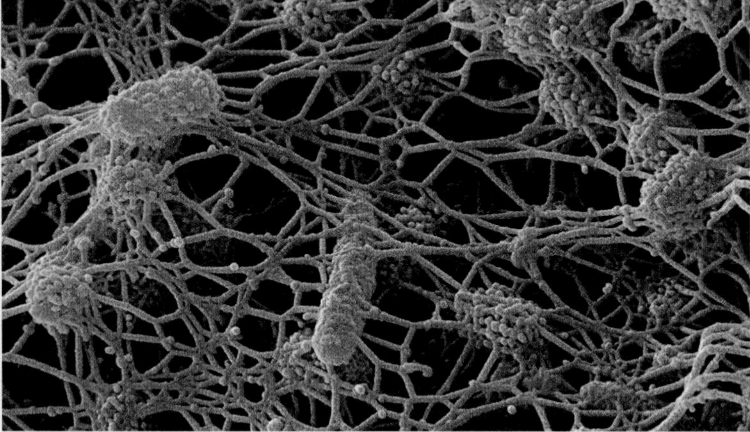

Below: A mining bee (*Andrena* species) feeds from a coneflower, an example of the food chain—the feeding order in a community of organisms, with each organism gaining energy from the organism preceding it.

Right: A researcher examines stem cell lines through a microscope, which are displayed on a computer screen, at the Burnham Institute, California, USA. Stem cells have the potential to develop into many different types of cells.

In most animals, for example, the ovum from the female begins to grow and develop only when fertilized by the sperm of the male. Obvious to us today, such a notion was thought improbable or even obscene a few centuries ago. When sperm were first discovered in human semen through the use of the microscope, popular opinion regarded them as parasites.

Many crucial discoveries began in the nineteenth century. The insights of Gregor Mendel, gained after exhaustive experimentation but forgotten for 50 years, can be summarized as follows. Through the process of fertilization, each parent passes on packages of genetic information, later called "genes," which helps to control how the offspring appears and behaves. At the same time, studies of the nucleus in the heart of every cell were beginning to uncover the structures that held that information.

Nearly a century later, after the contributions of many brilliant minds, these enquiries climaxed in the unraveling of the architecture of the vital chemical DNA, showing how inherited information is coded and passed on during cell division. This led to the development of technologies to manipulate inherited characteristics, so-called "genetic engineering."

These very powerful techniques are among the most important practical outcomes of the growth of biological knowledge in recent decades, with significant implications for agricultural production, industry, and human health. They are also among the most controversial. In the eyes of many critics, interference in the mechanism of inheritance is equivalent to "playing God."

There are undoubtedly more controversies to come. Through the Human Genome Project, a huge international collaboration, which climaxed in the early years of the twenty-first century, we can now read the genetic message that controls the way each of us appears and behaves, including our potential to suffer particular diseases. Cloning and the use of stem cells are other advances for which the scientific understanding remains clearer than the moral implications.

Perhaps most controversial all is the possibility that biologists may be able to create life artificially. That remains an immense challenge, both morally and scientifically.

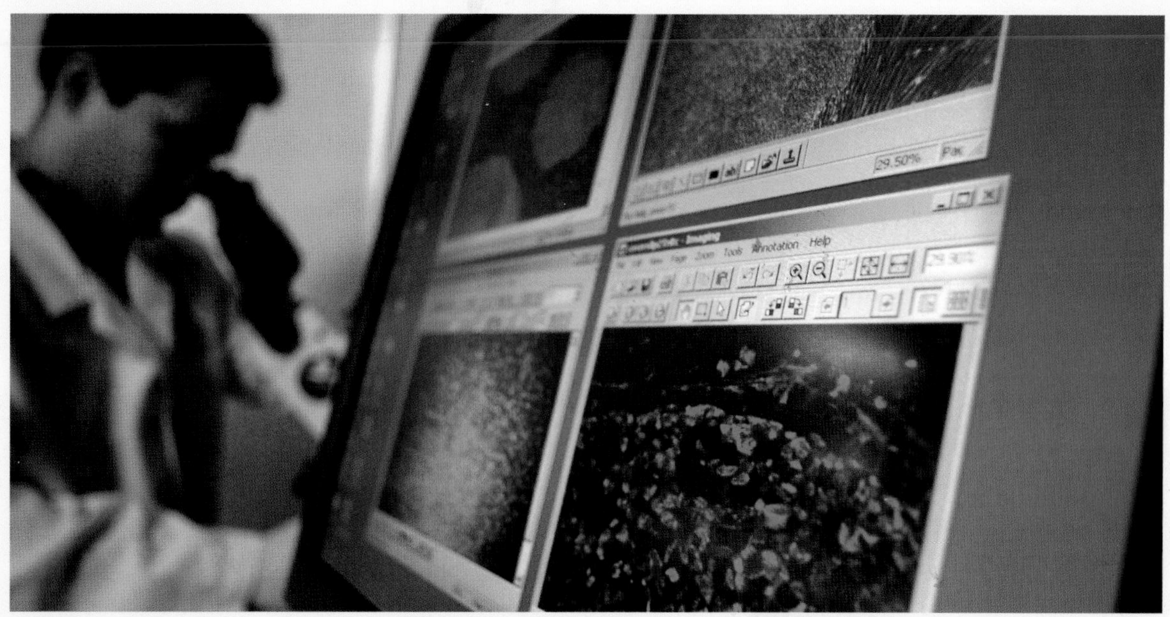

The value and utility of any experiment are determined by the fitness of the material to the purpose for which it is used …

Gregor Mendel, geneticist, 1822–1884

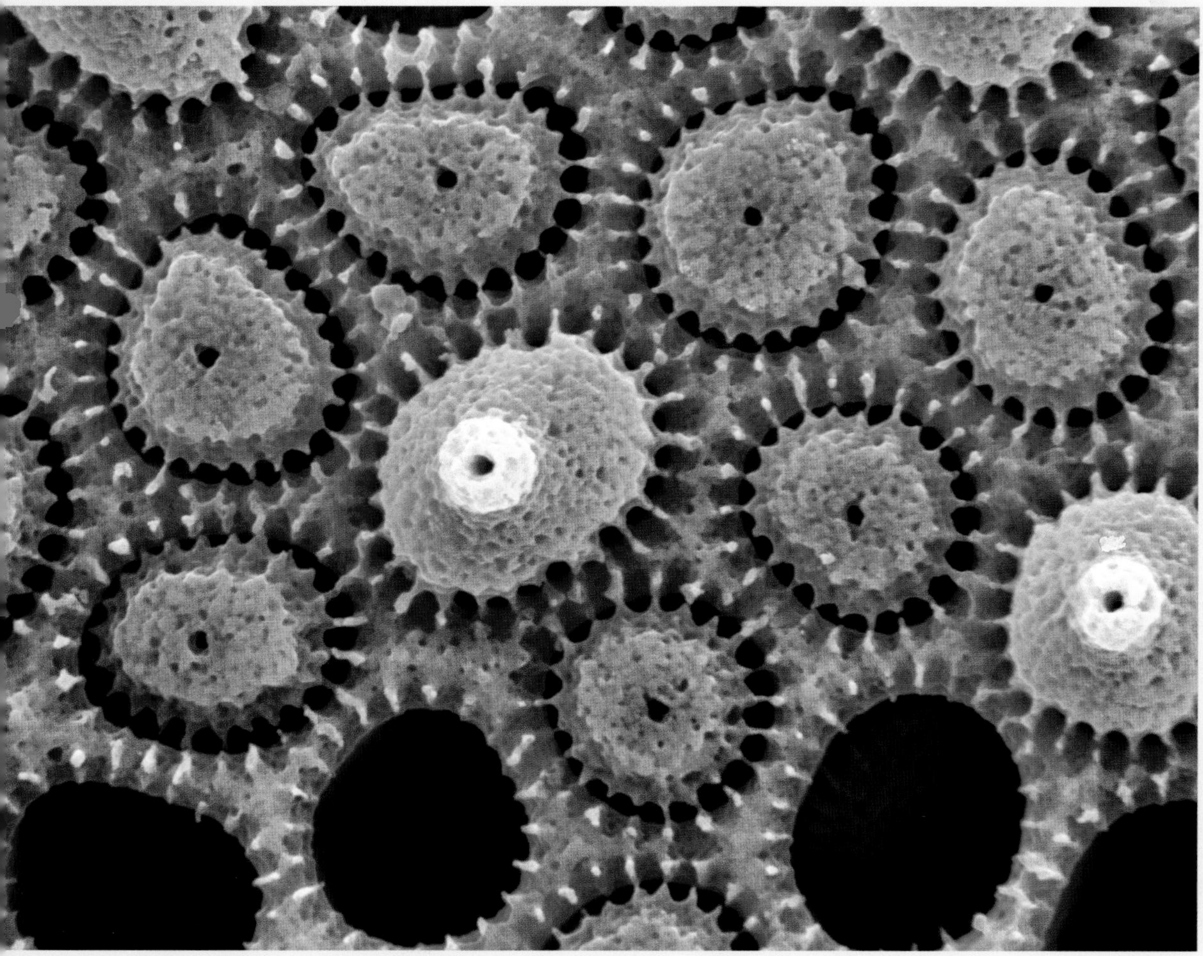

Left: Bacillariophyta are diatoms, unicellular aquatic microorganisms. This image has been taken by a scanning electron microscope, therefore it is a true image. However, the microscope does not acquire color information, so color has been added artificially.

Below: Human sperm cells-move via their flagella—long thin appendages resembling tails that allow them to move. Produced in the testes of the male, sperm swim up the female uterus with the aim of fertilizing the ovum.

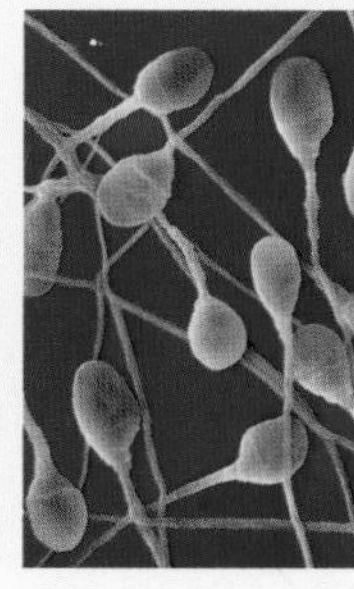

Some four billion years ago life appeared on Earth. From those first, simple, single-celled organisms, an incredible diversity of life forms has evolved, filling habitats from the deepest oceans to the highest mountaintops. Keeping track of this diversity is no easy task, but by naming and classifying organisms we can more easily identify and understand the life around us.

Life and the classification of living things

The biosphere is the broad collection of all life on Earth. The huge diversity and complexity of life forms, not to mention the sheer numbers of individual organisms, is overwhelming. Although around 1.7 million life forms on Earth have been identified and named, the actual number could be anywhere up to 100 million! There are many life forms that we'll never know about—they'll become extinct before we identify them, or we simply won't find them. To determine the full extent of life that exists on Earth, we need to be able to identify and name what we do know in an organized, consistent way.

Above: Carl Linnaeus (1707–1778) developed the fundamentals of the system of scientific classification that is still in use today. The system assigns all organisms a two-part name, and then groups them in a series of progressively broader categories. This marked the birth of modern taxonomy, and revolutionized the identification and classification of life.

Cataloguing life's diversity

Humans have been classifying the life around them since Aristotle's time, and possibly even earlier. Before the mid-eighteenth century however, there were no clear or fast rules for naming life, and attempts to classify organisms were sometimes very confusing. This all changed when Carl Linnaeus, a Swedish botanist and physician, proposed a system that assigns all living things a two-part name. The first part, the genus, identifies a group that the organism belongs to. The second part of the name, the species name, distinguishes the organism from other organisms in that group. A tiger, for example, is *Panthera tigris*. Its name identifies it as belonging to the genus *Panthera*, but being distinct from lions (*Panthera leo*), leopards (*Panthera pardus*), and jaguars (*Panthera onca*). This approach to naming life was broadly adopted, and the Linnaean bionomial system is the foundation of scientific classification today.

Taxonomy and the tree of life

The identification and naming of organisms is called taxonomy. Within taxonomy there is a specific hierarchy that allows scientists to name species in such a way that also provides information about their relationship to other species.

Species that have similar characteristics are grouped into genera (singular, genus), genera into families, families into orders, orders into classes, classes into phyla (singular, phylum) for animals or divisions for plants and fungi, and finally phyla into kingdoms.

The broader the groupings, the less like each other the members of that grouping are. For instance, all members of the genus *Equus* (horses, zebras, donkeys, and asses) are much more closely related than all members of the class Mammalia, which encompasses all the mammals on Earth. This means that the common ancestor of all horses occurred more recently in evolutionary time than the common ancestor of all mammals.

The relationships of organisms to each other are often depicted on a "tree," which not only shows where different groupings sit in relation to each other, but when groups diverged in evolutionary history. At the base of the tree are the earliest life forms—simple, single-celled organisms. As life evolved, new branches formed on the tree. The branches can be aligned with the various taxonomic groupings—phyla are large branches, from which smaller classes, then orders, families and genera branch off, with species at the tips of the branches. What this means is that that all of the organisms on any particular branch of this tree are more closely related to each other than they are to organisms on other branches.

Left and above: Two majestic members of the genus *Panthera*. The lion, *Panthera leo*, shares many characteristics with the the leopard, *Panthera pardus*, but they are distinctly different species. The two-part, or binomial, naming system for all living things makes classification a much more straightforward and more informative exercise.

WHAT IS A SPECIES?

The species is the fundamental unit of scientific classification (for example, this butterfly is a species of the *Danaus* genus), yet defining exactly what constitutes a species is not always straightforward. Broadly, a species is a group of organisms with a unique set of characteristics that distinguishes them from all other organisms. The challenges arise when determining which characteristics are the defining ones. For a long time, species identification was based on physical characteristics. Organisms that looked different enough from others were classed as different species. Yet sometimes different species can look all but identical, so this morphological basis of defining a species has drawbacks.

The biological species approach defines a species as a population whose members have the potential to freely interbreed with one another, and produce fertile offspring. Having the potential to breed and actually breeding can be two very different things, so although widely accepted as a way of defining a species, this approach has limitations.

With the introduction and growth of population genetics, and improving molecular and biochemical analysis, species can now be determined on a cellular level, adding more complexity to the issue. Irrespective of the challenges associated with defining what a species is, the existence of species as natural groups of organisms cannot be debated. The many recorded instances of local taxonomy, assigned by people with no formal biological training, corresponding well, if not exactly, with the identification of species by university-trained taxonomists, is a wonderful illustration of this.

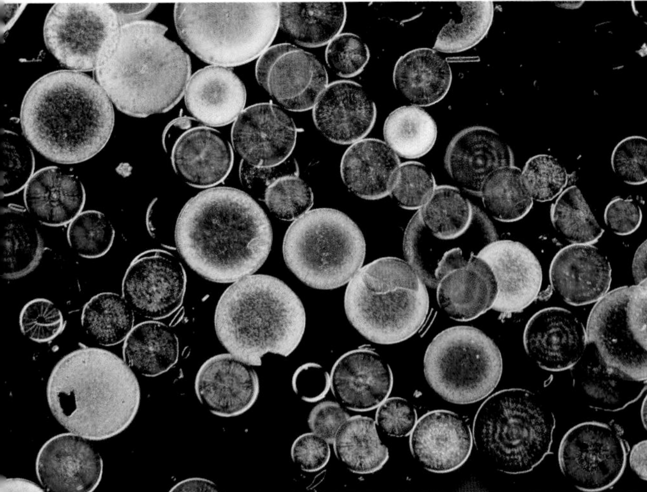

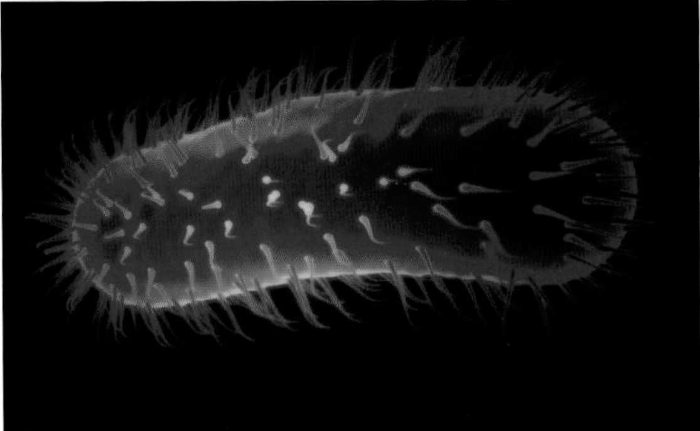

Top: Diatoms are microscopic unicellular algae found in aquatic environments. Their cell walls are made of silica and display delicate and beautiful symmetrical markings. Diatoms are a food source for many aquatic animals.

Above: A microscopic view of *Helicobacter pylori*, a spiral-shaped bacterium recognized as one of the major causes of gastric ulcers. This microorganism is also associated with stomach cancer.

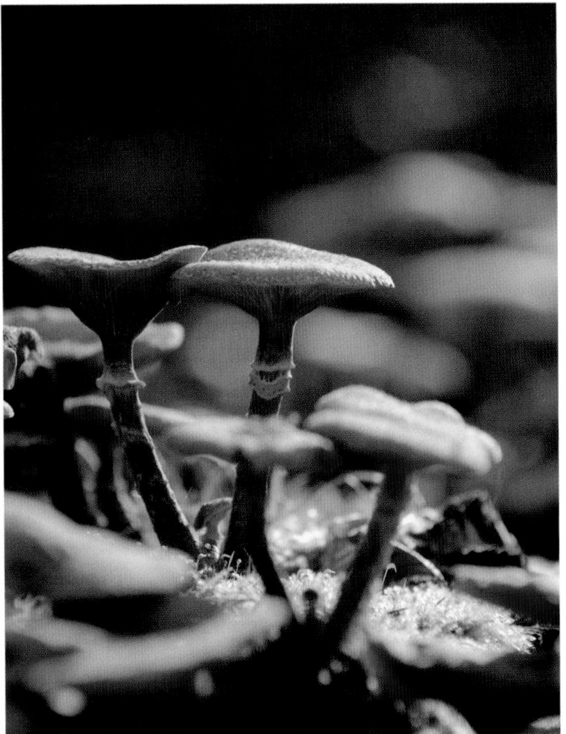

The totality of life, known as the biosphere to scientists and creation to theologians, is ... so internally complex that most species composing it remain undiscovered.

E. O. Wilson, biologist, b. 1929

Left: Mushrooms belong to the Kingdom Fungi. Fungi come in an amazing array of shapes and colors. They play an important role as decomposers, as they recycle dead organic matter into nutrients used in various parts of the ecosystem. The specialized study of fungi is called mycology.

Broader classifications

While kingdoms are incredibly broad classifications in terms of the life they encompass, there are broader classifications still that are used to describe life on Earth. The broadest classification consists of two groups, into which all life is divided based on cell structure. Those organisms with relatively simple cells with no cell nucleus are termed "prokaryotes." Those organisms with more complex cells that do contain a nucleus are termed "eukaryotes." These groups can be split into three domains, Archaea and Bacteria, containing prokaryotes, and Eukarya, containing eukaryotes.

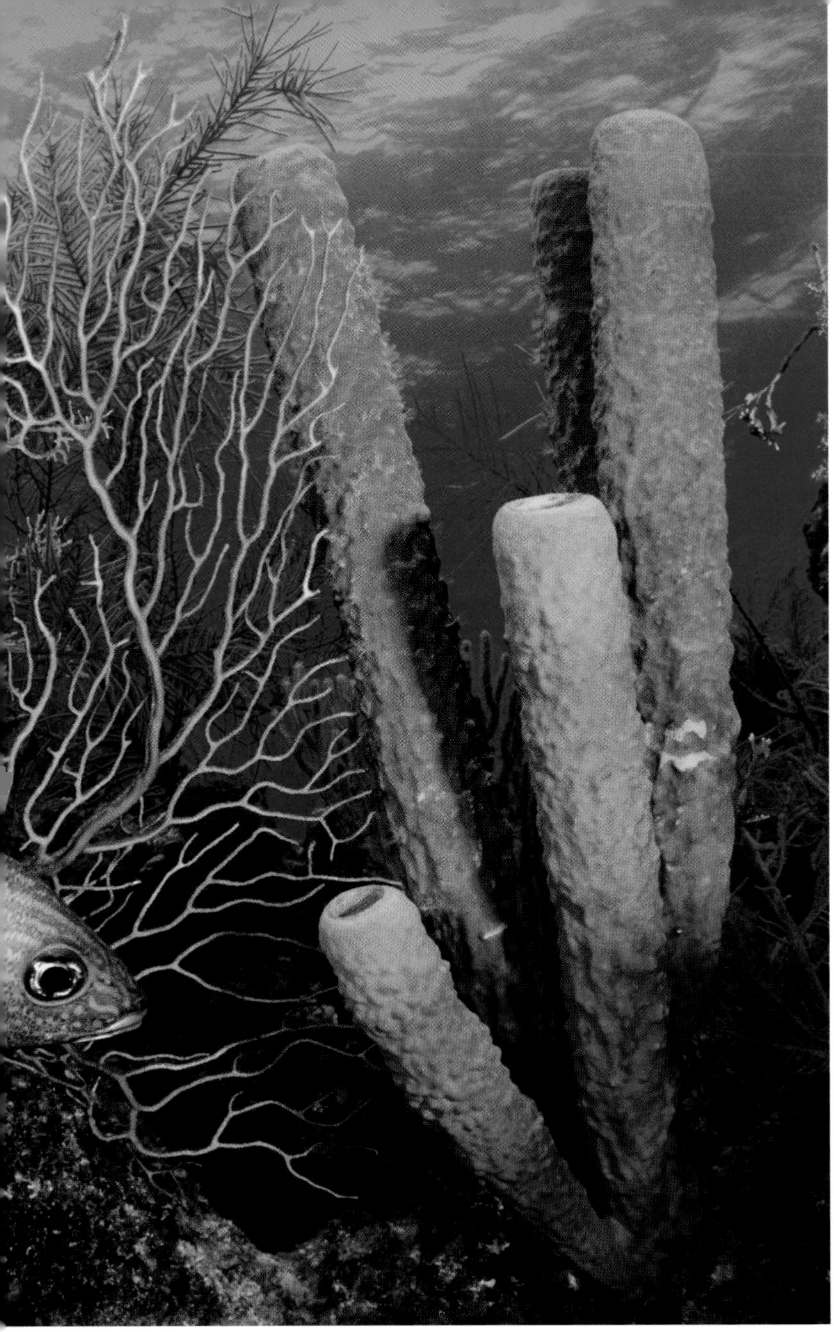

The Archaea domain contains some of the oldest forms of life on Earth. Generally single-celled, these prokaryotes thrive in environments that resemble early Earth, such as in or around deep sea vents and hot springs. Some do not need oxygen and live very well in atmospheres that are rich in carbon dioxide and hydrogen; others have been found in extremely alkaline, acidic, and salty environments. Although Archea look similar to bacteria under the microscope, they are so distinctive biochemically and genetically that they have been assigned their own domain.

The Bacteria domain contains the simplest life forms on Earth. These single-celled prokaryotes are also the most abundant—it has been estimated that just a single gram of soil can contain up to one billion bacteria cells, consisting of at least 5,000 species. They can be found in every habitat, including other organisms. Bacteria provide essential services to the biosphere, breaking down organic matter and recycling its components for use by other organisms. Some can cause disease in higher organisms. See pages 272–273.

Eukarya—life as we know it

The final domain, Eukarya, contains all the eukaryotes. The life forms we are most familiar with are grouped in this domain—the animals (kingdom Animalia), plants (kingdom Plantae) and fungi (kingdom Fungi). Eukarya also contains a fourth kingdom, Protista, which consists of any eukaryotes that are not animals, plants, or fungi.

Yeasts, molds, mildews, and mushrooms belong in the kingdom Fungi. Fungi obtain the energy they need to grow and reproduce by absorbing nutrients. There are around 70,000 identified species of fungi, although it has been estimated that there could be around 1.5 million species. Along with bacteria, fungi play an important role as decomposers, recycling nutrients through the ecosystem. Fungi are also important for the relationships they form with plant roots as "mycorrhizae" (fungus roots). These fungi supply essential minerals and nutrients from the soil to over 90 percent of plants. Fungi also used by humans for food (mushrooms and truffles), medicine (penicillin), and in commercial processes (yeast).

Kingdom Plantae contains all the land plants. More than 250,000 species of plants have been identified so far, and there are countless species yet to be discovered. Plants produce their own nourishment by using sunlight to convert carbon dioxide and water to oxygen and carbohydrates in a process called photosynthesis. As a result, plants shape the world we live in by having a significant impact on our atmosphere, and providing nutrition for other organisms. Plants also provide habitats for other organisms. See also pages 300–305.

The animal kingdom contains an incredible amount of diversity—sponges, snails, spiders, sharks, dogs, elephants, and humans all fall into this category. So far more than one million animal species have been named. While we are probably most familiar with a small subset of this kingdom known as vertebrates—animals with backbones including fish, amphibians, reptiles, birds, and mammals (the group humans belong to)—more than 95 percent of all species are invertebrates, animals without backbones.

Of the invertebrates, the largest group by far is the arthropods, which includes insects, spiders, scorpions, and crustaceans (such as lobsters, crabs, shrimp, and barnacles). Unlike plants, animals are not able make their own nourishment. Instead they need to obtain it from other organisms.

Left: French grunts (*Haemulon flavolineatum*) swimming over a coral reef. Found in sub-tropical waters around Bermuda, the West Indies, and from the north of the Gulf of Mexico down to Brazil, these fish generally travel in large schools. They feed primarily on species of small crustacean.

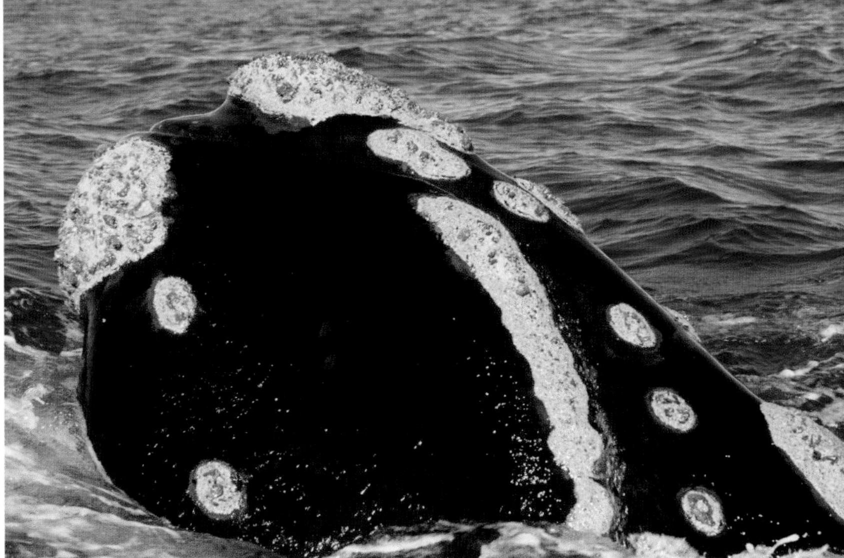

Below: The southern right whale (*Eubaleana australis*) is a baleen whale. Baleen whales do not have teeth. Rather, they have long flat bristles called baleen that hang across the mouth. When the whale takes a mouthful of water, prey is collected on the baleen plates, while the water flows out again. This is filter-feeding.

Charles Darwin was a British naturalist, who specialised in geological fieldwork. His work with fossils prompted the idea that living organisms had changed form and evolved over time to become the species we see today. In 1859, when he published his controversial theory of evolution, the scientific community was deeply shocked.

Evolution and natural selection

At the core of Darwin's argument was the suggestion that all species of life, including humans, evolved from a common ancestor through a process he named natural selection. This theory, now widely accepted, has become the cornerstone of modern biology, and is the central premise of evolutionary theory.

Mechanisms of evolution

A "population" is a community of organisms of the same species that is able to interbreed freely. Within a population, individuals generally display a number of genetic variations that may result in slight differences in appearance or behavior, but which do not prevent successful interbreeding within the population. For example, in humans, eye color reflects a particular genetic variation within the population, but this characteristic does not prevent individuals with different eye colors from successfully producing offspring.

The collective variation within a given population is called the gene pool. The gene pool is determined by the sum of all genes, and their allelic forms, and is an important determinant in identifying that evolutionary change has taken place.

Fundamental to the entire process of evolution is the understanding that a population needs to demonstrate genetic variation, upon which selective forces are able to act. This can occur according to a number of mechanisms.

Genetic mutation can result in the introduction of new traits within a population, which are subsequently inherited by future generations. Migration between two populations may also lead to the

Page 2.
Autograph of Charles Darwin

I have now recapitulated the chief facts and considerations, which have thoroughly convinced me that species have been modified, during a long course of descent, by the preservation or the natural selection of many successive slight favourable variations. I cannot believe that a false theory would explain, as it seems to me that the theory of natural selection does explain, the several large classes of facts above specified. It is no valid objection that science as yet throws no light on the far higher problem of the essence or origin of life. Who can explain what is the essence of attraction of gravity? No one now objects to following out the results consequent on this unknown element of attraction; notwithstanding that Leibnitz formerly accused Newton of introducing "occult qualities & miracles into philosophy." —

Charles Darwin

p. 514 3d Edit of "Origin"

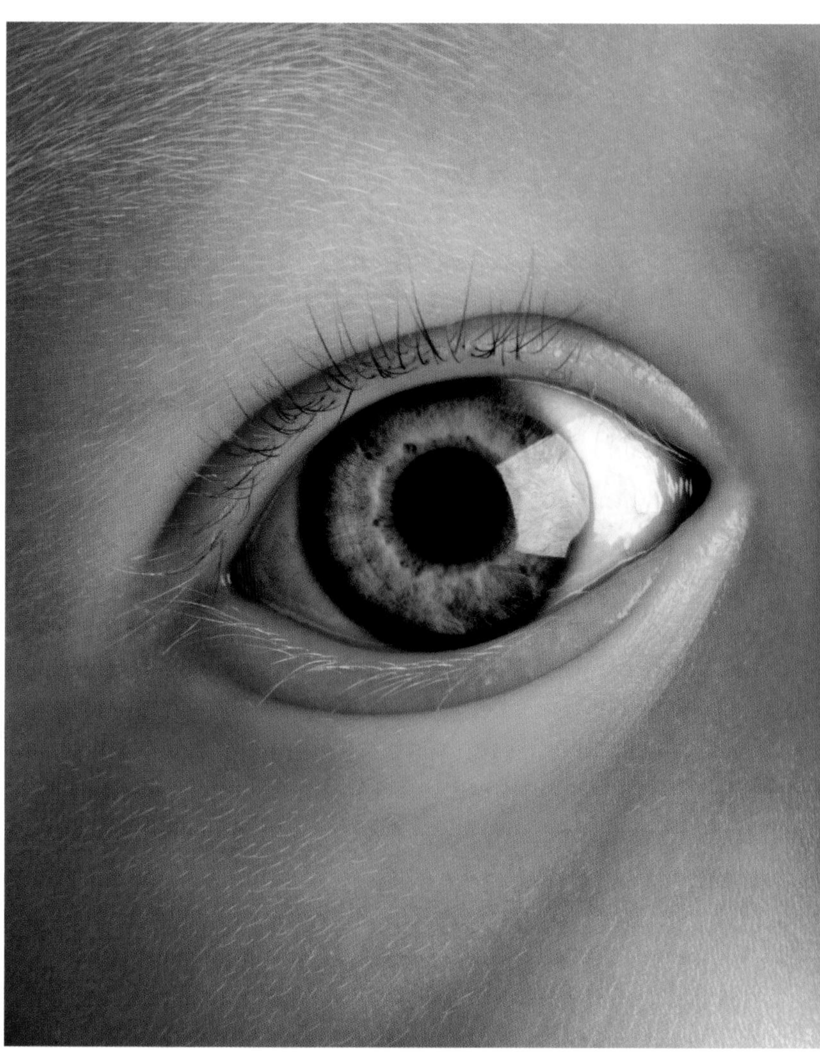

Above: Our eye color is determined by the combination of genes we receive from our parents. Generally speaking, brown is dominant and blue is recessive, but two brown-eyed people can still produce a child with blue eyes.

Left and far left: British naturalist Charles Darwin and an extract from his paper dealing with his theory of natural selection, dated c. 1865. Darwin's theories transformed the way that scientists thought about the natural world.

My mind seems to have become a machine for grinding general laws out of large collections of facts.

Charles Darwin, naturalist, 1809–1882

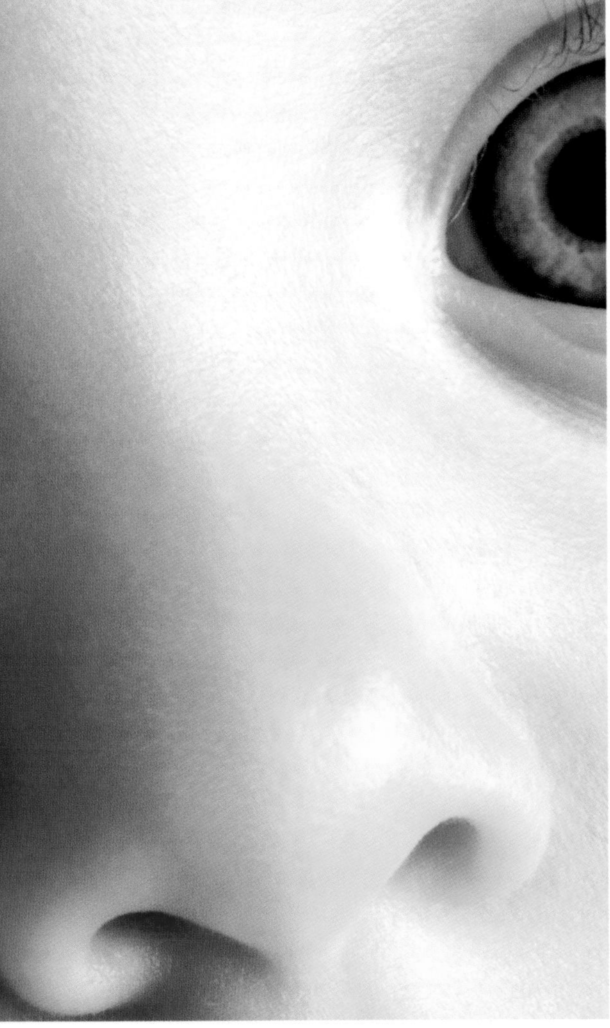

THE CASE OF THE PEPPERED MOTH

One of the best examples of natural selection and evolutionary fitness is the study of the peppered moth *Biston betularia*, pictured below, during the Industrial Revolution. In England prior to 1800, the peppered moth had a light colored pattern on its wings. Dark moths were very rare, and highly sought after as collectors' items.

However, during the course of the nineteenth century, a steep rise in the numbers of dark moths was observed. A likely explanation is the rise in soot and industrial waste that was being produced, with dark residues found on surrounding trees and wildlife. In Manchester, dark moths made up 98 percent of the peppered moth population.

It is suggested that over time, the lighter colored moths became less common, as local birds preferred to eat the light-colored moths over their darker, more camouflaged, counterparts. Populations of light colored moths were only found to exist in local pockets far away from industrial areas.

Some scientists have warned against attributing the population changes solely to the ability of birds to preferentially identify the lighter colored moth. Despite this, the steep rise of dark moths is a very compelling example of natural selection at work.

Below, left and right: With over 350,000 known species, beetles make up more than one-third of all Earth's insects. Found in most parts of the world, they are great survivors, their outer skeletons acting as effective armor. Natural selection has allowed beetles to adapt to various environments; color acts as both camouflage and, in some species, to attract prey.

introduction of new traits, as newcomers mate with the population, and pass on their traits to their new offspring.

Genetic drift refers to random events, which may cause a change in the number of variants within a population. For example, a population of beetles might display variations in color—green and brown. If a human were to accidentally step on a population of these beetles, and more brown beetles than green were killed, this would alter the frequency of that particular trait within the gene pool.

Finally, Charles Darwin proposed that populations could become better adapted to their environment from one generation to the next through a process he called natural selection.

Natural selection and survival of the fittest

Natural selection describes the process of favorable traits becoming more prevalent within successive generations of a population. Natural selection occurs when individuals of differing genotypes differ in their fitness, and make disproportionate contributions to the gene pool. This concept of evolutionary fitness is one of the most critical principles of evolutionary biology.

In essence, fitness describes the capability of an individual genotype to reproduce, and pass its genes onto the next generation. An individual's genotype is expressed as a phenotype, and natural selection acts on an individual's phenotype within a particular environment. Because of this, differences in fitness can arise due to differential survival and reproduction of particular genotypes.

Mathematically, fitness can be stated as the product of viability (proportion of adults that survive to mate) and fertility (ratio of offspring to fertile parents).

Conditions for natural selection

For natural selection to take place, Darwin specified that four conditions had to be met. First, variation in traits should exist

THE EVOLUTION OF MAN

In his 1871 publication, *The Descent of Man*, Darwin controversially suggested that humans and apes had evolved from a common ancestor. The proposition that humans could be descended from some primitive ape-like form shocked the public.

Today, all humans belong to the species that is named *Homo sapiens sapiens*. The first *Homo* species was believed to have existed in east Africa over 2 million years ago. It is now believed that modern humans have evolved from this original species.

Fossil evidence indicates that *Homo habilis* co-existed alongside another primitive hominid life form called *Australopithecus* ("southern ape") (below left), however the main difference between the two species was *Homo*'s significantly larger brain size, and his ability to make tools.

As *Homo* species have evolved, science has recorded a progressive increase in brain size. *Homo erectus* (below right) evolved approximately 1.5 million years ago, and was found in southeast Asia. *Homo erectus* is believed to have been one of the more successful *Homo* species, surviving over one million years, until approximately 300,000 years ago.

Anthropologists seem to agree that the modern form of *H. sapiens* evolved out of *H. erectus* approximately one million years ago. However, much debate still exists around whether *H. sapiens* evolved in Africa, and migrated to other regions (Out of Africa model), or whether *H.sapiens* evolved simultaneously in different regions of Europe and Asia (Multiregional model).

Australopithecus

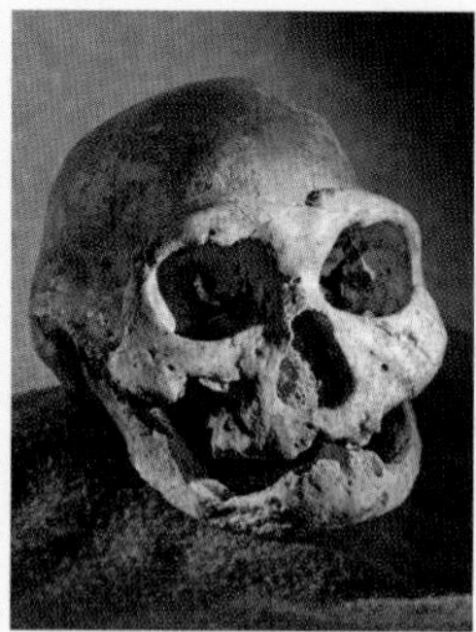

Homo erectus

Bottom right: The dyeing dart frog (*Dendrobates tinctorius*), found in northeastern Brazil and parts of French Guinea, is a species of poison dart frog. It has no need of camouflage, as its beautiful bright colors warn predators away.

Below: Native to eastern North America, this male spring peeper frog (*Pseudacris crucifer*) is calling for a mate from among bog grasses. These frogs feed on various insects, and in turn are preyed upon by birds, spiders, and snakes.

between individuals in a population, and some of this variation results in differences in fitness. For example, tree-dwelling green beetles may be less likely to be eaten by birds than brown beetles, due to their ability to hide in the leaves. This trait makes the green bugs "fitter" than the brown bugs in that particular environment.

Second, Darwin specified that there should be reproductive differences between the variants. Using the previous example, if the brown beetles are eaten more often than the green ones, they are less likely to reach their full reproductive potential, and have fewer offspring.

Third is the assumption that the genetic trait is inherited, thereby passing on fitness to the subsequent generation. Finally, Darwin suggested that fitter individuals would reproduce more often, and make a relatively greater contribution to the next generation. Over time, this means that the fitter trait will grow in number, and the less fit trait will decrease in number. If, and only if, all of these conditions are met, then evolutionary change by natural selection will occur.

Outcomes of evolution

There are two main processes that define the outcomes of evolutionary biology—microevolution and macroevolution.

Above: A finch sits on the tail of a marine iguana (*Amblyrhynchus cristatus*) on the Galapagos Islands, Ecuador. Charles Darwin called them "imps of darkness," but this amazing animal has adapted well to its environment. It is the only lizard that lives near and off the sea.

Microevolution is concerned with the changing composition of the gene pool, and refers to the changing frequency of particular alleles over time. The changes observed during microevolution can be explained by a number of evolutionary forces, including mutation, migration, inbreeding, genetic drift, and natural selection. Population genetics is a branch of mathematics dealing directly with the outcomes of microevolution.

By contrast, macroevolution looks at evolutionary forces on a grander scale, and leads to the creation of new species. The process, called speciation, generally spans long periods of time, and involves the cumulative effects of evolution over that period. To provide a definition of a species is a complicated one; however a good starting point is to think of a species as a group of organisms capable of interbreeding and producing fertile offspring.

While the process of microevolution can be easily observed and confirmed, that of macroevolution is much more difficult to confirm, due to the lengthy timeframes associated with speciation. However, it is the process of macroevolution that provides the framework for Darwin's proposition that all species of life evolved over time from one common ancestor.

Organisms have several features in common. They are made up of individual, microscopic units known as cells; they are chemically complex and often highly organized; they sense, respond to, and exchange material with their environment; they are capable of independent growth, development, and reproduction; and over time, they exhibit changes that may be adaptive.

Organisms

Being alive means that an organism is able to move, respond, grow, and reproduce. For example, a web-spinning spider moves as it constructs its web. It responds to the vibrations resulting from prey tangled in this web. It grows as it feeds, periodically shedding its skin and eventually, after mating, the female spider builds an egg sac within which it deposits its eggs. The tree that supported the spider's web began life as a seed deep within the soil. With the addition of water, sunlight, and nutrients, this seed sprouted roots and shoots that pushed through the substrate, either toward or away from the light. The topmost shoots eventually emerged from the ground and over time, the seedling grew to become a tree. Finally, the tree blossoms. The flowers are fertilized to produce fruits that themselves, house new seeds. And so, the cycle continues.

All organisms eventually die. The spider will die, falling to the ground or forming the first meal for its newly emerged offspring. Without adequate water and nutrients, or as a result of the impact of disease or pests, the tree may lose its leaves, becoming dry and brittle, before itself, succumbing to death.

Organisms can be large, such as an elephant, whale, dinosaur, or California redwood, or too small to be seen by the naked eye, such as bacteria. These tiny organisms are microorganisms (see pages 268–269). Organisms can be single-celled such as bacteria, many algae, and protozoans, or multicellular such as jellyfish, reptiles, human beings, shrubs and trees.

Above: An orb web spider (*Argiope minuta*) waits patiently for prey to come to her. Some orb web spiders also build wide bands shaped like the letter X into the web. This reflects light, attracting insects to it.

Left: African elephants (*Loxodonta africana*) are the largest land animals on Earth. They are distinguished from Asian elephants (*Elephas maximus*) by their larger ears. These magnificent mammals are herbivorous.

SUPERORGANISMS

Some scientists extend the organism classification to include the term "superorganism." Social insects such as ants (right) and bees form colonies made up of many individuals, each of which plays a role in maintaining the entire colony. In essence, the colony acts as a single individual and the activities of the individuals may be considered insignificant unless viewed as part of the host superorganism.

Human beings too have been described as superorganisms. The human body contains more than 500 species of bacteria whose cell count vastly outnumbers the population of human cells. These microbes play a vital role in maintaining the host organism. Bacteria within our alimentary tract break down our food, produce vitamins for our use, and help remove toxins.

Society also has been described as a superorganism made up of individual units, similarly to the way a multicellular organism is made up of individual cells. Each individual human is vital to the existence of the superorganism, which in turn, maintains each and every one of us. This has led to the Gaia theory, the concept of Earth as a planetary superorganism, regulating all life upon it.

> A seed hidden in the heart of an apple is an orchard invisible.
>
> Welsh proverb

Left: The General Sherman Tree in Sequoia National Park, California, USA, is the largest tree in the world. This sequoia (*Sequoiadendron giganteum*) is 275 ft (84 m) high with a circumference at the ground of 102 ft (31 m).

The body of an organism also exhibits some form of symmetry. Marine organisms such as sea anemones, starfish, and jellyfish are radially symmetrical because their body parts "radiate" out from a central point, giving them more than one axis of symmetry (they can be divided into two symmetrical halves by any of several planes). Worms and vertebrates are bilaterally symmetrical—they can be divided into two symmetrical halves through one axis of symmetry (imagine a line "splitting" you from head to groin, passing through your nose and navel). Bilaterally symmetrical organisms thus have a front (anterior) and rear (posterior) end, and both a dorsal (back) and ventral (belly) surface. Flowers can be either radially or bilaterally symmetrical. Organisms that have no form of symmetry are termed asymmetrical. The freshwater protozoan, ameba, with its constantly changing body shape, is asymmetrical.

All organisms are assigned to various groups on the basis of shared characteristics. This form of classification, known as taxonomy, is based on a naming system devised by Carl Linnaeus (see page 258). The major taxonomic groups are kingdom, phylum, class, order, family, genus, and species. Using this form of classification, each organism is given a unique scientific name. The scientific name of of a dog is *Canis familiaris*, that of the California redwood *Sequoia sempervirens*.

Right: A yellow anemone flower (*Epizoanthus scotinus*), one of the countless organisms living in the ocean. *Epizoanthus scotinus* is a cnidarian, which is a member of a phylum of soft-bodied aquatic creatures with radial symmetry.

In 1674 Dutchman Anton von Leeuwenhoek used a tiny hand lens to study a drop of pond water and his description of the little "animalcules" that he saw was the first recorded observation of microorganisms—living creatures too small to be seen with the naked eye.

The marvelous world of microorganisms

The existence of microorganisms was not known until the development of the light microscope enabled scientists to observe the microbial world. Leeuwenhoek's discoveries were largely ignored until the nineteenth century when Louis Pasteur showed that bacteria living freely in the environment were responsible for decomposition and souring of milk and wine. The role of microorganisms in infectious diseases was established by Robert Koch who in 1876 proposed the "germ theory of disease"—that each infectious disease is caused by a particular microorganism.

Environmental importance of microorganisms

Microorganisms have existed on Earth for many billions of years. It has been estimated that the number of microorganisms present in a handful of soil is greater than the total number of people who have ever lived on Earth.

Microorganisms occupy all kinds of habitats and play a vital role in preserving our ecosystem. Microorganisms form the basis of the food chain in rivers, lakes, and oceans. They are essential for decomposition, breaking down waste materials to release chemicals, and trapping nitrogen from the air for plant growth to provide food for animals. Bacteria in the gut of animals assist digestion of cellulose in their diet. The human intestine is populated by millions of beneficial bacteria that aid in digestion and produce essential vitamins that contribute to our health.

Kinds of microorganisms

The microbial world comprises bacteria, viruses, protozoa, algae, yeasts, fungal spores, and the microscopic stages in the life cycles of some parasites. They range in size from viruses a few nanometers in diameter to large algae which may be up to 50 micrometers in size and even larger protozoa. Most microbes are harmless or essential for life. Few cause disease.

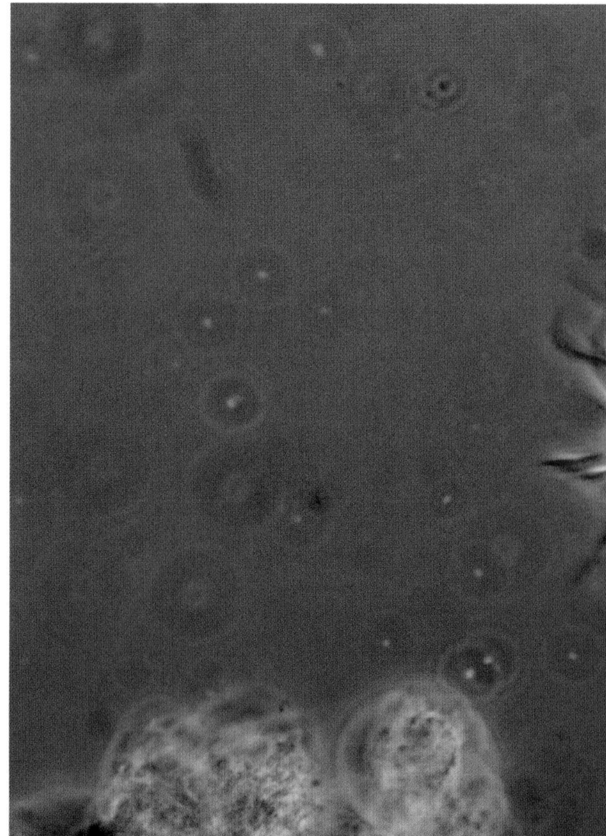

Right: Protista is the biological kingdom that is made up of unicellular eukaryotic organisms, such as this protozoan, *Euplotes patella*, which has hair-like protuberances that help to propel it along.

Above: There are countless billions of microorganisms in a handful of soil, indeed in most environments, including our homes, our bodies, and on the food we eat. Too small to be seen by the naked eye, microorganisms play an essential role in biological processes.

Left: A scanning electron microscope works by aiming a beam of electrons at the microscopic sample under study. The sample is coated with a thin layer of a heavy metal, such as platinum. The electrons deflect from the sample at various angles while a detector gathers information and produces a three-dimensional image.

Microorganisms are divided into two main groups: Prokaryotes (bacteria) and Eukaryotes (protozoa, algae, and fungi). Viruses require another cell in order to reproduce, and so do not fall into either of these groups. Prokaryotes are mostly free living, have a cell wall, but do not have a defined nucleus. Eukaryotic cells have a nucleus and other defined cell structures. Medical science is mainly concerned with bacteria and viruses.

Protozoa

Protozoa are single-celled eukaryotic organisms found mainly in watery environments. Protozoa are divided into groups depending on the way they move. Some swim using tail-like flagella, others have cilia (small hair-like appendages) or pseudopods (flowing extensions of their cytoplasm). Non-motile protozoa live in animal hosts and may be transmitted by insects. Most are harmless but some can cause serious diseases such as malaria, toxoplasmosis, amebic dysentery, giardiasis, and cryptosporidiosis.

Algae

Algae are photosynthetic aquatic plants that use energy from the Sun and carbon dioxide from the air to synthesize carbohydrates and release oxygen. They are unicellular, filamentous, or multicellular and their color may vary from green to red, brown or yellow, depending on the nature of the photosynthetic pigment present. Unicellular

In the fields of observation chance favors only the prepared mind.

Louis Pasteur, chemist and bacteriologist, 1822–1895

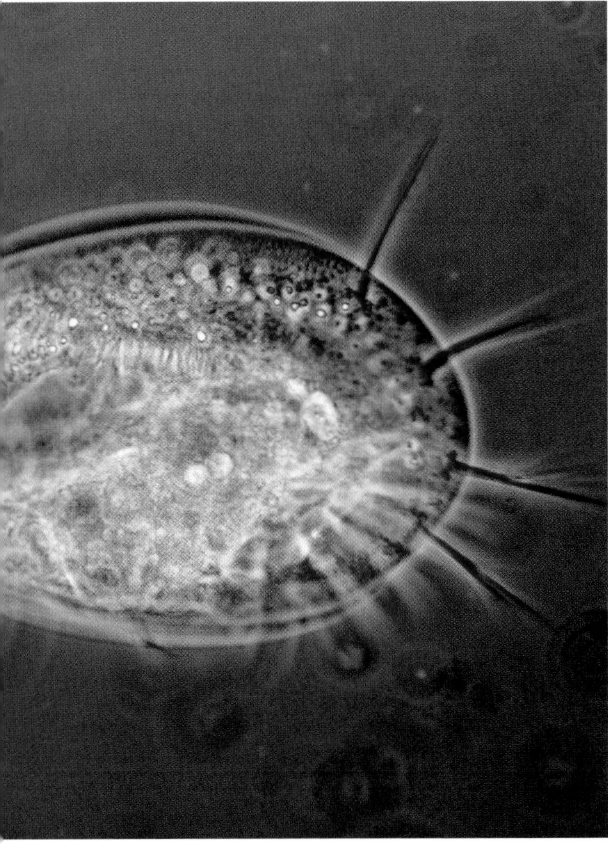

algae or phytoplankton form the basis of the food chain in aquatic environments. They also play a very important role in the return of oxygen to the environment.

Fungi

There are thousands of different kinds of fungi, including unicellular yeasts and larger multicellular molds. They are important for decomposition and as a source of some antibiotics, such as penicillin). Very few fungi cause disease in humans.

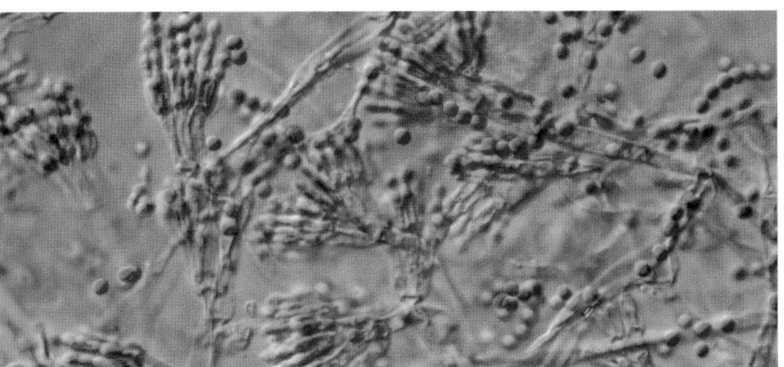

PASTEUR DISPROVES THEORY OF SPONTANEOUS GENERATION

Many people used to believe that life could arise spontaneously from non-living matter. Louis Pasteur (depicted below in his laboratory) believed that it was actually microorganisms that live freely in the environment that were responsible for decomposition and rotting of food. To disprove the theory of spontaneous generation, Pasteur devised an ingenious experiment. He poured nutrient broth into a flask with a long thin neck, then heated the neck and bent it into an S-curve, a "swan-necked" flask. He then boiled the broth in the flask, killing all the bacteria present in the flask, and expelling all the air. On cooling, air re-entered the flask, but the bacteria were trapped in the S-bend in the tube. Even after a long period of time, no growth of microorganisms occurred in the flask, conclusively showing that new cells could only arise from pre-existing living cells.

Right: Chlamydomonas is a genus of unicellular green algae that swim by using two flagella. It also contains a chloroplast for photosynthesis. Chlamydomonas has nothing to do with the disease chlamydia.

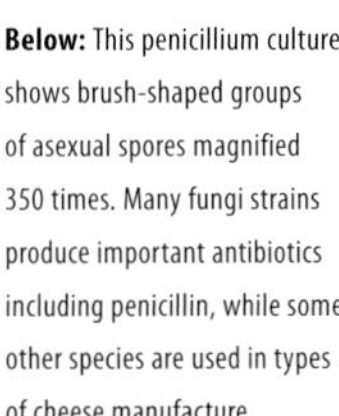

Below: This penicillium culture shows brush-shaped groups of asexual spores magnified 350 times. Many fungi strains produce important antibiotics including penicillin, while some other species are used in types of cheese manufacture.

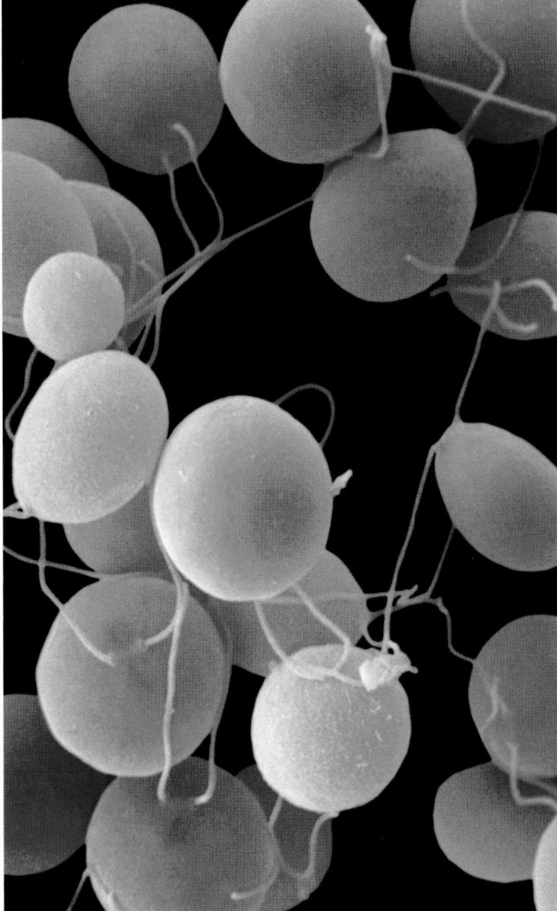

Viruses are responsible for many infectious diseases in humans, ranging from AIDS to the common cold. There are many different kinds of viruses and they cause disease not only in humans but also in animals, plants, and even bacteria.

Viruses

Viruses are among the smallest known infectious agents. They range in size from the polio virus (20 nm diameter) to the large pox virus (400 nm diameter). Viruses are so small that they require the high resolution of electron microscopy for their structure to be visible. Each virus has a distinctive shape and structure. They contain only one type of genetic material (DNA or RNA) that is enclosed in a protein coat called a capsid. Some of the more complex viruses have a protein core enclosing the nucleic acid. Others have an additional outer layer called an envelope.

Viruses are not cells and cannot carry out any metabolic functions without the involvement of a host cell. They use the host cell DNA and some host cell enzymes to produce new viral particles. During replication the host cell is usually damaged or destroyed, so viral infection always causes disease.

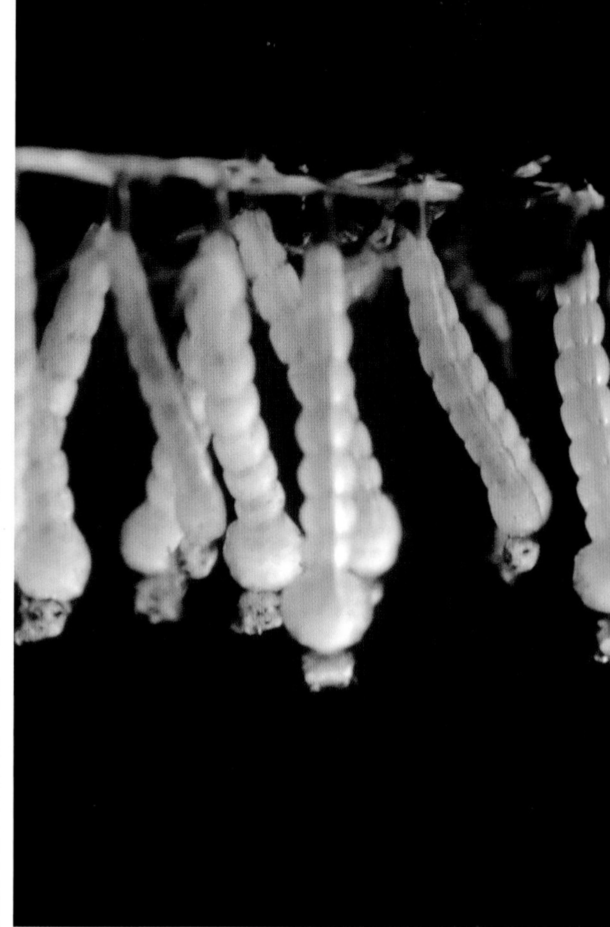

Right: Dengue fever mosquito larvae. Dengue fever is a virus carried by the striped *Aedes aegypti* mosquito. It flourishes during the rainy seasons but can survive in rain-filled flower pots, plastic bags, and other receptacles year round. This virus is endemic in the tropics and subtropics.

Host range and specificity

Viruses are highly selective for the range of hosts and the type of cells they infect. Specific structures on the surface of the virus recognize and attach to corresponding sites on the outside of the target cell. In general, plant viruses infect only plants, bacterial viruses infect only bacteria, and animal viruses infect only animals. Within these broad categories, most viruses will attack only one type of animal or plant, and may even be specific for only one kind of cell or tissue within that organism (for example, polio only attacks nerve cells in humans and rhesus monkeys.)

A few viruses, such as rabies, can infect humans as well as animals and there are some viruses that are able to cross the species barrier and infect birds as well as animals, such as influenza. In some cases an animal may be a reservoir for a virus that only causes disease symptoms in humans, such as hantavirus.

> It is perfectly obvious that no one … can save the worldfrom the horrors of tsunamis, hurricanes, earthquakes and … influenza.
>
> Richard Reeves, writer

Transmission of viral diseases

Viruses are unable to survive for long outside their host cell so most are transmitted from person to person by close direct contact. They are transmitted as aerosols in respiratory secretions (colds and flu), by contact with infected lesions (chickenpox, herpes), or in bodily substances such as feces (gastroenteritis, polio), blood, and semen (hepatitis B, HIV/AIDS). (See also pages 350–351.)

Arboviruses are transmitted by biting insects such as mosquitoes. These include West Nile virus, and viruses that cause yellow fever, dengue fever, Ross River fever, and some forms of encephalitis. In most cases these viruses have a reservoir in a wild animal or bird and are transmitted to humans by an insect bite.

INFLUENZA

In 1918 an influenza epidemic swept the world causing the deaths of around 40 million people. Although it was known how the disease was spread, the virus responsible had not been identified, so the medical profession had no weapons to fight it, although gauze masks were used to prevent its spread (left). Today we face the threat of another influenza epidemic—avian influenza or bird flu, but this time we are better prepared.

We now know that the influenza A virus has two "spikes" on its outer envelope—hemagglutinin (H), which is involved in the attachment of the virus to its target cell, and neuraminidase (N) which allows the release of new viral particles from the infected cell. Genetic changes produce new virus types with variations in the spikes which affect the infectivity of the virus. The avian flu (H5N1) is not very infectious to humans but there is concern that if it mutates it might cause a pandemic (see pages 360–361). Health authorities are tracking the spread of the disease and the type of virus and are ready to manufacture a vaccine if the virus mutates.

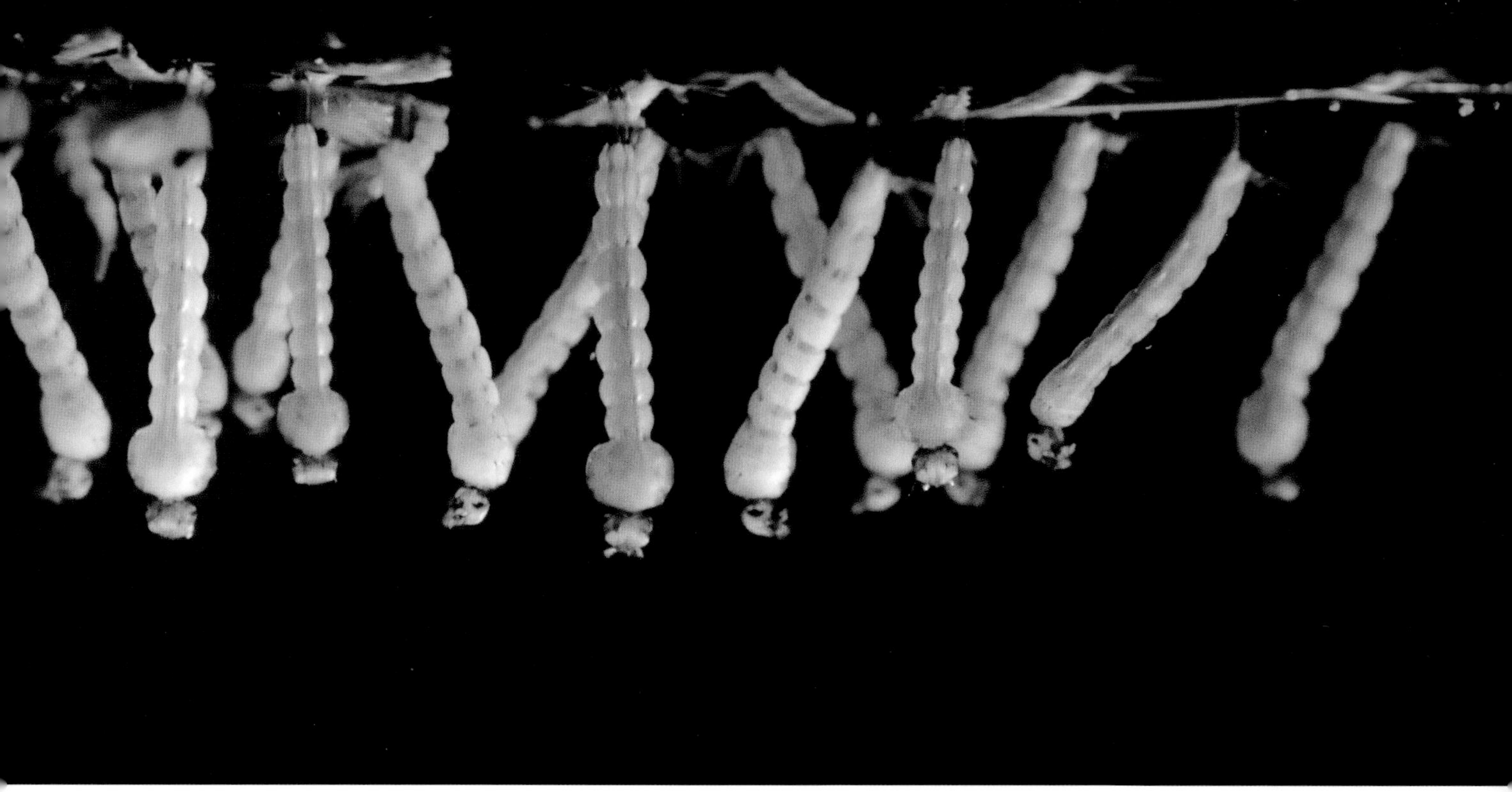

Treatment of viral diseases

Because viruses use the host cell to replicate, any drug that targets the virus will also damage the host cells, so there are few effective antiviral drugs. Vaccination is the most effective way to stop the spread of viral diseases. A successful vaccination campaign has led to the worldwide eradication of smallpox and the World Health Organization is now targeting measles and polio.

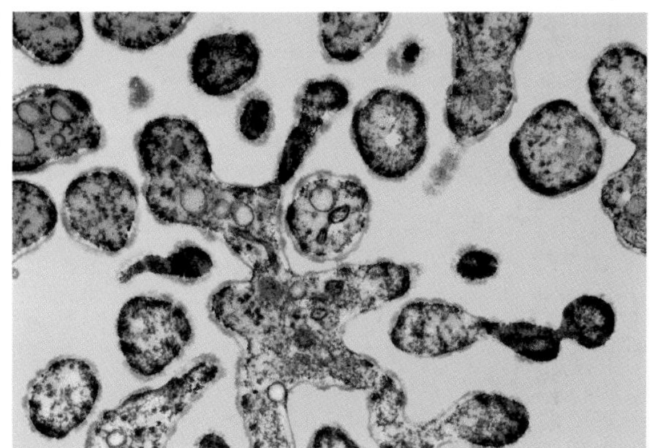

Right: A color-enhanced transmission electron micrograph image of measles. Measles is caused by the Morbilli virus which is highly contagious, and is spread in respiratory secretions. This virus is formed by a budding process and released as membrane-bound particles from an infected cell.

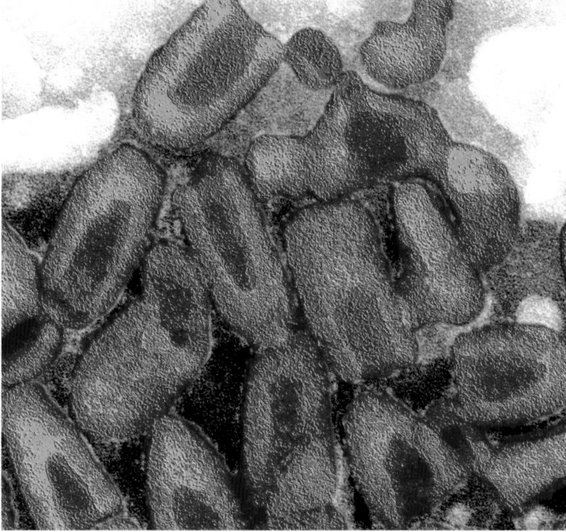

Right: The rabies virus, magnified 75,000 times. This virus is bullet-shaped and has single-stranded RNA. Mainly a disease of animals, rabies can be transmitted to humans via a bite from an infected dog.

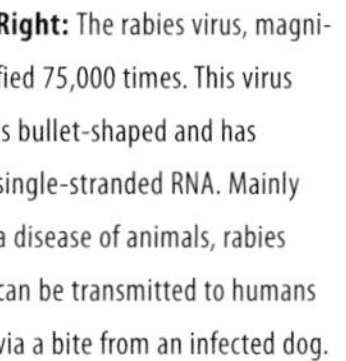

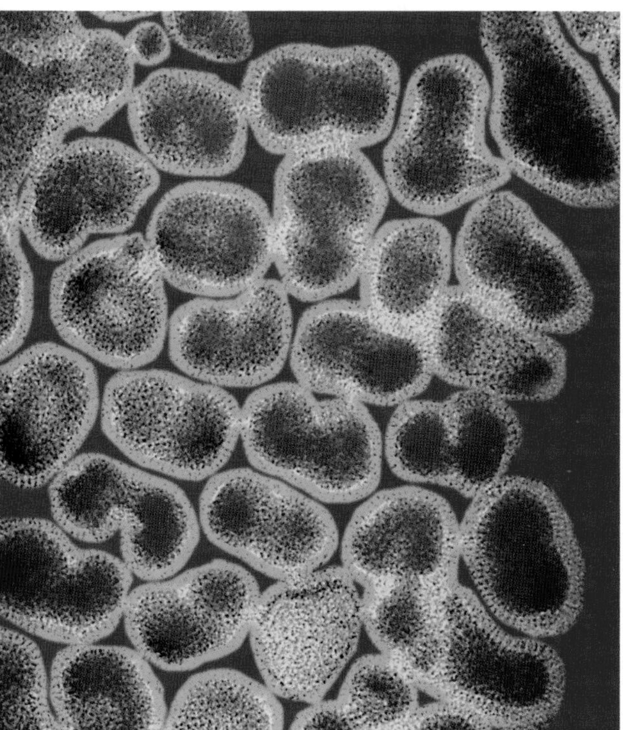

Left: The influenza virus, here magnifed 100,000 times, often mutates, making it hard to develop vaccines suitable for all strains, so epidemics or occasional pandemics of influenza occur. The flu virus enters the body when miniscule respiratory secretions from an infected person are inhaled.

Most people think of bacteria as germs that cause disease. In fact very few of the thousands of different kinds of bacteria that inhabit Earth cause disease. Most of them are beneficial, even essential, for the maintenance of life on Earth.

The benefits of bacteria

Right: Although it is often referred to as blue-green algae, cyanobacteria are in fact bacteria. Cyanobacteria are a major pollutant of waterways and the toxins they produce can affect animals that drink the polluted water.

Bacteria are microscopic single-celled organisms surrounded by a characteristic cell wall. Most can survive and reproduce freely. They are classified as prokaryotic because they do not possess a defined nucleus or other membrane-bound sub-cellular organelles. However, bacterial cells are capable of carrying out all the functions of more complex higher organisms and are used extensively for research because they are easy to grow and handle in a laboratory. Studies with bacteria have provided an understanding of cellular metabolism, the structure of DNA and the nature of genetic inheritance, and were instrumental in breaking the genetic code.

Bacteria are found in a wide variety of habitats. Only the disease-causing Rickettsiae and Chlamydiae exist as intracellular parasites inside animal cells. All others are free living.

Some bacteria live in extreme conditions, such as in icy regions of Antarctica, thermal hot springs, or in areas of high salinity. Some bacteria of medical importance form resistant cells called endospores in order to survive adverse conditions. A small group called cyanobacteria are photosynthetic.

Characteristics of bacteria

Bacteria range in size from 1or 2 up to10 microns. They have quite distinct shapes—round (coccus), rod-shaped (bacillus), or spiral (comma-shaped vibrio, spirochete or spirillum). The cells are enclosed by a membrane and a thick cell wall that maintains their shape. They may contain vacuoles, but have no defined intracellular structures. Many bacteria use one or more flagellae to propel themselves. Some have fimbriae—tiny hairlike structures used for attachment. They reproduce by splitting into two identical daughter cells, a process called binary fission. For some bacteria this occurs every 30 minutes.

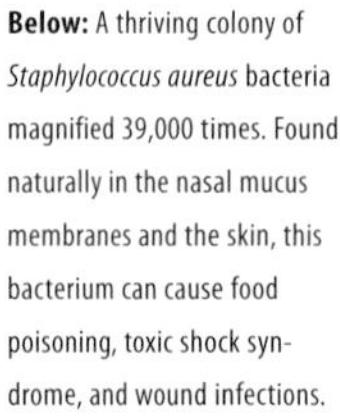

Below: A thriving colony of *Staphylococcus aureus* bacteria magnified 39,000 times. Found naturally in the nasal mucus membranes and the skin, this bacterium can cause food poisoning, toxic shock syndrome, and wound infections.

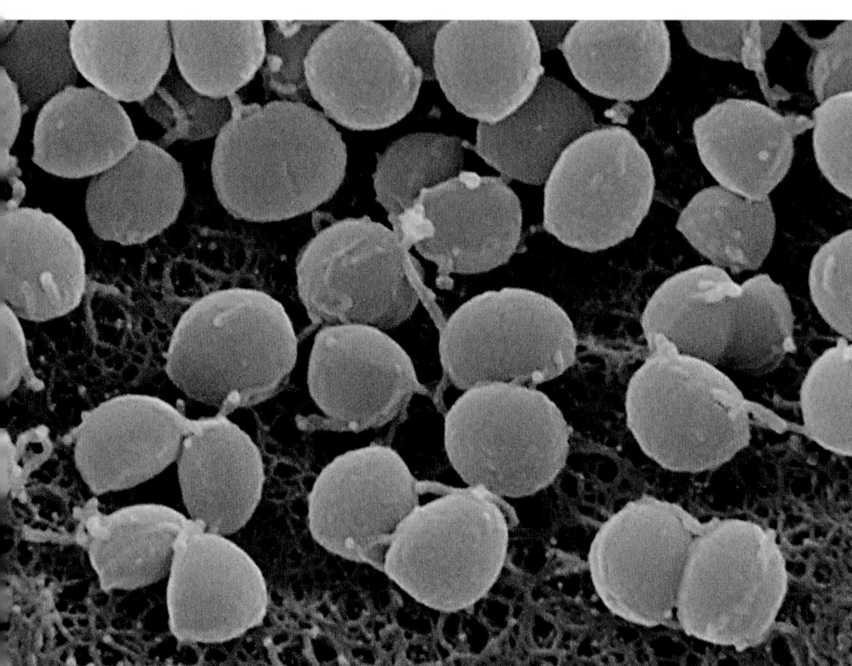

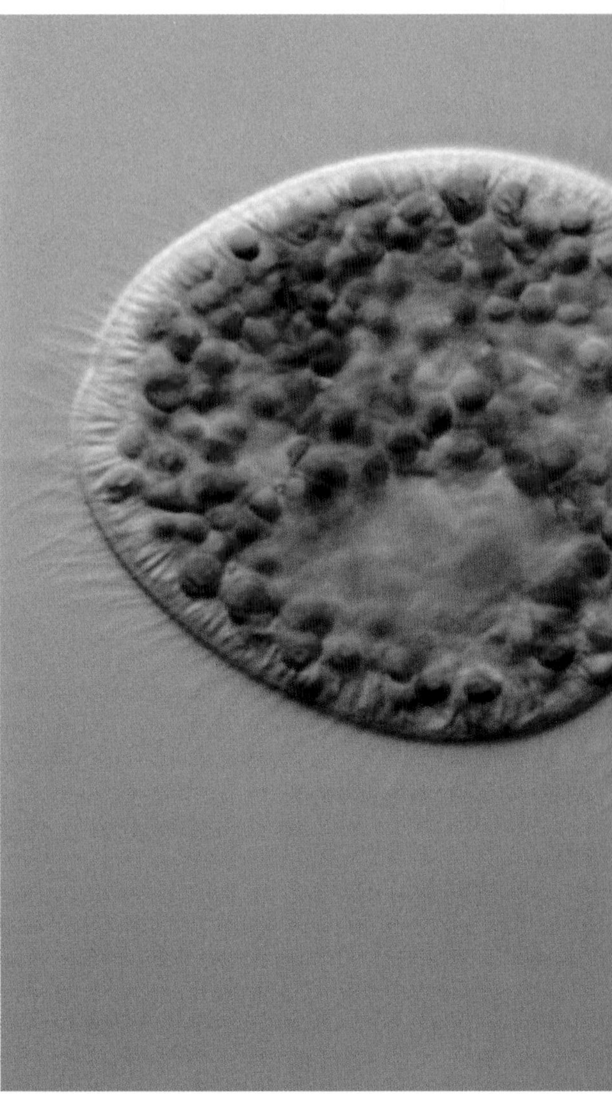

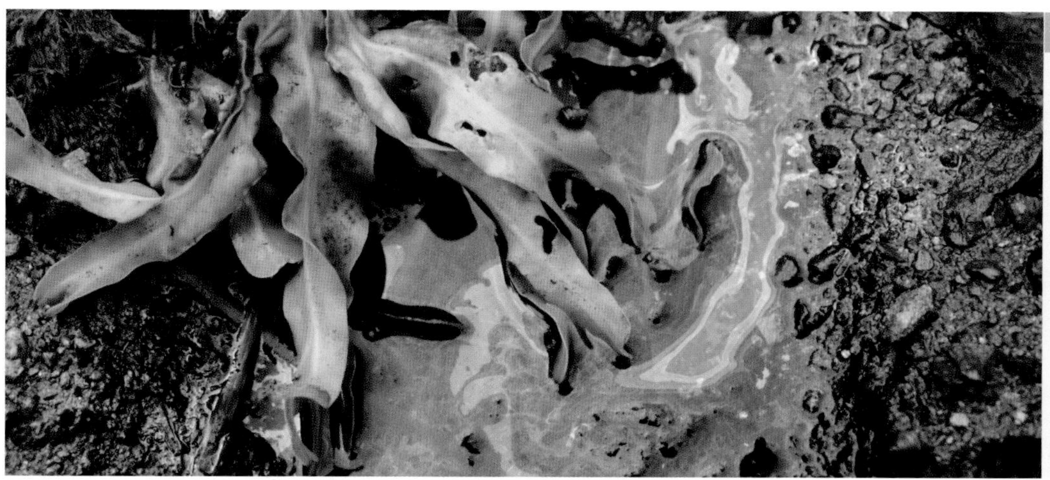

CLEANING UP OIL SPILLS

Certain bacteria, e.g. Pseudomonads, use hydrocarbons as an energy source. This ability has been used in bioremediation projects, to clean up polluted industrial sites or oil spills that threaten ecosystems. When the oil tanker the *Exxon Valdez* ran aground in Alaska in 1989, over 11 million gallons of crude oil was washed ashore (left) causing an environmental disaster and huge loss of wildlife. The clean-up involved the use of bacteria to break down most of the oil, and the area has gradually recovered.

For the first half of geological time our ancestors were bacteria. Most creatures still are bacteria, and each one of our trillions of cells is a colony of bacteria.

Richard Dawkins, biologist, b. 1941

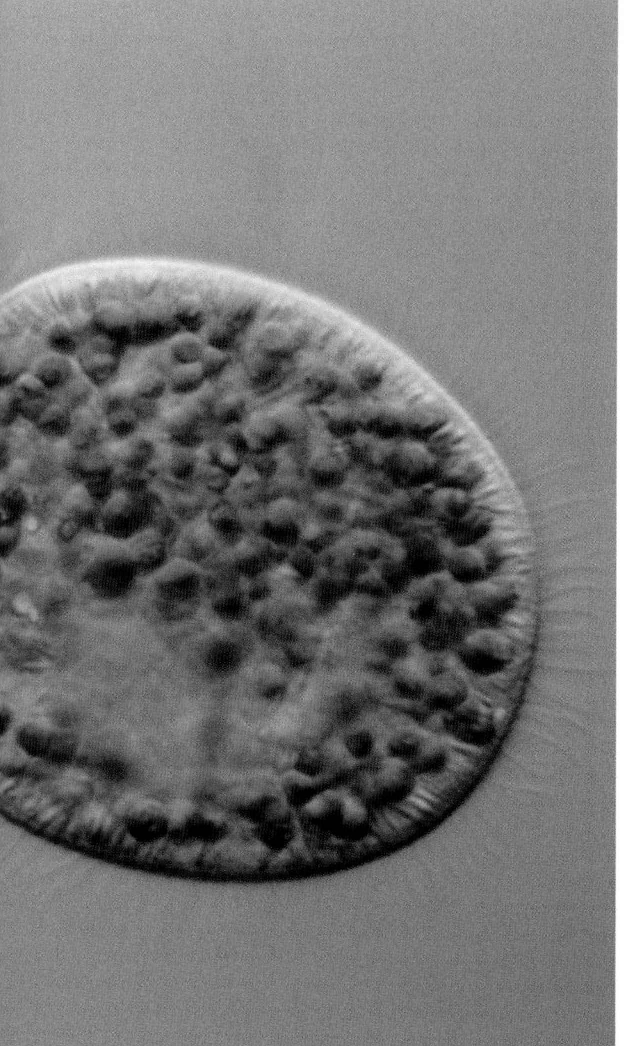

Left: The cells of *Paramecium bursaria*, a so-called slipper animalcule, are dividing by fission, an example of asexual reproduction that results in identical cells or "clones."

Staining properties

Danish bacteriologist Hans Christian Gram developed a method of staining bacteria in order to observe them under the microscope. Most bacteria fall into one of two groups: Gram-negative or Gram-positive. The staining property is related to the structure of the cell wall and has proved useful in diagnostic clinical laboratories. Gram-positive cell walls contain peptidoglycan whose synthesis is prevented by certain antibiotics such as penicillins and cephalosporins. Gram-negative bacteria are not very sensitive to treatment with these antibiotics. A few bacteria do not stain with the Gram stain.

Mycoplasmas do not have cell walls. Mycobacteria have a waxy layer outside the cell and are identified by the Ziehl-Neelsen or Acid-Fast stain (e.g. Mycobacteria tuberculosis).

Bacteria of medical importance

Very few bacteria cause disease in humans, but those that do can cause serious illness, even death. Among the most serious diseases caused by bacteria are tuberculosis, whooping cough, diphtheria, gastrointestinal diseases, pneumonia, Legionnaires' disease, meningococcal disease, gonorrhea, syphilis, and wound infections caused by staphylococci and streptococci. Diseases such as botulism, tetanus, gangrene, and anthrax are caused by bacteria that form resistant endospores under adverse condions. Bacteria affect the body by the production of enzymes and toxins that damage various organs. Before the discovery of antibiotics many of these diseases were often fatal. Antibiotics changed that, but there is growing concern that many bacteria have now become resistant to antibiotics.

Bacteria in the environment

Bacteria play a vital role in decomposition of organic matter, releasing nutrients into the soil used for plant growth. They break down sewage into water and methane. One group of bacteria live on the root nodules of leguminous plants and absorb nitrogen from the air, converting it into nitrogenous compounds that enrich the soil. Another group, the Actinomycetes, are a valuable source of antibiotics.

Bacteria are important in food production. Lactobacilli are responsible for fermentation of milk to produce cheese and yogurt. Others are used in the production of vinegar, citric acid, and other food additives.

Left: Lactobacilli produce lactic acid through the fermentation of lactose and are the reason why milk sours in order to make yogurt. These beneficial bacteria are used extensively in the food industry in the production of yogurt and cheese. Lactic acid works as a food preservative.

The cell is a fortress and houses many busily working parts. Every living organism is made up of individual cells that replicate, produce energy, communicate with other cells, and make proteins that are the workforce of the cell.

The incredible machinery of the cell

Cells are generally divided into two main types—prokaryotic and eukaryotic—based on the organization of their interiors. Bacteria are prokaryotic cells—they have a simple structure without inside compartments (see pages 272–273). Eukaryotic cells are the type of cell found in all other organisms.

The intracellular organization of cells

Eukaryotic cells contain a number of organelles enclosed by membranes, which compartmentalize the various cellular activities so they can be performed more efficiently. The nucleus stores and replicates genetic information in the form of DNA. The endoplasmic reticulum (ER) is a network of membranes that has a variety of functions. It acts as a support for the ribosomes where synthesis of proteins takes place; enzymes in the ER recognize toxic proteins that need to be disposed of to prevent cell damage, and it is also involved in the secretion of proteins and hormones. There are even two "trashcans" in the cell, proteasomes and lysosomes, that literally chop toxic proteins into pieces so that they do not cause cell damage.

The cytoplasm of the cell consists of fluid containing salts, sugars, and structures that bathe the organelles. The cytoplasm also houses the microtubules which are involved in many processes, one of the most important being cell division (see pages 276–277). To allow all these activities to happen, energy is needed, so the cell has designed an energy-producing powerhouse, the mitochondrion. In plant cells, chloroplasts carry out photosynthesis, trapping energy from the Sun into the synthesis of energy rich sugar.

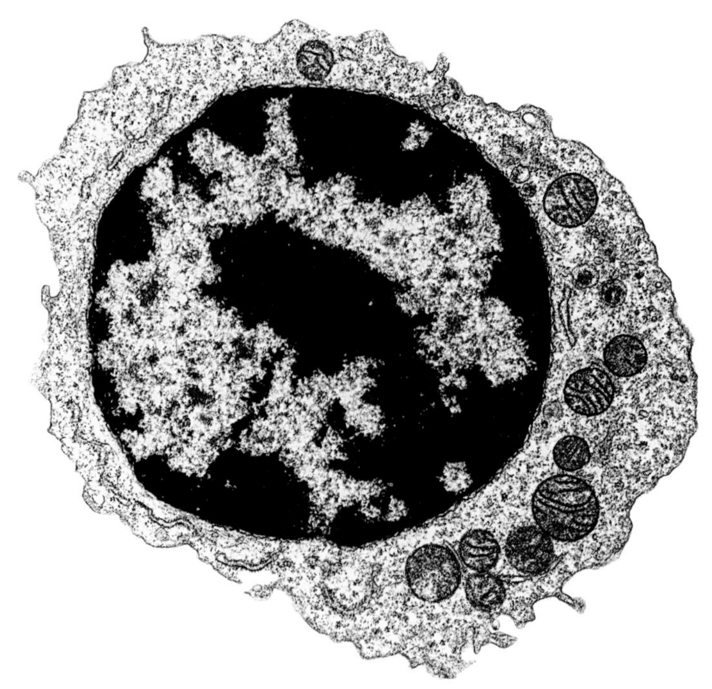

Above: This lymphocyte, or white blood cell, magnified 20,550 times, is a eukaryotic cell. It shows numerous major organelles such as a large nucleus (orange) and multiple mitochondria (blue).

When cell parts fail to function

What complexity, what diversity, for such a small organization of even smaller parts! If all these parts do not perform their jobs, and too many toxic proteins and damaged DNA is generated, the cell is brave enough to martyr itself. This is an effort by the cell to prevent copies being made of its damaged self, and in so doing to protect the human from diseases such as cancer. This process is known as programmed cell death or apoptosis. Obviously sometimes things go wrong and cancer develops, due to mutations in the DNA in the nucleus of the cells. Apoptosis limits the damage and acts, in more cases than we are even aware, to protect us.

> Man, like other organisms, is so perfectly coordinated that he may easily forget … that he is a colony of cells in action …
>
> Albert Claude, biologist, 1899–1983

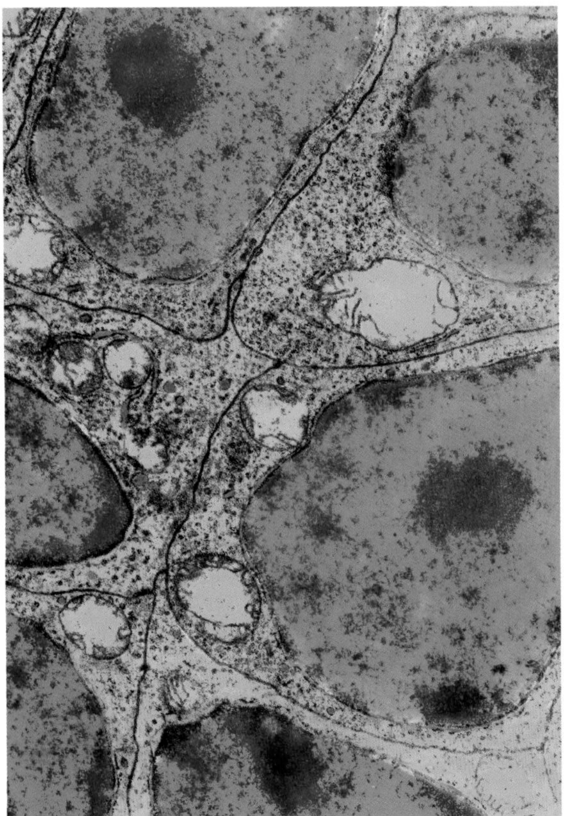

Left: A transmission electron micrograph of mammalian cells, magnified 250 times, showing cell walls, nuclei and mitochondria. Mitochondria function primarily to convert energy from food into a form that the cell can use.

Cellular organizations

There is further internal complexity depending on the role of the particular cell. A liver cell needs to be able make enzymes to break down toxins such as alcohol. A brain cell needs to be able to receive growth signals and talk to other brain cells in order to survive. A stem cell quietly awaits signals from the environment telling it what kind of cell it is destined to be. A cancer cell has lost control of the cell cycle and is replicating too many times.

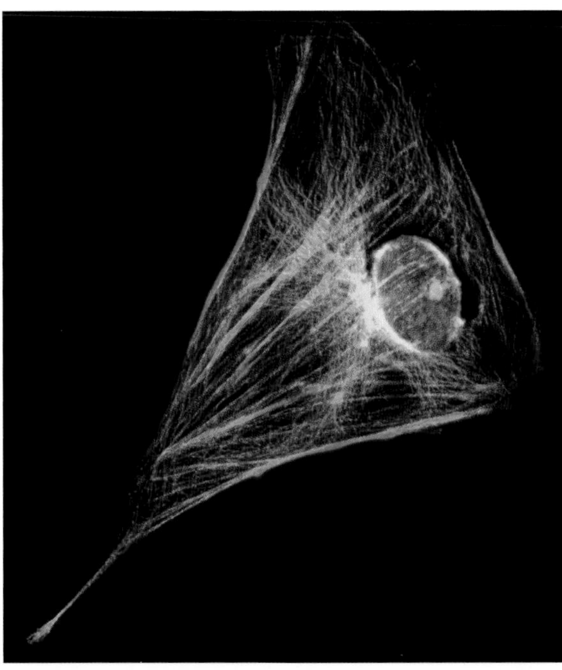

Left: A single cancer cell from a tumor cell line. The cell is derived from a fibroblast cell tumor. Fibroblasts are connective tissue cells that may differentiate into osteoblasts (in bone), chondroblasts (in cartilage), or collagenoblasts (in collagen).

THE CELL'S DISPOSAL UNITS

The main trashcans in the cell are the lysosome and the proteasome. Lysosomes are bags of hydrolytic enzymes that digest and dispose of large unwanted macromolecules, without causing damage to the rest of the cell. Proteasomes recognize and dispose of unwanted or defective proteins. When this does not happen, diseases such as Alzheimer's result. Imagine the trashcans as the waste disposal unit in your sink. When things that need to be degraded are fed into the waste disposal in an orderly manner, everything gets degraded. When too much is fed in at the same time there is a backlog. Sometimes mutations in our genes allow us to make lots of toxic proteins, overwhelming the waste disposal machinery, leading to backlog in the cell. This backlog leads to disease. Understanding which, where, and when proteins are degraded becomes important as we try to understand why some protein pieces that should be removed, are not, and thus why diseases such as Alzheimer's occur.

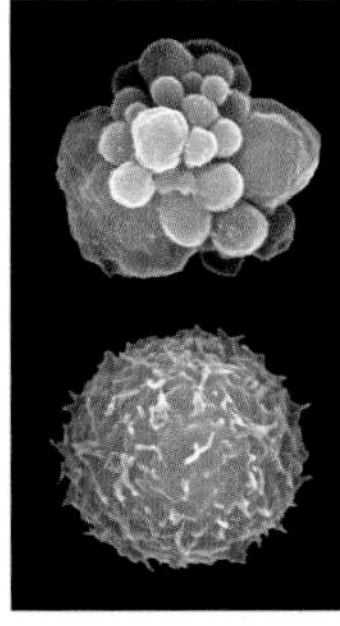

Left: Apoptosis or cell death, of a white blood cell is shown on the top, while a normal white blood cell is shown on the bottom. Apoptosis is a process of programmed cell death that occurs when the cells of the human myeloid cell line are deprived of growth factors.

Below: A magnified conceptual image of a generic human cell. Visible are the nucleus, golgi, centrioles, and mitochondria enclosed in a plasma membrane. This membrane is semi-permeable—pores allow the passage of some chemicals through.

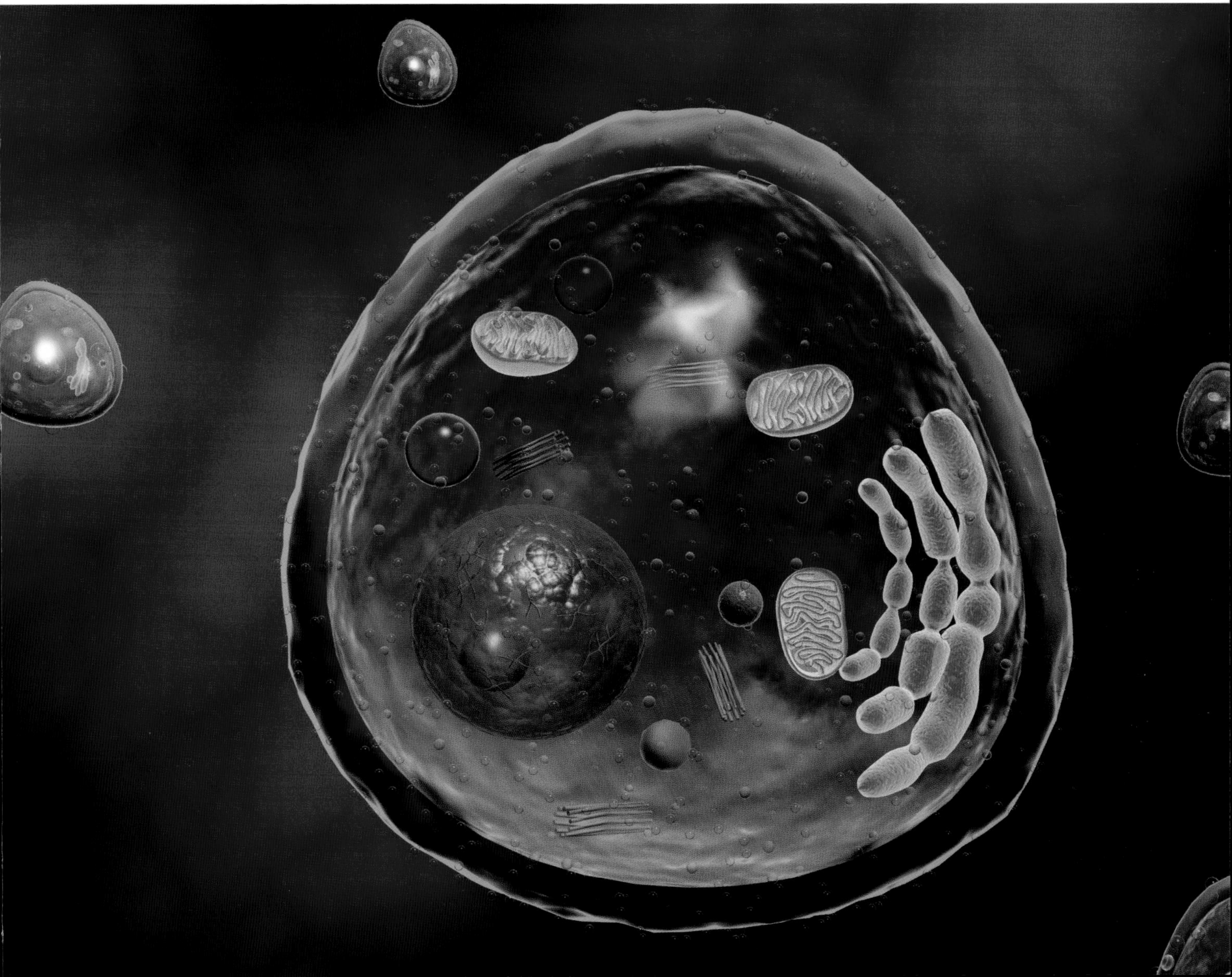

The basic characteristic of life is the ability to reproduce. All living organisms are in a constant state of renewal and repair. When animals and plants grow, their cells divide to form larger cell masses. During the development of a human embryo, the stem cells develop into specialized structures. When sexual reproduction occurs, two cells merge to produce a completely new organism.

Cell division

The timing and rate of cell division are crucial for normal growth and development to occur, and are regulated by special chemicals in the cytoplasm of the cell.

DNA and heredity

The genetic makeup of each cell is contained in its DNA, which is made up of a long strand of nucleotide bases arranged in a particular order. Two strands of DNA are coiled around each other to form a double helix and this is associated with protein in a structure called a chromosome. Bacteria contain only one chromosome, which is located in its nuclear region. In higher organisms the chromosomes are enclosed in a membrane to form a nucleus. The number of chromosomes varies in different species. Humans have 23 pairs of chromosomes that carry all the genetic information required to create another human being. During the process of cell division, it

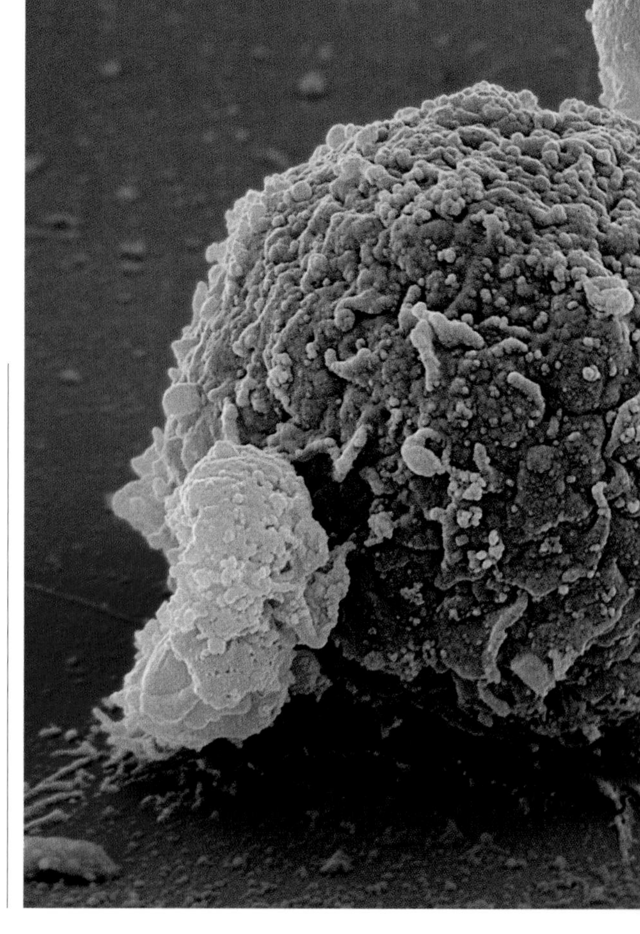

Right: Shown here magnified 14,300 times, two human embryonic stem cells have just completed the process of mitosis. Mitosis is the process whereby eukaryotic nuclei divide.

HOW GERMS MULTIPLY

Ever wondered why you can feel well one day but be sick with a raging fever the next? Many of the bacteria that cause infection, such as *Pseudomonas aeruginosa*, below, only take about 30 minutes to undergo simple cell division. Since one cell makes two, and two divide into four, and four make eight and so on, this means that by simple cell division, each cell multiplies into a million cells in about eight hours. No wonder you feel bad!

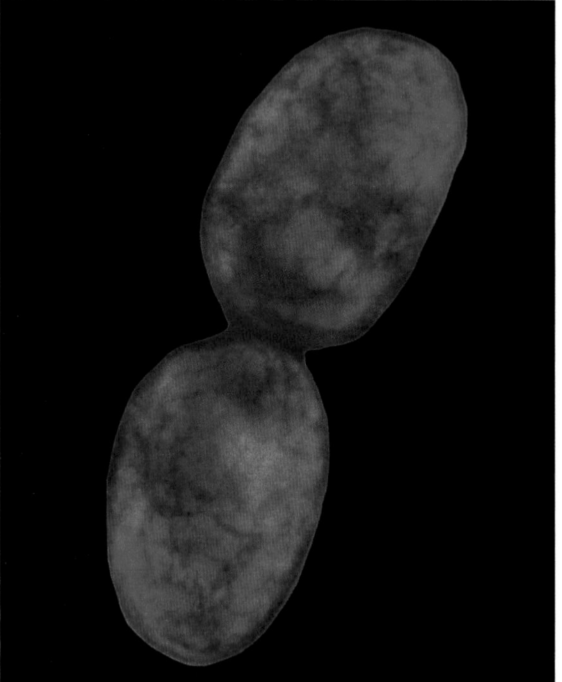

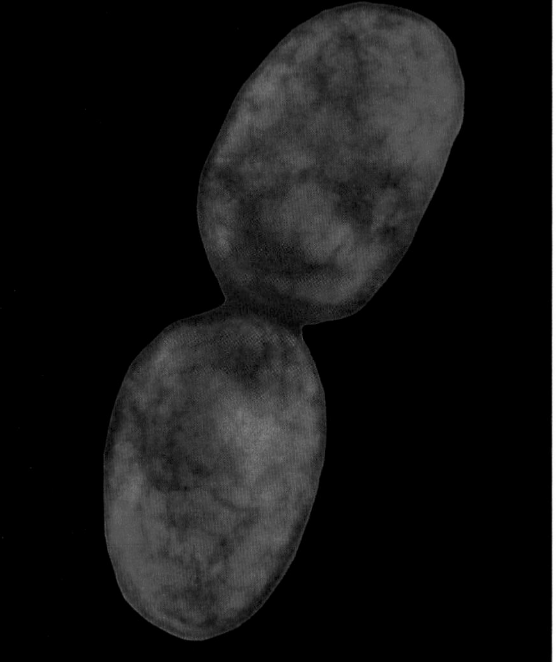

... I reasoned that study of the cell cycle responsible for the reproduction of cells was important and might even be illuminating about the nature of life.

Paul Nurse, biochemist, b. 1949

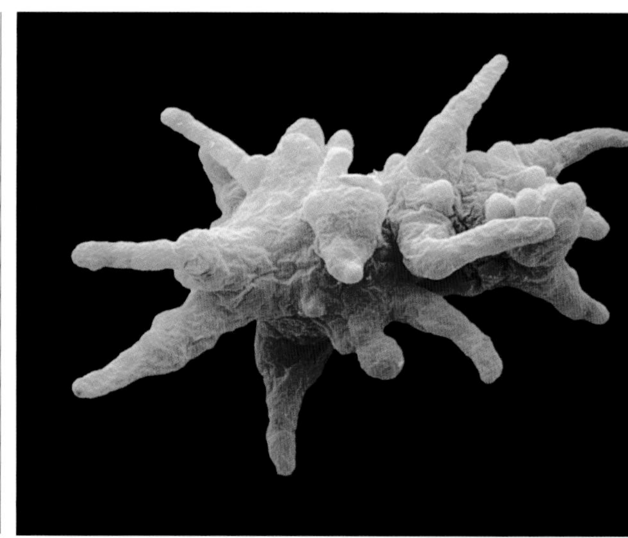

Right: An ameba dividing. The ameba is a mass of cytoplasm that has an ever-changing shape. Before cell division can take place, cytoplasm must form around each new nucleus.

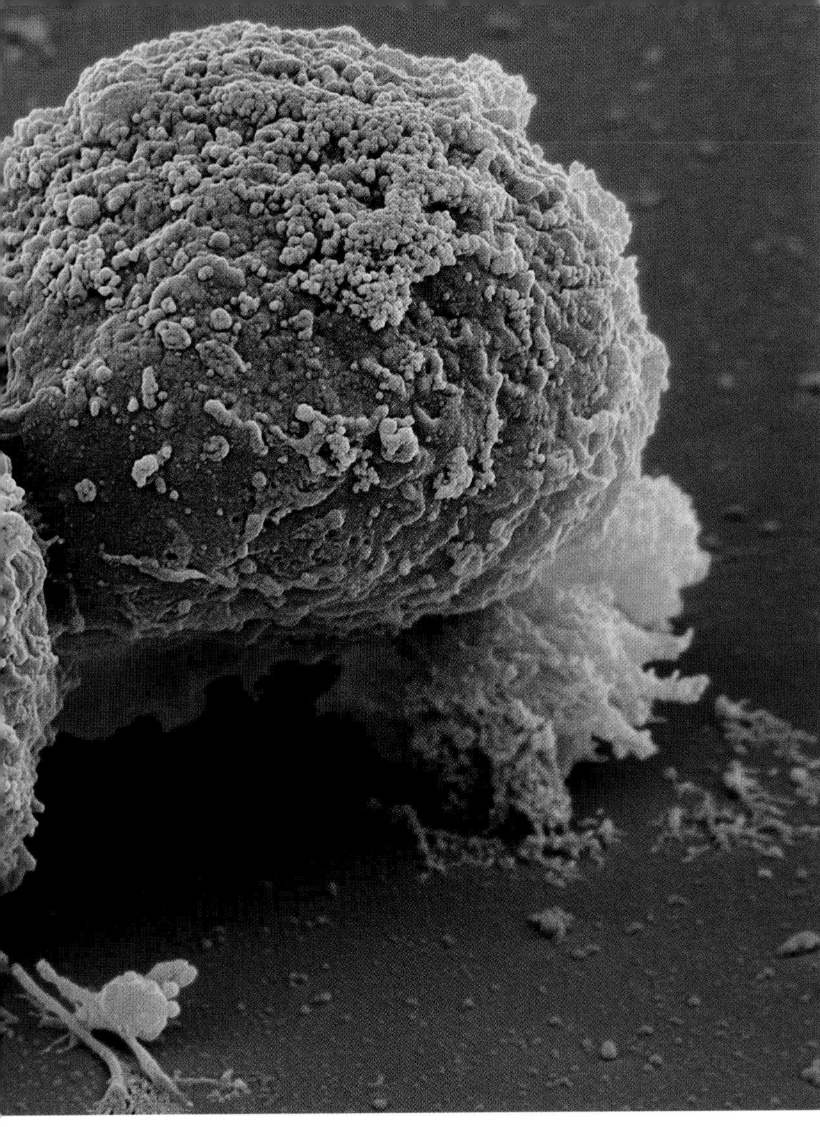

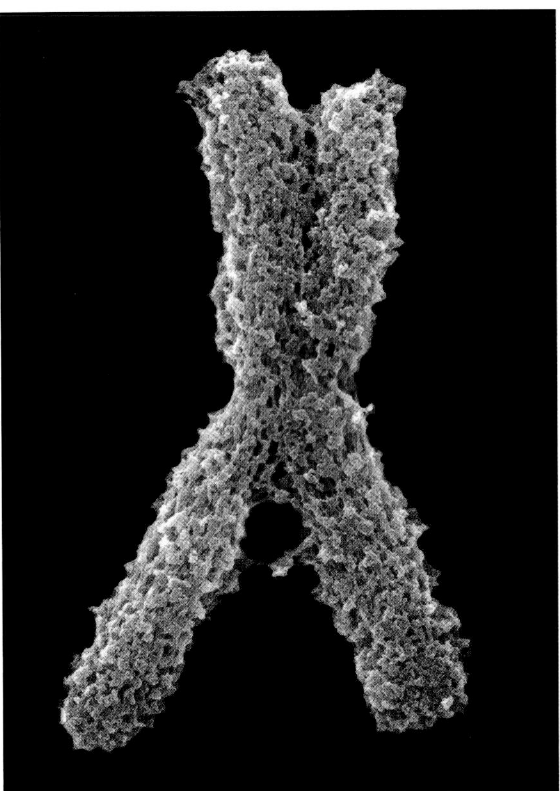

Above: A chromosome consists of a single molecule of DNA. This image was taken with a scanning electron microscope, so it is a true image. However, because the microscope does not acquire any color information, the color has been added artificially.

Below: Magnified 3,600 times though a scanning electron microscope, this image shows the telophase, or final, stage of mitosis. During this stage, the nucleoli reappear and the nuclear membrane begins to develop around the two groups of daughter cells.

is essential that an accurate transfer of genetic information from one generation to the next occurs, and this is achieved by the exact replication of the strands of DNA.

Simple cell division—binary fission

When bacteria reproduce, the cells merely split into two identical daughter cells. In order for these cells to be the same as the parent cell it is necessary for the DNA to first be copied so that there are two identical strands—one for each new cell. The double helix unwinds and a new DNA strand is built on the template or pattern provided, so that there are two new double helices formed. Each new chromosome then moves to opposite ends of the cell, and a wall is formed between them before the cell splits into two new cells.

Mitosis—asexual reproduction

When animals and plants grow, their cells undergo a process called mitosis that is similar to simple cell division but consists of several stages. Because the chromosomes are enclosed inside a nuclear membrane and the cells contain other structures in the cytoplasm, the process is more complicated than in bacteria. First each DNA strand is copied, then the nuclear membrane disappears and a spindle-like structure is formed. Microtubules in the cytoplasm help to pull the chromosomes apart so that one copy of the DNA is pulled to each end of the cell. A nuclear membrane is formed around each new set of chromosomes, the cytoplasm divides and new cell membranes are formed around the two new cells that are identical to the parent cell.

Sexual reproduction—meiosis

During sexual reproduction and cell division, the DNA from two different parent cells is combined to make a new organism. This

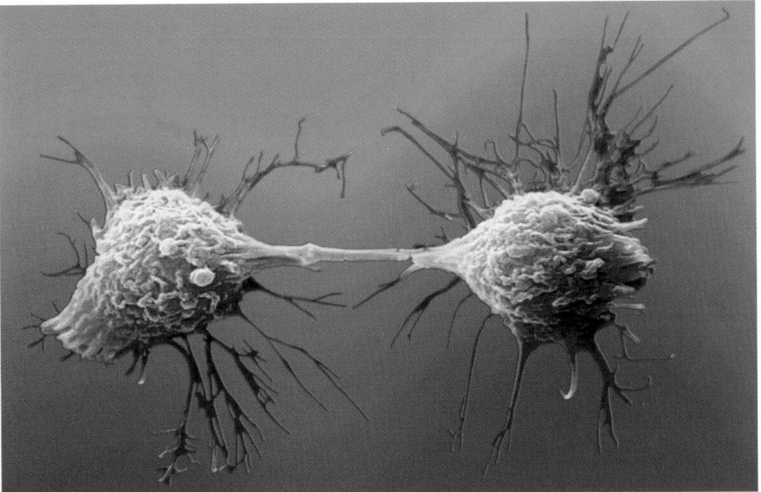

involves the formation of sex cells containing only half the DNA of the parent cells. First, each pair of chromosomes in the parent cells is separated. The cell then divides in such a way that each sex cell or gamete receives only one of each pair of chromosomes.

In animals, the female cell is called an egg and the male cell is called a sperm. When mating occurs, the egg and sperm fuse so that the offspring contains pairs of chromosomes again, but one of each pair is derived from each parent. The embryo then undergoes mitosis to produce more embryonic cells which then differentiate into specialized cells and organs.

Most of us think of proteins as steak, chicken, and eggs. This is not completely wrong as these are all made of different cells that have proteins in them, but when biologists talk about proteins they are thinking of something a lot smaller.

Proteins

Cellular proteins—shape relates to function

Proteins are the hard-working inhabitants of the cell. Because the shape of a protein is important for its function, proteins are constantly under quality control surveillance. They are formed and shaped and checked to see if they were made properly. If not, then it is off to the trashcan (proteasome or lysosome) for them and back to the drawing board for the protein machinery (see pages 274–275).

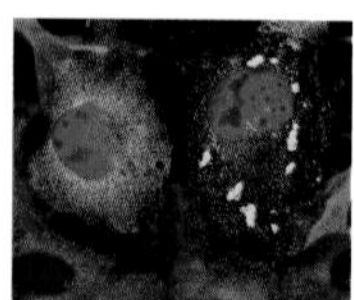

Above: Genetically engineering cells to over-express certain proteins triggers spontaneous assembly of stress granules, sites where the cell temporarily stores mRNAs it is no longer translating. One of the first proteins shown to assemble stress granules is TIA-1 (shown in yellow). The assembly of stress granules recruits other pre-existing proteins such as eIF3 (red), and causes the dispersal of other structures involved in mRNA decay (process bodies, blue).

Regulation of protein synthesis

Proteins are made in the cell with the help of many concerted working cell parts. As mentioned on pages 274–275, the nucleus of the cell houses all our genetic information. When a particular protein is needed—for example, an enzyme known as alcohol dehydrogenase is needed when we drink alcohol—our cells make this protein on demand. Making proteins on demand is best illustrated in someone who suffers from alcoholism. That person can break down so much more alcohol than a non-drinking individual exposed to alcohol for the first time. Why? Because the cell has made many copies of the alcohol dehydrogenase enzyme protein to help this person to clear alcohol from their system. So the alcoholic drinks, and sometimes even functions, perfectly well, while the cell works overtime to break down the alcohol they ingest. Eventually liver cells die and become fat-laden from having to work so hard, but this can go on for years due to the fabulous protein-generating cellular machinery.

How and where proteins are made

To make a new protein, the cell must select the gene with the information needed to make the required protein. Many proteins, known as transcription factors and polymerases, are involved in allowing this gene to be recognized and copied. The copy that is made from the gene of interest is called the messenger ribonucleic acid (mRNA). This message is then read by another part of the cell, the ribosome, to literally translate this mRNA message into the information needed to make the protein that is in demand. This newly formed protein has an important job to do. Since protein structure relates to function, the cell has many checks in place to make sure that the newly synthesized protein was properly assembled. If not it goes to the trashcan for that new protein and the machinery starts again.

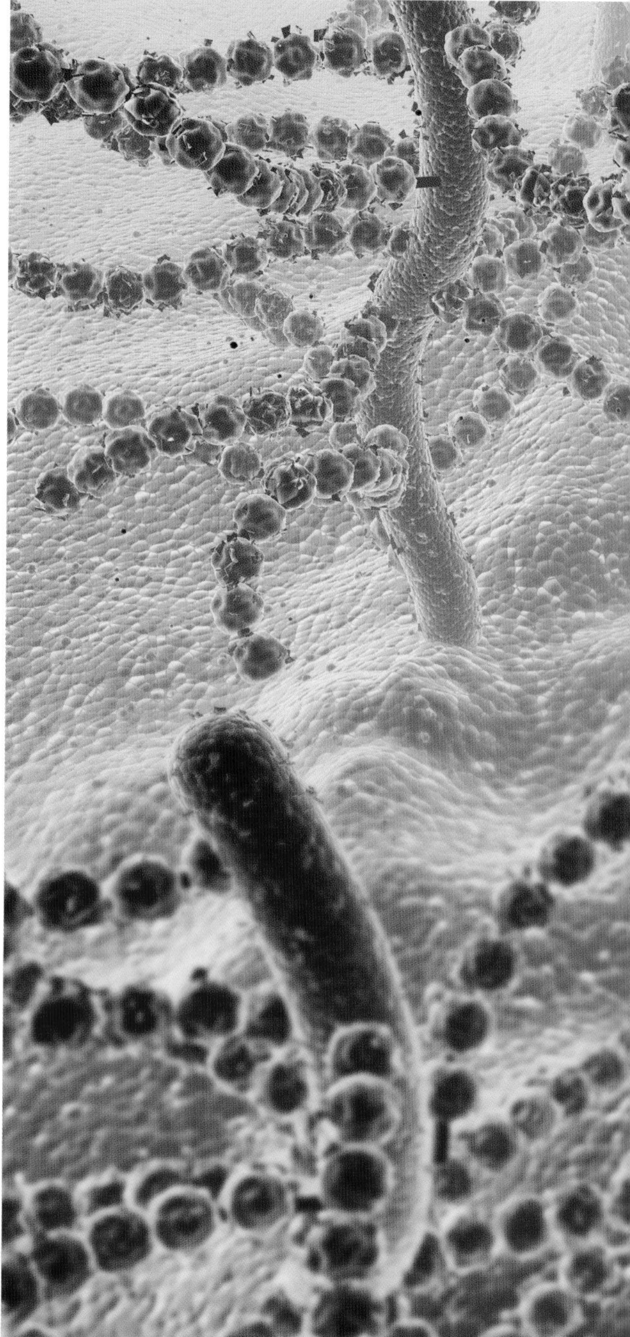

Right: A conceptual image of linear sugars on the surface of a cell. The sugars are attached to strands of protein.

Happy is he who has been able to learn the causes of things.

Virgil, poet, 70–19 BCE

Where do proteins perform their functions?

Some proteins are secreted, others are inserted into membranes, while others stay in the part of the cell in which they were made (endoplasmic reticulum) to carry out the work of that organelle. Some proteins are involved in helping things to get in and out of the cell (carrier proteins), others are involved in getting ions to pass across membranes (channels) and others still help other proteins to be removed when they need to be (chaperones) because they will be toxic to the cell and cause disease if they remain.

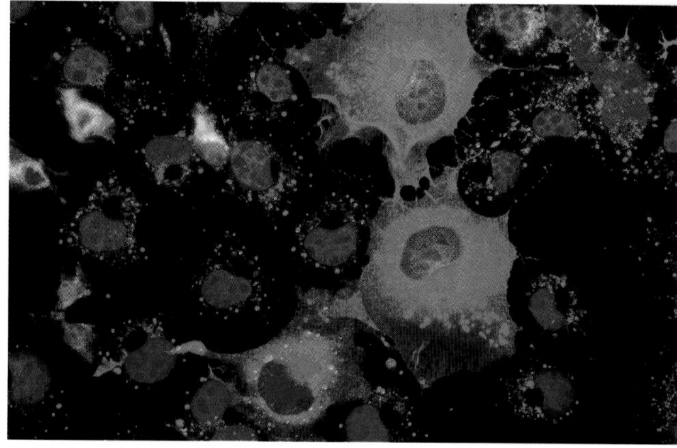

Right: Making more of specific proteins helps cancer cells to survive under conditions that kill normal cells. These cancer cells have been genetically modified to over-express the protein (shown in green) involved in apoptosis, or cell death.

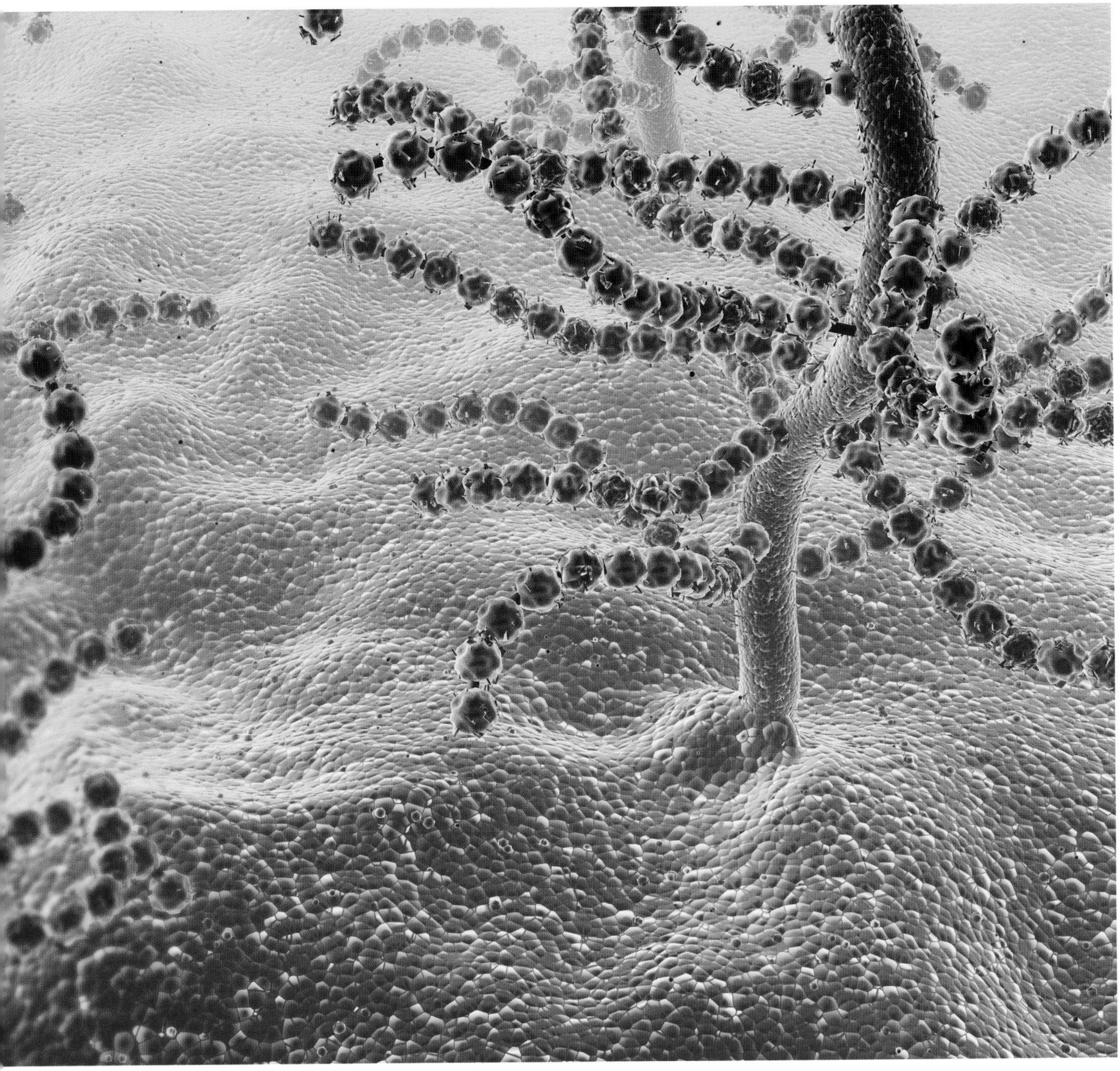

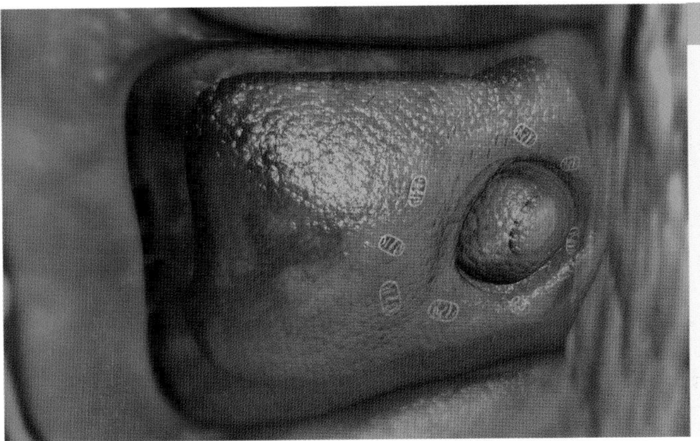

CYSTIC FIBROSIS—THE RESULT OF PROTEINS GONE BAD

The importance of proteins is best displayed when proteins that are supposed to perform functions such as the trafficking of ions across membranes do not perform their functions or are not working properly. For example, some people with cystic fibrosis have this disease because they have a genetic alteration in the gene that codes for the Cystic Fibrosis Transmembrane Conductor Regulator (CFTR). This channel transports chloride ions across membranes, but some mutations in this gene lead to the CFTR protein being trashed soon after it is made. This means that CFTR can no longer perform the function of a chloride channel. As a result the cell cannot pass chloride ions across its membranes and the thick mucus and infertility of the cystic fibrosis patients is the result. How incredible that one little protein not performing its function properly causes this massive effect on an individual. This highlights the importance of each and every protein in the cell being made properly, and performing its job properly, for us to live as healthy individuals. (At left is a microscopic view of a mucus cell.)

DNA, the genetic material found in every cell of the human body, carries the information that determines how we are put together and how we function. Since the discovery of its structure in 1953, many advances in DNA technology have been made including DNA sequencing, cloning, and understanding how DNA affects disease.

The building block of life

DNA, or deoxyribonucleic acid, is a long chemical structure made up of a sugar and phosphate group backbone. Coming off every sugar group in the DNA backbone, are molecules called nucleotide bases. Nucleotides can be thought of as letters in the alphabet. As words are made up of different orders of letters from the alphabet, genes are made of up different orders of nucleotide bases from the nucleotide pool. While the alphabet has 26 letters, there are only four different nucleotide bases in the nucleotide pool—(A) adenine, (T) thymidine, (G) guanine, and (C) cytosine. One strand of DNA can be so long that it can carry over a million nucleotides. This is just like having a sentence that is a million letters long! The entire set of nucleotides that a living organism has is its genome. The human genome has over 3 billion nucleotide bases. In comparison, the genome of a plant, for example rice, has a genome of 420 million bases, the genome of a mosquito only 27 million bases.

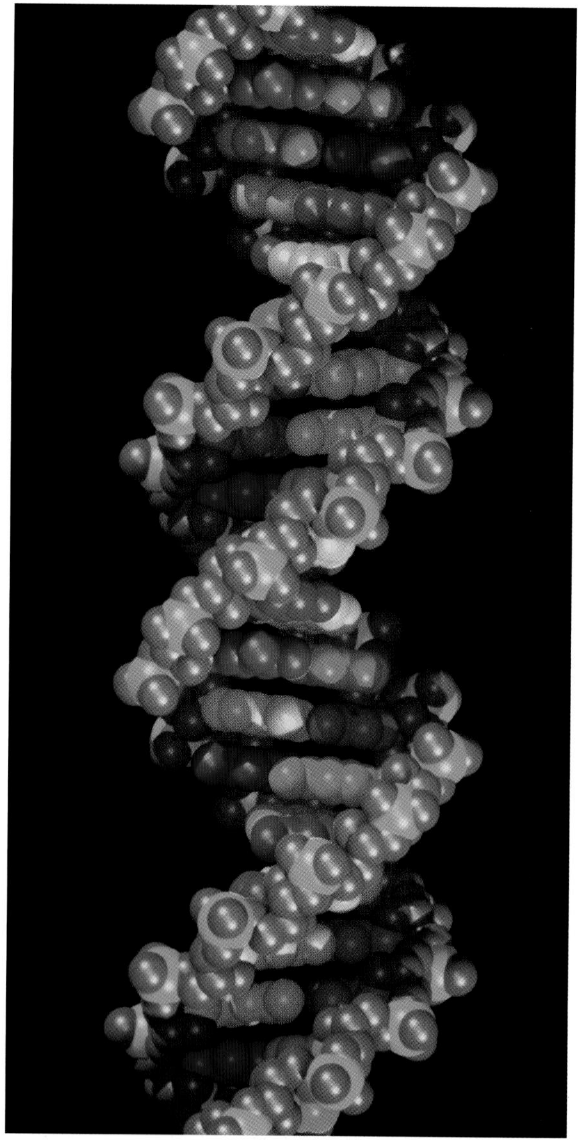

Right: A computer-generated model of a DNA molecule, which shows its double helix organization, with the sugar-phosphate backbone and the four different nucleotides—adenine, cytosine, guanine, and thymine—as well as the major and minor grooves of the molecule.

Where is DNA found?

DNA is found in every cell of a living organism. It is mostly found in a pair with another strand. The way that two strands of DNA meet and join is unusual, and was discovered in 1953 by James Watson, Francis Crick, and Maurice Wilkins. They found that two strands of DNA intertwine around each other forming a structure called the double helix. Imagine the double helix as a ladder—the sides of the ladder are the sugar phosphate backbone of each DNA strand, and the rungs are the nucleotide bases from each strand joining in the middle. This ladder-like structure is then twisted from the top like a ribbon, forming the double helix.

Joining up

What makes the structure of DNA even more complex is how the different nucleotides from each strand of DNA join. Nucleotides are complementary, so only adenine from one strand can pair with thymidine from the other, and only guanine can pair with cytosine. The complementary nucleotide bases are held together by hydrogen bonds, making the double helix stable. The DNA takes this structure to allow the nucleotide sequences to be replicated when cells divide.

If you were to take the DNA out of one cell and stretch it out, it would be 6 feet (1.8 m) long. And if you were to combine the DNA from every human cell, it would reach to the Moon and back 6,000 times. To fit this much DNA in our small cells, the human body has developed a way to tightly pack the genome into structures called chromosomes. Humans have 46 chromosomes (23 pairs) found in the cell nucleus. Each chromosome is made up of one strand of DNA carrying different nucleotide sequences, so different genes. When people reproduce, one of each chromosome pair is passed on to the offspring, resulting in a mixture of the DNA of both parents. This is how you acquire characteristics from each parent.

DNA was the first three-dimensional Xerox machine.

Kenneth Boulding, economist, 1976

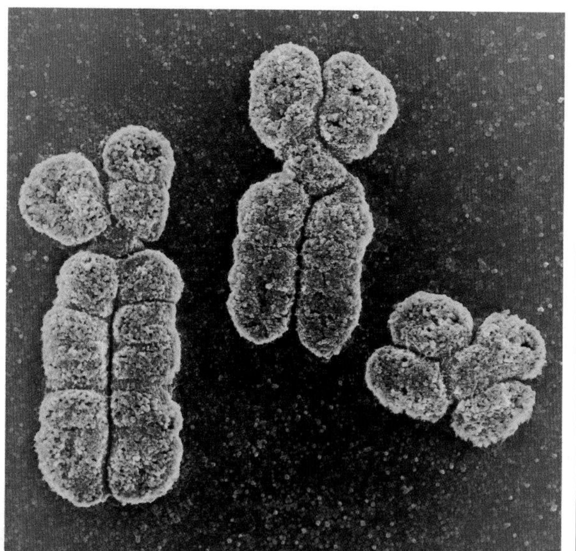

Left: A colored scanning electron micrograph image of human chromosomes, showing centromeres and chromatids. Chromosomes are the thread-like structures composed of DNA and proteins that are found in the cell nucleus.

Left: At the Nobel Prize ceremony in 1962. From left are scientists Maurice Wilkins, Max Perutz, Francis Crick, John Steinbeck (the novelist), James Watson, and John C. Kendrew.

Below: A medical technician examines DNA samples that are glowing pink under ultraviolet light. DNA analysis is the starting point for many molecular biological procedures including identification of individuals. Apart from identical twins, each of us has a different complement of DNA in our genes.

CHANGES IN YOUR DNA

When the nucleotide sequence of DNA changes, it is known as a DNA mutation. DNA mutations can include a deletion of a single nucleotide, a change from one nucleotide to another, or a deletion or repetition of large portions of DNA. Most DNA mutations do not have a negative effect, but if a mutation occurs in such a way that it alters what a gene encodes for, it can cause an illness. A medical condition caused by a DNA mutation is called a genetic disorder. Genetic disorders such as cystic fibrosis, Down syndrome, and Alzheimer's disease can be caused by a detrimental DNA mutation in one or many genes. Many forms of cancer are also due to genetic mutations. DNA can be damaged by environmental factors, like UV rays from the Sun, as well as things from inside your body, like oxygen free radicals. The human body has its own DNA repair mechanisms to help protect it from disease. These repair enzymes are constantly checking your DNA sequence and fixing any mistakes they might come across. However, they do not get to every mistake in time.

It was Gregor Mendel and his study of peas that initiated the identification of genes—regions of DNA responsible for inheritable characteristics. These first studies in the 1800s opened the door to understanding where genes sit in the genome, what they do, and how they are translated into proteins with the help of the genetic code.

It's all in our genes

The DNA sequence of the human genome contains 3 billion nucleotides. Within this long stretch of nucleotides, are smaller, stand-alone segments of DNA called genes. It is the order of the nucleotides within these smaller regions of DNA that spell out what each gene will code for. Every gene in the genome theoretically codes for one protein. These proteins make up the entire human body. So, ultimately, it is the genetic information found within each different gene that controls how our body is made. Every characteristic we possess, from eye color to shoe size, is in our genes.

Scientists expected that there would be thousands of genes in our genome. Yet when the Human Genome Project was completed in 2003 (see pages 284–285), scientists realized that humans had fewer than 30,000 genes, much fewer than they originally thought. They have since found that one gene can result in more than one protein being produced, through cellular processes like gene splicing. This in turn decreases the number of genes humans need.

"Junk" DNA

More study into the genes of the human genome found that they only accounted for 2 percent of its entire nucleotide sequence. This means that there are still billions of nucleotides found between genes that don't code for a specific protein product. Much of this spare DNA is closely linked to a specific gene and can be responsible for a gene's expression and regulation. Other repetitive regions of DNA that do not code for protein have been termed "junk" DNA. Although scientific research has tried to identify exactly what junk DNA does, scientists haven't yet determined its role.

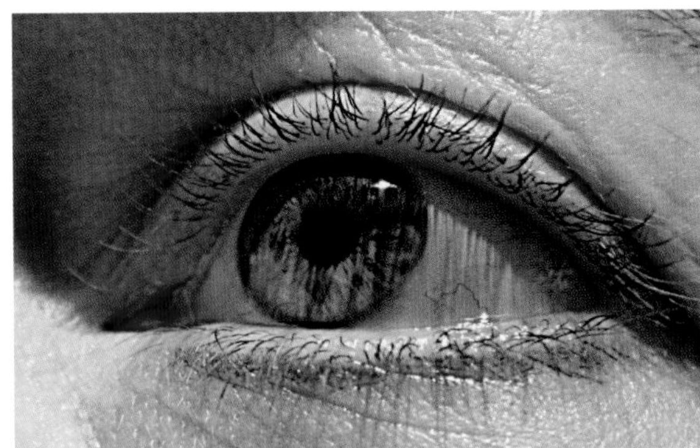

Right: Scientists have recently identified two genes that are responsible for macular degeneration, the gradual deterioration of eyesight in the elderly that can lead to blindness. Research also shows that smoking and being overweight carry a high risk of the condition, in which the central part of the retina degenerates.

Right: Genetics determines, among many other things, the color of our hair. Red hair is the least common natural hair color, and recent studies have suggested that only between 1 and 2 percent of the world's population are redheads.

> The genetic code describes the way in which a sequence of twenty or more things is determined by a sequence of four things of a different type.
>
> Francis Crick, biologist and physicist, 1916–2004

Left: This machine produces "synthetic" fragments of DNA by combining, in specific sequences, the bases adenine, cytosine, guanine, and thymine. The small fragments are often used as primers in polymerase chain reaction, a method used to make millions of copies of one specific gene sequence.

Opposite: A research analyst examining a DNA sequencing gel. DNA sequencing is the method of determining the exact order of the bases adenine, cytosine, guanine, and thymine in a piece of DNA.

Turning genes on and off

As every cell in an organism contains its entire genome, genes from the genome are usually regulated so that they are switched on and off at certain times or in certain places. This helps the cell to be more organized and to save energy. Some genes encode for proteins that the cell needs every day. These genes will be turned on all the time, while other genes may be switched on only when the cell needs them. There are other genes that encode for proteins that are only required in particular parts of the body, for example the brain. So only cells found in the brain will turn these genes on, while cells everywhere else in the body will have them turned off.

Similarly, other proteins are only needed at particular stages of development, for example puberty. Only at this time of development will these genes be turned on.

Most organisms tightly regulate when their genes are on or off like this—this is known as "gene expression."

THE GENETIC CODE

To convert the DNA sequence of a gene into a protein requires the help of the genetic code, a set of rules that the machinery of a cell follows in order to read the DNA sequence of the genome and convert the information it reads into new proteins. Unlike DNA, which is made up of many nucleotides, proteins are made up of many amino acids. To translate the nucleotide sequence of DNA into an amino acid sequence of a protein requires the use of codons. A codon represents three consecutive nucleotides, or a tri-nucleotide sequence, and every codon stands for one amino acid. There are 64 possible codons that can be made out of three consecutive DNA nucleotides, but only 20 amino acids that they need to code for. This means that more than one codon can code for each amino acid. There are also specific start and stop codons at the beginning and end of each gene, which also correlate with the beginning and end of each protein. The process of converting DNA sequence into amino acids is called translation. The genetic code is universal among all living things.

On June 26, 2000, a joint press conference by then-President of the USA, Bill Clinton and former British Prime Minister Tony Blair announced with much fanfare the completion of the first draft sequence of the entire human genome, an accounting of the three billion base pairs that make up the DNA sequence that spells out much of who we are.

The human genome project

The historic moment was the culmination of an idea that had its genesis some 15 years earlier, and started in earnest in 1990. When the idea was first proposed it seemed an impossible dream, but with the development of large automated sequencing machines, and new techniques for generating as much data as possible, and spurred on by a heated race between a public consortium of scientists led by Dr Francis Collins at the United States National Institute of Health and Sir John Sulston at the United Kingdom's Sanger Institute, as well as a privately funded effort headed by Dr J. Craig Venter at his biotechnology company Celera Genomics, the initial draft was completed two years ahead of schedule.

Right: A close up of DNA sequencing displayed on a computer monitor. The information has come from a DNA analysis machine. The results gained from DNA sequencing machines have a low percentage of errors.

Sequencing technologies

The two competing groups used different variations of sequencing technologies to achieve their common goal. Since its invention in 1975, sequencing has relied on a method known as Sanger, or chain-termination sequencing, which generates a series of overlapping sequences. The public consortium International Human Genome Sequencing Consortium (IHGSC) used a hierarchical approach, creating large ordered libraries of DNA fragments, which could be sequenced and then joined together to assemble the genome sequence. The Celera approach was to chop the entire genome up into random fragments and then assemble the data later using complex algorithms to identify overlapping sequences.

The human genome was declared largely complete in 2003, with an accuracy of one error per 10,000 bases and 99 percent coverage of the euchromatic, or gene-containing, portion of the human genome. This means that the only regions not sequenced are regions where repeated sequences and other structural anomalies make it impossible using current technology to determine the sequence. Since then, more gaps have been filled in and errors corrected.

One of the big surprises of the human genome sequence was the relatively small number of protein-coding genes it contained. Early

Without a doubt, this is the most important, most wondrous map ever produced by humankind.

Bill Clinton, 42nd US president, b. 1946

CRAIG VENTER'S GENOME

In 2007, the first diploid genome of a single individual was published. Fittingly, the sequence belonged to J. Craig Venter, pictured at left, who as head of the Celera Genomics genome project, became the catalyst for the completion of the human genome sequence years ahead of schedule. It was produced by researchers at the J. Craig Venter Institute, a non-profit institute focusing on applications of genomics.

Venter's genome sequence is, so far, unique. Unlike the genomes that are used to generate the human genome sequence for the human genome project, his sequence includes both copies of each chromosome, one inherited from Venter's mother, the other from his father, some six billion bases in total.

It provides valuable information about the variation in the genome, containing 4.1 million variations over 12.3 million bases of DNA, many of which have never been described before. It seems that there may be a lot more variation between individuals than has previously been thought.

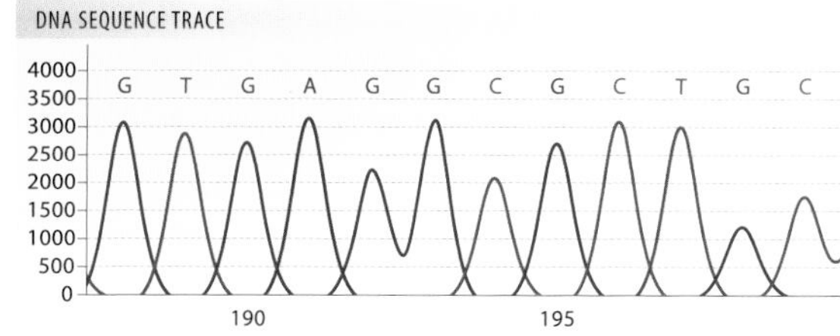

Above: DNA sequencing is the process that determines the sequence of nucleotides along a strand of DNA. One valuable application of DNA sequencing is testing for hereditary diseases.

estimates from the completed sequence, based on data from gene prediction programs, suggested 30,000–40,000 genes, and this has been dropped even further to between 20,000–25,000 genes—a far cry from the hundreds of thousands predicted before the project began. This means humans have only a few more genes than far simpler organisms, such as the roundworm *Caenorhabditis elegans*, which has 20,000 genes. However, each gene can encode more than one protein. The secrets of the non-protein coding regions of the genome are slowly emerging.

The availability of the human genome sequence has dramatically changed the way scientists identify genes, significantly speeding up the process of finding specific genes and determining their function.

What is the human genome?

The human genome contains essentially all of the hereditary information contained in a human cell's DNA, the so-called blueprint for a human. It consists of 46 chromosomes—22 pairs plus the gender-determining chromosomes X and Y. One of each pair of chromosomes comes from an individual's mother and the other comes from the father.

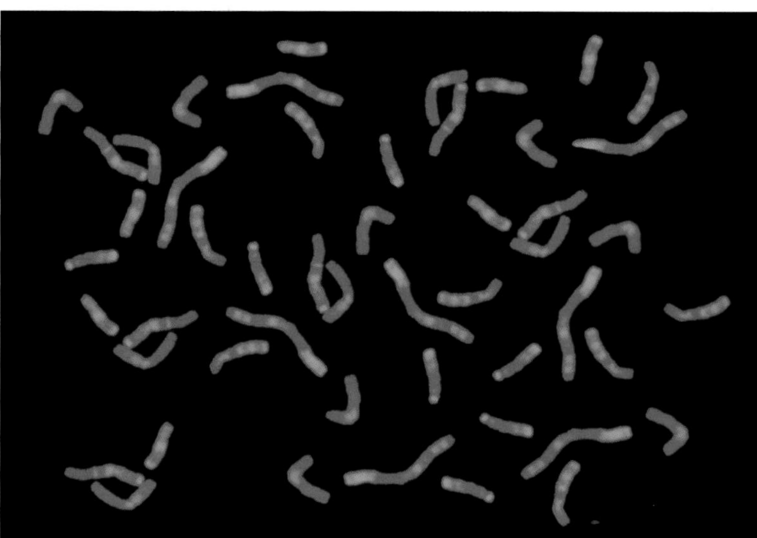

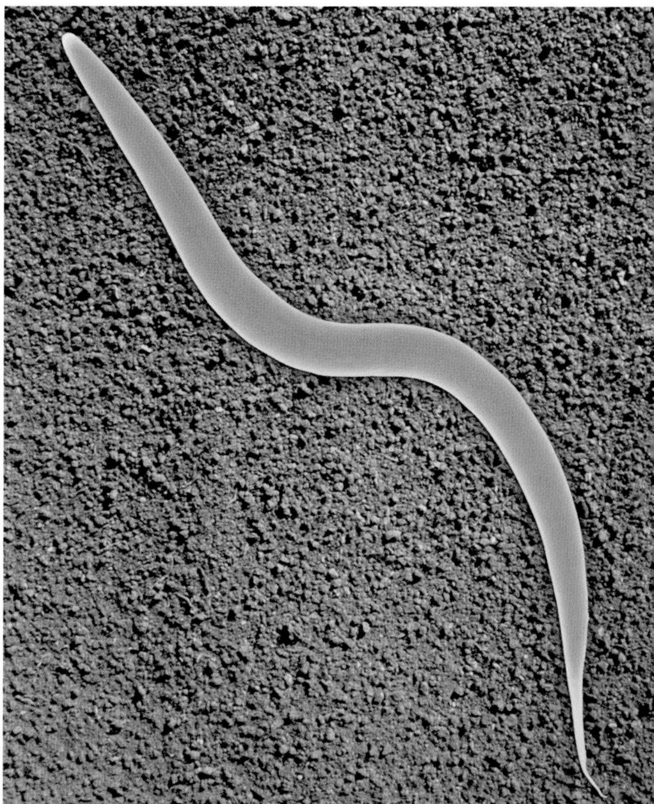

Above: Each human chromosome is formed from a single DNA molecule containing many genes. A chromosomal DNA molecule contains three specific nucleotide sequences which are needed for replication to be able to occur. We have 23 pairs of chromosomes.

Right: The free-living nematode worm (*Caenorhabditis elegans*) is used extensively in genetic and biological research. It was the first multicellular organism to be have its genome sequenced. Amazingly, humans have only a few more genes than this worm.

The field of stem cell biology is one of the most rapidly evolving and exciting in the area of medical research due to its potential to offer a new horizon in the treatment of some of the most severe clinical conditions, for which appropriate therapies are largely unavailable.

In the spotlight—stem cells

This potential lies in the fundamental capacity of stem cells to develop into any of the multitude of cell and tissue types in the body. Coupled with the natural anti-inflammatory properties of some adult stem cells, this represents a new paradigm for cell-based therapies.

Stem cells, which are found in all multicellular organisms, are characterized by their fundamental ability to self-renew by indefinite, yet tightly controlled, divisions, and their capacity to differentiate into a variety of new cell types. This latter capacity, to form multiple cell types, is known as pluri- or multi-potentiality. These abilities enable stem cells to develop into the multitude of tissues in the embryo, and to regenerate damaged tissue in the adult.

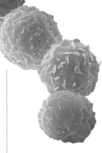

Above: Hematopoietic stem cells isolated from adult human bone marrow. These stem cells are part of the hematopoietic/reticulo-endothelial system and give rise to specialized blood cells such as macrophages and histiocytes.

Types of stem cells

Stem cells are broadly divided into two categories based on their potential for differentiation and the stage of life they are found. These are Embryonic Stem Cells (ESCs), which are able to differentiate into any cell type, and Adult Stem Cells (ASCs), which are mostly tissue-specific, such as hematopoietic or neural stem cells (blood or brain stem cells respectively). Although the study and clinical use of hematopoietic stem cells (HSCs) has been carried out for decades, it is only recently that our concept of stem cells has broadened significantly to include other tissue-specific ASCs, but more so ESCs.

Embryonic stem cells

ESCs were first identified as a consequence of research into reproduction and development, notably through the technology of *in vitro* fertilization (see pages 364–365) where it was discovered that a single cell could differentiate into the plethora of cell types found in the body. Conceptually it was assumed that the very earliest cell cluster (inner cell mass) developing in the blastocyst post-fertilization would form into the developing embryo and all cells and tissues and organs therein. By extracting such cells and developing appropriate culture conditions for expanding them as undifferentiated cell lines, it was shown that they could form into entire embryos when they were transplanted into receptive wombs. Such cell lines were given the name "embryonic stem cells" and were first isolated in a mouse by Martin Evans and Matthew Kaufman in 1981, and subsequently in humans in 1998 by Jamie Thomson.

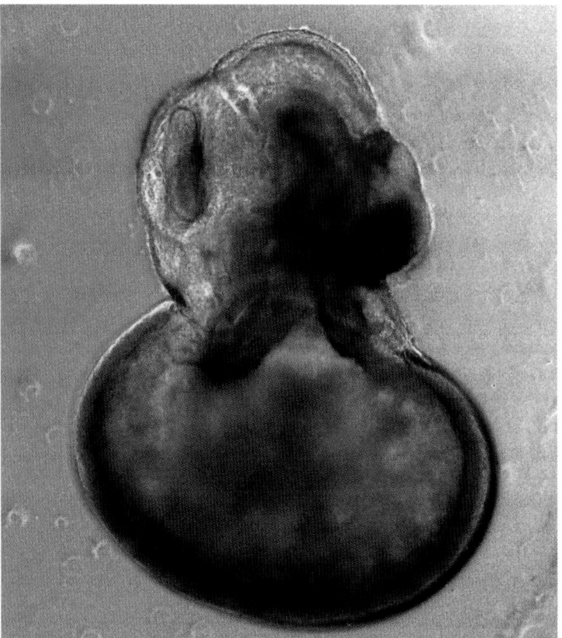

Left: An embryoidic body, a stem cell that has differentiated in suspension and which has the ability to become any kind of human cell, growing in a solution in a research laboratory.

Right: A researcher holds up human embryonic stem cell cultures in a laboratory at the Burnham Institute, California, USA. Many scientists believe that that stem cell research will lead to effective treatments for such diseases as Parkinson's disease and diabetes.

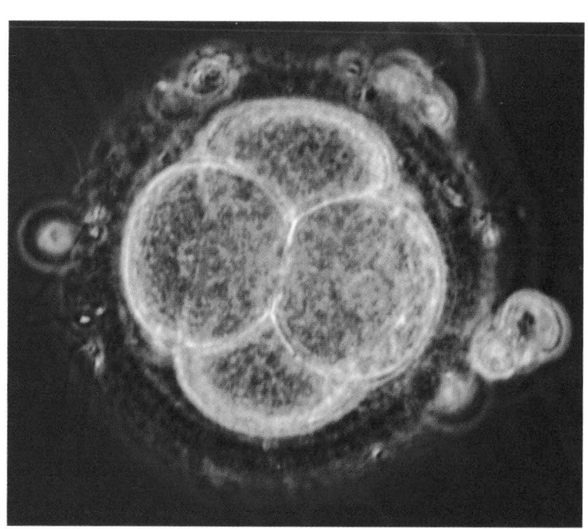

Left: A four-celled human embryo. Ethical objections have been raised into embryonic stem cell research. ESCs are derived from the inner cell mass of a blastocyst (one of the earliest masses of cells after fertilization). For research reasons, discarded IVF embryos, eggs that had been fertilized, then frozen, were used to extract and culture ESCs, thereby establishing stem cell lines. The debate over the use of these discarded embryos centered on the fact that, if implanted in a womb, these embryos may have developed into healthy babies.

What makes ESCs so special is their capability of developing into the three lineage-directions of development, known as germ layers—mesoderm (which contains developing organs such as the heart and kidney, as well as the blood system and muscle), endoderm (which contains the respiratory and gastrointestinal tracts as well as organs such as the pancreas and thymus) and the ectoderm (which contains the skin and nervous system, including the brain).

It is this broad-spectrum differentiation capability of ESCs that provides promise. Indeed significant progress has been made toward guiding ESCs down desired lineages in the hope of creating cells and tissues for clinical use—a process known as directed differentiation. However, as the immune system is structured to combat foreign organisms, one of the main drawbacks to ESC research is that any clinical treatment involving their therapeutic derivatives will need to contend with potential immune rejection. Traditionally this involves a protracted regime of immune suppressive drugs and their incumbent high degree of morbidity. However, a very recent discovery may negate this need by using the patient's own cells. This discovery came out of genetic profiling of stem cells relative to other cells

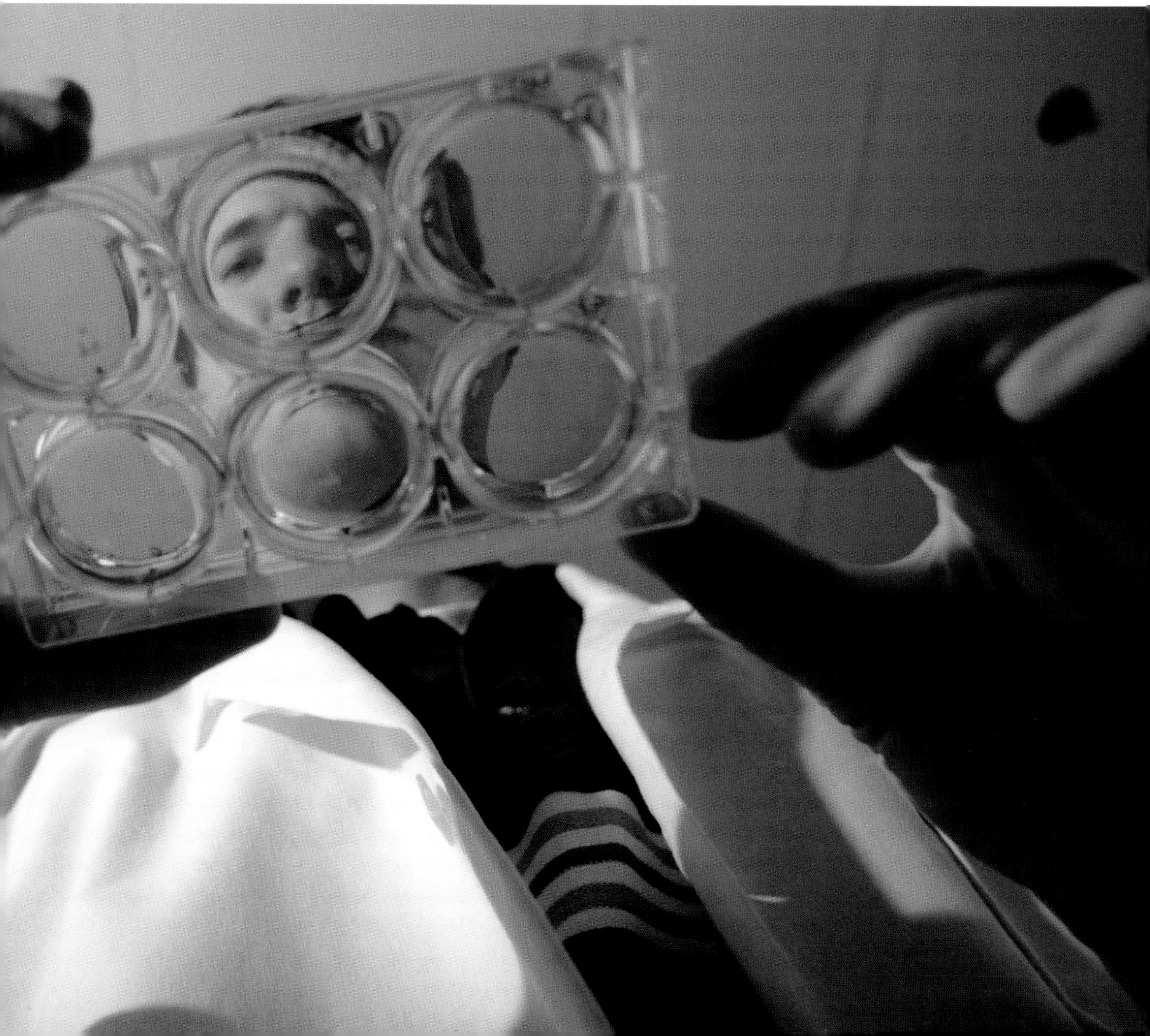

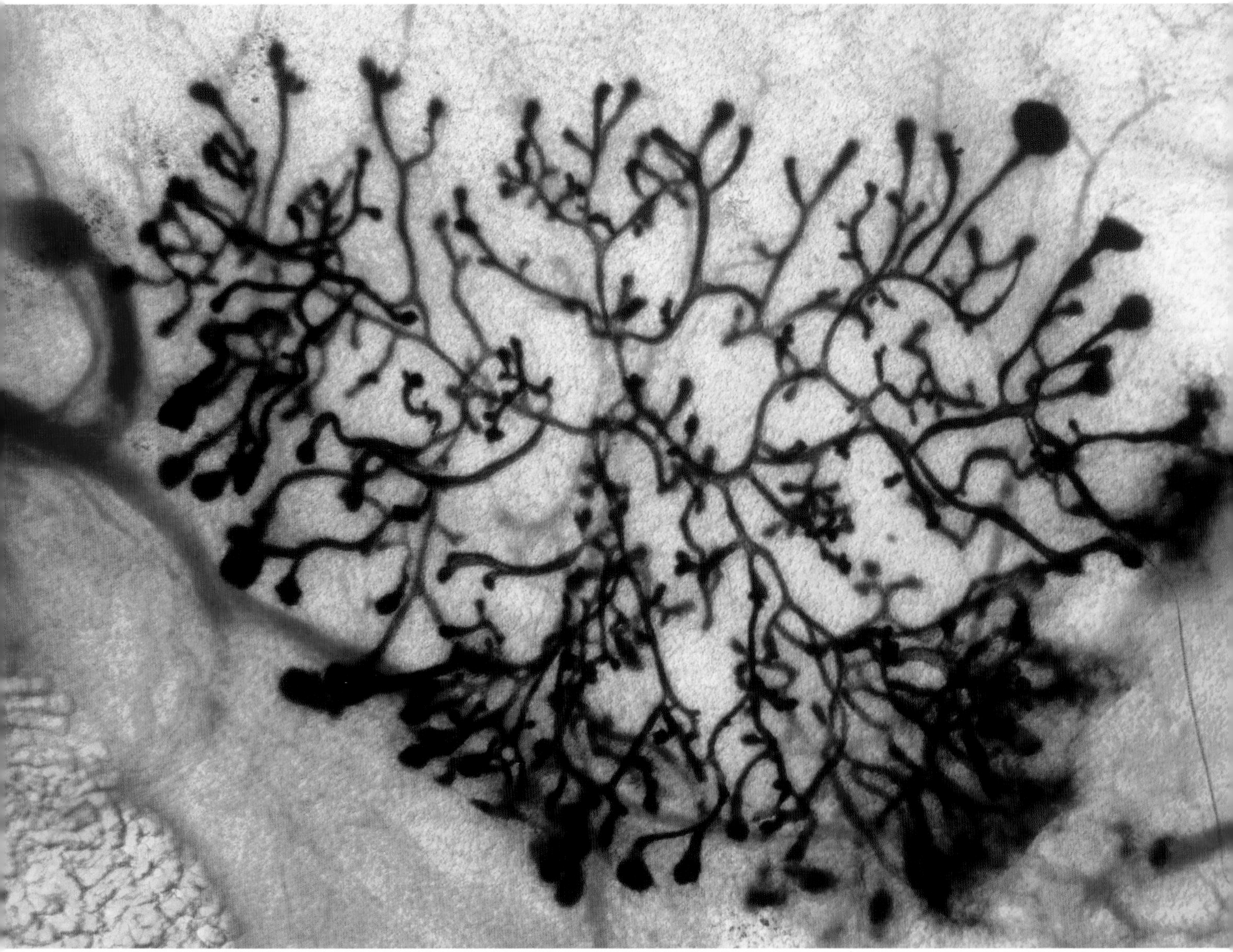

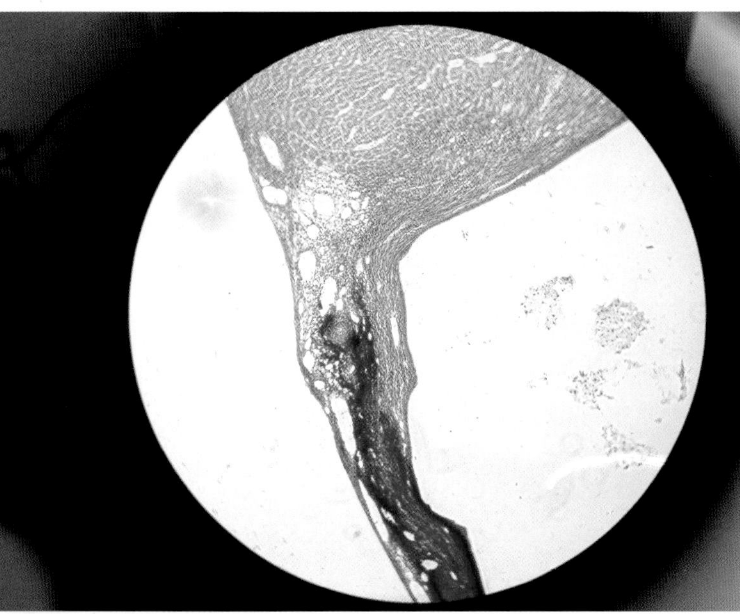

Left: This view through a microscope shows human heart stem cells (colored blue) repairing rat heart cells that have been damaged from a heart attack (colored pink). Researchers hope that stem cell research will help people who suffer chronic heart disease.

Above: This photograph shows a branching network of milk transport ducts, typical of development within the mammary gland, derived from a single stem cell. Researchers discovered the rare breast stem cells which could lead to new drugs or therapies to treat breast cancer.

and focused on what actually makes a stem cell a stem cell, in the hope that more mature cells can be "reprogrammed" to functionally resemble ESCs. This was achieved by inserting four specific stem cell genes into a mature skin cell. These reprogrammed skin cells, known as induced pluripotent stem (iPS) cells, were able to differentiate down multiple cell lineages as well as self-renew—mimicking the fundamental characteristics of ESCs. Taken together with work on the directed differentiation of ESCs down a particular desired cell lineage, this use of a patient's own cells would overcome the issue of immune destruction of the tissue as it recognizes the graft as "self," and provide a feasible means of tissue regeneration in the clinic. However, such cells are still a long way from clinical utility because of safety concerns. It is not yet known whether these cells will be susceptible to cancer formation, especially given that one of the inserted genes (c-Myc) is a common oncogene.

PERSONAL STEM CELLS

With the advent of potential clinical uses of stem cells has arisen the practice of stem cell banking. Stem cells can be harvested and stored from umbilical cord blood (for hematopoietic stem cells) as well as amnion and Wharton's Jelly in the newborn, and bone marrow and fat in the adult (for mesenchymal stem cells). These cells can all be stored indefinitely for future clinical use. Furthermore, recent findings on induced pluripotent stem cells, taken from mature adult cells and reprogrammed to take on a stem cell-like function, promise that stem cells can be derived from the patient's own tissue and used for clinical treatments. The use of banked cells negates the need for any suppression of the immune system as there will be no rejection of the tissue, seeing that the treatment essentially uses the patient's own cells. Adding to the promise of using banked stem cells for clinical regeneration, is the use of somatic cell nuclear transfer (SCNT) technology to test drugs specifically against a patient's own cells *in vitro*, thereby testing if treatments will work on disease and patient-specific stem cells, before trying treatments on the patient. This is done by harvesting diseased tissue and using SCNT to produce disease-specific and patient-specific stem cells.

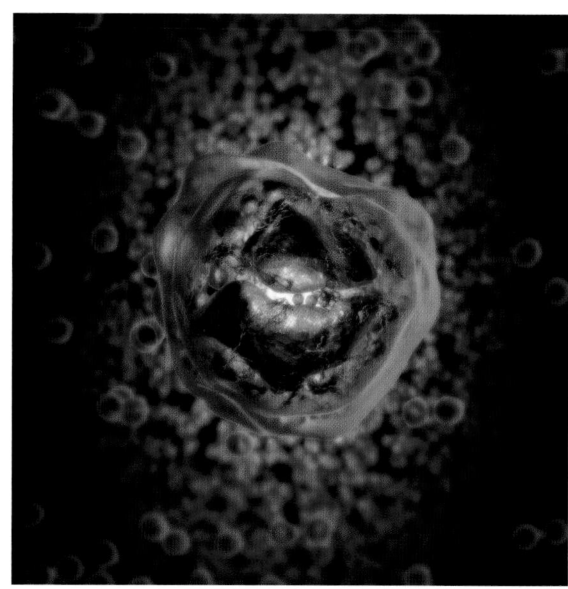

Right: A magnified conceptual image of a neutrophil granulocyte. Cells that are derived from hematopoietic stem cells include the multiple white blood cells of the immune system (such as T cells, B cells, neutrophils and macrophages) used to fight infections.

Below: This photograph shows neurons (red) and astrocytes (green), both of which can be made from neural stem cells. It is hoped that this research will lead to breakthroughs in treating neuro-degenerative genetic disorders such as Huntington's disease.

Stem cells are like toenail clippings with a better career plan.

Scott Adams, cartoonist, b. 1957

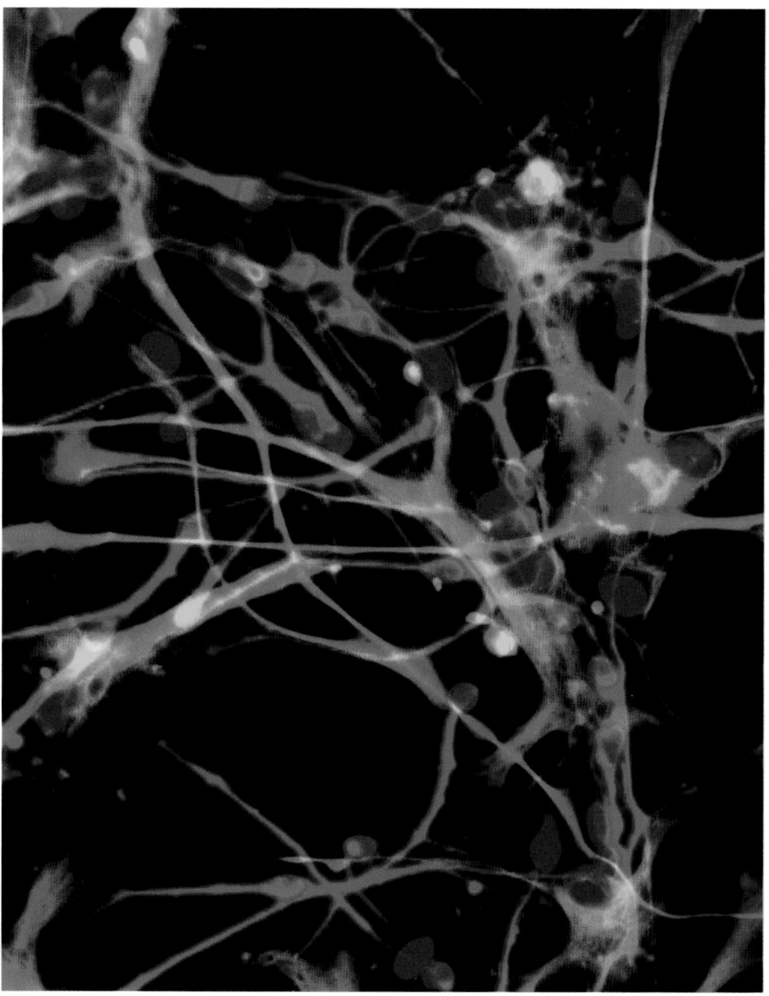

Adult stem cells

The most widely investigated stem cell to date is the hematopoietic stem cell (HSC). HSCs have been widely used clinically in bone marrow transplants to restore blood cells (hematopoiesis) following destruction by chemotherapy and radiation therapy as standard treatments for many leukemias and lymphomas. HSCT is also used for anemias (red blood cell disorder) and to attempt normalizing the immune system in some autoimmune diseases (diseases where the body's immune cells attack healthy tissue), such as multiple sclerosis. HSCs reside in the central region of bone, the bone marrow. Within this region, HSCs, with the assistance of signals from the supporting microenvironment, differentiate and develop into all blood cell lineages, a process that occurs on a continuous basis for the duration of life. Cells that derive from HSCs include red blood cells for oxygen carriage; platelets for clotting of blood; and multiple white blood cells of the immune system (such as T cells, B cells, neutrophils and macrophages) used to fight infections.

Types of ASCs

Recent research into other bodily systems has led to the discovery of many other types of ASCs. These cells provide daily replenishment of tissues and organs as part of natural "turnover." They have been identified in tissues such as pancreas, lung, heart, liver, muscle, brain, skin, hair, gonads, and endothelium. Two in particular have recently received much attention: Mesenchymal and Neural Stem Cells (MSCs and NSCs respectively). MSCs are found in many tissues of the body but have been primarily isolated from the bone marrow and fat, and enriched and expanded by *in vitro* culture. These cells are capable of differentiating into fat, muscle, cartilage, and bone cells, as well as bone marrow stromal cells (involved in supporting blood and immune cell development).

Along with their differentiation ability, MSCs also have very potent anti-inflammatory and immunosuppressive properties. This not only protects them from being rejected by the host (patient) immune system but also allows them to very effectively inhibit inflammatory reactions and block immune responses. This has led to research into their potential uses in autoimmune diseases. In addition to inhibiting some disease processes, MSCs also facilitate tissue regeneration from endogenous stem cells at the site of injury.

Neural stem cells are able to differentiate into the nervous system in both embryonic development and adult regeneration. They exhibit the same self-renewal potential as other stem cells and can be found in both the fetal and adult brain and spinal cord. Research has been focusing on the use of NSCs in the repair of spinal cord injuries as well as in neurodegenerative conditions such as Parkinson's disease.

An embryo is the progenitor from which an organism develops. It begins as one cell and develops into multitudes of cells—diverse cell types that make the body's complex tissues and organs such as the skin, liver, and brain. Embryogenesis is the route by which early cells develop an identity and organize into a new, unique individual.

From one cell to many: Embryology

How does one cell eventually become a community of varied cells? To do this, the cells of the embryo communicate, transmitting signals that dictate what each cell will ultimately become in the adult. Even though some of these signals have been discovered, a complete picture is yet to emerge.

Studying embryos

In humans, an embryo begins at fertilization until around eight weeks of pregnancy, when organs form. After this, it is called a fetus. Many animal eggs and embryos are very small and are studied microscopically in the laboratory. However, eggs in the animal kingdom come in all sizes. Reptile and bird eggs are relatively large with a protein-rich, jelly-like coating that sustains the embryo. In all animals, embryos form after the father's sperm incorporates into the mother's egg. Called fertilization, this sets in motion events that cumulate in a new, individual genome—events facilitated by factors within the egg.

By the time they unite, the egg and sperm have half their number of genes, so each contributes to the embryo. Thus, we inherit half our genes from our mother and half from our father. Following this genetic reconfiguration, the one-cell embryo starts to divide. Mice, one of the most studied laboratory animal, provide much of our knowledge of embryos. Experiments with mice show that at these earliest stages, all the embryo's cells are pluripotent—that is, they can become any type of body cell. After a few days, some cells decide what to become and change into trophoblast cells, forming the placenta that nurtures the baby during the pregnancy (see illustration at right). Where a pluripotent cell takes on specific characteristics, is called differentiation. The remaining cells have been harvested to generate human embryonic stem cell lines.

Eggs are used for cloning technology, whereby genetic material from an adult cell is re-set, making it behave embryonically. Being its clone, the embryo generated using this procedure has the same genome as the adult cell.

Sending special signals

In humans, the embryo implants into the mother's uterus and the cells change. To do this, they strategically move around, allowing exposure to signals that dictate their ultimate position and identity. This orchestrated strategy is so successful that most animals—frogs, fish, mice, and humans—follow it, dramatically restructuring the embryo. This forms distinct layers, foundations from which all of our body cells derive. Cells of one layer, positioned at the outer embryo become skin, brain, and nervous system. At three weeks of pregnancy they push inwards, forming the neural tube, which generates the brain and spinal cord. The developing heart then begins to circulate blood around the embryo. Further signalling and rearrangement generates organs and tissues. Thus, embryogenesis involves cells developing an identity, maintained throughout an organism's life.

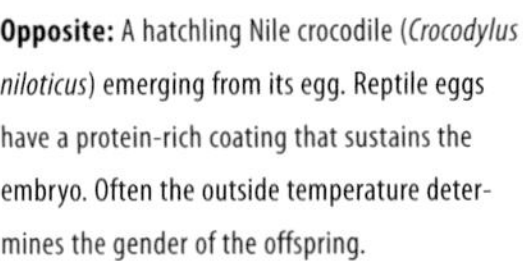

Opposite: A hatchling Nile crocodile (*Crocodylus niloticus*) emerging from its egg. Reptile eggs have a protein-rich coating that sustains the embryo. Often the outside temperature determines the gender of the offspring.

Below: The blastocyst is the structure that is formed in early mammalian embryogenesis, after the formation of the blastocele, but before implantation. The outer cell eventually forms the placenta. Here implantation has occurred.

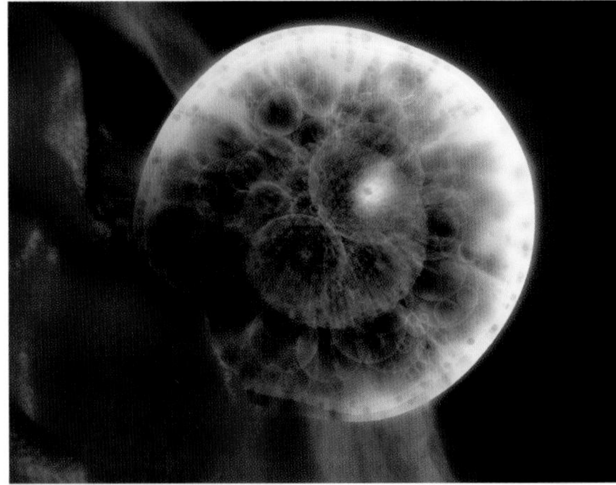

... two cells unite to form the fertilized egg cells. That is a marvel ...

Erwin Schrödinger, physicist, 1887–1961

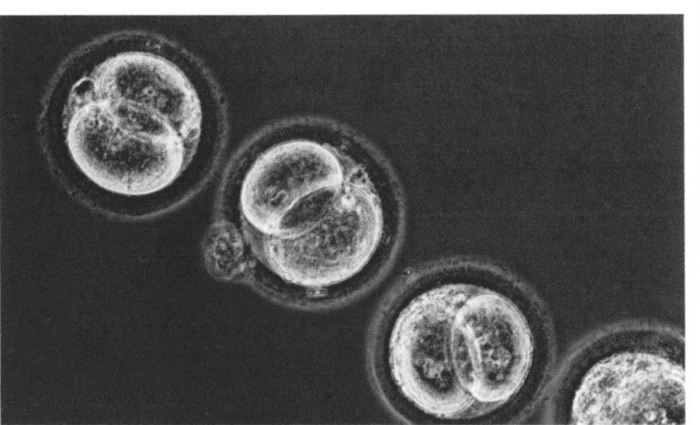

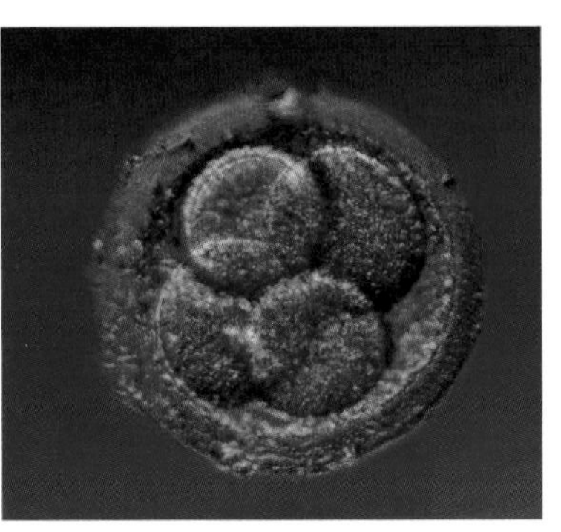

Left: A two-day old human embryo pictured at the four cell stage of development, magnified 260 times. Once an egg has been fertilized by a sperm the resulting zygote's nucleus is activated and it begins to grow by dividing.

Far left: A group of two-celled human embryos. A zygote results from an oocyte being fertilized by a single sperm; embryo refers to the next stage of development after the zygote has divided. Many believe the embryo begins at fertilization or at syngamy, when the parents' genomes join. Others believe that an embryo starts after fourteen days of growth. Our beliefs of what an embryo is can alter over time, a reflection of society's altered views about family, women's rights, reproduction, and medical technology.

SEARCHING FOR ANSWERS

Christiane Nusslein-Volhard was fascinated by how one cell becomes a complex, coordinated, living animal. In 1995, she, along with Edward Lewis and Eric Wieschasfor, was awarded a Nobel Prize for "discoveries concerning genetic control of early embryonic development." This was a great achievement, one uncommon for women scientists. Born in Germany in 1942, she became interested in biology at an early age. While working at the European Molecular Biology Laboratories in Heidelberg, she and Eric Wieschasfor studied fly embryos. They developed a way to alter genes, which are inheritable parts of DNA. By doing this, the scientists revealed the gene's role in embryo growth.

They identified many genes, essential for normal embryo development and rightly predicted that they influenced each other. They named the genes flamboyantly, *oskar*, *gurken*, and *hedgehog*. The "hedgehog" gene is important in the understanding of human birth defects, critical for brain and spinal cord development.

In 1997, the world was stunned when Scottish scientists at Roslin Institute created the cloning of the much-celebrated sheep "Dolly." However, this amazing breakthrough generated uncertainty over the meaning of "cloning"—an umbrella term traditionally used by scientists to describe different processes for duplicating biological material. This was an astounding development as most scientists thought that it was impossible to clone a mammal from an adult cell.

Carbon copies

When the media reports on cloning, they are usually only talking about the type called reproductive cloning. Yet there are different types of cloning. Cloning technologies can be used for purposes other than producing the genetic twin of another organism. Here, three types are discussed—DNA or gene cloning, reproductive cloning, and so-called therapeutic cloning.

DNA or gene cloning

This cloning technique refers to a process whereby a DNA fragment of interest from one organism may be cloned to produce multiple copies of the gene for further study. Bacteria are most often used as the host cells for a DNA fragment, but yeast and mammalian cells are also utilized in the laboratory.

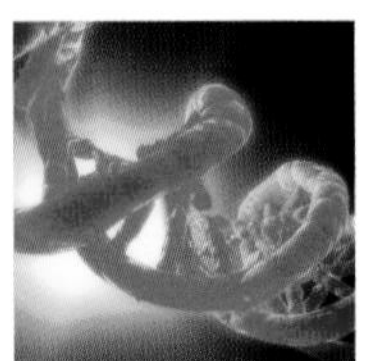

Above: The double helix strands of DNA. DNA contains the genetic instructions for the development and functioning of all living organisms. Within cells, DNA is organized into chromosomes. Chromosomes within a cell make up a genome.

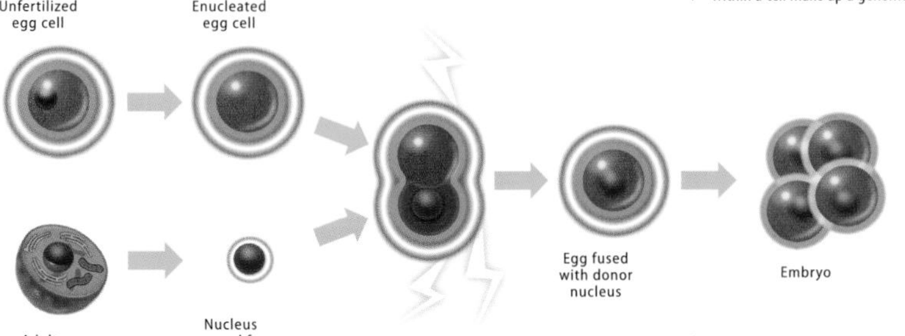

Above: To create a clone by nuclear transfer, an ordinary egg cell has its nucleus removed (enucleation). At the same time, the nucleus of the donor cell is obtained, ready for insertion into the egg. Electricity is used to initiate chemical reactions that will fuse the two together and cause the egg to act as though it has been fertilized. The egg cell, now containing the complete DNA of the adult donor, begins to divide and develop into an embryo.

Left: Scientists at South Korea's Seoul National University present three genetically identical female Afghan hounds, produced a year after the nation's first cloning of a dog.

Reproductive cloning

Reproductive cloning is a process used to generate an animal that has the same nuclear DNA as another pre-existing animal. Dolly the sheep was cloned by a process referred to as "somatic cell nuclear transfer" (SCNT). A single cell was taken from the udder of a mature sheep (the donor) and the nucleus of that cell, carrying the complete DNA of the donor sheep, was removed. The nucleus was then placed in an egg cell that had had its nucleus removed. The reconstructed egg containing the DNA from the donor cell was then treated with chemicals or an electrical current to stimulate cell division. Once the cloned embryo reached a suitable stage it was transferred into the uterus of another sheep, where it developed normally. Strictly speaking, Dolly is not truly an identical clone of the donor sheep. Only the clone's nuclear DNA is exactly the same as the donor, as some of a clone's genetic makeup comes from mitochondria in the cytoplasm of the enucleated egg.

Dolly's success is quite remarkable because it showed that the genetic material from a specialized adult cell, such as an udder cell programmed to express only those genes needed by udder cells, could be reprogrammed to generate an entire new organism. Prior to the birth of Dolly, it had been taken for granted that once a cell became

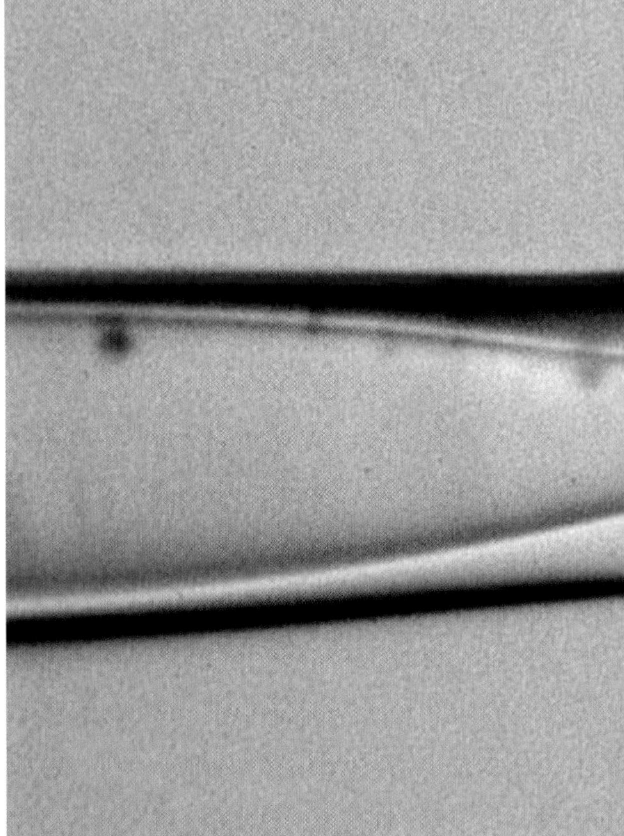

What we want is to stimulate an informed public discussion of the way in which the techniques might be misused as well as used and to ensure legislation was put in place to prevent misuse.

Ian Wilmut, embryologist who created Dolly the sheep, b. 1944

Left: Dolly, the cloned sheep, at home at the Roslin Institute in Edinburgh, Scotland. Ther research team there was the first to achieve the cloning of a mammal from am adult cell.

specialized as a liver, heart, udder, bone, or indeed any other type of cell, the change was irreversible and other unneeded genes in the cell would become inactive.

It appears that errors or incompleteness in the reprogramming process cause the high rates of death, deformity, and disability observed among animal clones. The somatic cell nuclear transfer cloning (SCNT) process is expensive and inefficient—276 attempts were necessary to produce Dolly. Since Dolly, many other animals have been cloned by somatic cell nuclear transfer, but attempts to clone certain species have been unsuccessful.

Below: A micropipette is used to inject a nucleus into an enucleated egg. Researchers at the University of Hawaii, USA, have used this technique to clone mice from adult cells, repeating the success of Dolly.

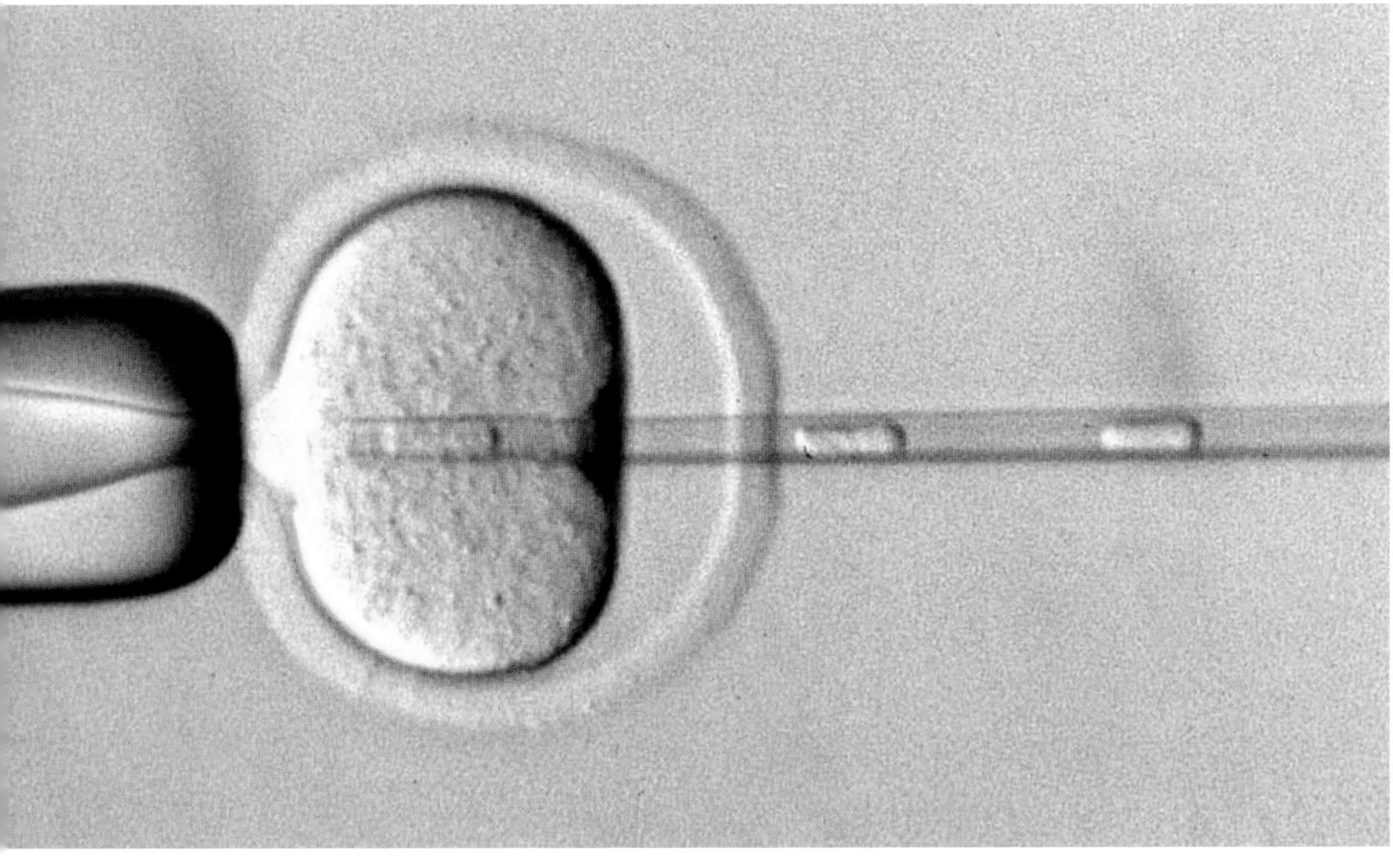

THERAPEUTIC OR EMBRYO CLONING

Therapeutic or embryo cloning is the production of human embryos for research. The same process (SCNT) that was used to produce Dolly could be used to produce human embryos. In therapeutic cloning, the resultant embryo is not transferred to a human female uterus to develop.

The aim of therapeutic cloning is to harvest stem cells to study human development and to treat human disease. (Stem cells (see pages 286–289) are unspecialized cells that exist with the specialized cells of a particular organ. They are used by the body to repair or replace damaged specialized cells.) The stem cells are harvested five days after the insertion of a donor cell into an enucleated egg. The extraction process destroys the embryo and thus raises serious ethical concerns.

Stem cells promise a cure for many diseases and traumas, but to date no successes have been reported from stem cells derived from embryo cloning. On the other hand, stem cells derived from human sources other than from the destruction of an embryo have proved more promising and have cured or alleviated many human conditions. These "adult" stem cells have been found in various parts of the human body, for example cord blood, uterus, bone marrow, and fat cells.

The human body is a remarkable structure containing a complex set of systems composed of organs, tissues, and cells working together to maintain a stable internal environment in response to a variable external environment. This requirement to maintain a relatively constant set of optimal conditions (homeostasis) is essential for all physiological processes to occur.

The remarkable human body

Furthermore, the body gathers information about its surroundings, responds, and adapts to any given situation allowing it to move, feed, or escape from potentially harmful situations.

In order to sense its surroundings, the human body relies on input from its various sensory organs, which convert the collected information into electrical impulses for interpretation by the central nervous system. The central nervous system then responds appropriately from generating thoughts to activating movement.

Information about the external environment reaches the brain through the five major senses—sight, hearing, smell, taste, and touch.

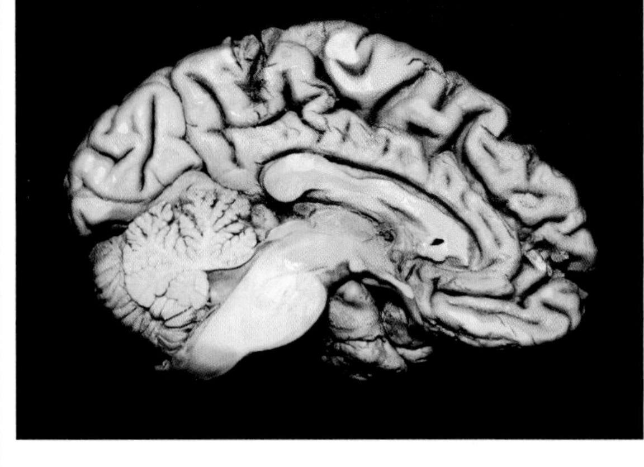

Right: The human brain is capable of continually adapting itself to new situations. Even in old age, new neurons can be grown. As well as controlling all our physical actions, the brain is closely associated with intelligence and emotions.

The brain

The central nervous system is made up of the brain and spinal cord. Through a complex network of nerves that carry electrical signals to and from the brain, the nervous system gives the body the ability to respond to stimulation, allowing us to walk, dance, drink, love, hate, think, and remember. One part of the nervous system allows the body to respond to external stimulation (somatic nervous system); another regulates the involuntary actions of the body's internal organs (autonomic nervous system).

The brain is the control center of the central nervous system and is made up of several major parts, each with its own specific function. The major components are the brain stem, cerebellum, and cerebrum.

The brain stem is the upper, enlarged, lobe-like end of the spinal cord as it enters the brain. It controls many autonomic functions of the body such as the actions of the heart and lungs, as well as reflex actions such as coughing, sneezing, swallowing, and vomiting. The brain stem is the pathway for the millions of nerves that transport impulses back and forth between the brain and the spinal cord.

Located at the rear of the brain, the cerebellum constitutes approximately 10 percent of brain weight. Acting on instructions from the cerebrum, the cerebellum is involved with the coordination of voluntary muscle movements, posture, and balance. As such, it coordinates input from muscles, tendons, joints, the ears, and inner ear. The brain's "automatic pilot," it enables the body to carry out the precise movement involved in everyday activities from walking to playing the piano. It is also thought that the cerebellum may play a role in learning and acquiring language.

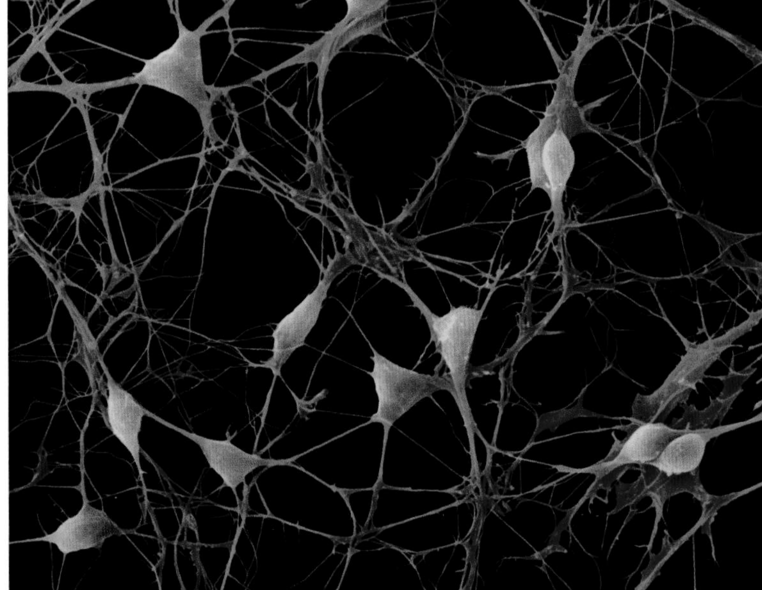

Below: Neurons have a large cell body with several long processes extending from it, usually one thick axon (red) and several thinner dendrites. The axon carries nerve impulses away from the neuron. Its branching ends make contact with other neurons and with muscles or glands. The highly branching dendrites receive information from other neurons. This complex network forms the nervous system.

The cerebrum is made up of two cerebral hemispheres; these are the familiar large wrinkled swellings of the forebrain. The cerebrum contains billions of nerve cells or neurons and a fine layer of gray matter that is linked with human intelligence. Each hemisphere is composed of many lobes, which correspond to a specific function. For example, the occipital lobes receive and interpret signals from the eyes. The cerebrum regulates and controls voluntary muscle actions as well as mental activities, for example, thinking, emotion, intelligence, memory, and reasoning.

Within the cerebrum are other specialized areas of gray matter that perform vital actions. These are the basal ganglia, thalamus, and hypothalamus. The basal ganglia are involved with the fine control and coordination of body movements, while the thalamus gathers sensory information from the viscera and special sense organs before passing it to the cerebrum. The hypothalamus helps to maintain a constant internal environment and is intimately involved with hormonal control, thermoregulatory behaviors such as sweating and shivering, thirst, appetite, and emotional reactions. The cerebrum also contains regions that are involved with thought, personality and speech. See also pages 330–331.

Sight

Light rays pass into each eye and form an image on the retina, the lining of the back of the eye. Each eye acts like a camera, the two eyes working together to provide stereoscopic vision so that objects can be seen in three dimensions including depth of field and distance. Thus, objects are viewed as standing out from their background as opposed to a flat two-dimensional image on paper.

The major components of the eye include the cornea, pupil, lens, iris, and the retina.

Light passes through the transparent cornea and is refracted by the lens prior to entering the pupil, the small opening of the eye. The pupil is surrounded by a muscular diaphragm, the iris, that opens and

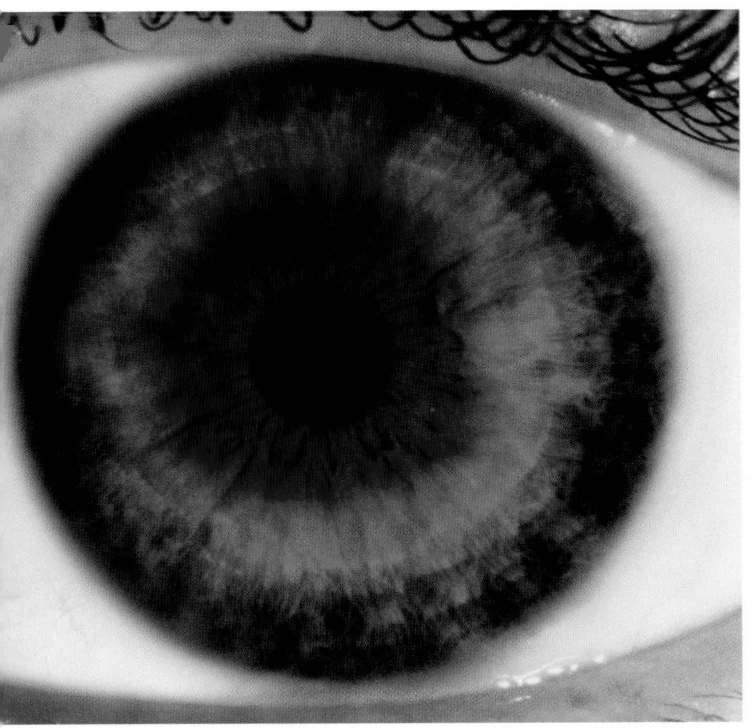
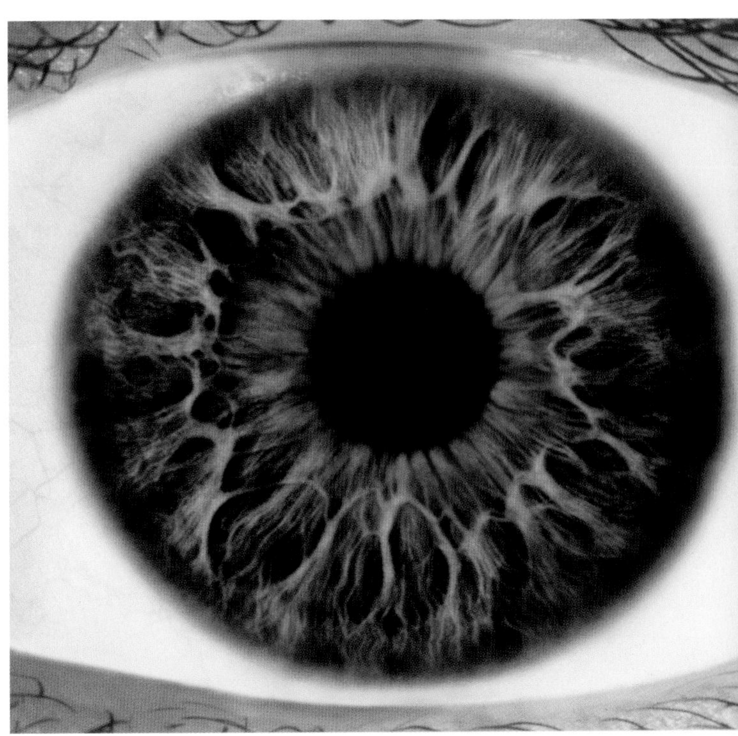
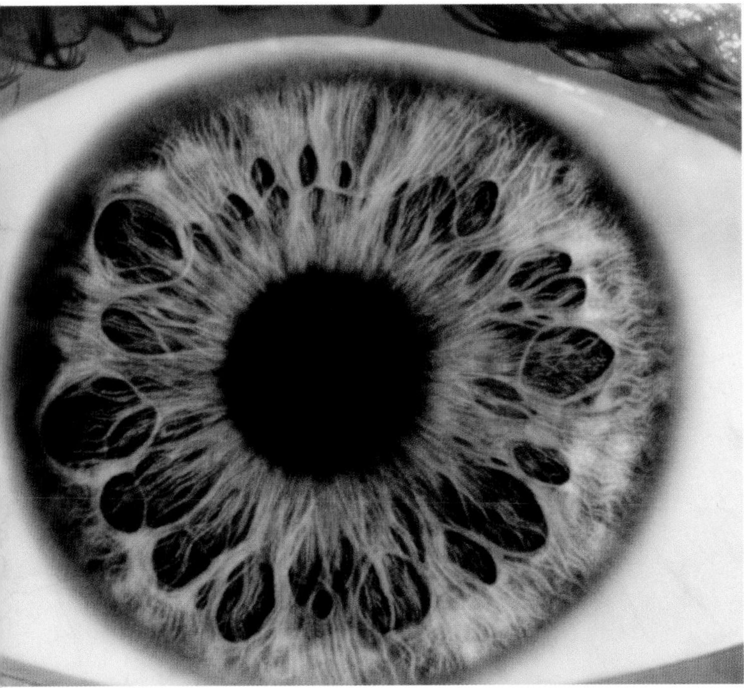
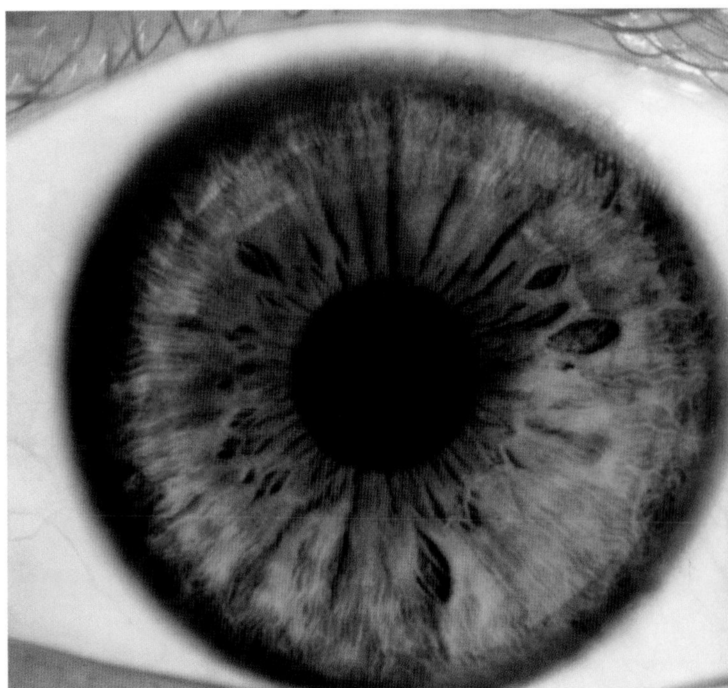

closes in response to the level of surrounding light. In bright light, the iris contracts, reducing the size of the pupil and limiting the amount of light reaching the retina; in conditions of low light, the iris relaxes, opening the pupil and permitting the entry of more light.

In addition to the refractive property of the cornea and lens, light rays are further refracted by the humors of the eye. The aqueous humor is watery and located beneath the cornea in front of the lens, while the more gelatinous vitreous humor, located posterior to the lens, helps the eye maintain its shape. Light rays ultimately fall upon the inner lining of the back of the eye, the retina.

The retina carries a rich blood supply and contains specialized light sensitive cells known as rods and cones. Rods far outnumber cones and are able to detect shadows and movement. They are also

Above: Four eyes, each of a different color. The iris is the colored part of the eye, and it controls the amount of light that passes through the pupil. In bright light, it closes up the pupil, while in weak light, it opens it up wide.

TRAVEL SICKNESS

Travel sickness is a common affliction associated with riding on or within moving vehicles. The nausea that is experienced is a result of confusion within the brain's sensory center and the messages it receives from the vestibular or balance system of the inner ear. This system tells the brain that the vehicle is moving. However, the eyes do not necessarily provide the brain with the same message. If the eyes detect little or no movement, as when a book is being read, then this is the message delivered to the brain. It is the combination of these conflicting messages that lead to the feelings of nausea and dizziness.

BALANCE

The vestibular system of the inner ear is responsible for balance. It is made up of a series of semi-circular canals that are perpendicular to one another. Movement of the head causes the fluid of the semi-circular canals and vestibular system to flow, which then stimulates the hair cells. This movement is converted into electrical impulses that are transmitted to the brain via vestibular nerve fibers. However, the sense of balance involves the integration of a number of systems including vision and mechano-receptors in muscles and the feet.

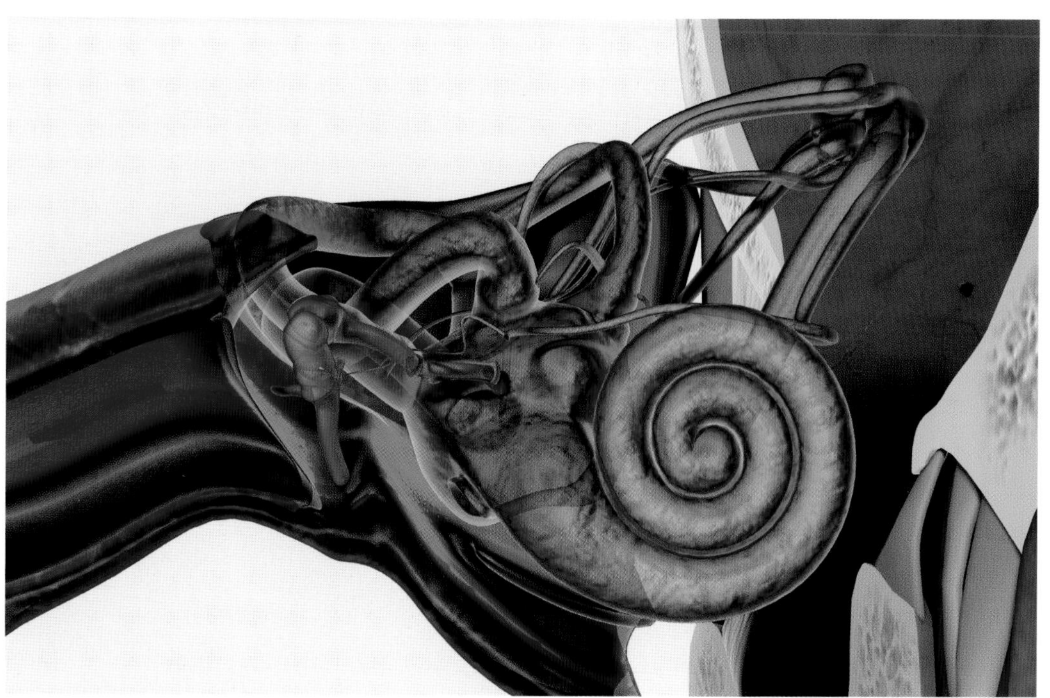

Opposite page: A mammal cochlea, magnified 6,100 times, showing the first row of inner hair cells with bodies of the cells exposed. The cochlea contains cells and fluid and when vibrations travel through the fluid, the hairs move, and the brain translates this as sound.

Left: An anterior stylized view of a cross-section of the ear in the head showing its inner anatomy. The shell-like part is the cochlea; the blue part is the eardrum or tympanic membrane. On average, people can hear sounds in frequencies ranging from 20 to 20,000 Hertz.

able to work effectively in low levels of light. Cones are involved with the detection of color and require bright light to function optimally. Images that fall on the retina affect the light-sensitive pigment within the rods and cones, triggering electrical impulses that travel along the optic nerve to the brain. The region where the optic nerve enters the eyeball is devoid of light sensitive cells. Thus, images that fall upon this area are not seen. This is the blind spot.

Hearing

The ear is designed to receive sound waves and convert these to electrical signals that can be interpreted by the brain. The ear consists of three major components—the outer ear, middle ear, and the inner ear. An additional fourth component is the transmission pathway of sound impulses to the brain.

The outer or external ear is composed of the fleshy pinna or auricle and the external auditory canal and provides directional sensitivity for hearing. The external ear gathers sound waves from around the surrounding environment and channels these through the auditory canal to the eardrum or tympanic membrane. As the auditory canal narrows, the sound waves are amplified as they reach and cause vibrations of the eardrum.

The eardrum separates the external auditory canal from the middle ear. The middle ear is an air-filled space, which is sometimes called the tympanic cavity and is located within the temporal bone of skull. The portion of the temporal bone above and below the auditory tube is called the mastoid and is filled with air filled spaces or cells. The middle ear communicates with these "mastoid cells" via small connections and also with the nasopharynx through the auditory or Eustachian tube. The auditory tube enables equalization of air pressures on both sides of the eardrum.

The middle ear also contains the three smallest bones of the human body, the auditory ossicles. These are the malleus, incus, and stapes, more commonly known as the hammer, anvil, and stirrup. The vibrations of the eardrum are transmitted to the auditory ossicles, resulting in sound being carried to the inner ear.

The inner ear is made up of two main parts—the cochlea, a spiral tube that contains fluid and a series of sensory hair cells that contain auditory detectors, and the vestibular system, which is responsible for balance. Vibrations transmitted via the auditory ossicles disturb the hair cells of the cochlea, which then send impulses to the brain via the auditory nerve. These impulses are then interpreted as sound.

> The human body is the best picture of the human soul.
>
> Ludwig Wittgenstein, philosopher, 1889–1951

Above: Specialized sensory cells in the nasal passages give us our sense of smell. It has been estimated that humans can distinguish more than 10,000 different smells.

Smell

The nose contains the nostrils and sensory cells embedded in the lining of the nasal cavity. These sensory cells possess numerous rod-like projections each of which is covered by minute filaments or hairs. The entire nasal cavity, including these hairs, is constantly kept moist by secretions of the mucous membrane that lines the nasal cavity. This moisture is vital for the sense of smell as chemical receptors can only detect odors that are dissolved in water. The olfactory receptors transmit signals to the olfactory bulbs that project from the brain. Once a smell has been processed by the brain, it is retained in memory, so that if the same smell is encountered at a future date, the brain quickly registers the familiar odor.

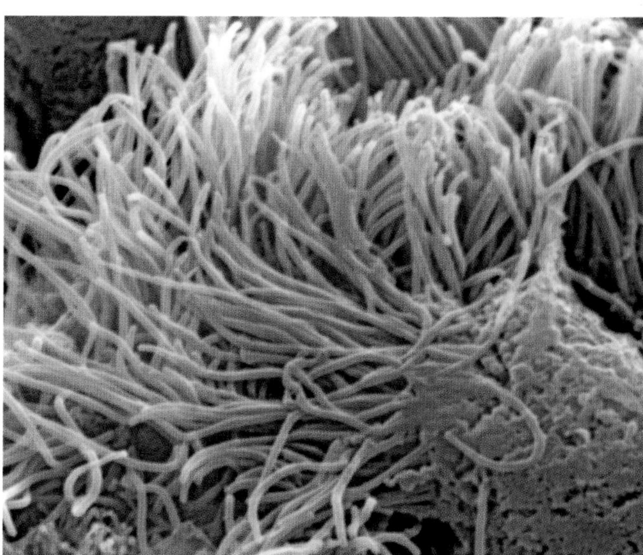

Right: Cilia, magnified here 14,000 times, are the tiny hairs lining the nasal passages. Cilia act as a filter and remove pollen, bacteria, dust and other particles from the air and trap them in mucus where they can be removed from the body.

Taste

In a similar way to smell, taste is a form of chemoreception, so the chemical must be dissolved in water. However, unlike the sense of smell, the object to be tasted must make contact with the tongue. Sometimes, taste becomes combined with smell because of the close connection between the mouth and the back of the nose (also known as the nasopharynx).

The tongue contains minute receptor organs called taste buds that are concentrated on the upper surface of the tongue. The majority are found on small projections of the tongue called papillae. In order for a substance to be tasted it must diffuse into a taste pore, dissolve in saliva, and contact minute hairs located on the taste buds. From here, signals are transmitted through cranial nerves to the brainstem and into the cerebrum. Typical taste sensations include sweet, salty, sour, and bitter. A common misrepresentation is that each of the taste sensations is associated with a specific area on the surface of the tongue. It is now known that taste buds for all taste sensations are located all over the tongue.

Touch

Touch involves the detection of pressure, temperature, and pain. It is a complex sense that enables the body to feel objects and sense its external environment. The skin contains hundreds of thousands of sensitive nerve endings, particularly in areas such as the lips and the fingertips. These nerve endings form sensors that detect touch. Receptors, such as those that detect touch, are bulb-shaped. Other nerve endings are arranged in complex networks located around the base of hairs, forming sensors that respond when the hair is moved or bent. In addition, throughout the body, there are free nerve endings that register pain as well as touch. Pressure and pain can be registered by several different kinds of nerve. All of these types of sensors are called mechanoceptors because they transmit signals when they experience a physical stimulus such as pressure. The skin also contains other sensors called thermoreceptors that are triggered in response to temperature changes. One form of thermoreceptor detects heat and the other, cold.

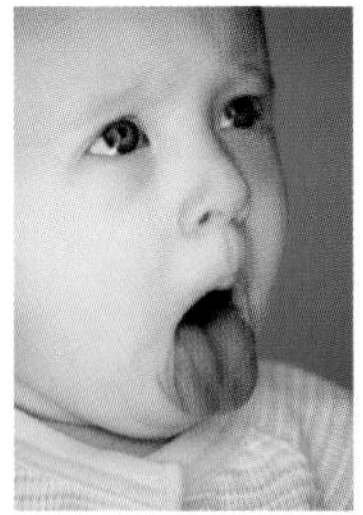

Above: The small projections on the surface of the tongue are called papillae and contain the taste buds. When we taste something, the brain registers the flavor as sweet, salty, sour, or bitter. When we have a cold, the upper part of the nose is not receiving the chemicals that trigger the receptors that inform the brain and provide us with the feeling of flavor. That's why when we're sick, our sense of taste is diminished.

Right: A high power view of the surface and taste buds of the tongue. We have about 10,000 taste buds and they renew themselves roughly every two weeks, although as we get older, fewer taste-buds are replaced.

THE SKIN

Skin— the largest organ of the human body—is a tough, flexible layer that protects the body from disease and injury, but also helps retain what is essentially a watery environment. In addition to moisture retention, the skin plays a key role in water balance and thermoregulation. For example, during hyperthermia, when body temperature rises above optimal levels, sweat glands in the skin secrete water onto the skin surface, which cools the body by the process of evaporation. The skin has three primary layers of tissues—the epidermis, dermis, and subcutaneous tissue. The epidermis, the outermost layer, is the skin's first protective barrier. It is from this layer that cells are continually shed or sloughed off and replaced by new cells. The human body replaces its skin approximately once three to five weeks. More than half a million skin particles are shed every hour; the average person loses approximately 100 pounds (45 kg) of skin by the age of 70 years. The epidermis also contains cells that produce melanin, the pigment that gives skin its color and helps protect it from harmful ultraviolet rays. The dermis, the skin's thickest layer, contains structural tissues such as protein fibers that give the skin its strength and elasticity, and houses hair follicles (left). It is from this layer that wrinkles arise but also fingerprints—minute ridges of dermis that project into the epidermis. The subcutaneous layer, the innermost layer of the skin, consists primarily of fat-storing cells within connective tissue. This layer helps cushion the body against injury and also acts to insulate the body against extremes of temperature.

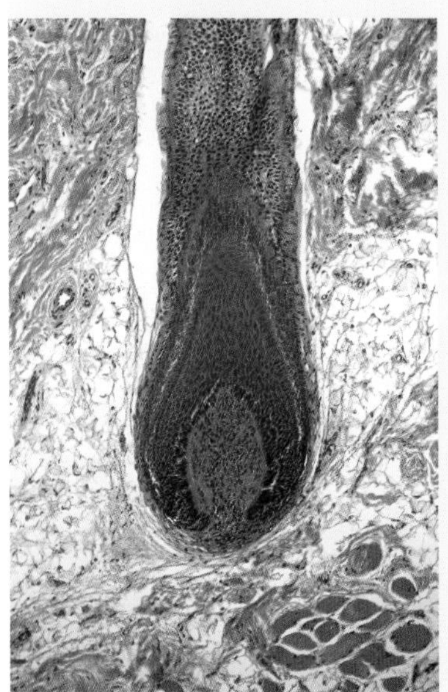

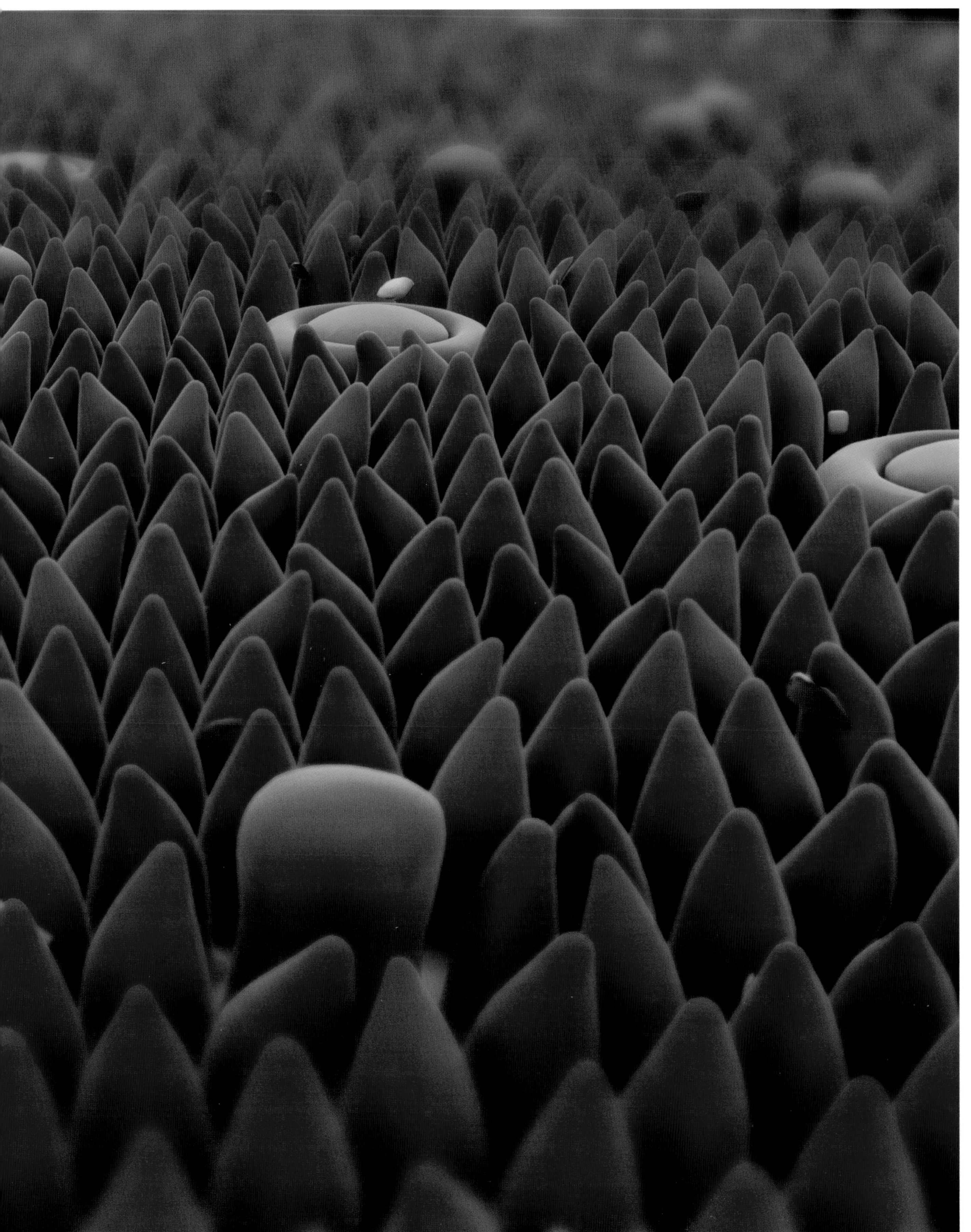

Plants fill almost every ecological niche on the planet. They can survive in water and on land, on the highest mountains, the coldest tundra, and the hottest deserts. Their importance to life on Earth is incalculable; plants provide us with food, oxygen, shelter, clothing, paper, fuel, and medicines.

Botanical beauties

Plants can be divided into two main groups—non-flowering plants, such as mosses, ferns, and conifers; and flowering plants, of which there are at least 300,000 species.

Non-flowering plants

Mosses are part of a group of non-flowering plants called bryophytes. They are mostly small, flattish plants that prefer moist regions such as stream edges, although some species live on hot rocks in deserts, and others in freezing polar regions. Ferns belong to a group described as seedless vascular plants, which have defined root and shoot systems. Unlike all other plants, bryophytes and ferns need water if they are to reproduce successfully; they produce free-swimming sperm that must travel through water to find the egg.

Conifers, cycads, and the maidenhair tree, *Ginkgo biloba*, are known as gymnosperms. They produce seeds, an evolutionary innovation that gave a massive advantage over their simpler, seedless relatives. The tallest known plant, the redwood, is a conifer that can reach up to 370 feet (113 m) in height.

Many conifers, like animals such as crocodiles, are known as living fossils; they are ancient plants that haven't changed for millions of years. The Wollemi pine, discovered in Australia in 1994, is a member of a family that became extinct in the Northern hemisphere around 65 to 100 million years ago.

Above: A leaf of *Gingko biloba*. The ancestors of this seed-bearing plant first appeared some 300 million years ago. The maidenhair tree, which is native to China, can grow to 80 ft (24 m) high. Male trees have pollen sacs, while female trees bear the seeds.

> What a desolate place would be a world without flowers. It would be a face without a smile; a feast without a welcome.
>
> Clara L. Balfour, writer, 1808–1878

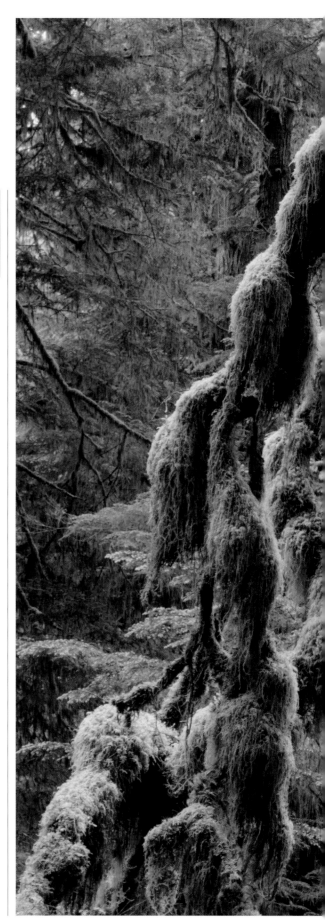

Right: A moss-covered branch in old growth rainforest in Pacific Rim National Park, Vancouver Island, British Columbia, Canada. Rainforests are home to about two-thirds of the world's plant species, some of which have not yet been named.

Left: The Wollemi pine (*Wollemia nobilis*) is an ancient conifer discovered in 1994 in the Blue Mountains of south-eastern Australia. Botanists have named this exceptionally rare plant a "living fossil."

Above: *Rafflesia arnoldii*, the world's largest flower measures some 3 ft (0.9 m) across. It takes months for the flower to develop and it lasts only a few days, so it is a rare treat to see one of these in the wild.

Flowering plants

Flowering plants are also known as angiosperms. Like gymnosperms, they produce seeds; but unlike gymnosperms, they also produce flowers and fruits. Angiosperms are incredibly diverse. Grasses, lilies, palms, cacti, daisies, wheat, potatoes, bananas, oak trees, and roses all look very different from one another, and yet are all flowering plants and share some basic similarities.

The world's largest individual flower, produced by a plant called *Rafflesia arnoldii*, can be up to 3 feet (90 cm) across. Found only in undisturbed forests of Borneo and Sumatra, *Rafflesia* has a second dubious claim to fame as one of the world's smelliest flowers, with its rotting flesh aroma giving it the nickname of the corpse flower.

DYES AND KILLERS

There are many ingenious uses for plants. Plant-based dyes, for example, have been used for thousands of years. Indigo leaves give a blue color to fabric; henna leaves are used as a hair, nail, and skin dye (see the hennaed hands pictured below); safflower petals provide a red tint for makeup such as rouge, and saffron's yellow dye colors food and fabrics.

Many plants produce deadly poisons to serve as chemical weapons against possible predators, which humans have taken advantage of. Some of them make life-saving medicines—foxgloves contain poisons called cardiac glycosides that can cause heart attacks, but when used medicinally, are a vital drug that slows and strengthens the heart beat.

Other plant poisons include cyanide, produced by the cassava plant; and ricin, one of the most toxic plant poisons, which comes from the castor bean plant. A single seed contains enough ricin to kill a child. Eating just three berries from deadly nightshade, *Atropa belladonna*, can kill an adult, but this didn't stop medieval women from using its poison, atropine, as cosmetic trick to dilate their pupils. Today, atropine is used as a cardiac stimulant and an antidote to nerve gas poisoning.

Left: The dramatic, colorful bracts of *Heliconia rendulata* have given it, and other members of the genus, the nicknames "lobster claw." Native to tropical America, southeast Asia, and parts of the Pacific, heliconias are pollinated mainly by birds.

Below: *Crocus tommasinianus* in flower. Grains of pollen are located on the anthers, the yellow part of the flower. When insects or birds come into close contact with the anthers, they unwittingly become pollinators of that plant.

Opposite: The long, narrow, probing bill of the broad-billed hummingbird (*Cynanthus latirostris*) is well suited to pollinate a range of flowering plants, particularly those that are red in color. The hummingbird links plant populations in different locations.

When is a plant not a plant?

Although algae are closely related to plants, they are different enough to be classified within a completely different kingdom—the Kingdom Protista. Just like plants, algae make their own food, but instead of living on the land, they take up residence in water. They vary enormously in size, from microscopic single cells to the brown algae called kelp that forms underwater forests up to 100 feet (30 m) tall. Some algae live in hot tropical waters, others can survive months of darkness under polar sea ice.

Fungi were once considered plants, but it is now known that they are actually more closely related to animals. Like algae, they belong to neither the animal nor the plant kingdom, but in the Kingdom Fungi. Instead of photosynthesizing as plants do, they obtain food by decomposing organic matter. The ultimate recyclers, they release carbon dioxide back into the atmosphere and return nitrogen to the soil, allowing these molecules to be re-used by plants and animals.

Flowers and reproduction

As far as plants are concerned, flowers are for one thing, and one thing only—sex. Flowers will go to extreme lengths to get it, tailoring their look, smell, and even taste to appeal to potential pollinators.

Pollen and pollinators

Flowering plants reproduce in basically the same way as members of the animal kingdom—sperm cells fertilize an egg cell to produce an embryo. The plant embryo develops within a protective seed, which must be released from the parent plant and germinate elsewhere to grow into a new plant.

The sperm cells of flowering plants are contained within grains of pollen. Depending on the species, pollen may fertilize an egg in the same flower as the pollen, a different flower on the same plant, or a flower on another plant altogether. Some plants can accomplish pollination without outside help, but most need either a good gust of wind, or to attract a pollinator of some kind. Bees, beetles, butterflies, moths, bats, and birds pick up and transfer pollen grains when they visit flowers for a meal of nectar.

Animal-pollinated flowers try to be as attractive as possible to their respective pollinators, in order to increase their chances of fertilization. They may ooze seductive scents, or shape themselves to exactly fit a hummingbird's beak or a butterfly's mouthparts. The distinctive spots on foxglove flowers serve as "honey guides," signalling to insects exactly where they should land. Some orchid flowers look and smell exactly like a certain female wasp, thus fooling the male wasp into pollinating its flower.

The main event

Pollen grains are located on anthers, which are the flower's male organs. Pollinating insects and animals brush against the anthers, and then transfer this pollen to the female organs of another flower. After a pollen grain lands on the stigma, a tiny platform sitting atop the female organs, it extends a tiny tube that penetrates the stigma and grows down through the female organs until it reaches the ovary, which is usually well out of sight and protected within the main body of the flower.

Sperm cells are carried along in the pollen tube and are then released into an ovule—the future seed. One sperm fertilizes an egg cell, which becomes the embryo, another fertilizes cells that will become the endosperm. The endosperm is a nutrient-rich food source that nourishes the developing embryo, and in many species, also feeds the seedling while the seed germinates.

While the ovule develops into a seed, the ovary develops into a fruit. Its walls thicken into a variety of different layers depending on the fruit. They may be soft and fleshy, like tomatoes; fibrous, like coconuts; or tough, like apples and pears.

The final stage: seed dispersal

Flowering plants use an incredible variety of methods to disperse their fruits and seeds as effectively as possible. Although some plants rely on passive dispersal, merely dropping their seeds or fruits to the ground, others have developed fruits that can be carried for very long distances by wind, water, and animals.

Right: From the humble acorn, the mighty oak tree grows. *Quercus* species, a seedling of which is pictured here, are usually very long-lived. They provide a home for a wide variety of birds, insects, and animals.

HUNGRY PLANTS

When times get tough, it pays to plan ahead. Some plants have become excellent at trapping food while staying rooted to the spot. The Venus flytrap (*Dionaea muscipula*), at left, catches unwary insects with its spine-edged leaves. When an insect, such as this wasp (*Vespula vulgaris*) sets off a fine trigger hair on the leaf's surface, the two halves of the leaf snap together, and the spines interlock like a zipper to prevent escape. The flytrap soaks its prey in digestive chemicals to extract the goodness, and opens a few days later to let the hard shell of the insect fly away in the wind.

Sundews set a different trap—a very sticky one. Their club-shaped hairs are tipped with drops of liquid that smell like dinner to a passing insect. The sticky ooze traps the prey, and the hairs bend inwards to surround it. The insect is slowly dissolved, giving the sundew a tasty meal. Pitcher plants use their funnel-shaped leaves to trap insects and digest them in a similar way. Carnivorous plants usually live in poor soils. They rely on catching insects to provide mineral nutrients, such as nitrogen and potassium, which are lacking from the area the plants live in.

Dandelions produce tiny, light fruits that are easily carried along by the wind. Maple fruits also rely on wind, with two thin wings that help them be blown about. Tumbleweed scatters its seed as the whole plant bounces along in the breeze.

Plants that grow near water often produce air-filled fruits adapted for floating. The coconut has adapted to long ocean journeys, and can delay germination for up to two years. A few plants rely on their own power to spread their seed; the dwarf mistletoe, for instance, forcibly shoots its seeds out of the fruit at approximately 62 miles per hour (100 km/h), sending them up to 50 feet (15 m) away.

Colorful sweet fruits aren't only enjoyed by humans. Many birds and mammals play a significant role in dispersing the seeds of many plants, simply by eating the fruits and passing the undigested seeds out in another location. Other plants cover their fruits with hooks, spines, or sticky substances that lodge into fur and feathers, allowing the fruits to travel long distances before the unknowing travel agent dislodges them at a new location.

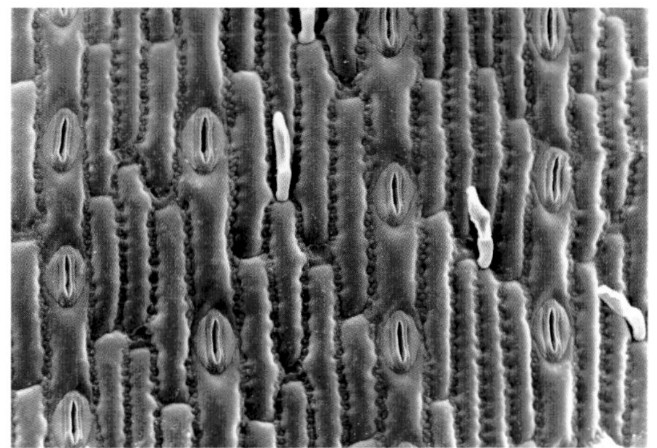

Right: Stomata (yellow) are microscopic pores bordered by guard cells found on the epidermal surface of leaves and stems. Gas exchange (oxygen, carbon dioxide, and water vapor) occur through these tiny openings. Three trichomes (light brown) are appendages to the epidermis that protect the plant from insect pests.

Photosynthesis

Plants use energy from sunlight to make their own food. Apart from algae and some bacteria that can also do this, the rest of life on Earth cannot—we all ultimately rely on plants for our survival. Plants not only form the basis of the food chain, but they also soak up carbon dioxide from the atmosphere, factories, and cars, while at the same time providing us with fresh air to breathe.

Above: The common dandelion (*Taraxacum officinale*) disperses its seeds with the help of the wind. This is very successful, with this plant declared a weed in many parts of the world.

The power of the Sun

Plants make their own food using the process of photosynthesis. Essentially, photosynthesis uses light energy to power the breakdown of carbon dioxide and water into sugar and oxygen.

Leaves take carbon dioxide up from the air through tiny holes called stomata. Within the leaf cell, the Sun's light energy is absorbed and used to split water molecules, producing oxygen gas, hydrogen atoms (protons), and electrons. Known as the light reaction, this is the first half of the two stages of photosynthesis.

The oxygen diffuses out of the same holes that the carbon dioxide came in by. The movement of electrons and protons drives the production of a compound called ATP (adenosine triphosphate), which the cell uses as an energy source to power many chemical reactions.

Left: A close-up photograph of a leaf, showing the rib structure and veins. The veins carry water and inorganic compounds right into the leaf. They then carry organic compounds produced by photosynthesis away from the leaf, to other parts of the plant.

In the second stage of photosynthesis, which is called the carbon-fixation reaction, ATP's chemical energy is utilized to join, or "fix" the carbon from carbon dioxide gas to a simple five-carbon sugar, creating a six-carbon sugar.

The six-carbon sugar immediately splits into two three-carbon sugars, which are converted to either sucrose or starch. The sucrose is transported out of the leaves and around the rest of the plant, thus effectively acting as the plant's food source. Starch, a major storage carbohydrate, is kept in reserve until such time as the plant needs energy for growth and development.

The key to efficiency

Plants can use up to 90 percent of the light that hits them, compared with less than 30 percent for today's average commercial solar panel. However, the efficiency of photosynthesis is greatly affected by the levels of carbon dioxide and oxygen within the leaves.

The stomata in a plant's leaves, which permit the passage of carbon dioxide and oxygen in and out, can open and shut depending on the environmental conditions. When the stomata close, the carbon dioxide supply is cut off and oxygen levels build up, which reduces photosynthetic efficiency. Some plants avoid this problem by conducting photosynthesis in different ways.

Maize, sugarcane, and many tropical plants physically separate the two stages of photosynthesis into different cells. Carbon from carbon dioxide gas is initially "fixed" to an intermediate molecule that rapidly converts to a form that can travel into a different cell. The carbon is then released and re-formed into carbon dioxide gas for use in the carbon-fixation process previously described. Preventing the loss of carbon dioxide in this way makes photosynthesis about two to three times more efficient than in plants like wheat, rye, and oats, that carry out the whole process within the same cell.

Plants such as cacti, which have evolved to survive in hot, dry environments, use a different trick. They separate the photosynthesis reactions by time. They take up and temporarily fix carbon dioxide at night, then carry out the standard carbon-fixation during the day. By keeping their stomata tightly shut during the heat of the day, they prevent loss of carbon dioxide and conserve water at the same time.

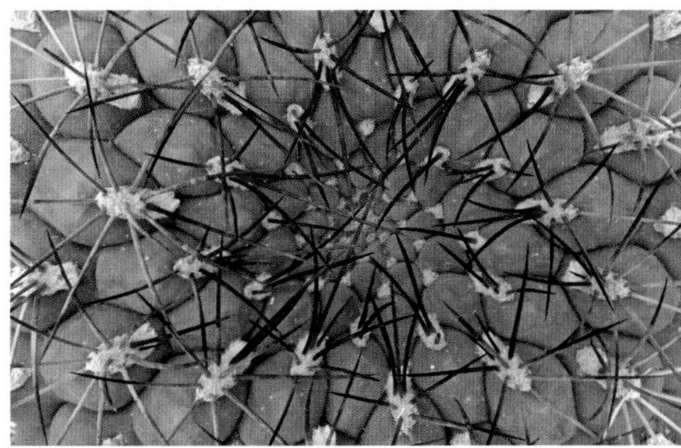

Right: By keeping its stomata closed during the hot daytime hours, the giant chin cactus (*Echinocactus saglionis*), like other plants of hot arid regions, manages to conserve water.

Today, the variety of animals on Earth seems almost endless. They swim in water, dig through the soil, and fly through the air. They even live inside the bodies other animals.

The amazing animal kingdom

Scientists classify animals into two main groups—animals without backbones, called invertebrates, and animals with backbones, called vertebrates. Invertebrates, which make up over 90 percent of all animal species, include sponges, worms, mollusks, and insects. Vertebrates include fish, amphibians, reptiles, birds, and mammals.

Plants or animals?

You might be surprised by some of the so-called simple invertebrates. Corals might look like plants, but they are actually colonies of tiny animals called polyps. They are responsible for constructing the world's largest living thing—the Great Barrier Reef—and their importance to our oceans cannot be understated. Coral on the ocean floor provide shelter from strong currents, protection from predators, nurseries for juveniles, and breeding areas for marine life. Sponges are classified as animals despite the fact that they don't have any of the

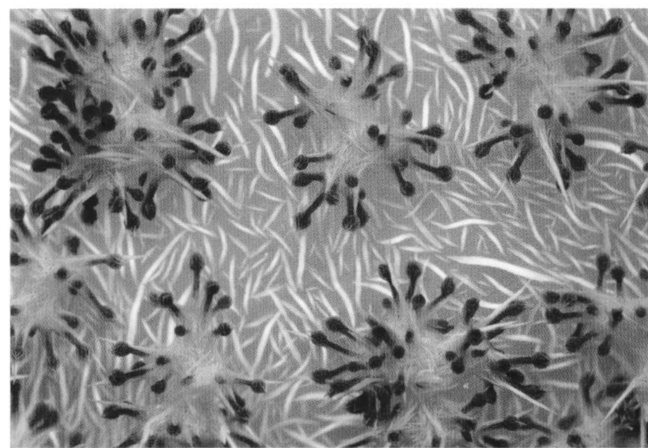

Right: The soft carnation coral (*Dendronephthya* species), showing detail of the polyps. These brightly colored coral feed almost exclusively on phytoplankton, microscopic aquatic plants.

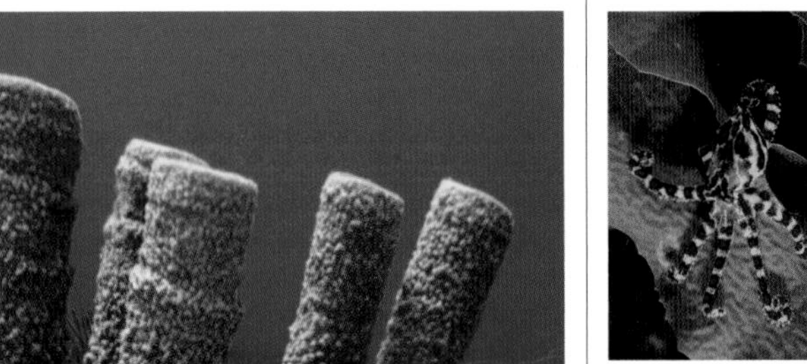

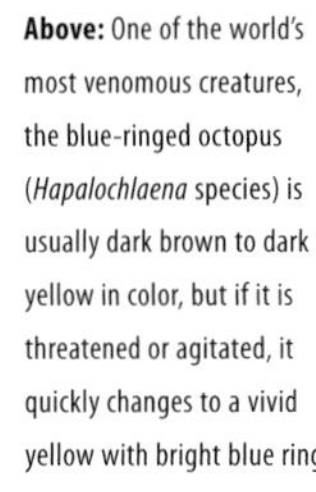

Above: One of the world's most venomous creatures, the blue-ringed octopus (*Hapalochlaena* species) is usually dark brown to dark yellow in color, but if it is threatened or agitated, it quickly changes to a vivid yellow with bright blue rings.

Left: Tube sponges grow over coral reefs effectively killing the polyps. Sponges resemble plants, but are in fact part of the animal kingdom. Interestingly, they have no true tissues or organs.

body parts, inside or out, that we expect an animal to have. They don't even move. Instead, most of these moss-like animals stay attached to an underwater rock or reef.

Mollusks

No group of animals is more diverse than the mollusks. Some of them have arms, others eyes, some have both, and others have neither. The smallest live in tiny shells as big as your fingernail, and the largest, such as the giant squids, weigh around 4,000 pounds (1800 kg)!

Most mollusks, including snails, clams, and oysters, have shells. A slow walk along most coastlines will reveal the remains of washed up mollusks—from delicate, patterned shells and fan-like scallops, to flat oyster shells, coated inside with an iridescent sheen of mother-of-pearl. However, the beauty of their appearance belies their main function—to protect the soft bodies of their owners.

Octopuses are considered the most neurologically sophisticated of the mollusks. The exact extent of their intelligence is still being debated among biologists, but problem-solving and maze experiments have shown that they do have both short-and long-term memory. An octopus at a German zoo learned to open a jar of shrimps by copying zoo staff.

Cuttlefish also have clever ways. Males can simultaneously adjust the color of one side of their bodies to show a dominant display toward other males, while the other side of their body shows a calm display towards a potential mate.

Right: Tridacna clams live in the warm shallow waters of the Pacific and Indian Oceans. One species, the giant clam, *Tridacna gigas*, can weigh up to 500 pounds (227 kg). These clams obtain their nourishment from the water and from algae cells within the mantle.

Below: Cuttlefish, such as this *Sepia* species, are not fish but mollusks. Masters of disguise, their skin is covered with pigmented cells that reflect light in different colors, providing this animal with camouflage for any occasion.

When thou seest an eagle, thou seest a portion of genius; lift up thy head!

William Blake, poet, 1757–1827

Arthropods

Arthropods include arachnids, crustaceans, and insects, and form the largest group in the animal kingdom, numbering more than all the other groups combined. They are characterized by the possession of a segmented body with appendages on at least one segment.

The success of arthropods is partly due to their ability to adapt rapidly to changing environmental conditions. They are also excellent hunters. The most varied and remarkable methods are exhibited by the order of arachnids, spiders. The bolas spider, for example, captures its prey by deception. At night it spins a short line of silk with a sticky globule of silk at the free end, hanging from its leg. The sticky globule may contain pheromones that mimic the scent of a certain female moth species, attracting unwary male moths within range. Not all spiders catch food in a web, however; jumping spiders and wolf spiders chase and pounce upon their prey, and brightly colored crab spiders await their nectar-loving prey in flowers.

Above: The hunting spider (*Cupiennius salei*) is also known as the banana spider, because it is often found on banana trees in its native Central America. Hunting spiders have big eyes to help them spot their prey. In comparison, the eyes of web-building spiders are small—instead, they rely on vibrations in the web to sense their prey.

Left: Wasps are usually social insects and live in nests that can contain anything up to 10,000 workers. Some wasps, however, are solitary. Wasps are arthropods and belong to the order Hymenoptera.

Scaly citizens

Water covers over 70 percent of Earth's surface, offering an enormous variety of habitats. Fish can be found from inter-tidal regions to deep sea trenches and at all levels in between.

Fish show an amazing variety of body forms and behavioral traits. Catching a meal is tricky for the deep-sea dwelling anglerfish, so they carry their own "fishing rod" with a bioluminescent lure on the tip (see page 419). Female seahorses lay their eggs into the male seahorse's pouch. He fertilizes the eggs, and at the end of the "pregnancy," gives birth to between 100 and 250 fully-formed baby seahorses.

The idea that all fish have gills, fins, and scales, and live in water is not entirely correct. All fish breathe using gills, but many species, such as the seahorses, lack scales; others, such as some eels, have no fins. Some fish, such as lungfishes, can even spend time out of water.

Amphibians: making a splash

Early amphibians were the first animals to leave the water and come ashore to breathe the atmosphere. Today, they lead double lives—one on water and one on the land. Frogs, toads, and salamanders belong to this fascinating group.

Most amphibians breathe through their skin, which is often moist and sticky. Almost all of them go through a tadpole stage in their metamorphosis into adulthood.

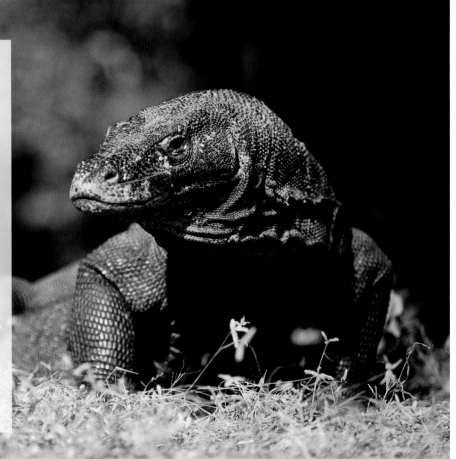

REPTILES: LIVING FOSSILS

The dinosaurs may have died out, but some of their relatives are still alive today. The reptiles are represented by about 7,000 species of lizards, snakes, turtles, and crocodilians.

The ancient Crocodilia order includes crocodiles, alligators, and gavials, which have narrower jaws. Crocodiles are the largest living crocodilians, and the animal most likely to eat a human. Classic opportunistic predators, they lurk patiently below the surface near the water's edge waiting for potential prey to stop and drink, before launching out of the water and using their powerful jaws to kill a potential meal.

The largest lizard living today is the Komodo dragon (*Varanus komodoensis*), right. It can grow up to 10 feet (3 m) long and weigh up to 366 pounds (166 kg). Komodo rely on camouflage and patience. The dragon springs on its victim, using its sharp claws and serrated, shark-like teeth to slice into its prey. Their saliva is alive with strains of fierce bacteria, and within 24 hours, the stricken creature dies of blood poisoning. Dragons will calmly follow any escapees for miles, waiting for the bacteria to take effect.

These prehistoric lizards have thrived for millions of years on a few islands in Indonesia. However, poaching, human encroachment, and natural disasters have driven the species to endangered status.

Above: The slimy salamander (*Plethodon glutinosus*) is so-called because it secretes a slimy glue-like substance from its skin glands, an effective form of protection against predators. Native to eastern and central United States, its diet consists of ants, beetles, and worms.

SUPER SENSES

Many animals have developed particularly keen senses to match their specific survival needs. Bats, such as the little brown bat (*Myotis lucifugus*), below, avoid obstacles and nab insects by emitting ultrasonic squeaks and interpreting the echo that sound waves make after bouncing off surrounding objects. This biological sonar, called "echolocation," is also used by dolphins and whales in waters where vision is limited.

Boa snakes and pit vipers have temperature-sensitive organs located between their eyes and nostrils that allow them to sense their prey's body heat. Snakes also use their tongues to "sniff" out the presence of food, enemies, or a mate. Chemical particles collected by a snake's flicking tongue are dipped into a special pit, called Jacobsen's organ, in the roof of its mouth. It's there that the odors are processed and translated into electrical signals that are sent to the brain.

Moths are able to detect love signals, called pheromones, that are emitted by the opposite sex up to 6 miles (9.5 km) away. Some studies have suggested that humans can also detect pheromones, although at much closer quarters.

Animals that are active at night tend to have a particularly well-developed sense of touch. Rats and cats, for example, use their whiskers in the same way that blind people use canes. By whisking the hairs across objects they come across, they are able to form a mental picture of their environment, even in the dark.

Right: Seahorses (*Hippocampus* species), such as this one swimming among Gorgonian sea fans (*Gorgonacea* species), are the only species where the male becomes "pregnant." This happens after the female deposits her eggs into the male's pouch.

Birds of a feather

Birds have many features in common with other animals, but they have one particular feature that makes them unique—feathers. A bird depends on its feathers for its power of flight, though some species have lost this ability.

Birds inhabit ecosystems on all continents, and several bird species have adapted to life on and in the world's oceans. Some seabirds only come ashore to breed, and some penguins have been recorded diving to almost 1,000 feet (300 m). Birds can also reach great speeds and travel enormous distances. The arctic tern, for instance, travels the farthest in its annual migration, flying a round trip between the North and the South Pole each year.

Mammals

Despite their size differences, the pygmy possum and the blue whale have something in common: They are both mammals. They belong to the diverse group of warm-blooded, air-breathing animals with backbones. The term mammal comes from the Latin word mamma, which means breast. Every female mammal has special glands called mammae, which are used to nourish their young with milk. Most mammals are born, rather than hatched, but there are some exceptions. The exclusive monotreme group of mammals has only two members, the echidna and the platypus, which both lay eggs!

Above: Red squirrels are found in Europe, North America, and parts of Great Britain. Squirrels are classified in the order Rodentia, mammals with long incisors that they try to keep short by gnawing. Rats, mice, and gophers are also rodents.

Movement

The ability to move is a sign of life in the animal world. Animals crawl, walk, hop, run, glide, or fly. The underlying theme of most of these forms of transport is the contraction of muscles.

Jetting about

The way an animal moves depends on its size, shape, and the environment in which it lives. Animals that live in water, such as octopuses and squid, move by jet propulsion—taking in water and ejecting it forcibly through a funnel as their muscular mantle contracts.

Winging it

Flight across the animal world varies dramatically, from the clumsy patterns of moths, to the acrobatic maneuvers of dragonflies, and the

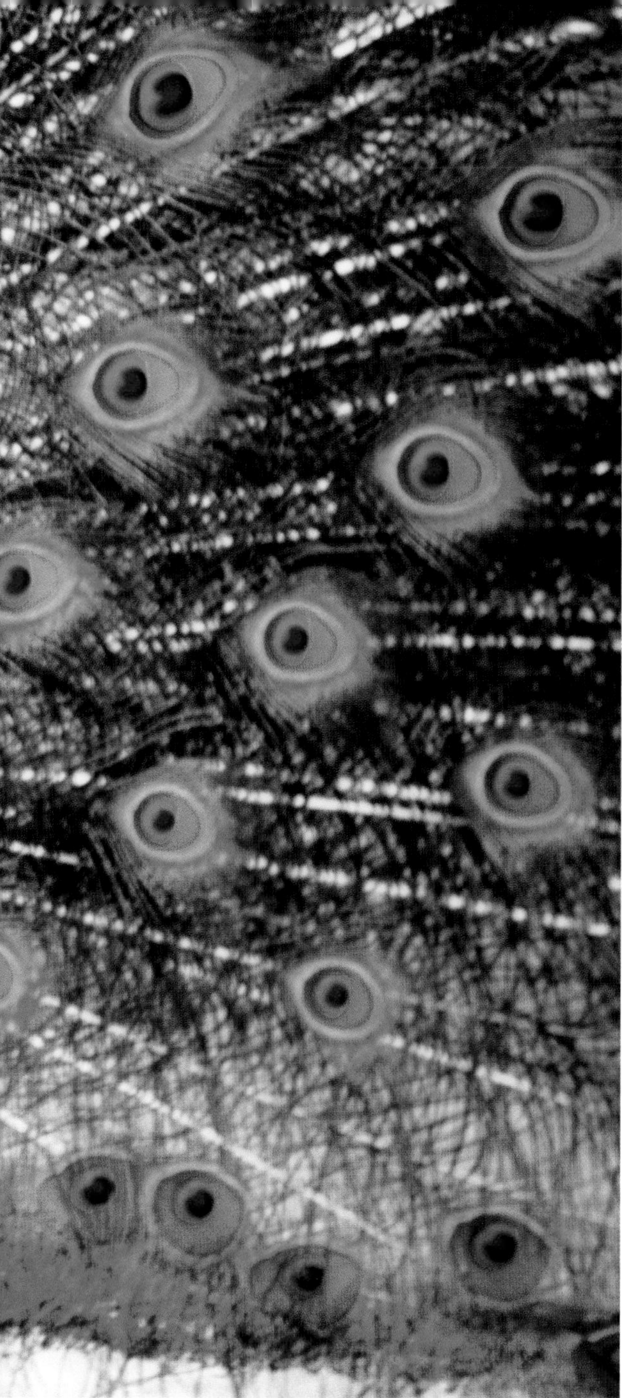

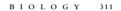

Left: There's nothing quite as glorious as the iridescent plumage of a peacock (*Pavo cristatus*), the national bird of India. The male displays his tail coverts in mating season. Females prefer males with plenty of spots on their feathers.

Below: Native to eastern Australia, the duck-billed platypus (*Ornithorhynchus anatinus*) is a monotreme, an egg-laying mammal. The male of the species has a venomous ankle spur, used to kill smaller animals.

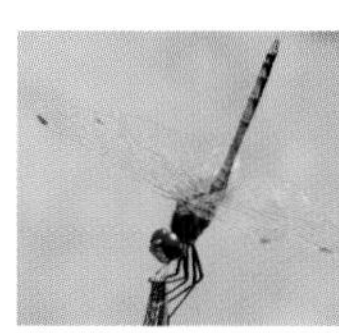

Above: Dragonflies (*Anisoptera* species) have two pairs of wings, which are usually held out in a slightly downward position. The wing veins are fused at the bases, so the wings cannot be folded over the body when the dragonfly is relaxing.

graceful soaring of the albatross. Dragonflies, which have two coordinated pairs of wings, were among the first insects to fly. Several of the insect orders that evolved later than dragonflies have modified flight equipment; bees, for example, hook their wings together and move them as a single pair.

Almost every part of a bird's anatomy is designed in a way that enhances flight. Their wings are specially curved in a shape called an airfoil that helps to produce an upward lift, the same principles of aerodynamics that are used in airplanes. Birds use energy to flap their wings and push against gravity so that they can stay airborne. Some birds, such as eagles, have wings adapted for soaring on air currents and flap their wings only occasionally, whereas others must flap continuously to stay aloft.

Thermal soaring is one of the most effective ways of traveling. Birds use the rising hot air of a thermal to climb to enormous heights. They then convert this height advantage into distance by gliding, wings outstretched, to the base of another thermal and then repeating the entire process.

The majority of birds that take advantage of thermal soaring, such as the birds of prey—eagles and hawks—have long, broad wings with separated primary feathers for delicate directional control.

Minute maneuvers

Some of the most impressive movements in the animal kingdom come from its very smallest representatives. The flea, scourge of pets and their owners, can at least claim to be one of the world's greatest jumpers. Their high-jump record is over 6½ inches (17 cm), and in the long jump, they can cover a massive 12½ inches (32 cm). While this might not sound like very much, it is equivalent to a human jumping 450 feet (137 m).

The hummingbird family contains the smallest of all birds, named because of the humming sound made by their wings. They are capable of rapid forward flight, of hovering like a helicopter in mid-air while feeding on flowers (see page 303), and even backward flight for short periods. Hummingbirds can hover because they have short rigid wings that can turn like propellers in virtually any direction.

FEATHERS IN FOCUS

Some mammals and reptiles have become adapted for flight, but in their case, the bones of the hand became very long and skin stretched out between the bones to act as wings. Birds evolved differently, reducing the size of their hand bones and growing feathers. In being both extremely light and strong, feathers are among the most remarkable of the vertebrate adaptations.

Feathers are composed of keratin, the same protein that forms human hair, fingernails, and also the scales of reptiles. At the center of a feather is a rigid hollow shaft, the quill, from which the vanes radiate. The vanes are made up of barbs, which in turn bear even smaller branches called barbules. The barbs are hooked strongly together, for they must not part or allow air to pass through them when the bird is in flight.

Stiff contour feathers shape a bird's wings and body. Their barbules have hooks that cling to barbules on neighboring barbs. When a bird preens, it runs the length of a feather through its beak, engaging the hooks and bringing together the barbs into a precisely shaped vane.

Beneath the outer surface of the body feathers is a layer of down. Because downy feathers lack hooks, the free-form arrangement of barbs produces a fluffiness that is used to trap air. This helps to insulate the bird's body and keep it warm.

Seventy percent of Earth's surface is covered by ocean and in some places it is deeper than the highest mountains. From the icy water of the poles to the warm tropical waters around the Equator, its vastness affects many global processes; it is home to an array of extraordinary life forms and has long fascinated humans.

Biological processes in the ocean

Virtually all life on Earth is sustained by the energy from the Sun. In the ocean, light penetrates down to around 600 feet (180 m). This is called the photic zone, where literally billions of microscopic plants (phytoplankton) convert light energy into food by photosynthesis—a process of harnessing the Sun's light energy to produce carbohydrates (sugars). Phytoplankton are consumed by tiny plant-eating animals (zooplankton) which are, in turn, preyed upon by other animals, from those not much bigger than themselves to planktivorous (plankton-eating) whales and sharks. Thus, phytoplankton is the basis of almost all the life in the ocean.

Life in the depths

When the plants and animals at the surface die, they sink, creating "marine snow." This shower of organic matter provides food to the organisms below and delivers carbon to the ocean depths. The average depth of oceanic basins is around 10,000 feet (3,000 m), and life in the deep often relies on life from above.

Deeper in the ocean, as the light diminishes, the twilight zone begins, and deeper still, at around 1,000 feet (3,000 m), where no light can penetrate, is the dark zone. Here, animals tolerate cold temperatures, high pressure, and the dark, and possess specialized physical adaptations. Many are thin, transparent, or red or black in color, which helps them to hide from predators; others have big eyes, and large mouths and teeth for hunting. Some even produce their own light, called bioluminescence—an internal chemical process—to attract prey or to find a mate.

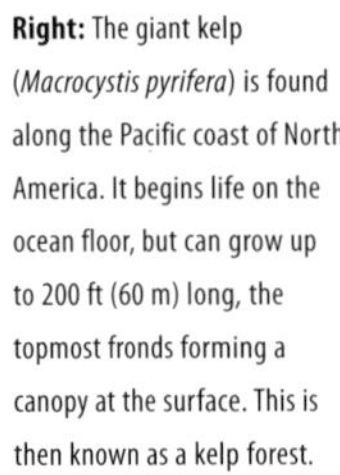

Right: The giant kelp (*Macrocystis pyrifera*) is found along the Pacific coast of North America. It begins life on the ocean floor, but can grow up to 200 ft (60 m) long, the topmost fronds forming a canopy at the surface. This is then known as a kelp forest.

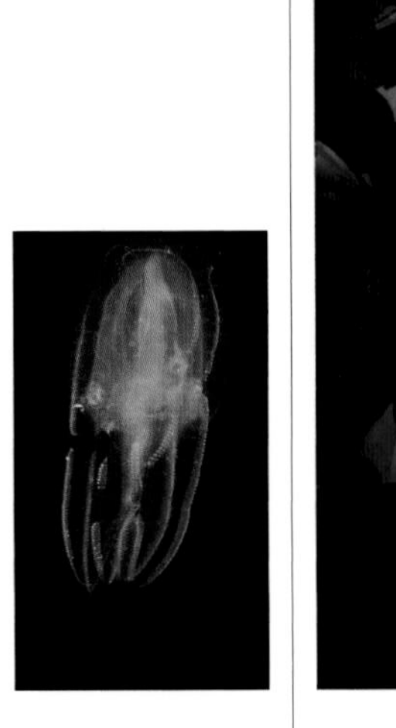

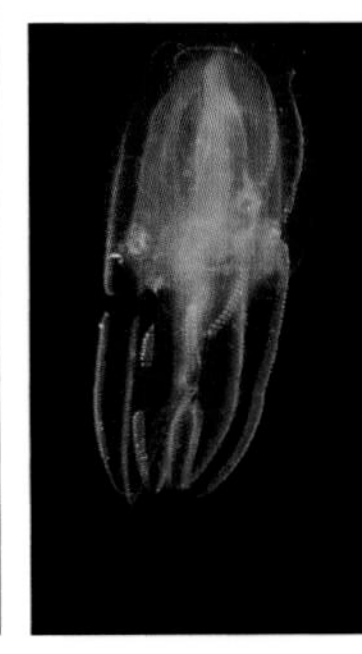

Right: Plankton range in size from microscopic organisms to jellyfish, and play a critical role in the marine food chain. Plankton is made up of plants and animals. The name comes from the Greek "planktos," which means "drifting."

Below: By studying the biological processes in the ocean, as well as the physical and chemical processes involved when air and water interact, scientists hope to come to a better understanding of global warming.

The sea, once it casts its spell,
holds one in its net of wonder forever.

Jacques Cousteau, undersea explorer, 1908–1997

Ocean circulation: Currents and carbon

As well as providing a habitat for a multitude of plants and animals, the ocean is also a vital part of Earth's cycling processes, upon which all life depends. Interactions between the ocean and the atmosphere drive global ocean circulation, affecting heat transfer, productivity, and the movement of carbon, all of which influence the global weather patterns (see also pagess 412–415).

We tend to think of ocean currents as moving horizontally across the surface, but large masses of water also move vertically, sinking slowly and spreading across the ocean floor. Winds create currents, moving warm water from the Equator towards the poles, where it cools—often freezing—leaving the water dense with salt. As the water becomes heavy it gradually sinks, pushing on the water below.

These deep-water masses move away from the poles until they return to the surface via a process called upwelling, usually at the Equator and along coasts where colder water replaces warm shallow water.

Upwelling is one of the most important processes to life on our planet, because it supplies nutrient-rich water to the surface in which phytoplankton flourish.

Carbon naturally cycles between the ocean, atmosphere, biosphere, and lithosphere, entering the ocean mostly as carbon dioxide gas. As phytoplankton use the carbon dioxide to photosynthesize, carbon moves from the surface to the ocean depths, hence any changes in atmospheric carbon dioxide will effect concentrations in the ocean and consequently, all life within it.

Above: A tarpon (*Megalops atlanticus*) invades a school of silversides. The small fish eat zooplankton, and are eaten by larger fish such as the tarpon. Large fish are prey for seals, which in turn are prey for killer whales. This biological interaction is known as the food chain.

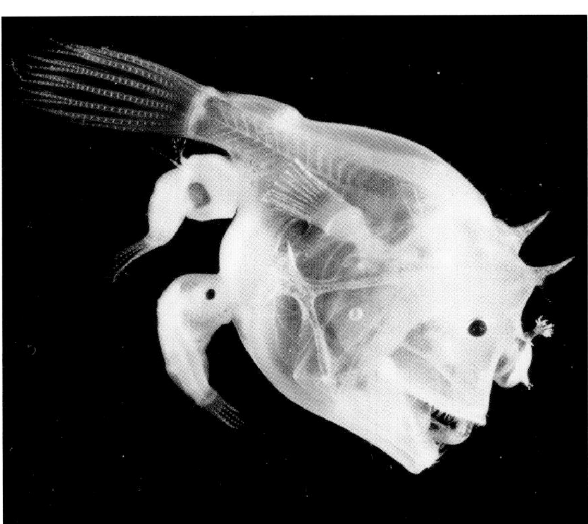

Left: There are almost 100 species of deep sea anglerfish, all living in the photic zone of the ocean. These fish rely on bioluminesence and sensitive feelers to locate their prey.

EXPLORING THE OCEAN

The ancient Greeks and Romans greatly contributed to early knowledge of the ocean, while the voyages of the Arabs, Vikings, Spanish, and English added to what is now well-known oceanographic knowledge.

Although marine scientists had collected animals from the ocean floor since the early 1800s, it was thought that no life existed in the deep ocean due to the high pressure and absence of light. In 1934 the first manned submersible, created by zoologist William Beebe, pictured below in the center, was lowered from a ship to a depth of 3,027 feet (920 m). Later development of manned and unmanned deep-sea submersibles has enabled depths of 35,000 feet (10,600 m) to be reached and the secret life of the ocean to be revealed. For example, complex biological communities have been discovered living in hot chemical-infused water around deep-sea thermal vents (see page 418).

We are all in contact with water every day: It is all around us and inside us. Land animals originally evolved from ocean-dwelling species and today water is still a major component of their bodies. Water is needed in almost all biological processes and there would not be any life without it.

The biological importance of water

Water is important in biological processes in all living organisms. We know that water covers three-quarters of Earth's surface and that the majority of it is stored in the oceans. However, only one percent of the world's water is the pure water that can be used in biological processes. Because water is a remarkable solvent, pure water rarely occurs in nature. Water exists naturally in three physical states, solid, liquid, and gas, and all three states are vital in biological processes. Water is the most abundant component of any organism; for example, a jellyfish is 98 percent water (that is why jellyfish are transparent), and the human body consists of around 65 percent water, in the blood, lungs, and brain. Constant movement of water between the atmosphere and Earth (known as the hydrological cycle) is essential for biological, ecological, and chemical processes: It regulates global and regional climate, allows us to experience rain, snow, and clouds and enjoy an impressive rainbow.

Right: Jellyfish are marine invertebrates, not fish, made mostly of water, which is why they are transparent. Jellyfish have stinging tentacles around their mouths, used to paralyze and kill their prey.

Left: A rainbow would not be possible without water, and can only happen when there is moisture in the air, such as after rainfall. Sunlight is then refracted through the raindrops creating a stunning arch of colored light. (See page 182.)

Water is the blood in our veins.

Levi Eshkol, 3rd Israeli prime minister, 1895–1969

NATURAL WATER CYCLE

The natural water cycle (the hydrological cycle) is the major force on our planet: It is an endless process that replaces the water on the planet continuously. Water slowly evaporates from oceans, rivers, and lakes, and then rises into the atmosphere where it condenses and forms clouds. When the atmosphere can no longer support the moisture it falls to Earth in the form of rain (or snow, at right) and runs back into the oceans, rivers, and lakes. During this cycle water is in constant movement and generates local climate.

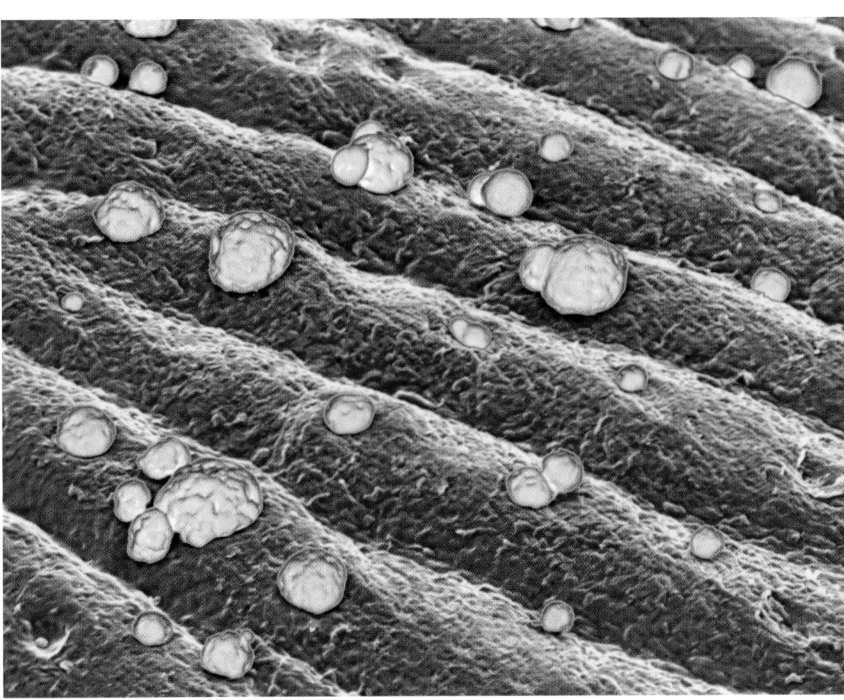

Right: A human fingertip surface, magnified 120 times, seen through a scanning electron microscope, showing the fingerprint ridges and droplets of sweat. Perspiration is the process of removing fluid (mainly water and salt) through the sweat glands in the skin. It is the human body's natural way of regulating its internal temperature. When we are feeling hot, we sweat. As the sweat evaporates on the skin's surface, it removes the excess heat and cools the body down.

The role of water in living organisms

Water is a good solvent because of its structure (H_2O). It is slightly ionized, is a polar molecule, and most elements and compounds can dissolve in it. For example, blood consists mostly of water, thus essential substances (oxygen, hormones, minerals) easily dissolve in it, making them ready to be transported to tissues and various organs. Water is also involved in removing the waste material of metabolic processes from an organism.

Because water is not easily compressed, it can provide support in various organisms. For example, earthworms possess a hydrostatic skeleton containing liquids (mostly water) under high pressure that helps them to keep their form stable. Plants remain upright and stiff because water fills up the cells and makes them turgid. Water also surrounds many internal organs, reducing friction and allowing free, easy movement. Here water plays the role of lubricant.

The three forms of water

Water has three forms. The most common and accessible for us is the liquid form of water: We use it for drinking, we swim in rivers and the seas. It also exists beneath Earth's surface as groundwater. Water provides a wonderful habitat for many organisms—from the smallest (bacteria) to the largest (whales, sharks, corals).

Water reaches its maximum density at 39.2°F (4°C) and becomes less dense as the temperature decreases until it reaches 32°F (0°C). At this temperature water becomes solid (ice) and can float on the surface. This property is vital for aquatic organisms, especially those living in polar regions, as it enables them to survive throughout the year and protects them until the temperature gets extremely low.

The thermal properties of water are also very important in biological processes. It takes a lot of time for water to reach its boiling point of 212°F (100°C) and therefore a lot of time to lose energy before the temperature starts to fall. This is important in maintaining body temperature in mammals, the transpiration process in plants, and evaporation in the environment (gaseous form of water). For example, mammals sweat when their body temperature is too high. Because sweat mainly consists of water, energy is released in the process and the organism's body temperature returns to normal, thus ensuring their survival. The evaporation process in nature increases when the air temperature gets higher: Water in gaseous form increases and causes humidity.

Above: Earthworms have no bones, and are supported by a hydrostatic skeleton, which is made up of water in their cells. This creates a pressure that gives the worm its shape as well as its ability to burrow through the earth.

Right: Beneath the ground of the El Tatio geyser field in the Atacama Desert in Chile lie channelways of boiling water. Heat-loving microorganisms called extremophiles thrive in many geothermal areas.

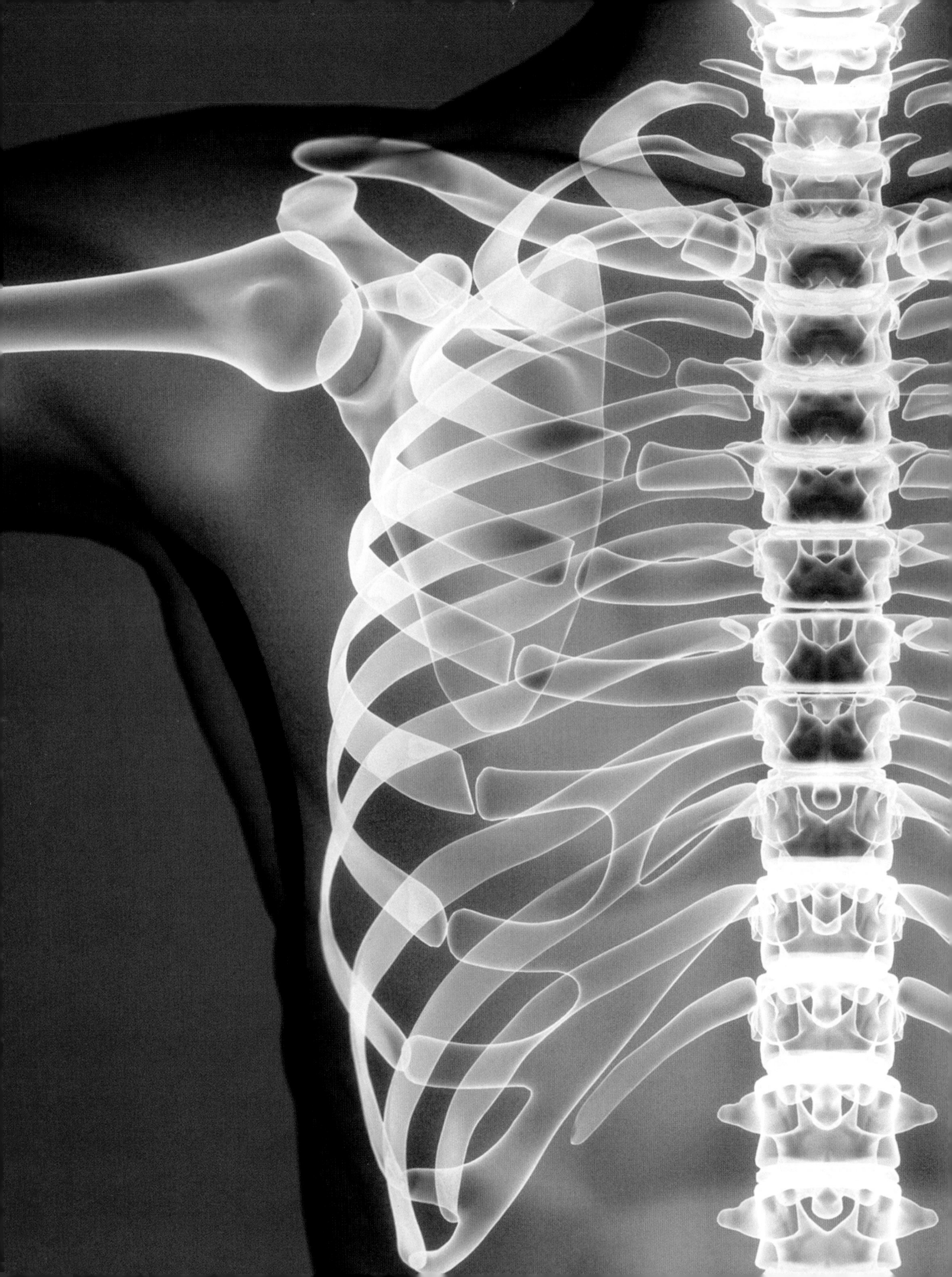

MEDICINE

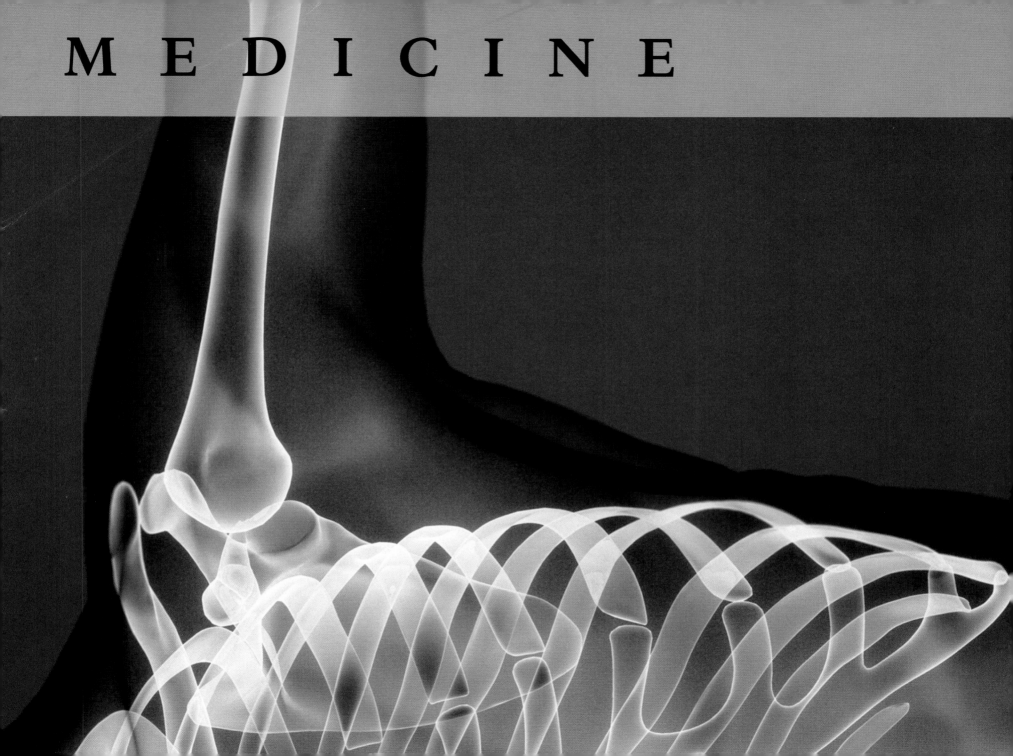

For most of us, nothing matters more than our own health and that of those dear to us. So it's not surprising that finding ways to keep us well and to cure us when we get sick has been a major concern for thousands of years. Before we knew much about the causes of illness, we tried to safeguard our health through a mixture of folk remedies and religious practices.

Introduction

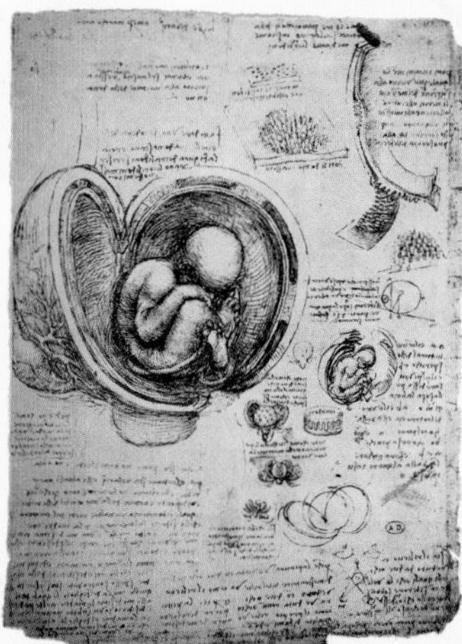

Above: Leonardo da Vinci (1452–1519), who was required to learn anatomy as part of his training as an artist, made one of the first scientific drawings of a human fetus in utero. Leonardo initially learned topographic anatomy, drawing muscles, tendons and other visible anatomical features, but as a successful artist he was later allowed to dissect human corpses in hospitals in Florence, Milan, and Rome. From 1510 to 1511 he collaborated with the doctor Marcantonio della Torre on a study of anatomy for which he produced more than 200 drawings.

Many old folk remedies were quite efficacious. They would not have persisted so long had they been worthless. But we had no idea why they worked, if they did, and therefore how we might enhance their effects. Any improvements over time came mostly by trial and error. Replacing guess-work with knowledge has been the task of medical science, and the outcomes have often been startling, especially in recent decades.

From anatomy to physiology

Medical science began with anatomy, since it was relatively easy to dissect animal and even human corpses (though the latter was frowned upon by religious authorities) and take note of the structures in the tissues and organs laid open to study. The great anatomists such as Andreas Vesalius and Leonardo da Vinci combined patience, acuity, and skill in remarkable amounts, and left us strikingly detailed drawings of many of the human bodily structures. Even more detail was revealed once the microscope came into use 400 years ago.

It was a greater challenge to move from anatomy to physiology, from describing body parts to working out exactly what they did and how they did it. Often the originators of new ideas had to fight tradition and prejudice, or lacked vital pieces of information. The circulation of the blood forms a powerful example.

For 1,500 years, doctors in Europe followed Galen's teachings. Galen believed that the two sorts of blood, dark red blood flowing in veins and bright red blood in arteries, originated in different organs—venous blood in the liver, arterial blood in the heart. Carried to various parts of the body they were consumed by tissues.

However, seventeenth-century English doctor William Harvey had a different view. He argued that the blood circulated around the body in great loops driven by the heart, to which it kept returning. One loop took bright red blood to muscles and organs, returning it darkened. The other looped passed the dark red blood through the lungs which restored its bright color.

Right: Edward Jenner, as depicted by French painter Gaston Melingue (1840–1914), performing the first vaccination against smallpox in 1796. Jenner's patient was an 8-year-old boy called James Phipps. To test his observation, drawn from folklore, that milkmaids who suffered the mild disease of cowpox never contracted smallpox, Jenner inserted pus from a cowpox pustule into an incision on the boy's arm.

Harvey's views were never accepted by the medical establishment of his day. And there was some problem with evidence. Before the widespread use of the microscope we could not see the tiny blood vessels, capillaries, in the muscles and lungs necessary to complete the two loops. But in the long term Harvey was proved correct. His explanation is still accepted today.

> Don't live in a town where there are no doctors.
>
> Jewish proverb

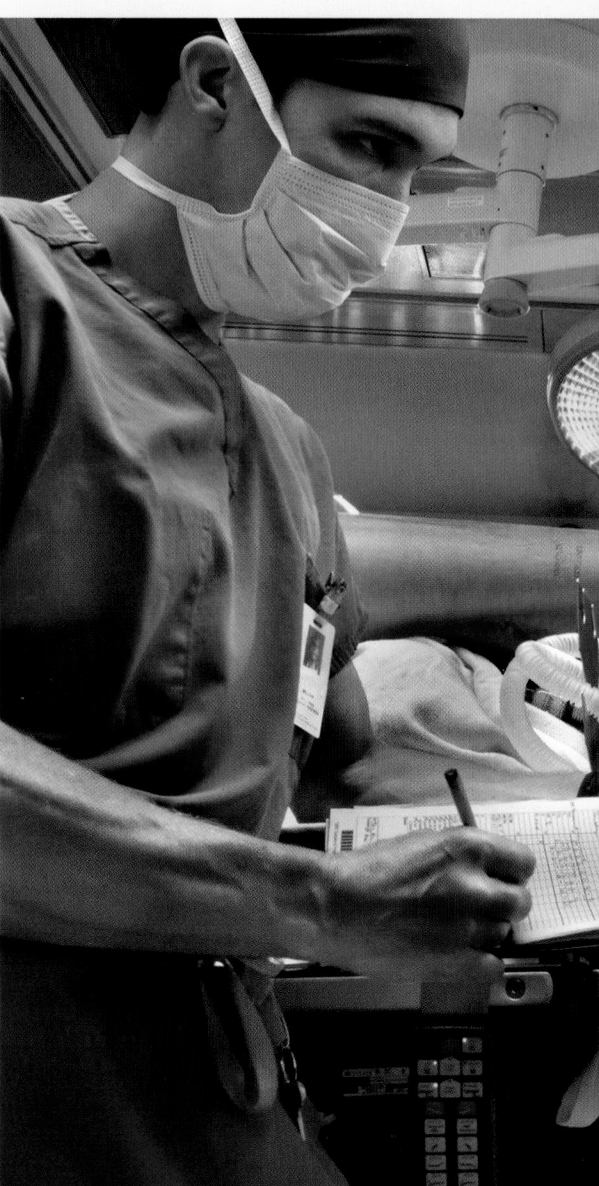

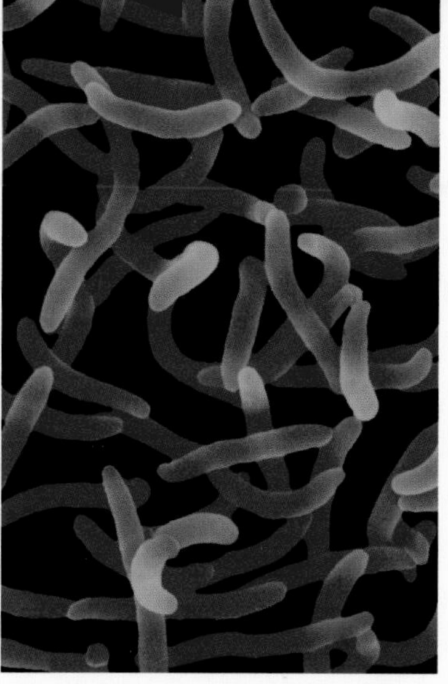

Right: Small, beautiful, and deadly, *Vibrio cholerae* is a gram-negative, facultatively anaerobic, curved rod bacterium that we now know causes cholera in humans. Isolated in 1855 by anatomist Filippo Pacini, it was identified by physician Robert Koch in 1885. That same year, Spanish doctor Jaume Ferran i Clua developed the first cholera vaccine.

The causes of disease

Equally intriguing is the long-running argument as to the possible causes of disease. For centuries outbreaks of the plague or cholera were blamed on "miasmas" of infected air, or seen as expressions of God's wrath. We now know that most diseases are due to microorganisms, but the "germ theory" had two false starts before being finally accepted in the nineteenth century due to incontrovertible evidence assembled by Louis Pasteur.

In the sixteenth century, Veronese physician Girolamo Fracastoro unsuccessfully promoted the idea that disease was spread by tiny "seeds" passed on by direct contact, contaminated clothing and utensils, or even through the air. The same idea was proposed 200 years later by Markus Plenciz ("every disease has its organism"), but again failed to win support, even though by then the microbes later implicated had been observed through microscopes.

The march forward

Broadly speaking, medicine has three arms—prevention, diagnosis, and therapy. Advances in each of these areas require the development of new ideas and new technologies. The nineteenth century saw many major technological advances which revolutionized medical practice—the introduction of anesthetics and antiseptic methods which vastly improved the outcomes of surgery; new diagnostic tools ranging from the stethoscope to X-rays; the development of vaccines against a wide range of diseases, continuing Edward Jenner's pioneering efforts in the control of smallpox.

Medicine continues to advance at an often breathtaking pace. Recent decades have seen an increased use of monoclonal antibodies to treat many conditions, and we may soon see stem cells similarly employed. But a great many challenges remain, including the increasing incidence of microorganisms unaffected by our most powerful antibiotics and the threats posed by new epidemics such as AIDS and avian influenza.

Above: English naturalist, geologist, paleontologist, and physician Edward Jenner (1749–1823), who developed the first smallpox vaccine.

Below: Coined by Oliver Wendell Holmes, Sr. in 1846, the word "anesthesia" comes from a Greek phrase meaning "without sensation." Modern medicine has advanced far beyond the painkillers and soporifics of earlier times.

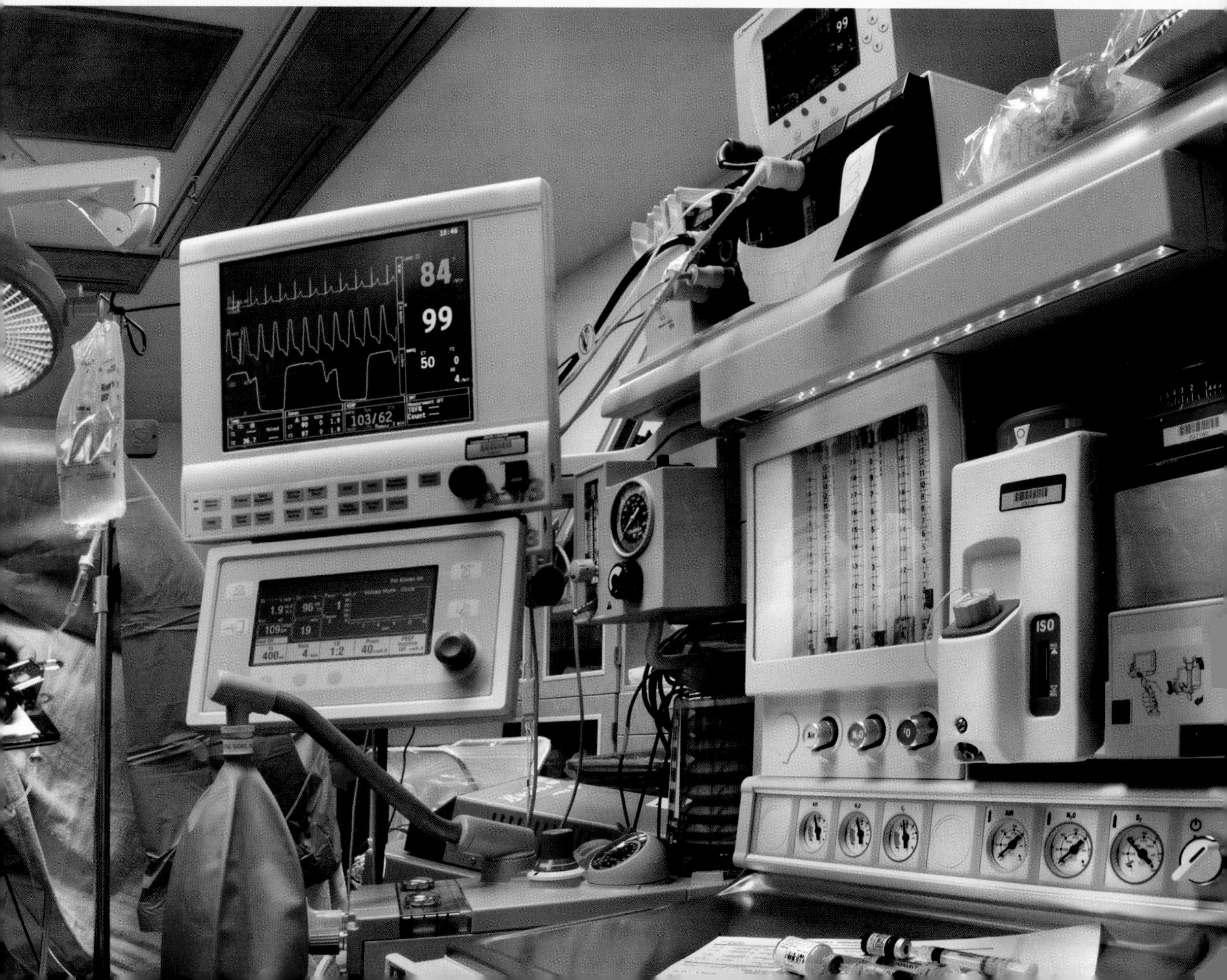

Many consider modern medicine to have been founded in ancient Greece, developed in Turkey, refined in the Middle East, and matured in Europe. The work of Hippocrates, the "father of medicine," Trotula, the "world's first gynecologist," and others, evokes dedication to healing, accuracy, and the development of ethical guidelines. Their work influenced the changing association between medical practices, spirituality, and the Church.

A dose of early doctors

Hippocrates

The Greek physician Hippocrates of Cos (*c.* 460–380 BCE), often called the "father of medicine," devoted his creativity to medicine and healing. Unlike many of his contemporaries, he believed illness to have rational causes, rather than demonic, religious, or supernatural ones. Although Hippocrates argues that ethics and morals are not related to the causes of physical ailments, they were integral in his treatment of patients, as reflected in the multi-authored *Corpus Hippocraticum* that contains the Hippocratic Oath. This compilation of biomedical methods and code of practice indicated healing as the paramount concern of the physician, with financial and social gains secondary reasons for practicing medicine.

One physical explanation Hippocrates gave for illness involved the concept of humoral pathology. He understood the body to consist of four humors: blood, phlegm, choler (yellow bile), and melancholic (black bile). An imbalance, or isonomia, specifically an excess of one or another humor, resulted in ill health. To prevent this, Hippocrates advocated good nutrition, rest, hygiene, and adequate ventilation. If someone fell ill, Hippocrates insisted that observation, investigation, examination, and record-taking be conducted prior to making any suggestions for recovery. Thus, the institutions of prognosis, record-taking, and case-by-case treatment began.

Above: Few of the facts of Hippocrates' life are known with any certainty, but the Hippocratic school of medicine he founded revolutionized medicine in ancient Greece, establishing it as a discipline in its own right, distinct from theurgy and philosophy.

Cure sometimes, treat often, comfort always.

Hippocrates, physician, *c.* 460–*c.* 380 BCE

Avicenna

During the European Dark Ages, the Middle East underwent a period of major scientific and medical advances, from which the West later benefited greatly. Significant contributions to the refinement of medical knowledge were made by the Persian medico Abu 'Ali al-Husayn ibn 'Abd Allah ibn Sina, or Avicenna (*c.* 980–1037). Avicenna began his medical studies in Bukhara at the age of 16, then went to practice at the Samanid court at Bukhara. When the court was overthrown a year later, he began work as scientific *wazir* or vizier in Esfahan.

Avicenna advised a measured and holistic medical approach. His remedies for pain built upon the understanding that *mukhaddirat*

Below: Abu Bakr al-Razi Rhazes (867–925) also known as Razi, a Persian physician and alchemist depicted here in his laboratory in Baghdad. An early proponent of experimental medicine, Razi is considered the father of pediatrics. He was also a pioneer of neurosurgery and ophthalmology.

(analgesics) were not the primary method of relief. He favored rest, relaxation, massage, hot compresses, or distractions. If these methods proved insufficient, relief was sought via cooling or toxic analgesics. Coolers included ice and cold water; toxins included opium, mandrake, poppy, henbane, hemlock, black nightshade, and lettuce seeds.

By the age of 21 he had written the *Kitab al-Qanun fi al-tibb (Canon of Medicine)* which superseded Baghdad doctor Al-Majusi's only treatise *Kitab Kamil al-sina'ah al-tibbiyah (The Complete Book of the Medical Art)*. In five sections the *qanun* (canon) covers general practice, therapeutic substances, diseases of body parts, general ailments, and methods of preparing remedies.

Trotula

Trotula de Ruggiero (d. 1097) trained, and is believed to have taught, in the famed medical school at Salerno, Italy, the first university to admit women. Dedicated to healing, she approached medicine with an open mind, incorporating a mix of pagan, Muslim, and Christian practices. Staunchly committed to women's health, she developed new techniques for episiotomy, uterine ruptures, pain management, and attenuation of bleeding and infection (including both herbal and animal treatments). Further, she was unafraid to defy Church guidelines that women should suffer through childbirth as punishment for original sin, and promoted opiate use to alleviate labor pains.

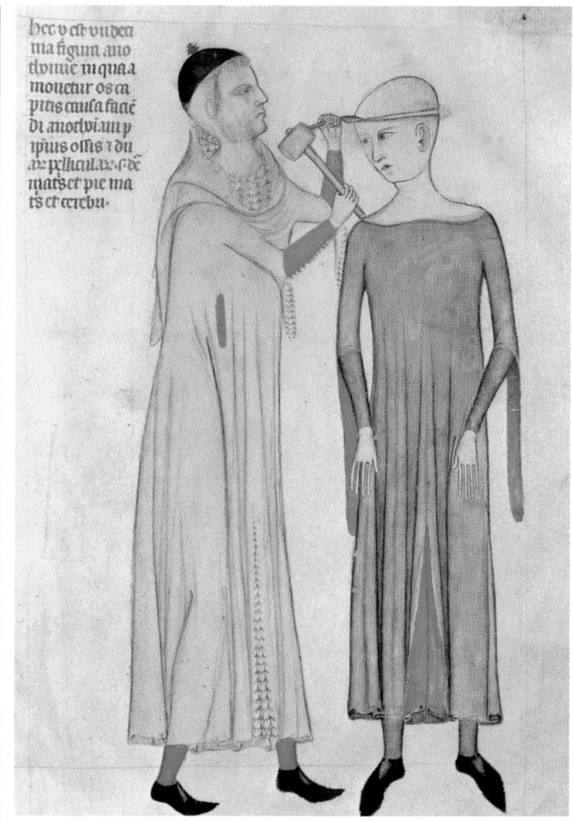

Right: Trepanation (also known as trepanning), in which a hole is drilled in the skull to relieve pressure on the brain, is the most ancient form of surgery for which evidence exists, going as far back as Neolithic times. Hippocrates and Galen both refer to the procedure, which was also used during the Middle Ages and the Renaissance as a cure for various ailments, including seizures and skull fractures. The survival rate was high and the infection rate low.

Below: *De Chirurgica* by Master Rolando (fourteenth century) shows plants used in medicine: angalles against tapeworm and mandrake to look after wounds.

GALEN

Galen, often considered the world's first experimental physiologist, lived in Pergamum, in modern Turkey. At the age of 17, he began medical studies in Alexandria and Smyrna; once they were completed, he traveled to Greece, Asia-Minor, and Palestine, developing his medical skills. At age 28, he returned to Pergamum to practice medicine. His fame reached Emperor Marcus Aurelius, who summoned him to Rome as his personal physician. Through tending to gladiators' wounds Galen furthered his anatomical knowledge and initiated his surgical practices.

Galen extended Hippocratic investigative methods to structure and function experiments. He tied ureters and observed the resultant renal swelling, thereby demonstrating kidney involvement in urine production. Through severing nerves, and observing lost motor or sensory control, he demonstrated nerve function, concomitantly discovering afferent and efferent neurons. Galen complemented experimental and medical work with the manufacture of vegetable- and animal-based pharmaceuticals; he also compiled treatises, providing details on their composition and correct dosage. Like the Hippocratic works, Galen's treatises discuss medical ethics, which remain as important today as they were in the first century CE.

In times gone by, there was little distinction between medicine, magic, and religion. The ancient Egyptians believed exogenous forces such as the gods caused illness. In the case of divine punishment or curse, healers including priests, magicians, and physicians attended to the spiritual causes prior to treating symptoms.

From folk healers to modern medicine

Pharmaceuticals and surgery treated symptoms, as evidenced by the *c.* 1552 BCE Ebers papyrus of magical and folk remedies for optical, dermal, gynecological, and other conditions. The Edwin Smith papyrus, *c.* 3000 BCE, directs physicians to take a patient's history, conduct an examination including an inspection of bodily fluids and secretions, then make diagnoses and decisions on treatment and surgery.

Roman medicine

At the beginning of the Roman Empire, 753 BCE, most Romans held similar beliefs to the ancient Egyptians regarding disease and illness. This began to change, however, when Gaius Marius organized the Roman army into a highly regulated, specialized defense force in the first century BCE. Medical innovations flourished as a corollary of the need to deal with battlefield wounds, and army camp diseases, that were much more pressing concerns than supernatural issues.

Of particular note is an innovation that was made by the Emperor Augustus. After the civil war that followed the assassination of Julius Caesar in 44 BCE, Augustus instituted medicos as a professional group within the army, insisting at the same time that they attend medical school with civilian doctors. This led inevitably to medical advancement. Everything the army doctors learned, devised, and improvised on the battlefield—such as sterilizing surgical implements, treating open wounds antiseptically with acetum, and amputating if necessary—they recorded and shared with others at the medical school. Further, the preventative measure of placing army camps at a distance from insect-infested wetlands benefited the health of both soldiers and civilians alike, as did the installation of sewerage systems.

Above: Bas relief, *c.* 100 CE. An ancient Roman doctor's tool kit included forceps, scalpels, catheters, and arrow-extractors, as well as a range of painkillers and sedatives, including extracts drawn from opium poppies (morphine) and henbane seeds (scopolamine). Folk remedies were tried and tested in battle.

Right: A nineteenth century English engraving of the "Death Cart" making its rounds during the Great Plague of London, an outbreak of bubonic plague that killed over 70,000 people in 1665–66. Though far smaller in scale than the earlier, more virulent outbreak known as the "Black Death" in Europe (1347–1353), it was remembered for centuries afterwards as the Great Plague because it was one of the last widespread outbreaks in England.

Left: An unguent box in the form of a double cartouche, from the tomb of Tutankhamun (*c.* 1370–1352 BCE). Tutankhamun's unguent jars were all made of calcite (alabaster), since it was believed such containers were good preservatives. Some of the perfumes and unguents used by the ancient Egyptians (such as spikenard, myrrh, and galbanum) now seem rare and exotic, while others, such as fenugreek (used as a remedy for bronchial infections) still play a role in medicine today.

Early medieval medicine

Following the fall of the Roman Empire in 476 CE, the West went through what we now call the Dark Ages, a time when scientific and intellectual advances stalled. Medical attitudes strongly reconnected with the church, astrology, and superstition. Under the influence of the church, illness was often regarded as the manifestation of divine displeasure; treatment required repentance.

During the mid-medieval period, the Crusaders brought medical knowledge from the Middle East to Europe. Despite the advances

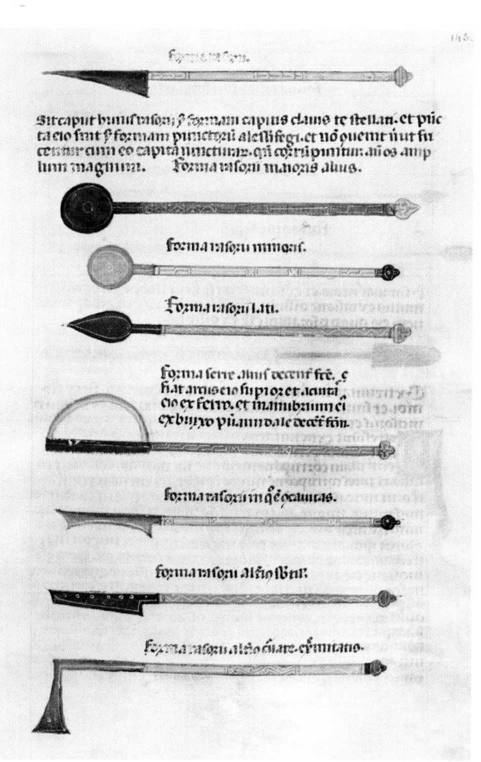

Left: Though Hippocrates is usually considered the "father of medicine," the Andalusian-Arab physician, surgeon, chemist, cosmetologist, and scientist Abu al-Qasim al-Zahrawi (936–1013), known in the West as Abulcasis, is often called the "father of surgery." His thirty-volume *Kitab al-Tasrif (The Method of Medicine)* included illustrations of his renowned collection of surgical tools, totaling 200 pieces, and its final volume, "On Surgery," was not only the earliest elucidation of this subject, but remained the best until relatively recent times.

Below: Western anatomical and surgical knowledge was slow to develop in the Middle Ages. For the most part surgery was limited to the removal of cysts and cataracts, and minor operations to cure war wounds.

available in Arabic medicine, however, the religious core of European medical practices remained, to the detriment of the West when the bubonic plague swept through Europe in the 1300s.

Medieval European medicine

During the Middle Ages, the medical influences of the Greek and Semitic peoples arrived in Europe via the port town of Salerno, near Naples. Through the dedicated accumulation of medical information, the school at Salerno contributed significantly to the medical canon. Along with information about anatomy, it introduced a staunch insistence on proper training, a system of certification for medical practitioners, and the admittance of women to medical school.

In the twelfth century, Sicilian Prince Roger passed a law that required aspiring doctors to sit examinations based on Hippocratic, Galenic, and Avicennic writings. The following century, his grandson Frederick II of Naples decreed that these examinations had to be conducted publicly by the masters of Salerno. These requirements were later relaxed, and Western medicine has only returned to such stringency in the past century or so. Thus, the Salerno School of Medicine provided a benchmark of industry standards, through which it became famous for producing exceptional practitioners.

In England, the enmeshment of religion and medicine resulted in close links between hospitals and churches. For example, the twelfth century St Thomas Hospital, London was attached to the priory of St Mary Overie. Further, due to church–state interaction, politics affected hospitals and medicine; St Thomas Hospital closed in 1540 following Henry VIII's dissolution of the Catholic Church, then reopened in a more conducive political climate in 1561.

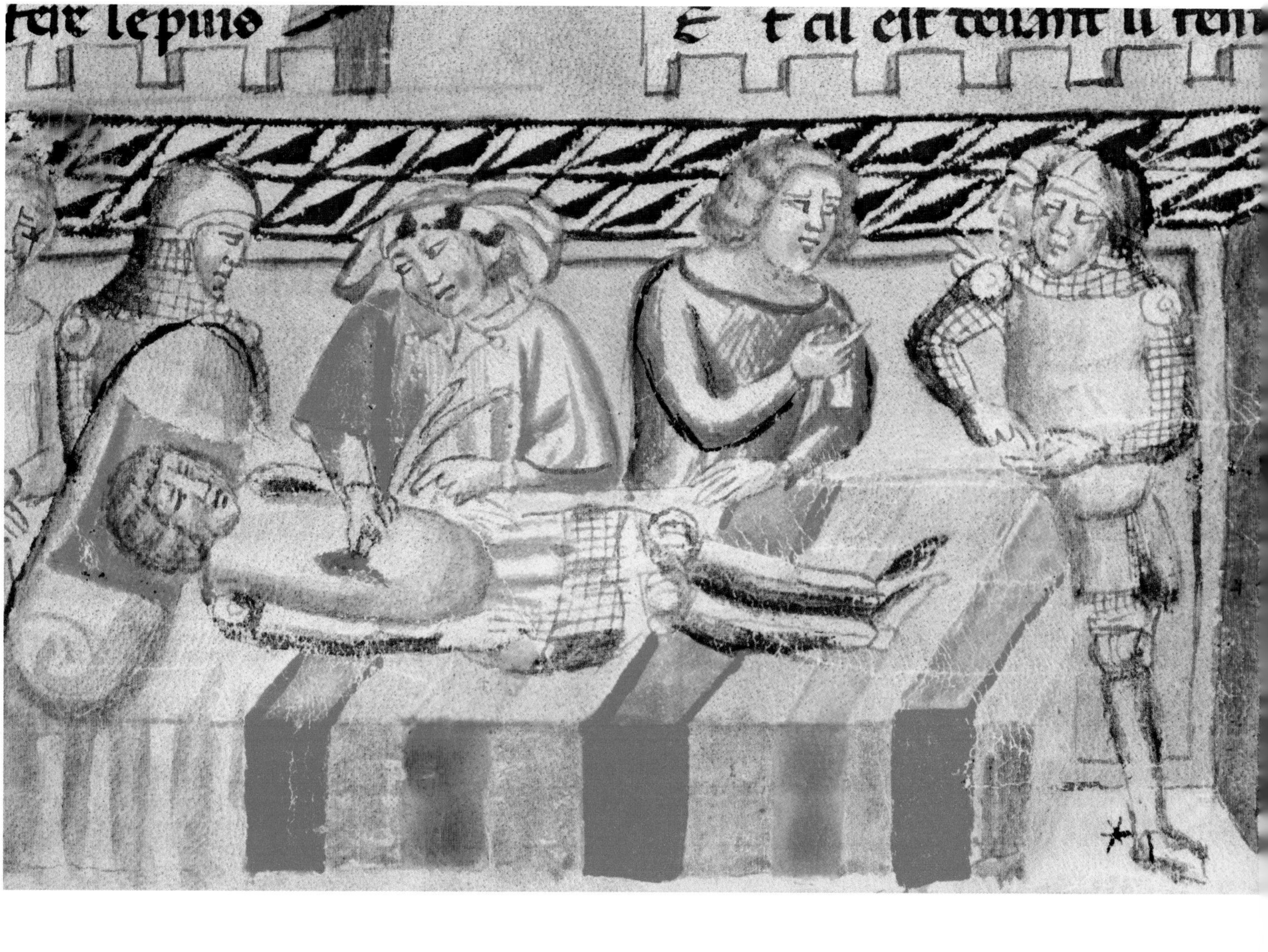

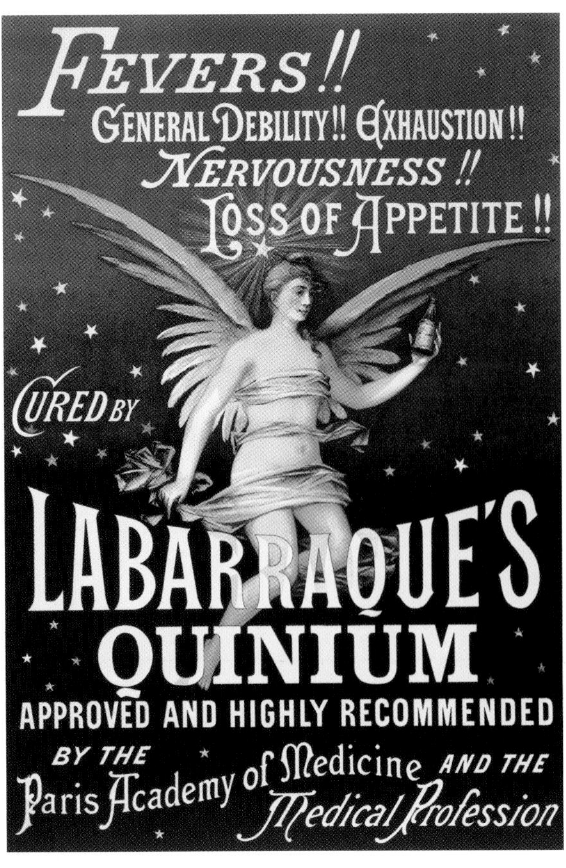

Left: Antoine-Germain Labarraque (1777–1850) was a French chemist and pharmacist who did much to improve the sanitary conditions in hospitals and army barracks. In 1825 he pioneered the use of sodium hypochlorite as a disinfectant. His quinium pills, taken with wine, were said to be effective against intermittent fever.

Victorian medical theory and miasma

The Industrial Revolution brought higher standards of living, but also the filthy, smoke-filled air, and respiratory illnesses that were part of the cost of attaining industrial wealth. When heavy fogs in December 1873 and January 1880 coincided with increases in weekly deaths of 268 and 692 respectively, the conclusion drawn was that industrial wealth borrowed from public health, via miasma: unpleasant, unhealthy vapors that spread illness, disease, and decrepitude.

During the nineteenth century, smell became an effective means of class distinction. Those emitting vile stenches, of a kind that acted as agents for disease, included the poor, domestic servants, prostitutes, convicts, rag pickers, pederasts, sailors, immigrants, tramps, vagabonds, and the working classes. Many considered the stench of the poor, who could not afford the luxury of space, to be reflexive, further infecting the vectors of disease. As if to support this, the poor were most likely to fall ill. Despite the flaws in miasmic theory, its popularity improved health and lowered morbidity and mortality rates, through the health reform acts, and the hygiene actions taken in attempts to remove foul, disease-causing fumes.

> Formerly, when religion was strong and science weak, men mistook magic for medicine; now, when science is strong and religion weak, men mistake medicine for magic.
>
> Thomas Szasz, psychiatrist, b. 1920

Below left: A nineteenth-century medical chest, showing a variety of glass containers. The use of glass vessels for medicinal storage declined after the fall of the Roman Empire, and was not revived for nearly a thousand years.

Enlightenment

During the late seventeenth and eighteenth centuries, philosophers and medicos engaged in an active process of replacing superstition with rationalism, religion with science, and emotion with intellectualism. This process, or era, has been termed the Age of Reason, or the Enlightenment. Experimentalist scholars such as Robert Boyle (1627–1691) tried to create a greater sense of certainty regarding the generation of matters of fact, and to bolster popular faith in science. Science, technology, and progress in the form of human advancement dominated Western civilization. During this period, the critique of religion and superstition was a slow, surreptitious development, which often utilized existing cultural and religious traditions; the expansion of scientific and medical knowledge was a means of increasing the sense of certainty in the production of knowledge.

Because the church considered the contributions of Galen (see page 321) compatible with doctrine, medicos maintained a humoral theory of health throughout the Renaissance and the Enlightenment, until the nineteenth century when the Industrial Revolution created a need for contemporary notions of public health.

NINETEENTH CENTURY CHOLERA AND BIOGEOGRAPHY

In healthy individuals, water resorption from the large intestine prevents the body from dehydrating, resulting in solid feces. Contemporary medical knowledge is that *Vibrio cholerae*, the pathogen responsible for cholera, dehydrates the body by causing water to be pumped into the gastrointestinal lumen, and feces are watery. During the nineteenth century, Britain suffered cholera epidemics. Various theories were proposed as to their cause. These included poison emanating from Earth (telluric), atmospheric electricity (electric), insufficient ozone (ozonic), choleric patients' putrefied bodies producing ammonia, and a miasma that spread the disease (zymotic). Most popular was the zymotic theory. To the anti-contagionist mind, the stench of the poor generated disease and choleric airs. They believed inadequate ventilation in crowded abodes facilitated the spread of cholera.

When the epidemic of the 1850s devastated London, John Snow doubted the popular miasmic understanding of the disease, and investigated with biogeographic mapping. He traced the source to a water pump, culminating in the removal of the pump handle. Snow's map and the isolation of *Vibrio cholerae* by physician Robert Koch (pictured) provided evidence against miasmic theory. However, other maps and data showed that a greater number of deaths occurred at lower altitudes, higher temperatures, and lower winds that allowed foul smells to develop. This, coupled with class anxieties, seemed to support miasmic theory.

It is now evident that biogeography both supported and refuted miasmic theories, at least in the case of cholera. This does not mean that it was futile; on the contrary, biogeography remains an important aspect of medical enquiry and theorizing.

Left: German orderlies dress the wounds of an injured British soldier, 1916. Front-line units were able to provide only the most superficial medical care, but with isolated exceptions, captured soldiers on both sides reported fair treatment.

Below: Modern surgery often requires skilled individuals to interact effectively in a variety of complex situations. Surgeons, anesthesiologists, perfusionists, and nurses must work as a team to ensure a successful outcome. Any breakdown in communication could spell disaster for the patient.

Modern medicine

As the world entered the twentieth century, medicine began to advance at a rapid rate. Karl Landsteiner increased the likelihood of surgical operations having positive outcomes by identifying different blood types, and their significance, in 1901. This was of fundamental medical importance during World War I, when the concept of blood banking for soldiers was first introduced.

The treatment of soldiers in World War II was improved by the discovery of antibiotics. Ernest Duchesne first reported penicillin in 1897, yet it remained unnoticed until 1928, when it was rediscovered by Alexander Fleming, and later medically applied by Howard Florey.

In concert with antibiotics, vaccines made a significant impact upon global health. Emil von Behring, the recipient of the first Nobel Prize for Medicine, helped develop serum therapy against diphtheria, and vaccines against tuberculosis. Further work curbed the spread of polio, influenza, hepatitis A and B, and smallpox. More recently, Ian Frazer developed a vaccine that protects women against several strains of the human papilloma virus, a precursor to cervical cancer. To the minds of many medico-scientists, this is significant, as they focus on anti-cancer developments as the future of medicine.

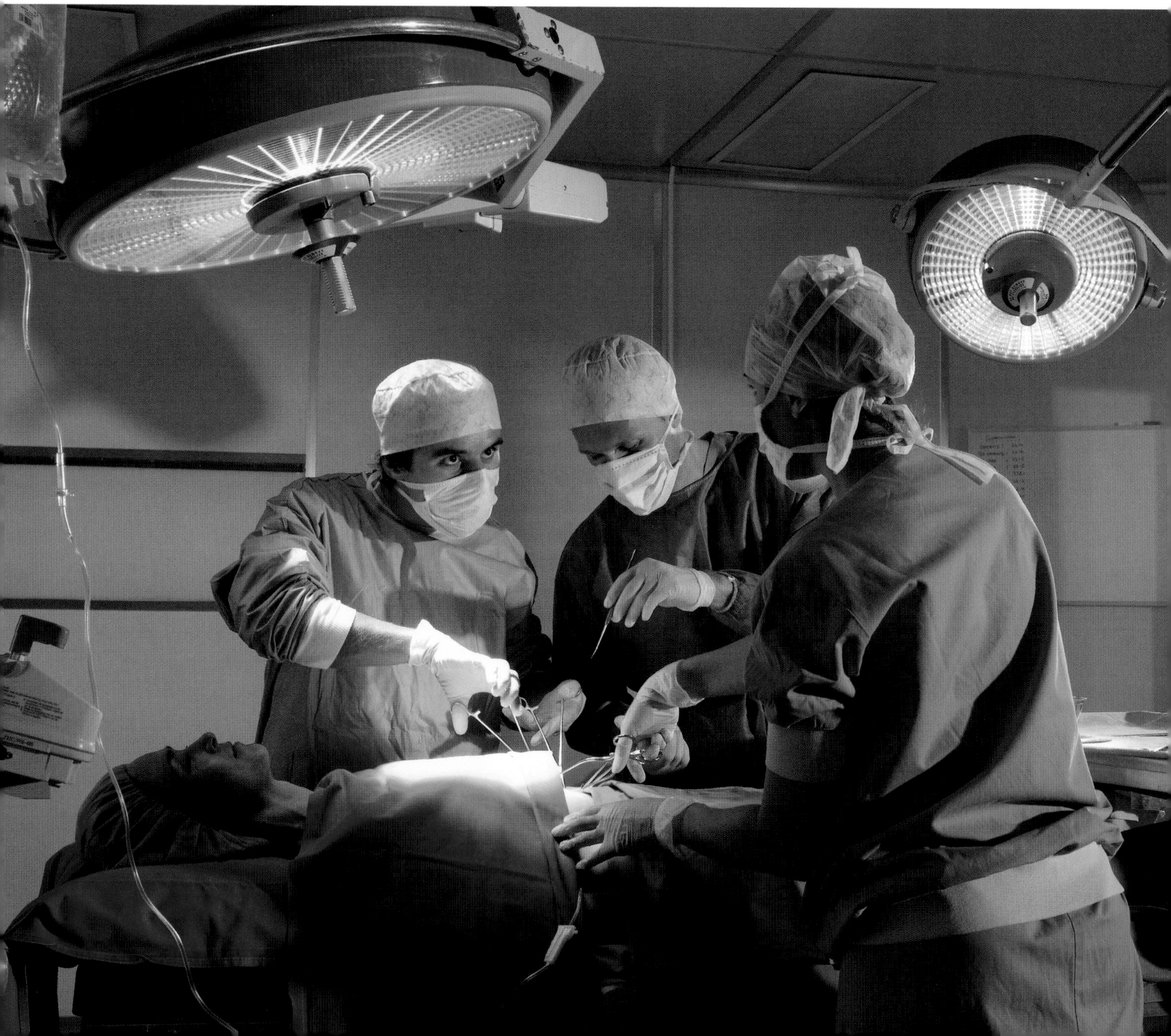

Most of us are aware that the skeleton is the structure that supports our movements and protects our body parts. It consists of bones, joints, cartilage, and ligaments, accounting for about 20 percent of our body mass. But many may not know that the skeleton is an engineering marvel which, in some parts, such as the femur (thigh bone), is stronger than reinforced concrete.

The human framework

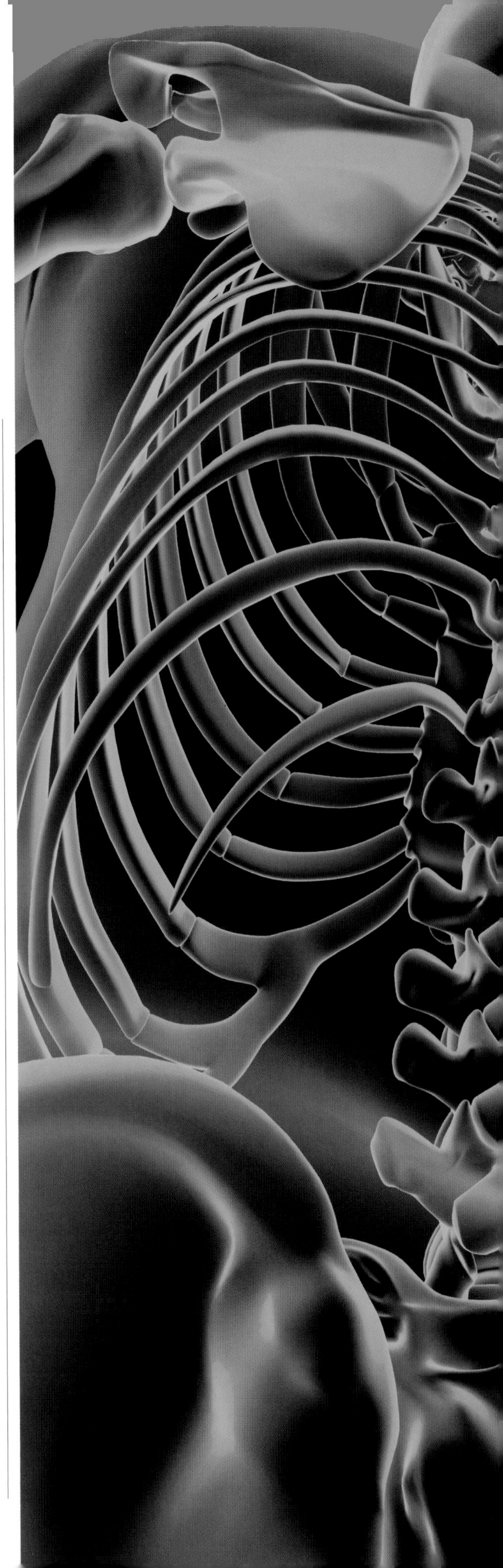

Axial and appendicular skeleton

The skeleton is divisible into two parts: The axial and the appendicular skeleton. The axial skeleton comprises the skull, the vertebral column (spine), and thorax (rib cage), forming the long axis of the body. These bones help protect, support, or carry other body parts. The skull protects the brain and supports the face. The vertebral column has 24 movable vertebrae plus the sacrum and coccyx. Between each vertebra is a fibrocartilage intervertebral disc, which acts as a shock absorber and gives the vertebral column flexibility. The thorax has 12 pairs of ribs, the sternum, and the thoracic vertebrae. It protects the organs of the thoracic cavity, such as the lungs and heart.

The appendicular skeleton comprises arm, forearm, hand, thigh, leg and foot bones, and the girdles (shoulder and hip bones) that attach the limbs to the axial skeleton. The appendicular skeleton is used for locomotion and manipulating the environment.

Bones

There are 206 bones in the adult human skeleton. Bones are classified according to their shape—long, short, flat, or irregular—and according to the proportion within them of compact to spongy bone.

Bones make up most of the skeleton, giving the body its shape, protecting and supporting organs, providing levers for muscles to pull on, storing calcium and other minerals, acting as the site of blood cell production, and being composed of living cells and matrix. Matrix includes both organic substances and inorganic components. The organic substances, secreted by osteoblasts (bone-creating cells), give the bone tensile strength. The inorganic components, hydroxyapatite (calcium salts), make the bone hard.

Bone is a dynamic living structure, one that is continually being remodeled. New bone replaces old bone in response to hormonal and mechanical stimuli (such as physical activity).

Joints and mobility

Joints are the junctions between the bones. At the joints, ligaments connect bones, helping to stabilize joints and allow certain movements. Also, at the joints, cartilage is usually found at the ends of the bones, allowing the bones to glide over each other. Cartilage is also

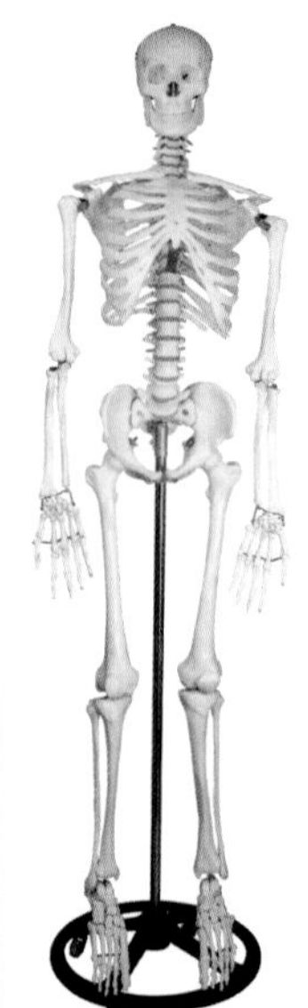

Above: Beginning with Galen, physicians saw the structural necessity of the skeleton, but were usually more interested in other parts of the body. It was Avicenna who first suggested that the best way to gain knowledge of the skeleton was to see it separated from the rest of the body. This idea became so widely accepted in the Renaissance that, by the late sixteenth century, anatomy theaters were filled with articulated skeletons.

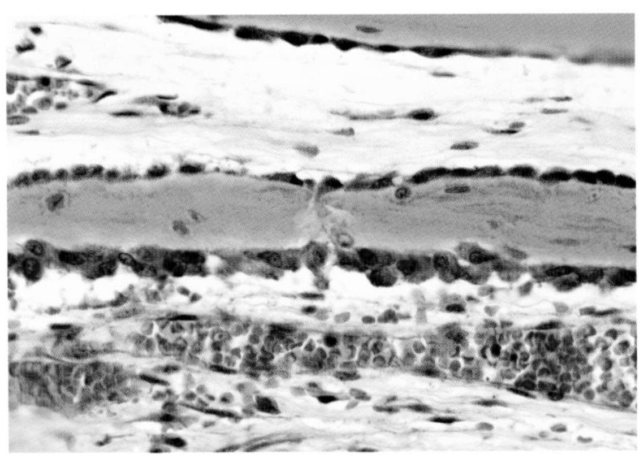

Left: Two types of cell of great importance in bone formation are osteoblasts and osteoclasts. Osteoblasts (shown here in developing bone membrane) are epithelial-like cells that form a monolayer over sites of active bone formation, depositing calcium into the protein matrix. Osteoclasts dissolve stored calcium and carry it to tissues when needed. Collagen, a flexible, gelatin-like matrix, comprises one-third of the bone's components.

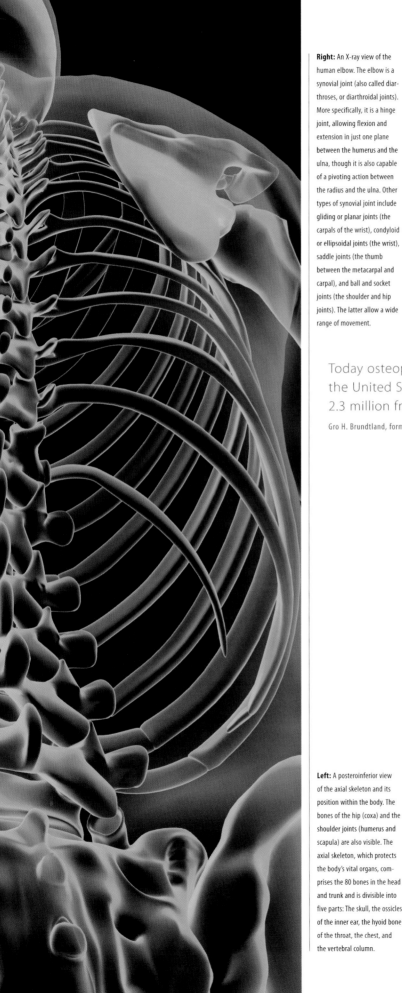

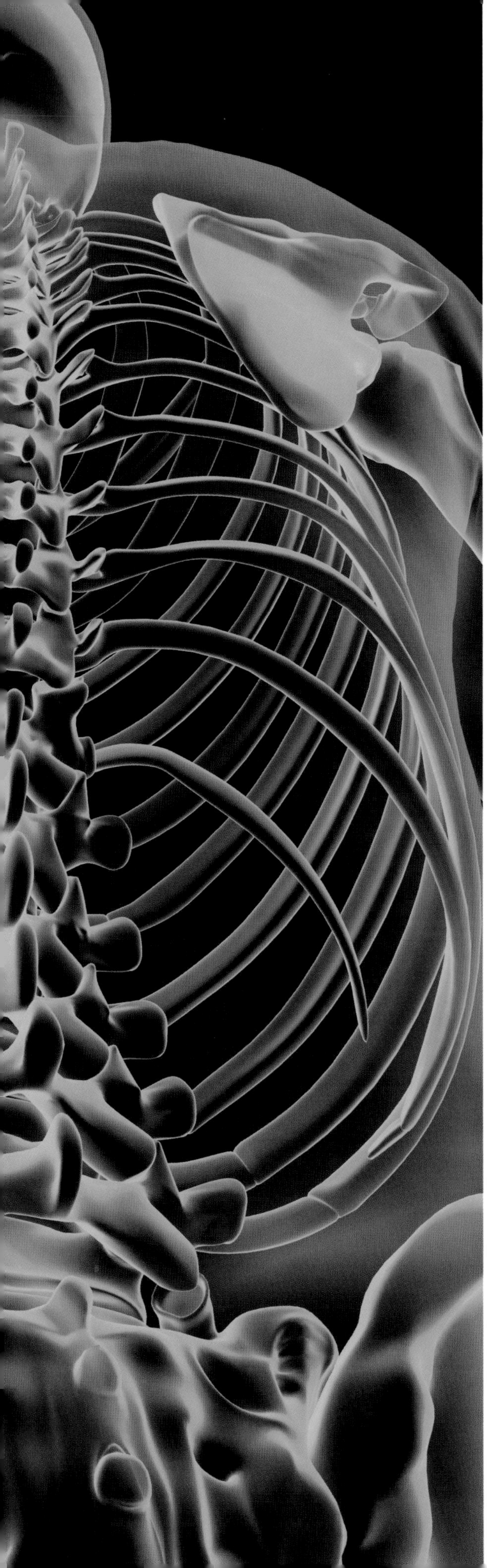

Right: An X-ray view of the human elbow. The elbow is a synovial joint (also called diarthroses, or diarthroidal joints). More specifically, it is a hinge joint, allowing flexion and extension in just one plane between the humerus and the ulna, though it is also capable of a pivoting action between the radius and the ulna. Other types of synovial joint include gliding or planar joints (the carpals of the wrist), condyloid or ellipsoidal joints (the wrist), saddle joints (the thumb between the metacarpal and carpal), and ball and socket joints (the shoulder and hip joints). The latter allow a wide range of movement.

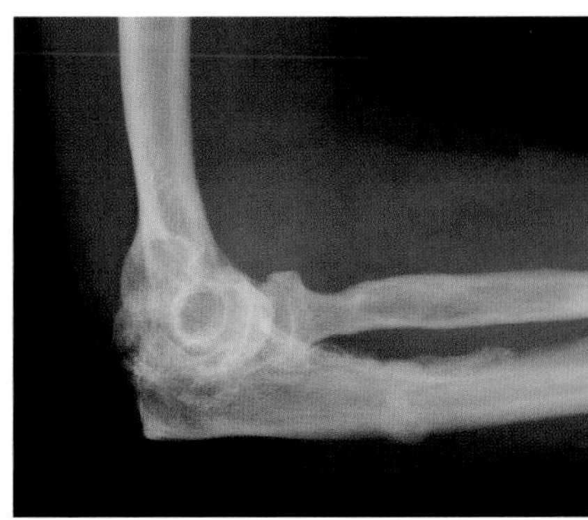

found in other areas of the body, such as the nose and parts of the ribs.

Joints are classified as fibrous, cartilaginous, or synovial. Fibrous joints occur where bones are connected by fibrous tissue and no joint cavity is present. Cartilaginous joints occur where bones are united by cartilage and no joint cavity is present. Synovial joints make up most of the joints in the human body and allow for the greatest range of movement. They have a joint cavity enclosed by a fibrous capsule lined with a synovial membrane and reinforced by ligaments; the articulating bone ends are covered with cartilage, and synovial fluid is present in the joint cavity. Some synovial joints, such as the knee, the largest joint in the body, contain fibrocartilage discs that absorb shock.

> Today osteoporosis affects more than 75 million people in the United States, Europe and Japan and causes more than 2.3 million fractures in the USA and Europe alone.
>
> Gro H. Brundtland, former Prime Minister of Norway, b. 1939

WEAR AND TEAR

Excluding traumatic injury, joints usually function well until middle age, at which time, symptoms of connective tissue stiffening and osteoarthritis (pictured below) begin to appear. As we age, the intervertebral discs become thinner; if the effects of osteoporosis are added as well, the result can be a gradual loss of height and an increased risk of disc herniation. Loss of bone mass increases the risk of fractures, and thoracic cage rigidity can induce breathing difficulties. Regular exercise can delay the onset of these effects, but excessive exercise may promote the early onset of arthritis. When joints fail due to wear and tear from sports, old age, or disease, it is unlikely that they can be regenerated.

Prosthetic (artificial) joints may be used to restore the function of degenerated and damaged joints. Prosthetic hip joints have been made from metals, plastics, and ceramics in many different designs, all of which have aimed to make incremental improvements to joint mobility, implantation, retention, and biocompatibility. The first hip prosthesis was implanted in 1939. Today's prosthetic hips are advanced medical devices, sometimes containing composite materials, designed to mimic the superior strength-for-weight properties of natural bone.

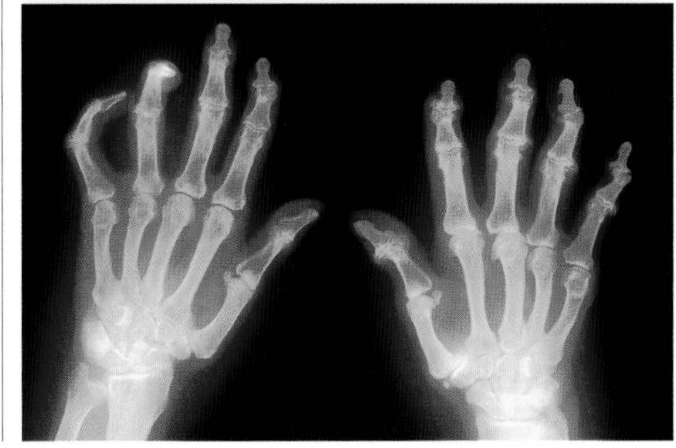

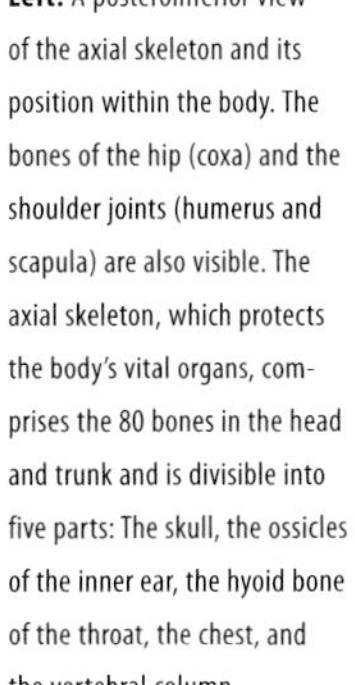

Left: A posteroinferior view of the axial skeleton and its position within the body. The bones of the hip (coxa) and the shoulder joints (humerus and scapula) are also visible. The axial skeleton, which protects the body's vital organs, comprises the 80 bones in the head and trunk and is divisible into five parts: The skull, the ossicles of the inner ear, the hyoid bone of the throat, the chest, and the vertebral column.

Bulging biceps are usually the first thing that comes to mind when we think about muscles. But there is more to muscles than a well-toned appearance. There are three types of muscle: Skeletal, cardiac, and smooth. Each performs a different function and together they make up nearly half our body weight.

Taut and terrific

Skeletal muscle

Skeletal muscle is the muscle that we can voluntarily flex (contract) and, through exercise, grow bigger, for example, the biceps muscle in the arm.

There are over 600 different skeletal muscles in the human body. The heaviest is the gluteus maximus, which forms the buttocks, and the smallest is the stapedius (about 1/20 of an inch (1.3 mm) long), which is attached to the smallest bone in the body located deep in the ear.

Although skeletal muscle tires easily and must recuperate after short periods of activity, it is powerful—with reports of people lifting cars to save loved ones in a road accident. The remarkable power of skeletal muscle, however, can be modulated to suit different activities. Your hand, for example, can gently hold an egg without breaking it or grip a heavy textbook with a force of about 70 pounds (32 kg).

Skeletal muscle is made up of long, thin fibers (cells) that have tendons on each end which attach to bones and connect two bones together. As a result, contracting and relaxing skeletal muscles enables us to move our limbs. For example, contracting your biceps enables you to bend your elbow.

Cardiac muscle

Cardiac muscle is found only in the heart. It is responsible for making the heart contract like a pump, which helps move blood around the body. You cannot control your heart beat because the heart beats automatically under the influence of a natural pacemaker. Highly resistant to fatigue and capable of continuous aerobic respiration, the heart can beat continuously throughout the entire span of your life without tiring—beating 100,000 times per day.

HEART ATTACK

The heart, which is made up of cardiac muscle, pumps blood around the body. The heart itself needs a blood supply, so that the cardiac muscle receives the nutrients and oxygen it needs to keep working.

Blood comes into the heart from the coronary arteries (pictured). If the heart is not receiving enough blood from these arteries because they have been blocked by a build-up of fat, the cardiac muscle fibers will die. This can result in heart attack (myocardial infarction).

If the supply of oxygen to the cardiac muscle is cut off temporarily this leads to angina (chest pain) but the cardiac muscle fibers survive.

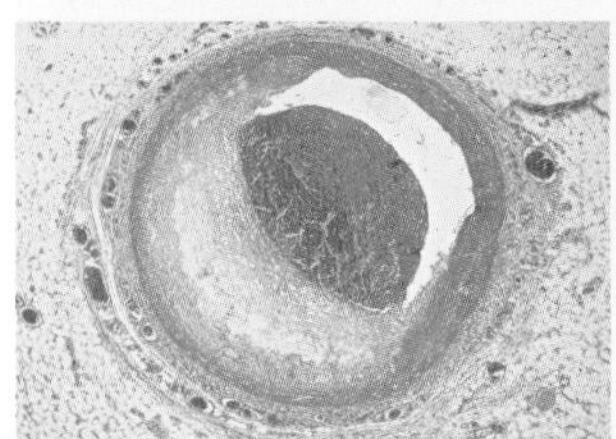

Unlike many tissues in the body, cardiac muscle fibers do not regenerate. So when heart attacks cause these fibers to die, they are replaced by scar tissue which does not beat. This often leads to heart failure or death.

Left: Whereas the skeleton gives the body structure, the muscular system allows it to move. In vertebrates, the muscular system is controlled through the nervous system: Before a skeletal muscle fiber can contract, it has to receive an impulse from a nerve cell. Generally, an artery and at least one vein accompany each nerve that penetrates the epimysium of a skeletal muscle. The skeleton and muscles work together as a lever system, with the joints acting as fulcrums to carry out instructions from the nervous system.

Opposite: Skeletal or "voluntary" muscle (main picture), makes up roughly 40 percent of an adult's body mass. Anchored by tendons to bone, it is striated, consisting of specialized cells (muscle fibers) that are multinucleated, with the nuclei located just beneath the plasma membrane. Each skeletal muscle fiber is a single cylindrical muscle cell. Cardiac muscle (left inset) is striated in structure like skeletal muscle, but "involuntary" like smooth muscle. The latter is found in the walls of organs and bodily structures such as the esophagus (right inset), stomach, intestines, bronchi, uterus, urethra, bladder, blood vessels, and skin (where it controls the erection of body hair).

Right: The biceps is one of the better-known human muscles, and is often well-defined even in non-athletes. Its full name, *musculus biceps brachii*, a Latin phrase meaning "two-headed muscle of the arm," refers to the fact that it consists of two distinct bundles, each with its own origin but with a common insertion point near the elbow.

Smooth muscle

Smooth muscle is found in hollow organs, such as the stomach, the urinary bladder, the respiratory airways, and the reproductive tract. The role of this type of muscle is to force fluids through organs by squeezing them. For example, the smooth muscle found along the walls of the digestive tract contracts and relaxes to move food from the mouth to the anus. Smooth muscle is also important for a mother pushing a newborn baby through her uterus when giving birth.

Unlike skeletal muscle, we cannot control smooth muscle.

> It takes 17 muscles to smile and 43 muscles to frown.
>
> Popular, yet mistaken, saying

Muscle fiber structure and function

Muscles are made up of many long and thin cell fibers bundled together. They are called fibers because they are like threads—being 0.04 to 1.6 inches (1 to 40 mm) long, but around 10 to 100 μm (micrometers) in diameter. For example, the calf muscle is made up of about a million of these thread-like fibers.

Muscle contracts (becomes shorter) when signals travel from the brain via nerves to the muscle fibers. Muscle contraction not only produces movement, such as running, facial expressions, and heart beat, it also helps maintain posture. Skeletal muscle, for example, is always contracting to a certain degree and never fully relaxes. When you are standing up, sitting down, and even when you are asleep, a constant, partial muscle contraction holds the body in position. Were this not so, the body would be limp and flop to the ground.

Muscles are also responsible for stabilizing and strengthening joints, and generating heat, which is crucial for maintaining core body temperature at around 98.6°F (37°C).

Weighing 4½ pounds (2 kg) and making up approximately 3 percent of the body's total weight, the nervous system is the most complex of the body's systems. An intricate, highly organized network of billions of neurons, its structures, including the brain, spinal cord, spinal nerves, and sensory receptors, make up the nervous system.

The nervous system

Dividing the nervous system

The main subdivisions of the nervous system are the central nervous system (CNS), which consists of the brain and spinal cord, and the peripheral nervous system (PNS), which includes all nervous tissue outside the CNS. The PNS can in turn be divided into the sensory (afferent) division and motor (efferent) division. The sensory division consists of nerve fibers that convey impulses to the CNS from sensory receptors located throughout the body. These sensory receptors include those involved in touch, taste, hearing, sight, smell, and all the other senses (there is no firm agreement among neurologists as to their number). The brain processes these sensory signals and decides on an appropriate response. The motor division of the PNS transmits signals from the CNS to effector organs such as muscles and glands. The signal can be for a muscle to contract (for movement) or for particular glands to secrete (e.g. sweat). Some muscle contractions can be voluntary, but the involuntary (autonomic) nervous system is vital for the body to maintain itself automatically. For example, the body self-adjusts imbalances in blood pH, pressure, and temperature.

Right: The human brain and cervical spinal cord. The more advanced the animal, the more convoluted the brain. These convolutions provide a larger surface area while keeping the volume of the brain compact enough to fit inside the skull. The folding allows a greater number of neurons to fit into a smaller volume. The human brain contains about 100 billion neurons, each linked to as many as 10,000 other neurons.

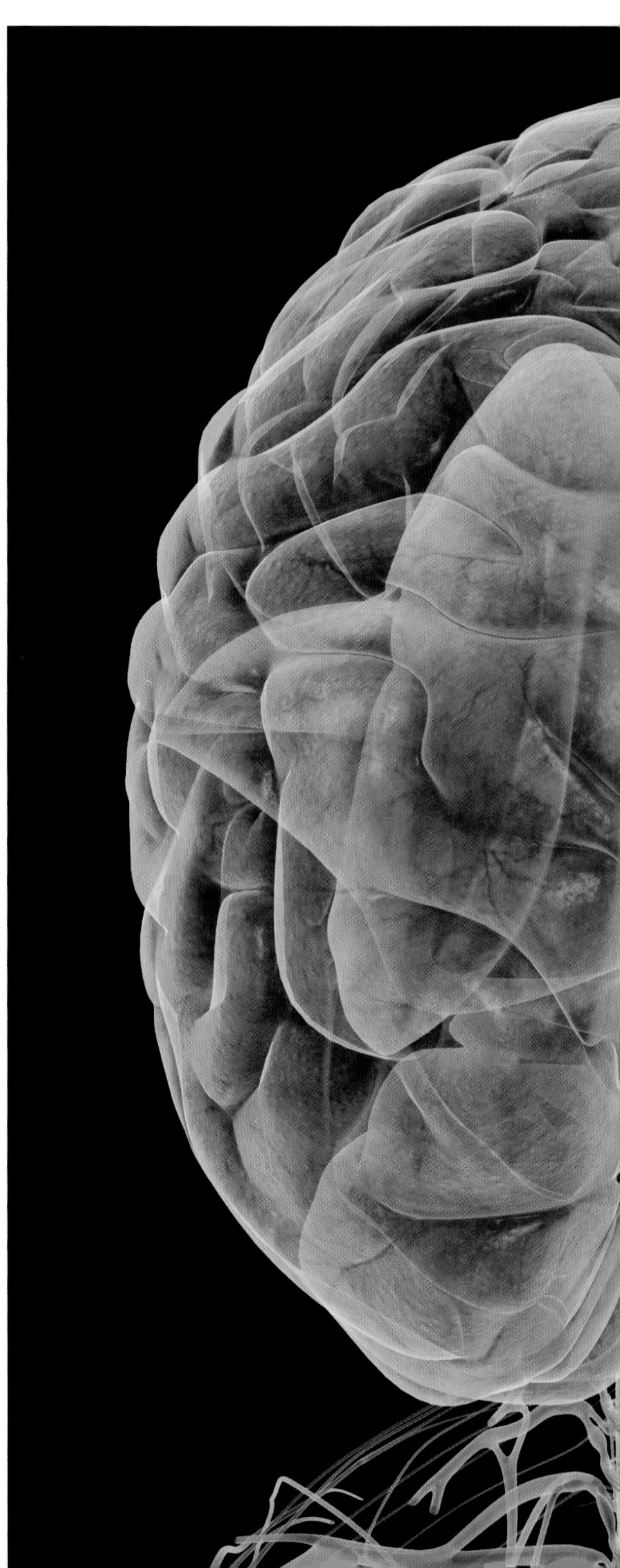

> The brain is a wonderful organ; it starts working the moment you get up in the morning and does not stop until you get into the office.
>
> Robert Frost, poet, 1874–1963

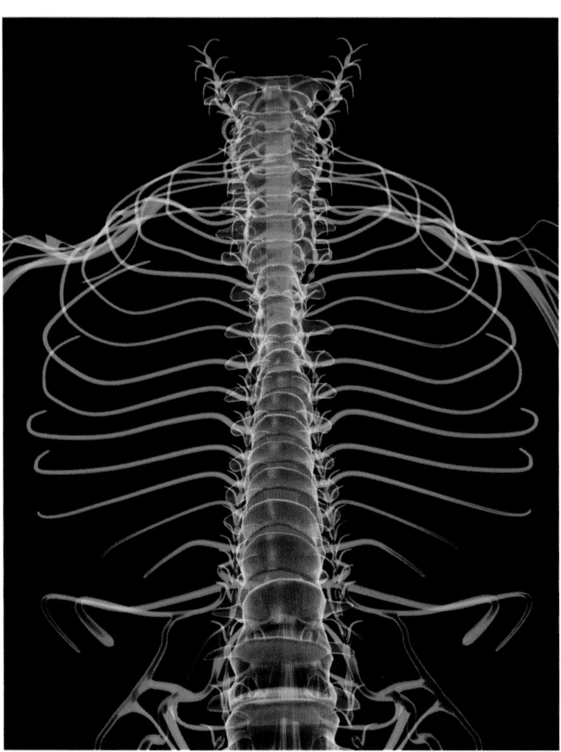

Left: Posterior stylized view of the central nervous system emanating from the spine. The spinal column is not visible within the spinal cavity, but the nerves radiating from the spine are shown. Together with the peripheral nervous system, the central nervous system plays a fundamental role in controlling behavior.

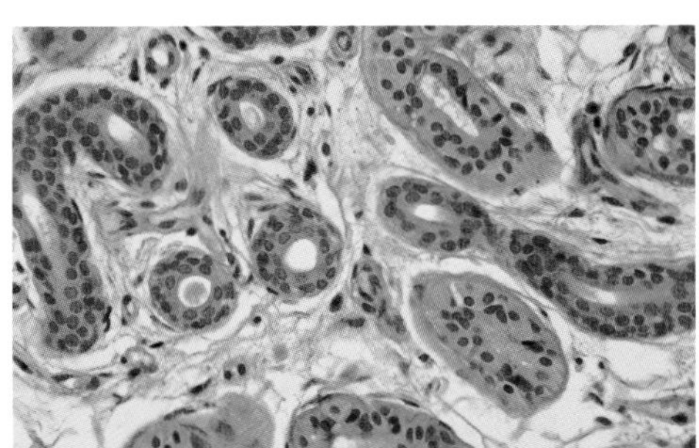

Left: Human skin showing eccrine sweat glands, which are coiled tubular glands extending from the outer to the inner layer of the skin. They play an important role in regulating body temperature, and are distributed over almost the entire surface of the body, but they are most abundant on the palms of the hands, the soles of the feet, and the forehead.

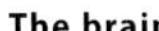

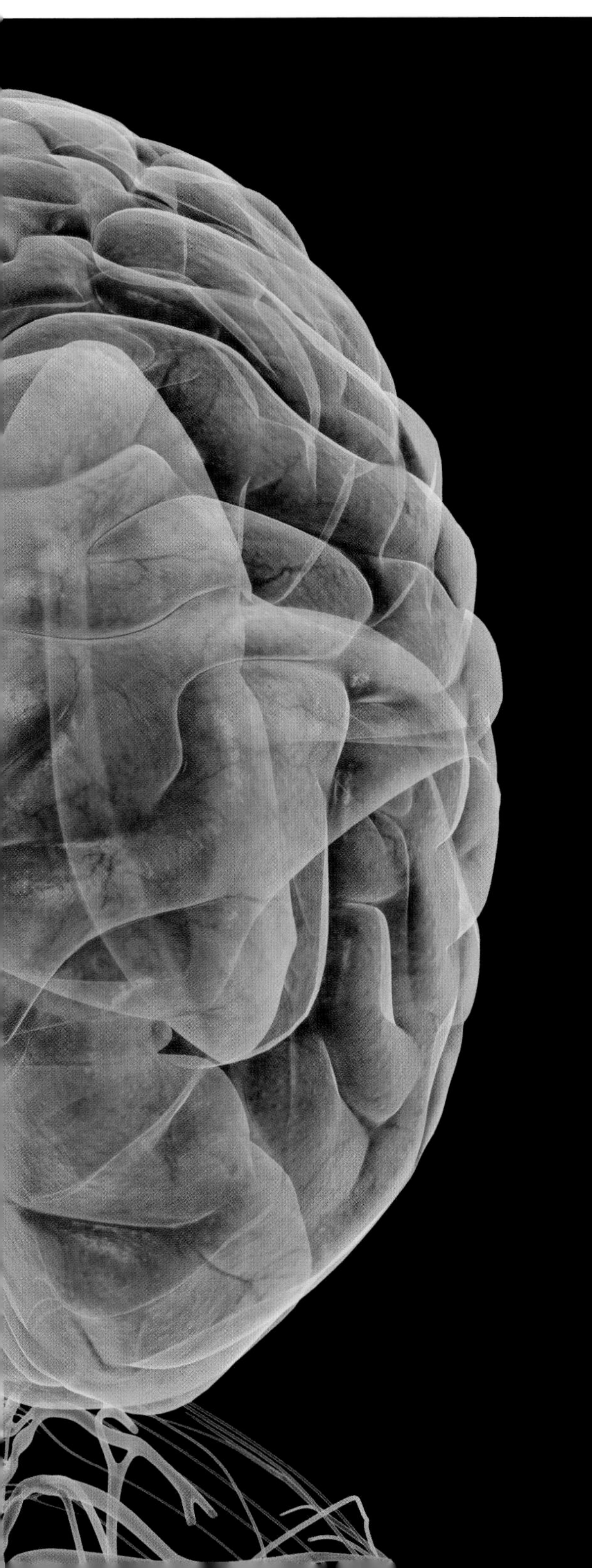

The brain

The brain—"control central"—can be divided into four main parts: cerebrum, cerebellum, diencephalons, and brain stem. The cerebrum, which is the largest division and the most obvious part of the brain, is divided into left and right hemispheres (left and right halves). The right hemisphere controls the left side of the body and vice versa. The cerebrum is involved in thought, memory, vision, hearing, touch, language, and movement. The cerebellum sits under the cerebrum and is involved in balance and fine muscle control. The diencephalon contains structures that maintain body homeostasis (equilibrium) and regulate appetite. Finally, the brain stem connects the rest of the brain to the spinal cord and thus controls the most basic functions of life such as breathing, heart rate, and blood pressure.

The brain is protected inside the skull, and the spinal cord is encased inside the vertebral column. The CNS is bathed in cerebrospinal fluid which cushions it and serves in the exchange of nutrients and waste between the blood and the CNS. The blood does not have direct contact with the CNS. (See also page 290.)

Right: Photomicrograph of nerve cells. Nerve cells have the same basic structure as other body cells, with a surrounding membrane containing the nucleus and cytoplasm, but they have a special, elongated shape. A typical motor nerve, carrying instructions from the brain to the muscle, has a tuft of short, rootlike projections, called dendrites, at one end. At the other end is a long, thin projection called the axon, which may split and divide up to 150 times and be attached to numerous muscle fibers. The function of these cells, which are often grouped together like the strands of a rope, is to carry information from one area of the body to another.

Cells of the nervous system

Nervous tissue consists of two cell types: neurons and neuroglia. Neurons carry electrical impulses (signals), allowing them to be transferred from the outer parts of the body to the brain and back. Some neuron cells are tiny; others are the longest cells in the body. The motor neurons that enable you to wriggle your toes extend from the spinal cord at waist level, to the muscles in your foot. Neuroglia support, nourish, and protect the neurons. Some wrap around neurons like the insulating rubber around electrical wires.

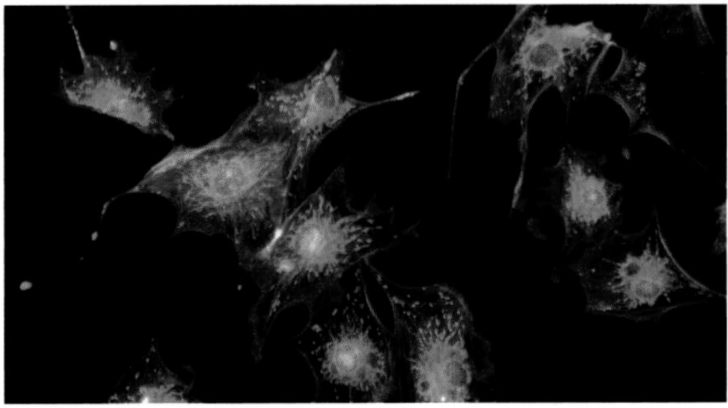

DISORDERS OF THE NERVOUS SYSTEM

The science that deals with the normal functioning and disorders of the nervous system is called neurology. Since the nervous system is complex and fragile, faults can occur during formation in the fetus or with ageing, or the nervous system can be permanently damaged from injuries. Epilepsy is a nervous system disorder affecting about one in 25 people in which uncoordinated torrents of electrical signals are sent within the brain, causing seizures, ranging in severity, to occur. (The photographs below show a brain before and after surgery to attempt to cure epilepsy.) Other disorders include Alzheimer's disease, which is associated with a shortage of a signaling chemical in the brain called acetylcholine, and structural changes in the brain which cause memory loss. If the spinal cord is broken through injury or accident, sensory and motor signals cannot be relayed from body to brain, so the patient is unable to feel any sensation or move the limbs below the break in the spinal cord. If the break occurs in the spinal cord below shoulder level, then paraplegia results and the legs are paralyzed. If the break occurs in the neck, then both the legs and arms are paralyzed—this is quadriplegia.

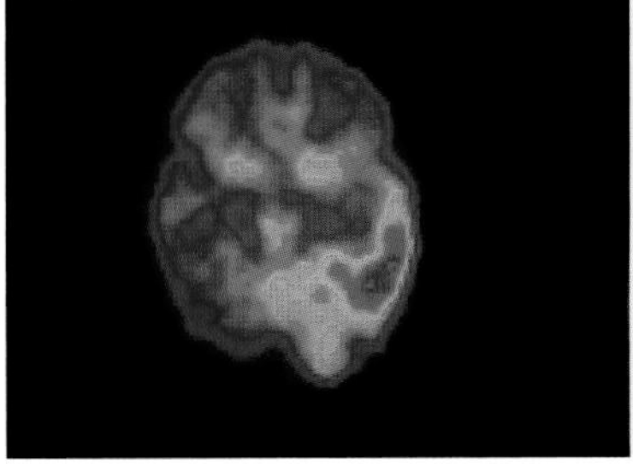

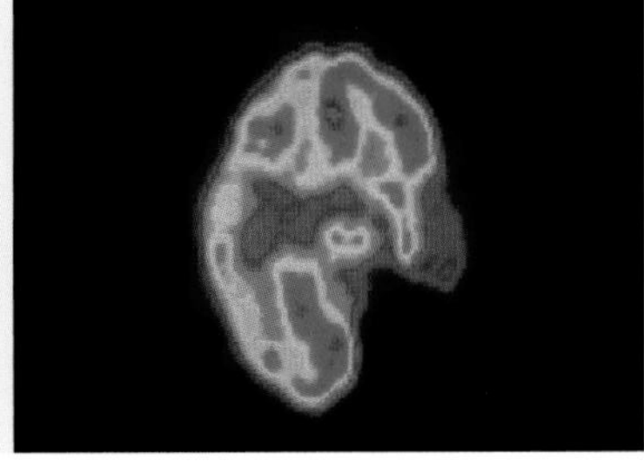

Our survival depends on the oxygen in the air we breathe. Our respiratory system allows us to breathe in the oxygen we require and breathe out the carbon dioxide waste produced by our body. An average adult breathes in 2.3 gallons (10 L) of air per minute while standing still, and this increases to over 12 gallons (50 L) per minute when running.

A breath of fresh air

What is the respiratory system?

The respiratory system is made up of the nose, nasal cavity, pharynx (back of the throat), larynx (voice box), trachea (windpipe), and the lungs. Inside the lungs is a network of air tubes structured similarly to a tree, starting with the larger diameter bronchi, branching off into smaller diameter bronchioles, and terminating with alveoli sacs (which are shaped like bunches of grapes).

If you cut a lung in half, the inside would look similar to a pink sponge. The internal surface area of a spongy adult lung is almost equal to the singles area of a tennis court.

Respiratory function

The major function of the respiratory system is to supply the body with oxygen and remove carbon dioxide from the body. When air is inhaled through the nose, it is filtered by the hairs in the nostrils, humidified, and warmed before entering the lung. In the deepest part of the lung (the alveoli), oxygen diffuses from the air into the blood circulation, while carbon dioxide diffuses in the opposite direction, from the blood circulation into the lung. As we exhale, this carbon dioxide that has diffused from the blood circulation moves through the lungs and is released into the environment.

Within the body, the transport of oxygen and carbon dioxide is facilitated by red blood cells. These cells circulate in the bloodstream around the entire body, providing oxygen to all the body's cells and removing excess carbon dioxide which can be toxic.

Protecting the respiratory system

Thick hairs inside the nose block dust and floating air particles. Particles and microbes are also caught in the mucus lining the nose. This mucus moves to the back of the throat, where it is swallowed (mostly subconsciously), and digested by the stomach.

The body also has a safety mechanism for clearing the airway should it become blocked or irritated. The inner surface of the nose is covered with nerves that sense irritating particles (for example, dust, pollen, and allergens) which trigger the sneeze reflex. During a sneeze, a violent burst of air rushes through the nostrils and expels whatever irritant has landed in the nose.

In much the same way, the throat, trachea, and lung are lined with nerves to sense irritants or bits of food that can initiate the cough reflex. A cough does the same for the lung, trachea, and throat as a sneeze does for the nose: It expels foreign objects.

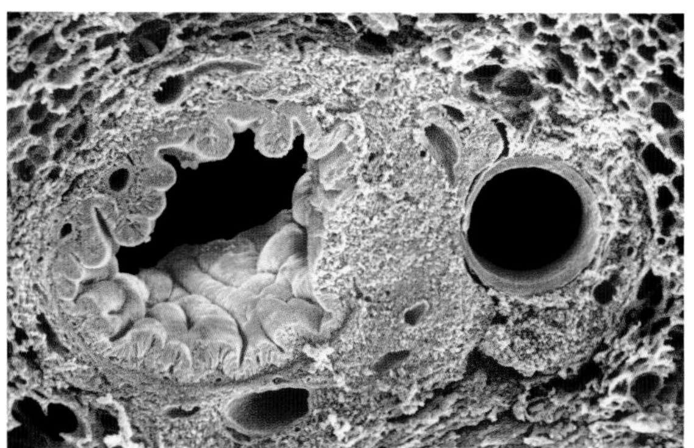

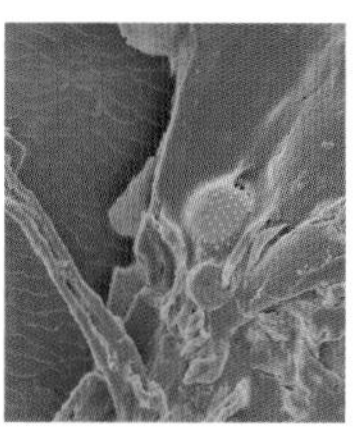

Above: Though sometimes considered unsightly, nose hair does an important job. This scanning electron microscope (SEM) image shows a nose hair with mucus and a trapped grain of pollen. The pollen has been prevented from entering the nasal passages and making its way to the lungs.

Opposite: Male torso showing the respiratory system and rib-cage. Ventilation is primarily controlled by the autonomic nervous system from the medulla oblongata and pons in the brain. Interconnected brain cells within the lower and middle brain stem coordinate respiratory movements. This region of the brain is so sensitive during infancy that the neurons can be destroyed if an infant is dropped or vigorously shaken. Destruction of these neurons can in turn result in the death of the infant.

Left: Lung section showing bronchus (blue) and pulmonary artery (red), plus many alveoli. Blood flows from the right side of the heart through the pulmonary arteries, then back to the left side of the heart through the pulmonary veins. The lungs add and remove many chemicals from the blood as it flows through the pulmonary capillaries. Almost all the body's blood travels through the lungs every minute.

Factors influencing breathing rate

An increase in the blood's carbon dioxide levels, which concurrently causes a drop in blood pH levels, will stimulate an increase in the breathing rate. This "flushes" the excess carbon dioxide out of the body and restores normal pH.

A decreased blood concentration of carbon dioxide causes the opposite effect: It slows the rate of breathing.

Exercise causes an increase in carbon dioxide levels in the blood and an increased demand for oxygen in the cells to create energy, which partly explains why breathing rates increase during exercise.

> There's so much pollution in the air now, that if it weren't for our lungs there'd be no place to put it all.
>
> Robert Orben, magician and comedian, b. 1927

THE MECHANICS OF BREATHING

As you breathe in, the diaphragm (a dome-shaped sheet of muscle below the lungs) contracts and flattens out, drawing downwards. The intercostal muscles (muscles between the ribs) also contract. Both these actions stretch the lungs and decrease the pressure inside the lungs relative to the air pressure in the outside environment. As a result, air gets sucked into the lungs until the air pressure inside and outside the lungs is equal.

As you breathe out, the relaxation of the diaphragm and intercostal muscles, and the natural elasticity of the lungs, compress the inflated lungs. Since the compression causes the air inside the lungs to come under pressure, this forces air out of the lungs and into the environment.

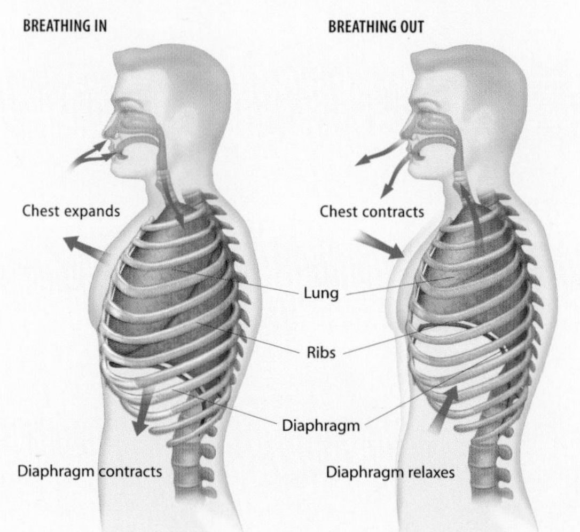

The human lymphatic system is vital for our health. It is crucial to maintaining our cardiovascular and immune systems, and for transporting fats. The lymphatic vessels return fluids that have "escaped" from the circulatory system back into the blood, and they also transport fats from the intestine to the circulatory system. The lymphatic organs house white blood cells essential for immunity.

The lymphatic system

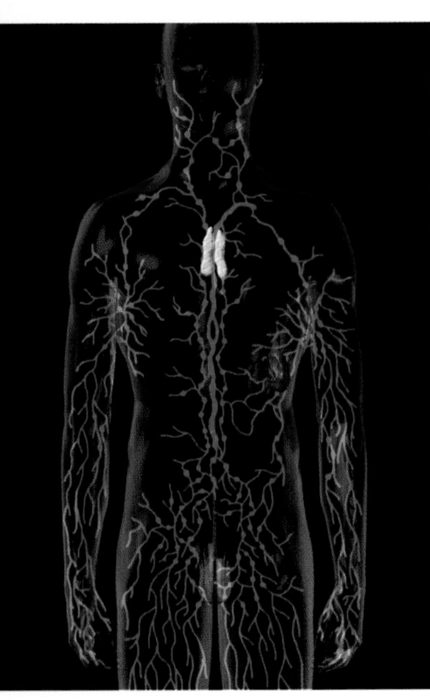

Channeling fluids

The lymphatic system drains excess interstitial fluid from tissue spaces, filters harmful things out of the fluid, and transports it back into the cardiovascular system.

Interstitial fluid is the fluid that surrounds the internal cells of the body. Composed of water, salts, amino acids, proteins, sugars, and fatty acids, among other things, it provides the medium through which nutrients and cellular wastes travel. Since blood is pumped under pressure through the circulatory system and through all the small blood capillaries, plasma is lost to the interstitial fluid by diffusing through the capillary wall or oozing through the cracks. This fluid must eventually be returned to the cardiovascular system, for without it there would not be enough fluid in the blood circulatory system to pump around the body.

The fluid transported in the lymphatic system is referred to as lymph, a clear, pale-yellow fluid, with an appearance and composition similar to plasma—lymph is basically plasma that has escaped from the capillaries. About 0.8 gal (3L) of lymph enters the lymphatic system every day for transportation back into the blood circulation.

Structure of lymphatic vessels

Lymphatic vessels are distributed throughout most of the body. They only allow one-way transport, since their role is to ferry lost fluids back into blood circulation. When the lymph reaches the heart, it becomes part of the blood and circulates around the body again under the pumping action of the heart.

Generally, lymphatic and blood capiliaries lay side by side. But unlike the blood circulatory system, the lymphatic system does not have an organ that pumps lymph through the lymphatic system. Instead, the movement of lymph relies on the "milking" action of adjacent skeletal muscles, pressure changes in the chest cavity during breathing, and pulsations from nearby arteries to slowly push the lymph through the lymphatic system. Vessels containing valves that only open one way prevent lymph backflow—the flow of lymph is always directed toward the heart.

Lymph nodes

Lymph nodes are small bean-shaped organs, less than 1 inch (2.5 cm) long, that are scattered in hundreds throughout most parts of the human body. Their numbers are mostly concentrated, however, near the surface of the armpits, breast, groin, and neck.

Above: Dendritic cells are membranous, spiny antigen-presenting cells which process foreign proteins into a peptide-MHC complex that can be recognized by T-cells. They are present in the interstitium of most organs and are highly concentrated in T cell-rich areas like lymph nodes and the spleen. Here, a dendritic cell communicates with a T Cell.

Top left: The human lymphatic system, shown here in stylized form, is a complex, decentralized network of organs, nodes, ducts, tissues, capillaries and vessels. It was first identified in 1652 by Olaus Rudbeck (1630–1702), a Swedish university dean, natural scientist, and archaeologist, who pointed to it as the likely source of production of white blood cells (leukocytes).

Opposite: Cross-section of a lymph node (main picture), showing its honeycomb-like structure. Lymph nodes are filled with lymphocytes that collect and destroy bacteria and viruses. When the body is fighting an infection, lymphocytes multiply rapidly and produce a swelling of the lymph nodes. Lymphatic tissue (inset), found in the lymph nodes, spleen, tonsils, adenoids and thymus, is similarly rich in lymphocytes and accessory cells such as macrophages.

Before lymph is drained and transported back into the blood circulation, it is passed through several lymph nodes where it is cleansed of pathogens. These nodes contain concentrated numbers of macrophages and lymphocytes that engulf and wage war on invading pathogens. Cleansing is important because if pathogens were allowed to gain entry into the blood circulation through the lymphatic system, they could easily multiply and cause serious illness, as they would then have access to every part of the body.

The spleen

The spleen is an organ in its own right, but it is also the largest single mass of lymphatic tissue in the body. Located behind the stomach, it is about 5 inches (12 cm) long. It is concentrated with white blood cells, which provide the body with immunity.

As blood (not lymph) circulates through the spleen, the spleen performs many functions, including removing microbes, and any old or defective blood cells, and storing blood-clotting platelets. During fetal life, the spleen produces blood cells.

> The lymphatic system is the body's garbage collector.
>
> Harvey Diamond, health author

Transporting fats

The lymphatic system transports fats and fat-soluble vitamins, such as vitamins A, D, E, and K that are absorbed by the gut, back to the cardiovascular system, where these fats and vitamins can then be circulated and used by cells, or stored for future use.

Lymphatic vessels are present in the intestines, where their role is to receive the absorbed fats which form fatty, creamy, white lymph called chyle, also delivered to the blood circulation.

SWOLLEN TISSUE

Damage to lymph vessels or lymph nodes that impairs or blocks tissue fluid drainage and causes interstitial fluid to accumulate in the tissue can result in edema—where the affected body part appears inflated. This condition is referred to as lymphedema.

In tropical and subtropical countries, filariasis—a disease caused by parasitic, microscopic, thread-like worms, most commonly *Wuchereria bancrofti*, that block the lymph vessels—can lead to a chronic and severe form of lymphedema called elephantiasis, where the skin hardens and limbs become grotesquely enlarged (left).

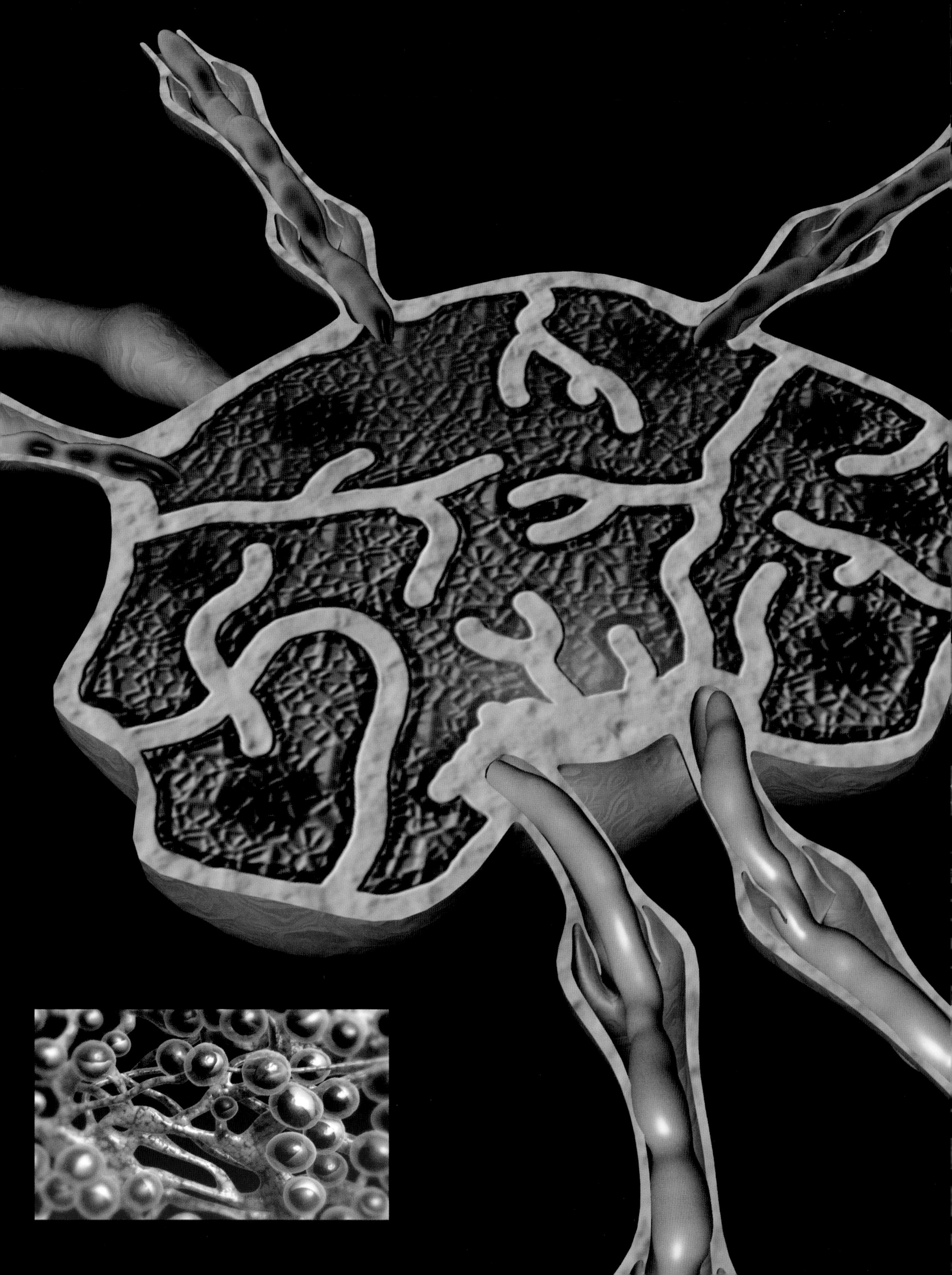

The circulatory system is the body's nutrient and communication highway. It supplies many of the other bodily systems with oxygen from the respiratory system, provides energy and nutrients from the digestive system to all other systems, and carries out a host of other important functions for maintaining a healthy body. Also known as the cardiovascular system, it consists of the heart, blood, and blood vessels.

Round and round—the circulatory system

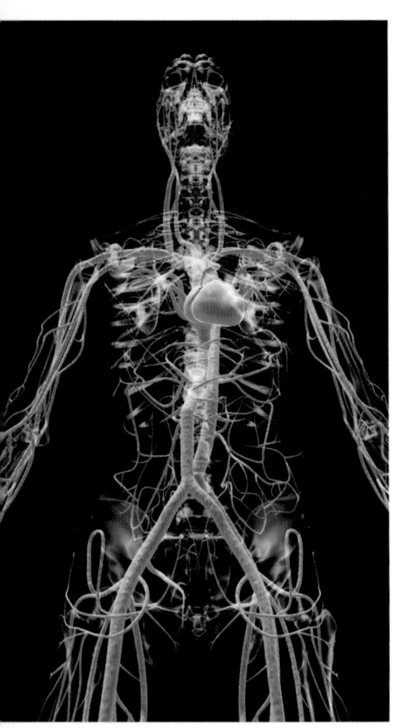

The circulatory system is divided into pulmonary and systemic circuits. The pulmonary circuit supplies blood to the lungs, where oxygen moves from the lungs into the blood and carbon dioxide moves out of the blood into the lungs. The systemic circuit supplies blood and nutrients to all other organs and tissues in the body. Oxygen and nutrients move out of the blood and into the tissues, while carbon dioxide and waste products move out of the tissues and into the blood.

The heart

The heart is a pump, circulating blood around the body. Cardiac muscle, found only in the heart, is responsible for its pumping action. Cardiac muscle contracts on a regular basis due to the electrical activity initiated by specialized pacemaker cells concentrated in certain regions of the myocardium (the heart's muscular wall). This electrical activity is what is recorded by electrocardiograms (ECG).

The heart consists of four chambers: Left atrium, left ventricle, right atrium, and right ventricle. The left atrium receives oxygenated blood from the lungs. It pumps this blood into the left ventricle, which in turn pumps the oxygenated blood to the body. The right atrium receives deoxygenated blood from the body, then pumps it into the right ventricle, which in turn pumps the deoxygenated blood back to the lungs to be oxygenated.

Blood

Blood is made up of plasma, red blood cells, white blood cells, and platelets. Adults have approximately 1.32 gal (5L) of blood circulating in their bodies, and plasma makes up 55 percent of its volume. Plasma

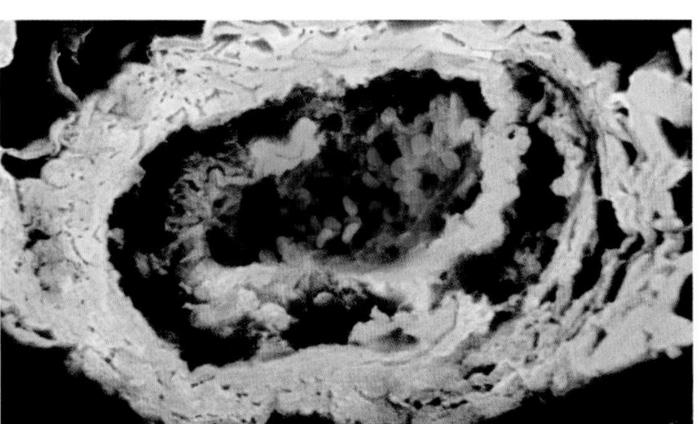

Left: Anterior stylized view of the cardiovascular system, with the skeleton also visible. As a blood distribution network, the cardiovascular system is a closed system (blood never leaves the network), but it works in tandem with the lymphatic system, an open system sometimes seen as part of a broader circulatory system which helps transport nutrients, gases, and wastes to and from cells, fight diseases, and stabilize body temperature and pH to maintain homeostasis.

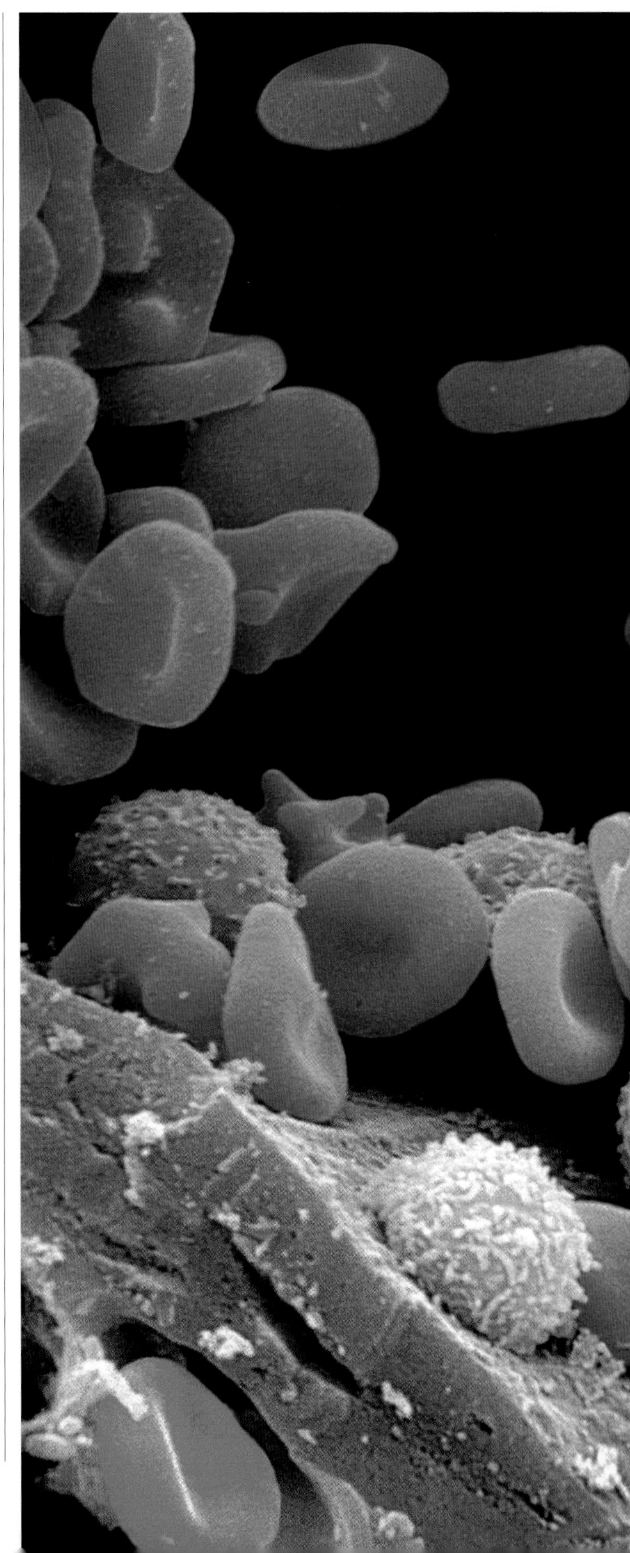

Right: Red and white blood cells jostle one another in the close confines of an arteriole. Red blood cells are filled with the iron-rich protein hemoglobin, which enables them to carry large amounts of oxygen and carbon dioxide around the body. Red blood cells do not have nuclei. Nor do white blood cells, which come in many varieties that play specialized roles in immune system actions. Some kill and eat germs and cancer cells, some produce antibodies, some stick to invaders and signal to other white cells to attack them. Others make the chemicals that cause inflammation at sites of infection. White blood cells can leave the bloodstream to enter lymph nodes or wage war on germs.

Left: Arterioles branch out from arteries to deliver blood to the capillaries. Arterioles are small in diameter and have thin walls, making them a primary site of vascular resistance, which affects blood pressure. This cross section of an arteriole in the calf shows red blood cells in the lumen. Smooth muscle surrounds the vessel making up the arteriole wall. Neural tubes are also present.

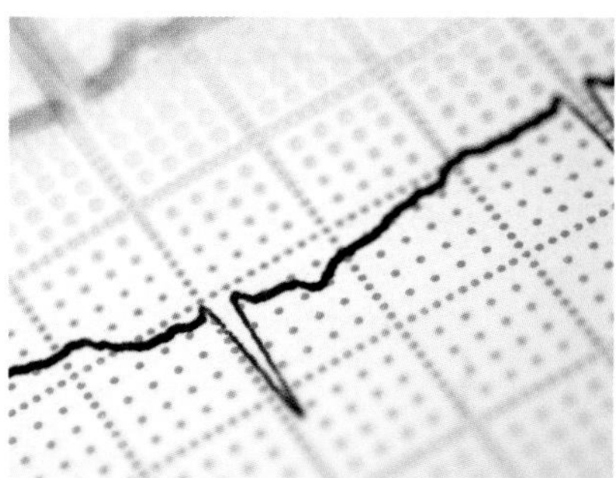

Left: An electrocardiogram (ECG or EKG) charts the electrical activity of the heart over time. Electrodes on different sides of the heart measure the activity of different parts of the cardiac muscle. The voltage between electrodes measures muscle activity, allowing doctors to determine the overall rhythm of the heart, and locate any weaknesses in specific parts.

THE STENT

A stent is designed to temporarily or permanently maintain or increase the diameter of a blood vessel—if, for example, it is being blocked by cholesterol-laden lumps. Stents were first introduced in the 1960s, and come in four traditional designs. Spring-like stents expand when a constraint is removed; thermal memory stents change shape in response to heat; balloon-expandable stents are inflated after implantation; and stents made of biodegradable polymers may also serve as drug delivery devices. Stents are usually designed to have radial and torsional flexibility, biocompatibility, reliable expandability, and visibility on X-rays.

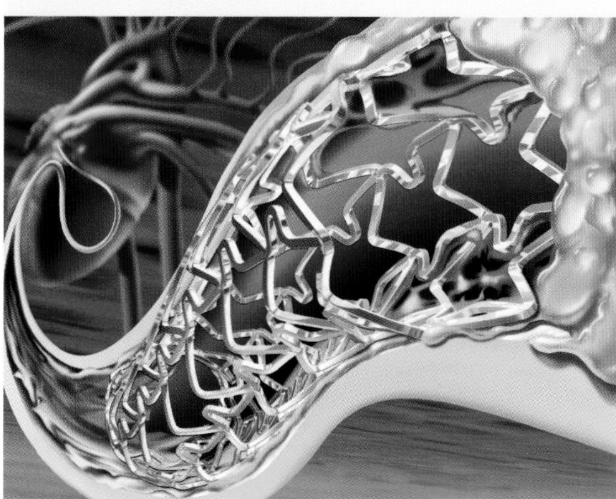

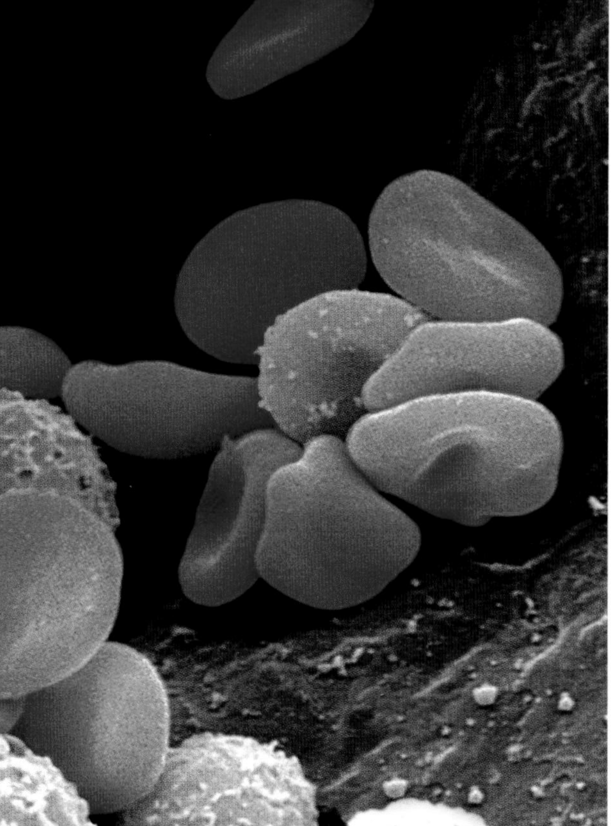

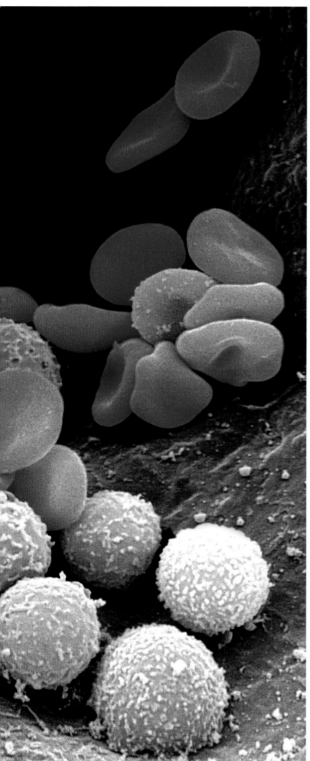

Varicose veins are the result of an improper selection of grandparents.

William Osler, physician, 1849–1919

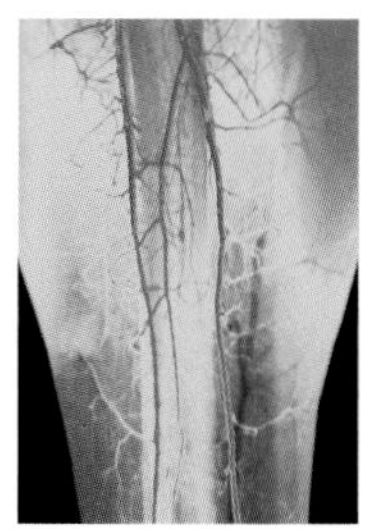

Above: This color-enhanced venogram shows the veins in a patient's leg. Veins have thick outer layers made of collagen wrapped in bands of smooth muscle, and interiors lined with intima (endothelial cells). Their location is much more variable from person to person than that of arteries. A venogram takes X-rays of a person's veins after they have been injected with radioactive dye. The procedure is usually performed to explore possible problems with the flow of blood back to the heart.

is 90 percent water, eight percent proteins, and a mixture of waste products, organic nutrients, electrolytes, and respiratory gases.

White blood cells and platelets make up less than one percent of blood volume. However, these cells play fundamental, necessary roles in defending the body against invading microorganisms and other foreign materials, and in enabling blood clotting.

Red blood cells make up around 45 percent of blood volume. They are the most abundant cells in the blood, numbering roughy 81,935 million per cubic inch (5 million per cubic mm) of blood. They provide a means of transport for oxygen and carbon dioxide, which they carry through the circulatory system.

Blood vessels

Arteries and smaller arterioles carry blood away from the heart to the capillaries. The aorta is the largest artery. Arteries have large diameters and thick walls, which, coupled with elastic tissue within the wall, gives them both stiffness and the ability to expand and contract.

Capillaries receive blood from the arteries and arterioles, and are the smallest and most numerous blood vessels in the body. Ranging from 5 μm to 10 μm (micrometers) in diameter, they have thin walls, allowing the exchange of materials between tissue cells and the blood.

Veins and venules carry blood away from the capillaries toward the heart. Veins have roughly the same diameters as arteries, but have walls about half as thick—though the largest veins, the venae cavae, are larger than the aorta, with diameters of 1.2 inches (30 mm). Unlike other blood vessels, veins are equipped with one-way valves that prevent blood pooling and only allow it to flow towards the heart.

The word "hormone" was coined in 1905 by the English physiologist Ernest Starling and is derived from the Greek word hormone that means "setting in motion." This is appropriate given the function of hormones: A signaling medium within the body that enables cells close by and far away to communicate with each other.

Hormones and the endocrine system

Hormones are secreted by endocrine glands and hormone-secreting organs, such as the hypothalamus, pituitary, thyroid, parathyroid, and adrenal glands, and the pancreas, ovaries, and testes. These glands and organs constitute the endocrine system. The study of endocrine function, anatomy, and disease is called endocrinology.

What are hormones?

Hormones are chemical messengers that travel through the blood and, when received by another cell, cause a metabolic effect. Once hormones start traveling in the blood, they can take anywhere from seconds to days to cause an effect but, once initiated, the effect tends to be long-lasting. Hormones influence reproduction and growth, mobilize the immune system, regulate water, electrolyte, and nutrient balance in the blood, and influence metabolism.

How are hormones grouped?

Hormones can be grouped by their function: Tropic hormones target other endocrine glands to stimulate their growth and secretion; sex hormones target the reproductive organs; anabolic hormones stimulate anabolism (a metabolic process whereby smaller molecules are synthesized into more complex molecules to build up body tissue); and there are other groups as well.

Another method of grouping hormones is based on a hormone's chemical structure. Water-soluble, amino acid-based hormones are the most common class. These hormones are derived from single amino acids, but can also comprise chains of amino acids linked together to form a protein—epinephrine (adrenaline) is a classic example. Another less common class consists of the fat-soluble, steroid hormones (such as testosterone) derived from cholesterol.

EFFECTS OF HORMONES

Typically, hormones circulate in low concentrations because only very small amounts of most hormones are required to achieve their effect. Once a hormone is released into circulation, it travels in the bloodstream around the body until, by chance, it finds a complementary receptor on a cell and binds to it. The recipient cell contains many signaling molecules which convert the hormone's message into a biological action, such as creating more proteins, undergoing cell division, or destroying itself. Each hormone will only bind to its complementary receptor—one that is exactly the right shape and size. A single hormone can impact on multiple areas in the body because its receptors are located in different organs or tissues. For example, thyroid hormones such as thyroxine (pictured), secreted by the thyroid gland, act on almost every cell in the body to increase the rate of metabolism. Sometimes more than one hormone has the same action; for example, GH and glucagon will both cause blood sugar concentration to rise.

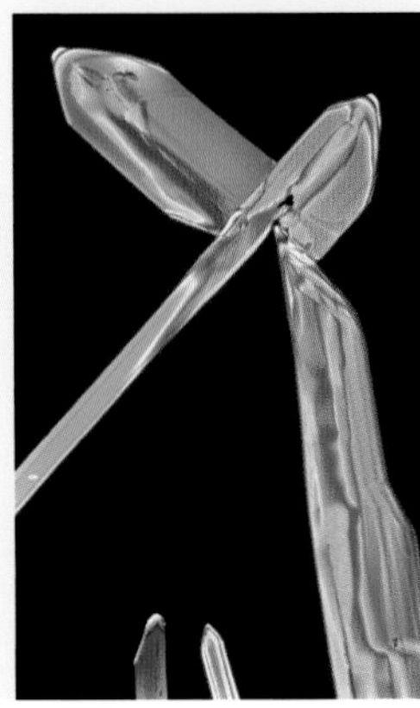

Right: Thyroid gland section. One of the largest endocrine glands in the body, the thyroid gland is found in the neck below the thyroid cartilage (the Adam's apple in men). By producing thyroid hormones, such as thyroxine and tri-iodothyronine, it controls the rate at which the body burns energy and makes proteins. It also controls the body's sensitivity to other hormones and produces the hormone calcitonin, which plays a role in calcium homeostasis.

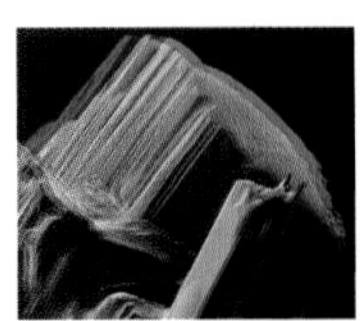

Above: Crystallized adrenaline molecules. Secreted by the adrenal glands, adrenaline is sometimes called the "fight-or-flight" hormone because it prepares the body for action.

Opposite: Progesterone (main picture) is a steroid hormone involved in the female menstrual cycle. It is crucial to pregnancy and embryogenesis in humans and other species. Testosterone (inset), also a steroid hormone, is secreted mainly in the testes of males and in the ovaries of females, though small amounts are also secreted by the adrenal glands. Progesterone levels tend to be relatively low in children and postmenopausal women. Adult males have levels similar to those in women during the follicular phase of the menstrual cycle, but usually have about forty to sixty times more testosterone than an adult female.

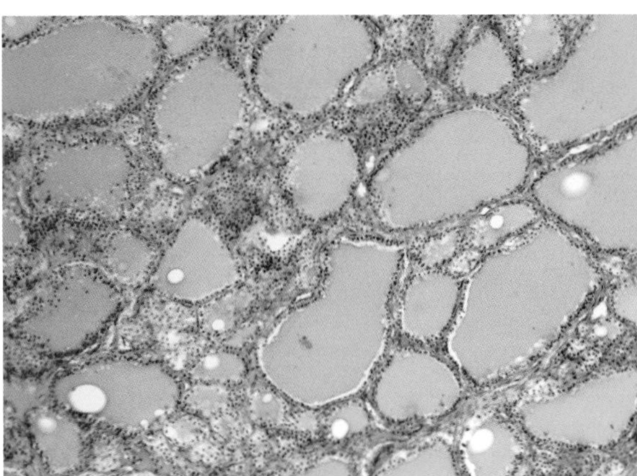

Components of the endocrine system

The endocrine system is coordinated by the hypothalamus, which is located in the brain, and translates electrical neural messages into chemical (hormonal) signals. The hypothalamus releases "releasing" hormones which stimulate the nearby pituitary gland to release "tropic" hormones. These "tropic" hormones then circulate to the primary hormone-producing organs, such as the pancreas, which secrete the hormones that have a metabolic effect on the body.

There are many endocrine tissues in the body, and they secrete a multitude of different hormones. For example, the adrenal gland releases aldosterone to regulate the electrolyte balance in the blood, and epinephrine to modulate the cardiovascular and metabolic response to stress. The pancreas releases insulin, which regulates the blood's sugar level. The ovaries release estrogen and the testes release testosterone to control reproductive function and gender characteristics.

> Everything in excess is opposed to nature.
>
> Hippocrates, physician, *c.* 460–*c.* 380 BCE

Regulating hormones

The release of hormones is regulated by a variety of factors, such as exercise, stress, and the amount of other hormones and nutrients in the body. For example, when the blood levels of growth hormone (GH), blood sugars, or fats are low, or just after exercising, the hypothalamus releases the growth hormone-releasing hormone (GHRH). GHRH targets the nearby pituitary gland. In response, the pituitary gland secretes GH which targets the liver, muscle, bone, and other cells. This induces anabolism and growth, and fats and glucose are mobilized into the bloodstream from storage as an energy source for anabolism. Typically, GH secretion operates on a daily cycle, with the highest levels generally being released at night during sleep. The total daily amount of GH secretion peaks during adolescence, then declines gradually as the body ages.

We all begin fetal development as females. If the fetus's developing gonads secrete androgens (hormones), male genitalia will develop. If not, the fetus will develop as a female.

In the beginning

The male reproductive system

During puberty, the testes begin to produce mature sperm cells called spermatozoa, and the testes and penis develop. The testes are located within the scrotum, which maintains a constant temperature of 96°F (35°C) to protect the spermatozoa. The penis, located in front of the scrotum, expels semen and urine.

During sexual stimulation, blood vessels in the penis become dilated and veins constrict. Consequently, more blood enters than leaves the penis, causing it to distend or swell. If sexual stimulation continues, spermatozoa will move from the epididymes (a bunch of tubules behind the surface of each testis) and through the ductus (vas) deferens, which connect to the ejaculatory ducts. An alkaline fluid is secreted into the ejaculatory ducts by the prostate (a gland at the base of the urinary bladder). Attached to the ejaculatory ducts, behind the prostate, are small glands called seminal vesicles, which secrete alkaline fluids with prostaglandins. These secretions increase spermatozoa

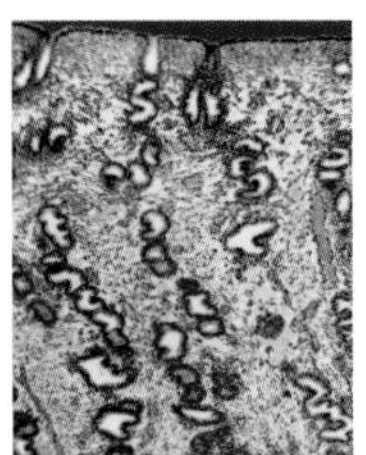

Above: Human endometrium (the inner membrane of the uterus) in the early secretory phase of the menstrual cycle, showing sections of the long, tortuous endometrial glands.

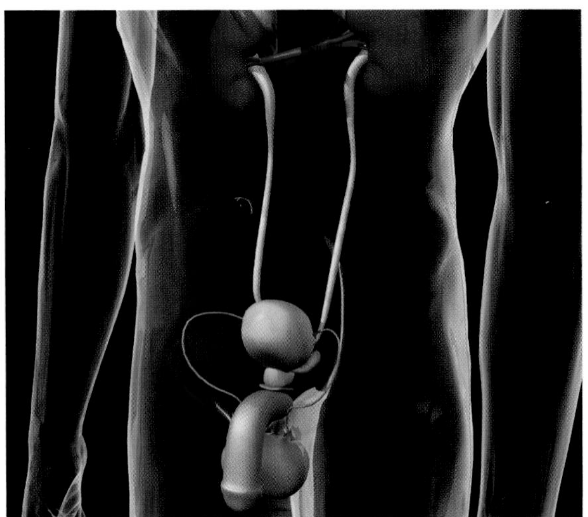

Left: The male genitourinary system. In anatomy, the genito-urinary system includes both the reproductive organs and the urinary system, which have a common embryological origin. The urinary tract is like a plumbing system, with special pipes carrying water and salts. Some parts of the male reproductive system are used for sperm production and storage (testes, scrotum, epididymis), some produce ejaculatory fluid (seminal vesicles, prostate, vas deferens), and some are used for copulation and the transfer of spermatozoa (penis, urethra, vas deferens, Cowper's gland).

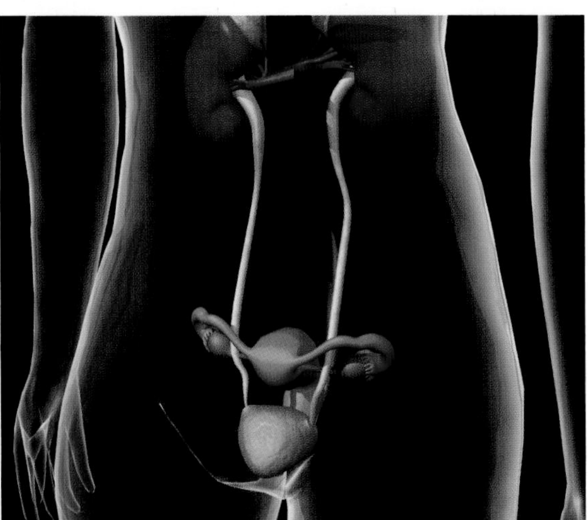

Left: The female genitourinary system. The female reproductive organs comprises the vagina, which acts as a receptacle for sperm; the uterus, which holds the developing fetus; and the ovaries, which produce the female's ova. The uterus is attached to the ovaries via the fallopian tubes. The vagina is attached to the uterus through the cervix and emerges from the body at the vulva, a region of the body that also includes the labia, clitoris, and urethra. During intercourse this area is lubricated by mucus secreted by the Bartholin's glands.

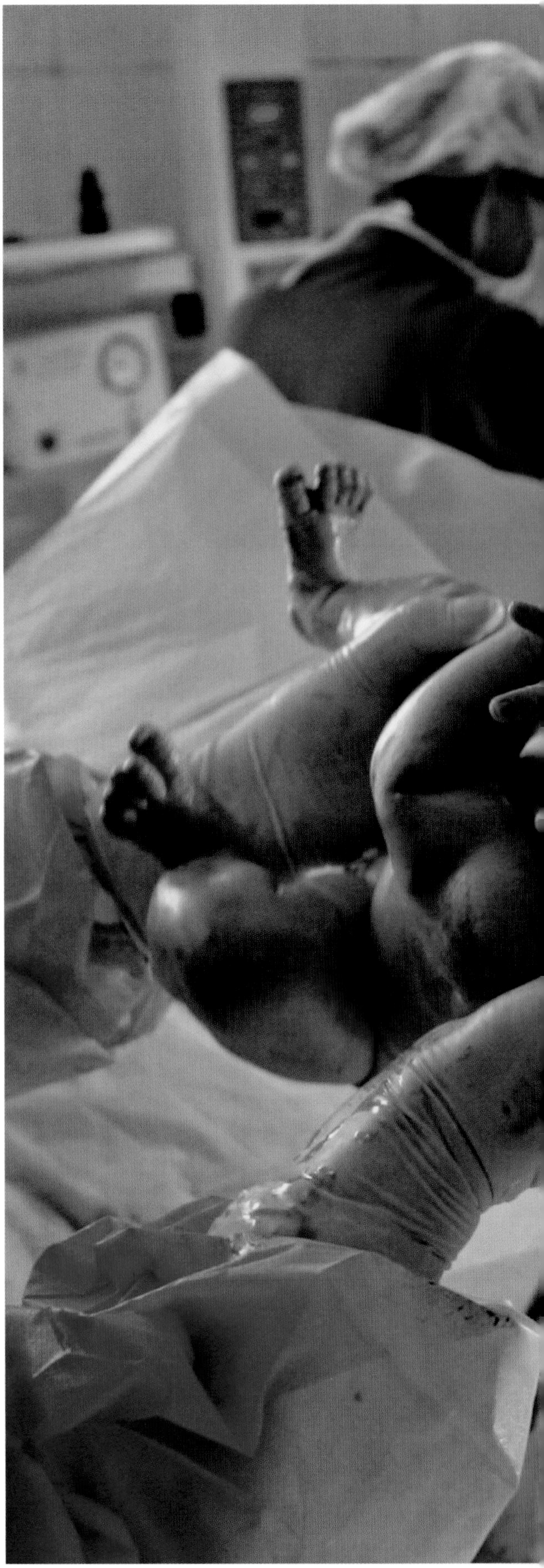

Left: Humans have a life cycle similar to that of other placental mammals, but childbirth is more dangerous than it is for other species. This is due to the relatively large fetal head circumference and the mother's narrow pelvis, a trait required for successful bipedalism. The chances of a successful birth increased significantly during the twentieth century.

motility, and together with spermatozoa they form semen. Finally, ejaculation occurs, when semen travels through the urethra—a tube in the penis that carries semen and urine—and is expelled outside the body. During sexual intercourse, the alkaline fluids, with prostaglandins that are found in the semen, neutralize the acidic pH of the vagina.

The female reproductive system

Unlike male genitalia, female genitalia are mostly internal. The ovaries (a pair of organs analogous to testes in males) are located in the upper pelvic cavity to the right and left of the uterus. They are held in place by ligaments. The outer edge of the ovaries consists of cells called primary follicles. Each ovary contains maturing ova (eggs).

Babies are such a nice way to start people.

Don Herold, humorist, 1889–1966

During puberty, ovarian hormones increase the size of mammary glands and lactiferous ducts in the breasts. Under the influence of a follicle-stimulating hormone, the follicle develops, then enlarges at the surface of the ovary, and ruptures, liberating the ovum (single ova or egg) into the uterine tube connecting to the uterus. This process is ovulation. The newly emptied follicle fills with blood and fat-rich luteal cells, forming the corpus luteum. If pregnancy does not occur, the corpus luteum degenerates, and hormone levels diminish, causing the blood vessels of the uterus to constrict. Because the inner uterine wall (endometrium) lacks a blood supply, the result is endometrial disintegration. Blood released from the uterine wall travels along the vagina—a hollow muscular organ between the rectum and urethra—and is expelled from the body. This is called menstruation. A menstrual bleed occurs approximately every 28 days. The first menstrual bleed in a female is known as menarche.

Two pairs of skin folds sit outside the vagina: the labia majora and minora. They protect the openings of the urethra and vagina. In addition to carrying menstrual flow, the vagina provides a passage for uterine secretions, an erect penis, and semen during sexual intercourse. Beneath the skin of the vaginal opening are vestibular glands, which secrete lubricating fluid into the vaginal opening to facilitate sexual intercourse. The clitoris, found at the upper section of the pudendal cleft, provides pleasure during sexual stimulation.

FERTILIZATION AND BIRTH

If semen enters the vagina, spermatozoa may move into the fallopian tube. If ovulation has occurred within the previous 24 hours, or will occur in the following five days, the spermatozoa may encounter an ovum (egg) in the fallopian tube, leading to fertilization (pictured). The zygote (fertilized egg) replicates itself over the next three days then enters the uterus and remains there free-floating prior to implanting in the endometrium. The corpus luteum, formed at ovulation, is retained, causing hormone secretions that influence the further development of mammary glands and lactiferous ducts in the breasts, and help in the maintenance of the uterine lining, as well as performing other functions that facilitate pregnancy.

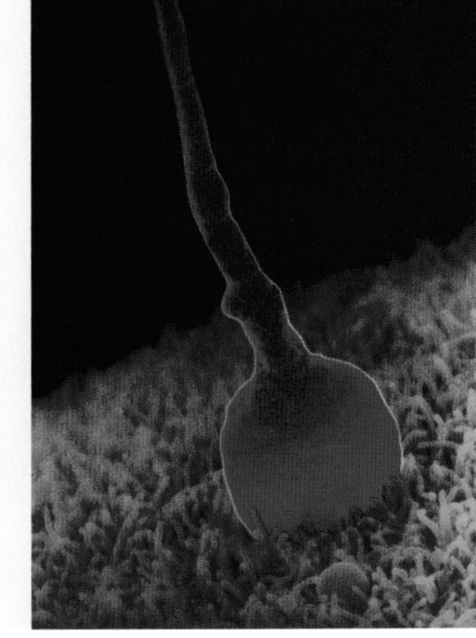

After a gestation period of nine months, labor is brought on by hormonal stimulation. Cervical dilation occurs; the amniotic sac ruptures and amniotic fluid is released; uterine and abdominal contractions begin to move the baby through the birth canal; the baby is born; and subsequently, the placenta is expelled. Following birth, hormone levels readjust to facilitate lactation.

The human body is made up of many kinds of complex organic molecules that enter the body as food. In this form they are too large to be absorbed directly, so they must be reduced by the process of digestion. Digestion results in the production of molecules small enough to pass through the cells lining the alimentary tract and into the bloodstream, to be carried around the body to wherever they are needed.

Making food work—digestion

Digestion involves both mechanical and chemical processes, through which food is reduced to its constituents. Of the major food groups, proteins are converted to amino acids, carbohydrates to simple sugars, and fats (lipids) to fatty acids and glycerol. These conversions occur with the assistance of specific chemicals called enzymes, which speed up the rate of chemical breakdown. Before it can be broken down chemically, however, food must be broken down physically. Mechanical breakdown increases the surface area upon which enzymes and other chemicals can act, thus allowing them to be more effective.

The process of digestion occurs in the alimentary tract. In mammals, the alimentary tract is essentially a long tube divided into regions that perform specific functions. At a broad level these regions are the mouth, esophagus, stomach, small intestine, large intestine, and anus.

Mechanical actions

Digestion begins in the mouth, where food is broken into smaller pieces by the mechanical action of teeth. Each type of tooth has been designed for a specific purpose. The front, chisel-like incisors cut off small portions of food, the canines stab and tear, while the broad and flattened cheek teeth (premolars and molars) are custom-built for crushing. Within the mouth, the tongue rolls the food into a "bolus" and mixes it with saliva before swallowing. Saliva contains a range of substances, including the enzyme salivary amylase which initiates the process of carbohydrate digestion. Thus, the chemical process of digestion also begins in the mouth.

Going down

When the bolus of food is swallowed, it enters the esophagus. During swallowing, it is prevented from entering the trachea (windpipe) by the epiglottis, a cartilaginous flap of tissue. The bolus is moved to the stomach by waves of muscular contraction called peristalsis. Involuntary contractions of smooth muscle are the primary means by which food travels through the different compartments of the alimentary tract. Its passage is smoothed by mucus, which is produced in large quantities by the specialized mucus cells lining the tract's surface.

Food enters the stomach, where it is partly stored before being moved into the small intestine. The stomach is the primary site of protein digestion. Proteins are broken down by the action of gastric juices secreted by the various glands and cells lining the stomach wall. This fluid contains the protein enzyme pepsin along with the large quantities of hydrochloric acid required by pepsin to function. Hydrochloric acid also helps destroy potentially harmful bacteria. With a series of coordinated muscular contractions, the stomach mixes its contents with these gastric juices.

Periodically, food leaves the stomach and enters the small intestine in small quantities known as "chyme." At this stage it is acidic, but chyme is neutralized by secretions from the pancreas and liver when it enters the first portion of the small intestine, the duodenum. The small intestine also comprises two other regions, the jejunum and the ileum. Throughout its passage, chyme is acted on by a combinaction of intestinal secretions and a coordinated series of mixing actions.

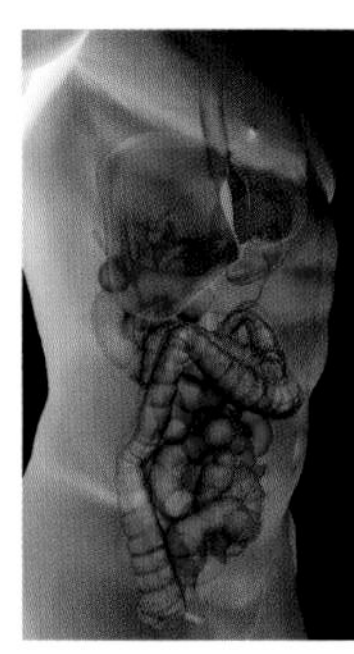

Above: An anterior, stylized view of the gastrointestinal system in a male torso. The stomach has been removed from this image to highlight the other organs. The digestive process comprises four main stages: Ingestion; mechanical and chemical processing; absorption; and egestion or the removal of waste.

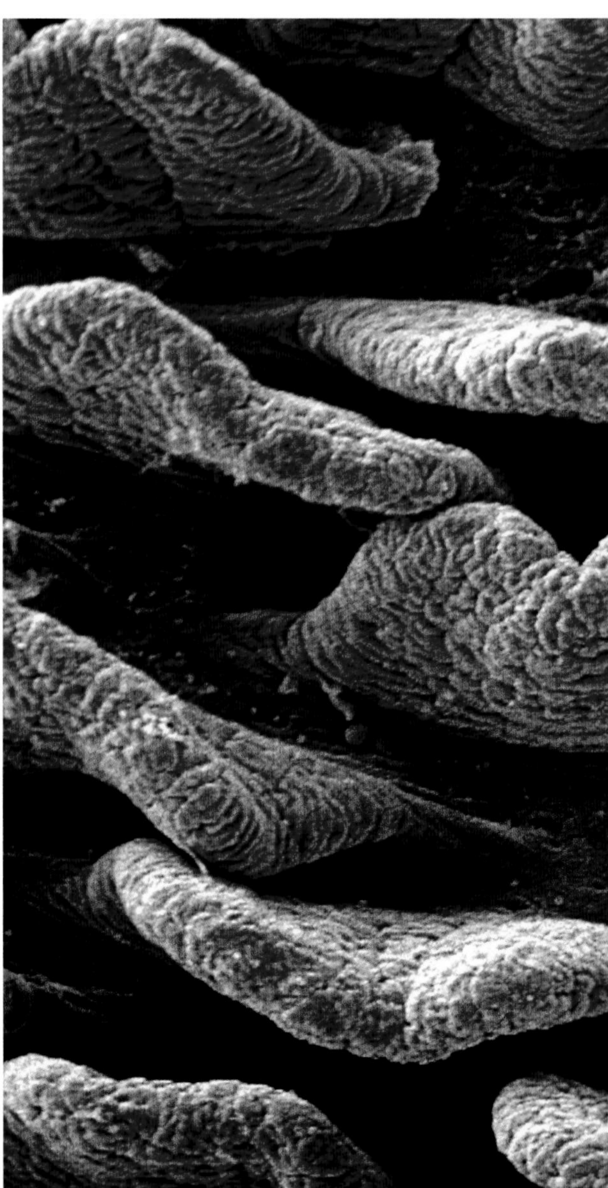

Right: The surface of the small intestine is lined with large, circular folds of epithelium on which villi cluster. Each villus is directly supplied by an extensive network of capillaries. The tongue-shaped villi of the jejunum region allow passive transport of sugar fructose and active transport of amino acids, small peptides, vitamins, and most glucose into the body.

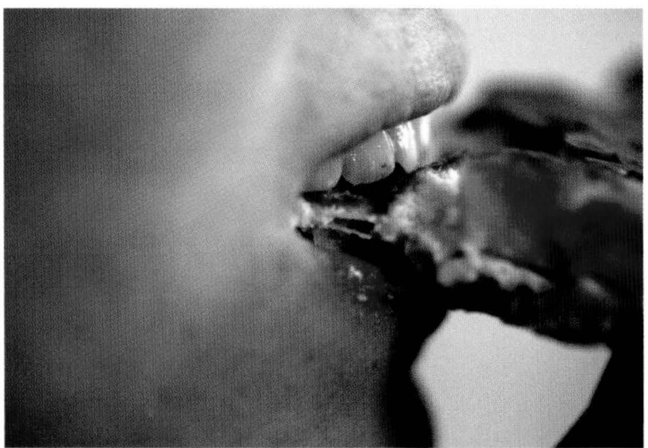

Left: Teeth can be used to tear, scrape, milk or chew food; each kind of tooth has its purpose. The incisors (from the Latin *incidere*, "to cut"), located in the premaxilla and mandible, are adapted to the action of shearing sharply, just as they are in other herbivorous or omnivorous mammals, such as horses. There are normally eight incisors in humans, two of each type: Central and lateral maxillary incisors, and central and lateral mandibular incisors.

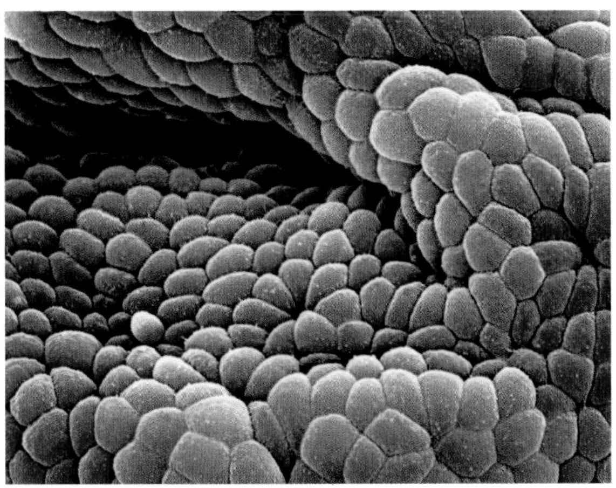

Left: Stomach or gastric epithelium. "Epithelium" is the general name given to tissue that lines the cavities and surfaces of structures throughout the body. In some places, it is as thin as a single layer of cells, but in others it is multi-layered and tougher. Epithelium is the type of tissue many glands are formed from, and it lines the outside of the body (the skin) as well as its inside (cavities and lumen). The outermost layer of the skin comprises dead stratified squamous, keratinized epithelial cells.

THE END OF DIGESTION

The process of digestion utilizes a great volume of water that must be reclaimed. Much of this occurs in the large intestine or colon (below). Once chyme has reached the colon, digestion is nearly complete. The remaining material, including roughage, dead cells from the alimentary tract lining, various salts, and water, then moves through the colon where it is acted on by bacteria. These bacteria produce the vitamins and enzymes required to break down fibrous vegetable matter, converting it to feces, which pass from the body via the anus.

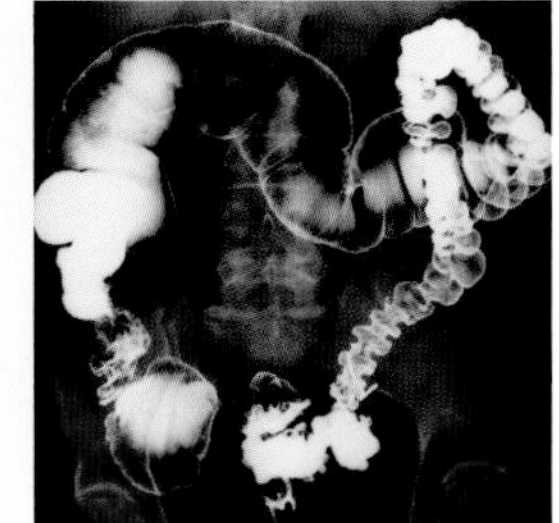

During this process, most proteins, fats, and carbohydrates are broken down into forms that can be absorbed into the bloodstream. In order to maximize absorption, the small intestine is lined by foldings and projections of its surface, termed "villi." Microscopically, each villus is in turn lined by microvilli. In the human small intestine, the combination of foldings, villi, and microvilli increase the total surface area to roughly the size of a tennis court.

Shall I refuse my dinner because I do not fully understand the process of digestion?

Oliver Heaviside, mathematician and physicist, 1850–1925

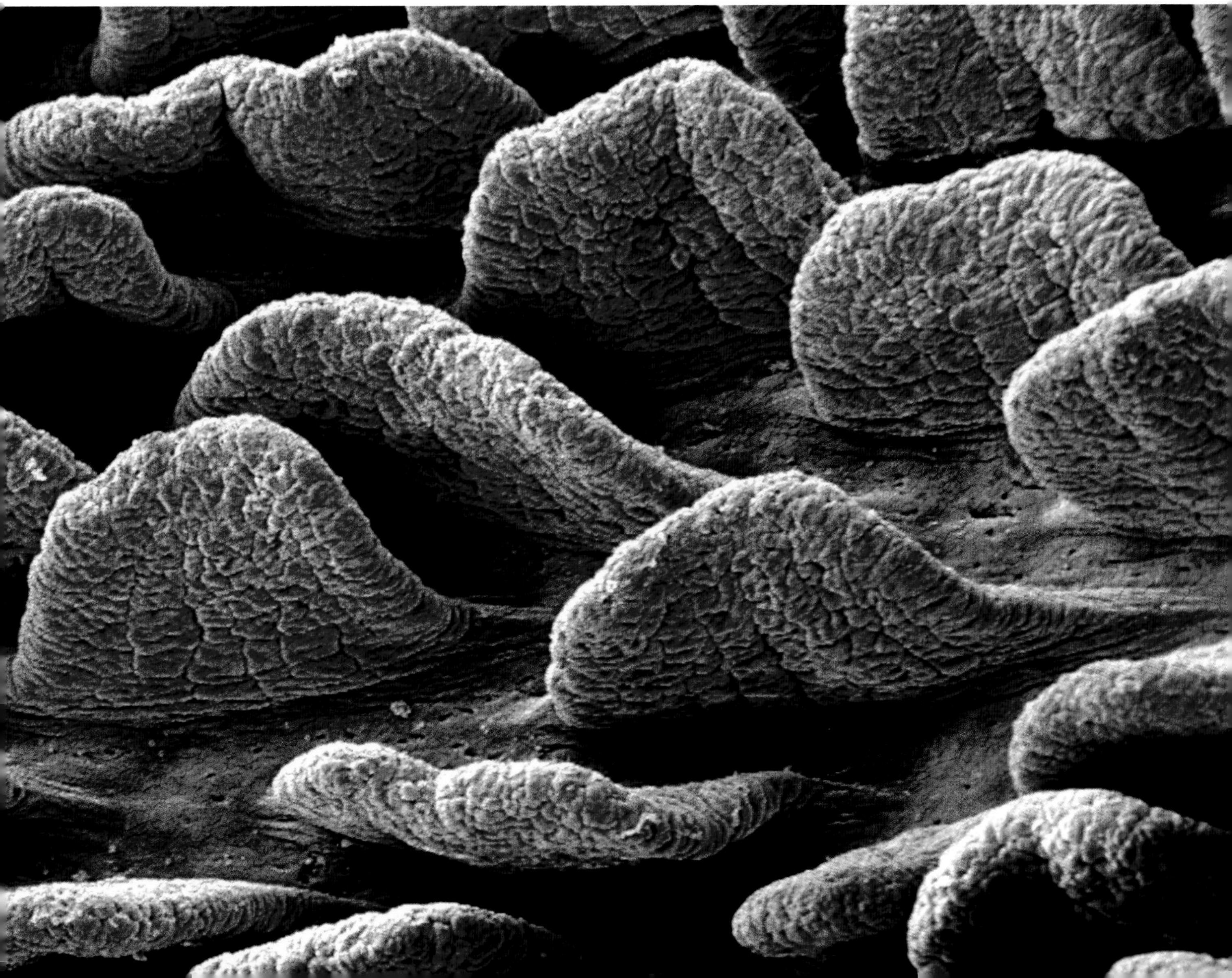

The immune system is a dynamic, microscopic barrier that actively prevents the human body from being overwhelmed by infection. It consists of a sophisticated network of cells, organs, and tissues that cooperate to protect the body. Immune cells are collectively called leukocytes, meaning "white blood cells," and they constantly survey tissues of the body for evidence of infection.

Immunology

Immunology studies the interaction of immune cells and antigens, an antigen being anything the immune system can react against. Antigens include parts of germs such as bacteria and viruses, but they are also present in human cells, where they sometimes become the focus for vigorous immune rejection of transplanted organs, and in otherwise harmless environmental phenomena, like cat hair or pollen, which sometimes cause the immune response known as allergy.

Working against disease

While general leukocytes, known as macrophages and neutrophils, engulf and destroy bacteria, T cells and B cells are more specialized killers, capable of remembering a particular microbe and responding more strongly to it upon subsequent encounter, often resulting in specific lifelong immunity to that disease. The response to the measles virus is an example of "immunological memory," since a person who has been infected once cannot be infected again.

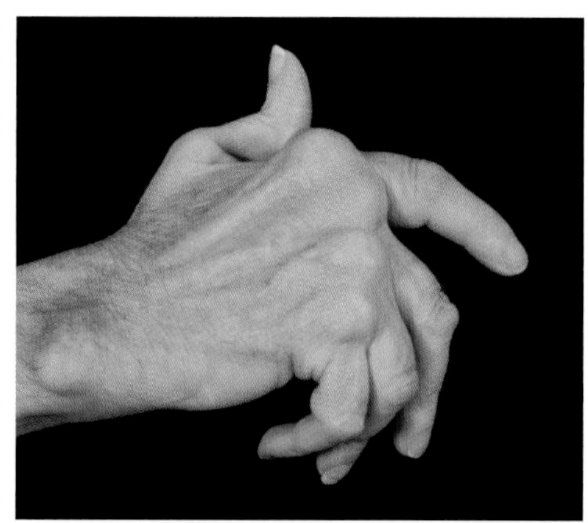

Right: Rheumatoid arthritis, a chronic systemic autoimmune disorder, attacks the joints and sometimes leads to painful and disfiguring inflammation and ossification. Its incidence increases with age. Women are affected three to five times as often as men, and smokers four times as often as non-smokers. Its cause remains unknown.

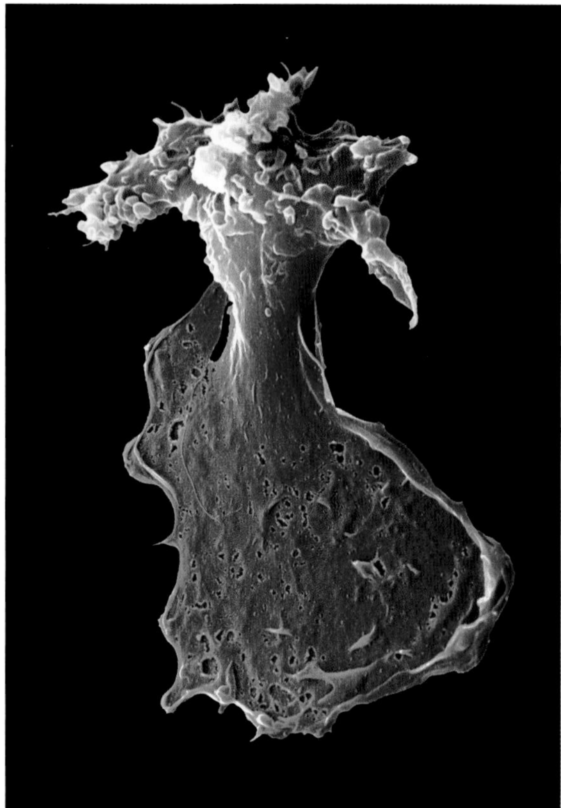

Above: Macrophage with extended pseudopod, or "arm." Macrophages get their name from the Greek phrase for "big eaters." Their role is to engulf, then digest, cellular debris and pathogens, and to stimulate lymphocytes and other immune cells to respond to any pathogens.

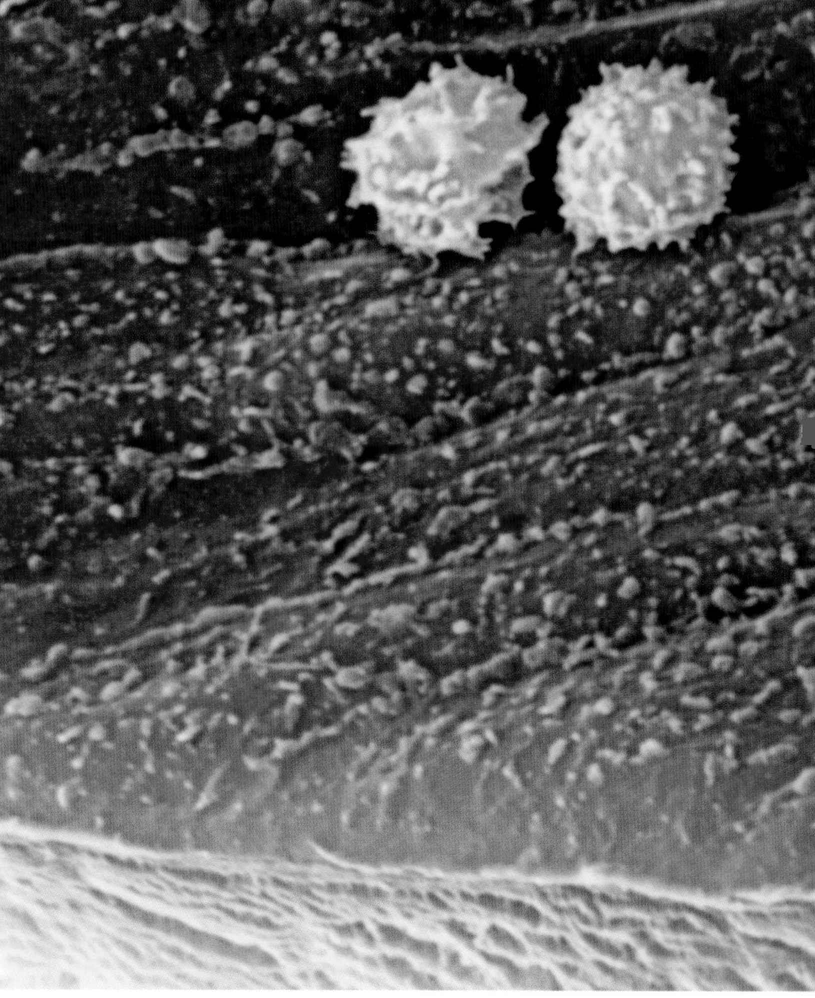

Right: The inside of a vein showing white blood cells, or leukocytes. A normal adult white blood cell count is approximately 4,500–10,800 white blood cells per cubic milliliter of blood. When an intruder such as a pathogen is present, the number of white blood cells increases.

As each T or B cell develops, its genes are randomly recombined and it generates a specific antigen receptor on the cell surface. This process is capable of generating 1011 unique B cell receptors, and 1016 unique T cell receptors, resulting in an enormous repertoire of immune cells, each specific for a different antigen. Thus the immune system maintains an incredibly diverse range of specificities and can react against virtually any disease-causing organism.

The B cell receptor, also known as an antibody, is shaped like a Y, with two sites to bind to antigen and one end anchored to the surface of the B cell. When a B cell receptor recognizes its antigen, the B cell starts manufacturing and secreting a great many antibodies into the blood, each one identical and able to bind to the same antigen. When they encounter the antigen, they bind tightly, blanketing it in a mesh. The antibodies block further infection by not allowing the bacteria or virus to bind to the cells it is trying to infect; they also signal to other leukocytes to destroy the germ.

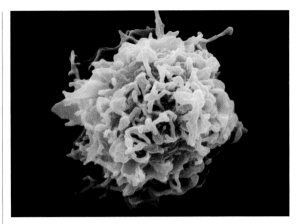

Right: A microscopic view of a T cell. The T is for thymus, the organ in which T cells mature. There are a number of different types of T cell, each with its own role, but they all differ from other lymphocyte types, such as B cells and NK cells, because of the special receptor on their surface that allows them to recognize antigens bound to major histocompatibility complex (MHC) molecules. When the receptor detects antigen and MHC, the T cell activates, triggering an immune response.

T cells can identify which of the body's myriad cells are infected by viruses. When a T cell receptor recognizes antigens, the T cell releases toxic particles to kill infected cells, removing the source of infection. T cells also secrete chemical messengers and express receptors that help B cells to make more antibodies. The components of the immune system thus work together to remove infection.

> The human immune system is remarkable in its capacity to respond to millions of different antigens.
>
> Neal A. Halsey, pediatrician

Self-antigens

The destructive power of T and B cells is enormous. Accidental immune system activation can result in a destructive response against "self-antigen"—a person's own tissues and organs. Juvenile diabetes is caused by immune destruction of the cells in the pancreas that produce insulin; multiple sclerosis is an immune response against particular nerve cells, resulting in progressive loss of body control. Other autoimmune diseases include rheumatoid arthritis (an immune response against antigens in the joints) and lupus (a response against the body's own DNA). Paradoxically, although many researchers aim to boost immunity, those studying autoimmunity and allergy are finding ways to turn off specific immune responses.

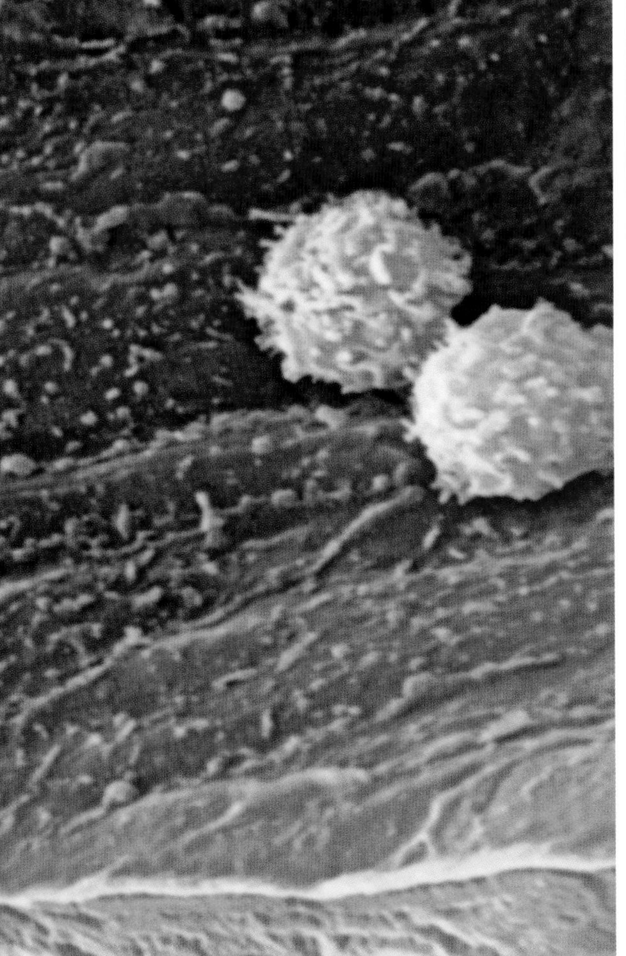

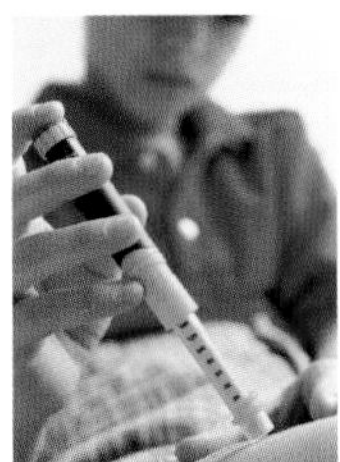

Above: Diabetes is a disorder of carbohydrate metabolism. Due to a lack of the pancreatic hormone insulin, sugars are not oxidized into energy; instead they begin to accumulate, and appear in blood and urine. Symptoms include excessive thirst and excessive production of urine. Since the discovery of insulin in 1922 by Frederick Banting and Charles Best, Type I diabetes has become a much more manageable disease.

THE AMAZING TOLERANCE OF THE HUMAN BODY

The body's ability to tolerate its own tissues while specifically destroying germs, or vigorously rejecting an organ transplant, is astonishing. It is extraordinary that T cells, able to initiate or carry out almost every immune response, are all derived from a single organ—the thymus (section below). For every 100 T cells produced, nearly 95 are destroyed before entering the blood stream, as the thymus screens for self-reactive T cells and deletes them before they can exit, thus inducing "self-tolerance." This process is crucial to the functioning of a healthy immune system.

The two cell types responsible for screening developing T cells are resident thymic epithelial cells and immigrant dendritic cells that develop within the thymus from bone marrow-derived hemopoietic stem cells (HSC). If this process could be harnessed clinically by incorporating donor epithelial cells or dendritic cells into the thymus of an organ transplant recipient, it might be possible to delete developing T cells reactive against donor tissue as well as self tissue, creating an immune system that would not reject cells, tissues, or organs from that particular donor. This possibility is just one of the reasons why immunology is now at the forefront of medical research and clinical delivery.

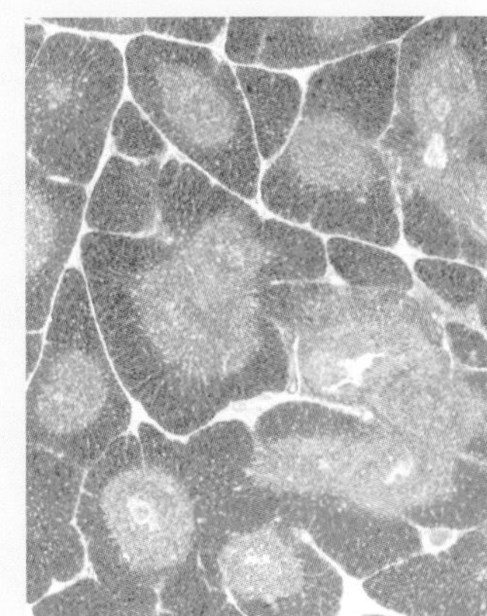

Vaccination is one of the major medical achievements of the twentieth century. It is based on inoculating a person, usually by injection, with part of a bacterium or virus, or a special preparation of a whole bacterium or virus, making the person immune to the disease that the bacterium or virus causes.

Injecting immunity

Thanks to the effectiveness and widespread use of vaccination, smallpox no longer occurs in humans, polio is close to eradication, and the incidence of many other diseases has reduced.

Above: English physician Edward Jenner (1749–1823) discovered vaccination by investigating folklore.

Vaccination breakthrough

During the eighteenth century, smallpox was one of the most prolific killers throughout the world, with an estimated 10 percent of the European population dying from the disease each year. That changed, however, when Edward Jenner, a country doctor from Gloucestershire, England, decided to test his theory that having the disease cowpox would protect people from subsequently contracting smallpox.

Jenner developed this theory when he noticed, having been alerted to the fact by popular folklore, that milkmaids who had experienced cowpox did not contract smallpox. In 1796, in the first ever reported vaccination experiment, Jenner took the pus from a milkmaid's cowpox lesion and injected it into a young boy who had never had cowpox or smallpox. The boy fell ill a few days later but soon recovered. Six weeks later Jenner infected the boy with fluid from a smallpox pustule. The boy did not contract smallpox, either then or several months later when the attempted infection was repeated.

> Vaccination is the medical sacrament corresponding to baptism.
>
> Samuel Butler, novelist, 1835–1902

Although Jenner is attributed with the discovery of vaccination in the eighteenth century, a similar procedure, now called variolation, was being practiced in China around 700 years earlier in 1000 CE.

The Chinese made people immune to smallpox by having them snort powdered scabs from smallpox victims, or by rubbing the same powder into scratches on their skin. By the seventeenth century, the success of variolation saw it become common practice in the Middle East and Europe in the fight against smallpox.

Variolation, however, was not without its risks. There was always the chance of contracting smallpox rather than becoming immune to it, and death occurred in around 1–2 percent of people. Variolated people sometimes also passed smallpox to others.

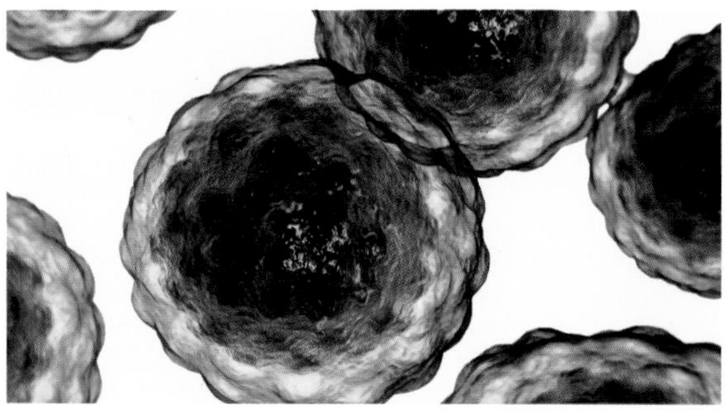

Left: A microscopic view of the polio virus, which existed quietly for millennia as an endemic pathogen. The disease was first noticed In the 1840s; in the 1880s, major epidemics began to occur in Europe, then the United States. The virus, which targets the central nervous system, infecting and destroying motor neurons, was identified by Karl Landsteiner in 1908.

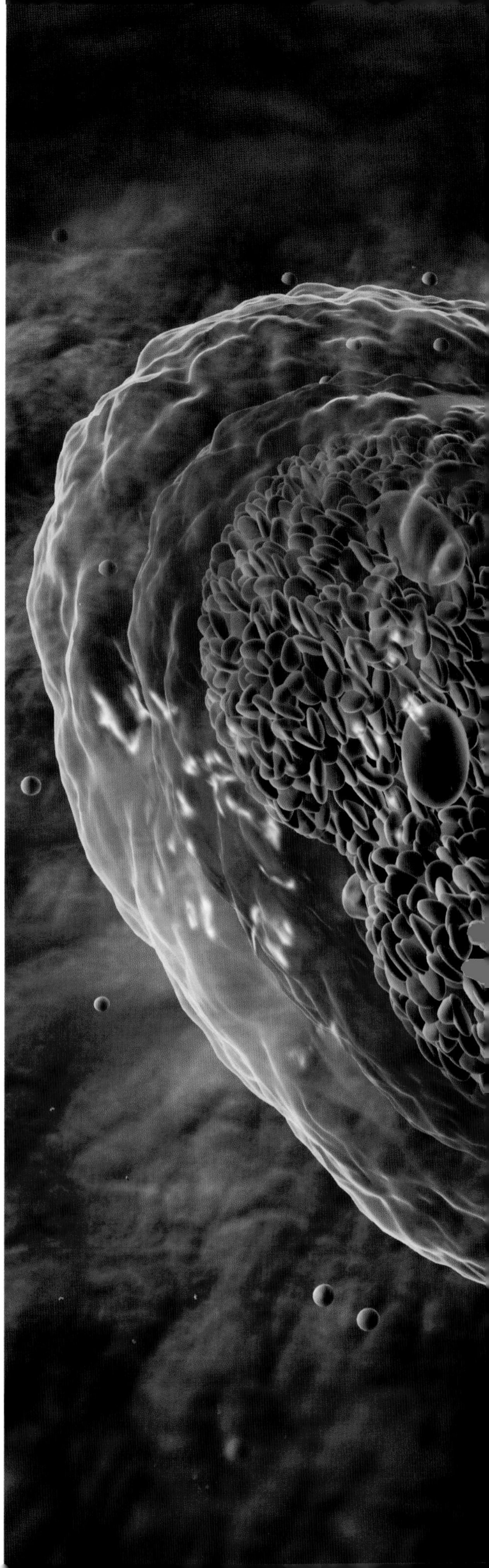

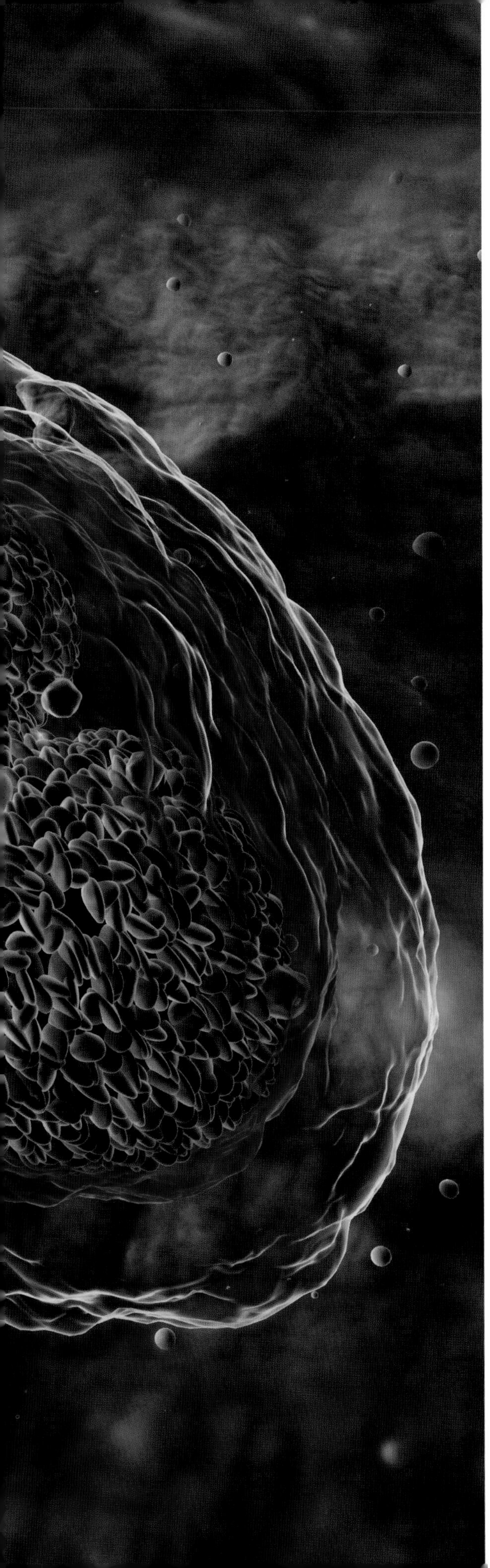

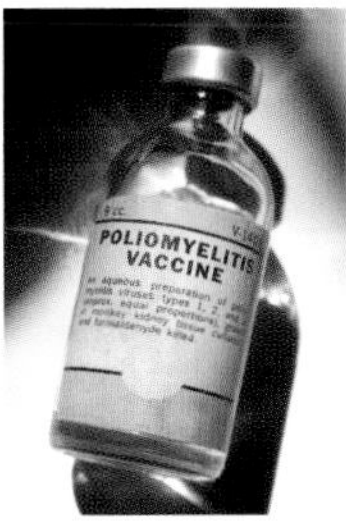

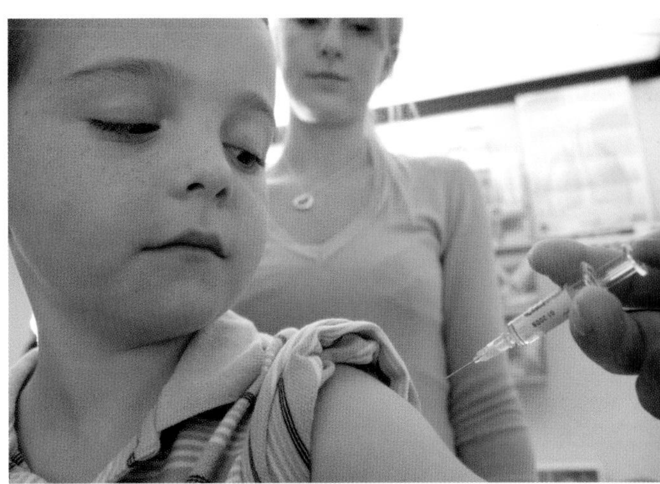

Above: Two vaccines have been used worldwide since the 1950s to dramatically reduce, and in some places eradicate, the incidence of polio. The first, developed by Jonas Salk, uses an injected dose of inactivated (dead) poliovirus. The oral vaccine developed by Albert Sabin uses live attenuated virus.

Above right: Vaccination in childhood can prevent subsequent health problems in both individuals and communities. Vaccination efforts have been controversial since their inception, but acceptance has grown steadily in light of their general success.

Left: The smallpox virus comes in two variants, *Variola major* and *Variola minor*. It is believed to have emerged in humans about 10,000 BCE. Having killed an estimated 400,000 Europeans each year in the eighteenth century and 300–500 million people in the twentieth century, it has now been eradicated by means of vaccination.

Weighing up the benefits

The World Health Organization (WHO) estimates that worldwide more than two million lives per year are saved through vaccination. The main groups to benefit are the very young and the elderly, both of whom are highly susceptible to infectious diseases. People with certain health, working, or living conditions may be at a higher risk of contracting particular diseases and are also targeted for vaccination. For example, health care workers are more likely to be exposed to diseases such as influenza and hepatitis B, and veterinarians and other animal handlers are at higher risk of contacting rabies. Vaccinations are also beneficial for travelers, offering protection from diseases such as yellow fever, typhoid, or Japanese encephalitis.

Even though many millions of lives have been saved by vaccination, concerns remain about the possible risks associated with the procedure. Sometimes mild side effects are experienced, such as pain, swelling, and redness at the injection site, fever, and irritability. Yellow fever, measles, mumps, rubella, and polio vaccines pose special risks for people with weakened immune systems, such as those on immunosuppressive drugs. Immunosuppressed people may contract the disease from these vaccines, and they are not used in pregnant women due to the risks they pose to the fetus. There has also been some concern expressed that the measles vaccine is associated with autism, though most of the published research does not support a causal relationship. While a severe reaction to vaccination rarely occurs, many more permanent side effects and deaths are avoided through vaccination.

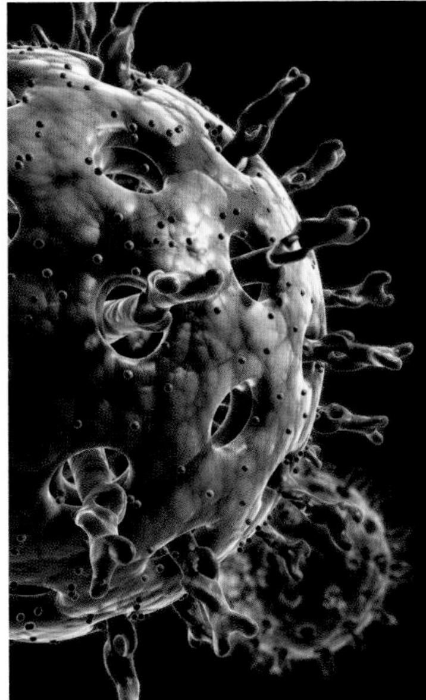

WHERE TO FROM HERE?

New vaccines for serious infections are constantly being sought, but require enormous amounts of time and money for research, development, and clinical trials. Two recently licensed vaccines which are now part of national immunization programs are designed to protect against the rotavirus (pictured) and the human papilloma virus. Rotavirus is the most common cause of severe gastroenteritis in young children and occasionally results in death. A vaccine against the four types of human papilloma virus (HPV) that cause the majority of genital warts and cervical cancer was also added to the immunization programs of many countries in 2007. While the HPV vaccine is highly effective, a regular pap smear for women is still essential since the vaccine does not protect against all cervical cancers. Other vaccines are currently being developed for diseases like malaria and human immunodeficiency virus (HIV) infection.

The success of modern medicine relies on quick and correct diagnosis of disease. Radioactive material in measured doses has been used for many years to identify and treat diseases. Another very important diagnostic advance made last century was the development of magnetic resonance imaging (MRI).

Diagnostic imaging

Revealing radiation

Radioactive material is often regarded as a cause of disease. However, nuclear medicine (a branch of radiology) uses radioactive material when taking images of what's happening inside a person's body, and small amounts of radiation are used to treat diseases such as cancer.

Nuclear medicine imaging reveals more about what is happening inside a person's body than computed tomography (CT scanning) because it shows tissue physiology as well as tissue structure. Nuclear imaging can tell the difference between scar tissue and a tumor, whereas CT scanning cannot.

Positron emission tomography (PET) is the principal nuclear medicine imaging technique used today. In PET, radioactive material administered to a person releases positively charged particles. These particles, positrons, quickly bump into negatively charged particles, electrons, in tissues. When a positron and an electron collide they annihilate each other, producing gamma rays. The PET camera uses sensitive crystals to detect these gamma rays; a computer assembles the signals into a visual representation of positron activity.

IMAGING SOMEONE'S ATOMS

An MRI (magnetic resonance imaging) scanning system has four principal components: An extremely strong magnet, a number of less powerful magnets, coils that transmit radio frequencies, and a powerful computer. The extremely strong magnet is the biggest part of the system.

The most common form of MRI is called hydrogen imaging because it targets the hydrogen atoms common throughout the body due to the high water content of its soft tissues.

The process is called "magnetic" because the large magnet in the MRI machine creates a magnetic field running straight down the person lying inside it. When a person is placed in this magnetic field, the nuclei of hydrogen atoms within their body no longer spin randomly, but in one of two directions: The direction of the magnetic field (parallel), or the opposite direction (anti-parallel). There are always slightly more hydrogen nuclei spinning parallel to the magnetic field than there are spinning anti-parallel.

The "resonance" part of the MRI scanning system utilizes the fact that a slightly greater number of hydrogen nuclei will be spinning parallel to the magnetic field. Special coils in the MRI machine generate radio-frequency pulses that make these nuclei swap their spinning direction. When the pulse stops, the nuclei relax, returning to their original alignment. In doing so, they release energy.

The "imaging" part of the MRI process relates to the detection and representation of this released energy. Data about it is sent to a computer, which transforms the data into a picture. Bones contain little hydrogen, and appear black in an MRI picture. The brain, right, contains an abundance of hydrogen, and appears bright.

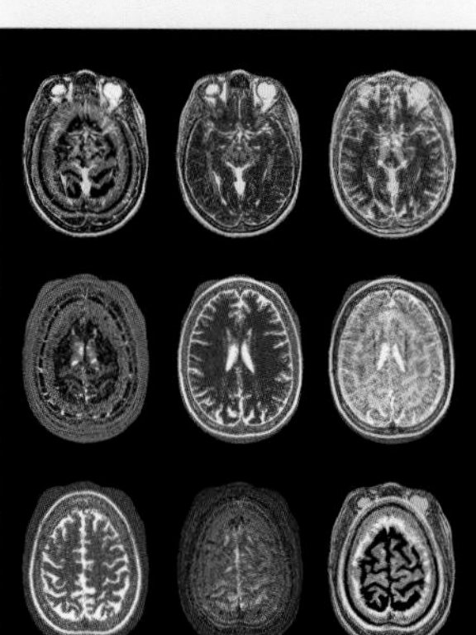

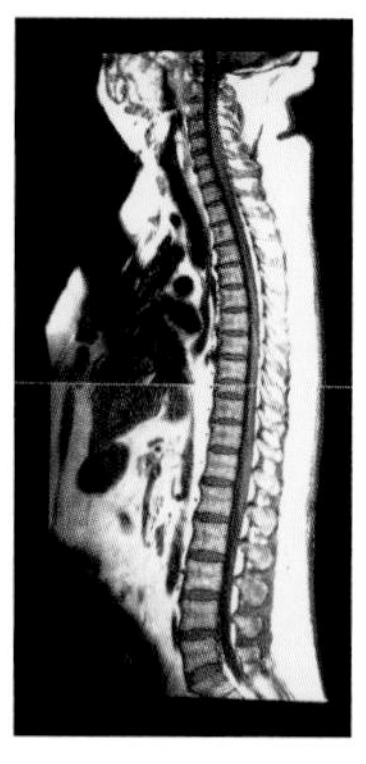

Above: An MRI (magnetic resonance imaging) scan of a spine. The greater soft tissue contrast provided by MRI scans makes them ideal for examining neurological, musculoskeletal, cardiovascular, and oncological structures in the body. MRI is mainly used to investigate neurological conditions and disorders of the muscles and joints, to evaluate tumors, and to reveal abnormalities in the heart and blood vessels.

Right: A PET (positron emission tomography) scanner is a precise diagnostic camera giving a view of the body's biochemical systems. It produces the kind of image seen in the foreground: A full body scan, in this case showing a cancer growth. Unlike CT and MRI scans, which detect bodily structures, PET scanners can detect and map bodily functions. They are increasingly used alongside CT and MRI scans because the combination gives both anatomic and metabolic information about the structures and biochemistry of the body.

Restoring radiation

Ionizing radiation damages DNA, so it seems strange that radiation can help us. However, cells generally don't complete their division process and multiply if their DNA is damaged. The same goes for cancer cells. Many cancers display rapid cell replication, and if their DNA is damaged by ionizing radiation, they have difficulty trying to repair the DNA before they can divide. Exposing a cancer cell to radiation turns the cancer's own rapid cell growth against it, causing its cells to self-destruct rather than replicate.

Unfortunately, radiation therapy also has a destructive effect on normal cells in the human body that naturally divide quickly, such as those found in the hair, stomach, intestines, blood, and skin. For this reason, people undergoing medical treatment with nuclear medicine commonly experience hair loss and nausea.

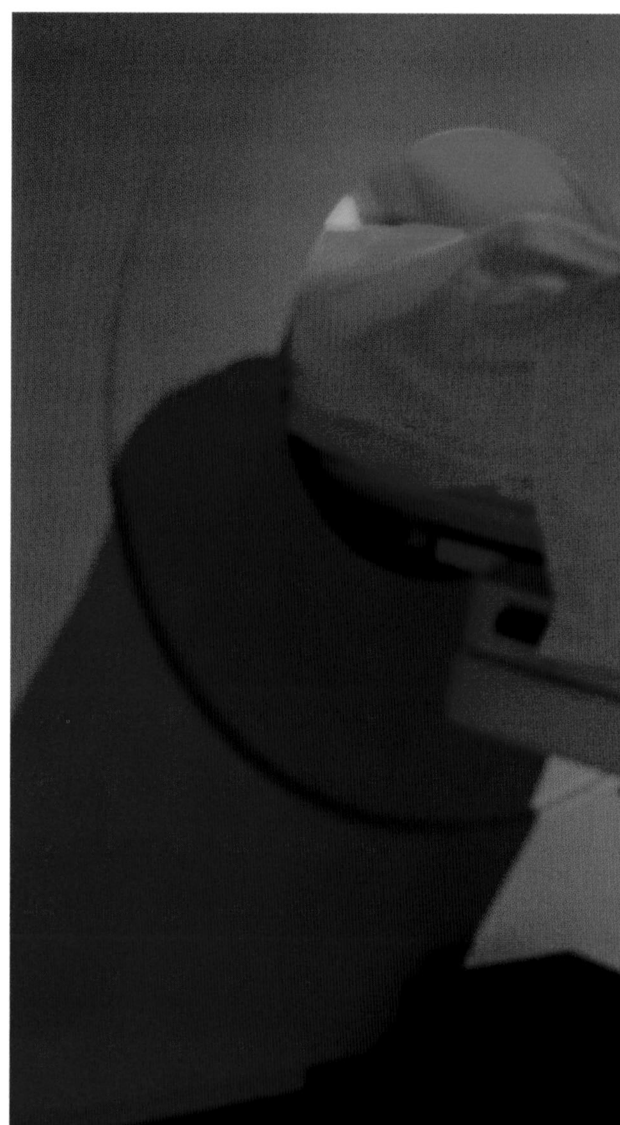

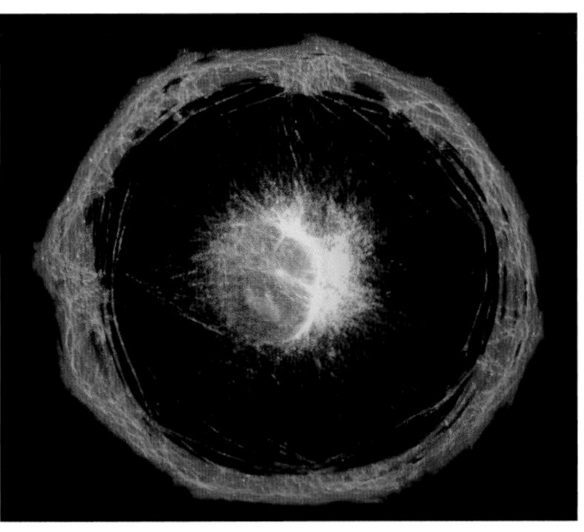

Left: A cancer cell taken from a tumor. This cell is from the fibroblast (connective tissue) cell tumor line. The blue color shows its nucleus; the red and green show actin and vimeutin, both parts of its cytoskeleton.

Right: MRI scanners only track protons from one "slice" of the body at a time, but together these slices give a three-dimensional picture. Contrast agents are sometimes injected to enhance blood vessels, tumors or inflammation.

Every year ... in the United States, seven out of 10 people undergo some type of radiologic procedure.

Charles W. Pickering, judge, b. 1937

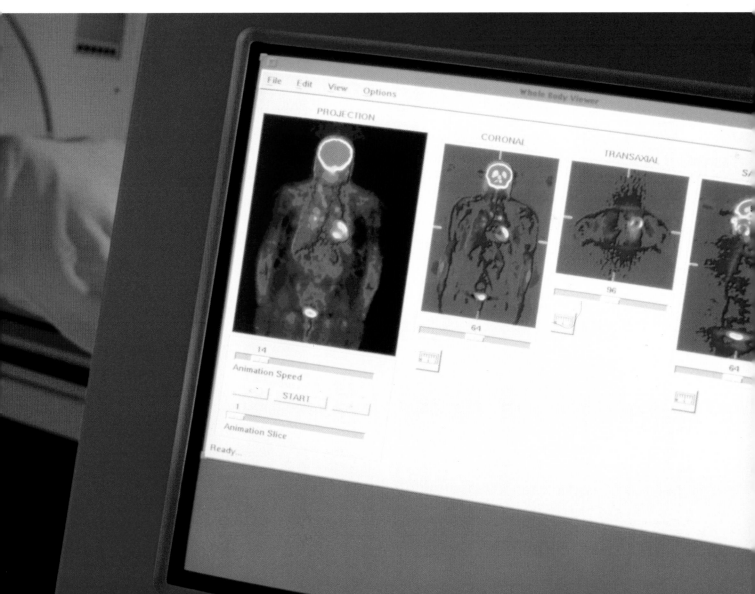

Picking up good vibrations

Magnetic resonance imaging (MRI) uses magnets and radio waves to provide an extremely detailed "look" inside human bodies, without causing side effects and without the use of scalpels. The results of an MRI scan can tell medical professionals whether someone has a problem and, if so, exactly where it is. People commonly undergo MRI scanning when their doctor suspects the presence of injuries, infections, unexplained alterations, or masses/tumors in the brain, spine, joints, abdomen, pelvis, or blood vessels.

An MRI scan can generate images of the structures and functions of the body in any plane, creating photograph-like slices that can be combined to form a diagnostic map of the body.

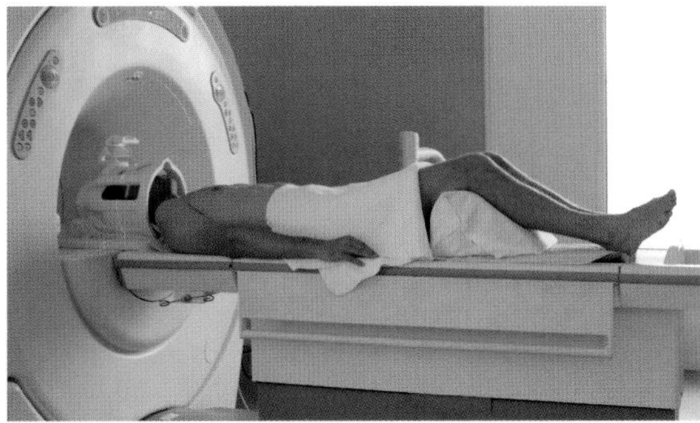

During the course of our normal daily activities, we come into contact with thousands of different types of microorganism, but only some of them (pathogens) cause disease. In order to contain the spread of a disease, it is necessary to identify not only the organism responsible for it, but also its reservoir or source (where it is found), and how it is transmitted.

How infectious diseases are spread

Reservoirs and sources of infection

A reservoir is a habitat in which microorganisms can survive for a long time. Microorganisms that have a permanent reservoir are referred to as "endemic" (always present) in a community. The reservoir may be a favorable environment, where microorganisms can survive or increase in large numbers, or it may be an unfavorable environment where they can survive in a resistant form. The reservoir may be a human, or some other animal, or even a non-living environment such as soil or water. Some pathogens (disease-causing microorganisms)—many kinds of bacteria, for example—can survive in the external environment. For some infectious diseases, humans are the only reservoir. Other pathogens infect animals as well as humans, in which case the animal acts as the reservoir for the pathogen.

The source of a disease is the individual or object from which the infectious agent is acquired. It may be contaminated food or water, or a fomite (an inanimate object that carries microorganisms—usually bacteria—on its surface). In the hospital environment, fomites are significant sources of infection.

The single most important reservoir and source of pathogens is the human body. An individual with a symptomatic infection acts as a reservoir for the pathogen as well as a potential source of infection. Sometimes people who are infected have no symptoms, but are still carriers of a disease (for example, hepatitis B) and can transmit the infection to another person. Humans with latent (asymptomatic) viral infections (for example, herpes) are also infectious.

Transmission of infectious diseases

The microorganisms that cause disease can be spread from person to person in many different ways. The method of transmission depends on the type of organism responsible for the infection (bacterium, virus, protozoan, or parasite). Each of these organisms has its own particular characteristics and requirements for survival, and these determine how it is transferred, how it gains access to a susceptible part of the body, and how it causes disease.

Above: An English plague doctor, *c.* 1656. Plague doctors assessed whether someone had the plague or not. None were qualified physicians as most real doctors had fled the city. If the plague doctor decided you had the plague, your home would be chained shut and a red cross painted on the door. Plague doctors wore protective clothing: A beak mask holding purifying spices, and a wand to avoid touching people.

Right: Scanning electron micrograph of a flea. Flea is the common name for any of the small wingless insects of the order *Siphonaptera*. They are external parasites, living by hematophagy off the blood of mammals and birds.

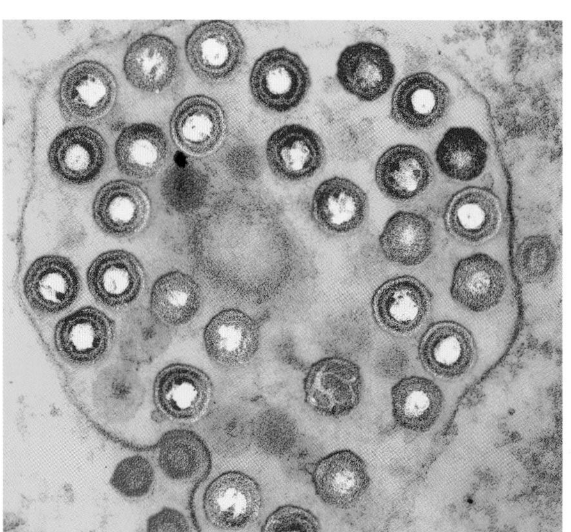

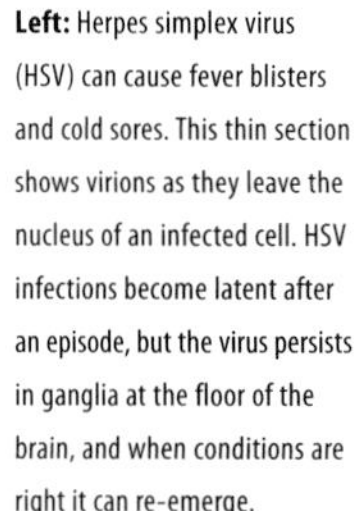

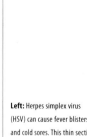

Left: Herpes simplex virus (HSV) can cause fever blisters and cold sores. This thin section shows virions as they leave the nucleus of an infected cell. HSV infections become latent after an episode, but the virus persists in ganglia at the floor of the brain, and when conditions are right it can re-emerge.

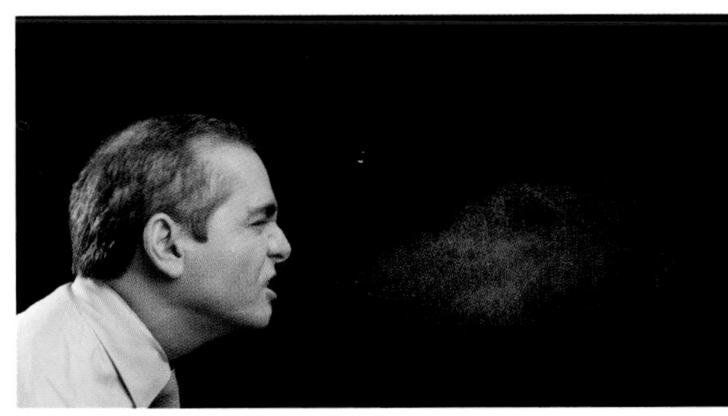

Left: Ancient Egyptian myths say that the first god, Atum, created the world by sneezing, producing air and moisture. Most sneezes, however, are less dramatic, and are harmless in healthy people. Usually caused by foreign particles irritating nasal mucosa, sneezes can also be caused by sudden exposure to bright light, or by a full stomach. They can help spread disease through potentially infectious aerosol droplets, about 40,000 of which can be produced by a single sneeze.

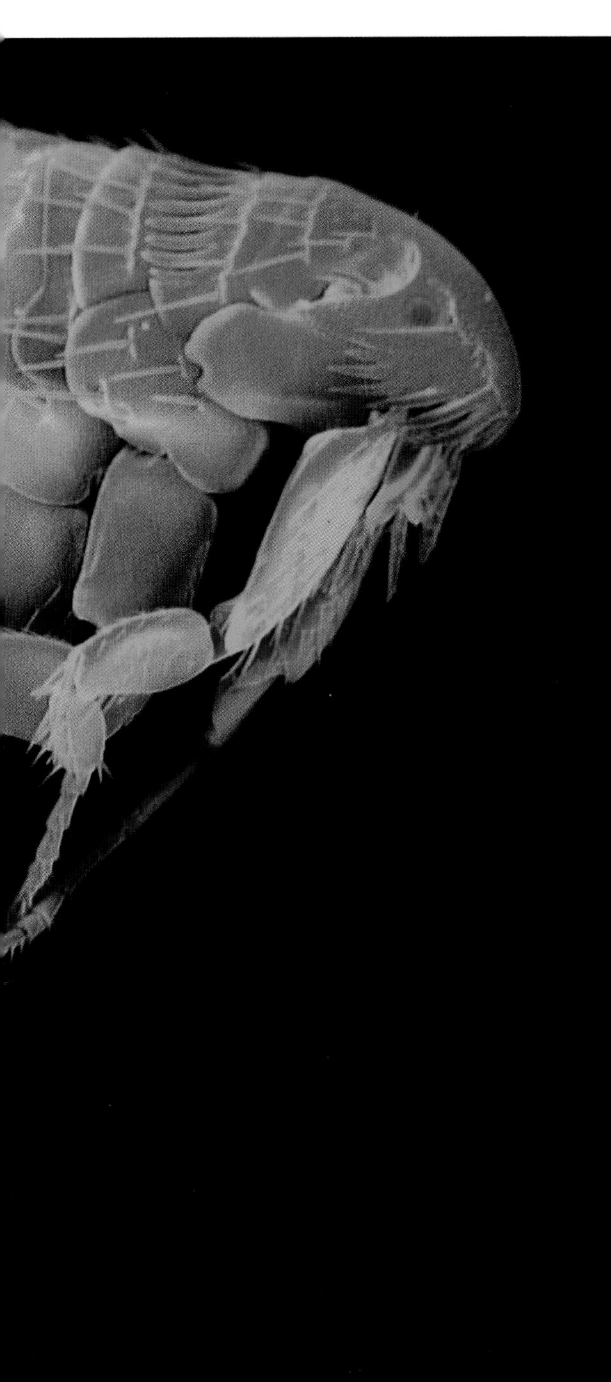

Transmission of microorganisms may occur by direct contact between infected individuals—carried on the hands, in respiratory secretions (colds and flu), in bodily substances (feces, urine, blood), or by intimate sexual contact. Direct contact is the main route of transfer for most viral infections. However, bacteria from the environment may be transmitted via indirect contact from contaminated surfaces, food, or water. Also, several important diseases, including malaria, are transmitted by insects that bite, such as mosquitoes, lice, and fleas.

Transmission in health care settings

Preventing infection is especially important in hospitals and other medical facilities, where there are large numbers of susceptible people gathered, together many with underlying illnesses or surgical wounds that could allow pathogens to enter the body. Here, hygiene is fundamental and there are strict sterilization and disinfection procedures to prevent transmission. Hand washing is particularly critical.

> HIV does not make people dangerous to know, so you can shake their hands or give them a hug. Heaven knows they need it.
>
> Diana, Princess of Wales, 1961–1997

Right: Kaposi's sarcoma, a tumor caused by human herpesvirus 8 (HHV8). Originally described by Moritz Kaposi, a Hungarian dermatologist, in 1872, it became known as an AIDS-defining illness in the early 1980s. Tumors are typically found on the skin, but can spread elsewhere, especially the mouth, gastrointestinal tract, and respiratory tract. Treatment of its underlying cause (immune system dysfunction) with antiretroviral therapy can slow or stop the progression of a tumor.

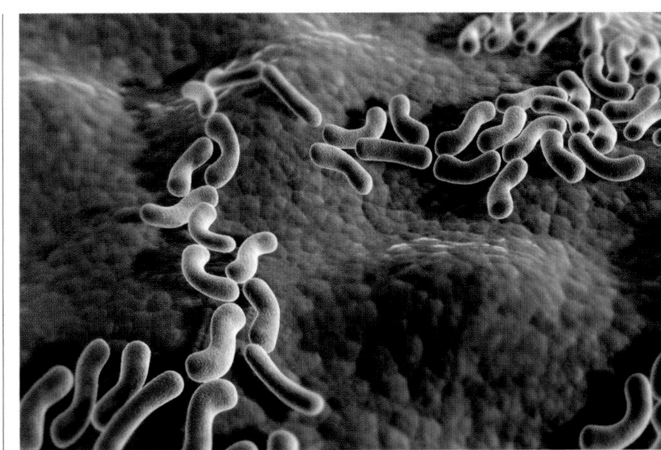

Below: Since the advent of the Industrial Revolution (*c.*1750–1850) and the discovery of germ theory in the second half of the nineteenth century, hygiene and sanitation have been at the forefront of the struggle against illness and disease. Puerperal fever was eradicated with cleanliness, and surgical mortality was reduced to acceptable levels.

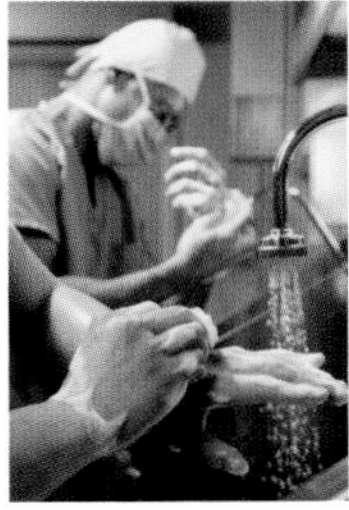

HIV/AIDS

In 1981 doctors in Los Angeles reported the occurrence of an unusual type of pneumonia in five young men. Three of the patients died. About the same time, in New York and California, patients were presenting with a rare cancer called Kaposi's sarcoma, pictured. This was the beginning of the HIV/AIDS (acquired immunodeficiency syndrome) epidemic as we now know it. At the time, nobody knew what caused the disease or how it was spread. The fact that it largely affected homosexual men suggested that it was sexually transmitted, but it was not until 1983 that the virus responsible for the disease was isolated, and subsequently shown to be spread in blood and body fluids. In the meantime, there was fear and hysteria surrounding anyone who tested positive for the virus. People were excluded from public places, and refused treatment at hospitals, until the method of transmission and risks of infection were established. The response to HIV/AIDS highlights the importance of epidemiology—the study of how diseases occur, are spread, and can be controlled.

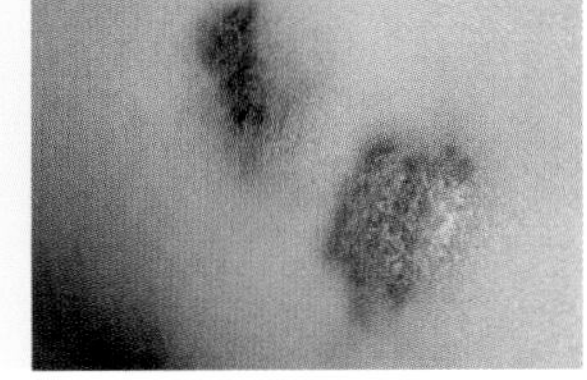

All of us at one point or another have visited a physician, complaining of an ailment. After a series of tests the physician usually offers a diagnosis, discusses treatment options, and sometimes prescribes medication. But how do physicians arrive at a diagnosis? Even with today's medical miracles, it is still not known how some diseases and illnesses are caused.

Diagnosing illness and disease

Who diagnoses illness and disease?

A physician has the difficult task of learning and knowing all that is normal about the human body and what could go wrong with it (pathology). Physicians need to know about the body's structure (anatomy), workings (physiology), and mind (psychology), but their primary role is to determine how far a person is from the normal state (homeostasis) by evaluating the disease or illness, then making a diagnosis. Other health professionals such as specialists, nurses, or medical technologists may also be involved in making the diagnosis.

How is a diagnosis made?

The diagnostic process starts the moment the patient walks into the physician's office. A thorough history is taken, including details of the patient's former health, work, and social status (for example, family situation) that may have contributed to the disease or illness. This is followed by a physical examination, where the physician may look for clues, such as a rash or swelling, and patient complaints such as headaches, and other pain. By this point, the physician may have drawn up a list of likely illnesses or diseases that may be causing the patient discomfort. If the physician is absolutely certain of the diagnosis, treatment will begin. But it is more likely that laboratory testing will be needed to confirm the diagnosis. This may include an analysis of biological samples, such as urine or blood, or taking medical images. Once all the results have been interpreted with respect to normal values, and other potential illnesses and diseases have been eliminated, the (differential) diagnosis has been made.

What diagnostic tools are available?

Laboratory testing has a vital role to play in the diagnostic procedure. A wide variety of laboratory tests are now available to investigate numerous parameters of the human body. Some of these tests are automated, others are conducted manually.

From a small blood sample, medical technologists and technicians are able to determine the state of a person's bodily systems, including that of organs such as the liver. Using a special analyzer to test the blood, technologists and technicians program the analyzer to run a list of biochemistry tests requested by the physician.

These days, most laboratory testing is performed in this automated way, but many manual procedures are still commonplace, such as the microbiological analysis of bodily fluids—sputum, feces, or urine.

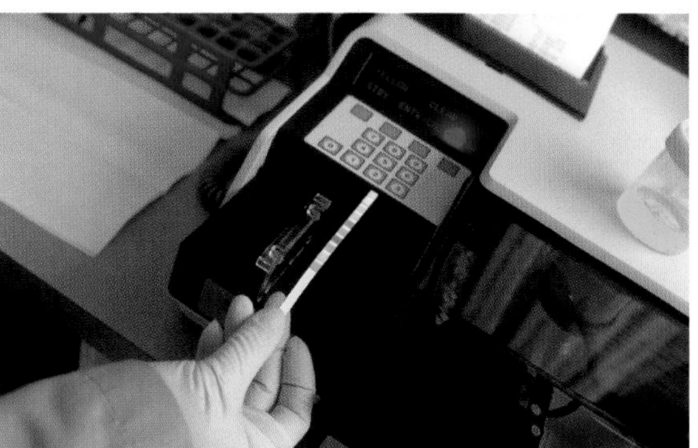

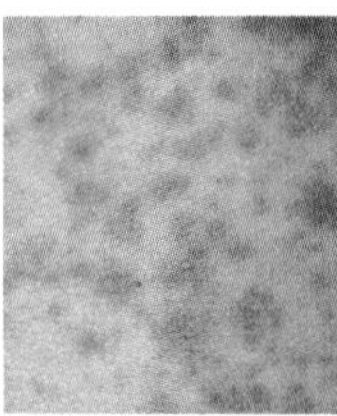

Above: A skin rash caused by an allergic reaction to laundry detergent. An allergic reaction is the body's way of responding to invaders such as bacteria and toxins. Its overreaction to a normally harmless substance (an allergen) is called a hypersensitivity, or allergic reaction.

Top right: Blood pressure (which is being measured here) refers to the force exerted by circulating blood on the walls of blood vessels. It constitutes one of the principal vital signs. The term generally refers to the pressure in the larger arteries, which take oxygenated blood away from the heart.

Below left: A laboratory technician testing a urine sample. Urinalysis, an array of tests performed on urine, is one of the most common methods of medical diagnosis. The number and type of cells and other material in the urine can yield helpful diagnostic information.

Opposite: A lab technician inserting a blood sample into a centrifuge, where test tubes containing samples are set rotating very rapidly in a slanted or a horizontal position. The heavier, solid parts are thrown toward the bottom, thus separating blood cells from plasma.

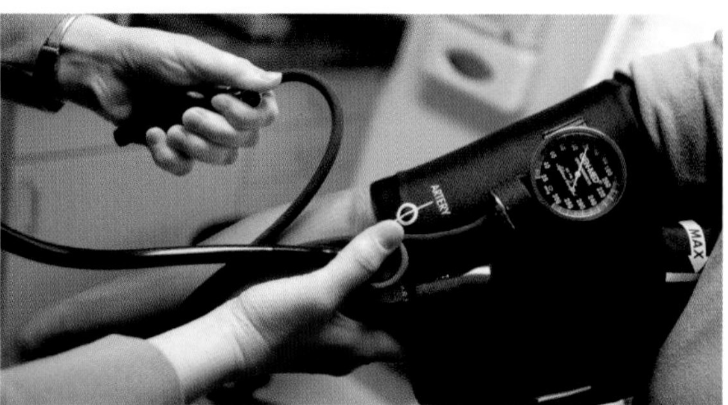

Technologists are required to prepare a microbial culture, which aims to simulate the conditions of the body on an agar plate by using media enriched with all the nutrients necessary for reproducing microbes. Within 24–72 hours, microbes causing a particular illness or disease may grow, greatly assisting with diagnosis and allowing doctors to prescribe the correct antibiotics. Other tools used in diagnosis include medical X-rays, ultrasounds, computed tomography scans (CT), magnetic resonance imaging (MRI), and biopsies.

Diagnosing illnesses or diseases of the mind is much more difficult as there are usually several causative factors, and differing theories as to what they are and how to treat them. Specialist physicians called psychiatrists are responsible for diagnosing mental conditions.

> Whenever a doctor cannot do good, he must be kept from doing harm.
>
> Hippocrates, physician, *c.* 460–380 BCE

WHAT FOLLOWS DIAGNOSIS?

Once a diagnosis is reached, the patient must work with the physician in following the prescribed treatment. This could range from a simple course of antibiotics for an infection, to months of chemotherapy to help rid the body of cancer. Monitoring the success of treatment is crucial to ensuring it is working. If after some time the patient does not respond to treatment, the physician will often refer the case on to a specialist who will hopefully find another, more successful way to tackle the illness or disease.

Common diseases affecting a cross-section of cultures include cancer, diabetes, and heart disease. Common illnesses include colds and influenza, depression, and viral and bacterial infections.

Parasites include a wide variety of organisms that live on, and derive nutrients from, another living organism, the host. Human parasitic diseases can be caused by protozoa, helminths (worms), and arthropods (insects). Parasites that live inside the host are called endoparasites; those that live outside the host are called ectoparasites.

Parasites: the ultimate freeloaders

The most successful parasites are those that maintain their own life processes without killing their host. Parasitic infections in humans can generate a variety of symptoms including malnutrition, weight changes, anemia, disturbed sleep, diarrhea, and skin lesions.

Diseases caused by parasites

Each year, parasitic infections occur in more than one billion people throughout the world and are responsible for several million deaths, mostly in developing countries. Some lead to major diseases such as malaria, sleeping sickness, schistosomiasis, and leishmaniasis, which cause severe debilitation and death among the populations they affect. Other parasitic infections, once considered minor, are becoming a much more significant threat because of their occurrence in immunocompromised individuals. These include the infections caused by the protozoa toxoplasma and cryptosporidium.

Parasite life cycles

Parasites often have complex life cycles involving, for example, the formation of resistant cyst forms, a passage through more than one host, or the use of a vector (insect carrier) for transmission. Many of the helminths (worms) have two or more hosts in their life cycle. The definitive host harbors the mature adult form of the parasite, while the intermediate host nurtures the immature or larval form. Humans frequently become "accidental" hosts, and acquire the parasite (the beef tapeworm, for example) by eating infected meat.

Above: Parasitic promastigotes that cause leishmaniasis (*Leishmania amazonensis*) in humans. This tropical disease, transmitted by bites from infected sand flies, is caused by parasites ingested by the insect vector during a blood meal from a mammal or human. It occurs in two forms: Cutaneous, which affects the skin and mucous membranes, and the "kala-azar" or visceral form which causes fever and liver damage, and can be fatal.

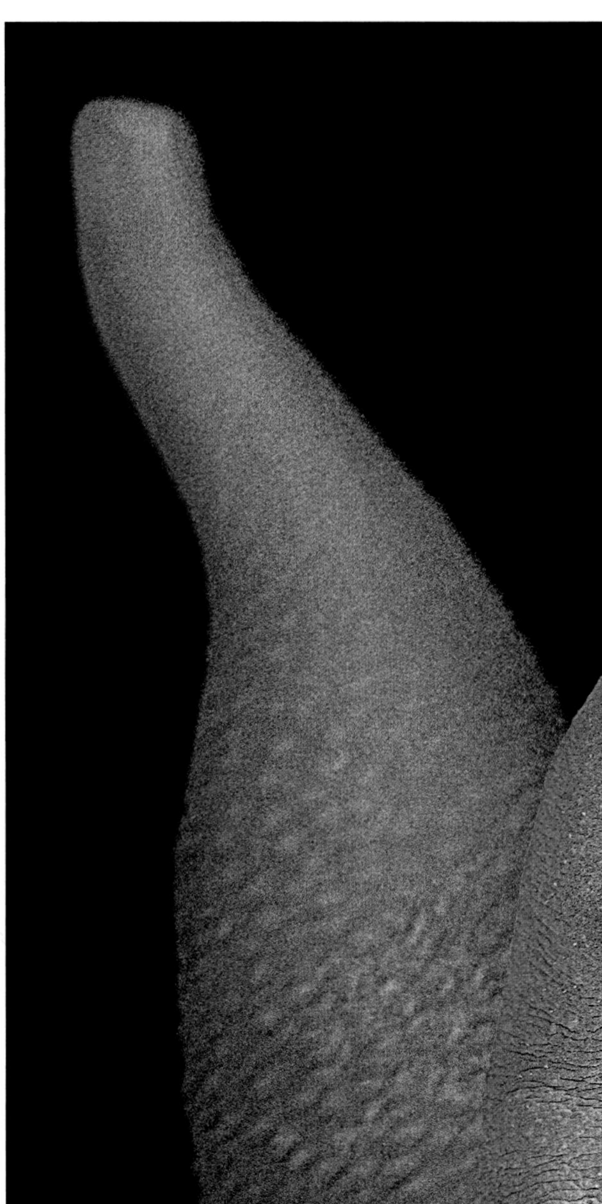

A child dying from malaria every 30 seconds is completely unacceptable ... insecticide-treated bed nets can reduce mortality by up to 25 percent.

Carol Bellamy, executive director of UNICEF, b. 1942

Left: Forests in Guatemala have produced thriving vegetation and abundant life of all sorts. In recent years, however, forest destruction and other forces have led to an increased incidence of diseases such as leishmaniasis. Cases occur more often in relatively rural areas, close to both forests and farmland, where parasite reservoirs and vectors reside together.

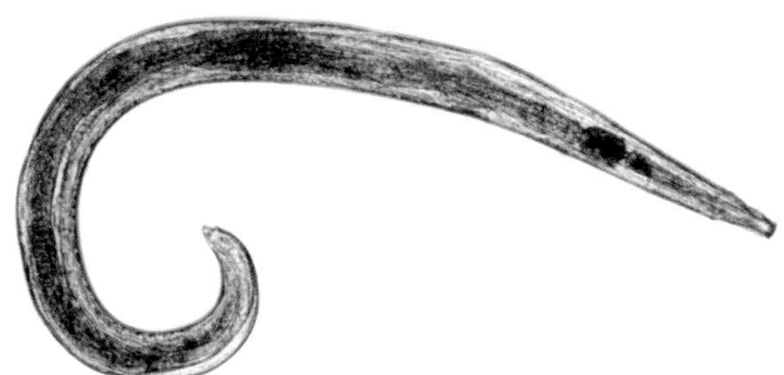

Protozoa

Protozoa, whose name means "first animals" in Greek, are single-celled organisms surrounded by an outer membrane. Some protozoa have hair-like appendages (cilia or flagella) for movement. They are mostly found in water habitats, but a number exist as intracellular or extracellular parasites in animals or insects. Protozoal infections may be acquired by the ingestion of the parasite in contaminated food or water, or via the bite of a blood-sucking insect carrier.

Worms (helminths)

The parasitic worms that may infect humans are *Platyhelminthes* or flatworms, including cestodes (tapeworms) and trematodes (flukes), and *Aschelminthes*, including the nematodes (roundworms). Parasitic worms are sequential hermaphrodites. Usually large organisms with a complex body structure, their larval stages may be quite small, only 100–200 μm in size, and the eggs are microscopic.

Cestodes (tapeworms) live as intestinal parasites in their definitive host. They consist of a head, or scolex, and a long body made up of segments, or proglottids. Trematodes (flukes), estimated to include about 18,000 to 24,000 species, have a flat, leaf-shaped body with muscular suckers that enable them to attach to host tissue. Their life cycle frequently involves more than one intermediate host.

Nematodes or roundworms are long, cylindrical worms with male and female sex organs located on different organisms. Human roundworms include the hookworm and threadworm.

Chronic worm infestations are common in children in developing countries. They can lead to serious problems, affecting many organs.

Ectoparasites

A number of arthropods (insects) are capable of existing in a parasitic relationship with the human body by inhabiting its outside surface, the skin. These are called ectoparasites. The most common are fleas, ticks, lice, and mites such as scabies. Some carry other diseases; for example, ticks may carry Rocky Mountains spotted fever. Scabies is a mite that burrows into the skin, creating itchy lesions. Scratching often leads to secondary bacterial infections.

Plant parasites

Sap-sucking insects and some types of fungi can live in a parasitic relationship on plants. And plants such as mistletoe, which grow on other plants, are usually considered to be parasitic.

Top left: The pinworm (*Enterobius vermicularis*) passes out of the human anus to lay its eggs, causing the infected person to feel itchy. Although common in temperate parts of the world, pinworm infection is not usually serious.

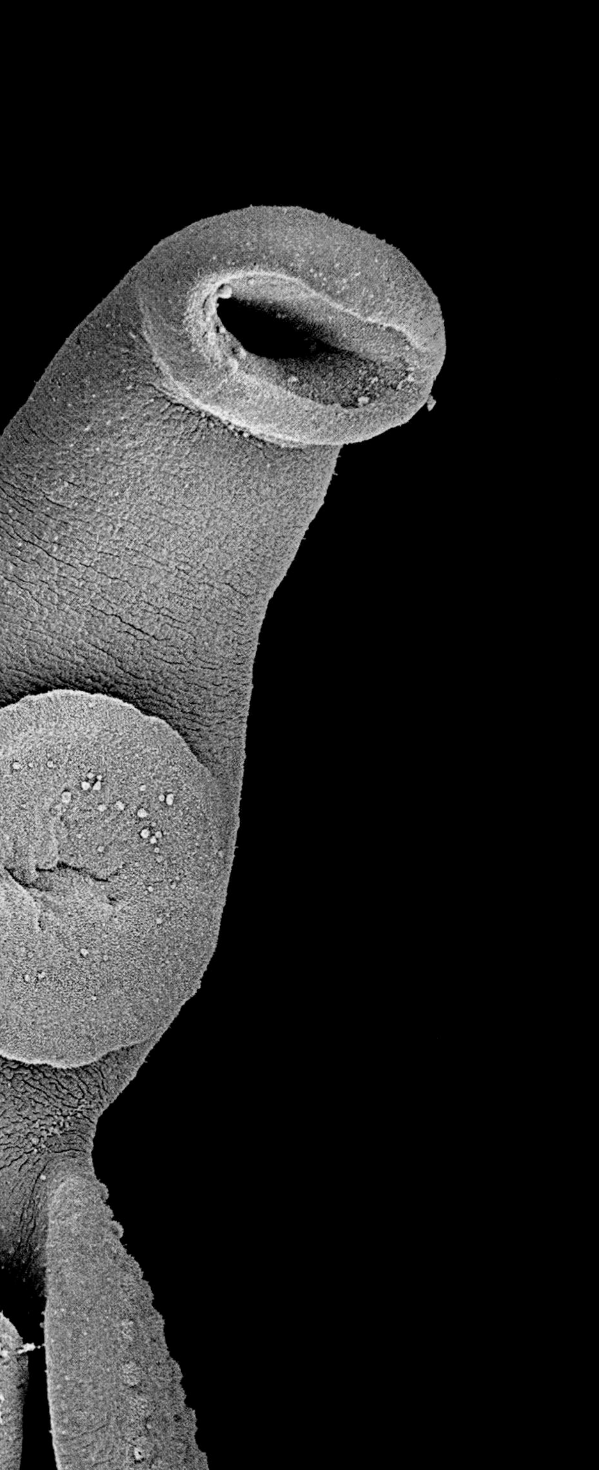

Left: Portrait of a male blood fluke (*Schistosoma mansoni*), the cause of schistosomiasis, a serious and widespread disease most commonly found in Asia, Africa, and South America. It has a low mortality rate, but schistosomiasis is an often chronic disease that can cause liver and intestinal damage, and can be very debilitating.

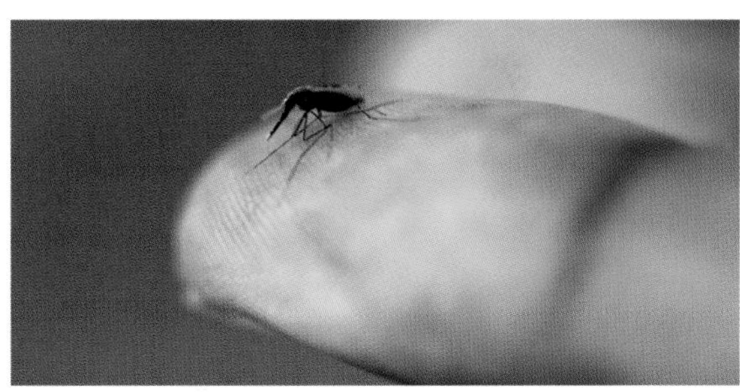

Right: Malaria parasites are carried by female *Anopheles* mosquitoes. These parasites multiply within the red blood cells of hosts, causing illness. There is no vaccine for malaria, but its transmission can be reduced by using nets and repellents, and by draining any standing pools of water where mosquitoes can lay their eggs.

HYDATIDS

Humans are the intermediate host for the dog tapeworm *Echinococcus granulosis* (pictured), which may lead to a condition known as hydatid cysts. Dogs and cats are the definitive hosts for this minuscule tapeworm, which is ¾–3 inches (2–8 mm) long. Eggs are shed in the feces and may be transmitted to humans from feces on the fur or tongue of the animal. The eggs hatch in the human intestine and migrate to various parts of the body, where they form large fluid-filled sacs of larvae, called hydatid cysts. Cysts may form in any tissue but those that form in the liver, lungs, and brain are the most common and the most serious. You can avoid hydatids by worming domestic animals regularly and making sure not to feed them raw meat.

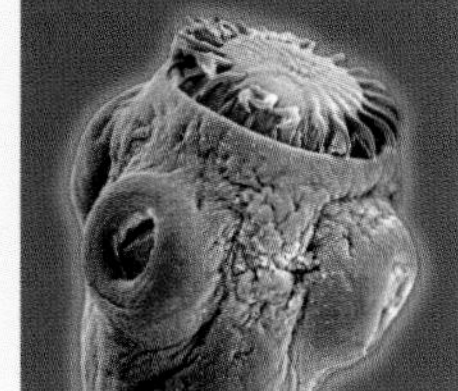

Our environment contains a multitude of microorganisms to which we are continually exposed, and the body harbors many billions of bacteria, especially on skin, in the intestine, mouth, and vagina. It is only thanks to the body's defense systems that we are not struck down continuously with disease.

Supplementing the body's defences

The human body is equipped with a vast arsenal of defenses that protect it against infection. The skin, the mucus membranes that line the mouth, and the respiratory, intestinal, and genital tracts, are very effective barriers against infectious agents. Internally, there are many chemical defences, including stomach acidity, which kills most microorganisms ingested in food, and enzymes in body fluids that destroy bacteria. There are also defensive cells, especially phagocytes, which ingest and kill foreign substances and microorganisms.

However, the body's ability to defend itself is not always completely effective and sometimes infection occurs. When this happens, antibodies are produced. These are molecules used by the immune system to identify and neutralize foreign agents, such as bacteria and viruses. When we become infected, it is the antibodies our body produces that are mainly responsible for our recovery. It takes between days and weeks to produce antibodies, which is why it takes that same amount of time to recuperate. These antibodies also protect the host from acquiring the same infection in the future, so that we suffer from many diseases, such as chickenpox, only once in our lives.

Antibodies are not always effective, and in some diseases, like AIDS and hepatitis B, they are ineffective. Also, there are certain times when people are more vulnerable to infection, due to immaturity of the immune system as with the fetus and young baby, or due to weakening of the immune system that comes with old age.

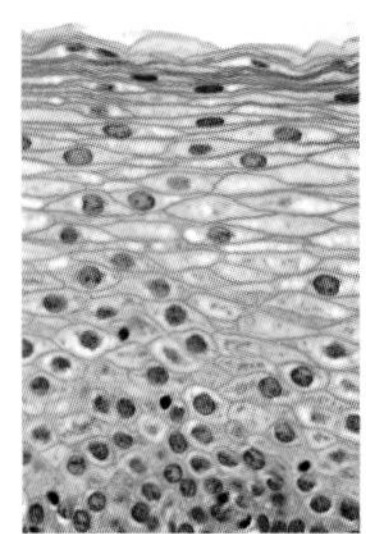

Above: Oral mucosa is the general name applied to the mucous membrane epithelium of the mouth. There are three varieties: Masticatory mucosa, found on the dorsum of the tongue, the hard palate and the attached gingiva; lining mucosa (pictured), found in most other parts of the mouth; and other specialized mucosa, such as those around the taste buds.

> In recent years science has learned that the human immune system is much more complicated than we thought.
>
> Philip Incao, doctor

Vaccination

Vaccination boosts the immune system of those who considered at a higher risk of infection. This has led to a substantial reduction in the instances of diseases that were once common and deadly, such as measles, mumps, polio, tetanus, and smallpox. Vaccines for rubella protect women and their unborn babies against German measles, an infection that can be devastating for the fetus. A recently developed vaccine now protects young adults from the human papilloma virus that is responsible for genital wart infection and cervical and penile cancer. Vaccination is discussed on pages 346–347.

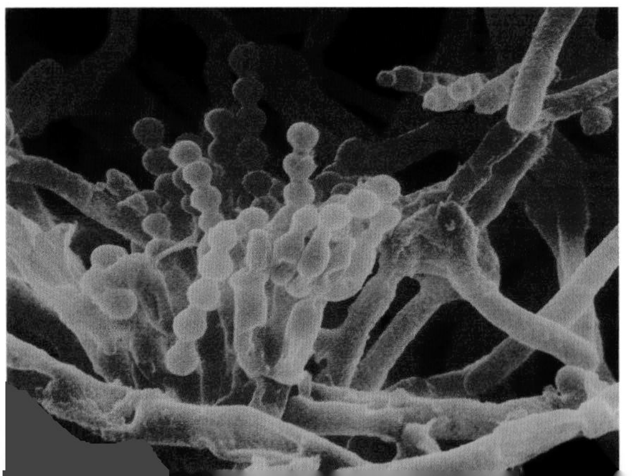

Left: A color-enhanced SEM (scanning electron micrograph) image of penicillium mold. Though Alexander Fleming was first to discover its antibacterial properties, and first to isolate the active substance he named penicillin, a number of ancient cultures had been using molds to treat infection for thousands of years. This worked, we now know, because of the antibiotic substances they produce.

Antibiotics

In 1928, a breakthrough in the treatment of infections came with Alexander Fleming's discovery of penicillin. He noticed that the growth of *Staphylococcus aureus*, a bacterium known as the "golden staph," was stopped by a common blue–green mold. Antibiotic is a term used for a natural product, like that discovered by Fleming, which is produced by a microorganism that can kill or inhibit other microorganisms. In common usage, the term also refers to synthetic or semi-synthetic antimicrobial drugs that have been chemically produced or modified to improve their effectiveness.

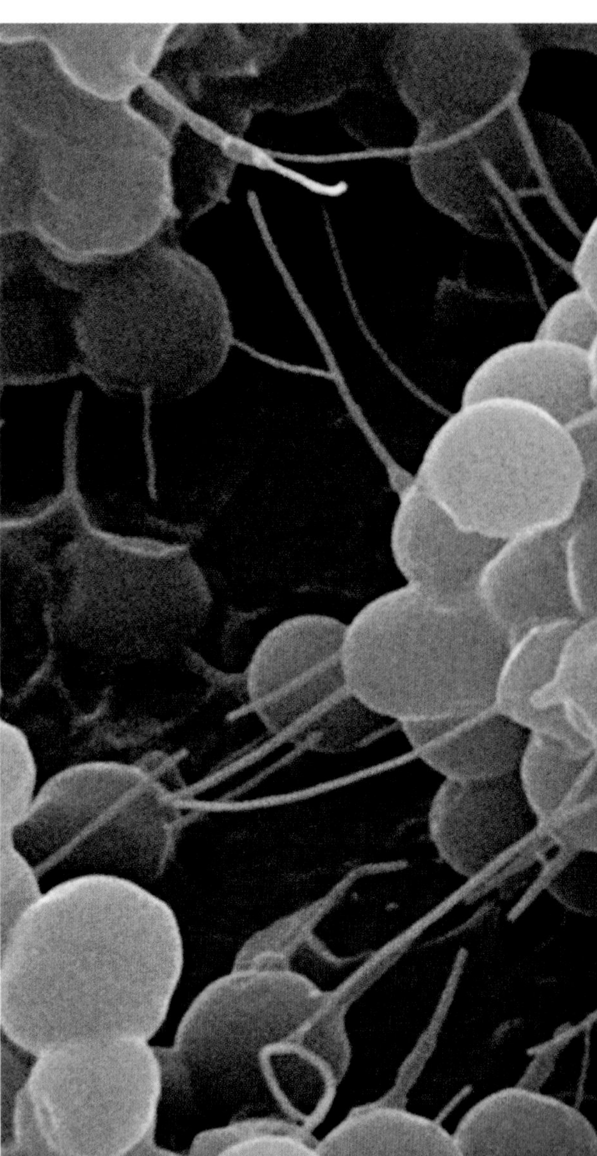

Some compounds, called broad spectrum antibiotics, are active against different types of microorganism; others have a limited range of activity. Some antibiotics, like penicillin, kill microorganisms, while others, like tetracycline, stop the growth of the organism without killing it. The latter type of drug is useful because, even though it does not kill the cell, it inhibits reproduction of the microorganism, allowing the immune system time to increase its natural defenses, such as antibodies, against the infection.

Left: Sir Alexander Fleming (1881–1955), the renowned bacteriologist famous for his discovery of penicillin, using the strong colors produced by certain germ cultures to paint pictures in his laboratory.

Below: *Staphylococcus epidermidis* bacteria are usually a harmless part of normal skin flora, but can cause infection on broken skin or wounds. A significant cause of infection in patients whose immune system is compromised, or who have indwelling catheters, the bacterium includes strains able to produce a biofilm, allowing them to adhere to the surfaces of medical prostheses.

IMMATURE IMMUNE SYSTEMS

The fetus and newborn babies have immature immune systems. As they do not make antibodies as effectively as adults and need time to build up their supply, they are more susceptible to infections. An unborn baby's defenses are naturally supplemented by the mother's own antibodies as they cross the placenta to the fetus. When the child is born, it receives more antibodies from its mother through colostrum, then breast milk. In the womb and then in the first weeks of life, the mother's antibodies protect the infant against a wide variety of microbes. Antibodies that the body produces itself last much longer than those that are transferred from mother to baby. Transferred antibodies last in the baby's circulation for only a few months, providing short-term protection from microbes to which the mother is immune. As the infant develops and grows, its own immune system matures and becomes more effective. Vaccination of the baby against the common childhood diseases also helps to supplement its defences.

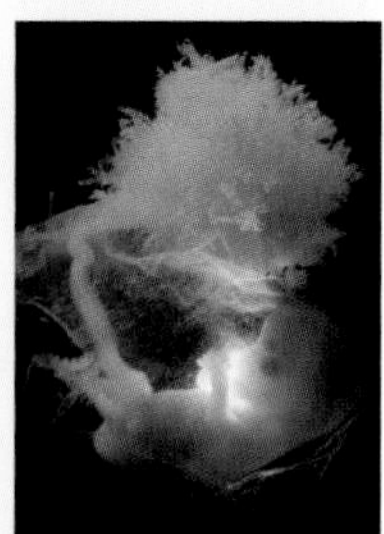

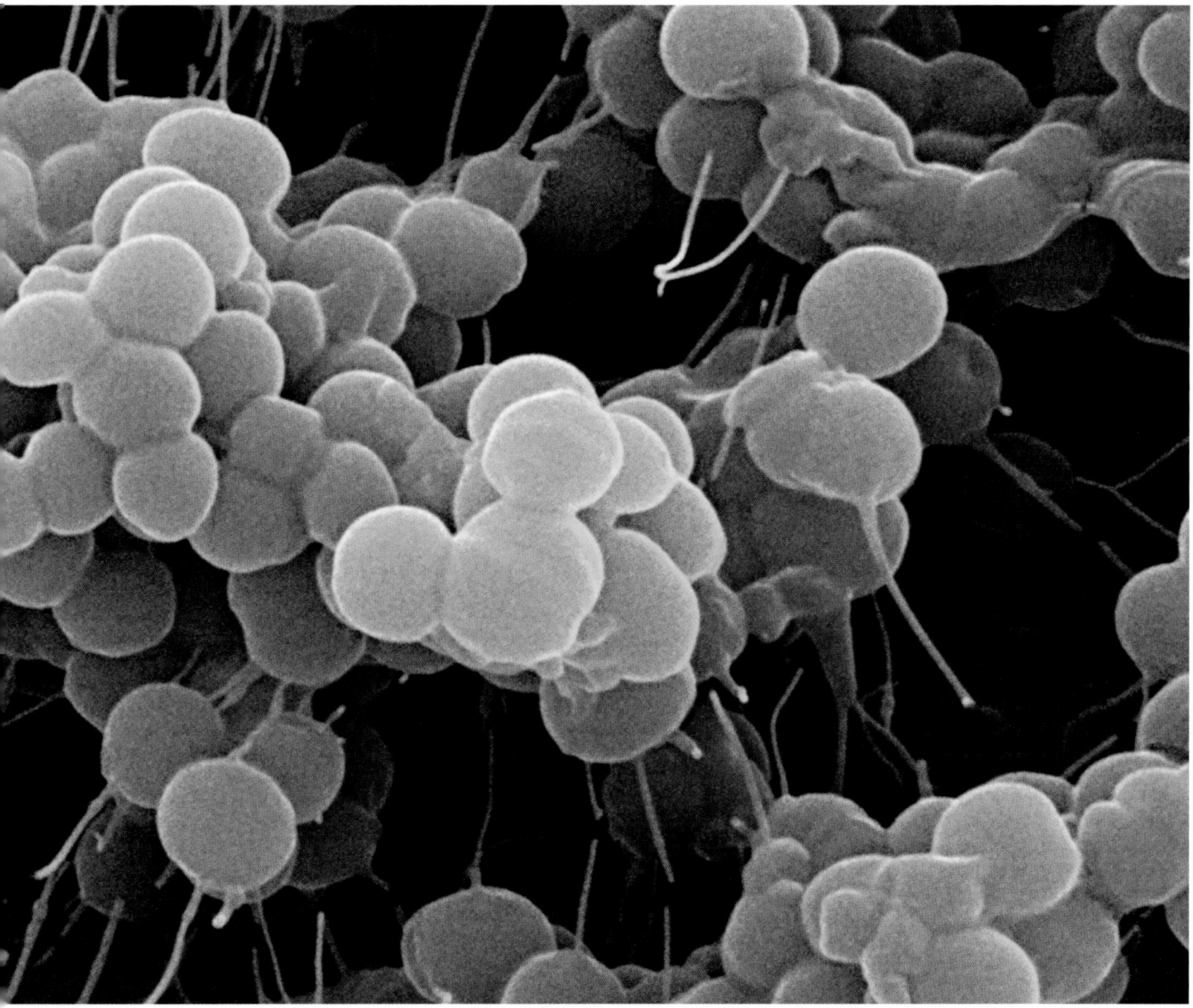

In the Middle Ages, the Black Death, or plague, killed millions of people across Europe. The bodies and clothing of those who had died were burnt, and sulfur and aromatic plants were burnt to purify the air in homes. These were crude forms of sterilization and disinfection. Unfortunately, these practices had little effect because the disease continued to devastate Europe for 400 years.

Sterilization and disinfection

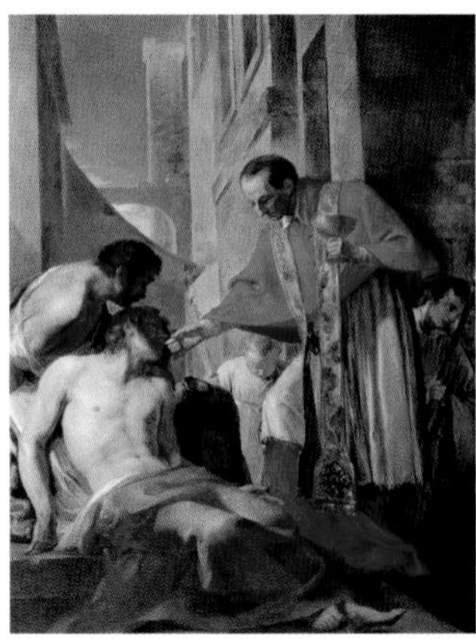

Left: St Charles Borromeo (1538–1584) administers the sacrament to a plague victim in Milan in 1576. Unlike most of his colleagues, Borromeo mixed with plague victims, so much so that 1576–77 is sometimes called the Plague of San Carlo. Worldwide, the "Black Death" claimed up to 75 million lives.

Despite vast improvements in health care over the past 50 years, serious outbreaks of infection still occur. They are usually found to be attributable to a breakdown in sterilization or disinfection procedures.

Sterilization

Sterilization is defined as the complete removal of viruses, bacteria, and all other living organisms from an object. Generally, sterilization is necessary in health care for objects that enter the body, such as surgical instruments and needles. There are a variety of methods that can be used to achieve sterilization, including heat, radiation, and lethal chemicals like hydrogen peroxide. Heat—either dry or moist—is the cheapest, most efficient, and most controllable method of sterilization, and is thus the most widely used.

Moist heat sterilization involves the use of steam at above boiling temperature, usually 249.8°F (121°C) or higher. In most health care practices this is achieved in an instrument called an autoclave, which is similar to a pressure cooker. It superheats steam under pressure to reach the higher temperatures required. While boiling water at 212°F (100°C) kills most microorganisms, it does not completely sterilize; some organisms, such as those that cause tetanus and botulism, can survive this temperature. The use of the term "sterilizer" for an instrument that simply boils water and creates steam is misleading.

Other methods of sterilization include dry heat, such as an oven at over 320°F (160°C); incineration; ionizing radiation, such as gamma rays; and chemicals, such as low temperature hydrogen peroxide.

Disinfection

Disinfection is the removal of large numbers or specific types of microorganisms from an object. Unlike sterilization, disinfection does not guarantee their complete removal, and is a much less certain process. Disinfection aims to reduce the number of microbes present, but to what level depends on the nature of the item being disinfected and how it is to be used. Disinfection is often used in hospitals when sterilization is not necessary or not possible, such as with trolleys, thermometers, floors, bedpans, and most importantly hands. The most commonly used disinfecting agents are chemicals, heat either at or below boiling temperature, and radiation.

Opposite: Sterilizing bandages in an autoclave, France, 1905. The autoclave, invented in 1879 by Charles Chamberland, uses very high heat to inactivate bacteria, viruses, fungi, and spores. Autoclaves are now standard features of medical settings, where they are used to sterilize equipment and other objects, or to sterilize medical waste.

Above: A disinfectant is an antimicrobial agent that can be applied to non-living objects to destroy microorganisms. Antibiotics are used to destroy microorganisms inside the body, antiseptics to destroy microorganisms on living tissue.

Left: Surgical masks are worn by health professionals during surgery and at other times to catch bacteria shed in liquid droplets and aerosols from the wearer's mouth and nose. The wearer is no less likely to catch a viral disease than someone not wearing a mask, but *is* less likely to pass on any infections. Masks are used only once.

Many chemicals can kill or inactivate microorganisms, but most are also toxic to humans and are therefore not safe enough to be used as disinfectants. Commonly used disinfectants include chlorine (household bleach), iodine, alcohol, and chlorhexidine. Alcohol and chlorhexidine are often used as antiseptics, that is, skin disinfectants.

> It may be a strange principle to enunciate as the first requirement in a hospital that it should do the sick no harm.
>
> Florence Nightingale, nurse, 1820–1910

The exposure of an object to sunshine might be described as a disinfection process because many microbes are killed by ultraviolet light. Not all microorganisms are killed by such light, but some are. Exposing wounds and ulcers to sunlight to keep them free from infection has been used as a treatment for centuries.

An increasingly vast array of soaps and cleaning agents containing disinfectants is being advertised and used in homes and other settings. However, there is a general lack of understanding of when and how to correctly use these products. In practice, most surfaces in the home can be rendered hygienically clean by being washed with soap or detergent, then thoroughly rinsed with water. In heavily soiled areas, such as toilets, careful use of household bleach provides very effective disinfection. Good disinfection of washed clothing can be achieved by drying the clothing in sunshine.

HOSPITAL INFECTIONS

Hospitals harbor disease-causing microorganisms, and hospitalized people generally have a higher risk of infection due to their underlying illness, or as a result of procedures or treatments they receive. Surgery, immunosuppressive drugs, treatments for cancer, and a host of other factors can make them more susceptible to infection. Once a person has an infection, they are then a potential source of infection for other patients and staff around them. Therefore, stringent cleaning, sterilization, and disinfection methods must be practiced at all times, for everybody's sake, in these institutions. When serious outbreaks of infection occur,they are usually due to breakdowns in sterilization or disinfection procedures.

Humans continue to be afflicted by new and old infections that are referred to by scientists as emerging infections. An emerging infection is either a new infection that has never appeared before or a known infection that has had a recent increase in incidence or geographic range.

Emerging infections

There are many diseases that fit this description, including avian influenza (H5N1 strain), severe acute respiratory syndrome (SARS), West Nile encephalitis, ebola hemorrhagic fever, human immunodeficiency virus (HIV) infection, variant Creutzfelt-Jakob disease, and multiply resistant *Staphylococcus aureus* (MRSA) infection.

Genetic flexibility of microorganisms

All living organisms have genetic mechanisms that help them adapt to their environment. Bacteria and viruses have a high mutation rate, which gives them an amazing capacity to rapidly change, and thus cause new or emerging infections. For example, different influenza viruses circulate each year, due to mutations. So that although someone may have immunity to last year's virus because they had the "flu" or were vaccinated against it, they are unlikely to be immune to this year's virus because it has changed. For this reason, an entirely new "flu" vaccine needs to be developed each year.

When penicillin first became available in the 1940s, its main use was for the treatment of infections caused by *Staphylococcus aureus* ("golden staph"). Within a few years, 50 percent of strains of *S. aureus* in hospitals were genetically different and resistant to penicillin. In the 1960s, more mutant strains appeared that were resistant not only to penicillin, but to other antibiotics as well. These multiply resistant strains of *S. aureus* (MRSA) are now the scourge of hospitals, causing infections that are difficult to treat and with a high mortality rate.

Human influences

Human activity has often played a critical role in promoting the development of emerging infections. Population growth during the Industrial Revolution led to extensive urbanization, and expansion of humans into virgin forests. This in turn promoted the spread of microorganisms from animals to humans, and between humans. For example, the natural reservoirs of the ebola virus, the cause of ebola hemorrhagic fever, are the rainforests of Africa and the Western Pacific. The intrusion of humans into these forests has led to the transmission of this virus from chimpanzees and gorillas to humans. Now the virus can be transmitted from human to human, which can result in explosive outbreaks of infection.

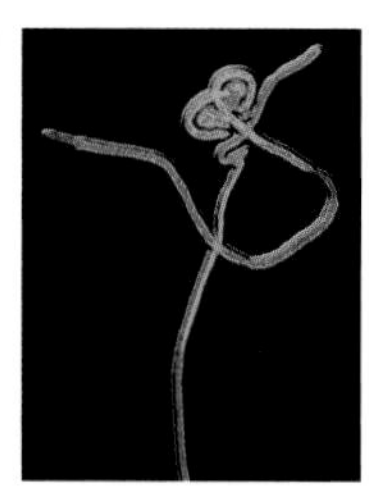

Above: The ebola virus, which causes a form of hemorrhagic fever, was first recognized in Africa in 1976. It is transmitted through direct person-to-person contact and is one of the most virulent viruses known, with a mortality rate usually over 50 percent. It typically spreads so quickly through the entire population of a small village or hospital that it runs out of hosts, burning out before it reaches a larger community.

Below: A microscopic image of human cells (darker) being infected with the severe acute respiratory syndrome (SARS) virus (lighter). The virus seems to have originated in Guangdong Province, China in November 2002, but did not attract public attention until February 2003, when an American businessman traveling from China developed pneumonia-like symptoms.

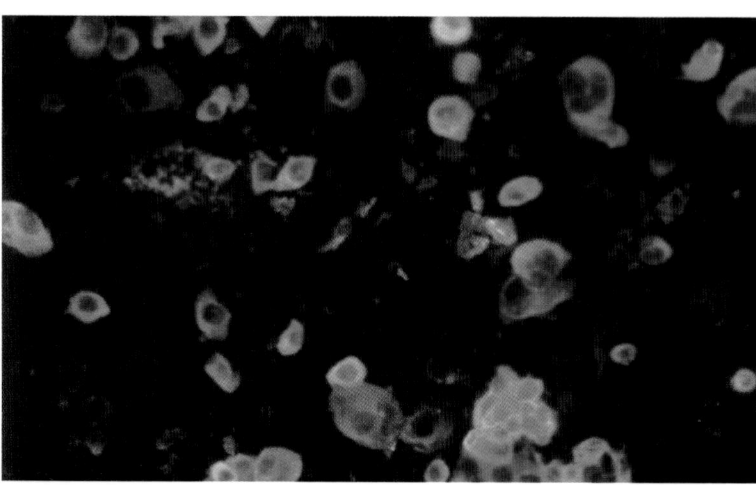

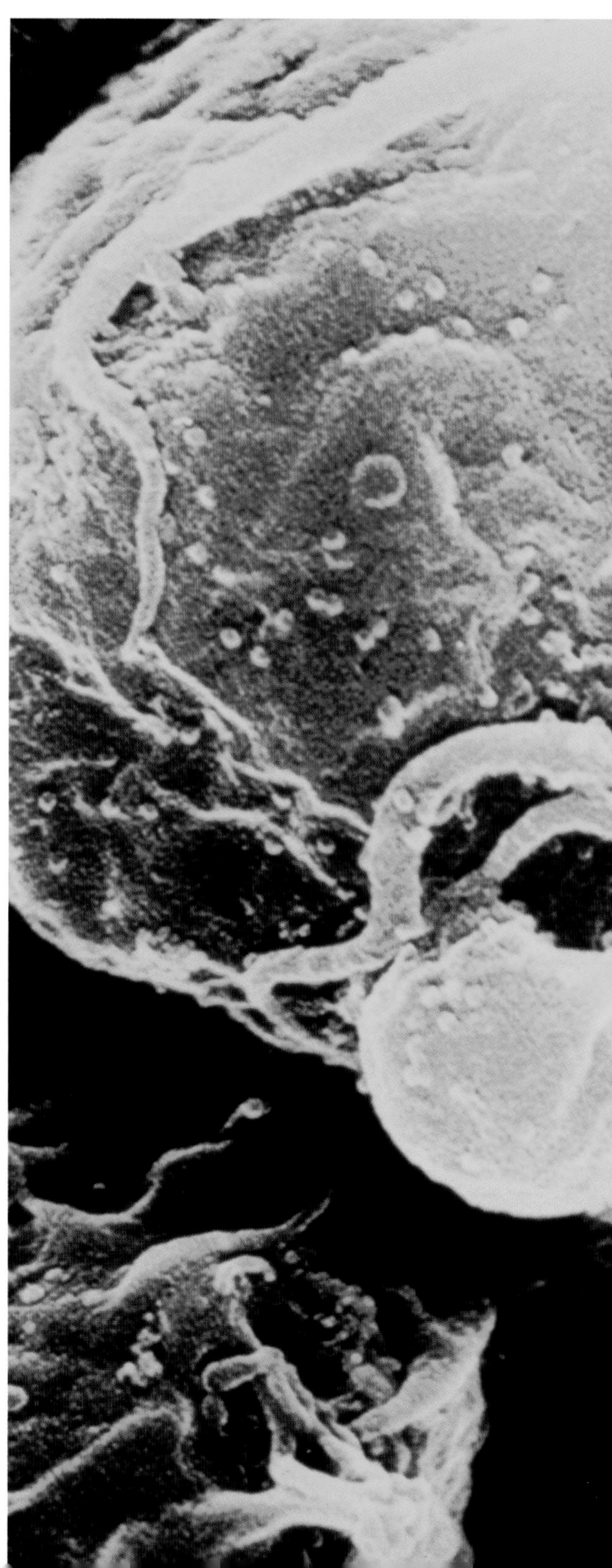

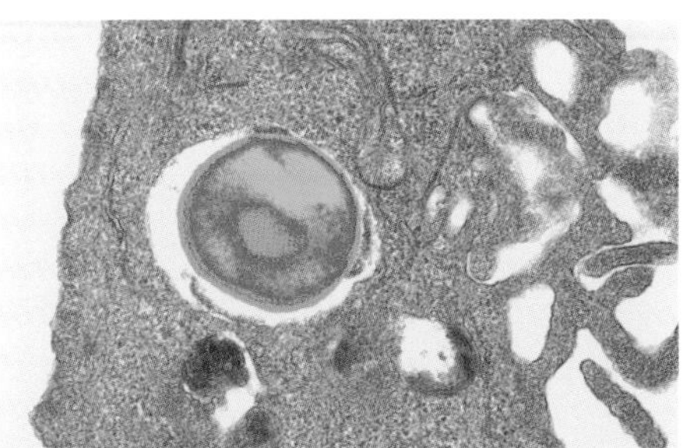

Left: A bacterial invasion of pulmonary epithelial cells by *Staphylococcus aureus*. Virulent and widespread, staphylococcus bacteria often occur in medical facilities, and have become resistant to many antibiotics.

The enormous traffic of humans around the world promotes the emergence and spread of new infections. Infected travelers who are yet to become symptomatic can reach any part of the world within 24 hours, taking their infection with them. For example, the first case of SARS occurred in China in November 2002; within five months, cases were reported in 30 other countries. The incredible speed with which the infection spread was clearly due to air travel by infected people. By June 2003 there were more than 8,000 cases with around 800 deaths reported. A similar mix of microbial and human factors will ensure that new infections continue to emerge and old ones re-emerge for decades, perhaps centuries, to come.

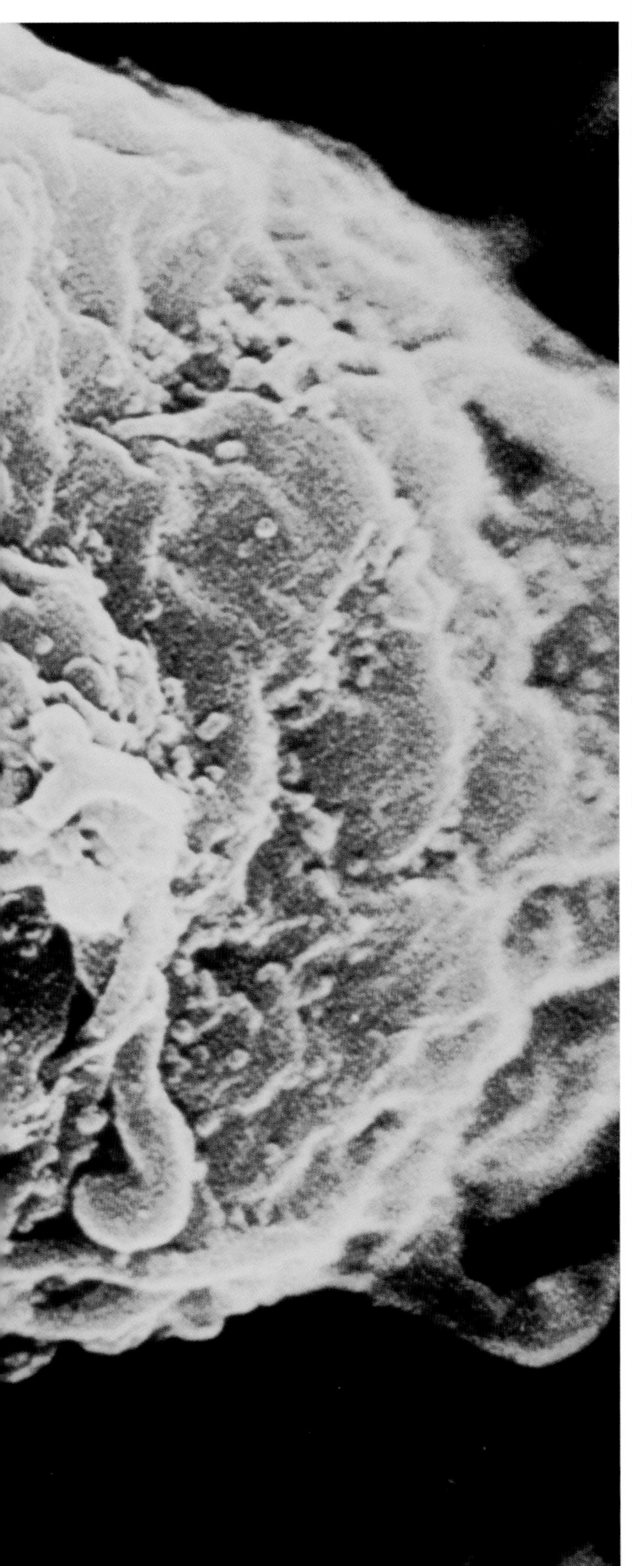

Left: Electron scan of the human immunodeficiency virus (HIV) infecting a cell. HIV is a retrovirus that can lead to acquired immunodeficiency syndrome (AIDS), a condition in humans in which the immune system begins to fail, leading to life-threatening infections. HIV has killed over 25 million people worldwide since 1981.

Animal influences

Avian influenza (strain H5N1) has the potential to cause a worldwide influenza pandemic, similar to the 1918 Spanish "flu" pandemic, which killed around 40 million people. The avian influenza virus originated in wild water birds. Once it spread to other birds, especially poultry, and possibly pigs, it had a direct route to humans. The only factor that has so far prevented a major pandemic is the fact that the virus has not yet learnt how to readily spread from human to human. Other animals are believed to be the sources of various emerging infections, including rodents (for some hemorrhagic fevers) and non-human primates (for yellow fever and possibly HIV).

> ... the human body is the product of many, many years of having fought various viral diseases, and has survived ...
>
> Fereydoon Batmanghelidj, doctor, 1931–2004

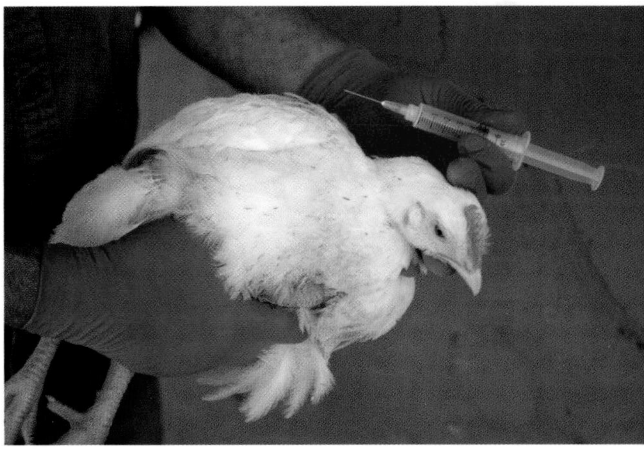

Right: The "bird flu" scares of recent years highlight the potential threat posed by emerging infections. Avian viruses had been spreading intermittently for some time between species, but before they began infecting humans in the 1990s, they were not considered important. Since then, they have been intensively studied. Scientific research has brought about changes in what is known about flu pandemics, as well as changes in poultry farming, vaccination research, and pandemic planning.

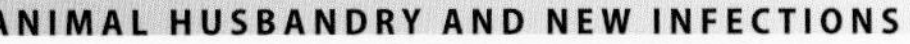

ANIMAL HUSBANDRY AND NEW INFECTIONS

Animal husbandry practices can cause the development of emerging infections. The outbreak of variant Creutzfeldt-Jakob disease (CJD) in the UK in the 1990s, was thought to have originated in sheep, which can suffer from a rare disease called scrapie. It was believed to have been spread to cows through the use of sheep carcasses in cattle feed, and then spread to humans through their consumption of beef. The disease is caused by an infectious protein, called a prion, and is one of a group of diseases that destroy brain tissue and are inevitably fatal. The slaughter and disposal of more than 4 million cattle in the UK helped to keep the epidemic under control.

Genes, the fundamental units of inheritance, are made of a biological chemical called DNA (deoxyribonucleic acid) (see pages 280–281). Every living thing on Earth has DNA in its cells and the genes carry information that shapes the characteristics of each organism. Long strands of linked genes are called chromosomes. Human cells contain 46 chromosomes, whereas most bacteria possess only one.

The medical significance of bacterial genetics

Bacterial genetics

Spontaneous mutations are thought to happen in bacteria roughly once every one million cell divisions. Because bacteria replicate so quickly—often as frequently as once every 20 minutes—they also have a high mutation rate (see pages 272–273).

There are three possible outcomes when a bacterial cell mutates. The first has no impact on the cell and the result is a silent mutation. The second is harmful, sometimes to the point where the cell cannot survive. The third is a beneficial mutation, which sees a cell emerge with characteristics that give it a better chance of survival and the ability to multiply. One important example of a beneficial mutation is the development of resistance to antibiotics in a bacterium. This is beneficial for the bacterium of course, not for humans, because it allows the bacterium to survive and continue to cause infection even in the presence of the antibiotic. Today, antibiotic resistance is of major concern in medical practice. Significant recent examples include the multiply resistant *Staphylococcus aureus* (MRSA) and the multidrug-resistant tuberculosis (MDR-TB).

Genetic engineering

Genetic engineering is a broad term: It refers to various techniques that have been developed to manipulate the genes of a living cell or organism. In one form of genetic engineering, scientists isolate a gene from one cell, then transfer it to another cell. It is possible, for example,

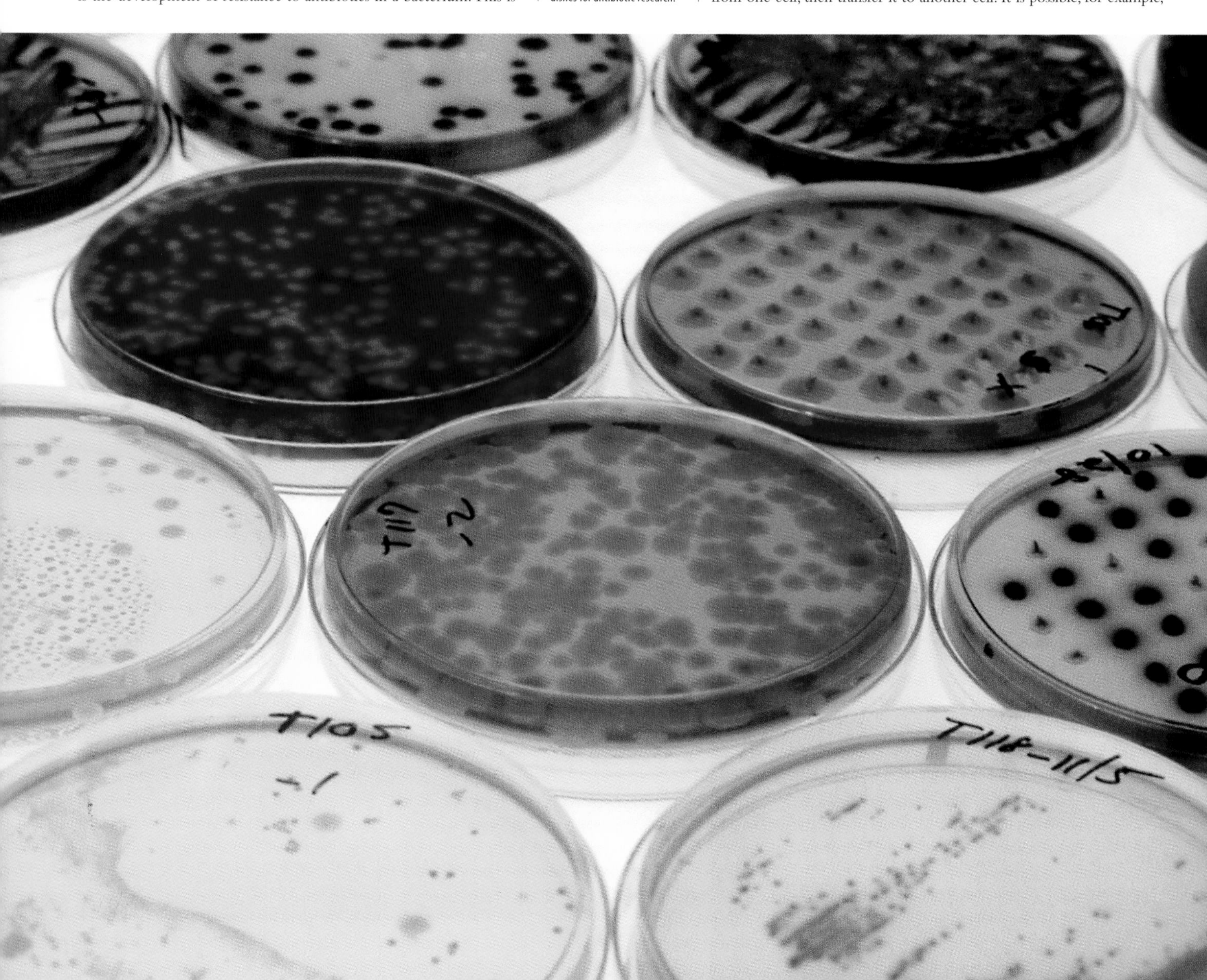

Below: Petri dishes are shallow glass or plastic cylindrical containers used by biologists to culture cells. Partially filled with warm liquid agar, plus nutrients, salts and amino acids, the dish is ready to receive a microbe-laden sample for testing after the agar solidifies. Here, a variety of bacteria are cultured in laboratory petri dishes for antibiotic research.

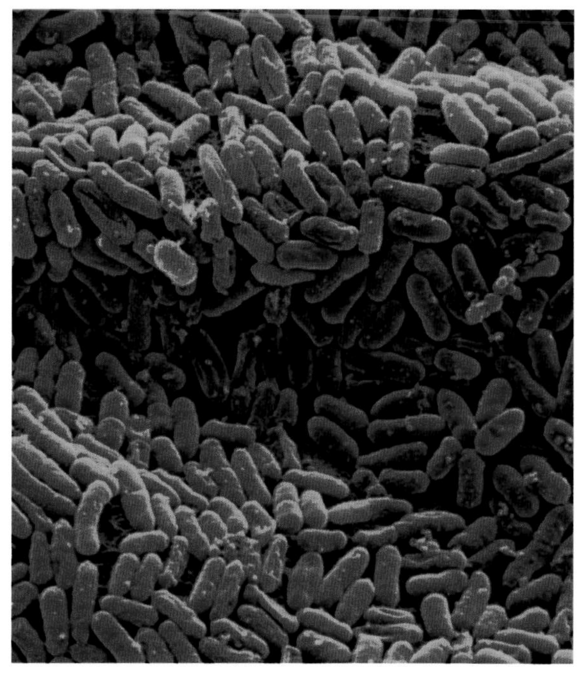

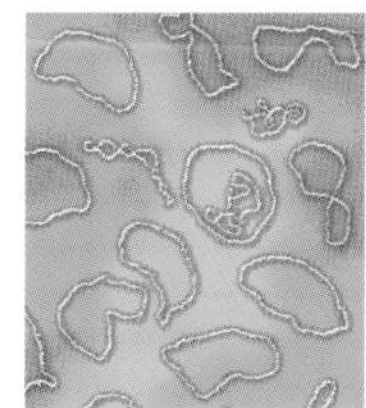

Above: Plasmids are extra-chromosomal DNA molecules capable of replicating independently. They can endow their hosts with DNA packages in times of severe stress.

Left: *Escherichia coli* (*E. coli*) is a rod-shaped bacterium that lives in the intestinal tracts of humans and animals. Some strains can cause cystisis and diarrhea, but it is usually not harmful. Its genetics are simple and easily manipulated, so it is very useful for research.

for a gene from a human cell to be inserted into a bacterial cell, and for the recipient bacterial cell to then be induced to make a human substance from this gene. One of the first uses of this technique was the production of human insulin by cells of the bacterium *Escherichia coli*. The gene for insulin production was taken from human pancreatic cells and inserted into the chromosome of the bacterium. The insulin produced can be used in the treatment of diabetes mellitus.

The use of genetic engineering techniques to force bacteria to produce substances of benefit to humans has enormous potential. Genes from humans, animals, or plants can now be inserted into a bacterium to produce a useful substance. Due to the rapid growth and reproduction rate of bacteria, and other organisms like yeasts, they can essentially be transformed into biological factories. This technique enables scientists to produce practically unlimited amounts of substances unavailable under normal circumstances or naturally present only in small quantities. So far, this technique has enabled the production of a range of products, including insulin, human growth hormone, interferon (a chemical that enhances the immune system), relaxin (used to ease childbirth), and biofuels.

Genetic engineering has also been used in agriculture. With genetically modified plants, farmers can have a successful harvest of fruit, vegetables, or grains despite what would normally be disastrous conditions for their crops, such as frost, drought, or insect plague. However, environmental groups have expressed concern over the widespread and uncontrolled use of such methods in agriculture.

Genetic engineers don't make new genes, they rearrange existing ones.

Thomas E. Lovejoy, biologist

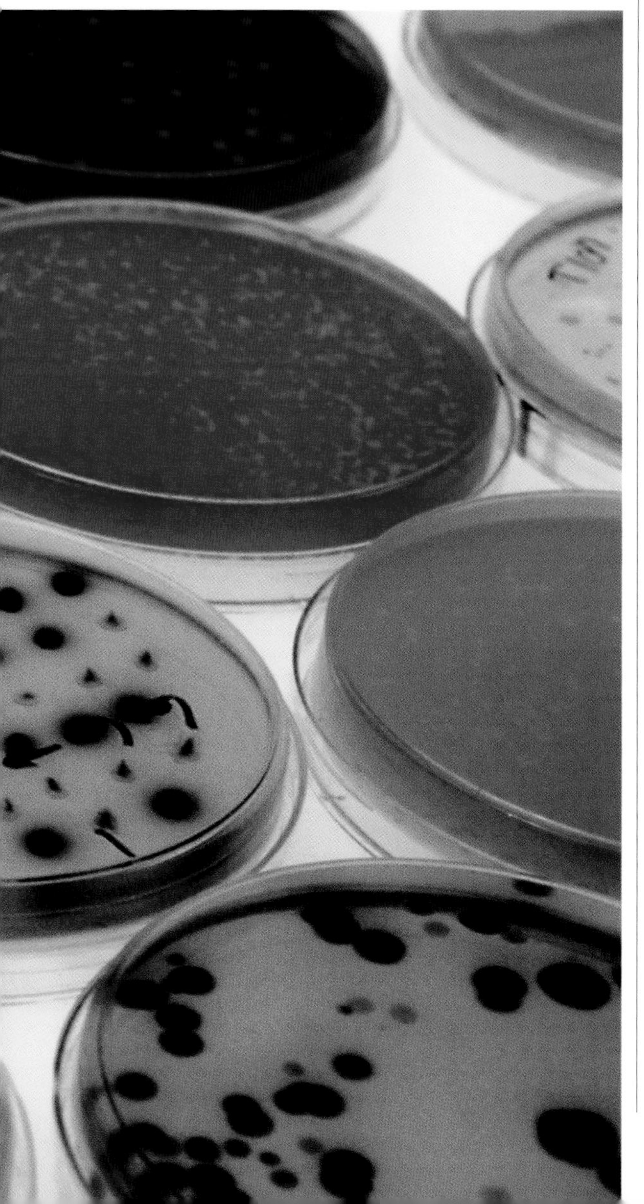

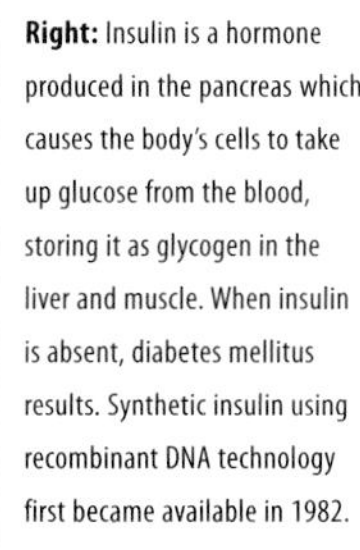

Right: Insulin is a hormone produced in the pancreas which causes the body's cells to take up glucose from the blood, storing it as glycogen in the liver and muscle. When insulin is absent, diabetes mellitus results. Synthetic insulin using recombinant DNA technology first became available in 1982.

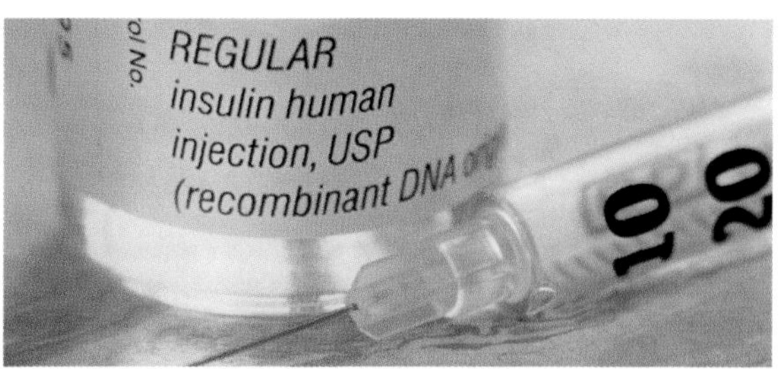

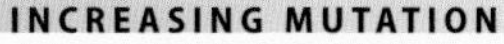

INCREASING MUTATION

The mutation rate of bacterial, human, and other cells can be increased if the cells are exposed to physical or chemical agents that affect their DNA. Known mutagens include ultraviolet light (sunlight), X-rays, benzene, asbestos, and cigarette smoke. The chemicals that cause human cells to mutate, and perhaps turn cancerous, are called carcinogens. The potential mutagenic properties of a chemical substance can be assessed in an Ames test, named after its developer, Bruce Ames (pictured), of the University of California. This method exposes a specific strain of bacteria to the chemical that is being tested for its mutagenic potential. If a particular change to a bacterium occurs, then the substance is deemed mutagenic, and possibly carcinogenic for humans. Further testing is then required to determine whether it is carcinogenic for human cells. This is done by exposing human or animal cell cultures to the suspected carcinogen. These methods have proven very useful for identifying chemicals that are mutagenic and ultimately carcinogenic for humans.

The world's first "test tube" baby, Louise Brown, was born on July 25, 1978 in England. Her birth unleashed a storm of controversy about the ethics of in vitro *fertilization (IVF). Nowadays,* in vitro *fertilization is just one of a number of assisted reproduction technologies used to help infertile couples fall pregnant.*

Making babies

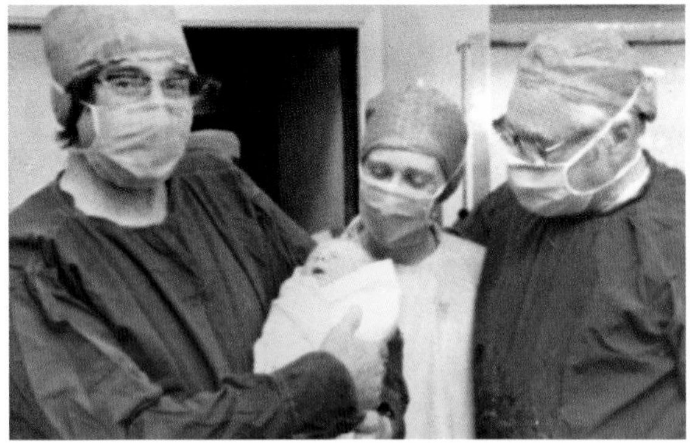

Left: Louise Joy Brown, the first ever "test tube" baby, was delivered by cesarean section on July 25, 1978. She was presented to the world by her mother and the team who pioneered *in vitro* fertilization: physiologist Dr Robert Edwards (left) and gynecologist Mr Patrick Steptoe (1913–1988) (right). IVF was originally developed to treat infertility caused by blocked or damaged fallopian tubes.

It is estimated that as many as one percent of babies born in Western countries are the result of assisted reproduction. It is also estimated that more than three million babies have been born as a result of assisted reproduction in the years since Louise Brown's birth.

At its simplest, assisted reproduction involves the use of hormone drugs to trigger ovulation. At its most complex, it involves testing embryos for the presence of specific genetic diseases before they are transferred to the woman's womb. Achieving pregnancy may involve the use of donor sperm, donor eggs, and even the donor womb of a surrogate mother willing to carry the pregnancy to term.

Baby-making basics

IVF success rates vary quite a lot, depending on the technique used and the woman's age. Generally speaking, however, 30–40 percent of IVF cycles will result in a pregnancy.

IVF is a multi-step procedure that involves removing ripe eggs from the woman's ovaries, combining the egg(s) with sperm from the father or from a donor, allowing fertilization to take place, and then transferring the fertilized embryo back into the woman's uterus where, with luck, the embryo will successfully implant into the wall of the uterus, resulting in pregnancy.

It's a lengthy and complicated process requiring the woman to take a series of hormones so that her ovaries will be encouraged to produce multiple mature eggs and her womb will be prepared to accept the embryo. Timing is everything. Eggs have to be surgically collected at exactly the right stage for fertilization to take place.

The eggs are fertilized in a laboratory dish with sperm provided by the partner or a donor and then allowed to grow and divide for up to five days before being transferred back into the womb. In the early days of IVF, multiple eggs were transferred back into the womb, and it was not uncommon for the result to be twins, triplets, or even more babies. But a multiple pregnancy is dangerous for both the mother and the babies, so the transfer of only one or two eggs is now routine. When more eggs are fertilized than are used, the remaining embryos are frozen and transferred at a later date.

> Once you put human life in human hands, you have started on a slippery slope that knows no boundaries.
>
> Leon Kass, bioethicist, b. 1939

DESIGNER BABIES

From the early beginnings of IVF and other assisted reproductive technologies, ethical debates have raged. Critics liken it to "playing God," with the Catholic Church being one of its most outspoken opponents. But nothing is more contentious than the potential to create designer babies. The development of techniques to test the embryo for the presence or absence of particular genes has made this a real possibility, even if it is unlikely for now.

Sometimes couples turn to IVF to prevent a severe genetic disease being passed on to their children. Preimplantation genetic diagnosis (PGD) is a technique that allows the removal of a single cell or two from the embryo for genetic testing. The presence of mutations responsible for specific genetic diseases, such as cystic fibrosis or muscular dystrophy, can be identified and those embryos discarded. Chromosomal problems can also be identified using this technique. Controversially, the test can also be used to identify the sex of the baby, and the potential use of the technology to select embryos with desirable attributes has led to much debate about designer babies. At left are the L'Esperance quintuplets, born in 1988. They were the first IVF quintuplets born in the USA.

Below: Dr Paul le Roux extracts eggs from a woman at the Cape Fertility Clinic in Cape Town, South Africa. The clinic has had such a high success rate that foreign women seeking pregnancy sometimes holiday in Cape Town while undergoing treatment at the clinic.

Opposite: Micromanipulation of a human metaphase II egg to remove a polar body (a small cell that buds off from the maturing egg) for genetic analysis. Controversially, pre-implantation genetic diagnosis (PGD) can be used to screen for genetic diseases.

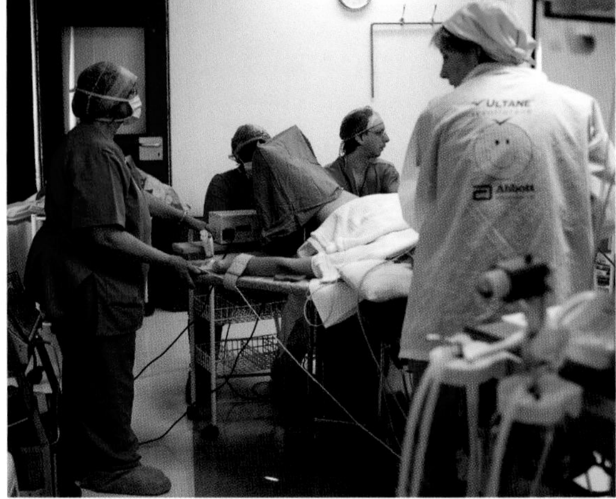

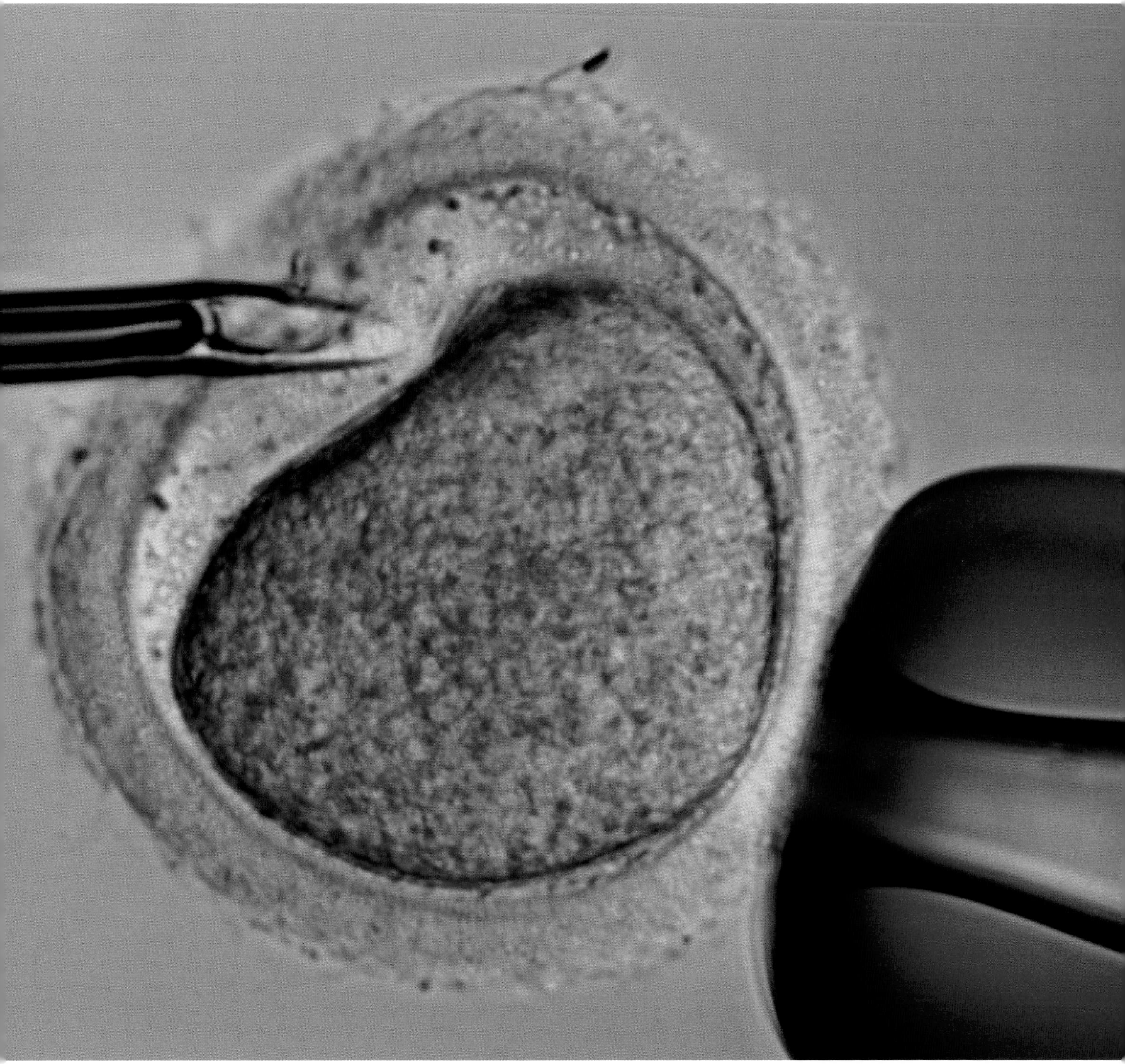

Other IVF

Another form of IVF is used when very low numbers of viable sperm is the cause of infertility. Known as intracytoplasmic sperm injection (ICSI), the technique is used to inject a single sperm into the egg using a very fine glass needle. The eggs are then cultured in the same way as for standard IVF procedures.

Other procedures

Other procedures include intra-uterine insemination, where sperm is injected into the uterus through the cervix to coincide with the release of eggs. There is also gamete intra-fallopian transfer, where the eggs are first removed from the woman, then placed back into her fallopian tube along with the sperm.

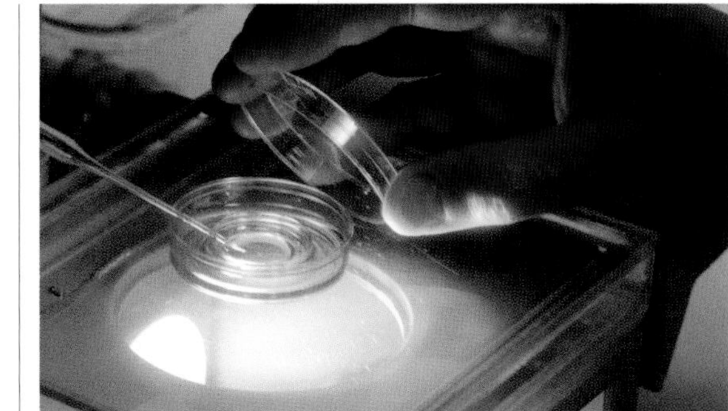

Right: A doctor fertilizes a woman's ovum as part of an *in vitro* fertilization prodecure. Early biological experiments used glass containers, hence "in vitro," which means "in glass". Today, however, the term refers to any biological procedure performed outside the organism it would normally be occurring in. An *in vivo* procedure is one in which the tissue remains inside the living organism.

Many famous writers and musicians of the eighteenth and nineteenth centuries, such as John Keats and Wolfgang Amadeus Mozart, died unexpectedly at young ages from infectious diseases. Had these distinguished individuals lived in more recent times, they probably would have enjoyed long lives and passed away from very different illnesses.

Deadly afflictions

Left: Composer Wolfgang Amadeus Mozart (1756–1791) died of "severe miliary fever" according to his doctors, but this only identifes a symptom (a rash that looks like millet seeds), it does not identify the cause. Many theories have been proposed: Trichinosis, influenza, mercury poisoning, and a rare kidney ailment. The likeliest cause is acute rheumatic fever; Mozart had had three or four attacks since childhood, and this disease tends to recur.

Opposite: Digitally enhanced chest X-ray of lungs showing cancer. Lung cancers are usually carcinomas derived from epithelial cells. The most common cause of cancer-related death in men and the second most common in women, they cause 1.3 million deaths annually. Non-smokers account for fewer than 10 percent of cases. With treatment, the five-year survival rate is currently 14 percent.

In 1900 the average life expectancy at birth in the USA was 47 years. Today, this has increased to a record high of 77 years. A century ago, death was most commonly caused by acute infectious diseases, such as tuberculosis, typhoid fever, syphilis, and pneumonia. However, due to improved nutrition, sanitation, and medical advances, we now live longer than ever and the major causes of death have changed. They are often related to ageing, and include cardiovascular disease, cancer, stroke, and chronic obstructive pulmonary disease. Lung infection remains the only leading infectious cause of death.

Cardiovascular disease

Cardiovascular disease is a broad term that is used to describe several heart conditions. The most common cardiovascular disease is called coronary heart disease, which can lead to a heart attack (myocardial infarction) and other life-threatening conditions. Cardiovascular disease is the leading cause of death in the USA and the UK.

Coronary heart disease is caused by a build-up of fatty deposits, called plaque, on the inner wall of the arteries that supply the heart muscle with blood. This plaque gradually clogs and narrows the inside channel of the arteries in a process termed atherosclerosis.

A myocardial infarction is usually set underway when an area of plaque is somehow disrupted. Blood cells and other components of the blood stick to the disrupted area and form a clot that suddenly restricts the blood flow to the heart muscle. This causes heart muscle cells to begin dying due to a lack of oxygen and can lead to acute heart failure, where the heart cannot pump blood at the rate and pressure required. It can also lead to arrhythmias, which are potentially fatal irregular or slow heart beats.

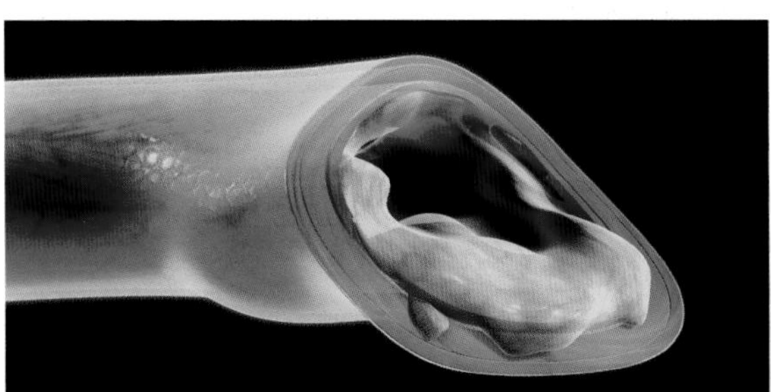

Left: Artist's representation of a partly blocked artery, showing plaque build-up. Plaque, made up of fat and cholesterol, reduces or stops the blood flow to parts of the body served by the artery. In peripheral vascular disease, for example, the supply of blood to the arms and legs is reduced, causing them to hurt or feel numb. Tissue death can result.

THE DEADLY CONSEQUENCES OF TOBACCO USE

Tobacco use is a major risk factor for cardiovascular disease, lung cancer, stroke, and chronic obstructive pulmonary disease. Despite this, its use is common throughout the world. Tobacco use is responsible for the deaths of one in ten adults worldwide and will kill approximately one-third of people who use it. If current trends continue, tobacco use will kill millions of people prematurely this century.

Tobacco products are made entirely or partly of leaf tobacco and are intended to be smoked, sucked, chewed, or snuffed. All contain the highly addictive psychoactive ingredient, nicotine.

A number of countries have developed legislation to restrict tobacco advertising and regulate who can buy and use tobacco products and where they can be used. However, the tobacco industry's history, its power and influence, and the nature of addiction make it a difficult public health issue worldwide.

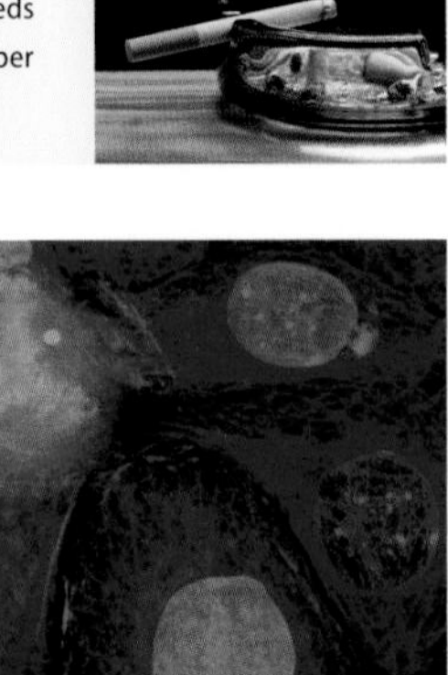

The harmful effects of tobacco use do not end with the user. Women who use tobacco during pregnancy are more likely to deliver babies of low birth weight, which is linked to an increased risk of infant death and a variety of infant health disorders. Environmental tobacco smoke has also been estimated to cause thousands of non-smoking Americans to die of lung cancer each year and hundreds of thousands of children to suffer from upper and lower respiratory tract infections.

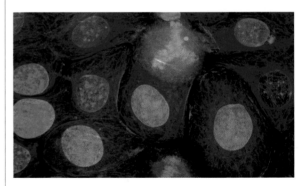

Right: Cancer cells from a human breast (laser scanning confocal micrograph), showing cell division occuring in some cells. The breast is a frequent site of tumors, both benign and malignant. Women are advised by the medical profession to examine their breasts regularly to check for any unusual lumps. Worldwide, breast cancer is the second most common type of cancer after lung cancer.

Cancer

Cancer is a general term used to describe diseases in which cells of the body begin to grow in an uncontrolled and purposeless way. In some cases, these cells move around and spread to other parts of the body. Cancer is the second leading cause of death in the USA and can arise from almost any cell type in the body, though it occurs most often in lung, breast, prostate, and bowel cells.

Cancer is caused by defects, called mutations, occurring in the genes of a cell. Some genetic mutations can cause the cell to begin producing faulty proteins. This can in turn cause the cell to become abnormal and begin to divide uncontrollably and, eventually, form a new growth known as a tumor.

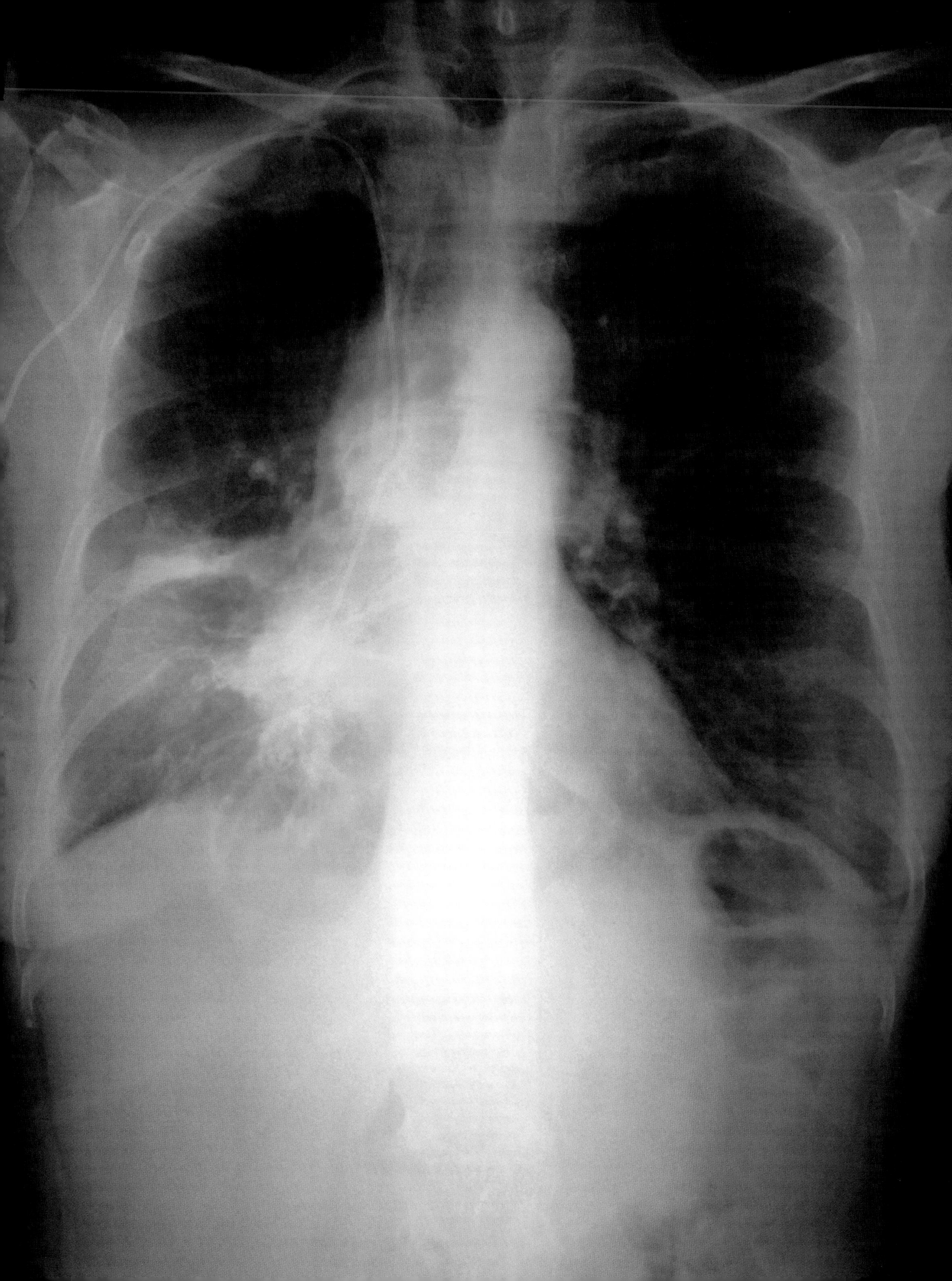

Tumors can be either benign or malignant. A benign tumor does not spread around the body and is unlikely to become life-threatening. A malignant tumor, on the other hand, grows into the surrounding tissues and eventually spreads to other parts of the body—a process known as metastasis. The malignant tumor cells then continue to divide and grow until they cause a critical organ to cease functioning, or until the enormous number of tumor cells becomes a burden the body can no longer sustain.

Although cancer is a disease caused by the mutation of genes, it is rarely inherited. Most mutations are caused by environmental factors called carcinogens. Hundreds of carcinogens have been identified, including ultraviolet radiation, tobacco smoke, arsenic, and asbestos.

Above: Asbestos Control Inc's Edward Zalig holds a pile of asbestos cleaned out from pipe insulation in Elk Grove Village, Illinois, USA. Asbestos, a group of minerals with long, thin fibrous crystals, was popular among manufacturers and builders in the late nineteenth century due to its resistance to heat, electricity, and chemical damage, its sound absorption and its tensile strength. However, the inhalation of asbestos fibers can cause serious illnesses, including mesothelioma and asbestosis, and since the mid-1980s many uses of asbestos have been banned in countries around the world.

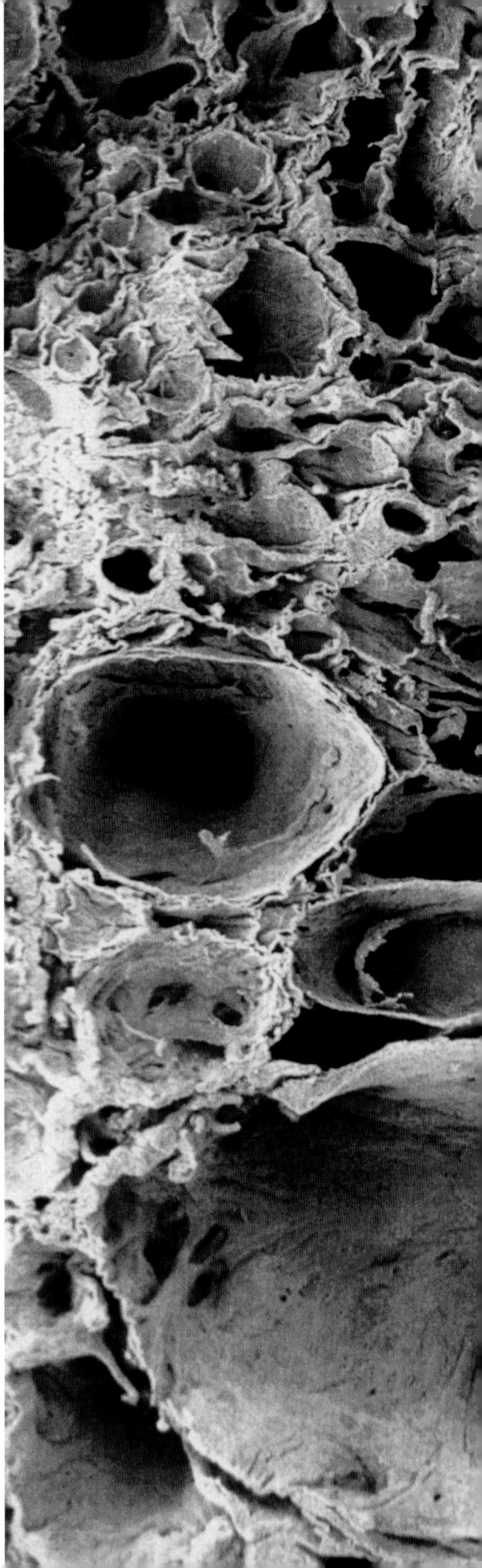

Stroke

Stroke, also known as cerebrovascular disease, occurs when an artery supplying blood to a part of the brain becomes blocked or bursts. As a result, that part of the brain becomes irreversibly damaged. Stroke is the third leading cause of death in the USA.

There are two types of stroke. An ischemic stroke is the most common type and results from blood clots in the arteries supplying blood to the brain. The clots tend to form in areas that are narrowed by a long-term build-up of fatty deposits in the arteries. This gradual clogging of the arteries, known as atherosclerosis, is the same process that can lead to myocardial infarction.

A hemorrhagic stroke, on the other hand, is caused by the sudden rupture of an artery in the brain, which in turn leads to bleeding and pressure on the brain tissue. High blood pressure is often the underlying cause of this type of stroke.

Both types of stroke cause brain tissue to be damaged or die. Interruption of a small artery may result in a speech impediment, weakness or numbness in one part of the body, or an unsteady gait. If a larger artery is involved, the result may be the total paralysis of one side of the body or, in severe cases, death.

Chronic obstructive pulmonary disease

Chronic obstructive pulmonary disease comprises two related diseases, chronic bronchitis and emphysema. In both these diseases the airways become obstructed, progressively diminishing the ability of the lungs to function properly. Chronic obstructive pulmonary disease is the fourth leading cause of death in the USA and is caused in most cases by air pollution or tobacco smoke.

Chronic bronchitis is caused by inflammation of the lining of the bronchi, which are the main airways that lead from the windpipe into

Right: Human lung showing the effects of emphysema, a condition in which the alveoli in the lungs become enlarged. This results from a breakdown of the walls of the alveoli, which causes a decrease in respiratory function, leading to symptoms such as shortness of breath and cough. The occurrence of emphysema is often associated with smoking.

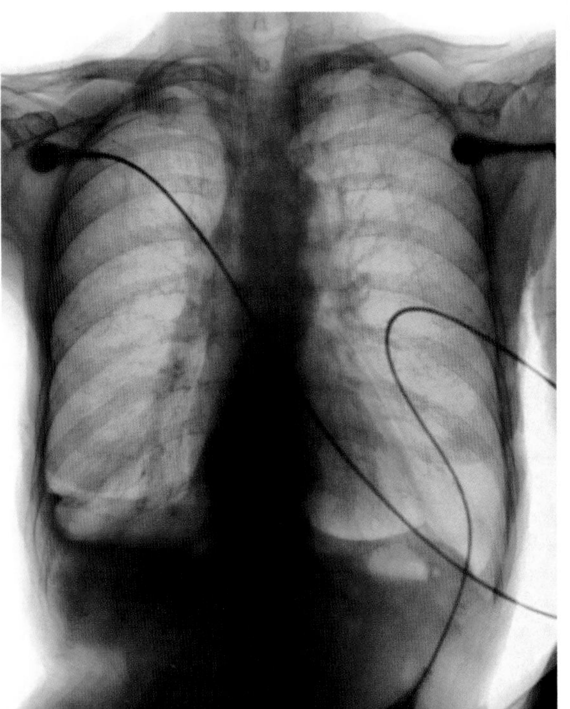

Left: X-ray of an elderly female smoker with chronic obstructive pulmonary disease (COPD) and a pulmonary effusion in the right side of the chest. Emphysema is seen as an increased lucency of the lungs, a change that is often more prominent in the upper lobes of the lungs of smokers. Hyperinflation is also present, with the chest vertically elongated and the diaphragms low and flattened. The heart shadow is also narrow.

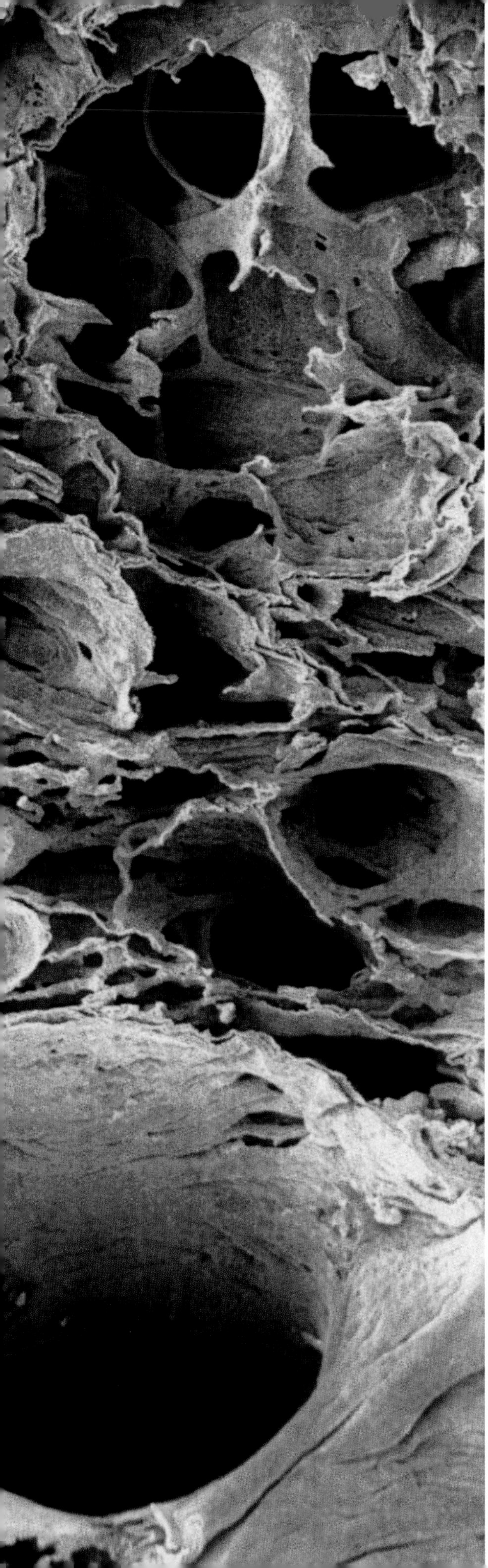

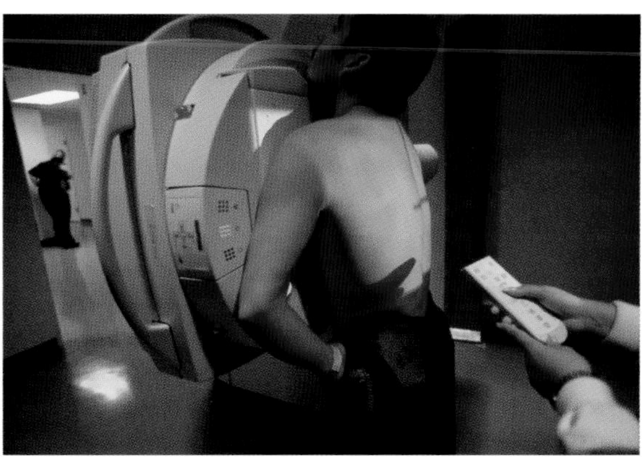

Right: An illegal immigrant is X-rayed by US authorities to check for TB (tuberculosis) infection. One-third of the world's current population has been infected by TB. A common and deadly infectious disease caused by mycobacteria, mainly *Mycobacterium tuberculosis*, it usually attacks the lungs, but can also affect the lymphatic system, the central nervous system, the circulatory system, the genitourinary system, bones, joints, and even the skin.

the lungs. This inflammation results in a narrowing of the airways, and also stimulates the production of mucus which can obstruct the airways even further. Eventually, chronic bronchitis results in reduced airflow and a severe shortness of breath.

In contrast, emphysema is characterized by enlargement of an important component of the lungs known as the alveoli. Alveoli are small sacs in the lungs that allow oxygen to enter the blood and carbon dioxide to be expelled into the air. This enlargement occurs due to the destruction of the walls between the alveoli and results in the formation of large air pockets that prevent air escaping and make breathing extremely difficult.

People with chronic bronchitis or emphysema eventually develop breathing difficulties that inevitably impair their quality of life. Ultimately, respiratory failure occurs, which prevents the body from supplying enough oxygen to tissues and cells to support life.

Each day we move closer to [scientific] trials that will not just minimize the symptoms of disease and injury but eliminate them.

Christopher Reeve, actor and activist, 1952–2004

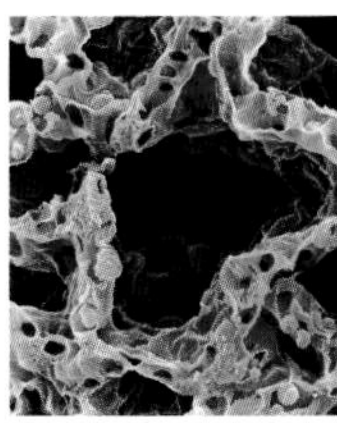

Above: Alveoli are tiny air sacs in the lung where the exchange of oxygen and carbon dioxide takes place. There are approximately 300 million alveoli in each lung. Though smaller than grains of salt, they have a very large surface area. During inspiration, oxygen diffuses through the alveoli walls into the blood; during exhalation, carbon dioxide diffuses in the opposite direction.

Major killers in low-income countries

In high-income countries, such as the USA, more than two-thirds of all people live beyond the age of 70 years and die of diseases that are often related to ageing, such as cardiovascular disease, cancer, stroke, and chronic obstructive pulmonary disease. Lung infection is the only leading infectious cause of death.

In contrast, less than a quarter of all people in low-income countries reach the age of 70 years and nearly one-third of all deaths are among children aged under 14 years. As is the case in high-income countries, cardiovascular disease represents the leading cause of death. However, a number of infectious diseases such as HIV/AIDS, lung infections, tuberculosis, diarrheal diseases, and malaria, together, claim more lives than cardiovascular disease.

Complications arising from pregnancy and childbirth continue to be another leading cause of death in low-income countries, often claiming the lives of both infants and mothers.

Currently, a relatively lower proportion of deaths are occurring each year in high-income countries than are occurring in low-income countries. Approximately one out of every six people in the world, or about 15 percent of people, live in high-income countries (mostly in North America and Europe), yet only 7 percent of all deaths occur annually in those countries.

In 1846, dentist W.T.G. Morton gave ether to a patient to enable surgeon John Warren to remove a growth from his jaw. So began the history of anesthesia at the Ether Dome, Massachusetts General.

No pain, no pain!

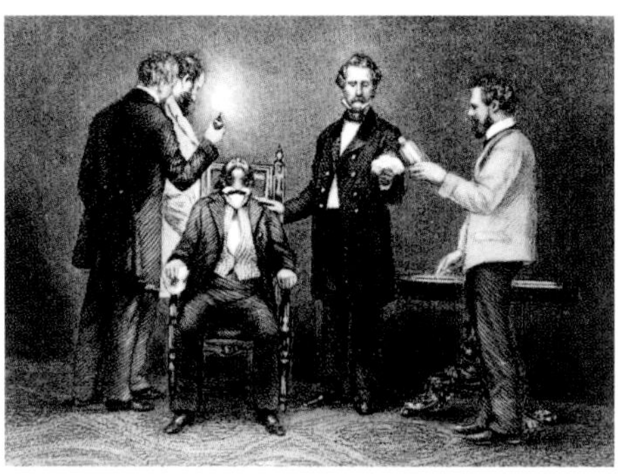

Its use travelled rapidly across the Atlantic and in 1853 John Snow gave chloroform to Queen Victoria for the birth of her eighth child, Prince Leopold. Chloroform "à la reine" made pain relief in childbirth fashionable in high society and acceptable for the masses.

Curare had been used to bring about muscle relaxation in cases of tetanus during the American Civil War, but it was not used until the mid-1940s for relaxing abdominal muscles during surgery. Thus was born the Triad of Anesthesia as we know it today: Amnesia, analgesia, and muscle relaxation. Subsequent advances in anesthesia often led to, or were a result of, advances in surgery. Eventually as it developed, anesthesia led to the formation of specialist intensive care, pain management, and perioperative centers.

Amnesia

It is not known definitively how general anesthesia works to reduce consciousness. The flow of sodium ions into the nerve cell is altered but the mechanism not clear, because anesthetic agents do not bind to receptors on the cell membrane nor do they alter transmitter substances at the nerve terminal. What is known is that when the flow of ions is stopped nerve impulses are not generated, and the patient loses consciousness. The brain does not store memories or register pain.

Analgesia

Pain relief during anesthesia may in part be provided by the use of nitrous oxide (laughing gas). Opiates such as morphine and pethidine are also used. Just as muscle relaxants combine with receptors at the nerve–muscle junction, the morphine molecule combines with special opioid receptors located throughout the body. The μ subgroup is present in the brain and gut. Combination with these leads to analgesia, respiratory depression, constipation, and constriction of the pupils.

Pain relief during and after anesthesia can also be provided by blocking nerve transmission through the use of local anesthetic agents such as lignocaine (xylocaine). These agents operate by blocking the opening of sodium channels along a nerve fiber. Epidural block is commonly used during childbirth.

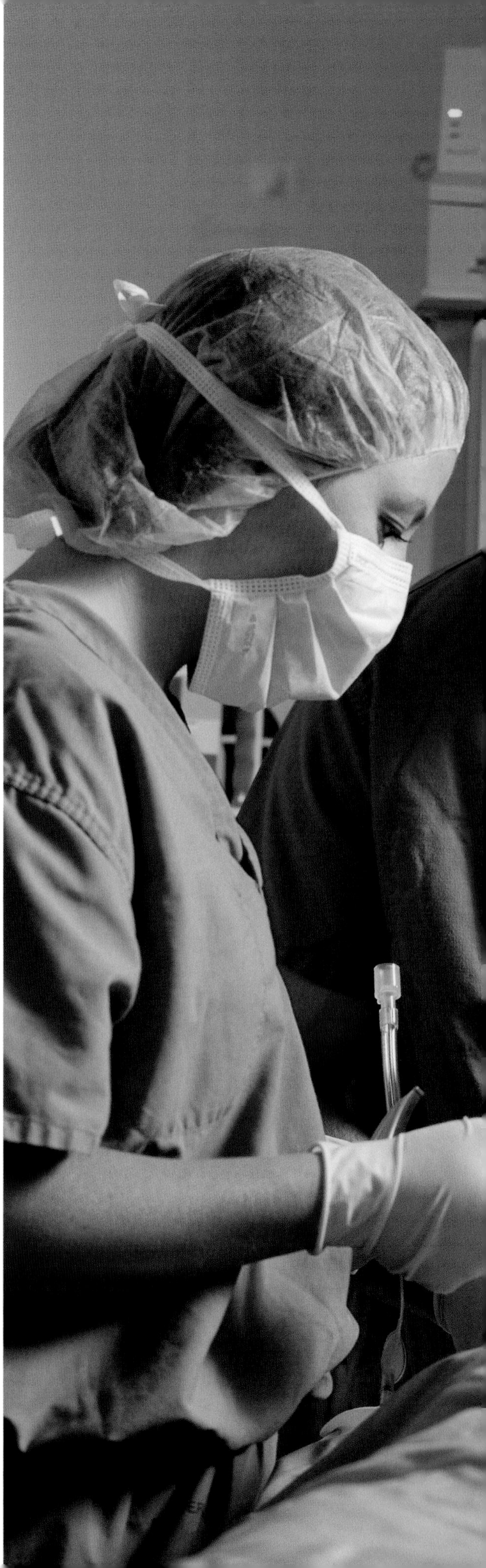

Above: Engraving depicting Boston dentist William Morton (1819–1868) administering ether for the first public demonstration of its benefits on October 16, 1846. Morton used diethyl ether (which was then called sulfuric ether) as an anesthetic agent.

Right: Coined (in a letter to William Morton) by Oliver Wendell Holmes, Sr in 1846, the word "anesthesia" means "without sensation." There are several forms of anesthesia, but all allow patients to undergo surgery with minimal pain and discomfort. Surgeons and anesthetists work hand in glove in today's operating theaters.

Left: The cultivation of opium poppies for food, anesthesia, and ritual purposes dates back to at least the Neolithic Age. Modern opium production has led to more potent methods of extraction and processing, as well as more sophisticated methods of consumption. Here, opium poppies produced for morphine approach ripeness in a field in the chalk downlands, Salisbury, England.

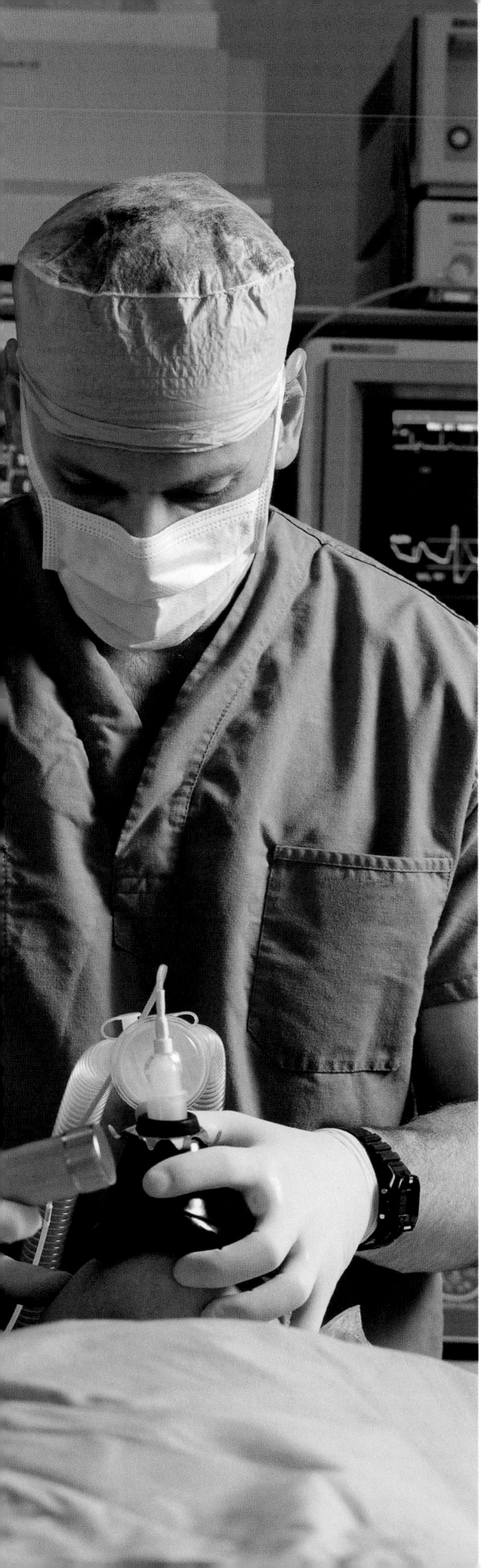

Above: German scientist Paul Ehrlich (1854–1915) developed "side-chain" or receptor theory to explain immunity and how antibodies are formed. All cells, he said, have receptors that act as gatekeepers, allowing certain substances to enter the cell.

Muscle relaxation

In the early 1900s, physiologists John Newport Langley and Paul Ehrlich developed receptor theory. This defined a receptor as a small region of a cell which, chemically combined with a drug, produces a chemical reaction leading to a biological response. To understand how drugs can combine with a receptor, think of the drug as a key and the receptor as a lock. Just as different keys fit into a lock, so different drugs fit with a receptor. The nature of the bond may vary—some drugs share electrons in their atoms with the receptor; other drugs combine as a result of electrostatic charges (opposites attract!).

Acetylcholine is released by the nerve fiber when a nerve impulse reaches the junction of the nerve ending and muscle fiber. It then crosses the junctional cleft and becomes attached to receptors on the muscle membrane. This chemical bond opens the sodium channels, allowing the inward flow of ions, depolarization, and the eventual contraction of the muscle. Muscle relaxant drugs such as curare combine with the muscle receptor, blocking the action of acetylcholine and thereby causing muscle relaxation.

> [Ether is] associated with such sweetness … they fall asleep for a while but waken later without harm.
>
> Paracelsus, alchemist and physician, 1483–1541

LAUGHING GAS

First prepared by Joseph Priestley in 1772, nitrous oxide inspired mirth in some quarters (see below), but was used by dentists for pain relief from about the mid-1800s. It was overshadowed by ether, however, for use in general surgery. The addition of oxygen to nitrous oxide led to a prolongation of anesthesia time and an increase in the popularity of nitrous oxide as an anesthetic.

Today nitrous oxide is also commonly used in general anesthesia. Anesthesia and analgesia are rapidly induced due to the chemical properties of nitrous oxide (low blood–gas coefficient), which replaces nitrogen in air-filled spaces at a faster rate than nitrogen can diffuse out. For every nitrogen molecule removed from air spaces, 35 molecules of nitrous oxide will pass in. This chemical property can lead to undesirable results, such as a rapid expansion of gas causing an increase in pressure if the space is closed, as in the case of a pneumothorax (collapsed lung). Nitrous oxide has a disinhibiting effect on consciousness and can cause the patient to giggle prior to becoming unconscious.

Society's attitude toward mental illness has changed dramatically over time, resulting in changes to the way mental illness is treated. Over the years, treatments have ranged from ritual, religion, and magic—including eradicating evil spirits, blood letting and spinning to tranquilize the "mad"—to probing the psyche and prescribing medication.

Understanding mental illness

Left: Austrian neurologist and psychiatrist Sigmund Freud (1856–1939) is considered to be the father of psychoanalysis. By putting his patients in a relaxing position, such as lying on a couch, he encouraged them to say whatever came into their minds. Known as "free association," this practice became the foundation of modern psychoanalysis.

Traditionally, treatment of mental illnesses has focused on a single factor, regardless of the perceived cause of the illness. But that is changing. Recent insights into the brain and mind reveal that our thoughts, feelings, and emotions are shaped by multiple factors, and that no one component can adequately explain the development of mental illness. Today, the predominant view is that individuals may inherit a susceptibility to a mental disorder, but whether or not it develops is determined by complex interactions between multiple genes, psychosocial events, and environmental stressors.

Current treatments reflect this view, with the two primary treatment categories being psychotherapy, which focuses on behavior, and somatic therapy, which is used to treat biological imbalances. For the majority of disorders, treatment tends to be most successful when psychotherapy and medication are used in combination. However, there is no cure for mental illness at present, and treatment aims to minimize symptoms and permit a normal lifestyle.

Psychotherapy

Psychotherapy is an interpersonal intervention, based on the assumption that the disorder stems from the way the patient reacts to and perceives the world. The therapist aims to create an empathetic and accepting environment that allows the patient to understand the basis of their problems, and to find solutions.

Psychoanalysis, once the dominant form of psychotherapy, is still practiced. However, psychotherapies focusing on interpersonal relationships, developing new ways of thinking and acting, and coping techniques such as relaxation, are now more common.

Left: Patients at an "insane asylum" or psychiatric hospital in Ohio, USA, 1946. Asylums did not become widespread in most parts of the world until the nineteenth century, though the first were built in the medieval Islamic world in the eighth century. Bethlem Royal Hospital (Bedlam) was the first in Europe. Founded in London in 1247, it became infamous for its cruel treatment of the insane. In 1700, "lunatics" were called "patients" for the first time, and by 1720 separate wards for the "curable" and the "incurable" had been established. Mental illness would eventually be regarded as a disease, to be diagnosed and potentially cured.

Somatic therapy

Somatic therapies include drug therapy and electroconvulsive therapy. Modern drug therapy began in the 1950s with the discovery of chlorpromazine and lithium to treat schizophrenia and bipolar disorder, respectively. Many thought that these "wonder drugs" would cure mental illness. Yet to date no "magic pill" exists.

The major psychotherapeutic drugs are generally categorized by their therapeutic application. Antidepressants primarily treat clinical depression; anxiolytics treat anxiety disorders and related problems such as insomnia; mood stabilizers are used for bipolar disorder, where they target mania rather than depression; and anti-psychotics treat psychotic disorders, such as schizophrenia.

Electroconvulsive therapy is used for severe depression that does not respond to other interventions.

Right: Throughout history people of different cultures have often explained deviant or abnormal behavior as the work of demons, evil spirits, and poisons. As a result, magical approaches to therapy evolved. With the development of the Christian church during the Middle Ages, exorcism, shrines, and saints became of great importance for the treatment of mental illness. In this panel from the Bernhardi Altar in Zwettl, Austria, St Bernhard is shown healing a demoniac.

Left: Vaslav Nijinsky (1890–1950) was one of the twentieth century's greatest dancers. He retired from the stage in 1919 at the age of 29 and, having been diagnosed schizophrenic, spent the rest of his life in and out of mental institutions. Before being committed, he wrote his *Diary*, combining elements of autobiography with appeals for compassion toward the less fortunate, and for vegetarianism and animal rights. Feeling, he argued, was just as important as reason.

The future

Continued progress enabling scientists to better understand brain function and genetics is likely to have a profound impact on the diagnosis and treatment of mental illness.

Mental illness is currently diagnosed by the symptoms a person experiences, but many disorders have similar symptoms, and this makes accurate diagnosis difficult. Identification of genetic variations related to specific disorders would enable clinicians to develop a more precise diagnostic system based on unique biological markers.

As details emerge about the brain circuits altered by mental illness, new targets for drug therapy will be identified, paving the way for medications with increased specificity. Ultimately, scientists are likely to pinpoint the genes that make certain individuals vulnerable to mental illness. This knowledge, together with the identification of modifiable psychosocial and environmental risk factors, could change the focus of psychiatry from treatment to prevention.

> The study of the human brain and its disease remains one of the greatest scientific and philosophical challenges ever undertaken.
>
> Floyd E. Bloom, neuroscientist, b. 1936

Above: English novelist and essayist Virginia Woolf (1882–1941) suffered from bipolar disorder, which affected many aspects of her life and work. In 1941 she took her own life by weighing down her pockets with stones and walking into the River Ouse. "I feel certain that I am going mad again," she wrote in a farewell letter to her husband. "I begin to hear voices, and I can't concentrate. So I am doing what seems the best thing to do."

MINDFUL MEASURES

Brain imaging allows scientists to visualize the brain and study the relationship between brain structure, function, and behavior. The images allow scientists to learn more about the structural and chemical changes that occur in the brain of someone with a mental illness, observe changes that occur as the brain processes information, and record responses before and after treatment to determine effectiveness.

Three commonly used procedures are positron emission tomography (PET), magnetic resonance imaging (MRI), and functional magnetic resonance imaging (fMRI). PET (below) answers questions about brain function by showing regional variations in brain activity in response to a stimulus. It measures changes in blood flow or metabolic activity, such as glucose utilization, by detecting positrons: Positively charged particles emitted by radioactive tracers that have been injected into the body.

MRI shows a detailed anatomical view of the brain. It uses magnetic fields and radio waves to produce high-quality, two- or three-dimensional images of brain structures without injecting radioactive tracers.

fMRI provides functional and anatomical views of the brain. It uses the magnetic properties of blood to detect blood flow, showing changes in brain activity as patients perform tasks or are exposed to various stimuli.

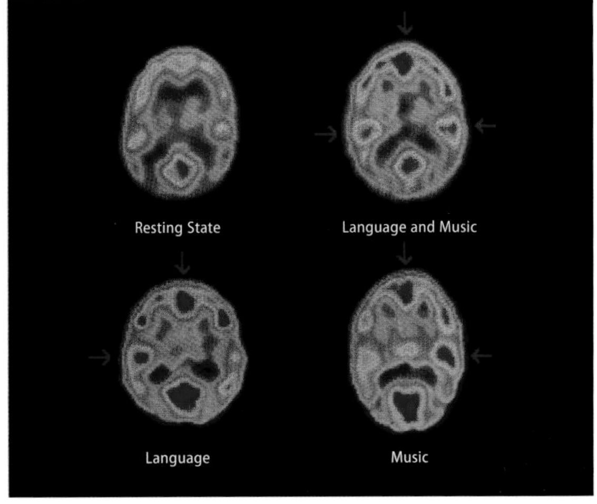

Many societies regard addicts and alcoholics as weak, morally corrupt individuals who have brought addiction on themselves. The science of addiction, however, supports the view that addiction is a brain disease with genetic as well as physical, social, and psychological risk factors.

Addiction

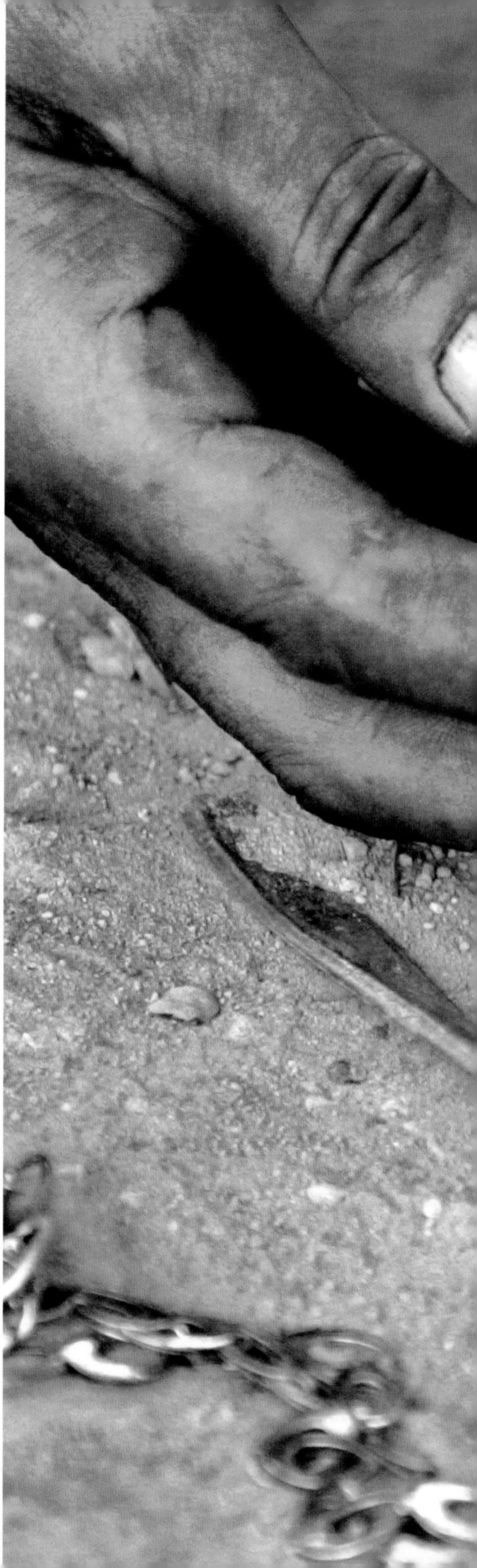

The science of addiction

Addiction is the continued and uncontrolled use of a substance despite the fact that it causes severe physical, psychological, or social consequences for the user.

With the diagnostic tools now available, including positron emission tomography (PET) scanning and functional magnetic resonance imaging (fMRI), we can look into the brain to see the structural and functional risks and effects of addiction. This, along with sophisticated human and animal studies conducted over many years, allows us to gain a deeper understanding of the disease.

Above: The *Journal of the American Medical Association* defines alcoholism as "a primary, chronic disease characterized by impaired control over drinking, preoccupation with the drug alcohol, use of alcohol despite adverse consequences, and distortions in thinking."

Reward, learning, and addiction

Deep in the primitive midbrain, below conscious thought, we have a nerve pathway (the mesolimbic dopamine pathway) that gives us a reward for doing those things that help us survive. Thus we get good feelings from food, sex (reproduction), and social interactions. This "go" pathway is linked to learned reward associations so that we can successfully repeat the process and have a higher chance of survival (for example, knowing where to find food again).

Euphoria-producing drugs or alcohol, when used repeatedly, overstimulate this nerve pathway. This "fools" the brain into thinking that the drug is by far the most important thing for survival. Subsequently, we learn and become highly sensitized in other areas of the brain to all things associated with acquiring and using the drug. As use continues, the reward pathway becomes less functional and, ultimately, the addict becomes compelled by learned associations to use the drug even though he/she is no longer getting rewards from it.

Decision-making and addiction

Nerve connections (mesocortical pathway) from the reward areas of the brain project to the decision-making prefrontal cortex of the brain, which is the "stop" area of the brain. As one continues to use the drug, alterations in brain chemistry in this area of the brain result in an impaired ability to plan behaviors in response to goals. As a result, the addict becomes unable to prioritize goals and adequately suppress inappropriate impulsive decisions.

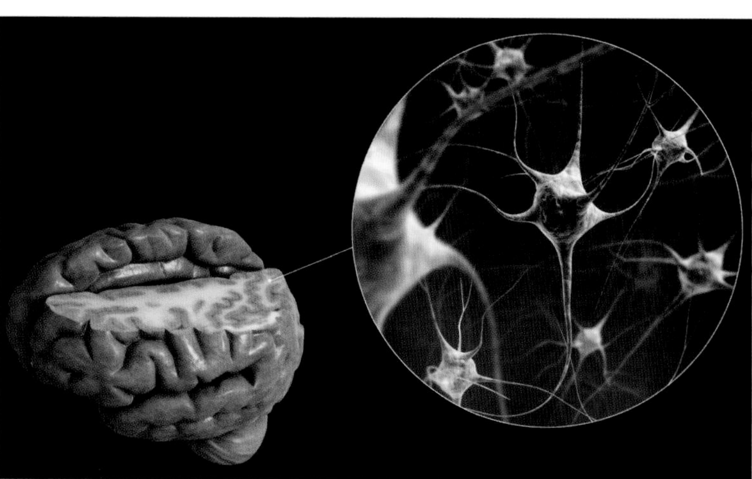

Below: Section of the brain with close-up of neuron tissue. In recent years, addiction has become a topic of considerable interest for neuroscience. The neurobiology of the learning and memory processes that characterizes addiction has a lot to teach science about the brain and how it works.

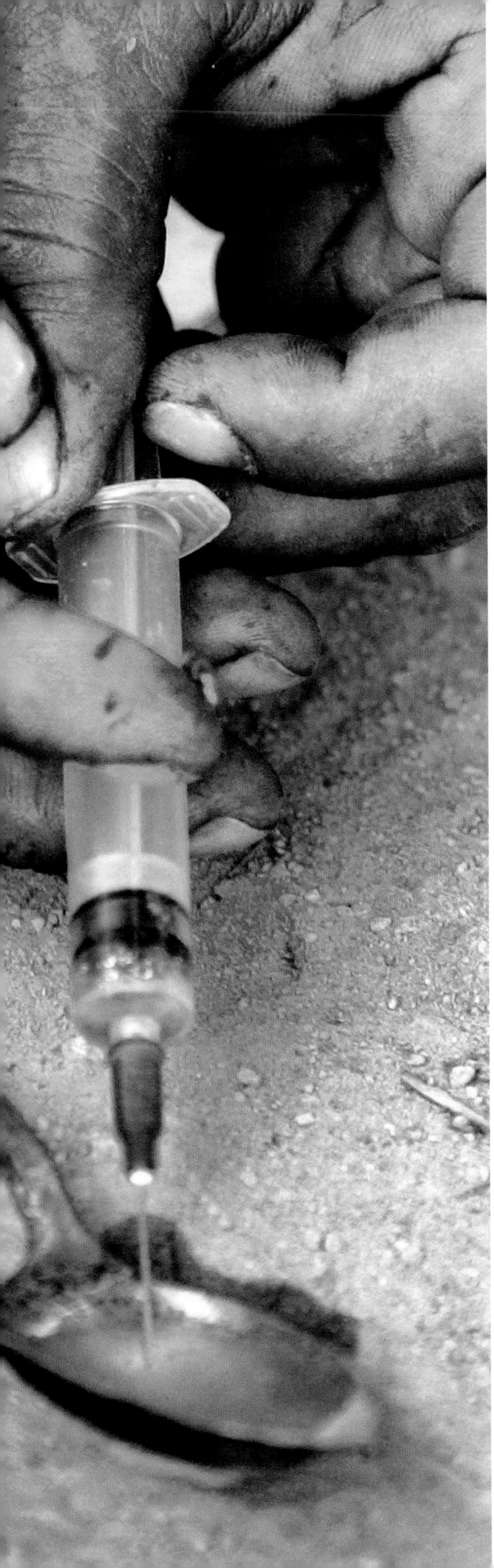

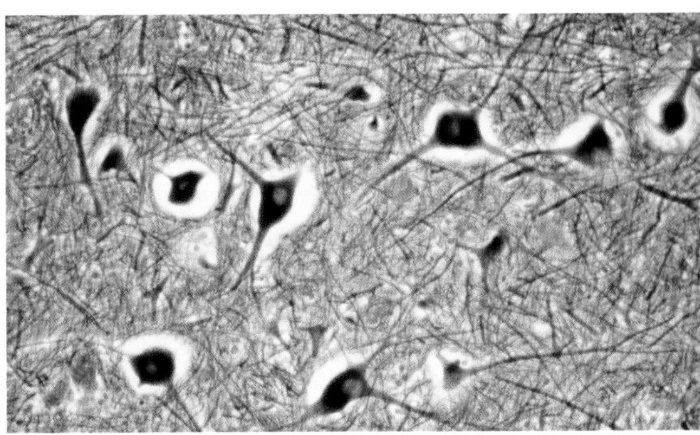

Right: Nerve cells (neurons) in a human brain stem. The nervous system, which comprises vast circuits of these delicate cells, all elaborately interconnected, is the body's control and communication network. Neurons sense changes both inside and outside the body, then interpret and explain them, and coordinate a response. Neuroscience has been studying the relationship between the brain's "reward" circuitry and addiction.

Although the addict may initially have made a decision to use the drug, continuing to use it leads to the hijacking of the "go" and "stop" areas of the brain. This results in overwhelming cravings to seek and use the drug, with simultaneous impairment in the ability to see the error of one's way and stop, despite the severe consequences. The addict is left wondering "why did I do this again?"

> I admire anyone who rids himself of an addiction.
>
> Gene Tierney, actress, 1920–1991

Other risk factors

Common risk factors for addiction include a dysfunctional family, or abuse in childhood or adolescence, parental use or attitudes, mental illness (for example, attention deficit, anxiety, or depressive disorders), and peer pressure. Further, there is evidence of a "developmental traumatology" in which many of the same brain areas hijacked in addiction develop abnormally in childhood and adolescence due to the stress of neglect or abuse.

Treating addiction

Although the changes in the brain associated with addiction may seem insurmountable, treatment can be quite effective. When needed, the brain has an amazing ability to change in structure and function (neuroplasticity). To effect change, treatment must be individualized, adjusted to the addict's changing needs, and must include treatment of coexisting mental disorders (dual diagnosis). Further, treatment must not be limited to detoxification from the drug (withdrawal), must be sufficient in length (a minimum of 90 days), have counseling and behavioral components, and include close monitoring and any required medical therapies. The goal of treatment is not just to stop drug use, but to return the addict to a life of productive sobriety.

Left: A drug addict prepares a dose of heroin before injecting himself. Statistics in the US suggest that in the 25–49 age group, drug overdose causes as many deaths as motor vehicle accidents. In recent years, the fear of infection (particularly HIV/AIDS) from sharing needles has led to an increase in the number of addicts who smoke or snort the drug.

THE GENETICS OF ADDICTION

Approximately 40–60 percent of addiction is considered genetic in origin. For example, a person who has a parent with an addiction has four to five times the risk of developing the disease.

There is evidence of a genetically determined "reward deficiency syndrome" in which a person under-experiences normal rewards in life. Related to decreased dopamine (pictured) functioning in the reward pathway, this syndrome may contribute to the initial decision to use a drug, as well as to the subsequent development of an addiction.

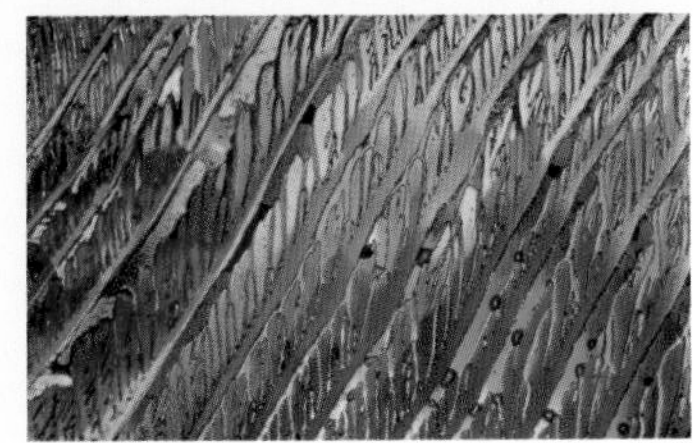

Frequently, addicts and alcoholics will report that when they first experienced the "high," they didn't just have a good time, but "found the answer to my life."

In the face of ageing, chronic disease, and social difficulties, geriatric medicine aims to maximize function. Although geriatricians don't usually cure disease or add years to older people's lives, they can improve the quality of people's lives.

Geriatrics: adding life to our years

Marjory Warren

In London in 1935, when elderly patients were essentially left to die in "Poor Law Infirmaries," Marjory Warren, the mother of geriatrics, took responsibility for the care of 714 patients—many of whom were old and infirm. She developed a rehabilitative, optimistic team approach, resulting in 35 percent of these people leaving the facility.

In response to Warren's work and her calls for a geriatric specialty, geriatrics developed first in Britain and later progressed to the United States after the American Geriatrics Society formed in 1942.

Geriatrics is now the largest specialty in Great Britain. But at a time when the "baby boomers" are ageing and the proportion of older people in the population is steadily growing, fewer physicians are training in the field in the USA.

Ageing

People aged 65–75 differ from younger adults due to ageing, chronic illness, multiple medications, and social needs which require a special understanding and approach. While ageing isn't a disease in itself, it does change outward appearance and reduces the ability to maintain self-regulated, stable physiologic functioning (internal homeostasis). As a result, the elderly are more susceptible to seemingly minor changes, such as changes in medications or the environment.

Ageing also alters the experience and presentation of disease in the elderly. An 80-year-old with acute appendicitis may experience mild abdominal discomfort, whereas an adolescent with the same condition may experience more severe pain. An elderly person with pneumonia may simply appear lethargic or disoriented, while a young adult with pneumonia may present with shaking chills, fever, productive cough, and shortness of breath. It's up to the geriatrician to look for and find the "hidden" disease.

THE ELDERLY DEFY CHRONIC ILLNESS

Maude always went to her doctor's office projecting a stately, confident, and kind demeanor. She answered questions appropriately, and expressed her sense of humor. She had 300 people at her 100th birthday party, which she thought was unnecessary, but enjoyed anyway. She became confused over the last months of her life, but was cared for at home, as are 95 percent of the elderly. She was always a pleasure to see, and despite her chronic illnesses she lived to over 103.

Andy was on many medications, including large doses of narcotic pain drugs, and would shout and throw things about his room. However, a couple of weeks after his doctor took him off many of the drugs, Andy became pleasant and conversational. Despite a hip fracture, a neck fracture, pneumonias, and multiple other illnesses, Andy was always comfortable and lived to the age of 100. (The picture shows a double hip replacement.)

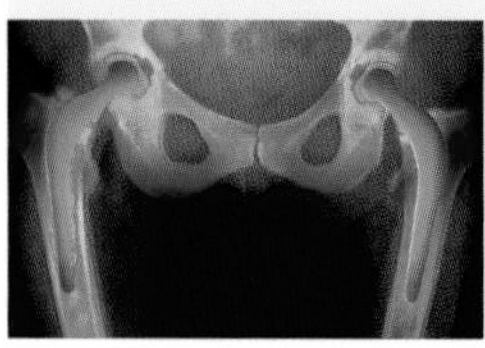

Left: As scientists discover more about the human body and how it works, the benefits of exercise and nutrition become ever more apparent and better understood. Here, a fit-looking senior citizen rehydrates after a gym workout.

Opposite: Variations in life expectancy around the world are due mainly to differences in public health, nutrition, and medicine between countries. Poverty also has a has a major impact. People live on average just over 32 years in Swaziland, 68½ years in India, and 81 years in Australia. Until the twentieth century, one in four women died in childbirth. Now, in developed countries, they outlive men by five years or more, though recently men have been catching up.

Chronic disease and excessive medications

Elderly people generally have multiple, incurable, but controllable chronic illnesses, such as heart, lung, or kidney disease to which they have adapted. It is commonplace for an 80-something person with congestive heart failure, diabetes, and renal failure to walk into their doctor's office and report that he or she is feeling fine.

Also, elderly people can have multiple, inappropriate, or excessive prescriptions, risking adverse interactions. Many well-intentioned physicians, having treated a person's various problems over the years, might be reluctant to stop a medication in case illness returns.

Death and dying

Death is a natural part of life, yet many of us, including physicians, are uncomfortable talking about it. Geriatricians commonly assist people who are dying, and their families. This can be a very rewarding aspect of their work, helping an older person die with dignity and without pain, and with their family. Educating the older person and their family about what to expect is a service that is highly appreciated, but one that requires sensitivity and good communication skills.

Planning in advance what an elderly person does or does not want—such as antibiotics, artificial feeding, and resuscitation—and determining who makes the decisions, if an elderly person is unable to make their own decisions, avoids conflict and uncertainty, and requires high ethical standards. Allowing elderly people and their families to make informed decisions (respect for autonomy), rather than attempting to promote the geriatrician's agenda (paternalism), is usually deemed the most appropriate approach.

> Old age is not a disease—it is strength and survivorship, triumph over all kinds of vicissitudes and disappointments, trials and illnesses.
>
> Maggie Kuhn, social activist, 1905–1995

Right: In fourteenth century Europe, great emphasis was placed on dying a "good death" according to Christian belief, a death that led to salvation. The Black Death was devastating Europe, and wars and other diseases tended to cut life short. Eventually, manuals were produced, telling the dying what to expect, and prescribing prayers, actions, and attitudes. Here, a dying man attends to one of the more practical duties, dictating his last will and testament.

Drugs that attack the root of a problem, treatments tailored to a patient's genetic profile, and artificial organs or, better yet, organs grown from a patient's own cells, are just some of medicine's latest innovations. As scientists uncover more detail about the way the human body works, new insights and new technologies are leading to new, innovative drugs, treatments, and devices.

Medicine's new face

Therapy just for you

Imagine going to the doctor and having your DNA profile checked before you are prescribed a personalized treatment for your ailment. Well, it's no longer science fiction; it's rapidly becoming a reality as scientists make astonishing advances in understanding just how our genes contribute to disease.

A great example is the use of a simple genetic test to determine whether an individual patient is likely to benefit from statin drugs, which lower circulating cholesterol levels in individuals at risk of developing cardiovascular disease. Genetic analysis has determined that there are two variants of the gene in the cholesterol-producing pathway targeted by these drugs, and one of the gene variants responds better to the drug treatment than the other.

Molecular solutions for molecular problems

Increasing knowledge about the fundamental pathways and processes of cells is also revolutionizing medicine. The first hormone to be produced by genetic engineering techniques (also known as recombinant

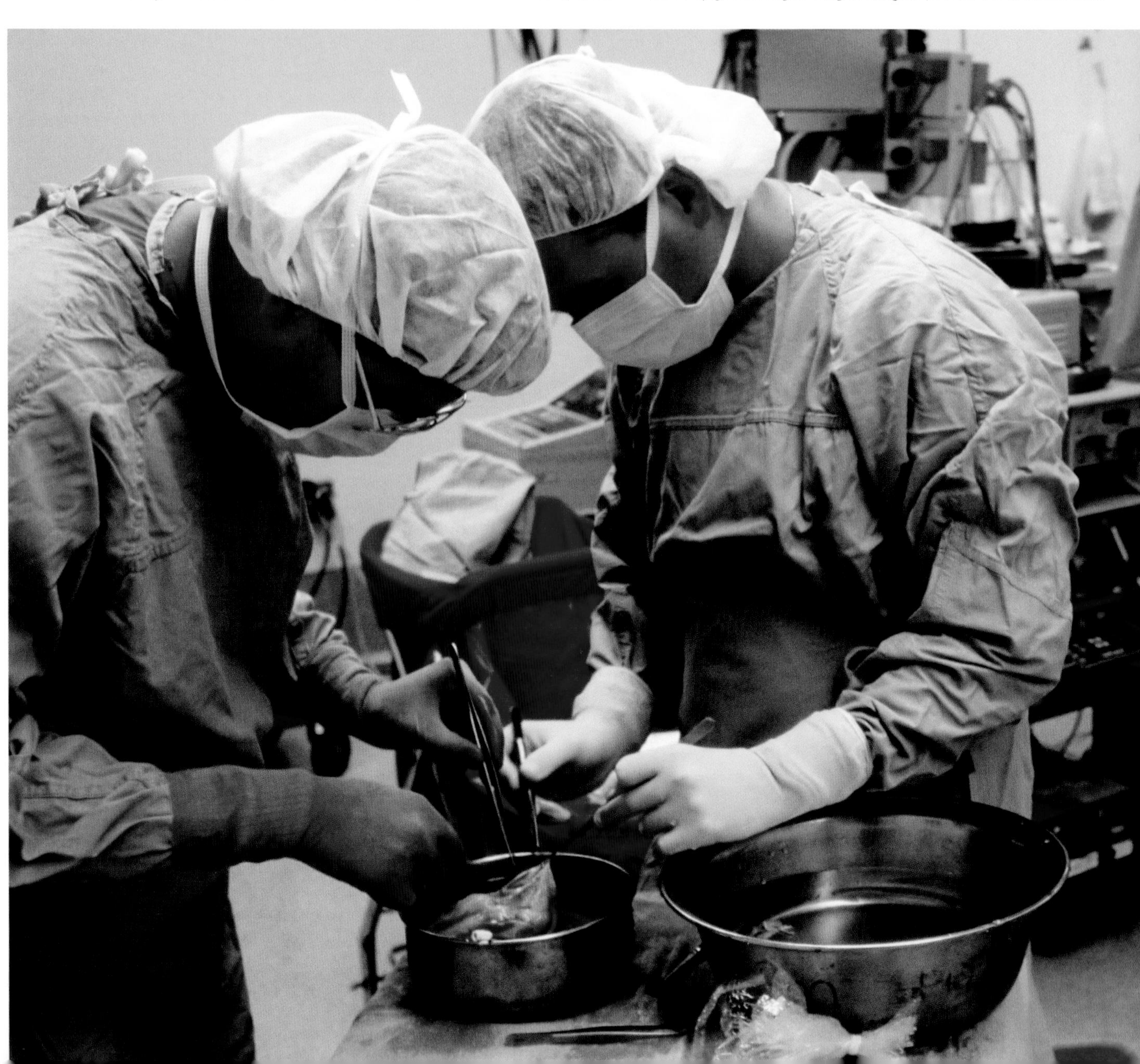

Below: Since 1967, human to human heart transplants have had a high rate of success and good postoperative longevity. A working heart is usually taken from a recently deceased human donor, but is sometimes man-made. The use of animal hearts is also being explored.

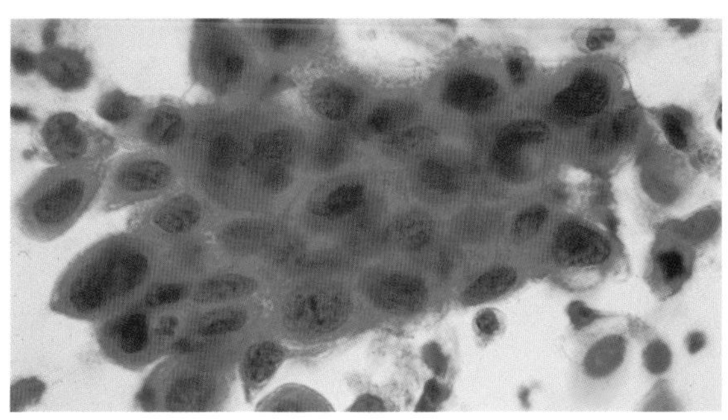

Left: Cancer cells in the cervix. In 2005, an effective vaccine against cervical cancer was developed by Dr Ian Frazer of the University of Queensland, Australia. It was soon approved by the US Food and Drug Administration and is now recommended for girls and women aged 11–26.

DNA technology) was insulin. Its release in 1982, as an alternative to animal insulin, was soon followed by the production of other small protein drugs, including growth hormones. Since that time, similar methods have been used to produce a growing number of other drugs, including the Hepatitis B vaccine, and antibody drugs for the treatment of cancer and other diseases.

A thorough, detailed understanding of the molecular structure of the influenza virus protein, neuraminidase, allowed researchers to design a small molecule called zanamivir. This molecule specifically inhibited the activity of neuraminidase in the first effective treatment and vaccine for the sometimes deadly influenza virus.

Since then, similar drug development methods have been used to design other types of drug, including antidepressants, anti-psychotics, and treatments for glaucoma, heartburn, and cancer.

Transplantable tissues

In the not-too-distant future, artificial tissues and organs grown from a patient's own cells or from a generic source of cells may replace donor organs in transplants. Already, skin cells can be harvested from an individual and grown in a laboratory to provide sheets of skin for treating burns and other severe skin damage. In the future, bone, cartilage, muscle, and blood vessels may be produced to repair damage sustained through disease or injury. Ultimately, a whole new organ—perhaps a heart or a kidney—may be grown for transplant.

> Medical science has proven time and again that when the resources are provided, great progress in the treatment, cure, and prevention of disease can occur.
>
> Michael J. Fox, actor and Parkinson's disease advocate, b. 1961

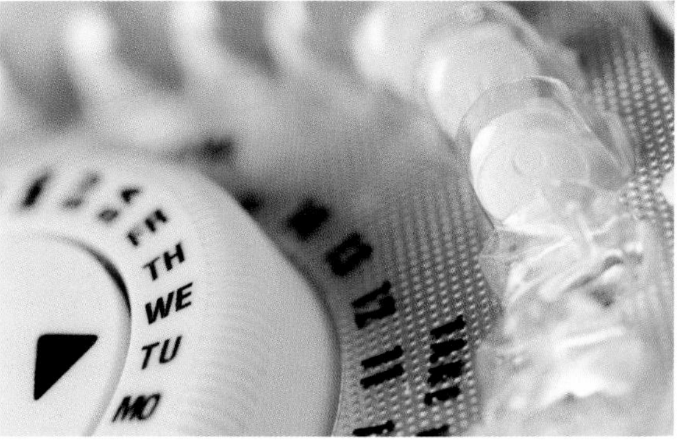

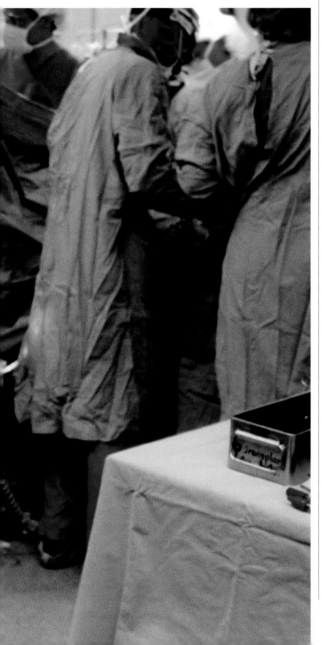

Right: Birth control pills, often referred to as "the pill," are a combination of estrogen and progestin taken by mouth to inhibit female fertility. Oral contraceptives were first approved for use in the USA in 1960, though their use was initially restricted to married women. It is sometimes said they put the swing into the sixties, but it was not until the early 1970s that their use became truly widespread.

BREAKTHROUGHS TAKE TIME AND MONEY

"Cure for cancer on the way," scream the headlines on a regular basis. But in reality, it is a long, slow, expensive process to get from an idea for a new drug to the market. And rightly so, because each new drug, treatment, or device must be proven to be safe and effective before it can be widely used.

Initially, studies are conducted in laboratories to understand what the drug is doing and to ensure it is targeting the correct process. Studies also determine if it can be taken orally or injected, how long it takes to clear from the body, and if there are any toxic by-products. Then, if the drug is safe and well tolerated by animals, it is given to a small group of people to see whether humans can safely tolerate it. If so, more studies with larger numbers of people are conducted. Typically, studies are randomized, double-blind placebo-controlled studies—groups of patients are randomly selected to receive the drug or a placebo or, in some cases, the current standard treatment for a disease. Double-blind means that neither the patient nor the medical practitioners administering the study know who is getting the new drug and who is getting the placebo until after the study is completed. This helps to minimize human bias affecting the results.

If, after all the studies have been done, a new drug appears to be safe and effective, or better than the existing drugs, the developer makes an application to regulatory agencies like the US Food and Drug Administration for approval to market the drug.

The field of pathology aims to understand the causes and progression of diseases in order to better treat or prevent their symptoms. Over the centuries, efforts have been made to alleviate pain and discomfort, but only recently have we begun to do this by understanding the fundamental causes of disease.

The road to modern pathology

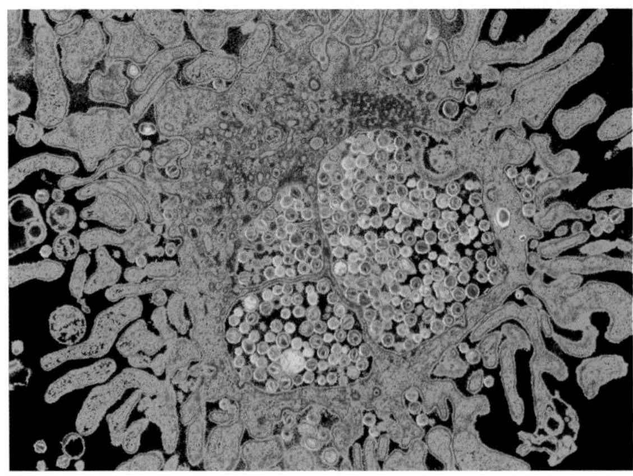

Left: This color-enhanced transmission electron micrograph (TEM) shows a cell that has been transformed into a factory producing HIV, the virus responsible for AIDS. HIV primarily infects vital cells in the human immune system such as helper T cells, macrophages and dendritic cells, replicating itself within them and making them targets for other immune cells. Highly active antiretroviral therapy (HAART), introduced in 1996, impedes its replication cycle.

Early morbid anatomy

Since the time of Galen (*c.* 129–216 CE), physicians have tried to understand the functions of our organs, though they were often impeded by social taboos, which prevented the dissection of bodies. However, the advent of universities removed this impediment by introducing the study of morbid anatomy—where sickness is viewed macroscopically (with the naked eye)—using dissection of bodies to study diseased organs and tissues.

The impact of microbiology

But it was only with the introduction of microscopy in the early nineteenth century that pathology as a science came into its own. Prior to that, illness was believed to be the result of a whole number of causes, ranging from the spontaneous generation of infective agents visible to the eye, such as worms or maggots, to the breathing in of poisonous gases or miasmas (bad air).

Often described as the patriarch of modern epidemiology, British physician John Snow (1813–1858) rejected the "bad air theory" in favor of "germ theory" (the theory that infectious microorganisms caused disease). He was famous for tracing a cholera outbreak to a water pump in Soho, London, in the 1850s. But his views on germs were not accepted until the latter part of the nineteenth century.

It was left to Louis Pasteur (1822–1895) to prove germ theory correct. He showed that contamination could indeed be caused by the transfer of bacteria (microorganisms) (see page 269).

However, the grandfather of modern experimental pathology is microbiologist Robert Koch (1843–1910). In the late nineteenth century he established a number of criteria for disease-causing agents. First, the agent must be found in all organisms suffering from the disease, but not in healthy specimens. Second, the agent must be taken from a diseased individual and grown separately. Third, the cultured agent should cause the same symptoms when introduced into another healthy organism. Last, the agent must then be re-isolated from the new host organism and shown to be identical to the original. These criteria remain the benchmark for declaring a microbe to be the cause of a particular disease.

German scientist Rudolf Virchow (1821–1902) had a similar impact on pathology by declaring that the cell was effectively the smallest component of an organism. His work with light microscopes took anatomical studies of disease down past the level of organs and tissues. Since then we've delved even deeper, describing diseases in terms of biochemistry and genetic inheritance.

> As it takes two to make a quarrel, so it takes two to make a disease—the microbe and its host.
>
> Charles Value Chapin, epidemiologist, 1846–1951

BEYOND THE DIVINE AND SUPERNATURAL

In ancient Greece, early non-mystical attempts to understand and treat illness involved the Greek physician Hippocrates (c. 460–380 BCE) (pictured by a fourteenth century artist below left, conversing with Hunayn ibn Isahq (808–73)). Hippocrates explained illness in terms of a system of "humors": Fluids which, if unbalanced, created physical and mental health problems.

Also during this time there was a dilemma: Whether to diagnose a disease based on the symptoms, called the Knidian approach; or focus on passive treatment and prognosis, and use general diagnosis, called the Koan approach. Hippocrates preferred the latter.

Although Hippocrates is considered the "father of medicine," our modern approach to disease and illness goes further than his view and is based on specific diagnosis, specialized treatment, and prognosis.

Below: German pathologist Rudolf Virchow (1821–1902), often considered the "father of pathology," also founded the field of social medicine and developed a standard method of autopsy still in use today. Virchow saw social reform as the road to good public health.

Modern pathology

The various studies of inheritance, from the work of Gregor Mendel through to the eventual description of DNA structure by James Watson and Francis Crick in the 1950s, have delivered to pathologists the tools necessary to analyze some of the most fundamental chemical building blocks in nature. For example, with advances in diagnostic tools such as fluorescence in situ hybridization (FISH), we can now examine whole chromosomes for indications of disease, well before symptoms have begun to make themselves felt.

Modern pathology is made up of a number of smaller disciplines: Hematology, which deals primarily with the constituents of blood; immunology, which deals with antigens and antibodies; microbiology, which deals with biological pathogens (disease-causing microorganisms); biochemistry, which deals with enzymes, hormones, and metabolic pathways; genetics, which deals with DNA coding and cytology; and histology, which deals with abnormalities in cells and tissues.

Above: *The Anatomy Lesson of Doctor Willem van der Meer in Delft*, by Pieter van Miereveld (1596–1623), is one of a number of paintings of the time that saw dissection as a new frontier of knowledge.

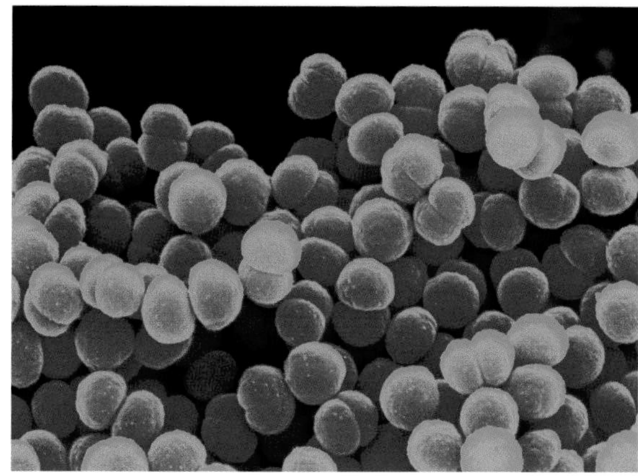

Right: *Neisseria meningitidis*, the bacterium that causes meningococcal meningitis. Infection on inhalation leads to inflammation of the meninges, the membranes that encase the brain and the spinal cord.

In a sense, every doctor is a detective trying to solve the mystery of what is making someone sick. Forensic medicine, one of the forensic sciences, requires highly specialized sleuthing skills that help police, litigants, or the courts find justice.

Medical detectives

Forensic pathologists are best known for performing autopsies to determine the time and cause of death: For example, it can be important to know whether the pilot died before or during an accident. However, they do not work alone. They are assisted by a range of forensic scientists with technical skills in areas such as fingerprinting, handwriting analysis, guns and other weapons, fires, aviation medicine, dentistry (think identifying the dead using dental records or fingering the murderer by analyzing bite marks), or entomology (the study of insects).

The role of forensic pathologists

Forensic pathologists require extensive training, lasting ten years or more, beginning with a basic medical degree. During their training they perform hundreds of supervised post-mortem examinations for their state coroner, including adult and infant deaths, unexpected natural deaths, accidents, suicides, homicides, and deaths during or soon after an operation or anesthetic. They need to know how to examine and dissect a body—for example, how to remove an eye—and how to collect organs, tissues, and fluids for microscopic, genetic, or toxicological examination. They must also be able to share their findings, sensitively and accurately, with the families and defend them, sometimes vigorously, in the courts. Unexcitingly, written reports take up many hours of a typical working day.

Left: Autopsy did not emerge as a means of exploring and discovering the cause of death until the eighteenth century, when it grew out of the practice of dissection and the study of anatomy. It's important that documentation such as toe tags or wristbands links each body to its proper identity.

Right: Forensic police officers in Manchester, England investigate the scene of a vicious attack on a 38-year old man, seeking evidence inadvertently left behind by his attackers. Targeted on Oxford Road, the victim was dragged to the subway under Mancunian Way, where he was beaten, struck on the head with a bottle, and had a cigarette stubbed out in his eye. The attack left the victim with a fractured skull and six broken ribs.

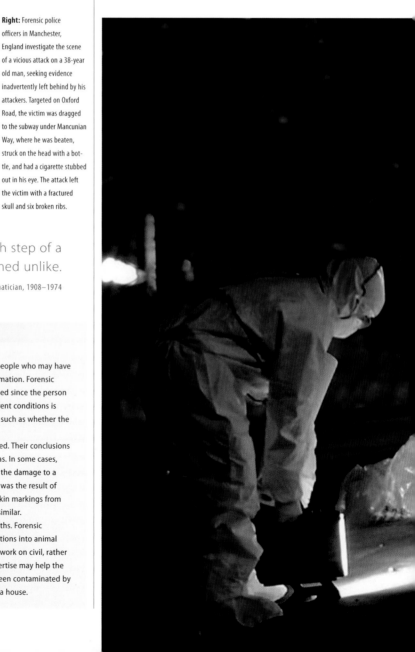

> The progress of science is the discovery at each step of a new order which gives unity to what had seemed unlike.
>
> Jacob Bronkowski, mathematician, 1908–1974

A BUG'S LIFE

Forensic entomology uses the study of insects to answer legal questions about people who may have committed suicide, been murdered, died because of neglect, or died from envenomation. Forensic entomologists can play an important role in determining how much time has passed since the person died. Detailed knowledge about insect breeding and feeding patterns under different conditions is required, as is the ability to identify insects. This knowledge can answer questions such as whether the body was buried soon after death or left exposed.

Forensic entomologists can also help work out whether a body has been moved. Their conclusions are based on a knowledge of which insects are found in particular geographic areas. In some cases, they are called upon to determine whether the damage to a corpse occurred before death or whether it was the result of insects feeding on the body. For example, skin markings from acid burns and from feeding ants can look similar.

Their work is not limited to human deaths. Forensic entomologists may be involved in investigations into animal cruelty or neglect, or they may be asked to work on civil, rather than criminal, cases. For instance, their expertise may help the courts to decide whether foodstuffs have been contaminated by insects or what sort of insect has damaged a house.

Solving puzzles

Forensic pathologists and other forensic scientists can play a role in a wide range of situations other than murder investigations. They may be able to determine how a particular pattern of injury occurred in someone who survived an assault; or reconstruct a road or workplace accident; or ascertain whether somebody was under the influence of a medication or illicit drug at the time of an incident. They may be asked to determine a person's identity or familial relationships, or to comment on whether someone died of natural causes or as the result of medical management gone horribly wrong.

Left: A forensic scientist prepares blood samples for DNA extraction for evidence in a sexual assault case. DNA evidence processed from the crime lab led to the identification of a suspected serial killer. Scientists from this lab are now trying to link evidence from other cases to the same killer. The lab deals with hundreds of cases at a time.

Right: One of the tasks of forensic dentistry is to identify individuals based on their dental characteristics. Most teeth have identifiable features that distinguish them from others, especially when X-rayed.

And forensic pathologists don't just work in the morgue. Correctly collecting evidence from the crime scene, then storing it and processing it in the laboratory may constitute vital steps in an investigation. A subsequent court case may succeed or fail depending on how well forensic scientists have done their job, so attention to detail and the ability to follow protocols meticulously is important. DNA testing is just one significant area of forensic medicine. It can be invaluable when a corpse is unrecognizable, especially following major disasters, such as investigating a mass grave or a jumbo jet crash site. It can also help a court to decide who the biological father is to determine whether child support should be paid; to decide whether someone really is a blood relative, and therefore eligible for a family reunion immigration program; or to confirm that genetic material at the crime scene really does belong to a suspect.

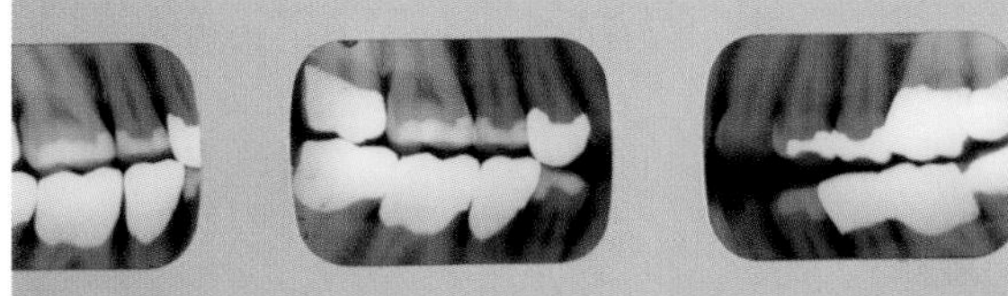

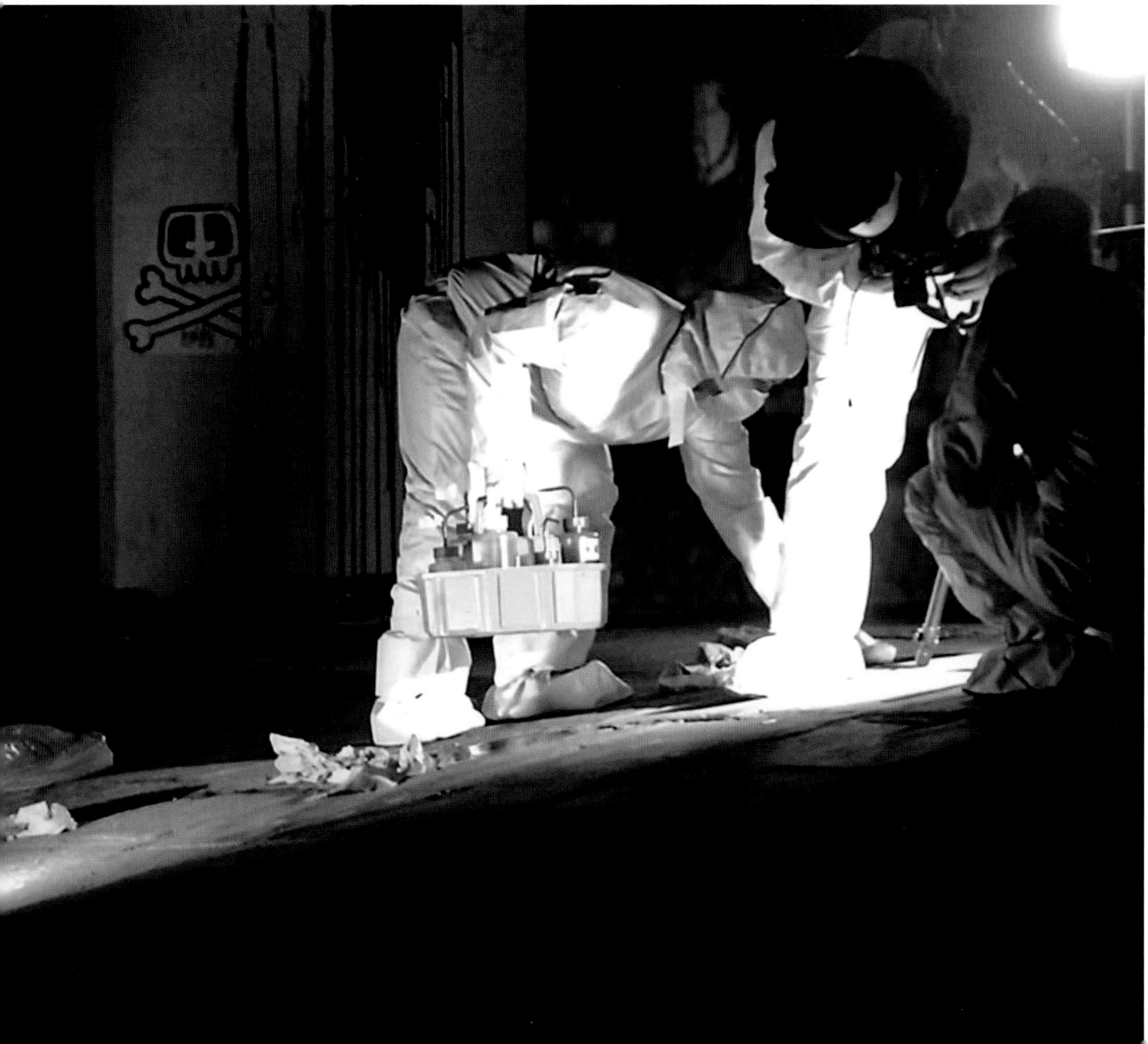

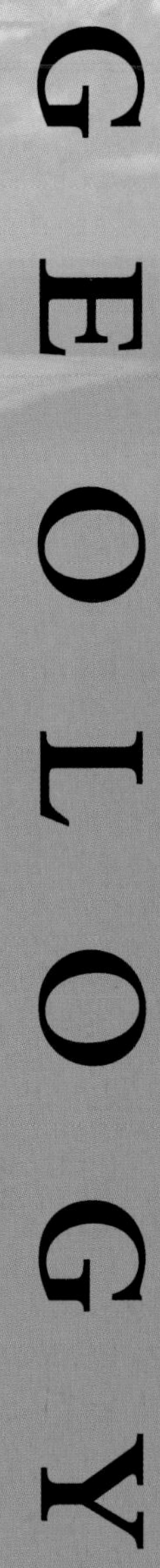
GEOLOGY

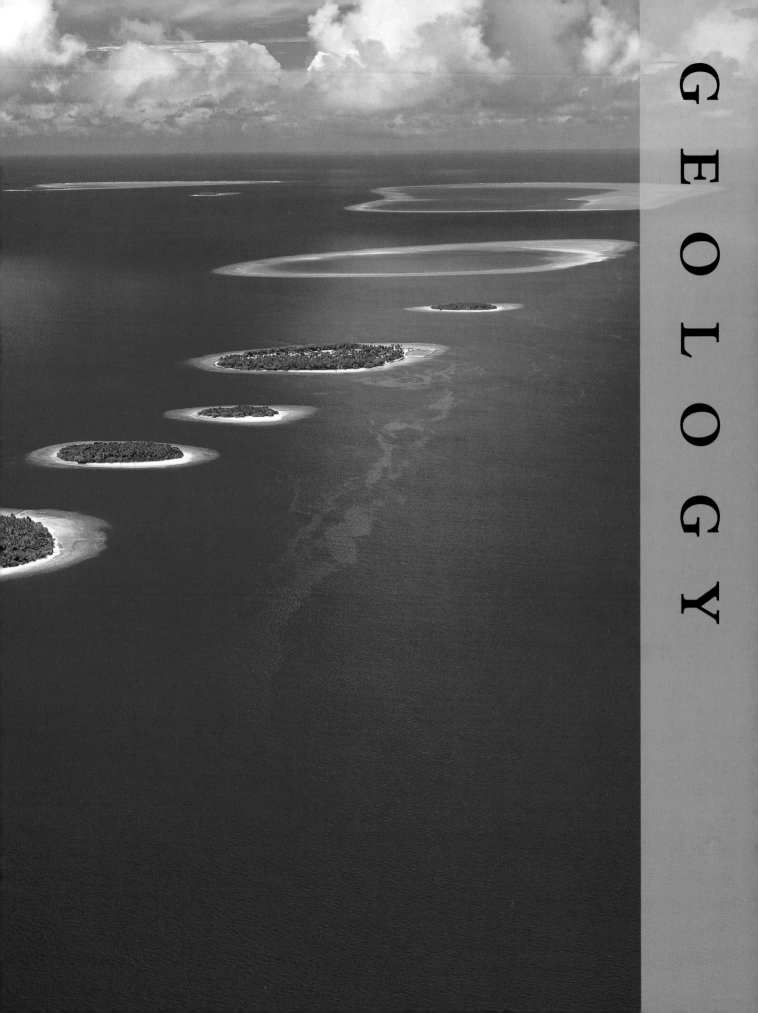

From early times, geology has been a practical business. The word means "knowledge of the Earth," and early observers were miners, wanting to find deposits of metal ores, distinctive rocks from which they could extract iron, copper, tin, lead, and other materials increasingly important in industry and daily living.

Introduction

Right: These prismatic rubies embedded in rock matrix are from Kashmir. Ruby deposits are also found in Tajikistan, Laos, Nepal, Afghanistan, Thailand, Sri Lanka, Myanmar, Pakistan, Vietnam, India, and parts of East Africa. Hues vary according to region, from light red in Sri Lanka to dark red in Thailand. The rubies of Mong Hsu, Myanmar, are purplish black at the core with a bright red periphery.

Later on, limestone was mined, crushed, and heated to yield mortar for building. Crushing the limestone together with certain clays produced hydraulic cement that would set underwater. Later still, coal deposits were prized as sources of energy. Rarer (and so more valuable) resources included precious metals like gold and silver, and gemstones such as diamonds, rubies, and emeralds.

Locating these treasures required careful observation of rock types, formations, and locations but not until the sixteenth century were more profound questions asked. How had the rocks been formed? Why were certain rock types found in certain locations? Exactly how old were the rocks? Indeed how old was Earth itself?

... the geology I plead for is that which states facts in plain words—in language understood by many rather than by few.

George Otis Smith, geologist, 1871–1944

How old is Earth?

For most people, religious texts provided answers to such questions. God had subjected Earth to a terrible flood during which the rocks and landforms familiar to us today had been formed. Adding together the periods of time mentioned in the scriptures indicated that Earth was at most 6,000 years old.

Leonardo da Vinci deduced that the rocks had been laid down underwater one layer at a time, with the oldest rocks at the bottom. That could be made to fit the scriptures, but there was more trouble with curious stony objects called "fossils." Da Vinci was perhaps first to suggest that fossils were the remains of once-living things. Yet if these had been destroyed in the great flood, why were so many found on the tops of mountains?

Da Vinci's questioning stimulated others. The sixteenth century German Georg Bauer (also known as Agricola) was arguably the first geologist (though the term was not used until several hundred years later). He set the pattern of relying on the evidence of his own observations rather than on scripture. He argued that wind and water had shaped Earth's surface and that volcanoes and earthquakes were evidence of Earth's internal heat.

Right: These dinosaur fossils at the Dinosaur National Monument in Utah, USA, are roughly 150 million years old. More than half of all the different kinds of dinosaurs that lived in North America during the late Jurassic Period are found in this quarry. It is thought their bodies were washed here by flooding, then buried under sediments that eventually lithified into sandstone and conglomerate rocks. Later, these rocks were lifted and tilted by mountain-building forces, and over subsequent millennia, the skeletons were gradually exposed by erosion, aided over the last hundred years by paleontologists.

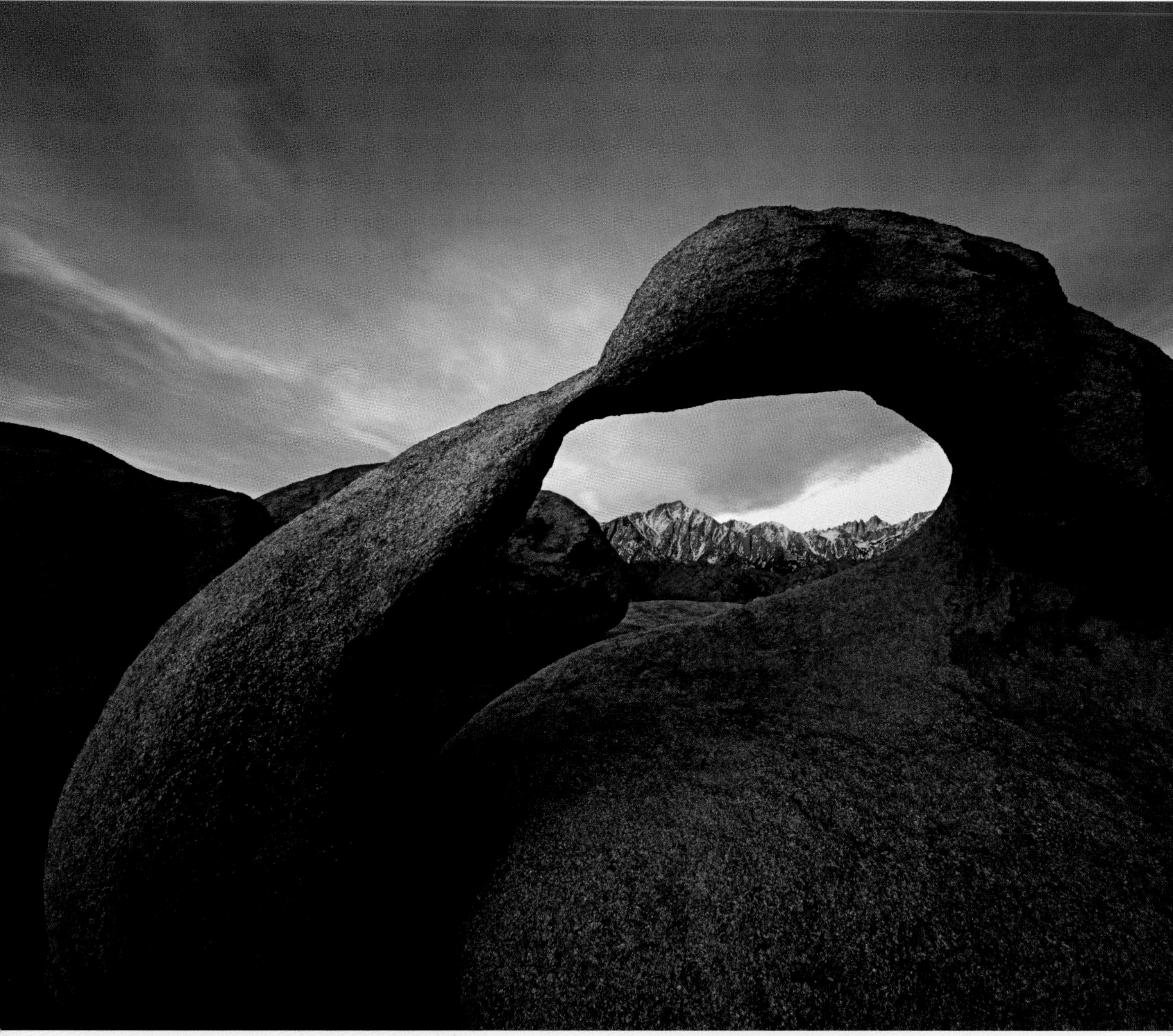

An abundance of theories

The history of geology has seen some great debates—the age of Earth was one. Evidence was accumulating that the 6,000 years allowed by the scriptures were not enough to permit the observed wearing-down of mountains by running water or the accumulation of salt in the oceans. French scientist Compte de Buffon argued that if Earth had once been a molten ball of rock flung off by the Sun it would have taken at least 75,000 years to cool. Church authorities responded to this challenge by burning Buffon's books.

That debate raged throughout the nineteenth century. English biologist Charles Darwin argued that biological evolution required many hundreds of millions of years at least. His countryman Lord Kelvin, arguing from a physics point of view, protested that Earth would have been stone cold long before then. At the end of the century, the discovery of radioactivity solved the conundrum from two directions. The decay of radioactive elements inside Earth liberated enough heat to provide Darwin with the time he needed, and careful study of those same radioactive elements provided clear evidence that Earth is in fact almost 5 billion years old.

A similar controversy lit up twentieth century geology. When German climatologist Alfred Wegener proposed in 1912 that the seemingly-fixed continents have drifted together and apart over millions of years, he was ridiculed. He had good evidence that continental drift had occurred, but could not explain what caused it. Nor could he explain how continents could plow through the solid rock on the beds of the oceans.

Fifty years later vindication came. We learnt that the ocean beds actually expand and contract over immense periods of time, pushing and pulling the continents around. Now broadened into the theory of plate tectonics, Wegener's once heretical ideas are mainstream, helping to explain, among other things, the distribution of volcanoes and earthquakes. Using lasers and satellites, researchers can measure the movement of the continents directly. As always in science, evidence settled the matter.

Above: The Alabama Hills, seen here at sunset, are only considered hills geographically. Geologically, they are the tip of a very steep escarpment in the Sierra Nevada Mountains. More rounded than their neighbors and featuring dozens of natural arches sculpted over eons by earthquakes and erosion, they consist mainly of orange metamorphosed volcanic rock around 150–200 million years old, and younger potato-shaped granite boulders, roughly 90 million years old.

Some 4.6 million years ago, during the Hadean Eon, Earth awoke. Starting as a pile of debris, Earth began to form in the Sun's spinning proto-planetary disk. As proto-Earth orbited, it cleared a swath by accreting numerous smaller bodies. It grew larger and hotter, then melted.

How Earth was formed

There were no surface rocks at this time, only a thin atmosphere of hydrogen and helium bleeding into space. The Hadean Eon stretched from Earth's beginning to about 3.8 billion years ago. There was no hard crust. Eventually, Earth became massive enough, and its gravitational field strong enough, to hold an atmosphere in place. It was a heavy poisonous mix of carbon dioxide, ammonia, methane, nitrogen, and water vapor. A thin crust began to form toward the end of the Hadean Eon and, even at these temperatures, water vapor was able to condense under high atmospheric pressure. Torrential rain poured into boiling oceans. There was no free oxygen around at this time, as it immediately became bound to hydrogen or other elements.

Core, mantle, and crust

During this early stage, Earth's own strengthening gravity pulled the heaviest elements, mainly metals such as iron, toward its center while the lighter ones, such as oxygen and silicon, rose upward toward its surface. Finally, Earth differentiated itself into three distinct layers of different density—the core, mantle, and crust.

At Earth's center sits a dense iron core, divided into the inner solid core, and an outer liquid core. It is only the incredibly high pressures toward Earth's center that maintain the inner core in solid form, otherwise it too would be molten. The mantle surrounds the core and is made mainly of the elements magnesium, silicon, and oxygen, which are combined as the mineral olivine. At times, fragments of the mantle, including the occasional diamonds that also form there, are carried to the surface by rapidly rising magma. A thin brittle low-density crust, akin to the cracked shell of a hard-boiled egg, lies on top of the mantle. The crust is made up almost entirely of the lightest elements—silicon, aluminum, and oxygen, combined mainly as the minerals feldspar and quartz.

Above: Quartz is a significant component of many igneous, metamorphic, and sedimentary rocks. A natural form of silicon dioxide, it can be found in numerous varieties and colors. This is an example of hematite quartz, or ferruginous quartz, so called because of its opaque reddish brown hematite (ferric oxide) coating.

Rock and roll

Earth was still very fluid during the Archean Eon, which followed the Hadean Eon, and lasted from 3.8 billion years ago to 2.5 billion years ago. The surface was very hot and continents were not yet forming—probably because heavy meteor bombardment was still shattering the crust, and vigorous plate tectonic activity was recycling the pieces at a very great rate. By the end of the Archean Eon, tectonic activity had slowed to near present-day levels.

The few traces of Earth's first rocks that still remain today are found in the most ancient centers of the continents of Africa, Australia, Greenland, and North America. The oldest mineral is a

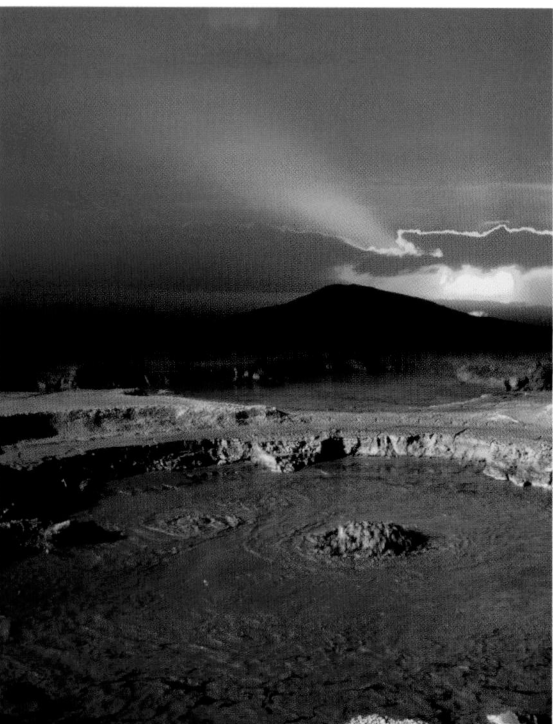

Left: Myvatn Reserve, Iceland, lies at the meeting point of the North American and Eurasian tectonic plates, which are drifting apart. One of the most geologically active areas on Earth, its formations include hot springs and mud pots that boil with Earth's inner heat and give off a sulfurous aroma.

Right: "The Wave" in Vermillion Wilderness (Arizona, USA) wears its geological history on its sleeve. Dune migration, fluid flows, sedimentation, and chemical reactions have all combined to produce the colored striations seen today. Although it follows the line of a natural fracture in a sandstone cliff, the bowl of "The Wave" has largely been shaped by wind erosion.

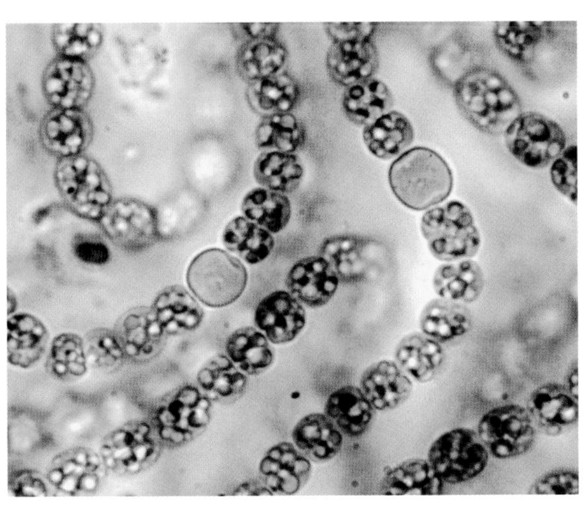

Left: *Cyanobacteria nostoc* are commonly found on moist rocks and at the bottom of freshwater streams and lakes. The large, transparent, thick-walled cells are heterocysts, which carry out the process of nitrogen fixation—taking nitrogen molecules from the atmosphere and converting them into nitrogen compounds such as ammonia, nitrate, and nitrogen dioxide.

The green pre-human earth is the mystery we were chosen to solve, a guide to the birthplace of our spirit.

Edward O. Wilson, biologist, b. 1929

zircon crystal, dated at 4.4 billion years old, but it was no longer part of a rock. It was found as a recycled grain in much younger rocks. The oldest dated complete rock formations were found in the Isua Greenstone Complex, in Greenland. These are 3.8 billion years old and are mainly altered metamorphic and igneous rocks.

Primitive life

Although the Archean Era atmosphere lacked free oxygen, temperatures had dropped to near normal levels, and there were now vast oceans. About 3.5 billion years ago, primitive cyanobacteria—simple prokaryote cells with no nucleus—appeared in the oceans. They began to grow hard mound-like structures called stromatolites by depositing concentric layers of calcium carbonate. These simple bacteria built huge reef complexes that became the first limestone rocks.

From about 2.3 billion years ago onward, oxygen that was being produced by photosynthesis began to build up significantly in the atmosphere. This oxygen increase is recorded in the world's banded iron formations—beautiful red and silver striped sedimentary rocks that also contain most of the world's iron ore reserves. Their tree-ring-like bands show that oxygen levels rose and fell in seasonal cycles. It is thought that the free oxygen poisoned the bacteria that produced it, before exhausting itself by oxidizing all the available iron. The cycle then began again with bacteria numbers increasing and oxygen levels building up in the atmosphere. As far as life on Earth goes, this was the beginning.

How can we possibly comprehend the notion of geological time? Earth was formed 4.6 billion years ago, and to appreciate the vastness leading up to the present day is conceptually difficult. We have no means to visualize such immensity of time.

Geological time

Geological time can best be understood by comparing it with something we can visualize. Teachers sometimes use a 1,000-sheet roll of toilet paper stretched out in a long corridor as a representation of our 4.6 billion-year-old Earth, to teach students about geological time. With each sheet representing 5 million years, the life of Earth stretches through 920 of them—practically the entire roll.

Visualizing geological time

Starting at the beginning of Earth time, we walk past 85 percent of the length of the roll before we can mark the first appearance of simple soft marine organisms at the beginning of the Ediacaran Period, about 635 million years ago. A little further on, there is an explosion of life forms at the beginning of the Phanerozoic Eon and the Cambrian Period, some 542 million years ago, when teeth evolved, along with other hard parts for attack, and shells for defense.

Fossils become common in the stratigraphic record from this time onward—usually only this part of the time scale is shown in geology textbooks. Marking all the following geological periods, and the arrival and departure of key animal and plant types, students realize they are quickly arriving at the end of the roll. The dinosaurs, together with other animals and plants, die out at the end of the Cretaceous Period in a massive extinction event just 13 sheets from the end (65 million years ago).

> Geological time is not money.
>
> Mark Twain, 1835–1910

The ancestral horse, which was about the size of a fox, appears 10 sheets from the end. Whales, bats, and monkeys appear five sheets from the end, and our own genus, Homo, evolves only a little less than half a sheet before the end. The earliest anatomically modern humans (Homo sapiens) appear in the last one-twenty-fifth of the last sheet (200,000 years ago), so our written history comes down to no more than the tangle of fibers at the very end of the last sheet.

Right: Caught in the act of devouring a small fish, this 50 million-year-old fossilized *Mioplosus labracoides* was found in the Green River Formation (Wyoming, USA). Such compromising geological snapshots, along with its teeth, suggest that *Mioplosus* was a voracious solitary predator.

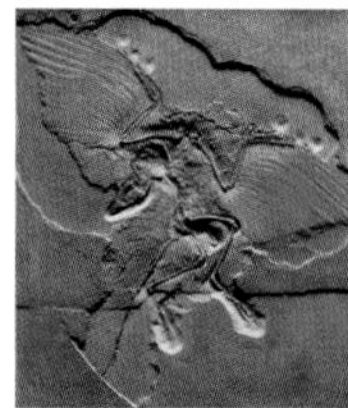

Below: This specimen of *Archaeopteryx*, discovered in 1876, has greatly enhanced our understanding of the evolutionary tie between dinosaurs and birds. It shows traces of feathers, but has the tiny sharp teeth of a meat-eating dinosaur. Sometimes called *Urvogel* ("first bird" or "original bird"), it lived in the late Jurassic Period around 155–150 million years ago, when Europe was an archipelago of islands in a shallow, warm tropical sea.

Rock clocks

Despite the development of a complex time scale relating the age of one region to another, we had no idea about the real age of Earth in absolute numbers of years until "atomic geological clocks" were discovered within different rocks. These "rock clocks" are based on the decay of radioactive elements (such as uranium) locked inside crystals of common minerals (such as zircon) that are found in many rocks. As we know the exact rate at which radioactive elements decay, we can determine the age of the minerals containing them simply by measuring the ratio of the parent element (say uranium) to its daughter decay product (say lead). The older the rock, the greater amounts of daughter elements it contains compared with parent atoms. The oldest known zircon crystal formed 4.4 billion years ago.

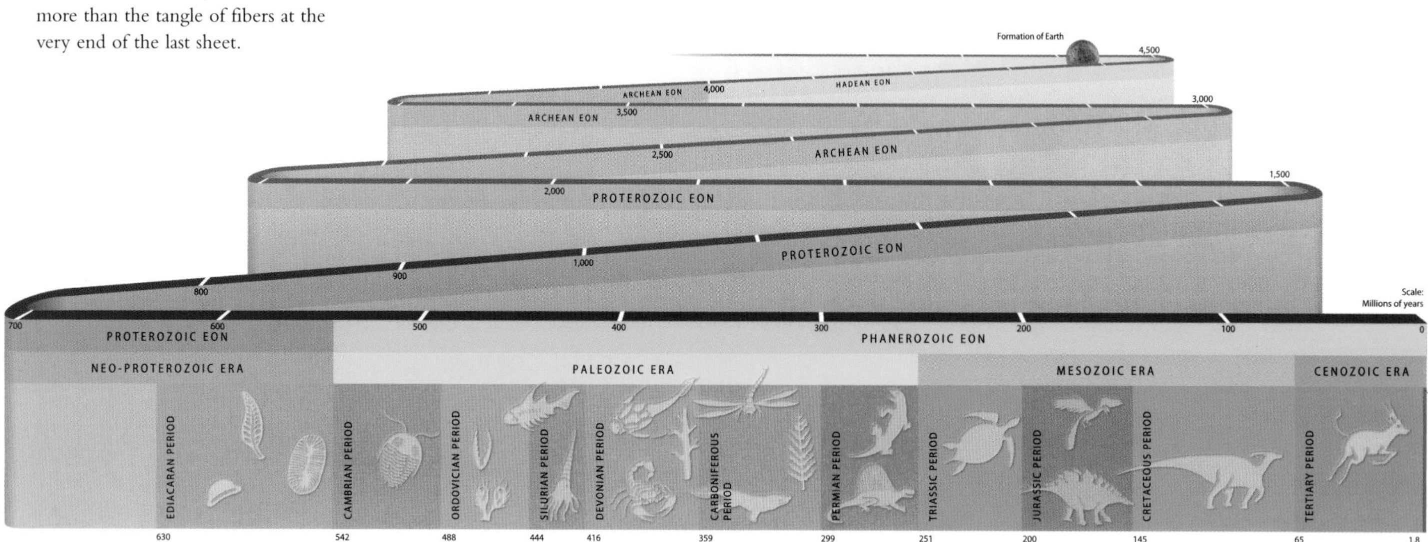

Above: An aerial view of the sawtooth peak of Ancient Wall in Jasper National Park, Canada. The limestone exposed in the Ancient Wall—formed in the shallow sea that covered the western part of North America 370 million years ago—may hold fossilized clues as to why large numbers of reef-building invertebrates died out during the late Devonian Period.

How geological time is divided

We divide geological time into eons, and smaller subdivisions called eras and periods, based on the appearance and disappearance of important life forms. At first the periods were arbitrary divisions, often named for regions of the United Kingdom where certain fossil types had been found. Boundaries were placed between one rock unit and another with different fossils. These fossil types were tracked over Europe, then linked with the rest of the world. It was only realized much later that many of these period boundaries represented major, often worldwide, catastrophes, known as a "mass extinctions", in which most life on Earth was killed, providing the opportunity for entirely different species to evolve.

Life evolves

In the beginning, single-celled life modified the environment by building up oxygen levels. Cells became more complex, developing a nucleus and paving the way for multicellular life forms. Predators evolved with sophisticated means of tracking and capturing prey, such as eyesight, teeth, and clawed appendages. Prey species countered with defense strategies such as plates, shells, and spines. Life filled the seas; plants and amphibians colonized the land, opening the way for reptiles.

After dominating for 200 million years, dinosaurs died out when the Chicxulub meteorite struck the Yucatan Peninsula, at the close of the Cretaceous Period. Mammals rose to power next, opening the way for the primates and finally our human ancestors.

The theory proposed by German climatologist Alfred Wegener in 1912 to explain the amazing jigsaw fit of Earth's continents is now a measurable certainty. Plate tectonics now provides our most up-to-date understanding of how our planet works. Today, anyone with an accurate global positioning system can monitor the slow movement of their home across the surface of the globe.

Tectonics

Wegener studied the pattern of scratch marks (striations) left behind by Permian Period glaciers on various continents. He noticed that, if those continents were all once joined together in a tight cluster around the pole, the otherwise random striations all pointed outward in a radial pattern from the South Pole. Furthermore, this also explained the perfect jigsaw fit of the Atlantic coastlines of Africa and the Americas. It all seemed so logical, but Wegener could not explain how or why such a movement of the continents would have happened. Then the following year, physicist Arthur Homes realized that the heat escaping from Earth's central furnace was driving a series of convection currents in the mantle, much like what happens in a pot of soup boiling on the stove.

Continents on the move

It is the slow-moving mantle convection currents that push the tectonic plates around the globe—at about the rate at which a fingernail grows. For decades, this concept was considered to be preposterous by the wider scientific community. Then in the 1960s, echo sounders, developed from ex-World War II submarine-tracking equipment, brought about an explosion in understanding of the topography of the ocean floors. Mid-oceanic ridges—huge submarine mountain ranges—were discovered and mapped.

More information came to light to support the theory when the patterns of Earth's magnetic field fluctuations recorded in the ocean floor basalts were found to be perfectly symmetric about the ridges. Samples of basalt collected from the ocean floor were also found to be youngest closest to the ridge and to grow progressively older toward the edges of the continents. It was realized Earth's crust is like the thin brittle shell of an egg. The plates grow and move apart at the mid-oceanic spreading ridges then eventually sink and disappear into the deep ocean trenches at a steady rate. The continents themselves, however, are thicker and lighter areas of the crust that float in the mantle at a higher level than the surrounding sea floor crust, and hence make dry land. Once created, the continental crust, like the froth on boiling soup, stays on top. It is never dragged down and recycled into the depths of Earth's mantle.

Right: Gemsbok visit streams delivering alkaline deposits to land near Lake Natron, a vast soda lake in Tanzania, in the heart of East Africa's Rift Valley. The lake, with its deposits of soda, salt, and magnesite, is the only known breeding habitat for flamingos.

Opposite page: Wild vicuna (a relative of the llama) roam the hills below Chimborazo, the highest mountain in Ecuador. Because Chimborazo lies just one degree south of the equator, and because Earth is not a sphere but an oblate spheroid with an equatorial bulge, the summit of Chimborazo is the point farthest from the center of the planet.

Below: Basalt columns and lava formations beside Skjalfandafljot River, near Aldeyjarfoss waterfall, Nordurland Eystra, Iceland. The Mid-Atlantic Ridge, which marks the division between the European and North American tectonic plates, runs across Iceland from the southwest to the northeast.

> It is a great philosophical breakthrough for geologists to accept catastrophe as a normal part of Earth history.
>
> Erie Kauffman, geologist, 1983

Rifting apart

Amazing landforms develop when a continent begins to break up. Initially a dome develops, creating a vast upland, before huge tensional cracks appear and open. Blocks of crust slide down along these faults creating deep flat-bottomed valleys, sometimes with their floors well below sea level. The great East African Rift and Lake Baikal are examples of rifting happening today. Basalt magma from Earth's mantle wells into the opening cracks—in places forming large volcanoes. Year by year, the rift slowly widens, with each valley wall eventually destined to become the shoreline of a distant continent. A very interesting event occurs when the land-locked rift valley finally breaches the edge of the separating continent—it is flooded by the sea, either gradually in stages, or as a raging torrent. The narrow seaway continues to grow into a giant ocean, like today's Atlantic Ocean, with the shape of the initial crack preserved in the shape of the mid-oceanic spreading ridge and the shape of the continental edges.

Trench warfare

It was ultimately realized by scientists that if the planet is not to expand like a great balloon, then the crust must disappear at a rate equal to that which is being created at the spreading ridges. The crust is being destroyed along lines of deep submarine trenches that ring the globe—it descends into these trenches and is melted.

Most of Earth's trenches occur in a line around the edge of the Pacific Ocean. It is a zone marked by active volcanoes and many violent earthquakes, appropriately named the Pacific "Ring of Fire." Therefore, as the continents separate, the widening of the Atlantic Ocean is being balanced by the closing of the Pacific Ocean along these zones of crustal destruction. The term "subduction zones" is used to describe how one plate, usually the oldest or most dense, slides beneath the other along an inclined plane. Numerous and often strong earthquakes mark this dipping subduction plane—sometimes also known as a megathrust.

Their source locations (foci) can be mapped to a depth of about 440 miles (700 km), at which point they die out, indicating that at this depth the down-going plate has begun to soften and melt. The molten products of the descending plate rise as huge buoyant bodies of andesitic magma and are responsible for the chains of active volcanoes that lie inland from, and parallel to, the deep trenches.

Above: Icebergs continually change their shape as they become eroded and grow imbalanced, then tip and roll into new positions. This iceberg shows the effects of both wind and water erosion.

Collision mountain ranges

Earth's crust cannot so easily disappear when two plates of buoyant continental crust are being pushed together. In this case, neither plate is able to descend beneath the other. The result is a collision of gigantic proportions, with the edges of both continents crushing together, folding and buckling upward. The Himalayas are the best example of a collision mountain range forming today—a result of India ramming into the Asian continent. This began about 50 million years ago, when the last remaining part of the ancient sea, the Tethys, was swallowed up and the two continental edges touched. Today, some of the Himalayan peaks reach over 5 miles (8 km) in elevation and rise at a rate of over 1 inch (25 mm) per year. As India is far too light and thick to subduct beneath Asia, the subduction process will eventually jam up and stop. Stress will continue to build, however, causing the subduction zone to "jump" to a new locality—most probably southward off the coast of India—and continue anew.

Understanding landforms

All Earth's landforms are a direct consequence of their geographic location and tectonic address on the planet's surface, as it spins on its axis and orbits the Sun. The forces driving the formation of our planet's physical features are gravity, pulling inward, and heat from Earth's interior, escaping outward.

The force of gravity works relentlessly to flatten out any topographical features. It drives rivers of water and ice to the sea, carving out landscapes and carrying and depositing huge loads of sediment in the process. The equally relentless forces of erosion remove giant mountain ranges grain by grain, leveling the land and gradually filling the ocean basins. Opposing the leveling, flattening force of gravity are the massive heat-driven convection cells within Earth's semi-plastic mantle. Transferring heat from the hot core to the cooler outer surface, these giant convecting plumes slowly rise through the mantle, cool, then sink again to begin anew.

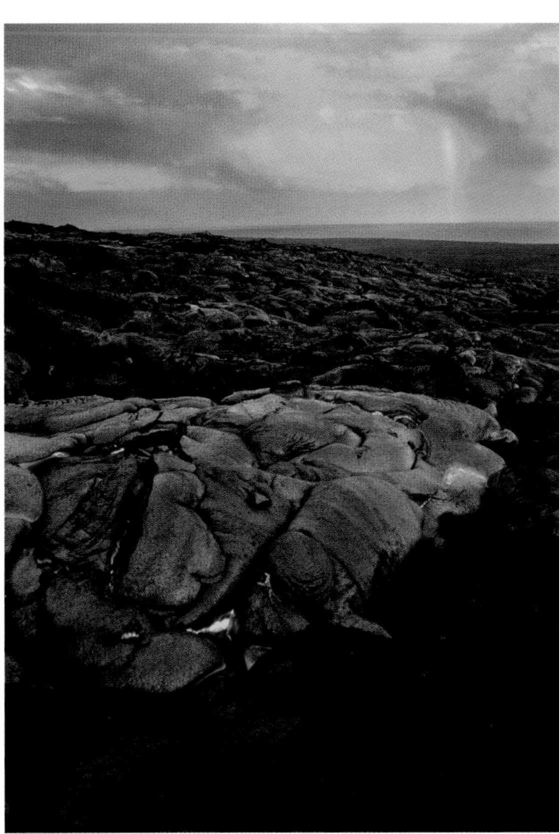

Right: Kilauea Volcano, Hawaii Volcanoes National Park, is the youngest volcano on the Big Island of Hawaii. Because it appears as a bulge on the southeastern flank of its larger neighbor, Mauna Loa, it was once thought to be a satellite, not a separate volcano. But in fact it has its own plumbing system, more than 37 miles (60 km) deep.

Left: Perhaps the most famous peak in the world, and certainly the highest, Mt Everest, shaped like a three-sided pyramid, is composed of multiple layers of rock folded back on themselves (nappes). Lower down are metamorphic schists and gneisses, topped by igneous granites. Higher up are mainly sedimentary rocks of marine origin, including the Yellow Band, a limestone formation just below the summit. Long revered by local peoples, Mt Everest's Tibetan name, Chomolungma, means "Goddess Mother of the World" or "Goddess of the Valley." Its Sanskrit name, Sagarmatha, means "Ocean Mother."

Earth's engine room

Earth's giant temperature engine is controlled by heat from Earth's still-cooling core escaping through the planet's surface into space. In places, thick continental crust acts as an insulating blanket causing heat to build up beneath it. This means that mantle convection is intensified beneath large areas of continental crust, much like boiling stew in a pressure cooker. These immensely powerful but incredibly slow moving convection currents push the plates of Earth's thin brittle crust around like plastic toys in a bathtub.

Major disasters

Catastrophic disasters, such as meteorite impacts or supervolcanic explosions, can dramatically change Earth's landscape and climate. Impacts have periodically wiped out nearly all life on the planet. For example, Chicxulub in Mexico's Yucatan Peninsula terminated the Cretaceous Period 65 million years ago. Life, however, has an amazing ability to rebound and those species resilient enough to survive the calamity evolved rapidly to repopulate the planet.

THE RISE OF MOUNTAINS

Over a period of hundreds of millions of years, continents are pushed halfway across the globe and rammed into each other. They are fused, and buckled upward into massive Himalayan-like mountain ranges reaching miles into the air, only to be eroded, then later rifted and pulled apart again. Marine fossils near the top of Mt Everest testify to the forces involved in this process.

Elsewhere, enormous slabs of ocean floor may push beneath adjacent plates of Earth's crust, forming deep oceanic trenches. Descending into the mantle, these slabs create earthquakes and melt to fuel long belts of explosive volcanoes along the overlying plate edge.

It is now understood that the continents move in enormous supercycles lasting hundreds of millions of years. Supercontinents form when all of Earth's landmasses are joined together in one huge continent. But they don't stay together for long before the single landmass becomes unstable.

Supercontinents

A supercontinent acts just like a gigantic thermal blanket, trapping Earth's internal heat underneath. This causes enormous plumes of rising mantle material to push upward against the supercontinent's base near its center. The convection currents gain strength and eventually rip the continent apart into jigsaw pieces.

We can only see dimly back about 1.1 billion years into Earth's continental history. At this time, a supercontinent existed, called Rodinia (Russian for "the motherland") surrounded by the Mirova Ocean (Russian for "global sea"). Earlier supercontinents probably existed before Rodinia, but the lack of fossils means that these cannot be pieced together. Rodinia broke apart about 750 million years ago. Its pieces reassembled to form Pannotia about 600 million years ago, which in turn broke up 50 million years later.

As well, large parts of today's Southern hemisphere continents formed one of the fragments from the break-up of Rodinia. These were grouped around a core landmass consisting of Australia, India, and East Antarctica. Then, around 520 million years ago, other fragments of Rodinia carrying Africa and South America that had moved eastward around the globe docked with the core landmass to form a new supercontinent called Gondwana.

The continental fragments of Rodinia came together again around about 275 million years ago to form Earth's last huge supercontinent known as Pangea. Our present continents are all fragments moving outward from the break-up of Pangea. At a future time, all of today's continental fragments will once again come crashing together to form a new supercontinent.

Biodiversity, too, reflects the cycle of supercontinental formation and break-up, with the number of different species becoming more abundant at times when the continents are fragmented and dispersed, and less so when the landmasses are joined. In recent geological time, for example, a different species of big cat has evolved on each of four major continents—the lion in Africa, the tiger in India and Asia, and the panther in South America. These were all the same species at the time of Pangea's break-up.

Right: *Nothofagus moorei*, or Antarctic beech (not to be confused with its South American relative, *Nothofagus antarctica*) is native to the eastern highlands of Australia. The pattern of distribution of the genus has fed speculation that it dates to the time when Antarctica, Australia, and South America all formed part of the supercontinent Gondwana.

Below: An aerial view of Livingstone Island, part of the South Shetland archipelago, Antarctica. Except for isolated patches, the land surface is covered by an ice cap, together with ash layers from volcanic activity on neighboring Deception Island.

Tectonic super cycles

Scientists suggest that cycles of amalgamation and break-up, known as tectonic super cycles, occur over periods of roughly 250 million years. There is evidence for at least six such cycles but there were probably many more. During these cycles there would have been extensive changes to the distribution of land and sea, greatly influencing biological and climatic conditions.

Earth's climate has alternated between extensive periods of warm tropical (greenhouse) and glacial (icehouse) conditions. The accumulation of continents at or near the poles could play a role in the development of large-scale ice sheets.

Forming today's continents

Gondwana remained intact until 180 million years ago when it started to break up in the Jurassic Period to form the continents of today's atlas. Africa, western China, and India were the first to detach from the remnants of Gondwana.

Australia was one of the last to separate. It started to move north when it split from Antarctica about 90 million years ago. However, it was India that took the record for tectonic "speed." It broke free from Gondwana about 135 million years ago, moving with a velocity of about 4 inches (10 cm) per year and crashing into the Eurasian Plate. The force of that collision produced the highest mountain range on Earth—the Himalayas. The impact was so colossal, in fact, that the uplift continues to this day.

When Gondwana finally broke up into the landmasses we know today, these newly formed continents took with them species of the flora and fauna of the time. Many of the southern continents of today share living representatives of these. For example, the Antarctic beech tree (*Nothofagus* species), which probably evolved during the Late Cretaceous Period, is found living on the former Gondwanan landmasses of New Guinea, New Calendonia, New Zealand, southern South America and southeastern Australia. It is not found in Africa or India, which means that *Nothofagus* probably evolved after these two continents separated from the rest of Gondwana.

Above: The golden jackal (*Canis aureus*), or Asiatic, oriental or common jackal, is native to northeastern Africa, southeastern Europe, and southern Asia. The largest of the jackals, it is the only one to occur outside Africa.

Left: Biodiversity is reflected across the continents. The Tasmanian tiger (*Thylacine* species) which was declared extinct in 1936, resembles the wild dogs of other continents.

A stone is ingrained with geological and historical memories.

Andy Goldsworthy, artist, b. 1956

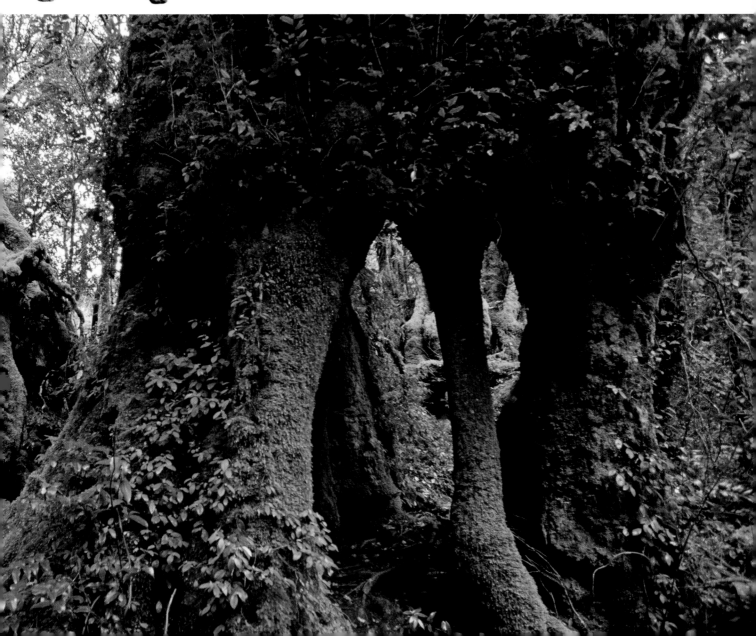

Just a short distance below Earth's surface, molten rock (magma) sits at high temperatures and pressures, awaiting the chance to burst forth, either in flows or explosions, wherever a weakness in the solid crust presents itself.

Vibrant volcanoes

Rift volcanoes

Rift volcanoes are the most abundant type of volcano. They lie mostly at the bottom of the oceans, forming a line of continuous eruption thousands of miles long between spreading plates. There are only a few places where these mid-oceanic rift volcanoes are visible on land, such as in Iceland, where the rift rises out of the ocean and cuts straight across the middle of the island.

Shield volcanoes

Shield volcanoes occur above specific "hot spots" in Earth's mantle where basaltic magma finds its way to the surface through a weakness in the crust. Flow after flow of runny low-viscosity lava builds up an enormous broad, flat, shield-like volcanic edifice, the entire process taking about one or two million years.

The Big Island of Hawaii stands 5 miles (8 km) tall, from ocean floor to summit. Activity ceases when the moving plate severs the shield volcano from its feeder pipe, stopping the lava flows. The immense weight of the shield volcano is so great that it begins to depress the sea floor and slowly sinks. After a few more million years the volcano disappears from sight beneath the waves. This mechanism is responsible for the many oceanic island chains, such as the Hawaiian and Tahitian chains in the Pacific.

Below: A composite volcano or stratovolcano, Japan's Mt Fuji is located above a subduction zone where the Philippine Sea tectonic plate is sinking beneath Japan. A tear or "hot spot" in the Philippine Sea plate directly beneath Mt Fuji permits enormous volumes of mantle material to regularly replenish its magma chamber.

Left: Popocatépetl, the second highest peak in Mexico, lies in the eastern half of the Trans-Mexican volcanic belt and is linked to the Iztaccíhuatl volcano to the north by a high saddle known as the Paso de Cortés. An active volcano, its name comes from the Nahuatl words popoca ("it smokes") and tépetl ("mountain"). More than 20 major eruptions are known to have occurred since the Spanish arrived in 1519, with the current cycle of activity beginning in 1947. In December 2000, tens of thousands of people were evacuated by the government based on the warnings of scientists. The ensuing display was Popocatépetl's largest in thousands of years.

Composite volcanoes

Composite volcanoes or stratovolcanoes grow at relatively closely spaced intervals along all of Earth's subduction margins, so it is possible to view several in a line from a single vantage point, such as along the edges of the continents encircling the plates of the Pacific Ocean; hence the name "Ring of Fire." Another place is the Indonesian archipelago. Deep trenches on the sea floor, such as the Japan Trench, Peru–Chile Trench, or the Java Trench, running parallel to the chains of volcanic peaks, mark the line along which plates are being forced beneath others and consumed. The magma that forms the volcanoes is of intermediate composition (andesitic), as it is the product of the descending and melting plate, as well as the seawater and ocean sediments. The magma tends to erupt explosively as ash and various-sized pieces of broken rock (collectively known as tephra), or more passively as lava flows. Over time, the alternating tephra falls and lava flows build up a distinctive, steep-sided, often symmetrical cone known as a composite cone or stratovolcano. Japan's Mt Fuji is a classic example.

A single stratovolcano can quickly wipe out generations of human development. Massive destruction took place in minutes when the 1902 explosive eruption of Mt Pelée totally obliterated Saint-Pierre, a city of 29,000 on the island of Martinique. The explosion that destroyed the summit of Mt St Helens in 1980 was even larger, but only 57 lives were lost due to the volcano's remote location and the successful management of the situation by the US Geological Survey.

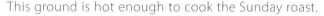

This ground is hot enough to cook the Sunday roast.

John Seach, volcanologist, 2000, before his boots melted on the ground at Lopevi volcano in Vanuatu

People living near a volcano are usually unaware of the danger. Today, Indonesian farmers work the fertile soils around lakes within the craters of their nation's numerous active volcanoes. There are entire Japanese cities spreading around the flanks of Mt Fuji, which has not erupted since 1707. Even Mexico City, one of Earth's largest urban populations, is located just 45 miles (70 km) from Popocatépetl volcano. This highly active volcano has had 15 eruptions since 1519 and is currently on Yellow Alert.

MT VESUVIUS, ITALY

Mt Vesuvius is perhaps Earth's most potentially dangerous volcano owing to its proximity to the vast metropolis of Naples, which has a population of 3 million as well as 600,000 in the immediate area of the cone. The volcano is best known for the catastrophic eruption in 79 CE that buried the towns of Herculaneum, Stabiae, and Pompeii beneath layers of cinders, ash, and mud. About 2,000 people died in Pompeii alone when a dense cloud of hot ash and gas surged down the mountainside. This eruption was the first in history to be documented in detail, by Pliny the Younger. Volcanologists now use the term "Plinian" to describe explosive volcanic eruptions that generate high-altitude eruption columns and blanket large areas with ash.

When a volcano ceases to erupt it is susceptible to erosion, and can be substantially reduced by rivers and streams flowing down its sides within a million years. Soft volcanic ash is removed first while harder parts of the volcano, like solid lava and magma, are more resistant. These spectacular volcanic remnants afford geologists a first-hand opportunity to study the internal workings of a volcano.

Volcanic remnants

When erosion breaches the crater or caldera walls, its lake water drains away into surrounding river systems. An Aleut village at Cape Tanak on Umnak Island, Alaska, was wiped out in 1870 by the sudden draining of the Okmok caldera lake. Nowadays, geologists and civil authorities watch such situations closely. Ten years after the 1991 eruption of Mt Pinatubo in the Philippines, lake levels had risen dangerously and there was a risk that the caldera walls would burst and flood communities surrounding the volcano with wet volcanic debris. A canal was quickly designed and constructed, and about a quarter of the lake was drained into the Maraunot River, relieving pressure on the upper caldera walls.

Above: Stacked stones on Bodmin Moor, a granite moorland in Cornwall, England, dating from the Carboniferous Period. Densely populated during the Bronze Age, the area's many prehistoric stone barrows and circles have been artfully polished by erosion.

Towering spectacles

A volcanic plug, or neck (also known as a spire or tower), is a nearly circular and vertical feeder pipe of a volcano that has been filled with solidified magma. Devils Tower, a possible example, rises above the Wyoming Plains, USA. Spectacular volcanic plugs are located near Rhumsiki in the Mandara Mountains of Cameroon; the largest is Kapsiki Peak, a spire standing 4,016 feet (1,224 m) tall.

Being difficult to climb and therefore well protected, volcanic plugs have sometimes become the sites of buildings such as castles, churches, monasteries, or temples. Many British castles, such as Edinburgh Castle, are located on isolated, craggy plugs, making them invincible to attack. The name "puy" comes from the volcanic plugs of the Avergne district in France's Central Massif, which stick out as steep, pillar-like hills. The twelfth-century St-Michel-d'Aiguilhe chapelle, in the town of Le Puy-en-Velay, is built atop a 280-foot (85-m) high volcanic rock pinnacle. Towering above a more modern part of this city, a bronze statue of the Virgin Mary stands on another 500-foot (152-m) tall bare rock plug.

> As in geology, so in social institutions, we may discover the causes of all past changes in the present invariable order of society.
>
> Henry David Thoreau, naturalist and philosopher, 1817–1862

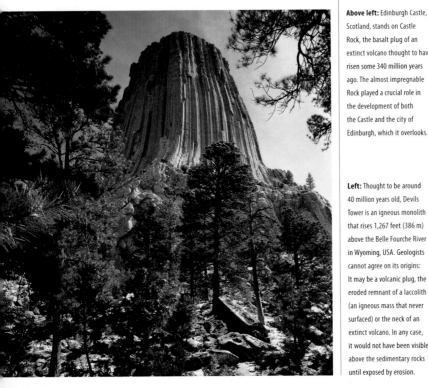

Above left: Edinburgh Castle, Scotland, stands on Castle Rock, the basalt plug of an extinct volcano thought to have risen some 340 million years ago. The almost impregnable Rock played a crucial role in the development of both the Castle and the city of Edinburgh, which it overlooks.

Left: Thought to be around 40 million years old, Devils Tower is an igneous monolith that rises 1,267 feet (386 m) above the Belle Fourche River in Wyoming, USA. Geologists cannot agree on its origins: It may be a volcanic plug, the eroded remnant of a laccolith (an igneous mass that never surfaced) or the neck of an extinct volcano. In any case, it would not have been visible above the sedimentary rocks until exposed by erosion.

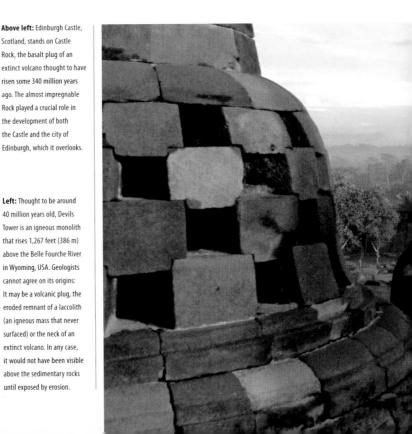

SHIPROCK, NEW MEXICO, USA

One of North America's most spectacular volcanic remnants, Shiprock is part of the Navajo Volcanic Field that erupted some 25 to 30 million years ago over an area of about 7,700 square miles (20,000 km^2) where the four states of Arizona, New Mexico, Colorado, and Utah meet. Highly gas-charged magma from Earth's mantle exploded onto the surface, creating numerous diatremes, vents, tuff pipes, and dikes. The violent nature of the eruptions is clearly evident in the sheer abundance of fragmented volcanic rock (breccia and tuff) in this area. Shiprock itself is composed of fractured volcanic breccia that has been injected with black dikes of an igneous rock called minette, a rare potassium-rich and silica-poor intrusive igneous rock believed to have come directly from small amounts of melting within Earth's mantle.

One of the most extensive Buddhist shrines is the temple of Borobudur in Java, Indonesia, where a broad plug of basalt has been turned into an ornately decorated stepped pyramid 115 feet (35 m) high and with a base 394 feet (120 m) square. Built during the Sailendra dynasty in the ninth century, it took several generations—about 80 years—to construct.

Other volcanic structures

Sills are sheet-like bodies similar to dikes, except that molten magma is injected along fractures parallel to or conformable with the surrounding bedding planes, rather than cutting across them. The Great Whin Sill of northern England is the primary example. Underlying much of south and east Northumberland and the Durham Coalfield, it reaches a maximum thickness of approximately 230 feet (70 m).

Under high magma pressure a sill can balloon out locally along its upper surface by lifting up the overlying strata. These structures—laccoliths—were first seen in the Henry Mountains of Utah, USA.

Batholiths are enormous bodies of igneous rocks that formed deep in Earth's crust, and have subsequently been exposed at Earth's surface by erosion. They form long linear belts that follow ancient subduction zones and other heat sources. Batholiths are usually made of rocks containing a high to moderate amount of silica.

Below: The ninth century Buddhist site of Borobudur (Java, Indonesia) is located between two twin volcanoes, Sundoro-Sumbing and Merbabu-Merapi. Sediment and pollen samples found near the site suggest a paleolake environment: Borobudur may have represented a lotus flower floating on the lake.

Boiling pools of strangely colored water, thermal springs with supposed curative powers, pots of bubbling mud, and sudden noisy spouting geysers are typical of geothermal zones—that is, those regions where the geothermal gradient, or the increase in temperature with depth, in the top few miles of Earth's crust is significantly higher th[illegible]

Hot spots: Ea[illegible]

Areas of high geothermal heat flow occur in di[illegible] which correspond to major boundaries betwee[illegible] Earth's main geothermal belts are found along t[illegible] and rift valleys, where the crust is young and thi[illegible] can more easily escape from the planet's interior. [illegible] coincides with Earth's major subduction zones w[illegible] volcanic activity is most common, and with collisi[illegible] heat is being generated by intense and ongoing cr[illegible] Localized geothermal fields are also found. These a[illegible] the isolated mantle hotspots responsible for building [illegible] volcanoes such as Mauna Loa, Hawaii.

Under the sea

The spreading ridge geothermal zones are mostly loc[illegible] the ocean where new basaltic ocean crust is being for[illegible] of rift volcanoes at the mid-oceanic ridges, and also in [illegible] Underwater geysers, or hydrothermal vents, in the deep oceans are visible only by using submersible craft. As the hot mineral-rich waters issue forth from these fissures and cool in the near-freezing waters of the ocean bottom, tiny black crystals of metal sulfides precipitate instantly as billowing, thick, black clouds. Hence these vents are known as "black smokers" and they sometimes build spectacular complexes of tall black chimneys. The first black smoker vent was only discovered as recently as 1977.

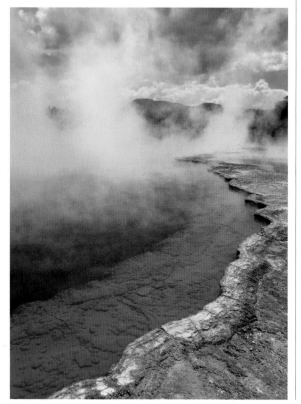

[illegible] ("gusher") that gave its name to all other geysers in the world is located in Haukadalur, but is not as active as some others, erupting only four or five times each day.

Left: The spectacular colors of the Champagne Pool, a hot spring in New Zealand's Waiotapu geothermal area, are caused by deposits of arsenic and antimony sulfides. The pool's name derives from the efflux of carbon dioxide it emits, similar to bubbles in a glass of champagne. Formed 900 years ago by hydrothermal eruption, the Champagne Pool is a relatively young system in geological terms. But it isn't only of interest to geologists and tourists. Recently, scientists examining its potential role as a habitat for microbial life forms have successfully isolated two novel bacteria and a novel archaeon. One of the bacteria, *Venenivibrio stagnispumantis*, which tolerates high concentrations of arsenic and antimony, represents a novel genus and species within the order *Aquificales*.

[illegible] ness and at an average depth of about 7,000 feet [illegible] has evolved near the black smokers, taking [illegible] ermal heat. Unusual marine creatures include [illegible] lams around the Pacific Ocean vents, while [illegible] n found at vents in the Atlantic Ocean.

[illegible] l zone

[illegible] hermal zone, running through the [illegible] atka, the Aleutians, North and South [illegible] Zealand, is supplied by heat from the vol-[illegible] zones along which the Pacific Ocean [illegible] nderneath the surrounding plates.

The primary role of the geologist is to recognize the existence of phenomena before trying to explain them.

B. M. Keilhau, geologist, 1828

The widening and opening of the Atlantic Ocean is driving this closure of the Pacific Ocean. Outside the Pacific, other geothermal belts fuelled by subduction heat include the Indonesian, Antilles, and South Sandwich island arcs.

Heat generated from the Collision Mountain Ranges

The Mediterranean–Himalayan geothermal belt follows the line of closure of the ancient Tethys Sea, where the Indo–Australian, African, and Arabian tectonic plates began to crunch into the southern edge of the Eurasian tectonic plate beginning approximately 50 million years ago. The collision, still going on today, is building some of the world's mightiest mountain ranges.

Geysers—spouting the news

The spectacular phenomenon of columns of boiling water spouting into the air at periodic intervals is extremely rare due to the specific set of geological conditions required for geysers to occur. There are only about 1,000 active geysers on Earth, and most occur at just two sites. North America's Yellowstone National Park has around 500

Below: El Tatio Geyser field, seen here at sunrise, is located high in the Andes Mountains of northern Chile. The largest geyser field in the southern hemisphere and third-largest in the world, El Tatio is also notable for having no significantly tall geysers. The tallest eruptions observed have been about 20 ft (6 m) high, with most being less than one meter. It is, however, a natural laboratory of great interest because of the geological and biological information to be gleaned from its sediments.

OLD FAITHFUL

Old Faithful in Yellowstone Park is a fountain geyser. Eruptions shoot between 3,700 and 8,400 gallons (14,000 to 32,000 L) of boiling water skyward to a height of about 100 to 200 feet (30 to 60 m). The duration of an eruption varies from 1½ to 5 minutes, and they occur about 65 to 92 minutes apart. Over 137,000 eruptions of Old Faithful have been studied and, most interestingly, a mathematical relationship exists between the size of an eruption and the interval between eruptions, with longer recharge times needed following larger eruptions. The reliability of Old Faithful is due to the fact that its subterranean plumbing system is not connected to the other thermal features of the Upper Geyser Basin.

geysers, and there are up to 200 in Dolina Geizerov on the Kamchatka Peninsula, Russia. The third-largest geyser field, El Tatio, Chile, contains just 38 geysers.

Three important geologic ingredients are needed for geysers to occur—a shallow source of volcanic heat, an abundant groundwater supply, and a unique natural underground plumbing system. The plumbing system is critical because it must be watertight and pressure-tight, much like a pressure cooker, in order to spurt out a jet of boiling water. Silica-rich volcanic rocks appear to be the best hosts. Rhyolite lavas and ignimbrites host most of the world's geysers. Silica dissolved from these rocks by the hot water is precipitated as mineral deposits called siliceous sinter, or geyserite, along the inside of the plumbing systems. Over time, these deposits form a watertight seal and strengthen the channel walls, enabling the geyser to work.

There are two types of geyser. Fountain geysers erupt from pools of water, usually as a series of sudden powerful bursts; cone geysers erupt from cones or mounds of geyserite, usually in steady jets that last anywhere from a few seconds to several minutes.

Turning on the works

Groundwater, superheated by contact with hot rocks, circulates into the geyser's plumbing system from below and meets cooler water flowing in from the surface. They mix and fill the plumbing system, and are gradually heated from below. This process may take minutes for small geysers, or hours, even days, for larger geysers. The water temperature reaches well above boiling point within the geyser's plumbing system, but it does not boil because of the pressure of the overlying water column. Eventually the water at depth becomes hot enough and starts to boil. Steam bubbles rise through the water column and physically push the overlying column of water upward. At the surface, this is visible as a sudden outpouring of water or a rising dome of water sometimes called a "blue bubble." At this moment, the loss of water causes the system to become unstable. Due to the reduction in pressure (which equals a drop in boiling point temperature) the superheated water in the plumbing system boils vigorously. Within seconds there is a sudden "flashing" of water to steam accompanied by a huge volume increase, so violent that it ejects the overlying water column into the air. Steam continues to roar from the vent until either the water is used up or the temperature within the plumbing system drops below boiling.

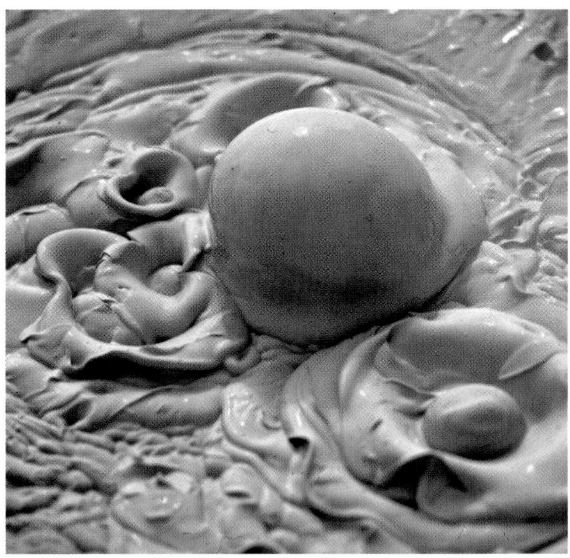

Right: Chalky, flower-shaped bubbles of hot mud in Waimangu Volcanic Rift Valley, near Rotorua, New Zealand. Waimangu, a hydrothermal system created by the volcanic eruption of Mt Tarawera on June 10, 1886, is the only such system in the world whose origin can be pinned down to an exact day.

Below: Many fumaroles and hot springs are found at Beppu in Oita Prefecture, one of the largest hot springs resorts in Japan. Eight of the area's thermal manifestations are referred to as jigokus, or hells. Bozu Jigoku ("Bonze Hell") contains thick, gray mud that boils and bubbles menacingly. Tatsumaki Jigoku ("Water-spout Hell") is a geyser that spouts water to a height of 82 feet (20 m) every 25 minutes. Chinoike ("Bloody Pond") has a vermillion color, and Umi Jigoku ("Sea of Hell"), pictured here, contains white particles that reflect the color of the sky.

The cycle of filling, heating, and boiling is then repeated. The duration of eruptions and times between successive eruptions varies greatly from geyser to geyser. Strokkur in Iceland's Geysir geothermal field erupts for a few seconds every few minutes, while Grand Geyser in the United States erupts for up to 10 minutes every 8 to 12 hours.

The impact of technology

The extraction of steam and taking of groundwater to run geothermal power plants often causes nearby geysers to cease being active. Of the five major New Zealand geyser fields in existence a century ago (Whakarewarewa, Rotomahana, Orakeikorako, Wairakei, and Spa), Whakarewarewa is the only one with a significant number of geysers still active. Fewer than 15 geysers are still erupting out of more than 130 that were active throughout New Zealand in 1950.

Another threat comes from mining, because thermal areas are often rich in valuable minerals, such as gold. Extraction may even require removing the geyser plumbing itself. A dramatic example was the cessation of activity in South America's second-largest geyser field (Puchuldiza, Chile) in May 2003, caused by the commencement of mineral exploration nearby.

More geothermal phenomena

Other geothermal occurrences include fumaroles (solfataras), hot springs, boiling springs, and mud pots. Fumaroles are steam vents that form when small quantities of water come in contact with an intense volcanic heat source. Hot pools and boiling pools, in turn, form where there is abundant groundwater present. The vibrant colors often seen in thermal pools, such as the Grand Prismatic Spring, Yellowstone National Park, are mainly produced by thermophilic (heat-loving) prokaryotes (bacteria) that live in the hot waters despite the harsh conditions. If the rock surrounding the springs is soft and friable, or if there is much soil, then the water turns into a thick muddy slurry and "mud pots" form.

Above: An aerial view of Grand Prismatic Spring, Yellowstone National Park, Wyoming. The vivid colors toward the edge of the spring are mainly caused by cyanobacteria. The gray-white color of the landscape around it is due to a precipitate called siliceous sinter, formed as the hot water dissolves silica in the spring's rhyolitic rock base.

Right: The center of Grand Prismatic Spring is sterile due to its extreme heat. Its deep blue color results from a light-absorbing overtone of the hydroxy stretch of water. This effect makes all large bodies of water blue, but is particularly intense at Grand Prismatic because of the high purity and depth of the water in the middle of the spring.

From the time Earth formed, it has been evolving. All its features—structure, atmosphere, oceans, continents, climate, and life—have undergone constant change. Once the planet had cooled enough for the crust and oceans to form, cyclic tectonic processes—over hundreds of millions of years—assembled then rifted apart supercontinents, changing the distribution of land and sea.

Climate

These changes to the crust undoubtedly had a great influence on the climatic and biological conditions, with climate varying between periods of icehouse-driven glaciation and more extensive periods that were greenhouse-driven, and were warm and tropical. Climate, in this context, means looking at global trends over geologic time. The cycles of glaciation, for example, can best be understood through paleoclimatology, the study of ancient climates, which uncovers evidence for the many climatic changes buried in the geologic record.

Time in the icehouse

Over geological time there have been at least four major glaciations when Earth was an icehouse—covered in massive polar ice sheets. The oldest dates back to the early Proterozoic Era, around 2.3 billion years ago. There was also a second icehouse period during the late Proterozoic Era, 780 million years ago, and another, between the Permian and Carboniferous periods, which started about 330 million years ago. The most recent began during the Tertiary Period, some 15 million years ago. Since each of the first three major glacial periods lasted 75 million years or more and the most recent started only 15 million years ago, it seems likely that today, although the climate is warming artificially, we are still under the influence of the icehouse conditions of the Tertiary Period.

When the roughly 600 million years for all four icehouse periods are expressed as a percentage of Earth's history, it is just 13 percent. So

Right: Dense tropical rainforest extends along parts of the Atlantic coast and into the uplands of Paraná, in southern Brazil. Other parts of the same region are treeless savanna or localized bushland.

Below: Lonnie Thompson with an ice core from the Quelccaya Ice Cap in Peru. Stored at Ohio State University at -4°F (-20°C), ice cores from tropical glaciers can yield climate data for up to 1,500 years into the past. Cores collected and analyzed over the last 20 years show that tropical glaciers are melting.

Above: Native to the Arctic, the polar bear (*Ursus maritimus*) is related to the brown bear, but has evolved to occupy a narrower ecological niche. No longer threatened by hunting, it faces a bigger challenge in global warming, as the melting of its sea ice habitat reduces its ability to find sufficient food.

Left: The Gobi Desert (Mongolian for "waterless place") is a continental desert extending for over 500,000 square miles (805 km^2) across southern Mongolia and northern China. The largest desert in Asia and the largest cold winter desert in the world, the Gobi encompasses mountains, rocks, grasslands, and salt flats, as well as sand dunes.

icehouse periods are "minor." They are the exception rather than the rule. Warmer climates with no permanent icecaps at the poles are apparently the norm. The warming that occurred between each of the icehouse periods would have been from the effect of greenhouse gases that caused the icecaps to melt and sea levels to rise significantly.

How climate changes

The changing configuration of the continents and oceans over cycles lasting millions of years has greatly affected climate. The positions of the continents in relation to each other, the circulation of the ocean currents around them, and their latitudes were all instrumental in determining climatic conditions. For example, when Australia finally separated from Antarctica about 45 million years ago, ocean currents started to flow between the two continents for the first time. This had the effect of producing cooler and drier climates on both continents. (See also the feature box on page 418.)

When continents are assembling into supercontinents, the collision of plates causes the crust to buckle through compressional forces, pushing up extensive mountain chains. These are often of an altitude that can interfere with the jet stream and so produce the cooling effect required to trigger the beginning of an icehouse cycle.

It also appears to be the case that when Earth's landmasses are concentrated near the polar regions, there is an increased chance for snow and ice to accumulate. But this is not always the case; there

FINDING OUT ABOUT ANCIENT CLIMATES

Science has discovered a vast number of clues to ancient climates. Some relating to past temperatures and weather patterns have been found in sea-floor sediments, which also reveal how much ice existed in the past. Ancient gas bubbles in cores drilled from polar ice caps can be made to yield detailed records of the atmospheric composition of the time, and also offer clues to past temperatures. No climatic event is too small to be recorded. Cycles of drought and rain are recorded in both living and fossil tree rings and coral, as well as in cave formations.

Left: Strange weather patterns signal the arrival of global warming. Here, sandbars are exposed as the Morris Reservoir water level in Azusa, California, USA, remains dramatically low near the end of the rainy season of 2007, when the San Gabriel River would normally have restored water reserves.

Right: Large mountains like Mont Blanc, on the French–Italian border, create their own weather. Blasting winds whip the summit, with temperatures as low as -58°F (-50°C). After the European heatwave of 2003, though not necessarily because of it, Mont Blanc lost over 29 inches (75 cm) in height.

have been warm periods in the past when polar landmasses were home to deciduous forests. So, the causes of the four major periods of glaciation are not fully understood.

Variations in solar radiation offer one possible explanation. For example, just small changes in temperature can tip the balance between snows completely melting over successive summers to snow remaining throughout the seasons. We understand cycles of solar activity over very short periods, but not over millions of years.

Other factors might include variations in the pattern of volcanic eruptions and in Earth's orbit, which can alter the amount of solar radiation received. These variations are cyclic and seem to tie in well with the known glacial–interglacial cycles during the Pliocene–Pleistocene Ice Age of the last 2.5 million years.

Climatic cycles

There are also climatic changes that take place over shorter periods of time. Over peiods of years and decades, Earth has cycles of drought and rain, and over periods of hundreds to hundreds of thousands of years, there is historical evidence of minor glitches like the Little Ice Age in the latter half of the seventeenth century. This abnormally cold period was possibly tied to the changing alignments of Earth, Moon, and Sun, which affected tidal pull and ocean currents. However, some believe it was linked to reduced sunspot activity.

Some climatic cycles over periods of thousands of years have been shown to relate to variations in Earth's orbit around the Sun. These have the capacity to alter the amount of solar radiation reaching Earth from the Sun and seem to work in recurring time intervals of 23,000, 41,000, 100,000, and 400,000 years. The figures for these cycles were calculated by a Serbian mathematician, Milutin Milankovich in the 1920s, and now bear the name Milankovich Variations. They match many of the glacial–interglacial cycles of the Pliocene–Pleistocene Ice Age over the past 2.5 million years.

> Geological change usually takes thousands of years to happen but we are seeing the climate changing not just in our lifetimes but also year by year.
>
> James Lovelock, scientist, b. 1919

CLIMATIC ZONES

It is possible to divide Earth's surface into climatic zones that broadly follow changes in latitude. Climate determines what plants will grow in a region and what animals will live there. These climatic regions coincide to a certain extent with patterns of temperature, rainfall and soils. Collectively, climate, plants, and animals create a biome.

A. Tropical climates experience high temperatures and rainfall, though in tropical monsoon and savanna regions, rainfall is very seasonal. Tropical rainforests like the Daintree in Australia are sometimes called "the world's largest pharmacy" because over one-quarter of modern medicines originate from their plants.

B. Arid and semi-arid climates are characterized by little rain and a huge daily temperature range. They include semi-arid or steppe, as well as arid or desert regions like Africa's Sahara. Over 14 percent of Earth's landmass can be classified as desert, with an average annual precipitation of less than 9.8 inches (250 mm).

C. Humid middle-latitude climates tend to have warm, dry summers and cool, wet winters. They include the Mediterranean and parts of western North America, Western and South Australia, southwestern South Africa, and central Chile. Plants in these areas have adapted to the difference in rainfall and temperature between seasons.

D. Continental climates are found in the interior regions of large land masses. Total annual precipitation is not very high and seasonal temperatures vary widely. Conditions can range from humid to cold, depending on factors such as proximity to large bodies of water, which makes temperatures colder.

E. Cold climates include both polar and tundra regions, which are characterized by permanent ice. Temperatures rise above freezing for only four months each year, when the top layer of the tundra regions' permafrost melts. The air above polar ice caps, the coldest places on Earth, is as dry as it is in most deserts.

Today's climates

"Climate" is more generally used to describe the average weather in a region over a period of roughly 30 years—a span of mere decades. But the climate that affects us from day to day is influenced by a number of factors. Climate changes with latitude or distance from the Equator, which affects the amount of solar radiation received. It is much hotter where the Sun's rays strike Earth's surface at 90° than it is at the poles, where the angle is very much less. Seasonal movements of air masses within the atmosphere also play a major role. These, in turn, influence two important climatic factors—air temperature and precipitation such as rain, hail, or snow.

Seasons result from the annual variation in the solar radiation reaching Earth's surface due to the tilt of its axis (23.5°) as it rotates in orbit around the Sun. When the Northern hemisphere tilts toward the Sun, it is winter in the Southern hemisphere and summer in the Northern. This is reversed when the Southern hemisphere is tilted toward the Sun.

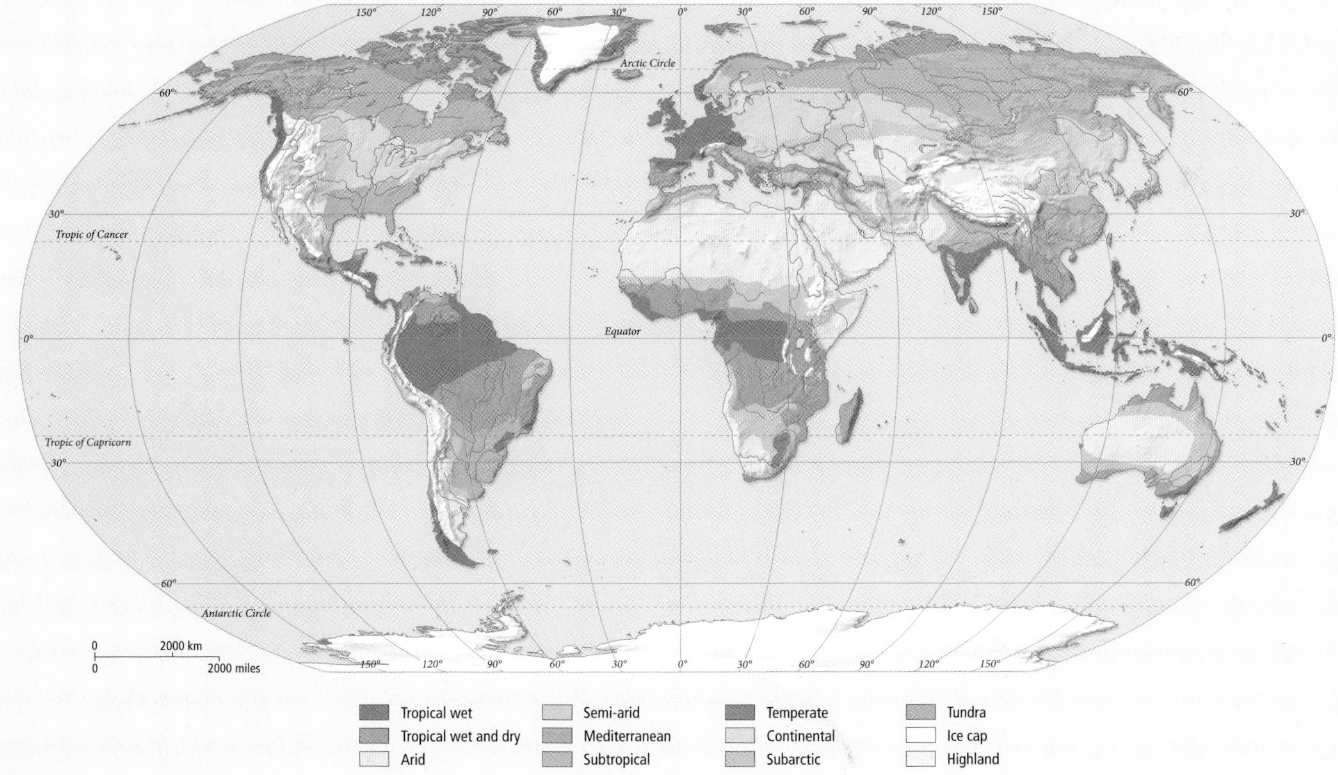

Left: Indian commuters make their way along a waterlogged street during a downpour of monsoonal rain in Amritsar, August 2007. The monsoon accounts for 80 percent of the rainfall in India, so agriculture and the economy depend on it. But it can have a downside, as evidenced by the Mumbai floods of 2005, which caused huge losses of life and property. In recent years, areas in India that used to receive scanty rainfall, like the Thar Desert, have been flooded due to prolonged monsoon seasons.

Our weather

Weather is the regional day-to-day experience of climate. There can be weather extremes, such as severe storms, hailstorms, hurricanes, droughts, and floods. Although distance from the Equator plays a role in a country's climate, not all countries at the same latitude experience the same weather patterns. Parts of some countries, such as India, might have very seasonal monsoons—warm, wet summer weather and dry, sunny winters—while others at the same latitude, such as the Sahara Desert, are arid. Still others, like the United Kingdom, have very changeable weather year-round. North America tends to have cycles of very extreme weather.

Altitude also influences weather. Mountain ranges can increase or decrease rainfall depending on the moisture content of prevailing winds. The climate of some coastal regions varies greatly from that found inland, such as the warm to cool temperate climate of Australia's eastern coast compared with the hot, arid and semi-arid interior regions. There is more about weather on pages 412–415.

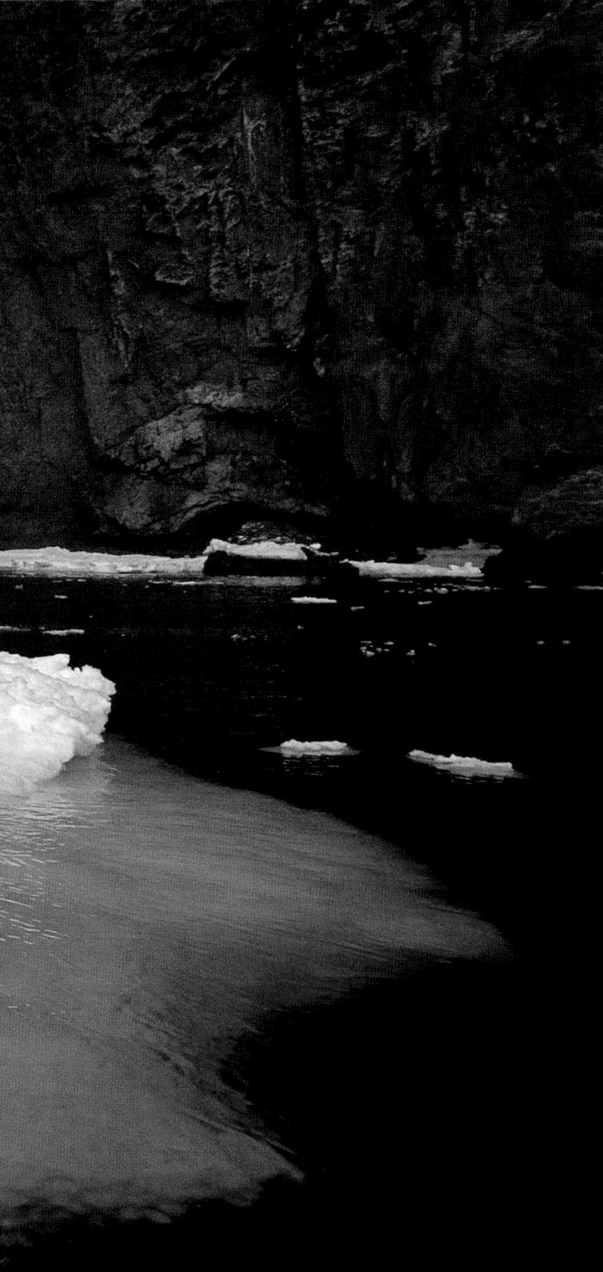

Above: The town of Ilulissat, Greenland, in August 2007. Disappearing ice caps could lead to higher sea levels all over the world, but Greenland's Inuit population are first to feel the effects of global warming. Air temperature has risen by 7°F (4°C) over the past decade and water temperature by half that amount. The ice in the fjords has become too thin to traverse safely, so fishermen and hunters avoid them. Warm water fish are becoming more common in their nets.

Left: Adélie penguins (*Pygoscelis adeliae*) crowded together on a melting summer ice floe, Possession Island, Ross Sea, Antarctica. Aside from the storm petrel, they are the most southerly distributed seabird. Their genus split from other penguins around 38 million years ago, and the Adélie species split off around 19 million years ago. With a reduction in sea ice and a scarcity of food, populations of the Adélie penguin have dropped by 65 percent over the past 25 years.

Right: A boy collects mussels from a dry pond in Yunyang Yunyang, Chongqing, China, August 2006. The worst drought in 50 years left over 10 million people without water. In China, more than 300 million people in rural areas are short of clean drinking water. The Ministry of Water Resources estimates that 40 percent of water in the country's 1,300 or so major rivers is fit only for industrial or agricultural use.

Climate change today

Relative stability of climate has persisted throughout 10,000 years of the Holocene Epoch, which started at the end of the last Ice Age. The very first modern humans, *Homo sapiens*, are thought to have appeared during the Pleistocene Epoch, around 200,000 years ago. In order to survive throughout that period of time, they would have had to endure several very bleak glacial cycles.

Evidence is clear that the climate is now changing at an alarming rate. We use records of long-term climate in order to understand how Earth's climate has changed in the past and how it might change in the future, and global warming is indicated. It is now agreed that this is the result of a build-up of greenhouse gases in the atmosphere, and that these gases are the products of human industrialization.

The greenhouse effect and global warming

The greenhouse effect is a natural phenomenon that has occurred throughout the planet's history. It is what keeps Earth's surface warm enough to be inhabited by life. It results from certain greenhouse gases in the atmosphere, such as carbon dioxide and water vapor, absorbing solar energy and radiating it back down to the surface. If this did not occur, this energy would be lost into space, in which case Earth's surface would be about 60°F (33°C) colder.

The volume of greenhouse gases in the atmosphere has varied over geologic time and this variation has caused the fluctuations between icehouse glaciation and greenhouse, warm, tropical climates. What we currently face, however, is the rapidly accelerating addition of greenhouse gases, such as carbon dioxide, into the atmosphere from human activities. This significantly increases average global temperature. This is not a measure of how hot or cold it is from day to day, but of the overall average temperature at the planet's surface.

Average global temperature is not unlike the average temperature of the human body in that a rise of only one or two degrees would be cause for concern, since it probably signals illness. So it is with our planet. Even a one-degree rise is a significant increase. The resulting global warming can cause dramatic changes to climate—polar ice caps and glaciers melt, and weather patterns change so that severe weather events occur more frequently. There can also be changes to climate in which desert areas begin to receive greater rainfall and regions of higher rainfall become more arid.

Earth's climate has never been stable. Climate change has been the norm. What is different about the climate change facing us today is that we are its cause. Never before has a single species of animal been seen in such vast numbers and with unprecedented power to alter the environment. *Homo sapiens* is in plague proportions.

If we are to survive, our only option is to change our ways. If we continue to allow carbon dioxide and other greenhouse gases to enter the atmosphere without sufficient checks and balances, temperatures will quickly rise. If we act now, however, we may still save ourselves and the biosphere on which we depend.

Weather affects our lives in myriad ways—from the clothes we wear to the jobs we have. Think about our homes—igloos in icy landscapes, white painted houses in hot places to reflect the sunlight, peaked roofs so snow can slide off. Jogging, sailing swimming, picnicking, flying a kite: Whatever the pastime, we love to know the weather.

Weather

From the very beginning, we have tried to predict the weather. Soothsayers were regularly engaged to predict good and bad years. Farmers believed that unsettled animals foretold storms, and sore knees meant cold weather. Sailors dependent on winds, and averse to storms, liked to be able to foretell the weather. It was not until information could be relayed quickly in the nineteenth century via telegraph that modern methods developed. A hundred years later the use of computers helped in data collection, improved predictions, and the dissemination of information.

Weather explained

Weather can be balmy or humid, sunny or rainy, turbulent, or calm. Weather is the state of the atmosphere at a given time and place, including its temperature, air pressure, wind speed, cloud cover, and moisture content. The Sun heats Earth's surface and the atmosphere above it. The warmed air, together with banks of clouds, circle the planet, producing the weather as we know it. Weather constantly changes as different masses of air interact. The study of weather is called meteorology, one of the oldest sciences in the world.

Temperature

Temperature, which is measured in degrees Fahrenheit or Celsius, is an indicator of the heat present in air. Cold is really the absence of heat. The Sun's rays warm the air at different rates according to how far they have to travel and whether they have to penetrate through a thicker blanket of air. So it is hotter at the Equator and colder at the poles. Temperature is also affected by altitude, distance from the sea, and prevailing winds. In extremely cold weather a strong wind makes it feel even cooler; this is called a "wind chill factor."

Above: Santorini, one of a small group of islands in the Aegean Sea formed by volcanic activity, is sometimes referred to as the "lost city of Atlantis." Like all the Cyclades Islands of Greece, Santorini has a Mediterranean climate—hot sunny summers and mild dry winters, with low average rainfall.

Sunshine is delicious, rain is refreshing, wind braces us up, snow is exhilarating; there is really no such thing as bad weather, only different kinds of good weather.

John Ruskin, critic and author, 1819–1900

Left: Igloos are ingenious structures built entirely of blocks of snow. Snow is a great insulator so the inside of an igloo can be quite toasty once it has been warmed by body heat or a small stove. In addition, the igloo's thick walls block freezing winds.

Air pressure

Air pressure is the weight of air pressing on Earth. When air is heated, air particles are more active, increasing the air pressure in that area. Because gravity exerts a pull on the air particles, air is denser closer to the surface. At higher altitudes there are fewer particles and the air pressure is lower. We measure air pressure in hectopascals or millibars using a barometer.

Air is also constantly moving. As the Sun heats a mass of air, it expands and rises in a process called convection, creating an area of low pressure. Air from high pressure areas rushes in to fill up low pressure areas to try to reach a state of equilibrium.

Left: Barometers are used to measure atmospheric pressure and are a good indicator of what the weather will be. High air pressure suggests fair weather, while low pressure suggests the chance of storms.

Below: Sophisticated hurricane warning systems operate around the world, thus saving lives and property. Florida, USA, has borne the brunt of many a destructive hurricane.

Wind

Wind is the flow of air from areas of high to low pressure. At the same time, air moves in spirals because of Earth's rotation. A 12-point scale called the Beaufort scale, devised by Francis Beaufort in the early 1800s, is used to measure wind speed—from below one for calm, 6 for a strong breeze that makes a whistling sound, and up to 12 for a hurricane. Hurricanes have their own scale as well.

Some areas of the world have recurring wind patterns which are given descriptive names. For example, the "mistral" is a destructive wind in the southeast of France; a "chinook" is a warm and gusty wind in the Rocky Mountains of the USA; the "southerly buster" brings a storm or cool wind at the end of a hot day along Australia's east coast; and a "sirocco" blows hot dusty sand from the Sahara across the Mediterranean.

Left: A thunderstorm seen from space. Thunderstorms occur when warm humid air rises and then cools suddenly. They happen all over the world, but are particularly prevalent in tropical regions.

Opposite page: Lightning splits the sky in Kagoshima Prefecture, Japan. Lightning is a sudden, dramatic, and visible electrical discharge from cloud to cloud or from a cloud to Earth.

Clouds and precipitation

Air rises due to convection currents because it is part of a front, or if it hits a mountain. When warm air rises, it carries water that has evaporated into vapor from lakes, rivers, streams, the oceans, and other surfaces. As it rises, the air cools until it reaches the point when the vapor is so cool it turns back into a droplet of water—it condenses, usually around a nucleus such as a speck of dust. At higher altitudes the water forms ice crystal clouds. When the water droplets become too heavy they fall as rain. Rain is only one type of moisture that falls from air. Other types of precipitation are snow, sleet, and hail, which fall when the temperature drops below freezing point.

Clouds are classified into types based on their shape and their height above the ground according to defined characteristics: Middle level (alto), low level or layered (stratus), wispy (cirrus), tall or heaped (cumulus), and rain-bearing (nimbus).

Lightning and thunder

When particles in clouds become electrified, they separate, with positive charges moving to the top of the cloud and the negative ones to the bottom. On the ground under the cloud a positive charge builds up too. The two charges are attracted to each other and the negative moves to the positive. A short but massive discharge of electricity shoots from cloud to cloud or from cloud to Earth. The process goes back and forth many times in the same lightning bolt. Lightning travels an average of 60,000 miles/second (96,000 km/second) and lasts for one-fifth of a second. A lightning flash sends out a whopping 100 million volts of electricity and heats the surrounding air to 54,000°F (30,000°C). Usually lightning hits the tallest object on the ground: a tree, a building, or lightning rod.

Thunder is the sound of the rapid expansion and contraction of heated air. Sound travels slower than the speed of light, so we see a flash of lightning before we hear the thunder. At any one time there can be thousands of thunderstorms happening in different places around the world.

Below left: A wind farm, such as this one in California, USA, is a group of wind turbines, placed close together and often in rows, that are used to produce electric power. They are considered environmentally friendly for the most part as they do not produce pollution.

Humidity

Water vapor is always present in the air. When there is very little water vapor, our skin feels dry and we get static shocks—it is dry. When there is a lot of water vapor in the air, it is humid. If the air is 100 percent full of water and cannot hold any more, it is saturated. Wind blowing across dry land and deserts is drier. Wind blowing across water will be moister because it picks up more water vapor, hence it is more humid near the coast.

Meteorologists utilize a complex method in order to measure humidity, which is dependent on the amount of water vapor in air at a specific temperature.

Modern forecasting

Meteorologists take thousands of observations around the world on Earth's surface, using balloons, and from satellite images. Local data gathered on the ground includes temperature, wind speed, wind direction, barometric readings, cloud cover, visibility, and humidity. Weather balloons have instruments to collect temperature and humidity at higher altitudes and send data back to Earth by radar. Snapshots taken by weather satellites can be carefully analyzed to create animations of the ever-changing weather patterns. These show the outline of countries, vegetation, cloud cover, hurricanes, floods, and thunderstorms. Specialized infrared images are taken of humidity and temperature zones. Information from these various sources is fed into computers using "forecast models" to generate weather maps. Meteorologists use the results to predict the weather outlook and to issue warnings about floods, storms, and hurricanes. Accurately forecasting the weather is still a tricky business, because it is limited by the complexity of the variables in the atmosphere, the type of data collected, and the models used.

Below right: A forecaster works on predicting the path and strength of Hurricane Lili as it approaches the Louisiana coastline on October 2, 2002 in Shreveport, Louisiana, USA. Lili was designated a Category 4 hurricane with sustained winds blowing at 135 mph (217 km/h).

HOW TO READ A WEATHER MAP

Weather maps in newspapers and on television are simplified for easier explanation. On a typical map, points of equal air pressure are connected with lines called isobars, forming concentric circles of high or low air pressure areas. The distance between isobars varies. When isobars are closer together it means that the wind is stronger. The concentric circles with H show high pressure systems and are associated with fine weather. An L indicates low pressure systems, which are usually rainy or cloudy. In the Northern hemisphere the air flows clockwise around highs and anticlockwise around lows. In the Southern hemisphere it goes in the opposite direction. (The map at right shows Greenland.)

The colored lines on weather maps show weather fronts, the boundaries between different masses of air. With a cold front—a blue line with triangles—cold air moves in to replace less dense, warmer air. It brings fresh winds, changeable weather, and possibly storms if the rising air develops into rain clouds. With a warm front—a red line with semicircles—warm air rises over the cold air, cooling as it ascends. Various cloud types are produced by warm fronts, which may lead to rain.

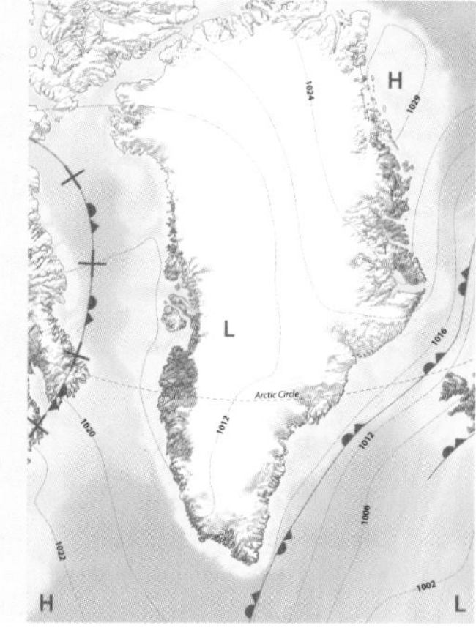

Imagine a forest and all the living organisms within it, from the smallest microbe through to the largest carnivore. Take into account the physical and chemical factors such as nutrient and energy flows and the water, air, rocks, and soil. By considering how all these things interact together, one can begin to appreciate the complexities of an ecosystem.

Ecosystems

There are many different types of ecosystem, and many complex factors that contribute to their functions and interactions. All are determined by the characteristics of their physical environment, such that the components and requirements for life are different from one environment to another. Yet there are some key characteristics and functions common to all ecosystems.

To survive and flourish, ecosystems need to transfer energy and recycle nutrients efficiently. The primary source of energy for almost all ecosystems is the Sun. Plants rely on the sun to photosynthesize, using sunlight to make chemical energy. This energy is transferred from one organism to another through flows of energy known as food chains. Plants, as producers of energy, are eaten by primary consumers, which in turn are eaten by secondary consumers and so on. Each level of the food chain is referred to as a trophic level. Carnivores are at the highest trophic level in most food chains. These energy flows are part of more complicated food webs, with different food chains interlinked to form a network of nutrient and energy transfer within an ecosystem.

The role of biotic interactions in ecosystems

Biological diversity is fundamentally important in maintaining the health of ecosystems. Biodiversity is defined as the variability among living things. It includes diversity within species, between species and between ecosystems themselves. Yet biodiversity is more than just the presence of a variety of animals and plants in a given environment. Biodiversity provides the conditions in which crucial biological processes such as pollination can take place. It supplies the world with natural products such as food, minerals, timber, and medicine—biotic products and components that are intimately interlinked with the physical and chemical processes of the ecosystems from which they come. Biodiversity impacts on the systems that purify water and air, and keep nutrients in the soil. When living matter decomposes, it returns to the ecosystem as nutrients and energy, thus contributing to and continuing the biochemical cycles. When biodiversity is reduced, all these products and systems are adversely affected.

While biodiversity is fundamental, some species are considered more ecologically valuable than others. This is because the removal of these "keystone species" can dramatically alter both the diversity and processes of an ecosystem. The earthworm is a keystone species because a number of other organisms rely on the physical changes worms make to the soil. The cassowary, a large bird that is found in Australian tropical rainforests, is considered a keystone species because of its vital role in ingesting and triggering the germination of a range of tropical plants. No other organism in this habitat is capable of playing this role. So without the cassowary, these tropical plants and the organisms reliant upon them could not survive.

Above: The Australian koala (*Phascolarctos cinereus*) has evolved to fill a very particular ecological niche. It lives almost entirely on eucalyptus leaves.

Right: The southern cassowary (*Casuarius casuarius*), found in southern New Guinea, eastern Indonesia and northeastern Australia, is a solitary forest dweller that pairs only in breeding season. It forages for fallen fruit, and distributes the seeds in its droppings.

Left: Two cheetahs (*Acinonyx jubatus*) feed on fresh kill at Masai Mara, Kenya, Africa. Unique even among cats in its speed and stealth, the cheetah is a carnivore that thrives in vast expanses of land where prey is abundant, though it can also survive in grasslands and some mountainous regions.

> All ... biological creatures are intimately connected with all of its physical systems, from the soils to the oceans to the atmosphere. Changes in any of these systems can affect everything else.
>
> David Suzuki, scientist and climate change activist, b. 1936

CAUSES AND CONSEQUENCES OF DAMAGED ECOSYSTEMS

As all ecological interactions are linked, a change to one component, either biotic or abiotic, has the potential to greatly alter an ecosystem's structure. So a number of different factors can impede a system's vigor. Since the industrial age, human activities have negatively impacted both directly and indirectly on ecosystems across the planet. Practices like land clearing, unsustainable fishing and agricultural methods, hunting, introduction of exotic species, and pollution have had direct impacts on the environment. Other problems such as salinity, erosion, acid rain (the effects of which are seen at right), changes to the water cycle, and climate change are symptoms of human activities that have negative global implications.

The consequences of sustained environmental pressures on ecosystems can be dire. The more obvious effects include extinctions, loss of diversity, destruction and fragmentation of habitats, reduction in the abundance and quality of natural resources, a drop in the productivity of crops, and the shrinking of wilderness corridors. More significant effects for humans are a reduction in the quality of air and the availability of clean water. Sometimes ecosystems can repair or modify themselves, but many are already beginning to crumble under the ongoing burden of human pressures.

Scientists believe that life began in shallow seas more than 3.8 billion years ago. Today with more than 70 percent of Earth's surface covered in oceans, we still depend upon on the riches within its waters. Oceans control and create our weather and yet its great depths are more mysterious to us than the surface of the moon.

Oceans

The oceans are one huge united system of salt water, with an average depth of close to 2½ miles (4 km). Containing an abundance of diverse habitats and life, more than 80 percent of the ocean is yet to be explored. Most living things are found in the sunlit zone at the top of the ocean, where a wide range of creatures live, from the mighty blue whale to the tiny plankton on which it feeds. Further down is the darker twilight zone, where only a small amount of light can penetrate. The animals here often have larger eyes and live on dead matter that falls from above. The deeper midnight and abyssal zones are quite devoid of light, with cold temperatures and extreme pressure. Around 90 percent of the ocean is in these dark zones. Many creatures found here produce their own form of chemical light, a phenomenon known as bioluminescence. The Hadal zone refers to water found in the ocean's deepest trenches such as the Mariana Trench, which is up to 36,160 feet (11,029 m) deep.

> We know that when we are protecting our oceans we're protecting our future.
>
> Bill Clinton, 42nd US president, b. 1946

Ocean properties and systems

One important property of ocean water is its density—it is 800 times denser than air. Consequently the pressure increases by 1 atmosphere for every 33 feet (10 m) of depth. This pressure determines the shape and functions of oceanic organisms. A further consequence of water's density is that sunlight can penetrate only a comparatively short distance, with plant life usually restricted to the top 164 feet (50 m).

A second important property of ocean water is its saltiness, with concentrations averaging from 3 to 3.7 percent. Salinity increases in areas of greater evaporation and decreases in cold water due to the diluting effect of melting ice. The more saline the water, the denser it is and the further it will sink.

Above: Said to be the largest animal ever to have existed, the blue whale (*Balaenoptera musculus*) eats up to 8,000 lb (3,600 kg) of krill every day, typically at depths of more than 330 feet (100 m). Numbers greatly declined last century.

Opposite: Named for its method of predation, the anglerfish (*Melanocetus johnsoni*) is a deep sea species equipped with a bioluminescent "fishing pole" which it uses as a lure to attract prey. Many varieties are found worldwide in both abyssal and continental shelf regions.

Below: The ice-covered Elephant Island, off the coast of Antarctica in the outer reaches of the South Shetland Islands, supports no significant flora or native fauna, but migratory gentoo penguins and seals are often found on its shores.

A third critical property of ocean water is its ability to store heat, with 1 gram of water increasing by 1.8°F (1°C) for every 4.18 joules (1 calorie) of heat it absorbs. One consequence of this is that the top 10 feet (3 m) of the ocean holds more heat than the entire atmosphere.

These properties determine the operation of ocean currents. There are two types of current—surface currents and deep ocean currents. Surface currents, which are largely driven by wind blowing over the ocean's surface, are usually fast moving and can be either warm or cool. Deep ocean currents, on the other hand, are much slower and form a more complex system. Deep ocean water can take hundreds of years to circulate then return to the surface. The surface and deep ocean current systems meet up in a process called the ocean conveyor belt. It transfers the heat of the planet and circulates nutrients from the depths of the ocean back to the surface. Through this complex system, the atmosphere and ocean work together to regulate the climate through the exchange of water, heat, and movement.

OCEANS AND CLIMATE CHANGE

Changes to the ocean's circulatory system could negatively contribute to the processes and impacts of climate change. A drop in salinity due to increased ice melting can disrupt or halt the deeper currents by reducing their ability to sink and transfer heat. This could result in more adverse weather events, such as longer and colder winters. It is believed that ice ages have occurred when this giant conveyor belt has been disrupted.

The ocean also plays a key role in the uptake of carbon dioxide. Alterations to the ocean's chemical composition and circulatory systems could greatly affect its ability to absorb CO_2, resulting in higher levels of this greenhouse gas in the atmosphere. Changes in salinity and water temperature also affect the biological processes of the ocean. This can result in a decrease in the abundance of plants and animals, a reduction in biodiversity and loss of habitats. Such impacts are already being observed with polar bears drowning in greater numbers because of shrinking icesheets. In the tropics, events such as coral bleaching are happening more frequently and are threatening the existence of coral reefs (below).

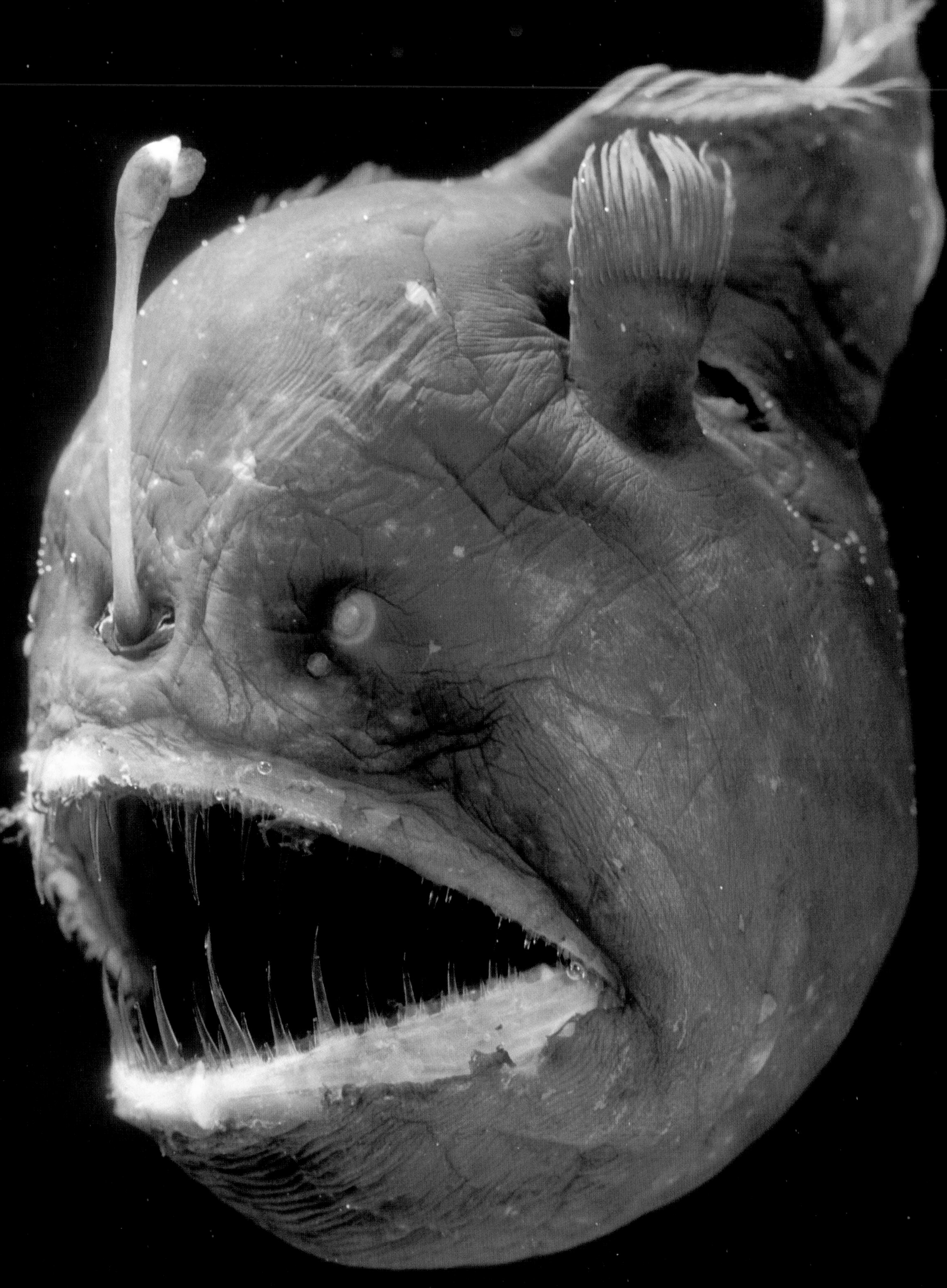

Deserts are places of contrast, with soaring daytime temperatures followed by freezing temperatures at night. Evaporation rates are so high that perspiration in humans seems minimal, yet even a short time without water is enough to create the light-headed euphoria of dehydration. Remote and hostile to human life, deserts remain one of Earth's last unpopulated frontiers.

Deserts

Deserts receive little precipitation—below 10 inches (250 mm) of rainfall per year. By this definition, the world's cold dry regions are also defined as deserts and are sometimes called polar deserts. With an area of 5,400,000 square miles (14,000,000 km^2), the Antarctic Desert, an ice desert, is the world's largest. The Sahara Desert is the world's largest hot desert, with an area of 3,320,000 square miles (8,600,000 km^2) covering Egypt, Libya, Chad, Mauritania, Morocco, and Algeria. At no more than 150,000 square miles (400,000 km^2), Australia's Great Sandy Desert and the Karakum Desert in Turkmenistan are the two smallest of the world's ten biggest deserts.

About one-fifth of Earth's land surface is desert. The major deserts, such as the Sahara and the Great Australian Desert (the general name given to all desert areas within Australia), occur in bands around the globe, generally outside of the tropics and between latitudes of about 20 to 35 degrees north and south of the equator. This is a zone of low winds and high pressure. Here, air masses from the equatorial and temperate convection cells descend and become dry and warm, so that generally speaking there is little chance of rain.

Deserts tend to be located toward the center and western side of large continental landmasses far from the reach of moisture-laden easterly trade winds (strong winds driven by Earth's easterly rotation). Furthermore, deserts will also form on the downwind or rainshadow side of mountain ranges because all moisture is extracted from the prevailing winds by uplift and cooling on the mountain's windward side. Sometimes an unfortunate combination of these factors produces exceptionally dry deserts, including some of the driest on Earth, such as the Atacama Desert of Peru and Chile, a virtually rainless plateau.

Above: The intermittent history of the Sahara Desert, which is almost as large as the USA, may go back about 3 million years. Its northern and southern reaches, like its highlands, include areas of sparse grassland and desert shrubs, with trees and taller shrubs growing around wadis, but the center is hyper-arid. The sand dunes there can reach heights of up to 600 feet (180 m).

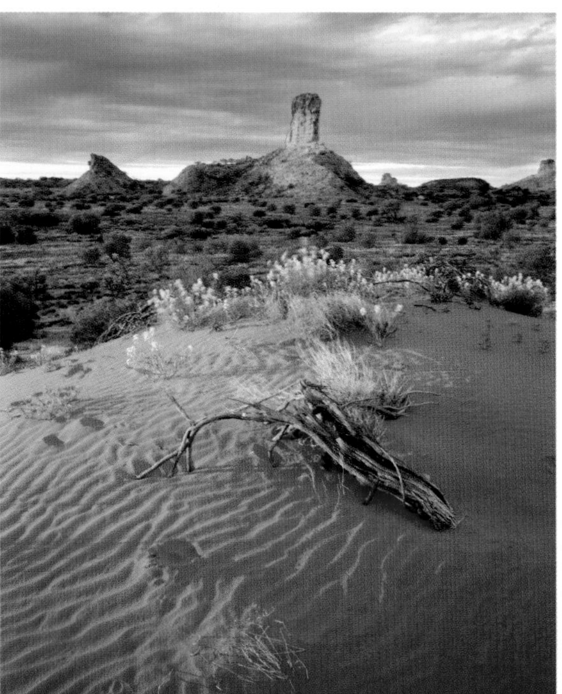

Left: Chambers Pillar, in Australia's Simpson Desert, is a red and yellow sandstone column rising 164 feet (50 m) above vast red plains. The desert around it contains the world's longest parallel sand dunes, held in position by vegetation. Dunes here vary in height from 10 feet (3 m) to over 100 feet (30–40 m).

Right: A crescentic dune in the Sahara Desert. Dunes come in a variety of simple shapes and in compound or complex combinations of these. Compound dunes are large dunes on which smaller dunes of similar type and orientation are superimposed, and complex dunes are combinations of two or more dune types. Compound and complex dunes are caused by variations in wind pattern.

Desert landscapes

Not all desert landscapes are vast oceans of wind-blown sand, or sand deserts, such as the Empty Quarter of Saudi Arabia. There are also stony deserts, such as the Tirari-Sturt Stony Desert in Australia, a vast plain of gravel and wind-polished pebbles or gibbers. The gibbers are so close together that they form a hard flat pavement and protect the underlying sand from erosion. Rocky deserts have shallow bedrock not covered deeply by talus or soil, or large pavements of bare rock swept clear of sand or gravel by wind. Extensive areas of this type of desert are found in the Libyan Sahara and are known as hammada. Plateau deserts occur on flat, bare tablelands, such as Golan Heights on the border of Israel, Lebanon, Jordan, and Syria. These rugged plateau lands are often deeply dissected by steep-sided narrow ravines or wadis. Mountain deserts feature bare and dry rock peaks such as those seen in the Tibesti and Ahaggar ranges of the Sahara.

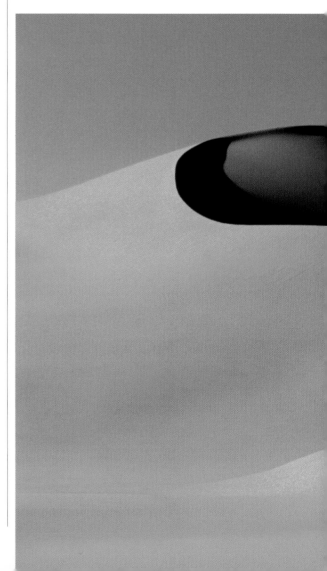

Above: The Ahaggar (or Hoggar) Mountains in the central Saharan highlands of southern Algeria. Climatically milder than much of the Sahara, this desert region, mostly comprising volcanic rock, is a major location for relict species and biodiversity.

Sand dunes

The lack of soil-stabilizing vegetation in deserts means that loose sand is common, and this can be blown into a variety of differently shaped sand dunes. The most common is the transverse dune, also known as a barchan or crescentic dune because of the crescent moon shape. A classic example, from North Africa's Sahara Desert, is pictured below. Barchan dunes form in unidirectional winds of moderate strength. Sand grains are blown up the windward slope of the dune and then roll down the downwind side. So the dune creeps forward, grain by grain. This can be a slow process, but barchan dunes have been measured in China traveling at about 300 feet (90 m) per year; they can bury trees, houses, or anything else in their path. In regions where the wind is regularly changing, the sand will not migrate in one direction and star dunes form. These dunes tend to pile upward, and in the Badain Jaran Desert of China, star dunes have been recorded reaching heights of around 1,600 feet (485 m).

Longitudinal or seif dunes form where winds are strong and unidirectional. These shoestrings of sand lie parallel to the direction of the wind. In the Sahara Desert they sometimes achieve lengths of 200 miles (320 km) and heights of 900 feet (275 m).

Parabolic dunes occur in the destabilized areas of vegetated dune systems, hence the name "blowout dune" is commonly used. The sand is pushed downwind into a parabolic arc in front of a wind-scoured corridor. Parabolic dunes are common in coastal areas and can be caused by any natural or human disturbance. For example, off-road vehicles cutting a track to the beach across a vegetated coastal foredune can initiate a blowout dune.

What makes the desert beautiful is that somewhere it hides a well.

Antoine de Saint-Exupéry, pilot and author, 1900–1944

The Namib Desert

The Namib, in Namibia and southwest Angola, is the world's oldest desert and one of the driest regions on Earth, with a history of aridity going back 50 million years. It receives less precipitation than the Sahara, yet supports a rich diversity of life, sustained by mild coastal temperatures and moisture-laden Atlantic fogs. A number of unusual plant and animal species are found only in this desert.

With an average width of just 60 miles (100 km), the Namib runs parallel to Namibia's Atlantic coastline for 1,200 miles (2,000 km), as it stretches ribbon-like from the mouth of the Orange River in South Africa's Northern Cape Province north through Namibia to Mossamedes in Angola. Located along a broad plain eroded into monotonously flat bedrock, it rises gradually on a gradient reaching inland from the coast to a height of 3,000 feet (1,000 m) at the Great Western Escarpment, which marks its eastern boundary. The escarpment's increased humidity, together with its annual rainfall of around 8 inches (200 mm), allows for thin coverings of annual grasses. Dry riverbeds etch their way across the Namib as linear oases, providing conduits for surface water, groundwater, and flood sediments.

In the central Namib, prominent steep-sided hills of granite called inselbergs rise from the surrounding low-relief plains. Spitzkoppe, a granite massif between Usakos and Swakopmund, is the country's most recognizable mountain. Made of granite more than 700 million years old, it rises roughly 2,300 feet (700 m) above the surrounding plains despite its summit being 5,835 feet (1,784 m) above sea level.

Salt deserts

Salt deserts occur in enclosed dry basins where drainage does not flow toward the coast but inland toward the center of the basin. Any rainfall that does occur carries soluble salts inward to temporary lakes or playa lakes that form at the basin's lowest points. These lakes may only have water in them once every few years or even decades. High evaporation rates mean that they very quickly dry out, leaving vast

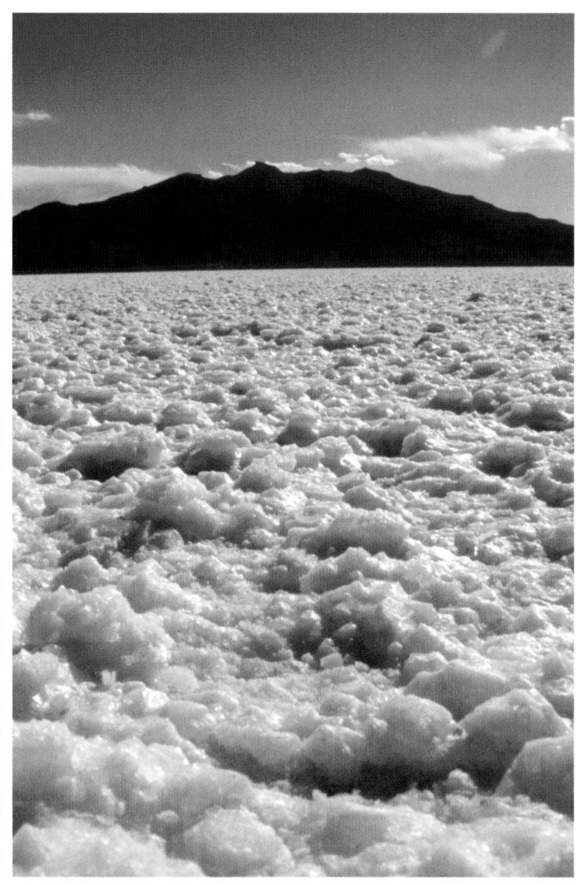

Right: Salar de Uyuni, in southwest Bolivia, is the world's highest and largest salt lake. Around 40,000 years ago, the area was part of Lake Minchin, a giant prehistoric lake. When the lake dried, it left behind two modern lakes (Poopó and Uru Uru), and two major salt deserts, Salar de Coipasa and the larger Uyuni. Salar de Uyuni is estimated to contain about 10 billion tons of salt, of which less than 25,000 tons is extracted annually. Every November, the lake becomes the breeding ground for three species of South American flamingo.

Below: The Sossusvlei salt pan lies deep in the dunes of the Namib Desert, Namibia. Sossus means "place of no return." Created by a river that flows through the Namib every five to 10 years before draining away between the dunes, Sossusvlei is surrounded by both dead and living trees, some of which are estimated to be 900 years old.

ANTARCTICA'S ICE DESERTS

Because Antarctica receives less than 10 inches (250 mm) of precipitation per year, it is classified as a desert. The interior of the continent receives less than 2 inches (60 mm) of rainfall annually, less than the Sahara. With almost no evaporation due to its cold climate, rainfall and snow have built up over millennia into the massive ice sheets we see today. Antarctica was not always a land of ice sheets and permafrost—300 miles (500 km) from the South Pole, sandstone beds lined with coal deposits have been found, laid down in marshy conditions under a cool moist climate. A series of dry valleys throughout Antarctica receive virtually no rainfall. These ice-free valleys can be up to 3 to 6 miles (5 to 10 km) wide between ridge crests and up to 30 miles (50 km) long, with permafrost patterns common to perennially frozen ground seen in their loose sand and gravel surfaces.

shimmering white flats of salt and other minerals such as gypsum and calcite. Some famous salt deserts include Salar de Uyuni in Bolivia, Dasht-e Kavir in central Iran, Australia's Lake Eyre Basin, and the Great Salt Lake of Utah, USA.

Desert vegetation

Desert plants have many water-conserving strategies. Xerophytes collect and store water and have features that reduce water loss, such as small waxy leaves, fewer surface pores (stomata), or no leaves at all. Long hairs on the leaf—trichomes—provide shade, break up wind flow over the leaf surface and hence reduce moisture loss, and act as water collectors during morning mists. The cactus closes its stomata during the day to reduce transpiration. Other xerophytes minimize water loss by shedding their leaves during the dry season. Many have either succulent fleshy stems such as cacti, or a large fleshy underground tuber, enabling them to store large amounts of water. Spines or prickly hairs discourage foraging animals from dining on the plant's succulent flesh in order to quench their thirst.

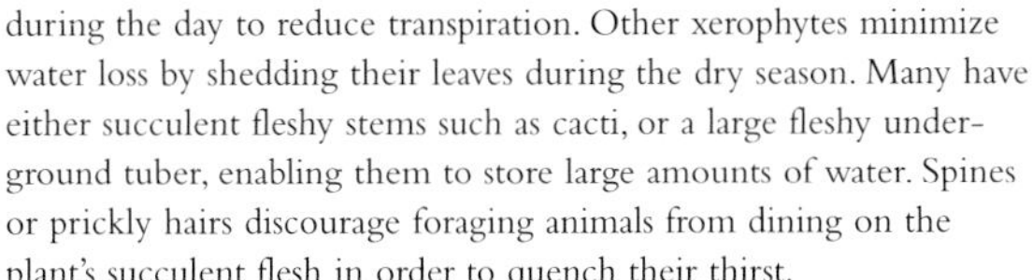

The root system of xerophytes can be extensive, shallow, and admirably adapted to maximize water uptake during brief sporadic rains, as in the case of the cactus. Or, they can be deeply penetrating, designed to reach a permanent subsurface water supply, as in the case of the acacia and the oleander. Desert palms have developed long taproots that enable them to survive.

Above: Saguaro National Park in Arizona, USA, gets its name from the saguaro cactus (*Cereus giganteus*) which is native to the region. Many other kinds of cactus, including barrel cactus, cholla cactus, and prickly pear, are also abundant, and the varied elevation within the park (which encompasses the Rincon Mountains) allows for a great variety of different species, such as ponderosa pine, oak, and Douglas-fir, to grow here.

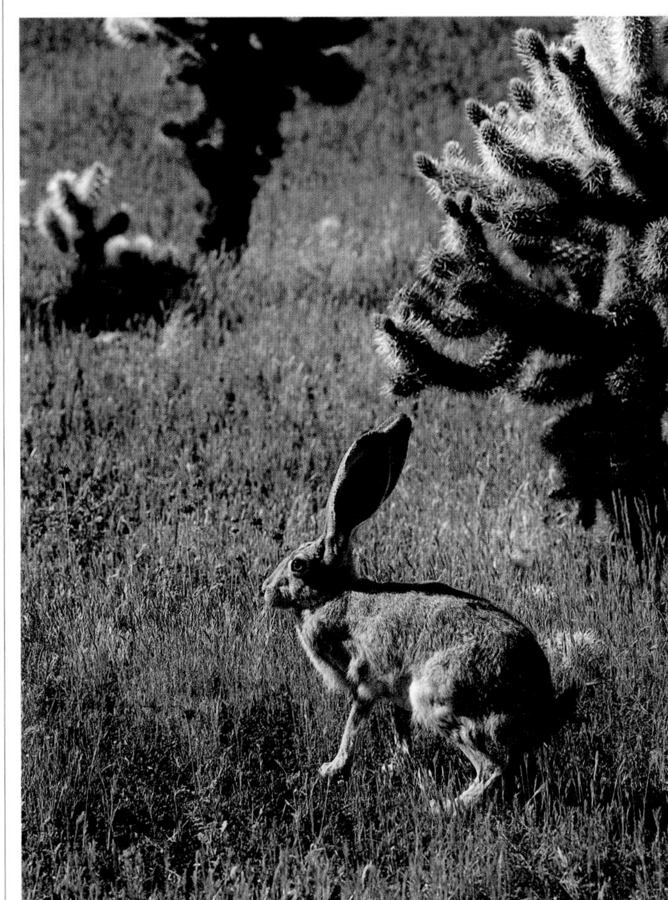

Right: A black-tailed jackrabbit (*Lepus californicus*) seeks the shade of a cholla cactus in the Sonoran Desert, which straddles large parts of Arizona and California, USA, and Sonora and Baja California, Mexico. One of the largest and hottest deserts in North America, the Sonoran Desert is home to many unique plants and animals adapted to its geology and climate. Of the three species of hare native to California, only the black-tailed jackrabbit is a desert dweller.

Without rivers, Earth would have no valleys, no waterfalls, no floodplains, and very little life on land. Gravity causes water to flow from high to lower elevations. Small trickles merge to form streams, then mighty rivers. Given a continuous water supply, a river will flow downhill until it reaches its mouth, usually at the ocean.

Rivers and waterfalls

Water, whether liquid or solid, is the most erosive force in nature, although the rate at which it erodes is controlled by a number of factors including the hardness of the bedrock, the volume of the water, and the gradient of the slope. Snowfields may compact into ice, forming "rivers of ice" or glaciers that also flow downhill. Glaciers are highly erosive, acting like giant rasps as they grind the underlying rock to form U-shaped valleys.

Cascades and rapids

Steep-sided V-shaped valleys are characteristically formed by faster flowing streams and rivers originating in higher altitudes, where water commonly drops over steep slopes and cliffs, forming rapids, cascades, and waterfalls. As the slope's gradient decreases and the rate of water flow drops, the valleys widen and erosion removes sediment from the banks and riverbed. But the slower flow velocity also results in load deposition, with the heaviest particles deposited first and the finest deposited in relatively still water where they might form floodplains, channel bars, and point bars.

Channels usually meander or flow along a winding course of looping bends. It is common for a river to meander across the entire width of its floodplain with the channels changing direction and migrating along the flats. In time, a loop of the meander might be totally cut off from the river itself to form an oxbow lake.

The oldest stage of a river has very low gradients and very wide floodplains over which meandering channels are restricted to sections of the floodplain. In this stage, deposition exceeds sediment supply, so there is a build-up or aggradation of surface materials as the river finally loses its capacity to transport sediment. It is at this terminal stage of the river that the sediment is mostly deposited at the river mouth as a delta, usually roughly triangular in shape.

Heaven above was blue, and earth beneath was green; the river glistened like a path of diamonds in the sun …

Charles Dickens, writer, 1786–1851

Below: Horseshoe Bend is the name given to a horseshoe-shaped meander of the Colorado River in Arizona, USA. Vegetation on the surrounding cliffs includes blackbrush, sage, yucca, and other desert plants. Conceivably, at some time in the future, the river could cut through the narrow neck of rock and abandon the horseshoe-shaped channel around it.

Left: Victoria Falls (Mosi-oa-Tunya, "Smoke that Thunders"), on the border between Zambia and Zimbabwe in Africa, is unusual in form: The Zambezi River drops into a deep, narrow chasm connected to a long series of gorges, allowing the whole falls to be viewed face-on. Victoria Falls arguably boasts the most diverse and easily observed wildlife of any major waterfall site.

Right: The Mekong Delta is a region in southwestern Vietnam where the Mekong River approaches and empties into the sea through a network of distributaries. The size of the area covered by water depends on the season. Thanks largely to the delta, Vietnam is the third largest exporter of rice in the world, after Thailand and India.

Below: The waterfall system that comprises Iguazú Falls extends along 1.67 miles (2.7 km) of the Iguazú River. The falls divide the river into the upper and lower Iguazú. The water eventually collects in a canyon that drains into the Rio Parana in Argentina.

When a channel becomes choked with sediment, it will break its banks and form a maze of numerous shallow, interweaving channels that tend to create a braid-like pattern with a distinctive appearance that gives rise to the term "braided stream." These distributaries, as they are known, are a common feature of alluvial fans, particularly those formed by glaciers and deltas.

Drainage basins

Drainage basins are areas that funnel the water coming into them into the streams that drain them. The shape and extent of a drainage basin depends on topography, rock type, and geologic structures within the watershed. The most remarkable is the Tibetan Plateau, the world's highest and largest plateau and watershed of six of the world's largest rivers: the Brahmaputra (China, India, and Bangladesh), the Indus (China, India, and Pakistan), the Salween (China, Myanmar, and Thailand), the Mekong (China, Myanmar, Laos, Thailand, Cambodia, and Vietnam), the Yangtze (China), and the Huang or Yellow (China). More than half of the world's population lives in nations served by the drainage basins of these six rivers.

IGUAZÚ FALLS, SOUTH AMERICA

This truly remarkable series of waterfalls forms where the borders of Argentina, Paraguay, and Brazil connect. Higher than Niagara Falls and wider than Victoria Falls, the Iguazú Falls are often said to make up the largest waterfall in the world. The source of the Iguazú River is the Serra do Mar near the Atlantic Ocean in Brazil. The river flows west for more than 800 miles (1,300 km) before cascading spectacularly over drops of up to 270 ft (80 m) high and 8,800 ft (2,700 m) wide. The flow rate varies seasonally, and the water splits into as many as 275 falls and rapids, the most impressive being the Garganta del Diablo, or Devil's Throat. Here, the water flows over a 500-ft (153-m) wide horseshoe-shaped cliff—it is this part of the river that forms the border between Argentina and Brazil.

The falls are formed by the Iguazú River as it cascades over a series of massive flat-lying to undulating basalt flows. These are a section of the extensive layers of flood basalts that cover parts of southern Brazil, Paraguay, Argentina, and Uruguay.

Earth has been free of ice and relatively warm, even at the poles, for most of its history. Yet the balance between solid ice and liquid water on the planet's surface is constantly changing. When ice builds up on the polar landmasses, average global temperatures drop, sea levels fall, and the planet experiences a glacial epoch.

Fjords and glaciers

Since Earth formed some 4.6 billion years ago, there have only been a few major glacial epochs. These are known as the Huronian, from 2.4 to 2.1 billion years ago, the Cryogenian (Sturtian–Marinoan) from 850 to 635 million years ago, the Andean–Saharan (Ordovician Period) from 450 to 420 million years ago, the Karoo (Carboniferous–Permian periods) from 360 to 260 million years ago, and the Late Cenozoic Era ice epoch that began about 30 million years ago—the epoch we are currently experiencing.

All the glacial epochs put together have only occupied around 10 percent of Earth's geological time. During major glacial epochs, great ice sheets form and thicken in the high latitudes and push toward the equator to cover as much as 40 percent of Earth's land surface. During glacial epochs, the temperature may drop as much as 25°F (14°C) in the mid-latitudes.

Below: Fox Glacier, on the west coast of the South Island of New Zealand, retreating for most of the last 100 years, has been advancing at the rate of about one meter every day since 1985. It drops 8,500 ft (2,600 m) on its 8-mile (13-km) journey to the coast, getting smaller as it goes. Together with the nearby Franz Josef glacier, it is one of very few glaciers in the world that come right down into a rainforest.

Interglacial periods

Within each major ice epoch there are Ice Ages, which alternate with shorter, warmer periods known as interglacials. Earth is currently experiencing an interglacial period that followed the last Ice Age. The climate began to warm around 18,000 years ago, then 15,000 years ago glacial advance halted and sea levels began to rise. Ice Age megafauna (mammals, birds, and reptiles) became extinct some 10,000 years ago, and then approximately 8,000 years ago the Bering Strait land bridge was submerged, cutting off human and animal migration routes into North America. The Holocene Epoch interglacial warm period peaked around 5,000 to 6,000 years ago.

Barely noticeable eccentricities in Earth's orbit may affect these glacial and interglacial cycles approximately every 100,000 years. These eccentricities include variations in Earth's elliptical orbit, tilt

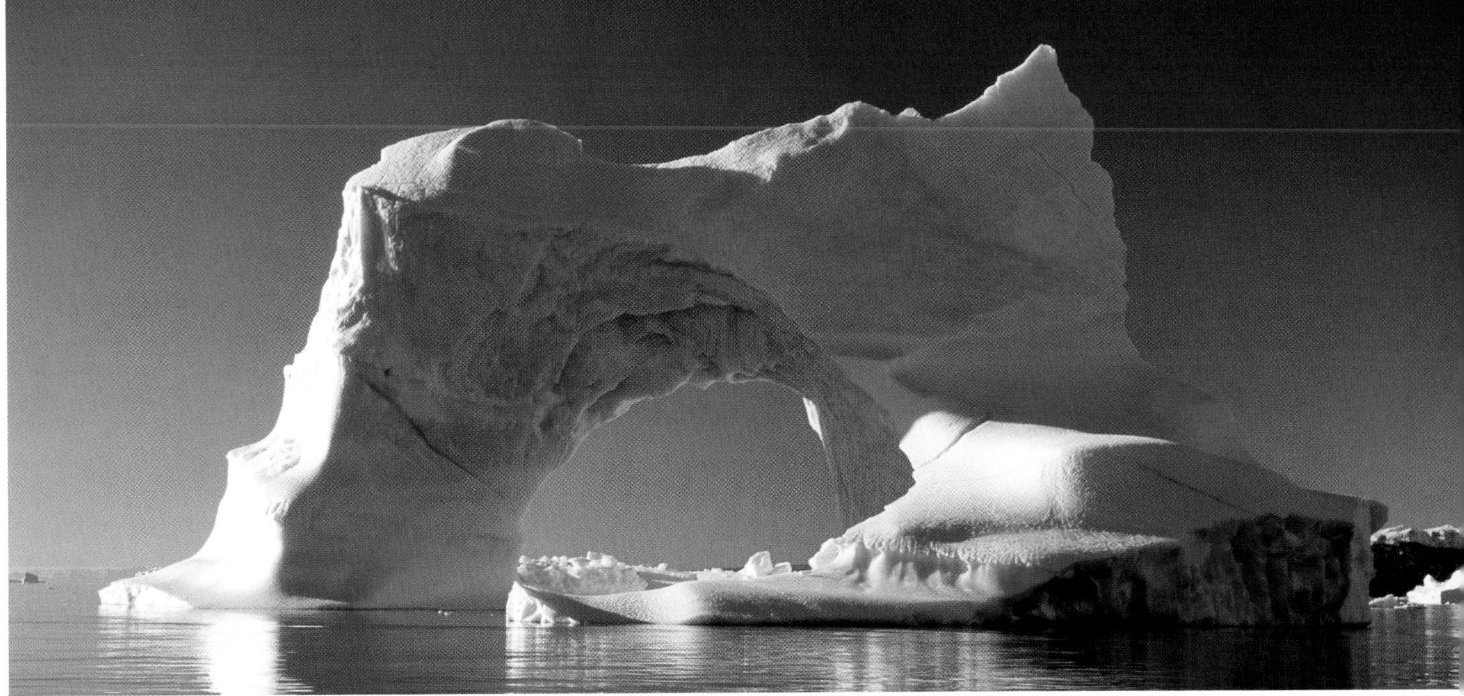

of its rotational axis, and a wobble or precession in Earth's axis. These are named Milankovitch cycles, after the twentieth-century Serbian civil engineer and geophysicist who first suggested that these fluctuations could account for the Ice Age–interglacial pattern.

Above: Icebergs, it is often said, always tell a story. This one, in Greenland's Scoresby Sund (the longest and one of the deepest fjords in the world), shows signs of wind and water erosion, and its old waterlines tell of its previous orientations in the water. Though sightseers are often tempted to row through such arches, it can be very dangerous to do so: Arches can collapse at any time due to their own weight, and as the iceberg can lose balance, and tip and roll into a new position.

Mini ice ages

Even shorter cycles of colder weather and warmer weather have been recognized within the current interglacial period, which started approximately 15,000 years ago. Some 5,000 to 6,000 years ago, Holocene Epoch interglacial temperatures reached an average of about 3½°F (2°C) warmer than present-day temperatures. This period was followed by a much colder spell during the Iron Age, around 1000 to 500 BCE. This cooling sparked re-glaciation of the coastal mountain ranges. Temperatures then rose and peaked between 1000 and 1350 CE, the Medieval Warm Period. They then fell again from 1400 to 1860 CE during the "Little Ice Age" and have been rising

> The glacier was God's great plough set at work ages ago to grind, furrow, and knead over, as it were, the surface of the earth.
>
> Louis Agassiz, geologist, 1807–1873

Left: The Tracy Arm–Fords Terror Wilderness area (near Juneau, Alaska, USA) includes two deep and narrow fjords, Tracy Arm and Endicott Arm, both over 30 miles (48 km) in length. The twin Sawyer Glaciers (North Sawyer and South Sawyer) are located at the end of Tracy Arm. The area supports a wide variety of wildlife: Black and brown bears, deer, wolves, harbor seals, and birds such as arctic terns and pigeon guillemots. Mountain goats, usually found at higher elevations, have occasionally been seen near the base of the glaciers.

HOW FJORDS ARE FORMED

Fjords are constructed by glacial ice build-up during an Ice Age.

A. Steep river valleys have been incised in the landscape, probably controlled by joints or planes of weakness in the rock. Such V-shaped valley profiles are typical of youthful, energetic rivers trying to cut down to sea level.

B. During an Ice Age, glaciers widen the valley into a broad U-shape. The cutaway at the base of the glacier shows how the moraine, or rocks imbedded in the ice, act as tools to carve, scour, and polish the walls and base of the valley.

C. When the glaciers retreat, the empty U-shaped valleys are flooded by the rising sea to form fjords. They are spectacularly beautiful, sheer-walled deepwater harbors with waterfalls often cascading from hanging tributary valleys.

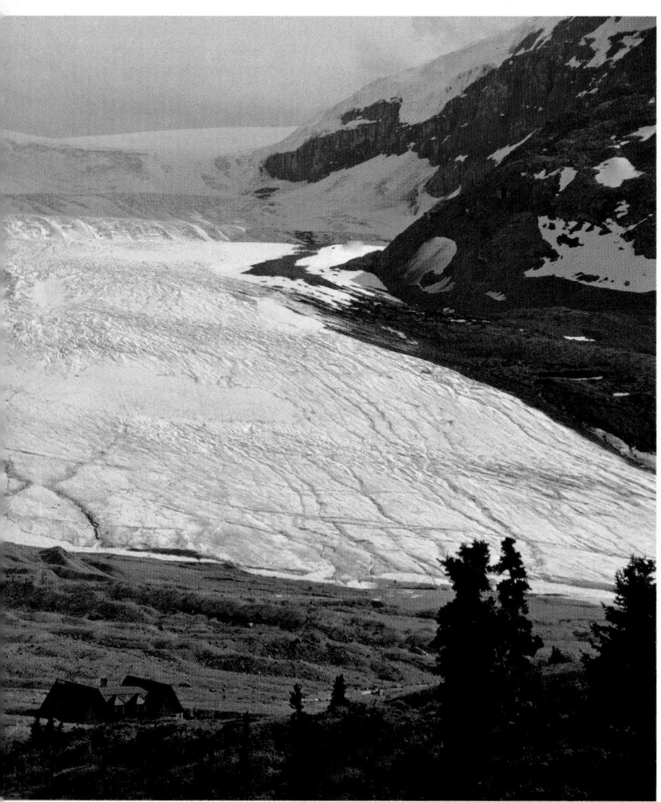

Left: Athabasca Glacier in Alberta, Canada, is one six principal toes of the Columbia Icefield in the Canadian Rockies. Due to global warming, it has receded nearly one mile (1.5 km) over the past 125 years and is now receding at a rate of nearly 9 feet (3 m) per year.

Right: Hubbard Glacier is a tidewater glacier in Russell Fjord Wilderness, Alaska, USA. From its source in the Yukon, it travels 76 miles (122 km) to the sea. Its open calving face is over 6 miles (10 km) wide. Calving occurs when glaciers meet large bodies of water, and involves the separation of portions of the glacier into the water. Many icebergs enter the world's oceans due to calving.

Below: Columbia Glacier, in Prince William Sound (Alaska, USA), is one of the fastest-moving glaciers in the world. Roughly 32 miles (51 km) in length, it has been retreating at an average rate of 0.4 miles (0.6 km) since the early 1980s, discharging icebergs as it goes.

since 1860 to the present day, known as the "Industrial Age" global warming cycle. These shorter cycles are probably driven by a combination of factors including ocean circulation, volcanic eruptions, meteorite impacts, solar variations, and the concentration of greenhouse gases in Earth's atmosphere.

It was long believed that these cycles came and went slowly on time scales measured in centuries. Current studies of ice cores taken from drill holes in Earth's icefields, however, reveal the potential for far more rapid temperature changes, known as "abrupt transitions," in which major shifts in Earth's climate occur on very short time scales. Abrupt transitions are explained by the recent discovery that large-scale ocean circulation in the North Atlantic can occur in one of two patterns, or stable states, that can switch suddenly. In one state, the warm Gulf Stream flows along the eastern coast of the United States and continues northward, past the British Isles to the Norwegian Sea, thereby keeping the climate of northwestern Europe relatively warm. In the second possible state, the northward extension of the Gulf Stream is "turned off," resulting in a much colder climate for all who live downstream, including the populations of Europe.

Glacial mechanics

Icefields begin to form in Earth's cooler regions whenever more snow has fallen over the winter than will melt during the following summer. In tectonic terms, both high mountainous regions and larger continents that have drifted into polar regions can act as zones of ice accumulation. The yearly snow build-up slowly compacts to form ice, which begins to move outward, or downslope, under its own immense weight. This is how glaciers are born.

Glaciers act as huge conveyor belts, moving ice and enormous amounts of rocky debris from the zone of accumulation above the snowline to the zone of ablation (melting) at lower elevations or at latitudes below the permanent snowline. Glaciers may move slowly but they nevertheless transform landscapes, crushing and pulverizing the rocks as they pass over. Bands of debris form starkly contrasting black stripes against the brilliant blue-white of the glacial ice, either lying within the glacier itself (medial moraine), along the edges of the glacier (lateral moraine), or as mountainous piles in front of it (terminal moraine). Glacial processing that occurred in the distant past has been of enormous benefit to humans. Glaciers are responsible for the thick, rich soils of North America and northern Europe, which in turn allowed for the development of numerous strong societies based on agriculture in these regions.

Fantastic fjords

Although glacial ice is continually moving down its U-shaped valley, the front edge or terminus of the glacier may either advance or retreat. The terminus moves forward or backward depending on how fast the ice melts away compared with how quickly new ice is brought to the front by the glacier. During warmer interglacial periods, glaciers shrink and retreat within their broad U-shaped valleys. All the extra meltwater from the ice causes global sea levels to rise, leading to the flooding of low-lying coastlines and coastal valleys.

The term fjord is used to describe a flooded U-shaped glacial valley. Fjords are characterized by smooth, near-vertical walls that rise majestically from deepwater estuaries. Their protected waters are often filled with abundant marine life. Waterfalls frequently cascade spectacularly from the hanging valleys. Fjords are commonplace along coastlines of once heavily glaciated lands such as Norway, New Zealand, Chile, and Alaska.

THE FJORDS OF NORWAY

Norway is renowned for its glacial scenery, such as that found in Jostedal Glacier National Park (right). This park surrounds one of mainland Europe's largest plateau glaciers, the Jostedalsbreen, and constitutes one of the largest wilderness areas in southern Norway. Spectacular scenic contrasts within a relatively short distance vary from luxuriant fjords to U-shaped valleys in ice-covered mountains. Lowland farms in traditional agricultural landscapes give way to peaks such as Lodalskapa, 6,834 ft (2,084 m) above sea level. Innumerable streams, rivers, and waterfalls lace the mountainsides during summer, fed by the ice cap and snow melt.

Sognefjord is Norway's longest fjord, stretching more than 120 miles (200 km) inland from its outer coastal islands to its inner precipitous cliffs. The region where this deep fjord meets up with the glaciers and Norway's highest mountains is considered one of the world's most beautiful travel destinations. Another fjord, Hardangerfjord, is 110 miles (179 km) long and lined by a number of picturesque villages. Tourists visit this fjord to enjoy the Hardanger spring blossoming, when the district's fruit orchards flower against the backdrop of snow-covered peaks.

Coastal landforms vary greatly. There can be broad, sandy, or cobbled (shingled) beaches; steep rocky sea cliffs, sometimes with small bays; tidal flats; or beach barrier islands with bars and lagoons. Waves and currents, produced by tidal activity and river discharge, interact with wind to create these environments.

Coastal landforms

Coastal landforms are grouped according to whether they are formed by depositional or erosional processes; for both, water is the main agent. The location of a coast relative to its position on a tectonic plate determines just which of these two processes dominates.

Tectonics and coastal landforms

The boundaries of Earth's tectonic plates are said to be convergent when plates collide and divergent when they move apart, or transform as the plates slide past each other. The continents carried by the plates have various types of continental margin, influenced by where they are situated in relation to plate boundaries. There are three distinct types: leading edge coastal margins; trailing edge margins; and the marginal seacoasts. The leading edge coastal margins are associated with convergent plates, such as those along parts of the Pacific coast of the Americas. Coastal landforms related to leading edge margins feature rugged and irregular topography, sometimes with high sea cliffs rising from the water line. Erosion dominates these coasts with small pockets of local deposition, such as small beaches in embayments. Large waves erode the cliff to form rock platforms, caves, and sea stacks. Trailing edge margins are relatively tectonically stable. The Atlantic side of North America is a typical example of a trailing edge margin. Here, wave action is moderate because of the gently sloping and wide coastal shelf. Large river systems contribute to the deposition of mud, silt, and sand that form a variety of landforms such as deltas, estuaries, barrier islands, tidal inlets, tidal flats, wetlands, and salt marshes. Marginal seacoasts are relatively stable because they are protected from open ocean processes, and smaller waves allow mud and silt to accumulate in the coastal zone. Their seaward plate boundary often has island arcs and volcanoes, such as those of Japan. Landforms associated with marginal seacoasts share some of the features of both leading edge and trailing edge margins, although the deltas that form along the marginal seacoasts are often the largest of all; for example, the Mississippi Delta in the Gulf of Mexico.

Below: The Twelve Apostles, in Port Campbell National Park on the southern coast of Australia, are stacks are made of rock formed 20 million years ago. The sea gradually eroded the softer limestone, forming caves in the coastal cliffs which then become arches. When these arches collapsed, rock islands as high as 148 ft (45 m) were left in the sea. The coast is still changing, with one large stack collapsing on July 3, 2005.

Change is the only constant.

Heraclitus, philosopher, *c.* 535–475 BCE

ENGLAND'S JURASSIC COAST

In 2001, a 100-mile (60-km) stretch of coastline between Old Harry Rocks in East Dorset and Orcombe Point near Exmouth in East Devon became Great Britain's first and only natural World Heritage Site. Known as the Jurassic Coast, its sea cliff exposures provide a continuous sequence of rock formations spanning 185 million years of geologic history along an unspoiled and accessible coastline of great beauty. Its landforms include stretches of discordant and concordant coastlines, examples of cove and limestone folding, tombolos, and natural arches. The region gained its name from the richness of its fossil record. Dinosaur footprints, a fossil forest, and huge ammonites such as those pictured below represent only a fraction of the preserved fossil evidence that can still be found.

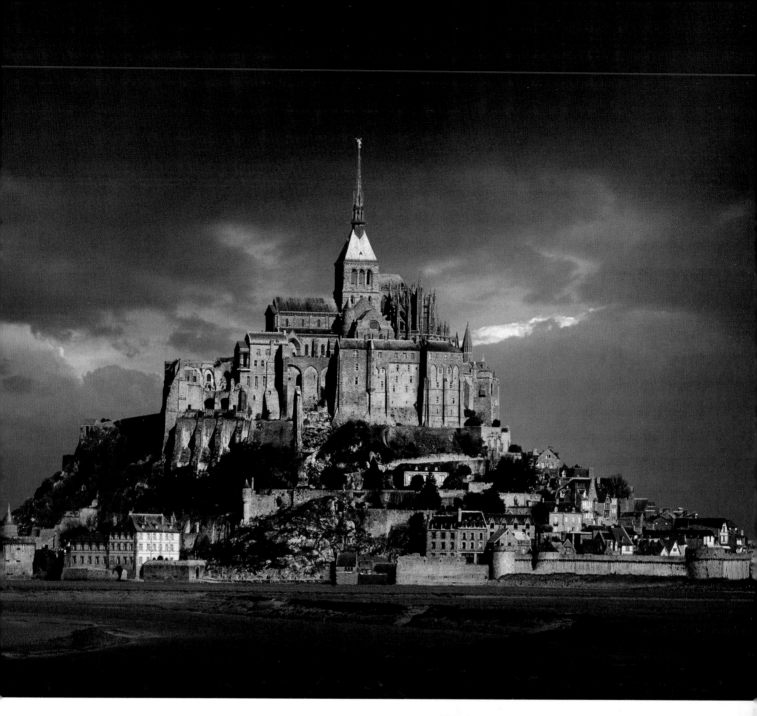

Changing coasts

Tectonic activity can dramatically alter coastal landforms, sometimes beyond recognition. Crustal deformation can cause uplift, raising beaches above sea level, or it can cause subsidence, so that the depressed land becomes prone to inundation by seawater. Associated seismic and volcanic activity can create new coastal landforms.

Tsunamis generated by seismic activity are highly erosive. They inundate low-lying land, destroying dwellings and often causing loss of life on an enormous scale. The earthquakes that generate tsunamis can also cause coasts to rise, leaving raised beaches and other coastal features sitting above sea level.

Climate change can cause sea levels to rise and fall around the world. These global or eustatic changes are due to the amount of ice held at the poles. Earth is currently undergoing a period of global warming which is associated with climate change, and this is melting an alarming volume of ice at the poles. Should the ice continue to melt, huge volumes of water will be released into the oceans, raising sea levels possibly some tens of feet (several meters), thus dramatically changing the existing coastlines worldwide.

Above: Mont Saint-Michel is a rocky tidal island in Normandy, France. Now located over half a mile (1 km) offshore, it was once part of the land. As sea levels rose, erosion whittled away the coast until several large blocks of resistant granite emerged, including Mont Saint-Michel.

Right: A popular sailing spot, the Gokova Gulf in Turkey is also of interest to geologists due to the submarine tectonic activity occurring there. Eastern Turkey is the youngest continent-to-continent plate boundary region on Earth. It is also, thus far, one of the least studied.

Inland landforms are often distinctive topographic features, having been created by a unique combination of tectonic activity, erosion, and climate. They include plains, plateaus, mesas, buttes, pinnacles, arches, folded and buckled rocks, domes, basins, different types of faults, cuestas, and razorback or hogback ridges.

Inland landforms

A large, flat, unbroken expanse of land is referred to as a plain, and an uplifted plain is called a plateau. Resistant geological strata often form a protective cap over weaker rocks beneath, creating an array of spectacular landforms. Once streams and rivers cut into the resistant top of a raised plateau, it becomes a series of flat tabletop mountains with steeply cliffed edges, known as mesas.

The weaker layers in a horizontal stack of sedimentary rocks allow weathering to undermine the stronger layers so that eventually the resistant layers collapse, forming a sloping apron of debris around the mesa's base called a talus apron. (An example can be seen in the photograph below, of the Mittens rock formation in Monument Valley, Arizona, USA.) Erosion takes place around the weakened sides of the landform in the form of parallel rather than downward retreat of its cliff faces, so mesas do not shrink in height but become aerially smaller. When the diameter of the flat tabletop becomes less than the height of the mesa above the plains, it is called a butte.

Below: Once a vast basin, the Colorado Plateau in Arizona, USA, comprises layers of sandstone and limestone deposited by meandering rivers. Over a period of tens of millions of years, the basin rose to become a plateau, then erosion went to work, wearing away the softer shales and leaving massive, vertically jointed slabs of sandstone like those shown below: The East and West Mittens buttes in Monument Valley. The vivid red color comes from iron oxide exposed by erosion.

Rumpled rocks

Sedimentary rocks can be slowly rumpled into a series of folds, much like a floor rug pushed up against a wall. The upper parts of the folds, or hills, are anticlines; the bottoms or troughs are synclines. Because the process takes place slowly, most of the sedimentary strata accommodate the bending. However, the very resistant layers generally develop a series of parallel tension cracks along the tops of the folds (at the point of maximum bending), like those you see if you bend a fruit candy bar. Because these cracks weaken the layers, the tops of anticlinal hills will quickly erode into a series of valleys that may eventually become deeper than the adjacent syncline valleys. This is referred to as topographic inversion.

More interesting landforms

When sedimentary strata are tilted at various angles by faulting, then exposed by erosion, a series of parallel ridges adjacent to a series of

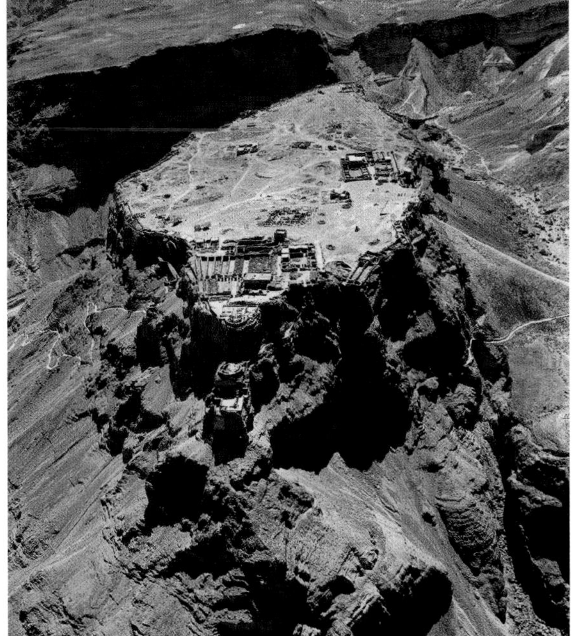

Left: From a geological point of view, mesas are transient phenomena, but on a human time scale they often appear to be permanent fixtures. So this one, not far from the Dead Sea in southern Israel, must have seemed to the Judean king who decided to build a fortress there in the late second century BCE. Masada, as it came to be known, takes advantage of the natural defense system provided by cliffs and escarpments that on its eastern side are 1,300 ft (400 m) in height. In the first century BCE Masada was further embellished by King Herod, who built a wall around the entire summit and turned it into a royal stronghold with spacious palaces and a state-of-the-art bathhouse. During the Roman siege of 66–74 CE, the site was defended by Jewish rebels, who committed mass suicide when their attackers blazed routes to the top.

long straight valleys can form. The Basin and Range Province, in southwestern USA and northwestern Mexico, is a typical example. Here, parallel, hot, dry valleys separate elongated north–south trending mountain ranges. Death Valley, California—one of the hottest, driest, and lowest places in the USA—is an exemplary basin and range valley.

If folding occurs in more than one direction, it creates closed dome and basin structures in sedimentary rocks. Oil exploration companies look for anticlinal domes below the land surface and sea floor, because these structures can trap and hold rising oil and gas. When these geologic structures erode, the relative positions of the harder and softer layers determine the appearance of the final landform. If the core of the dome is more resistant to erosion, an upland area will result with radial drainage flowing outward; if the central core is weaker, a central depression will form with drainage flowing inward. Resistant rock units can stand out as striking circular or elliptical natural walls that create completely enclosed valleys.

ARCHES

Arches are temporary geological structures that occur where closely spaced jointing has allowed erosion to carve thin fins or vertical slices of rock, which are then eroded through from one side to the other to form an arch beneath a more resistant cap rock layer. Water and wind are the main agents of erosion in inland settings.

Below: The landscape around Cappadocia, Turkey, has been described as "one of Mother Earth's stranger waking dreams." Volcanic tufa overlaid with basalt has been sculpted over time into narrow valleys crammed with cones. Some resemble giant drippings, sand castles, or "fairy chimneys" like those shown below, topped with large caps of basalt. To travelers' eyes they sometimes look like mushrooms, or huge ripples of soft ice cream, and the steep lunar-looking ridges are pitted with natural caves. People have been carving homes, cities, monasteries, churches, and more recently hotels, out of the landscape for 10,000 years.

> Civilization exists by geological consent, subject to change without notice.
>
> Will Durant, philosopher and historian, 1885–1981

Cappadocia, Turkey

Between 10 and 17 million years ago, Cappadocia, in the Anatolian highlands of central Turkey, was in the midst of a period of intense volcanic activity. Volcanic ash (soft tufa) fell onto the steppes of Cappadocia in depths of up to 500 ft (150 m) from Mt Erciyes and the surrounding volcanoes, Hasandagi and Golludag.

Then, during the mid-Pliocene Epoch, 2.9 million year ago, a large caldera formed in eastern Cappadocia and agglomerates such as basaltic lava began to accumulate alongside the soft tufa. These deposits have since been eroded and sculpted by steam, wind, and rain into the region's famous "fairy chimneys," pictured below.

Earth's mountain ranges play a major role in influencing climate and vegetation patterns. They are keys to help us understand the internal geologic workings of our planet, from the past to the present. Along the boundaries of Earth's actively moving tectonic plates, major mountain belts are still forming.

Mountains

There are two classes of mountain range that form at convergent plate margins: collision and subduction ranges. The most spectacular collision mountain belt in today's world is the Himalayas. Subduction mountain ranges encircle the Pacific Ocean basin. A third kind of mountain range is totally submerged beneath the oceans.

Right: A village in the foothills of the High Atlas Mountains, Morocco. Tectonic convergence between Africa and Europe is partly responsible for the formation of these mountains, though processes deep in the Earth's mantle may also have contributed. The Atlas ranges separate the Mediterranean and Atlantic coastlines from the Sahara Desert.

Collision mountains

The greatest mountain ranges form where the two plates coming together are thick continental landmasses. As they collide, neither landmass can go down, so they both push upward along the collision margin. On the geological time scale, the collision is relatively slow, only several inches per year, so the rocks are folded and deformed in

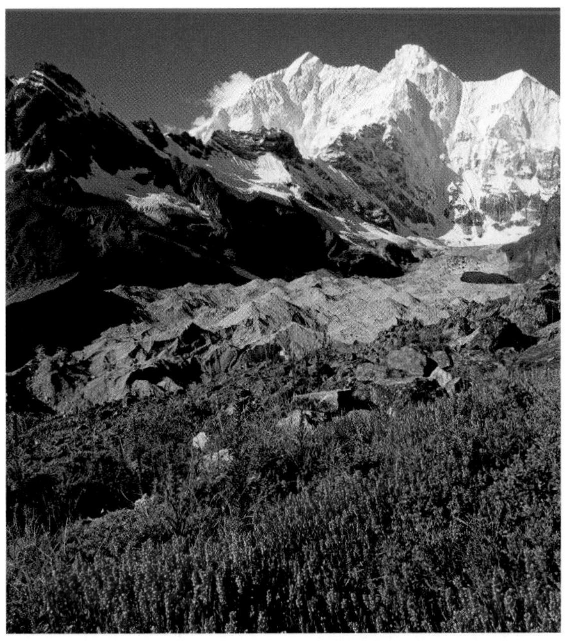

Left: Several glaciers cover the slopes of Mt Everest to its base. Kangshung Glacier, seen here with wildflowers growing on a moraine terrace beside it, extends from the eastern side of Mt Everest into China. Mt Chomolonzo, in the background, has three distinct summits. Its name means "bird goddess," and to some eyes, when looked at from this angle, it resembles a gigantic eagle with outspread wings.

a plastic or ductile manner, responding more like a firmly squeezed tube of toothpaste than something brittle that breaks or snaps. The rocks found in these mountains are mostly composed of shallow ocean sediments, such as sand, silt, and clay, which often teem with fossils. These have been crushed between the closing continents and then uplifted high above sea level.

Earth's active collision mountain ranges form a long continuous belt that includes the Himalayas, the Hindu Kush, the Zagros Mountains, the Makran and Sulaiman ranges, as well as the European Alps. This massive belt runs roughly east–west along the southern edge of the Eurasian tectonic plate, marking the line of closure of the ancient Tethys Sea. Here, the northern edges of the African, Arabian, and Indo-Australian tectonic plates began to crunch into the Eurasian plate some 50 million years ago, and today these ranges contain most of the world's tallest mountains.

Mt Everest—world's highest peak

Standing 29,035 feet (8,850 m) above sea level, Mt Everest in the Himalayan Mountain Range is located in one of Earth's most tectonically active regions, at the focal point of folding and upthrusting in a massive collision zone between two continental landmasses, the Indian and the Eurasian tectonic plates. Though the collision began about 50 million years ago, the Himalayas themselves started rising only about 25 to 30 million years ago. Today, India is still ramming northward and Everest's summit continues to rise by about $^1/_5$ to $^2/_5$ inches (0.5 to 1 cm) per year, so that in fact it is now a footstep higher than when Edmund Hillary and Tenzing Norgay famously first reached the summit on May 29, 1953.

> The highest of the world's mountains, it seems, has to make but a single gesture of magnificence to be the lord of all, vast in unchallenged and isolated supremacy.
>
> George Mallory, mountaineer, 1886–1924

Left: Mt Everest and Mt Nuptse, seen from Kala Pattar, Nepal. Tectonic folds in the region are very broad, with wavelengths of several kilometers. Structural studies suggest that of at least four generations of fold structures, two imparted a perpendicular disposition to the Himalayan trend.

Mt Everest is made up of three distinct rock units. The bottom layer is highly folded and metamorphosed schist and gneiss, although most of this part of the mountain is permanently covered with snow and ice. The middle layer is metamorphosed shale and sandstone intruded by granite, characterized by rugged dark cliffs and pinnacles. The top section of the mountain is gray limestone with a distinctly visible dipping yellow limestone bed, known as the yellow band, exposed in Everest's sheer three-sided pyramidal face.

Right: This aerial view of the Sulaiman and Makran mountain ranges in Pakistan shows how the northerly drift of India has caused the folded sedimentary Makran Range to bend from east–west to north–south. Much of the Makran region, which extends into Baluchistan and Iran, is a semi-desert coastal strip that rises very rapidly to mountains. The Sulaiman Range forms a barrier against winds that blow from the Indian Ocean, creating arid conditions across southern Afghanistan to the north.

Above: Aerial view of the Andes Mountains in Tierra del Fuego, Argentina. The longest continuous mountain chain on Earth, the Andes Cordillera is South America's main mountain system, running parallel to the Pacific coast for over 4,000 miles (6,400 km), from the Caribbean Sea to the Strait of Magellan. The Andes are folded mountains made of limestone, sandstone, slate, and granite, with large amounts of lava also present. Active volcanoes and frequent earthquakes add to the geological drama of the region.

Subduction mountains

Long continuous mountain ranges, studded with hundreds of active volcanoes, mark the tectonic plate boundaries where oceanic crust is colliding with continental crust or other oceanic crust. Along these margins, the denser oceanic crust slides (subducts) down deep trenches beneath the more buoyant continental crust. The subducting plate melts and produces the magma to drive the volcanoes. These subduction mountain ranges encircle the slowly closing Pacific Ocean basin. They form the backbone of New Guinea, Japan, the Philippines, and Kamchatka Peninsula along its western edge. To the north, the Aleutian Island chain is a largely submerged subduction mountain range with only the tips of the volcanic peaks emerging from the sea. Along the eastern edge of the Pacific basin, the mighty North American Cordillera and Andean Cordillera form the world's longest subaerial mountain range. Another major subduction mountain range stretches the length of the Indonesian archipelago.

Mountains under the sea

A third type of mountain range encircles the globe entirely beneath the oceans. Submarine ranges are, in fact, Earth's longest mountain ranges. Known as mid-oceanic ridges or rises, they occur along the spreading or divergent plate margins. These ridges are broad and low, rising about 1 to 2 miles (1.5 to 3 km) above the floor of abyssal plains but with a width of some 1,200 to 2,500 miles (2,000 to 4,000 km). The ridge crests, along which molten basalt is continually being emplaced, are held buoyantly in place by heat that is supplied by the upwelling convection currents in Earth's mantle. As the spreading plates carry the basalt away from the crests, it cools and subsides. If the planet's tectonic plate motor were to cease suddenly, these ridges would cool, subside, and eventually disappear.

Right: Mountain flights are one of the main attractions for tourists in Nepal, and offer the chance to view Mt Everest from unfamiliar angles such as this.

LIFE IN THE ANDES

The Andean or South American Cordillera is the longest mountain range in the world. The Andes run for some 5,000 miles (8,000 km) from the Caribbean Sea in the north to Tierra del Fuego, at the southernmost tip of South America. Unique life has evolved to survive in the often harsh, dry, and cold high-altitude conditions. Best known are the camel-like animals, known as guanaco (pictured below). Their long necks and surefootedness enable them to spot and avoid predators such as pumas and leopards, while their stomach structure and gastric secretions permit them to survive on the sparse and nutrient-poor grasses. Guanaco have been used by the native Indians for some 5,000 years, and domesticated into the alpaca and llama. These animals, renowned for their fine warm wool, are able to carry loads, and their dung is used as a cooking fuel.

Mountains and rivers

Orographic rainfall on the windward slopes of Earth's major mountain ranges means that they are also the sources of the planet's mightiest rivers. The Amazon and the Parana rise on the windward eastern slopes of the Andes and cross the South American continent to the Atlantic Ocean. Likewise, the Missouri and Mississippi rise on the eastern slopes of the Rockies. The Indus, Ganges, and Brahmaputra flow from the windward southern slopes of the Himalayas, and the Rhine, Rhone, and Po rise on the northern slopes of the European Alps.

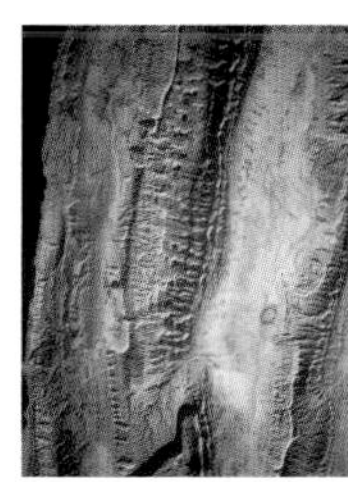

Above: The Zagros Mountains seen from the space shuttle *Atlantis*. The largest mountain range in Iran and Iraq, they were formed by collision of the Eurasian and Arabian tectonic plates. The folded sedimentary rocks were subsequently eroded: Softer rocks such as mudstone and siltstone were worn away, leaving linear ridges of limestone and dolomite, along with the trapped petroleum now mined in the region.

Below: The Verdon Gorge in southeastern France, named for the color of its waters, can trace its history back to the Triassic Period, though it was during the Jurassic that limestone deposits fractured, forming relief features. Glaciation during the Quaternary Period later remodeled the topography.

These rivers carry enormous amounts of sediment from the mountains to the sea. Sediment is deposited in immense deltas at the river mouths and as huge submarine sediment fans further out on the abyssal plains. The largest examples are the Indus Fan that accumulates in the Arabian Sea, and the Ganges Fan in the Bay of Bengal, fed by the short but powerful rivers draining the Himalayan ranges. The weight of sediment is so great that the ocean floor beneath these fans is being depressed into a basin shape into which more and more sediment is, in turn, being deposited. The river deltas themselves form some of the world's richest and most sought-after agricultural land.

Mountains and climate

Mountain ranges affect rainfall patterns, and thus influence climate. When wind travels over the sea it will evaporate water and retain the moisture up to its holding capacity (the higher the airmass temperature, the more water it can hold as vapor). Orographic rainfall occurs when warm moist air is blown against a mountain range's windward side. The airmass is forced to rise and cool and release its moisture as rain. The cool dry air then moves over the crest of the range and down the leeward side. Moving downhill, the air heats up rapidly, becoming even drier. As rain hardly ever falls here, this leeward side of the mountain is known as a "rainshadow."

The speed and direction of major airmasses traveling across the planet's surface are driven by Earth's spin and are therefore relatively uniform. As a result, it is common to find a well-watered forest on one side of a mountain range and a parched desert on the other. This rainshadow effect is clearly observed in many mountain ranges. Satellite images of the Himalayas show the verdant south-facing slopes of India, Pakistan, and Bangladesh, well-watered by monsoonal winds, in stark contrast to the dry cold deserts of China in the rainshadow on the Himalayas' north-facing slopes.

The Andes Cordillera creates another dramatic example of extreme climatic contrast between one side of a mountain range and the other. The well-watered windward eastern slopes in Argentina, Bolivia, and Peru feed the continent's major drainage basins, while the parched Atacama Desert of Chile and Peru lies on the leeward side of the range facing the Pacific Ocean.

Rift valleys are immediately recognizable by their flat floors and very straight steep walls formed by fault planes. Unlike most valleys, rift valleys have not been cut by erosion. Instead, their valley bottoms have dropped away, sliding downward along lengthy parallel fractures to create deep canyon-like slots that may be thousands of miles long.

Rift valleys

Rift valleys, also known as grabens, are collection areas for sediment and water. They contain Earth's largest and deepest freshwater lakes, such as Lake Baikal in Siberia, Lake Tanganyika in Africa, and Lake Superior in North America. Significant rift valleys and fault zones include Iceland's Mid-Atlantic Rift, the Red Sea–Gulf of Aden Rift, Turkey's North Anatolian Fault Zone, the Dead Sea–Jordan Valley Rift, faults and rift valleys of the North American west coast, and also the Spencer Gulf–Lake Torrens Rift in southern Australia.

The East African Rift Valley

The East African Rift is probably the best known great rift valley. This classic geologic structure is characterized by deep straight valleys bounded by sheer-walled fault escarpments. Voluminous outpourings of lava periodically issue forth from deep cracks opening along the rift. It stretches 3,700 miles (6,000 km)—halfway down the length of Africa—from Djibouti on the Gulf of Aden in the north to the Zambesi River delta in Mozambique. On its journey south, it passes through Ethiopia, Kenya, Uganda, Zaire, Rwanda, Burundi, Tanzania, Zambia, Malawi, and Mozambique. The valley resembles a huge arc-shaped bite-mark in east Africa, and continental drift will eventually carry this enormous chunk of land out into the Indian Ocean.

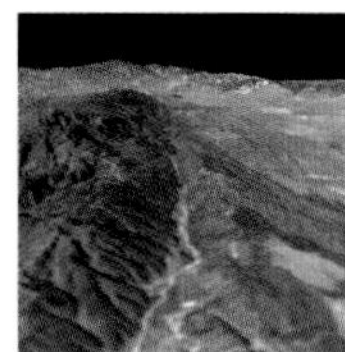

Above: This image of the San Andreas Fault is from the Shuttle Radar Topography Mission (SRTM), which has generated the most complete topographic database of Earth to date.

Below: The San Andreas Fault, which is in constant motion along the western coast of North America, marks the boundary between the Pacific and North American tectonic plates.

The East African Rift Valley is widening at about ⅕ inch (4 mm) per year, compared with a widening of ⅘ inch (20 mm) per year for the Red Sea–Gulf of Aden Rift. The lines of earthquakes and landforms indicate that the rift valley splits into two, with one branch running around the eastern side of Lake Victoria and the other around its western side. They rejoin on Lake Victoria's southern side. The incipient rift valley then divides again, branching in a southeasterly direction along Lake Malawi and southwest into Zambia and Botswana along the line of the Luangwa and Zambesi rivers.

Forming rift valleys

Rift valleys form as a result of uplift and tension in Earth's crust. The initial upward doming is caused by hot convection currents that rise within Earth's mantle and push against the base of the crust. These currents also create the strong tensional forces that eventually tear the crust apart. Initially, at the point of maximum tension, three rift valleys develop, oriented at 120 degrees to one another, and radiating in a star pattern, or triple junction. These tensional features usually begin to develop toward the middle of great continents, but eventually tear through to the edges and become flooded with seawater. Earth's textbook example of a triple junction can be seen where the three arms

of the Great Rift meet at the Afar Triangle of Djibouti—the East African Rift, the Red Sea, and the Gulf of Aden.

Tension in the arms of the triple junction continues, and the walls of the rift valleys very slowly move apart. Eventually, the divergence or spreading becomes concentrated along two of the rift valley arms (known as the active arms), while the third one slows down and eventually stops opening (known as the failed arm or aulacogen). In the case of the Afar Triangle of Djibouti, the active arms are the Red Sea and the Gulf of Aden; the failed arm is the East African Rift Valley.

Rift valley riches

Initially rift valleys fill with shallow seas; however, the process is not uniform. Alternate periods of flooding and drying out create alternating seas, salt lakes, and saltpans. Over time, massive sheets of salt and other water-soluble minerals, known as evaporates, build up on the floor of the rift valley. The minerals precipitate from the water as it evaporates into the atmosphere. Some of the most common examples include halite (common salt), anhydrite, gypsum, calcite, and potassium, as well as various magnesium salts such as sylvite, carnallite, kainite, and kieserite. Nitrate evaporate minerals are frequently mined for use in the production of fertilizer and explosives.

Eventually, when the warm shallow seawaters become more permanent in the rift valley, they begin to house a biological soup of minute aquatic algae and bacteria. These microorganisms live and die,

Above: Lake Magadi, the southernmost lake in the East African Rift Valley, lies in part of the graben just northeast of Lake Natron. An alkiline lake recharged mainly by saline hot springs, it is roughly 38½ square miles (100 km^2) in size. Frequented by the lesser flamingo (*Phoenicopterus minor*), it supports only one species of fish, a cichlid called *Alcolapia grahami*.

Top right: Lake Abbe, a salt lake on the Ethiopia–Djibouti border, is at the central point of the Afar Depression, where three pieces of Earth's crust meet. The ultimate destination of the Awash River, it is known for its 164-ft (50-m) high steam-venting limestone chimneys, and its flamingoes.

We learn geology the morning after the earthquake.

Ralph Waldo Emerson, poet and philosopher, 1803–1882

WHAT IS AN EARTHQUAKE?

Earthquake energy is released when stored stress in the crust overcomes frictional forces along a pre-existing fault, or ruptures unbroken rock. The point of rupture is known as the "focus" from which the energy radiates outward as compression waves (P-waves) or shear waves (S-waves). The magnitude of the earthquake is measured from seismographs and zones of intensity can be delineated based on the reported damage caused. The modified Mercalli Scale ranges from an intensity of one (no movement felt) to 12 (complete destruction with the ground moving in waves).

Left: A sliver of land emerges from alkaline deposits in Lake Magadi, Kenya. The water of the lake is a dense sodium carbonate brine that precipitates vast quanitites of the mineral trona (sodium sesquicarbonate). In places, the salt deposits are up to 130 ft (40 m) thick.

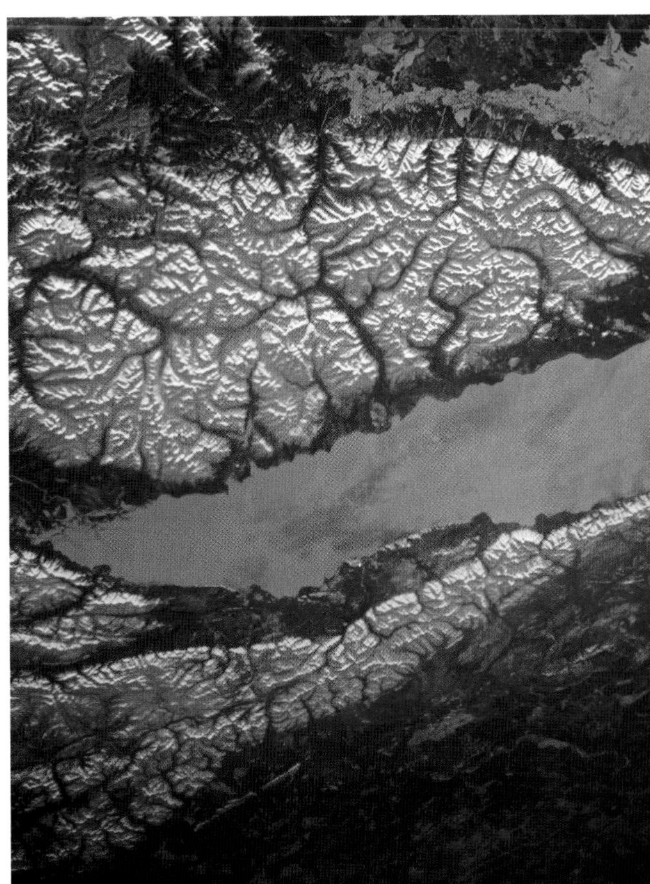

Right: The northern end of Lake Baikal in southern Siberia, as seen from the space shuttle *Discovery* in 1994. Long and crescent-shaped, the lake is sometimes known as "The Blue Eye of Siberia." Home to more than 1,700 species of plants and animals, two-thirds of which are found nowhere else in the world, it is of exceptional value to evolutionary science and was declared a World Heritage Site in 1996.

Above: The Baikal seal or nerpa (*Phoca sibirica*), a species of earless seal endemic to Lake Baikal, is the longest-lived of all seals and one of only three freshwater seal species in the world. How they originally came to Lake Baikal is not known, but they are thought to be related to the ringed seal of the Arctic, and may have come at a time when a sea passage linked the lake with the ocean.

Opposite left: Mt Kilimanjaro, an inactive stratovolcano in northeastern Tanzania, has fumaroles that emit gas in the crater on its main summit, and molten magma just 1,300 ft (400 m) below the crater. But it is known mostly for its ice fields, described by Ernest Hemingway as "wide as all the world, great, high, and unbelievably white in the sun." Since 1912, when they were first measured, these fields have lost 82 percent of their ice, and many scientists believe that it will all disappear in coming decades due to forest reduction and global warming.

accumulating on the sea bottom in vast numbers, where thick piles of younger marine sediments eventually bury them. Temperatures rise as the depth of burial increases and these organic-rich rocks eventually mature into the source rocks of oil and gas. These rocks are slowly cooked in Earth's giant pressure cooker until the longer organic chains of carbon and hydrogen break down into smaller lighter molecules of oil and gas that rise upward out of the source rocks.

An important factor in the formation of economic hydrocarbon reserves is the presence of impermeable geological structures that can trap the oil and gas, and prevent them from reaching the surface. Fine-grained sedimentary rocks, such as mudstones, will not let oil and gas pass, and if these have been folded into anticlines or dome shapes then they make perfect traps. Rising domes (or diapirs) of low-density salt from the underlying evaporate beds will also push up the overlying sediments into closed structures, which can act as hydrocarbon traps. More than two-thirds of all the world's 910 giant oilfields (those with more than 500 million barrels of ultimately recoverable oil or gas) are associated either with continental spreading margins facing open ocean basins or with continental rift valleys. The North Sea and West Siberian oilfields are important examples of hydrocarbon deposits that formed in a rift valley setting, while the Gulf of Mexico, Northwest Australian, and West African oilfields are classified as facing an open ocean basin.

Lake Baikal's youthful rift

Lake Baikal, also known as Dalai-Nor, or "Sacred Sea" in the Buryat and Mongol languages, lies in a classic young continental rift valley. The lake is surrounded by up-domed mountains, and stretches for 400 miles (630 km) across southern Siberia, though it only attains a width of 50 miles (80 km) at its widest point. The water is an amazing 5,370 feet (1,637 m) deep, making Lake Baikal the world's deepest lake, and also Earth's greatest natural freshwater reservoir. It contains over one-fifth of the world's liquid fresh water. The capacity of Lake Baikal is 5,500 cubic miles (23,000 km^3) of water, which is as much as the contents of North America's entire Great Lakes system put together. The solid rock bottom of the Baikal rift lies more than 5 miles (8 km) below sea level, and the rift valley is filled with some 4 miles (6.5 km) of sediment. This makes the rift valley associated with Lake Baikal the deepest continental rift on Earth.

The rift graben is actively widening at approximately ⅘ of an inch (20 mm) per year. Every few years, sizable earthquakes are recorded in the region indicating that slippage is still occurring along the graben-bounding faults. On August 29, 1959, during an earthquake of magnitude 9, the Baikal lake bottom was displaced for perhaps as much as 65 feet (20 m). Thermal spring activity in the area also demonstrates that hot volcanic rocks lie quite close to the surface.

Programs that involve the deep drilling of Lake Baikal's thick sediments are enabling scientists to map the history of the lake and rift, including its climate and ecology, as far back as 5 million years. The long and unbroken record that is now being uncovered presents a unique opportunity to understand how continents begin to break apart, and give rise to new ocean basins.

Canyons and gorges are carved by powerful rivers. In fact, they only exist wherever the land is rising quickly, giving water a vast amount of potential energy to erode rocks, carry sediment, and cut huge slots into the landscape.

Canyons and gorges

Canyon is the term generally used in the United States, while the word gorge is more common in Europe and Oceania. Canyons and gorges are different from a normal river valley in that they have steep to vertical sides, as opposed to more gently sloping V-shaped valley walls. Canyons are much more common in arid areas because the weathering of their surrounding rock walls takes place more slowly than it does in humid tropical environments.

Above: Part of the rugged George Gill Ranges, Kings Canyon, in Watarrka National Park, Northern Territory, Australia, features remarkable geological formations that date back 440 million years, when the interior of what is now Australia was sluiced by ancient lakes and rivers and covered by tropical forests, remnants of which can still be seen on the floor of the canyon, in the so-called "Garden of Eden." With walls towering 886 ft (270 m) above Kings Creek, the canyon has long been considered a sacred site by Indigenous Australians. The Luritja people have inhabited the area for over 20,000 years.

Where canyons and gorges are found

Canyons and gorges can be found wherever tectonic activity is at work creating active mountain ranges and volcanoes. These zones principally occur around colliding plate margins, where continents crush against one another, or are pushing over the top of the sea floor. The rising, volcanically active mountain ranges around the Pacific "Ring of Fire" mark the line along which the oceanic plates of the Pacific basin are being swallowed up by the surrounding plates. The Grand Canyon in Arizona, USA, and Copper Canyon in Chihuahua, Mexico, are just two examples of the often very dramatic gorges that cut into these active mountain belts.

The great collision mountain belts that stretch from the Himalayas of Asia to the European Alps host some of the world's deepest and most impressive canyons, such as the Yarlung Tsangpo Gorge in Tibetan China. Canyons also exist in places where Earth's crust is uplifted and torn open along tectonic plate boundaries, known as spreading margins. Here, plates move away from one another and create a gap. These fault-bounded gorges, such as Tanzania's Olduvai Gorge in Serengeti National Park, are more correctly called rift valleys.

Gorgeous gorges

The Grand Canyon, in Arizona, USA, is probably the world's best-known and most visited canyon. The Colorado River has cut this spectacular desert canyon through sedimentary layers, exposing an enormous span of geologic time; the 1.7 billion-year-old Vishnu schist is exposed at the very bottom. The Grand Canyon is often said to be Earth's largest canyon, but it is actually the Yarlung Tsangpo Canyon that takes that prize. Here the Brahmaputra River has carved a mighty gorge in the Himalayas to a depth of 3⅓ miles (5.3 km), which makes the Grand Canyon's depth of approximately 1½ miles (2.4 km) seem somewhat small in comparison. Other giant gorges in

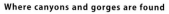

Left: Marble Canyon, a section of the Colorado River canyon at the northeastern end of Grand Canyon National Park in Arizona, USA, gets its name from its appearance rather than its geological content. Its walls, made of limestone and hermit shale, contain fossils more than 300 million years old, and indicate the presence of a shallow sea containing corals, sponges, crinoids, and brachiopods.

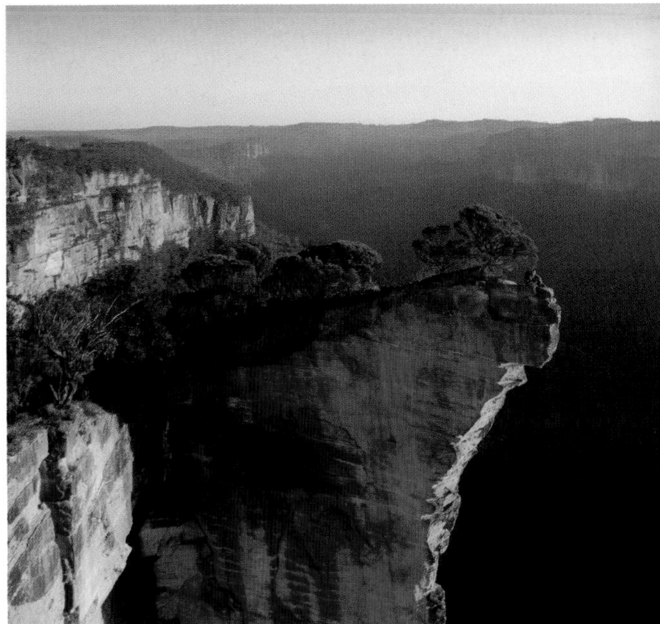

Left: The Colorado River carves a path through Marble Canyon. Some geologists believe that the river's ancient ancestor, which came into existence around around 60–70 million years ago, once entered the sea through Marble Canyon, but this point has been hotly debated.

Above: Hanging Rock, overlooking the Grose River Gorge in Blue Mountains National Park, New South Wales, Australia. The major rock outcrops in the park are sedimentary sandstones laid down in the Permian and Triassic periods, then uplifted and eroded over 90 million years.

PETRA GORGE, JORDAN

Petra is situated near the Dead Sea Rift, a fault that formed 30 million years ago when the Arabian and African tectonic plates began to move apart. Both plates are drifting northward but are moving at different speeds, causing geological stresses that create many earthquakes. Petra's gorges gradually formed along these stress cracks in the rock as they were widened and deepened by water erosion.

The multicolored walls of Petra's sandstone gorges were carved into colossal and beautiful buildings more than 2,000 years ago by the Nabateans—a powerful people who controlled the trade routes through Arabia. Petra's sandstone contains quartz grains with cement colored by the oxides of iron, aluminum, manganese, sodium, and lithium. These elements are responsible for the spectacular colors found in the sandstone. Oranges occur where there is a predominance of iron, yellows are produced by sodium and iron, pinks by the combination of sodium and lithium, and mauves are produced by manganese.

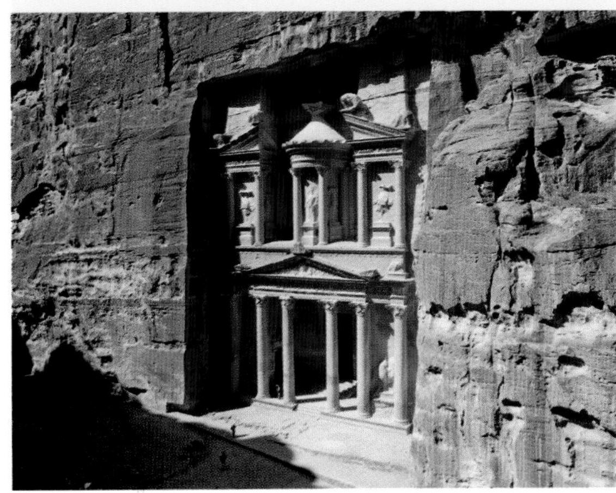

the area include the Kali Gandaki Gorge in Nepal, and the Polung Tsangpo Gorge in China. It is the phenomenal rate of uplift in this area, brought about by the ongoing collision between the Indian subcontinent and Asia, which is leading to the formation of these incredibly deep canyons. The land in this region is rising at a rate of more than 1 inch (25 mm) per year.

The Grand Canyon, USA

A World Heritage Site, the Grand Canyon encompasses 1,900 square miles (4,900 km²). The layers in its cliff were deposited when shallow seas flooded the eroded surface of Precambrian Vishnu schist about 600 million years ago. Thick sequences of sandstone were followed by sequences of shale and limestone (the remains of marine organisms) as the sea progressively deepened—followed by the reverse sequence as it shallowed again—all over a period of about 300 million years. The Coconino Sandstone layer, formed by desert sands, records a time about 275 million years ago when the sea retreated completely before returning again to deposit the uppermost sandstone and limestone layers. This cycle of oceans flooding, then waning, continued until about 70 million years ago, when the entire accumulated sequence of layers was lifted above sea level by the same tectonic activity that created the San Francisco Peaks and the Rocky Mountains. The Colorado River, formed by water draining from those mountains, began to cut through the rock. As the Colorado Plateau slowly rose, the river continued to cut downward. The plateau stopped rising 5 million years ago with the river entrenched its current course, continuing to erode and widen the canyon.

Entrenched rivers

Canyons and gorges are created by an entrenched river—a river no longer able to alter its course. A river will only etch its drainage pattern permanently onto the landscape when its ability to wear down

> The wonders of the Grand Canyon cannot be adequately represented in symbols of speech, nor by speech itself. The resources of the graphic art are taxed beyond their powers in attempting to portray its features. Language and illustration combined must fail.
>
> John Wesley Powell, geologist, 1834–1902

Below: The colossal uplift of the Colorado Plateau to heights of 5,000–10,000 ft (1,500–3,000 m) is responsible for the immense depth of the Grand Canyon. The uplift has steepened the gradient of the Colorado River, increasing its speed and enhancing its ability to cut through rock.

through the rocks matches or exceeds the rate of uplift in that area. In some cases, even the slow-flowing meanders, typical of an old river flowing across its floodplain, can become heavily entrenched.

Classic examples of entrenched hairpin meanders (meanders that loop back on themselves) have been carved by the Snake River in Snake River Canyon, Idaho, and also by the Colorado River in Canyonlands National Park, Utah, USA. Another, somewhat bizarre example can be seen on the Emu Plains, to the west of Sydney, Australia. Here, a meander of the Nepean River heads into the Blue Mountains, flowing around in a broad loop through a narrow canyon, before emerging again onto the plains.

Submarine canyons

Submarine canyons are very steep to vertical-sided gorges that cut into the sea floor of the continental slope. They are more closely spaced on the steeper continental slopes, and they cut into all rock types as well as loose sediment. They have only been discovered and explored by submersibles relatively recently. Some of the largest submarine canyons occur as offshore extensions of Earth's major rivers. The largest of these is Congo Canyon, which extends from the mouth of the Congo River. It is 500 miles (800 km) long, and 4,000 feet (1,200 m) deep. Other examples include the Amazon Canyon, the Ganga (Ganges) Gorge, the Hudson Canyon, and the Indus Gorge, which extend from the mouths of their respective rivers.

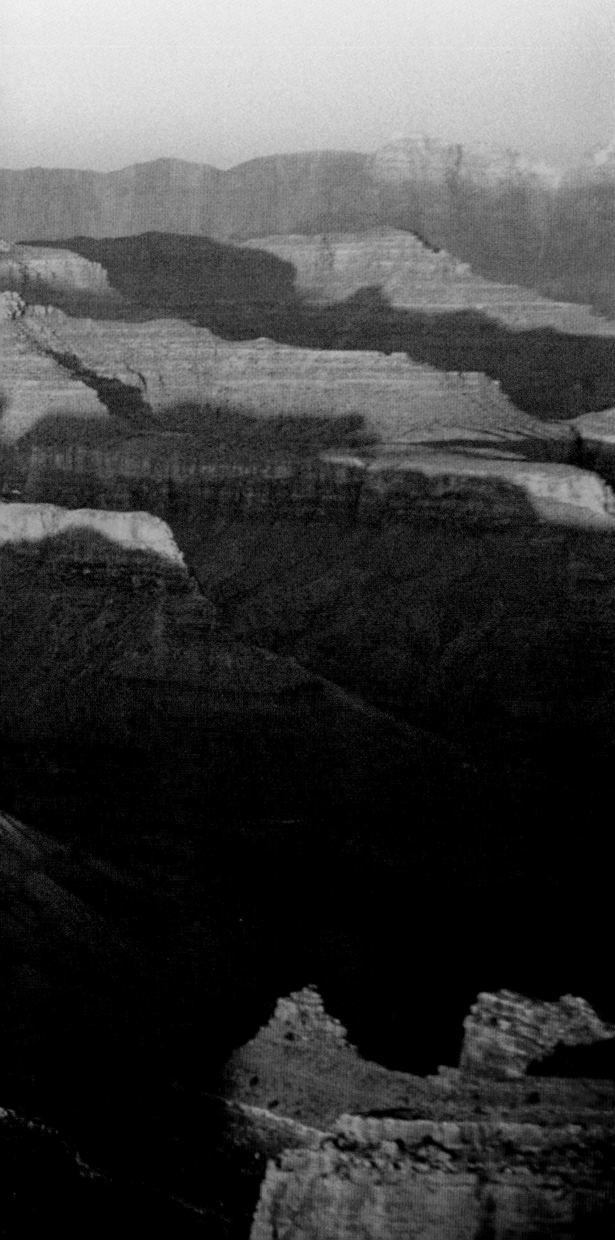

Above: The most famous part of Samaria Gorge is the narrow passage called "The Iron Gates." Part of the White Mountains region of western Crete, the gorge begins near the settlement of Omalos at an altitude of 4,100 ft (1,250 m) and emerges on the shores of the Libyan sea near Agia Roumeli.

Top: An ancient chapel, Ossia Maria, cut into the cliffs at Samaria Gorge, gave a version of its name (St Mary) to both the gorge and the village the chapel once served. The village was abandoned after 1962 when the National Park of Samaria was declared.

It is believed that submarine canyons are cut by dense, sediment-laden fluid flows that move through them like underwater rivers of sand. The canyons are continuously supplied by sediment that is brought to the deltas of the major rivers, or by sand that is carried along the beaches by long-shore drift and then fed into them. The movement of sediment through the canyons may also be sudden and sporadic, and some believe that this is due to collapsing masses of sediment triggered by earthquakes. These flows are known as turbidity currents or underwater landslides, and enormous sediment fans have been found on the abyssal plain at the mouths of submarine canyons as evidence of this mechanism.

Crete's abundant gorges

Along with its cultural marvels, the island of Crete boasts over 100 natural limestone gorges. The island was created about 10 million years ago, when the Mediterranean Sea rose up over the ancient landmass of Aegeis. Upheavals caused by the collision of the African and European tectonic plates raised the island further, and continuous uplift cracked the crust, bringing about volcanic eruptions. Glaciers, or possibly river torrents, brought about by climate changes eroded the limestone and marble, thus creating the gorges.

Of the many gorges in the Sfakia area of Crete, over a dozen run north–south into the Libyan Sea, including the 10-mile (16-km) long Samaria Gorge, Crete's longest canyon. With walls composed mainly of dolomite and covered in cypress trees, the gorge was created during the Quaternary Period through glacial action and karstification. Samaria Gorge is 1,640 feet (500 m) deep and, at some points, only 7 feet (2 m) wide. Other Cretan gorges are the 7-mile (11-km) long Imbros Gorge, the 2,300-foot (700-m) deep Ha Gorge, and the Kourtaliotiko Gorge, the cliffs of which are the roosting sites for the rare griffon vulture (*Gyps fulvus*).

Karst topography and caves occur wherever areas of limestone are exposed above sea level by tectonic uplift. Limestone is hard and tough, but is more soluble than other rock types, and is particularly susceptible to slightly acidic rain or groundwater. It does not take long for surface water to create a typical limestone landscape: a series of interconnected, often labyrinthine underground chambers and channelways, through which water can travel.

Caves and karst

Many places across the globe contain karst landscape, often vast expanses of it. The Caribbean Islands, for example, boast extensive networks of caves and passageways as well as sinkholes and karst formations. In North America, numerous limestone caves and karst systems are now protected within national parks. Australia's Nullarbor Plain is one of the largest areas of karst topography in the world.

Slight amounts of acid in rainwater react with calcium carbonate in the limestone to produce soluble calcium salts and carbon dioxide. Over time, countless raindrops create thousands of tiny pits and fluted rills on the rock surface that slowly become sharper and sharper. Eventually, the rock surface becomes a myriad razor-edged surfaces—almost too sharp to touch. These classic limestone dissolution features are referred to by geologists as karren or lapies.

Under the water table

Beneath the level of standing groundwater (the water table) in areas of high rainfall there is a flow of water that works its way outward and downward toward the sea. This happens in all rock types. Water picks the path of least resistance between the grains of a sedimentary

Right: A caver negotiates the waterfall that marks the entrance to Ora Resurgence Cave on New Britain Island, Papua New Guinea. Ora River, which feeds the falls, has been estimated to produce around 35 cubic ft (10 m^3) per second of white water rapids.

Below: Cavers light the entrance to Ora Cave. With a total length of just over half a mile (1 km), the cave is 1,040 ft (317 m) deep from the lowest point on the doline rim. The Ora system is believed to be one of the largest collapse dolines in the world.

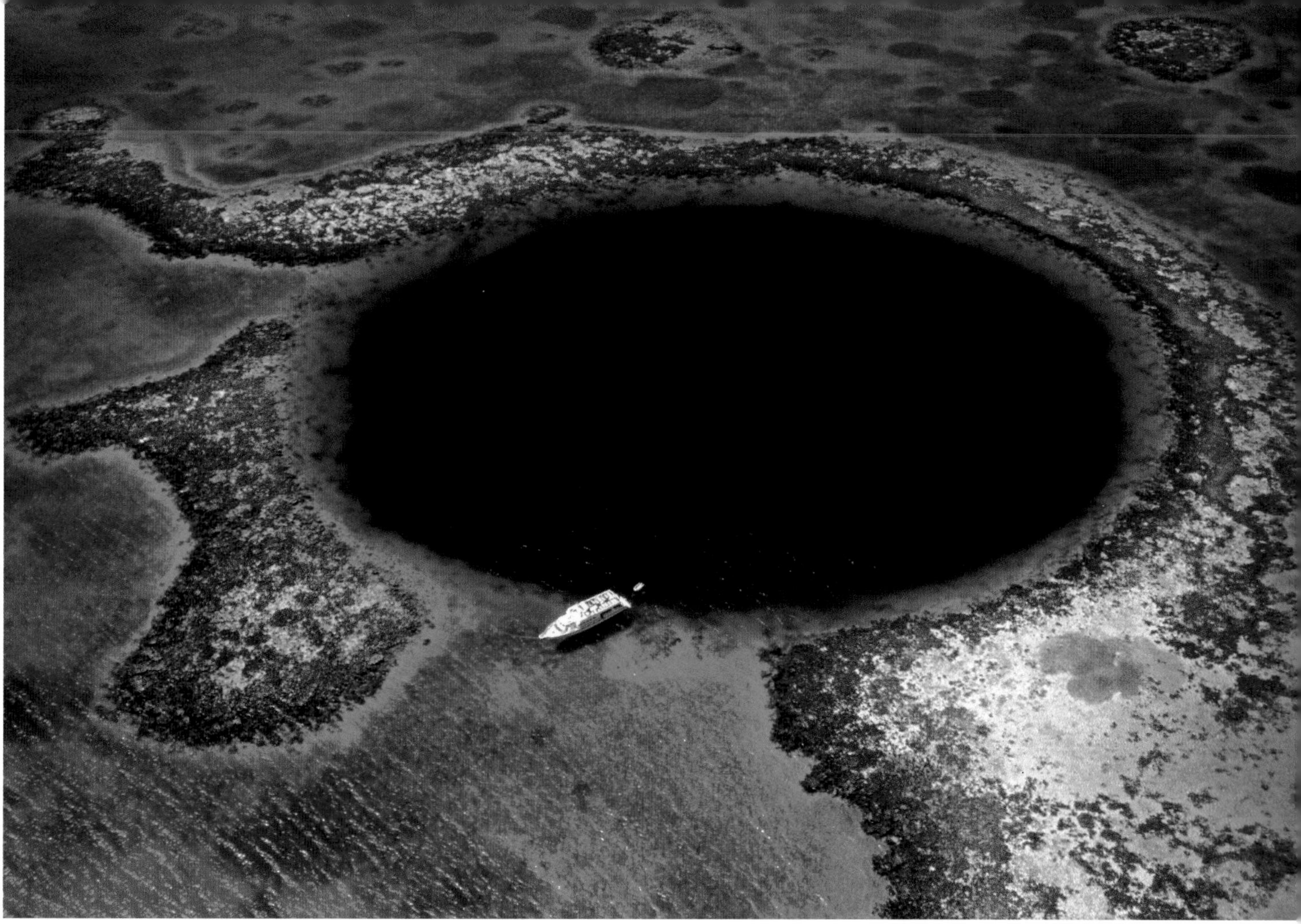

rock, between lava flows, through joints and fractures, or along fault planes. In limestone, however, the results are dramatic as the groundwater slowly turns initially tiny channelways into giant tube-shaped tunnels—huge underground river channels that carry fresh water, sometimes miles out to sea to emerge as underwater freshwater springs. These caves that form beneath the water table are known as phreatic caves. Often, such caves are eventually exposed after a general lowering of the water table, and it becomes possible to explore the caves without the use of scuba gear.

That sinking feeling

Individual caverns can grow to enormous sizes. If a chamber is close beneath the surface, its roof will become thinner and thinner. Eventually it either dissolves away or, if the cavern is large enough, collapses, creating an enclosed crater-like basin called a doline or sinkhole. These can range in size from small grassy depressions to enormous cliff-bounded wells containing entire forests or sometimes lakes. They are carefully explored as they often provide access points to larger cave systems. On New Britain Island, off the coast of Papua New Guinea, speleologists have explored the Ora Cave System by means of the Ora Doline. Here, a raging subterranean torrent emerges from the side of the heavily forested doline and crosses its floor before disappearing into a cave on the other side.

Blue holes

Blue holes are near-perfect circular sinkholes found particularly in the submerged karst of the Caribbean Sea. Approaching the edge of one of these structures, the light turquoise, shallow waters give way suddenly to a deep indigo blue as the sides drop vertically away into an abyss. The sight is even more dramatic from the air.

The Great Blue Hole at Lighthouse Reef, 60 miles (100 km) off the coast from Belize City, is about 1,300 feet (400 m) across and 480 feet (145 m) deep. Its walls are lined with ancient speleotherms, and

Above: The Great Blue Hole at Lighthouse Reef, off the coast of Belize, is regarded by many as one of the most spectacular diving locations in world. Its upper regions support an abundance of coral and tropical fish, but below a depth of 100 ft (30.5 m) these give way to an ancient world of stalagmites, stalagtites, and columns.

Right: Palau, an archipelago in Micronesia's Caroline Islands, is a chain of more than 200 knobs of rounded, weathered, forest-capped limestone, with steep sides plunging to the sea. The waters around Palau are home to 1,500 species of fish and 700 species of coral—as well as giant clams and incandescent starfish. This richness is due to Palau's geology and its location at the confluence of three major ocean currents.

the hole itself is the opening to a system of caves and passageways believed to connect through to the mainland. The stalactites indicate that the cave was air-filled during the last Ice Age, when the sea level was lower. The debris on the cave's floor suggests that the sinkhole formed when its roof collapsed. Distinct shelves and ledges carved into the limestone walls and running around their entire circumference show that the sea level did not rise in a single progressive movement, but rather in a series of distinct steps.

Above: Helictite formations in Cebada Cave, in the Chiquibul Cave System, Belize. Helictites, among the most beautiful and delicate of speleotherms, have been described in many types: Ribbons, saws, rods, butterflies, hands, curly-fries, and "clumps of worms." Made for the most part of needle-form calcite and aragonite, they typically have radial symmetry.

Sensational speleotherms

The very same limestone that is dissolved to create caves is often re-deposited as glistening formations all over their walls, ceilings, and floors, especially in humid, tropical climates. Speleotherm is the general name used for all such cave formations.

Speleotherms grow slowly; they are often deposited as microscopic crystals, drip by drip, as mineral-charged water falls from the ceiling or runs down the walls of a cave. Pointed stalactites grow downward from the ceiling toward fatter, rounder stalagmites building upward from the floor—once again, drip by drip. Eventually the two touch and fuse to form a single column.

Wavy or folded sheets of calcite known as draperies or shawls grow from sloping cave walls. These thin rock curtains are translucent and when lit from behind can show a beautiful alternating light-and-dark-colored growth banding. The colors vary from dark brown to white, caused by staining of calcite by impurities in the groundwater. Calcite can also build up some formations in the shape of rounded disks that project from a cave wall around a wall rock fracture. These are known as shields and grow at an upward angle, often with their undersides draped with stalactites.

> If one is sufficiently lavish with time, everything possible happens.
>
> Herodotus, historian, *c.* 484–425 BCE

Helictites are particularly enigmatic speleotherms because they appear to defy gravity by growing horizontally, neither up nor down. These twig-like growths are like twisted soda straws with a fine central canal. Water appears to travel to the end of the helictite by capillary action allowing it to grow longer in an apparently random direction.

Often, small walls or barriers of rock on the floor of wet caves, known as rimstone, trap the water in shallow basins referred to as rimstone pools. Changing conditions, such as turbulence in the water or the release of carbon dioxide, cause the precipitation of soluble minerals, and it is these minerals that form the rim.

Below: An explorer admires the selenite chandelier display in Carlsbad Caverns National Park, New Mexico. The caverns are part of the Delaware Basin region in West Texas and southern New Mexico, famous for its oil fields and fossilized reef. The Delaware Basin is in turn part of the larger Permian Basin.

CASTLES AND CAVES

Cave and karst features have been useful for castle builders. In Poland, ancient coral atolls have provided the foundations for castles such as Podzamcze. There, the solid limestone rim forms the walls and the lagoon forms the keep.

The most striking use of a cave for castle is at Predjama in Slovenia (pictured right). A fifteenth-century castle, Predjama Grad is constructed partly inside a large cave entrance. The cave provided water and a secret entrance for supplies during sieges. An older castle in the cave was home to the "robber baron" Erasmus, whose siege—and death by cannon fire—in 1484 has become legendary.

Above: Ha Long Bay, a World Heritage site in Quang Ninh province, Vietnam, features thousands of limestone karsts and isles topped with thick jungle vegetation that rise spectacularly from the ocean in various shapes and sizes. Several are hollow, with enormous caves. At least two are large enough to support permanent inhabitants, while others are the anchor points for floating villages of fishermen. The waters around them are home to 200 species of fish and 450 different kinds of mollusk.

Left: Stalactites and stalagmites in Carlsbad Caverns, New Mexico, USA. Explorer Jim White named many of the rooms in this cave system, including the King's Palace, the Queen's Chamber, the Papoose Room, and the Green Lake Room. He also named many of the caves' formations, such as the Totem Pole, Witch's Finger, Giant Dome, Bottomless Pit, Fairyland, Iceberg Rock, Temple of the Sun, and Rock of Ages. This photograph shows the Big Room, also called the Hall of Giants, where the stalagmites are six stories tall.

Rich pickings

Occasionally, rare colorful carbonate speleotherms, such as green malachite, blue azurite (copper carbonates), and pink rhodochrosite (manganese carbonate), can be found adorning caves. The caves of Catamarca Province in Argentina, with their delicately banded pink and white rhodochrosite stalactites and stalagmites, are a particularly striking example of this phenomenon. Unfortunately, most of these rare speleotherms have been removed for cutting and polishing into gemstones. In Mexico, thick cave deposits are mined for ornamental use. This flowstone material is known as cave onyx or Mexican onyx, and is a delicately banded, pale green to white translucent calcite.

Caves of gypsum

Caves formed in gypsum are less common than those in limestone, though the world's second-longest cave system, the Optimisticheskaya Cave in the Podolie region of the Ukraine, is made of gypsum. It has about 186 miles (300 km) of passageways. Secondary gypsum growths on the walls and ceilings of caves are known as gypsum flowers, while large, spectacular examples, such as those found hanging from the ceiling of the Lechuguilla Cave in New Mexico, are called chandeliers. Unlike speleothems, gypsum flowers grow from their point of attachment outward. Distortion in the crystal lattice causes the columnar crystals or fibers to curl as they grow, creating a flowerlike effect.

Older karst landscapes

As a limestone landscape continues to erode by the widening and deepening of existing sets of joints, a landscape of vertical-sided towers or conical peaks begins to emerge. These are called tower karst or cone karst landscapes. Some of the most spectacular tower karst are in the Guangxi Province of southern China. Here the magnificent cliffed towers of limestone are so abundant that the name fenglin, or "forest of stone peaks," is used to describe them. At Ha Long Bay on the coast of Vietnam, a rising sea level flooded a fenglin landscape to create the bay and its 1,600 towering islands. Around the base of each island at sea level there is a deep undercut, often big enough for boats to pass under at low tide. This marine erosion notch is caused by increased dissolution of the limestone at this level.

A science timeline

c. 4.6 billion BCE Our Sun is born (give or take a few millennia).

Egypt, 2800–2600 BCE Egyptians invent the first calendar of 365 days.

Egypt, 2700 BCE The first female doctor to have her name recorded is Merit Ptah. Her name is engraved in a tomb in the Valley of Kings.

China, c. 2700 BCE Acupuncture has its beginnings.

Egypt, c. 2600 BCE Imhotep wrote texts describing diagnosis and treatment of 200 diseases in Third Dynasty Egypt.

Babylon, Mesopotamia, c. 1600 BCE The Venus tablet of Ammisaduqa is among the first known Babylonian astronomical observations.

Ugarit, 1223 BCE The oldest eclipse is recorded on a clay tablet uncovered in the ancient city of Ugarit (in present-day Syria).

India, c. 500 BCE Sushruta wrote *Sushruta Samhita* describing over 120 surgical instruments, 300 surgical procedures, and classified human surgery into eight categories. He also performed cosmetic surgery.

Samos, Greece, c. 490 BCE Mathematician and philosopher Pythagoras dies.

Greece, 450 BCE Democritus coins the term átomos, meaning "the smallest indivisible particle of matter," from which we get the word "atom."

Greece, 420 BCE Hippocrates begins the scientific study of medicine by maintaining that diseases have natural causes. He also puts forth the Hippocratic Oath, thus marking the birth of modern medicine.

Left: American astronomer Edwin Hubble (1889–1953), pictured here at the Palomar Observatory in California, USA, proved that the Milky Way is not the entire universe when he discovered that Cepheid variables in spiral nebulae proved that they were far beyond the boundaries of our galaxy.

Previous spread: The skeleton of a dodo (*Raphus cucullatus*) is displayed opposite a reconstructed model of the extinct bird in the National Museum of Wales, Cardiff in 1938.

Greece, c. 400 BCE Aristotle deduces that Earth is round from the shape of Earth's shadow on the Moon during a lunar eclipse.

Greece, c. 360 BCE Eudoxus of Cnidus writes about planetary motions.

Greece, c. 300 BCE Euclid writes *Elements*, his treatise on geometry.

Greece, 280 BCE Herophilus studies the nervous system and distinguishes between sensory nerves and motor nerves.

Greece, 250 BCE Erasistratus studies the brain and distinguishes between the cerebrum and cerebellum. Around the same time, Eratosthenes measures Earth's circumference.

Syracuse, Sicily, 212 BCE Mathematician and inventor Archimedes is killed by a Roman soldier during the Siege of Syracuse.

Greece, c. 200 BCE Aristarchus of Samos, a Greek mathematician and astronomer, announces a theory of a Sun-centered universe. Around the same time, Eratosthenes devises a map of the world. He also estimates the circumference of Earth, and the distances to the Moon and Sun, as well as constructing a method for finding prime numbers.

Greece, 130 BCE Mathematician and astronomer Hipparchus catalogs over 800 stars.

Italy, 23 CE Pliny the Elder is born. His interests included astronomy, geography, botany, and pharmacology.

Rome, 50–70 CE Greek physician, Pedanius Dioscorides, writes *De Materia Medica*—a precursor of modern pharmacopeias that was in use for almost 1,600 years.

Greece, 100 CE Pioneering gynecologist Soranus gives women advice on contraception and abortion.

Egypt, 150 CE Ptolemy writes *Almagest*, a work on cosmology used for the next 1,000 years.

Rome, Italy, 180 CE Galen investigates the connection between paralysis and severance of the spinal cord.

Greece, c. 201 CE Galen, a pioneer of medicine, dies.

Alexandria, Egypt, 415 CE Greek mathematician Hypatia is murdered.

Persia, c. 780 CE Mathematician Muhammed ibn Musa al-Khwarizmi, who coined the word "algebra" is born.

Persia, 980 CE Physician Avicenna is born.

Cairo, Egypt, 965–1039 CE Muslim scholar Abu Ali Al-Hasan (Alhazen), invents the camera obscura, becoming the first person to use a device to observe the Sun.

World, 1006 History's brightest ever supernova is seen in the sky.

Cairo, Egypt, 1009 Ibn Yunus, famed astronomer, dies. His legacy includes the Hakemite Tables—astrological tables completed during the reign of Caliph al-Hakim, and named in his honor.

Baghdad, Iraq, c. 1010 Ibn Isa (Jesu Haly), physician, dies. His work focused on the field of optics, and he wrote *Tadhkirat al-Kahhalin* (*Notebook on Ophthamology*)—a comprehensive manual for ophthamologists.

Persia, c. 1010 Avicenna writes *The Book of Healing* and *The Canon of Medicine.*

Spain, 1013 Al-Zahrawi (Albucasis), famed Muslim physician, dies. He introduced a number of surgical techniques and invented several surgical instruments. He also wrote *al-Tasrif* (*The Method of Medicine*), a comprehensive encyclopedia of medical knowledge and various surgical techniques.

China, c. 1017 Inoculation with infected material from smallpox victims prevents serious cases of the disease in those treated.

China, c. 1026 Physician Wang Weiyi writes *Illustrated Manual of Points for Acupuncture and Moxibustion on a Bronze Figure with Acupoints*, a work on acupuncture.

Baghdad, Iraq, 1029 Al-Karkhi, mathematician, dies. His work includes *al-Kafi*, a book on arithmetic.

Spain, 1035 Ibn al-Saffar, astronomer, dies. He and Maslama ibn Ahmad al-Majriti introduced to Spain the theory of triangulating for surveying.

Hamadan, Persia, before 1037 Ibn Sina (Avicenna), Persian scientist, publishes *al-Qanun fi al-Tibb*, arguably the greatest work of medieval Islamic medicine.

Hamadan, Persia, June 1037 Ibn Sina (Avicenna), philosopher and physician, dies, aged 57. Among his scientific achievements are theories on the creation of mountains and the introduction of cubic equations.

Cairo, Egypt, 1040 Al-Haytham (Alhazen), mathematician and physician, dies, aged 57.

China, 1044 For the first time a gunpowder formula is recorded, in a manual on weaponry.

Ghazna, Afghanistan, 1048 Al-Biruni, scholar, dies, aged about 75. He was responsible for a number of important works on subjects ranging from astronomy and mathematics to geography and culture.

World, July 1054 Chinese and Arab astronomers record the supernova that creates the Crab Nebula.

Salerno, Italy, c. 1065 Constantine the African, a Tunisian scholar, visits the school of medicine, and begins translating his books on Islamic medicine into Latin.

Europe, April 1066 Halley's Comet is observed at the time of the Norman Conquest.

Samarkand, Uzbekistan, 1070 Omar Khayyam writes *Maqalat fi al-Jabr wa al-Muqabila* (*Treatise on Demonstration of the Problems of Algebra*).

Baghdad, Iraq, c. 1075 Al-Nasawi, Persian mathematician, dies. He is best known for his translation of Euclid's *Elements* for Arab readers.

Toledo, Spain, 1080 Botanist Ibn Bassal, who is in the employ of the Sultan of Toledo, writes *The Book of Agriculture*.

China, 1086 Shen Kua writes *Dream Pool Essays*, in which he discusses the use of the magnetic compass in navigation.

Toledo, Spain, 1087 Al-Zarqali (Arzachel), renowned astromoner, instrument maker, and inventor of an astrolabe, dies.

Bologna, Italy, 1088 The University of Bologna, the first university in the western world, is established.

Persia, c. 1088 Surgeon Zarrin Dast writes his treatise on ophthalmology, *Nur al-Ayun* (*The Light of the Eyes*). This amazing work includes descriptions of a number of groundbreaking treatments for eye problems and provides comprehensive information on surgery of the eye.

Italy, 1091 French-born Walcher observes an eclipse of the Moon and uses an astrolabe to determine the exact time of the eclipse.

Salerno, Italy, c. 1097 Physician Trotula of Salerno is born.

1100s

Kaifeng, China, 1100 Liu Ruxie, Chinese mathematician, publishes the first known version of "Pascal's triangle," a chart of coefficients of binomial equations.

China, 1117 Zhu Yu's *P'ingchow Table Talk* makes first mention of the use of the magnetic compass in maritime navigation.

Cairo, Egypt, 1120 An observatory is built for solar observations.

Kaifeng, China, 1120 Song dynasty generals use gunpowder flame-throwers, rockets, and bombs to slow the advance of the Manchurian Jurchen armies.

London, England, 1123 Rahere, a member of the court of Henry I, founds St Bartholomew's Hospital.

Artois, France, 1126 The first artesian well in Europe is drilled by monks—the wells are named after the town.

Worcester, England, December 8, 1128 John of Worcester makes the first known illustration of sunspots, with a drawing showing two dark spots on the Sun.

Kaesong, Korea, December 13, 1128 The Koryo-sa record the observation of a red light in the night sky (the aurora borealis), an event probably caused by sunspots five days earlier.

Provence, France, 1136 Abraham bar Hiyya, Spanish mathematician and scholar, dies, aged about 66.

England, 1144 *Liber de Compositione Alchimiae* (*The Composition of Alchemy*) is translated from Arabic by Robert of Chester.

Italy, 1145 Plato of Tivoli translates Savasorda's 1136 work on algebra. The translated work is published as *Liber embadorum*.

England, 1145 Robert of Chester translates al-Khwarizmi's work on algebra from Arabic to Latin.

India, 1150 Astronomer and mathematician Bhaskara writes *Siddhanta Siromani*.

England, 1150 Adelard of Bath, regarded as England's first scientist, dies, aged about 70.

Bingen, Germany, 1150 Theologian Hildegard of Bingen writes *Physica* and *Causae et Curae*, both books discussing her views on natural history and the use of plants and animals for healing purposes.

Damascus, Middle East, c. 1155 One of the first hospitals in the Arab world, the Bimaristan Nureddine, is built.

Seville, Spain, 1160 The mathematician and astronomer Jabir ibn Aflah, also known as Geber, dies, aged about 60.

Sicily, c. 1166 Great geographer and astronomer, al-Idrisi dies, aged about 67.

Canterbury, England, June 18, 1178 Gervaise of Canterbury records his observation of an event that is possibly a meteorite hitting the surface of the Moon.

China, February 1182 A supernova, observed since August 6 the previous year, finally disappears. It is also seen from Japan, first seen there one day later.

Novgorod, Russia, May 1, 1185 A description of a solar eclipse includes the earliest known record of a solar prominence.

Pisa, Italy, 1185 Burgundio of Pisa translates various texts by ancient Greek physician Galen from Greek to Latin.

Salerno, Italy, 1190 Matthaeus Platerius writes *Circa Instans*, which discusses the use of plants for medicinal purposes.

England, 1190 Alexander Neckham's work *De Naturis Rerum* (*On the Nature of Things*) is published.

Seville, Spain, 1190 The Tower of Seville houses Europe's first observatory.

Marrakesh, Morocco, 1198 Ibn Rushd, better known as Averroes, philosopher, lawyer, and scientist, dies, aged about 72.

1200s

Italy, 1202 Fibonacci's *Liber abaci* (*Book of Calculations*) is published, using Hindu–Arabic mathematics to revolutionize everyday use of mathematics.

Egypt, December 13, 1204 Spanish born philosopher and rabbi, Maimonides (Moshe ben Maimon), dies, aged about 70.

Diyarbakir, Turkey, 1206 Al-Jazari, the outstanding engineer, writes his *al-Jami bain al-ilm wal-amal al-nafi fi al-hiyal* (*The Book of Knowledge of Ingenious Mechanical Devices*). He is also a prolific inventor.

Persia, *c.* 1215 Mathematician Sharafeddin Tusi dies. He was responsible for an important treatise on cubic equations.

Paris, France, *c.* 1230 Mathematician and astronomer Sacrobosco writes his book on astronomy—*Tractatus de Sphaera*.

England, 1234 Michael Scott, scholar, dies, aged about 59. His many works covered a range of chiefly scientific subjects, including mathematics, physics, alchemy, astrology, and astronomy.

Syria, 1242 Ibn an-Nafis suggests that the right and left ventricles of the heart are separate and describes the lesser circulation of blood.

China, 1247 Ch'in Chui-Shao (Qin Juishao) writes his mathematical treatise *Shushu jiuzhang* (*Mathematical Treatise in Nine Sections*).

China, 1248 The first book on forensic science is published. *Hsi duan yu* (*The Washing Away of Wrongs*) describes how to detect the difference between death by drowning and death by strangulation.

China, 1248 Li Zhi (also known as Li Ye) writes his mathematical work, *Ce yuan hai jing* (*The Sea Mirror of Circle Measurements*).

Damascus, Middle East, 1248 Ibn al-Baitar, scientist, dies, aged about 60. His contributions to botany and medicine are remembered in his works *Kitab al-Jami fi al-Adwiya al-Mufrada* on botany, and *Kitab al-Mlughni fi al-Adwiya al-Mufrada*, his work on medicine that incorporated the use of herbal remedies.

England, 1250 Roger Bacon invents the magnifying glass.

Pisa, Italy, 1250 Leonardo of Pisa (Fibonacci), mathematician, dies, aged about 70.

Germany (?), 1250 Jordanus Rufus writes a veterinary manual.

Toledo, Spain, 1252 Commissioned by King Alfonso X and named for him, the Alfonsine Tables are the first European astronomical tables.

Buckinghamshire, England, October 9, 1253 Robert Grosseteste, scholar, dies, aged about 78. His interests included astronomy, mathematics, and optics in particular, and he counted many influential figures among his students, including Roger Bacon.

Maragha, Central Asia, 1259 Persian astronomer Nasir ad-Din al-Tusi builds an observatory for Mongol Great Khan Hulegu.

Paris, France, August 1269 Petrus Peregrinus (Petrus de Maricourt) writes his manuscript on magnets, *Epistola de magnete*.

Viterbo, Italy, 1270 Witelo's treatise *Perspectiva* (*Perspectives*) is published. It outlines his theories on optics.

Baghdad, Persia, June 26, 1274 Al-Tusi, scientist, dies, aged about 73. His interests covered many fields of science and mathematics.

Lombardy, Italy, 1275 William of Saliceto writes *Chirugia*, a record of his surgery and dissections.

China, 1276 Guo Shoujing, astronomer and engineer, develops instruments to help gather information for the development of the new calendar commissioned by Kublai Khan.

China, 1279 Li Zhi (Li Ye), mathematician, dies, aged about 86. He wrote many works, the most famous being *Ce yuan hai jing* (*The Sea Mirror of Circle Measurements*).

Beijing, China, 1279 Kublai Khan commissions the building of an observatory in the city to provide Guo Shoujing with more data.

Cologne, Germany, November 15, 1280 Albertus Magnus, great scholar, dies.

Italy, *c.* 1284 Eyeglasses are invented. The actual inventor is uncertain, with Alessandro Spina of Florence and Salvino D'Armante of Pisa both credited.

Italy, 1286 William of Moerbeke, scholar, dies, aged about 71. During his lifetime he translated numerous works from Arabic to Latin, bringing the discoveries and theories of Arab scholars in the fields of philosophy, medicine, mathematics, and science to a European audience.

Cairo, Egypt, 1288 Ibn al-Nafis, renowned physician, dies, aged about 75. He was the first to document the workings of the pulmonary circulation of the body, gas exchange in the lungs, and the structure of the lungs.

Oxford, England, 1292 Roger Bacon, scholar, dies, aged about 78. His interests ranged from theology to science, and he was particularly interested in the field of optics.

Viterbo, Italy, 1296 Johannes Campanus (also known as Campanus of Novara) dies, aged about 76. He is known for his works on mathematics and astronomy, including *Theorica planetarum*, which detailed the construction of a planetarium.

1300s

Cologne, Germany, November 8, 1308 John Duns Scotus, philosopher and theologian, dies, aged about 42.

Tunisia or Mallorca, 1315 Ramon Llull, Spanish philosopher and scientist, dies, aged about 83. (The place of his death is not certain.) His many works included mathematics, astronomy, and religious texts.

Padua, Italy, before 1316 Pietro d'Abano combines Arab medical science with Greek philosophy in his *Conciliator differentiarum*.

Bologna, Italy, 1316 Mondino di Luzzi publishes *Anathomia*, a book of his teachings on anatomy and methods of dissection.

Paris, France, 1320 Physician to the French king, Henri de Mondeville writes his *Cyrurgia*.

Tabriz, Persia, 1320 Al-Farisi, Islamic scientist and mathematician, dies, aged about 60.

France, 1321 Levi ben Gerson (Gersonides) writes his mathematical work *Book of Numbers*.

Marrakesh, Morocco, 1321 Ibn al-Banna, mathematician, dies, aged about 65. His works included *Talkhis amal al-hisab* (*Summary of Arithmetical Operations*).

France, 1329 Levi ben Gerson (Gersonides) completes his major work *Sefer Milhamot Ha-Shem* (*The Wars of the Lord*). This includes writings on many subjects, with his astronomical observations being particularly significant.

England, 1345 William Merlee, clergyman and scholar, dies. He kept detailed meteorological records, and attempted to predict the weather using these records.

Italy, 1348 Gentile da Foligna, physician, dies of bubonic plague, aged about 58. He was a pioneer in the study of kidney function.

Munich, Germany, 1349 William of Ockham, logician, theologian, and philosopher, dies, aged about 64. He is recognized for the theory known as Ockham's (also Occam's) razor.

England, August 26, 1349 Thomas Bradwardine, scientist and theologian, dies, aged about 59.

Europe, c. 1358 Jean Buridan, scientist and philosopher, dies, aged about 60. A student of William Ockham, he went on to publish a number of works, and ultimately argued against the theories of his mentor. He developed the concept of impetus and invented its name.

France, 1363 Guy de Chauliac publishes his *Chirurgia magna*, an influential surgical guide.

Padua, Italy, 1364 Giovanni di Dondi builds an amazing astronomical clock, which he calls an "astrarium."

Peking, China, 1368 A water-clock in the Mongol imperial palace is destroyed by order of the founding emperor of Ming Dynasty, who calls it a "useless contrivance."

India, 1370 Mahendra Suri writes his work on astronomy, *Yantraraja*.

Lisieux, France, July 11, 1382 Nicholas Oresme, scientist, dies, aged about 62. His wide range of interests included theology, mathematics, science, and music.

1400s

Venice, Italy, 1410 Ptolemy's *Geographia*, a comprehensive guide to mapmaking and geography, is translated into Latin and thus becomes available to European scholars and mapmakers.

Venice, Italy, 1410 Benedetto Rinio publishes his herbal (*Liber de simplicibus*), an illustrated guide to more than 400 plants of medicinal use.

Samarkand, Uzbekistan, 1420 Ulugh Beg, Timurid ruler, with a keen interest in both the arts and science, builds an observatory in the city.

Samarkand, Uzbekistan, 1427 Al-Kashi publishes his mathematical work, *The Key to Arithmetic.*

Samarkand, Uzbekistan, 1437 The star catalog *Zij-I Sultani* is published by Ulugh Beg.

Seoul, Korea, 1438 Astronomers complete the project to re-equip the royal observatory with the world's most up-to-date instruments.

Murano, Italy, 1450 Angelo Barovier invents crystalline glass.

Germany, 1451 Nicholas of Cusa invents concave lenses to correct near-sightedness.

Italy, 1452 Leonardo da Vinci is born.

Italy, c. 1454 Theodore of Gaza translates botanical works by Theophrastus (a pupil of Aristotle) into Latin.

Borgo Sansepolcro, Italy, 1470 Piero della Francesca, acknowledged not only as a great painter but also as a mathematician, publishes *Tratta d'Abaco*, a work on algebra.

Nuremberg, Germany, 1471 Johannes Müller (Regiomontanus) builds an observatory to conduct his studies in astronomy. He also installs his own printing press to enable publication of his findings in the fields of science.

Torun, Poland, February 19, 1473 Nicolaus Copernicus is born.

Rome, Italy, July 6, 1476 Johannes Müller (Regiomontanus), mathematician and astronomer, dies, aged 40.

Venice, Italy, 1482 Erhard Ratdolt prints a Latin translation of Euclid's *Elements*, a geometry text written in about 300 BCE.

Lyon, France, 1484 Nicolas Chuquet's *Triparty en la science des nombres* is the first mathematical work to use negative exponents.

Milan, Italy, 1488 Leonardo da Vinci combines his engineering skills with his concept of man taking to the skies when he sketches his famous flying machine.

Leipzig, 1489 Johann Widman writes the first text to use the plus (+) and minus (–) signs to indicate excess and deficiency, making it easier to express mathematics in symbols rather than in words.

Nuremberg, Germany, 1492 Martin Behaim is responsible for the first world map on a globe.

Venice, Italy, 1494 Fra Luca de Pacioli publishes his *Summa de arithmetica, geometrica, proportioni et proportionalita*, a synthesis of the mathematical knowledge of his time.

Milan, Italy, 1495 Leonardo da Vinci sketches his design for a parachute.

Leiria, Portugal, 1496 Abraham Zacuto publishes his *Almanach perpetuum*.

1500s

China, 1500 In what may be regarded as the first attempt to send a man into space, Wan-hu sat in a chair with 47 firework rockets attached. Sadly, when they were lit Wan-hu disappeared in a flash.

Switzerland, 1500 The first record of a mother and child surviving a caesarean birth is made after Jacob Nufer saves his wife and child.

Saint-Dié, Lorraine, France, 1507 Martin Waldseemüller's map of the world is the first to record the New World.

Lisbon, Portugal, July 29, 1507 Martin Behaim, German mapmaker and geographer, dies, aged about 48.

Rumpelmonde, Belgium, March 5, 1512 Gerardus Mercator is born.

Frombork, Poland, c. 1512 Polish astronomer Nicolaus Copernicus writes his *Commentariolus*, outlining his theory that the planets revolve around the Sun.

Zurich, Switzerland, March 1516 Konrad von Gesner is born.

Bologna, Italy, 1520 Italian mathematician, Scipione del Ferro, develops a method for solving cubic equations.

Tabriz, Persia, c. 1520 Persian mechanical engineer Hafez el-Esfahani invents two new types of water mill.

Vienna, Austria, 1525 Mathematician Christoff Rudolff uses the square root symbol for the first time in *Die Coss.*

Nuremberg, Germany, 1525 Albrecht Dürer publishes his *Treatise on Measurement.*

Europe, 1530 Physicians begin to use laudanum for pain management.

Germany, 1530 Physician Georg Agricola writes on mineralogy.

Louvain, Netherlands, 1530 Regnier Gemma Frisius describes a method of determining the longitude of a place, using clocks.

Europe, 1531 A comet, later known as Halley's comet, is observed.

Canada, March 1536 A native remedy called annedda made from brewing the bark of white cedar trees cures Jacques Cartier's crew of scurvy.

Basle, Switzerland, 1536 Paracelsus (Philip von Hohenheim) publishes *Der Grossen Wundertzney* (*The Great Surgery Book*).

Italy, 1539 Olaus Magnus illustrates the first map of Scandinavia, known as the "Carta Marina."

Europe, 1540 German scientist Valerius Cordus creates ethyl ether, a synthetic ether.

Fontenay-le-Comte, France, 1540 François Viète is born.

Paris, France, 1542 Jean François Fernel's book on physiology is based on his observations as a physician and anatomist.

Tubingen, Germany, 1542 Leonhart Fuchs writes *De historia stirpium commentarii insignes* (*Notable Commentaries on the History of Plants*).

Poland, March, 1543 Copernicus publishes *Die revolutionibus orbium coelestium* (*On the Revolution of Heavenly Bodies*), which contains his theory on heliocentric motion.

Frombork, Poland, May 24, 1543 Nicolaus Copernicus, astronomer and mathematician, dies, aged 70.

Basle, Switzerland, 1543 The structure of the human body is the subject matter for *De humani corporis fabrica* (*On the Fabric of the Human Body*), written by Belgian physician Andreas Vesalius.

Germany, 1544 Valerius Cordus, botanist and physician, dies, aged 29.

Italy, 1545 The method for solving cubic and quartic equations is published for the first time in *Ars Magna* (*The Great Art*)—the work of Girolamo Cardano.

Leuven, Belgium, 1546 Gerardus Mercator, famous for his work in mapping, declares that Earth has a magnetic pole.

Knutstorp, Denmark, December 14, 1546 Tycho Brahe, astronomer, is born.

Verona, Italy, 1546 Girolamo Fracastoro publishes his theory of contagion of disease.

Merchiston, Scotland, 1550 John Napier is born.

Kyoto, Japan, 1551 The first Western weight-driven clock mechanism is introduced into Japan.

Zurich, Switzerland, 1551–58 Naturalist Conrad Gessner publishes *Historiae animalium*, marking the beginning of modern zoology.

Venice, Italy, 1552 Gabriele Fallopius publishes *Observations anatomicae* (*Anatomical Observations*).

Rome, Italy, 1552 Bartolommeo Eustachio describes the Eustachian tube of the ear.

Chemnitz, Germany, 1556 Georg Agricola publishes his *De re metallica libri*, a comprehensive work on mining and smelting.

London, England, 1557 Replacing their long-hand versions, the addition sign, minus sign, and equals sign appear for the first time in English in Robert Recorde's *The Whetstone of Witte.*

Thrissur, India, c. 1560 Mathematician Chitrabhanu does groundbreaking work on integer solutions to series of algebraic equations.

Pisa, Italy, February 15, 1564 Galileo Galilei is born.

Zakynthos, Greece, October 15, 1564 Andreas Vesalius, surgeon and anatomist, dies after being shipwrecked, aged about 50.

London, England, 1565 Previously forbidden, the members of the Royal College of Physicians are given permission to perform dissections of the human body.

Duisburg, Germany, 1569 Gerardus Mercator creates a map using his own "Mercator projection."

England, 1570 John Dee writes a Mathematical Preface to an English translation of Euclid's *Elements.*

Italy, 1570 Giambattista della Porta perfects the camera obscura, commonly known as the pinhole camera.

London, England, 1571 *Pantometria*, by Leonard Digges, contains the first record of the theodolite.

Denmark, 1572 The astronomer Tycho Brahe details his sighting of a supernova in the Cassiopeia formation.

Scotland, 1574 Child prodigy James Crichton receives his first degree from St Andrew's University at the age of 14.

Switzerland, 1577: Jost Bürgi invents the first clock that incorporates a minute hand.

Folkeston, England, April 1, 1578 William Harvey is born.

Venice, Italy, 1579 The first glass eye is invented.

England, 1580 William Bourne publishes his description of the principles of a submarine.

Pisa, Italy, 1581 The sight of a swinging lamp leads Galileo to his theory on pendulum motion.

Paris, France, 1581 François Rousset's book *Traitte Nouveau de l'Hysterotomotokie* contains an account of a successful caesarean procedure (by Jacob Nufer, in 1500). It was Rousset who coined the phrase "Caesarean section."

Florence, Italy, 1583 Andrea Cesalpino's *De plantis* shows the first groupings of plants by structure and habit.

Macau, China, 1584 Matteo Ricci produces the first western-style world map in Chinese.

Antwerp, Belgium, 1585 Simon Stevin puts forward the decimal system in his work *De Theinde* (*The Tenth*).

Netherlands, 1590 Zacharias Janssen invents the compound microscope.

Paris, France, December 20, 1590 Ambroise Paré, surgeon, dies, aged about 80.

China, 1593 Li Shizhen (or Li Shih-chen), medical researcher, dies, aged about 75. He was the author of the *Compendium materia medica*, a comprehensive guide to the components used in traditional Chinese remedies.

Padua, Italy, 1594 The University of Padua builds an anatomical theater so that students can observe dissections.

Prague, eastern Europe, 1596 Georg Joachim Rheticus's *Opus palatinum de triangulis* (*The Palatine Work on Triangles*), a work on trigonometric tables, is published posthumously.

La Haye, France, March 31, 1596 René Descartes is born.

England, 1597 John Gerard publishes *The Herball of Generall Historie of Plants*, promoting medicinal use of plants.

Padua, Italy, 1597 Galileo Galilei invents the proportional compass.

Netherlands, 1599 The dodo bird is first mentioned in the 1598 record of Admiral Jacob Cornelius van Neck's visit to Mauritius.

1600s

England, 1600 Physician William Gilbert publishes his work on magnetics, *De Magnete* (*On Magnets*).

Italy, 1600 Priest and cosmologist Giodano Bruno is burnt at the stake by the Roman Inquisition for his heretical views on the nature of the universe. This included the view that Earth was not at the center of the universe.

Germany, 1603 Christoph Scheiner invents the first pantograph.

Rome, Italy, 1603 The Accademia dei Lincei is established by Federico Cesi.

Padua, Italy, 1603 Anatomist Hieronymus Fabricius discovers the venous valves in the lower limbs. He publishes these findings in his work *De venarum ostiolis*.

Bologna, Italy, c. 1605 Ulisse Aldrovandi, naturalist, dies, aged about 82. He is best remembered for his researches into the whole scope of natural history.

London, England, 1605 Francis Bacon publishes *The Proficience and Advancement of Learning*, which advocates that science and scientific knowledge should be based on experimentation and proof.

Padua, Italy, c. 1606 Galileo is credited with the invention of the first thermometer.

Middelburg, Netherlands, 1608 Hans Lippershey is credited with the invention of the first telescope.

Venice, Italy, 1609 Galileo refines and modifies Lippershey's telescope.

Prague, eastern Europe, 1609 Johannes Kepler (1571–1630) publishes *Astronomia Nova* (*New Astronomy*), in which he states his first two laws of planetary motion.

Europe, 1610 A number of observers (including Galileo and Thomas Harriott) observe spots on the Sun.

Rome, Italy, January 1610 Using his improved telescope, Galileo observes the four largest moons orbiting Jupiter, and the rings of Saturn (which he initially mistakes as ears). In the same year, he publishes *Sidereus Nuncius*.

Florence, Italy, 1612 Antonio Neri writes *L'arte vetraria* (*The Art of Glass*), in which he reveals the secrets of glassmaking, knowledge previously closely held only by master craftsmen.

Scotland, 1614 Scottish mathematician John Napier invents logarithms.

London, England, c. 1620 Cornelius Drebbel is credited with the invention of the submarine.

England, 1620 Francis Bacon publishes *Novum Organum*.

Linz, Germany, 1621 Johannes Kepler publishes *Epitome Astronomiae Copernicanae* (*Epitome of Copernican Astronomy*).

Netherlands, 1621 Willebrord Snell determines the law of refraction.

Tübingen, Germany, 1623 Wilhelm Schickard invents a calculating machine.

Ulm, Germany, 1627 Johannes Kepler publishes *Tabulae Rudolphinae* (*Rudolphine Tables*), combining the calculations of Tycho Brahe, with Kepler's own observations on planetary motion.

London, England, 1628 William Harvey publishes his findings *On the Motion of the Heart and Blood in Animals*.

Florence, Italy, 1632 Galileo publishes his controversial *Dialogo sopra i due massimi sistemi del mondo, tolemaico e copernicano* (*Dialogue Concerning the Two Chief World Systems, Ptolemaic and Copernican*).

Peking, China, 1633 Xu Guangxi writes *Nong-zheng Quanzhu* (*Complete Treatise on Agriculture*).

Jiangxi, China, 1637 Song Yingxing writes *Tiangong Kaiwu* (*Exploitation of Resources*), a pioneering encyclopedia of technology.

Amsterdam, Netherlands, 1638 René Descartes publishes his mathematical work *La Géométrie*.

Amsterdam, Netherlands, 1641 Descartes publishes *Meditationes de Prima Philosophia* (*Meditations on First Philosophy*).

Paris, France, 1642 Blaise Pascal invents a mechanical calculator.

Florence, Italy, January 8, 1642 Galileo Galilei, mathematician and scientist, dies, aged 77.

Lincolnshire, England, January 4, 1643 Isaac Newton is born.

Florence, Italy, 1644 Italian scientist and mathematician Evangelista Torricelli invents the barometer.

Paris, France, 1644 Marin Mersenne (1588–1648) releases his mathematical work on prime numbers, *Cogitata Physico-Mathematica*.

Amsterdam, Netherlands, 1644 Descartes publishes *Principia Philosophiae* (*Principles of Philosophy*), on a range of sciences.

Paris, France, 1647 Scientist and philosopher Pierre Gassendi publishes his work on astronomy—*Institutio astronomica*.

Bologna, Italy, 1651 Giovanni Riccioli publishes his *Almagestum novum*, including a map of the Moon (drawn by Francesco Grimaldi), which names many of the craters after scientists.

Copenhagen, Denmark, 1652 Thomas Bartholin discovers the lymphatic system.

France, 1654 Correspondence between mathematicians Blaise Pascal and Pierre de Fermat becomes the basis of probability theory.

Netherlands, 1655 Mathematician and astronomer Christiaan Huygens is the first to suggest that Saturn is surrounded by a ring.

England, November 8, 1656 Edmond Halley is born.

Netherlands, 1657 Christiaan Huygens invents the first pendulum clock.

London, England, 1660 The Royal Society is founded, formalizing the meetings of several scientists of the time. The Society gains the approval of King Charles II.

England, 1660 Robert Hooke discovers the relationship between a stretched spring and its tension.

Bologna, Italy, 1661 Marcello Malpighi becomes the first to view the capillary system and writes an article on his discovery.

London, England, 1662 Robert Boyle advocates the use of experimentation to provide proof of scientific theories in *The Sceptical Chymist*.

Paris, France, August 19, 1662 Blaise Pascal, mathematician and religious philosopher, dies, aged 39.

Cambridge, England, 1664 Thomas Willis publishes his work on the anatomy of the brain and nervous system—*Cerebri Anatome*.

Castres, France, January 1665 Pierre de Fermat, mathematician, dies, aged about 63.

London, England, 1665 Robert Hooke publishes a book on microscopes and his observations using them—*Micrographia*.

Amsterdam, Netherlands, 1665 Athanasius Kircher publishes *Mundus subterraneus*, which describes the geology and biology of the underground world.

Bologna, Italy, 1666 Giovanni Domenico Cassini discovers a polar ice cap on Mars.

France, June 1667 The first recorded blood transfusion is carried out by Jean-Baptiste Denys. He uses blood from a lamb to save the life of a young boy.

Seoul, Korea, 1669 Song Lyong and Yi Minch'ol construct a clockwork-driven armillary sphere modeling movements of heavenly bodies.

Europe, 1671 Though a number of calculators had been invented in previous years, Gottfried Wilhelm von Leibnitz improves the design with his Step Reckoner.

Delft, Netherlands, 1672 Regnier de Graaf publishes his findings on the human reproductive system and describes the ovarian follicles, which are named Graafian follicles in his honor.

Paris, France, 1673 Christiaan Huygens, Dutch astronomer, mathematician, and physicist, writes *Horologium Oscillatorium*.

Peking, China, 1673 Jesuit astronomer Ferdinand Verbiest re-equips the Imperial Observatory with European-style instruments.

Hamburg, Germany, 1674 Alchemist Hennig Brand discovers the chemical element phosphorus.

Delft, Netherlands, 1674 With no formal training or education in the sciences, Anton van Leeuwenhoek becomes the first to view bacteria through his own self-built lens.

Paris, France, 1675 Danish astronomer Ole Rømer uses his observations of Jupiter's moons to estimate the speed of light.

Peking, China, 1678 Li Shizhen compiles the *Bencao Gangmu* (*Compendium of Essential Herbs*), a massive encyclopedia of traditional Chinese medicine.

London, England, 1678 Robert Hooke publishes his findings that states the forces applied to a spring and its subsequent extension are in direct proportion. This law is subsequently named after him (Hooke's Law).

London, England, 1682 Edmond Halley follows the progress of a passing comet that would later bear his name, and calculates the date of its recurrence (1758).

Germany, 1684 Mathematician Gottfried Leibniz publishes his findings on calculus. He and Isaac Newton, who worked separately, are both credited with the invention of calculus.

England, 1686 Following his earlier volume on plant classification, *Methodus Plantarum Novum*, John Ray classifies around 18,000 British plants in his great work, *Historia Plantarum*.

London, England, 1687 Isaac Newton publishes a work, *Principia Mathematica* on his law of universal gravitation and several laws of motion.

Lucca, Italy, 1699 In an attempt to stop the spread of the disease, the body and possessions of anyone dying of consumption (tuberculosis) are burned, by government order.

1700s

Cambridge, England, 1702 Cambridge University establishes chairs for the sciences, with the endowment of a chair in chemistry.

London, England, 1704 Isaac Newton's *Opticks* is published, outlining theories of light and color.

Basle, Switzerland, August 16, 1705 Jacob Bernoulli dies, aged 50.

Boston, North America, January 17, 1706 Benjamin Franklin, inventor, scientist, and statesman, is born.

Rashult, Sweden, May 23, 1707 Carl von Linné (Linnaeus) biologist, is born.

Lichfield, England, 1707 Sir John Floyer measures the pulse beats of his patients as a means of diagnosis.

Leiden, Netherlands, 1708 Hermann Boerhaave, prominent medical professor and physician, publishes *Institutiones Medicinae*.

Hanover, Germany, November 14, 1716 Gottfried Leibnitz, mathematician and philosopher, dies, aged 70.

London, England, 1717 Nurseryman Thomas Fairchild pollinates one *Dianthus* species with another and produces Europe's first artificial hybrid plant.

Greenwich, England, 1718 Edmund Halley discovers the true motion of "fixed" stars.

London, England, 1718 Lady Mary Wortley Montagu promotes the Turkish practice of inoculation against smallpox.

Germany, 1724 Physicist Daniel Gabriel Fahrenheit, the inventor of the mercury thermometer, introduces his temperature scale.

Greenwich, England, 1725 John Flamsteed's star catalog, *Historia Coelestis Britannica*, is published posthumously.

London, England, March 31, 1727 Isaac Newton dies, aged 84.

London, England, 1728 James Bradley's star observations lead him to calculate the speed of light to be 183,000 miles (295,000 km) per second.

Paris, France, 1728 Pierre Fauchard publishes *The Surgeon Dentist*, the first dental textbook, and invents the term "dentist."

Jaipur, India, 1728 A new royal observatory is completed. It contains one of the world's largest sundials.

London, England, 1729 Stephen Gray demonstrates that electricity can travel along some materials and not others.

London, England, 1731 John Hadley invents the reflecting quadrant (the basis of the sextant). American Thomas Godfrey independently invents the same instrument.

Teddington, Middlesex, England, 1733 Stephen Hales publishes studies of blood pressure measurements.

Paris, France, 1737 Philippe Buache uses contour lines to show elevation on maps.

London, England, 1738 Instrument maker Benjamin Martin creates his first universal microscope, a new type of portable microscope.

Switzerland, 1738 Daniel Bernoulli publishes *Hydrodynamica*, including Bernoulli's Principle, that as the speed of a fluid increases, the pressure it exerts decreases.

St Petersburg, Russia, June 7, 1742 The mathematician Christian Goldbach conjectures that every number greater than 2 is a sum of 3 primes, in a letter to Leonhard Euler.

Sweden, 1742 Anders Celsius proposes that a temperature scale of 0 to 100 (based on the boiling and melting points of water) be adopted for all scientific measurements.

Philadelphia, North America, 1746 Benjamin Franklin's kite experiment proves that lightning is a form of electricity.

Bazentinele-le-Petit, Picardy, Erfurt, Germany, 1745–1746 Scottish monk Andreas Gordon describes a device that can be used to store electricity. Musschenbroek and Cunaeus build this "Leyden jar."

Plymouth, England, June 16, 1747 James Lind's experiments with scurvy show citrus fruits to be the cure.

Stockholm, Sweden, 1751 Axel Frederik Cronstedt discovers nickel.

London, England, 1752 The great Scottish obstetrician William Smellie publishes *A Treatise on the Theory and Practice of Midwifery*.

Uppsala, Sweden, 1753 Carl von Linné (Linnaeus) publishes *System plantarum*.

Halle, Germany, 1754 Dorothea Erxleben is the first woman to be awarded a medical degree in Germany.

Edinburgh, Scotland, 1754 Joseph Black discovers carbon dioxide, to which he gives the name "fixed air."

Cape of Good Hope, South Africa, 1754 French astronomer Nicolas Louis de Lacaille and his assistants complete their study of the southern sky, identifying about 10,000 stars.

Prohlis, Germany, December 1758 Johann Georg Palitzsch observes the return of Edmund Halley's comet.

Halle, Germany, 1759 Kaspar Wolff publishes his dissertation *Theoria generationis* on embryos.

Peking, China, 1760 The first explanation of the Copernican system in Chinese is presented to Emperor Qianlong by visiting Jesuit priest Michel Benoist.

Lyons, France, 1762: Claude Bourgelat establishes the world's first veterinary school.

Moscow, Russia, November 20, 1764 Christian Goldbach, mathematician, dies, aged 74.

London, England, 1766 Henry Cavendish discovers the gas known as "inflammable air;" it is later named hydrogen.

Eaglesfield, Cumberland, England, September 6, 1766 John Dalton, scientist, is born.

London, England, 1771 Joseph Banks returns from his Pacific expedition with many new plant specimens.

Edo (Tokyo), Japan, 1774 Sugita Gempaku produces the first Japanese translation of a European textbook on anatomy and medicine.

Leeds, England, 1774 Joseph Priestley experiments with air in sealed containers and isolates "dephlogisticated" air, later known as oxygen.

Lyons, France, January 20, 1775 André Ampère, physicist, is born.

Bologna, Italy, 1780 Luigi Galvani's experiments with frogs lead him to the theory that muscle and nerve cells contain electricity.

Bath, England, March 13, 1781 William Herschel discovers the planet Uranus.

Edo (Tokyo), Japan, 1782 Shizuki Tadao publishes first book in Japanese on Newtonian physics.

Annonay, France, June 4, 1783 Joseph and Étienne Montgolfier send a balloon to 6,562 ft (about 2,000 m); the first public demonstration of a hot air balloon.

Paris, France, August 27, 1783 Jacques Charles launches the first hydrogen balloon.

St Petersburg, Russia, September 18, 1783 Leonhard Euler, Swiss mathematician and scientist, dies, aged 76.

Birmingham, England, 1785 William Withering publishes his findings on treating dropsy patients with foxglove leaves.

Paris, France, January 1786 Pierre Méchain is the first to discover the comet Encke. It takes 3.3 years to complete a revolution around the Sun.

Paris, France, 1787 Antoine Lavoisier publishes the chemical nomenclature system devised with Antoine Fourcroy, Guyton de Morveau, and Claude-Louis Berthollet.

Berlin, Germany, 1789 Chemist Martin Heinrich Klaproth discovers uranium and zirconium.

Philadelphia, USA, April 17, 1790 Benjamin Franklin dies, aged 84.

France, July 16, 1794 An optical telegraph line using Claude Charré's semaphore system begins operating between Paris and Lille.

Berkeley, England, 1798 Edward Jenner publishes vaccination findings against scourge of smallpox.

1800s

Pavia, Italy, March 20, 1800 Alessandro Volta informs Sir Joseph Banks of his invention—an electric "pile," a battery that produces a steady stream of electricity.

Jena, Germany, 1801 Johann Wilhelm Ritter discovers radiation past the violet end of the spectrum, a year after William Herschel found radiation past the red end.

Madras, India, April 10, 1802 William Lambton establishes a baseline near Madras, beginning the Great Trigonometric Survey of India.

London, England, November 24, 1803 Thomas Young proposes a wave theory for light, thus supplanting the idea that light is made up of particles.

London, England, 1803 John Dalton presents his atomic theory.

Germany, 1805 Friedrich Wilhelm Sertürner isolates morphine from opium.

Paris, France, 1808 Joseph Gay-Lussac publishes his law of combining volumes, which relates to the ratio between the volumes of reacting gases.

Paris, France, 1809 Jean-Baptiste Lamarck publishes his *Philosophie zoologique*, outlining his theory of evolution.

Germany, 1810 Samuel Hahnemann publishes his *Organon of Rational Therapeutics*, based on the principle of the "Law of Similars," which is still used in homeopathy.

London, England, 1811 Surgeon James Bell publishes *An Idea of the Anatomy of the Brain*, in which he distinguishes between motor and sensory nerves; this was a major breakthrough in physiology.

Paris, France, 1812 Marie Boivin publishes her book on obstetrics and gynecology, *Memoire de l'art accouchements*.

Stockholm, Sweden, 1813 Jöns Jacob Berzelius develops a system of alphabetical symbols for the elements, based on their Latin names.

London, England, 1817 James Parkinson publishes *An Essay on the Shaking Palsy*, in which he describes the disease later given his name.

Paris, France, 1817 Pierre-Joseph Pelletier and Joseph-Bienaimé Caventou isolate chlorophyll.

Copenhagen, Denmark, 1819 Hans Oersted discovers that electricity and magnetism are related.

Paris, France, 1820 André Ampère makes discoveries in the new field of electromagnetism.

Bonn, Germany, 1820 Christian Friedrich Nasse discovers that hemophilia affects males only and is carried by and inherited from females.

Slough, England, August 25, 1822 German astronomer William Herschel dies, aged 83.

Dole, France, December 22, 1822 Chemist and microbiologist Louis Pasteur is born.

London, England, October 1823 The first issue of *The Lancet* is published by Thomas Wakley.

Nagoya, Japan, 1823 Yoshio Nanko attempts to reconcile Copernican and traditional Japanese cosmologies.

Nagasaki, Japan, 1823 Philipp Franz von Siebold, German physician, gives classes in Western science to Japanese scholars.

Jena, Germany, 1823 Johann Dobereiner produces fire by the interaction of air, hydrogen, and platinum. His "lighters" become popular accessories.

Denmark, 1825 Hans Christian Oersted, physician and chemist, successfully produces aluminum.

England, 1825 William Sturgess invents the electromagnet.

England, 1825 Michael Faraday isolates and describes benzine.

Germany, 1826: Georg Simon Ohm formulates the relationship between voltage, current and resistance (Ohm's Law).

England, 1830–33 Charles Lyell publishes his *Principles of Geology*.

England, 1831 Michael Faraday discovers electromagnetic induction, making possible the electric transformer and generator.

Canada, 1831 James Clark Ross locates the magnetic North Pole.

Germany, France, USA, 1831 German chemist Justus von Liebig, French pharmacologist Eugène Soubeiran, and American chemist Samuel Guthrie independently discover chloroform.

England, 1834 Charles Babbage conceives a design proposal for the Analytical Engine, which possesses the logical features of the modern computer.

Germany, 1836 Wilhelm Beer and Johann von Madler publish the first exact map of the Moon, entitled *Mappa Selenographica*.

Edo (Tokyo), Japan, 1837 Udagawa Yoan begins translation of chemical works of Antoine-Laurent Lavoisier into Japanese.

London, England, 1837 J. F. Royle's *Antiquity of Hindoo Medicine* introduces Indian Ayurvedic medicine to the western world.

England, 1839 William Robert Grove invents the first hydrogen fuel cell.

England, 1840 Astronomer John Herschel invents the blueprint process of copying architectural and engineering drawings.

England, 1840 Physicist James Joule explains the "Joule effect," that heat produced in a conductor by an electrical current is proportional to the resistance and to the square of the current.

Vienna, Austria, 1842 Christian Doppler describes his principal regarding the effect of velocity on sound and light waves.

Ireland, 1844 Francis Rynd, physician, invents the hypodermic syringe and administers the world's first subcutaneous injection.

USA, 1844 Samuel Morse transmits the first demonstration telegraph message. The message "What God Hath Wrought" travels from Washington to Baltimore.

France, 1846 Astronomer and mathematician Urbain Le Verrier predicts the existence of Neptune.

Italy, 1846 Chemist Ascanio Sobrero prepares nitroglycerine.

Ohio, USA, February 11, 1847 Thomas Edison, inventor, is born.

Edinburgh, Scotland, March 3, 1847 Alexander Graham Bell, inventor, is born.

Edinburgh, Scotland, November 15, 1847 Obstetrician James Simpson successfully anesthetizes a woman in childbirth with chloroform.

England, 1847 George Boole publishes the *Mathematical Analysis of Logic*.

England, 1848 William Thomson, Lord Kelvin, devises the absolute, or Kelvin, scale of temperature.

Berlin, Germany, 1850 Rudolph Clausius states the second law of thermodynamics.

Germany, 1851 Franz Neumann formulates the laws of electromagnetic induction.

England, 1852 Herbert Spencer, biologist, writes *The Developmental Hypothesis*, using the word "evolution" for the first time.

England, 1852 Chemist Edward Frankland establishes the theory of valency, the basis of structural chemistry.

Nigeria, Africa, 1854 For the first time, quinine is administered successfully to treat malaria.

France, 1854 Scientist Henri Sainte-Claire Deville synthesizes aluminum to create the first commercial process.

England, 1854 George Boole publishes *An Investigation into the Laws of Thought, on which are founded the Mathematical Theories of Logic and Probabilities*.

London, England, 1854 An epidemic of cholera kills 10,000 people. Dr John Snow traces the source of an outbreak in Broad Street to a single water pump, thus validating his theory that cholera is water-borne, and forming the beginnings of epidemiology.

London, England, 1854 John Tyndall demonstrates the principles of fiber optics to the Royal Society.

Austrian Empire, 1856 Gregor Mendel, an Austrian monk later known as the "father of modern genetics," begins his experiments in heredity and variation in pea plants.

Germany, 1858 Physicist Julius Plucker identifies cathode rays (electrons) and discovers that they are deflected by a magnetic field.

England, 1859 Charles Darwin publishes *On the Origin of Species by Means of Natural Selection*.

France, 1859 Scientist Gaston Plante develops lead-acid battery.

France, 1862 Léon Foucault determines the speed of light.

England, 1863 Thomas Huxley publishes *Evidence on Man's Place in Nature*, making the first claim to imply evolution directly to the human race.

Ireland, December 8, 1864 Mathematician George Boole, inventor of Boolean algebra, dies, aged 49.

Austria–Hungary, 1865 Gregor Mendel publishes *Experiments in Plant Hybridization*, discussing the laws of genetic inheritance.

France, 1865 By heating milk and wine to 133°F (56°C) for 30 minutes, Louis Pasteur destroys harmful bacteria. The process becomes known as pasteurization.

Canada, 1867 Emily Stowe, the first female doctor in Canada, practices medicine illegally after being declined permission to write the Canadian exams required for foreign practitioners.

Stockholm, Sweden, 1867 Alfred Nobel, chemist and industrialist, patents dynamite.

England, 1868 After observing an eclipse using electromagnetic spectroscopy, astronomer Joseph Norman Lockyer identifies and names the element helium.

St Petersburg, Russia, 1869 Dmitri Ivanovitch Mendeleev publishes his periodic table of the chemical properties of the elements.

England, December 21, 1872 HMS *Challenger* sails from Portsmouth on a three-year voyage of exploration. Chemists, physicists and biologists collaborate with navigators to map the sea.

England, 1873 James Clerk Maxwell, eminent Scottish physicist, publishes his *Treatise on Electricity and Magnetism*, expounding his ideas on field theory and magnetism.

Germany, 1873 Graduate student Othmar Zeidler synthesizes DDT when working in a laboratory at the University of Strasbourg.

Vienna, Austria, 1873 The International Meteorological Society is formed.

England, 1874 George Stoney estimates the unit of charge in electrochemistry and calls it the electron.

Philadelphia, USA, March 10, 1876 Scottish-American inventor Alexander Graham Bell transmits his voice via the first telephone, saying "Mr Watson, come here. I need you."

Grays, Essex, England, May 1876 Alfred Wallace publishes *The Geographical Distribution of Animals*, in which he divides the world into six fauna regions and also introduces the concept of biogeography.

Berlin, Germany, March 24, 1882 Robert Koch presents his discovery of the bacterium that causes tuberculosis.

Kiel, Germany, 1882 Anatomist Walther Flemming describes cell division, a process he calls mitosis.

Paris, France, November 17, 1884 Count Hilaire de Chardonnet produces artificial silk from cellulose, the first man-made fabric, later known as rayon.

Paris, France, July 6, 1885 Louis Pasteur's rabies vaccine is used on humans for the first time. After receiving the vaccine, nine-year-old Joseph Meister makes a miraculous recovery after being bitten by a rabid dog.

Karlsruhe, Germany, 1887 Heinrich Hertz begins his experiments to produce Maxwellian electromagnetic waves, the first radio waves.

Washington DC, USA, October 1888 The first edition of *National Geographic* magazine is released. The National Geographic Society was founded in January by 33 influential Americans seeking to advance geographical knowledge.

Baltimore, USA, October 1891 William Osler completes the manuscript for *The Principles and Practice of Medicine*, a text that codifies medical practice.

Scotland, August 4, 1894 Chemists William Ramsey and Lord Rayleigh discover the inert element argon.

Bologna, Italy, August 1895 Guglielmo Marconi transmits radio waves over a distance of about 1.3 miles (2 km). He embarked on study of the works of Heinrich Hertz last year, experimenting with the wireless transmission of messages using Hertzian waves.

Würtzberg, Germany, December 28, 1895 Physics professor Wilhelm Röntgen presents a paper outlining his discovery of X-rays. He produced the first of this new type of ray on November 8, and had spent the next eight weeks repeating his experiments.

Right: American inventor Thomas Alva Edison (1847–1931), pictured here in 1920 experimenting in his chemistry laboratory. Among his more than 1,000 inventions, Edison is probably most famous for the development of the phonograph and the light bulb.

Vienna, Austria, 1896 Sigmund Freud coins the term psychoanalysis or "free association," referring to the investigation of the psychological causes of mental disorders.

Stockholm, Sweden, 1897 Physicist Vilhelm Bjerknes develops the circulation theorems that bring together hydrodynamics and thermodynamics. His work leads to the development of weather forecasting.

England, 1897 Physicist Joseph John "J. J." Thomson discovers the electron. (In 1906, he is awarded the Nobel Prize in Physics for the discovery.)

Paris, France, July 1898 Marie and Pierre Curie coin the term radioactivity after they discover the elements polonium and radium.

Calcutta, India, 1898 After eight years of research, the Scottish physician Ronald Ross determines that malaria is transmitted by mosquitoes.

1900s

Friedrichshafen, Germany, July 2, 1900 Count von Zeppelin successfully launches an airship that he has been working on since 1891.

Vienna, Austria, October 14, 1900 Psychiatrist and neurologist Sigmund Freud publishes *The Interpretation of Dreams*.

Berlin, Germany, December 14, 1900 Max Planck's quantum theory is unveiled.

Canada, May 6, 1901 Niagara Falls is used for power generation.

Paris, France, June 12, 1901 Physicist Henri Becquerel, working with radium, proves that atoms have an internal structure.

Cuba, February 22, 1902 Dr Walter Reed determines that yellow fever is spread by mosquitoes.

West Orange, New Jersey, USA, May 28, 1902 Thomas Edison invents a new storage battery made of nickel and iron in an alkaline solution.

London, England, December 1902 Oliver Heaviside predicts the existence of a conducting layer in the atmosphere which could transmit radio waves.

Chicago, USA, February 16, 1903 A new method for filling teeth, porcelain inlay, is developed.

Stockholm, Sweden, December 10, 1903 Pierre and Marie Curie share the Nobel Prize in Physics with Henri Becquerel for their work on radioactivity.

Stockholm, Sweden, December 10, 1904 Russian physiologist Ivan Pavlov wins the Nobel Prize in Medicine. Through his experiments with dogs, he discovers conditioned reflexes, such as salivation, in relation to food.

France, April 31, 1905 Psychologist Alfred Binet develops the Intelligence Quotient (IQ) test, finding a method by which to measure the ability to think and reason.

Bern, Switzerland, December 1905 Albert Einstein publishes four papers—including articles on special relativity, Brownian motion and the photoelectric effect—providing the foundations of modern physics.

Germany, May 10, 1906 Microbiologist August von Wasserman devises a test for syphilis.

USA, December 24, 1906 Canadian Reginald Fessenden, discoverer of AM radio waves, broadcasts over 100 miles (160 km).

Stockholm, Sweden, December 10, 1907 Alphonse Laveran wins the Nobel Prize in Physiology for his discovery of the role of protozoa in causing malaria and leishmaniasis.

Salzburg, Austria, April 27, 1908 The inaugural International Congress of Psychoanalysis opens.

Manchester, England, December 1908 Hans Geiger and Ernest Rutherford develop a device for measuring radioactivity, the Geiger counter. Rutherford is also awarded the Nobel Prize in Chemistry.

Denmark, March 1909 Biochemist Soren Sorensen invents the pH scale of acidity.

Stockholm, Sweden, December 10, 1909 Guglielmo Marconi and Karl Braun share the Nobel Prize in Physics.

New York, USA, February 27, 1910 In a world first, an X-ray machine is used to guide doctors in removing a nail from a child's lung.

Paris, France, December 1910 In an attempt to develop an inexpensive method by which to produce pure oxygen, Georges Claude invents the neon light.

New York, USA, 1911 Thomas Hunt Morgan, studying the cross-breeding of fruit flies, proves that genes are carried on chromosomes.

Stockholm, Sweden, December 10, 1911 Marie Curie receives an unprecedented second Nobel Prize, this time in Chemistry, for her work on radium. Despite this, the Académie des Sciences refuses to admit her, as she is a woman.

Kent, England, February 10, 1912 Surgeon Joseph Lister, inventor of antiseptic surgery, dies, aged 85.

New York, USA, September 1912 Carl Jung delivers the lectures that shall form the basis of his Theory of Psychoanalysis.

Russia, December 1912 Polish biochemist Casimir Funk coins the term "vitamine" following his isolation of vitamin B1.

Germany, 1912 Climatologist Alfred Wegener puts forward his theory of continental drift, and that all the continents were originally part of one supercontinent.

Cambridge, England, December 1913 The third volume of Bertrand Russell and Alfred North Whitehead's *Principia Mathematica* is published. Its defense of logicism is greatly influential.

Munsterlingen, Switzerland, February 13, 1914 French criminologist, Alphonse Bertillon, inventor of scientific identification of criminals, dies, aged 61.

London, England, December 1914 Ernest Rutherford, 1908 recipient of the Nobel Prize for Chemistry for his work on the atom, is knighted. His current work is on detection of submarines by radio monitoring.

Strasbourg, Germany, December 1914 Seismologist Beno Guttenberg discovers the discontinuity between Earth's mantle and its core.

Berlin, Germany, November 25, 1915 Einstein's General Theory of Relativity is published.

Germany, December 1915 Meteorologist Alfred Wegner publishes his controversial theory of continental drift and the supercontinent of Pangea in *The Origins of Continents and Oceans*.

France, December 1915 This year physicist Paul Langevin invents the first active sonar-type device for detecting submarines.

USA, December 1916 Frederick Kolster develops the radio direction finder, allowing ships to take bearings out of sight of land. The US Navy realized the importance of this and continued work on it.

Berlin, Germany, September 1917 Einstein adds a cosmological constant to the General Theory of Relativity, in order to describe a universe that conforms to theoretical expectations.

Switzerland and Austria-Hungary, December 1917 This year Sigmund Freud releases *Introductory Lectures in Psychoanalysis* and Carl Jung, *The Psychology of the Unconscious*.

Stockholm, Sweden, June 1, 1918 German physicist Max Planck is awarded the Nobel Prize in Physics for his work on the establishment of the theory of elementary quanta.

Manchester, England, January 3, 1919 Ernest Rutherford successfully splits an atom.

Principe, Gulf of Guinea, West Africa, May 29, 1919 Arthur Eddington photographs changes in the stars' positioning that confirm Einstein's theory of relativity.

Tucson, USA, December 1920 A.E. Douglass creates dendrochronology, scientific dating based on tree ring growth.

France, July 18, 1921 The BCG tuberculosis vaccine is developed by Albert Calmette and Camille Guérin, and used on an infant.

Toronto, Canada, July 27, 1921 Frederick Banting and Charles Best isolate the hormone insulin from the pancreases of dogs.

Stockholm, Sweden, December 10, 1921 The Nobel Prize for Physics is awarded to Albert Einstein for his services to theoretical physics.

Switzerland, December 1921 Hermann Rorschach develops his inkblot test for studying personality.

Baddeck, Nova Scotia, Canada, August 2, 1922 Alexander Graham Bell, scientist and inventor of the telephone, dies, aged 75.

Queenston, Ontario, Canada, August 1922 The world's largest hydroelectric power station is completed near Niagara Falls.

Munich, Germany, February 10, 1923 Wilhelm Röntgen, the discoverer of X-rays, dies, aged 77.

Austria, December 1923 Sigmund Freud's psychology text *The Ego and the Id* is published.

Stockholm, Sweden, December 10, 1924 Willem Einthoven is awarded the Nobel Prize in Medicine.

London, England, December 31, 1925 Surgeon Henry Souttar performs a successful operation inside the heart of a young patient, a world first.

Auburn, USA, March 16, 1926 Physicist Robert Goddard's liquid-fuelled rocket is launched successfully.

Solo, Central Java, Indonesia, August 31, 1927 Scientists believe that *Pithecanthropus erectus*, "Java Man," may be the earliest human forebear.

Germany, 1927 Physicist Werner Heisenberg develops his Uncertainty Principle, which states that an observer cannot know both the position and momentum of particles at the quantum level at any given instant.

Belgium, 1927 Georges Lemaître presents his "hypothesis of the primeval atom," later to become known as the Big Bang theory.

London, England, September 15, 1928 Bacteriologist Alexander Fleming makes a discovery while studying *Staphylococcus* bacteria; *Penicillium notatum*, a mold growing on some specimens, has killed the bacteria.

Boston, USA, October 12, 1928 The Iron Lung, a machine enabling a person to breathe, is used for the first time on a child with infantile paralysis.

Washington DC, USA, January 17, 1929 Edwin Hubble observes that all galaxies are moving away from each other by an analysis of the light spectra emitted changing wavelength (red shift).

New York, USA, December 1929 Russian-born biochemist Phoebus Levene identifies the components of DNA.

Arizona, USA, May 24, 1930 Pluto is named as the ninth planet in our solar system.

Sweden, December 12, 1930 Karl Landsteiner wins the Nobel Prize for Medicine.

New Jersey, USA, October 18, 1931 Inventor Thomas Alva Edison dies in his home, aged 84.

Alabama, USA, November 8, 1931 A new radioactive halogen, later known as astatine, is discovered by Frederick Allison.

Germany, December 1931 Max Knott and Ernst Ruska jointly invent the electron microscope.

Cambridge, England, February 27, 1932 In the journal *Nature*, James Chadwick reports the potential existence of a new subatomic particle, the neutron.

World, 1932 Vitamin C is isolated by Hungarian scientist Albert Szent-Györgyi de Nagyrápolt and his research fellow Joseph Svirbely, and independently by American scientist Charles Glen King. Szent-Györgyi later won a Nobel Prize for his discovery, though controversy still remains over whether both men deserve equal credit for the breakthrough.

USA, April 28, 1932 A yellow fever vaccine is created.

USA, December 1932 Astronomer Theodore Dunham finds there is carbon dioxide in Venus's atmosphere.

New Jersey, USA, May 1933 Astronomer Karl Jansky detects radio waves coming from the center of the galaxy.

UK, July 7, 1933 Doctors announced they have been able to isolate the influenza virus.

USA, October 31, 1933 A scientific research vessel, the *Atlantis*, finds clues that life exists in our deepest oceans.

Savoy, France, July 4, 1934 The discoverer of radium, Nobel Prize winner Marie Curie, dies aged 66 from aplastic anemia, almost certainly due to exposure to radiation.

USA, December 10, 1934 Harold Clayton Urey wins the Nobel Prize for Chemistry for his 1931 discovery of deuterium, the heavy form of hydrogen.

New York, USA, June 27, 1935 Wendell Stanley is the first to crystallize a disease virus, showing it is infectious.

California, USA, August 20, 1935 Scientists isolate Vitamin E.

USA, December 1935 Charles Richter develops a new logarithmic scale to measure earthquake intensity.

St Petersburg, USSR, February 27, 1936 The father of classical conditioning, Russian physiologist Ivan Petrovich Pavlov, dies aged 85.

Germany, February 1936 Testing begins on a rocket with a 3,300-lb (1,498-kg) thrust.

England, 1936 Mathematician Alan Turing develops the concept of an abstract computer that can be adapted to simulate the logic of any computer that could possibly be constructed (later known as a Turing Machine).

Durham, North Carolina, USA, December 1937 Surgeon J. Deryl Hart reduces the number of post-operative infectious deaths this year by using ultraviolet lamps in operating rooms.

Chicago, USA, March 1937 Cook County Hospital sets up the world's first "blood bank" to store blood taken from living donors, expected to greatly increase survival from major surgery.

Berlin, Germany, December 1938 Radiochemists Otto Hahn and Fritz Strassmann unexpectedly discover nuclear fission.

USA, December 1938 A fourth kingdom, for bacteria, has been added to the three existing kingdoms in the taxonomy of life—plants, animals, and protista.

Netherlands, March 17, 1939 Mathematician Johannes van der Corput of the University of Groningen shows that prime numbers can be infinitely progressed.

Long Island, USA, August 2, 1939 Albert Einstein recommends to President Roosevelt that the USA develop an atomic bomb.

California, USA, February 27, 1940 Martin Kamen and Samuel Ruben discover an isotope, called "carbon-14," which decays at a set rate.

Europe, February 12, 1941 Rhesus negative blood is shipped to war-torn Europe as standard supply after it was discovered recently that blood types are negative or positive; wounded soldiers need a matching type when undergoing transfusions.

Oxford, England, November 26, 1941 Howard Florey makes a high-yield penicillin that kills bacteria. It is expected to increase survival rates from surgical infections.

USA, December 1941 Plutonium is synthesized by Glenn Seaborg and other collaborators.

Chicago, USA, December 2, 1942 The Manhattan Project creates the first successful self-sustaining nuclear chain reaction in uranium.

France, June 1943 Jacques-Yves Cousteau makes his first dive with the aqualung he and Emile Gagnan invented, to a depth of 60 ft (18 m).

Sweden, December 10, 1943 The Nobel Prize for Medicine is awarded to Henrik Carl and Peter Dam for the discovery of, and research into, vitamin K.

New Jersey, USA, December 1943 A cure for tuberculosis and meningitis has been found—the antibiotic streptomycin.

New York, USA, May 9, 1944 New York Hospital establishes the world's first eye bank.

Germany, September 7, 1944 Germany uses ballistic missiles called V-2 rockets as weapons. They are far more advanced than cruise missiles or the V-1 bomb.

Sweden, December 10, 1944 Otto Hahn is awarded the Nobel Prize for Chemistry.

USA, July 16, 1945 The world's first atomic bomb is exploded at "Trinity," a test site in New Mexico. The detonation was equivalent to the explosion of around 20 kilotons of TNT.

Hiroshima, Japan, August 6, 1945 An atomic bomb is dropped on the city of Hiroshima by US air forces. Three days later, the city of Nagasaki is also bombed. The consequences prove devastating.

Sweden, December 10, 1945 The Nobel Prize for Medicine is awarded jointly to Alexander Fleming, Ernst Boris Chain, and Howard Florey.

Pennsylvania, USA, February 14, 1946 The Electronic Numerical Integrator and Calculator (ENIAC) is unveiled. It weighs a massive 30 tons (30.6 tonnes).

Buffalo, USA, October 2, 1946 A medical symposium suggests that smoking may cause lung cancer.

Raroia, Polynesia, August 7, 1947 Norwegian marine biologist Thor Heyerdahl completes his 101-day journey across the Pacific Ocean in a balsawood raft.

USA, August 29, 1947 Scientists find that plutonium fission can be used to produce power.

California, USA, October 14, 1947 Test pilot Chuck Yeager becomes the first person to break the sound barrier in flight, achieving the feat in a rocket-powered XS-1 fighter plane.

USA, April 1, 1948 A letter by George Gamow in *The Physical Review* postulates a theory that the universe was started, and many elements within it created, with an event that became known as the Big Bang.

California, USA, June 3, 1948 The Hale reflecting telescope at the Palomar Mountain Observatory commences operation.

USA, May 1, 1949 Gerard Kuiper discovers a new satellite orbiting Neptune. He also finds the atmosphere of Titan (Saturn's satellite, in 1944) and Mars (1948, carbon dioxide).

UK, May 6, 1949 EDSAC, one of the first British computers, runs its first mathematical programs.

New York, USA, March 23, 1950 The United Nations sets up the World Meteorological Organization.

England, March 28, 1949 British astronomer Fred Hoyle coins the term "Big Bang" on a BBC radio program, *The Nature of Things*. Ironically, he doesn't support the theory.

USA, March 31, 1950 *Astrophysical Journal* reports new insights into the nature of comets by astronomer Fred Whipple. He proposes that comets are "dirty snowballs," consisting of ice mixed with rock particles.

Chicago, USA, June 17, 1950 The world's first kidney transplant operation is performed by Dr Richard Lawler.

Philadelphia, USA, June 15, 1951 Dr John Mauchly and J. Presper Eckert Jr. demonstrate UNIVAC, the first commercial computer.

Mexico City, Mexico, October 15, 1951 Dr Carl Djerassi develops a synthetic oral contraceptive, later known simply as "the Pill."

Pennsylvania, USA, March 8, 1952 The first artificial heart is implanted into a patient.

Washington DC, USA, May 7, 1952 Geoffrey Dummer publishes his idea for an integrated circuit chip.

London, England, 1952 Biophysicist Rosalind Franklin produces X-ray diffraction images of DNA, making inroads to the study of DNA.

Cambridge, UK, April 25, 1953 James Watson and Francis Crick, with Maurice Wilkins, solve the mystery of reproduction in their findings of the molecular model of DNA.

UK, May 14, 1953 Scientists report they believe jet planes can damage eardrums and houses.

England, November 21, 1953 The "Piltdown Man," a skull found by Charles Dawson in Sussex in 1912 believed to be of an ancient human, is proven to be a hoax.

Iowa, USA, December 1953 Frozen sperm is used to impregnate a woman for the first time.

Groton, USA, January 21, 1954 The USS *Nautilus*, the first atomic submarine, is launched by First Lady Mamie Eisenhower.

USA, June 16, 1954 The first official flight of a vertical take-off and landing plane takes place.

Boston, USA, December 24, 1954 A kidney is transplanted from Ronald Herrick to his brother Richard.

USA, January 11, 1955 Lloyd Conover patents tetracycline, an effective broad-spectrum antibiotic.

New York, USA, January 25, 1955 An atomic clock that is accurate to within one second every 300 years is developed.

Michigan, USA, April 12, 1955 The results of the field trials of Jonas Salk's polio vaccine reveal that the vaccine is effective.

New Jersey, USA, April 18, 1955 Albert Einstein dies, aged 76. Dr Thomas Harvey performs the autopsy and takes Einstein's brain home to study.

USA, April 26, 1955 Calvin Fuller and Gerald Pearson at Bell Laboratories develop a solar cell using a tiny sliver of silicon.

USA, July 24, 1956 Ernst Brandl and Hans Margreiter are granted a patent for oral penicillin.

USA, September 25, 1956 The first transatlantic telephone cable commences functioning.

California, USA, December 1956 The Palomar sky survey, a seven-year project to photograph the entire northern sky, is completed this year.

London, England, February 28, 1957 Cancer experts express concern about the health of Australian soldiers exposed to radiation during British atomic tests in Australia.

California, USA, September 1957 Oceanographers Roger Revelle and Hans Seuss reveal that oceans cannot absorb all carbon dioxide being released, which will eventually lead to the phenomenon known as global warming.

USSR, October 4, 1957 The Soviet Union stuns the world by launching the world's first artificial satellite, *Sputnik I*, into orbit, effectively beginning the Space Age.

USSR, November 3, 1957 The Soviet Union follows *Sputnik I* with *Sputnik II*, containing a dog named Laika, the first living animal to travel into space. Laika died a few hours after launch when the cabin became too hot.

USA, 1957 William Grey Walter invents the toposcope for brain EEG topography.

USA, September 1958 A prototype of an integrated circuit is created on a piece of silicon.

Washington DC, USA, October 1, 1958 Civilian space agency NASA (National Aeronautics and Space Administration) is established.

Cleveland, USA, October 29, 1958 Dr Mason Sones makes the first diagnostic coronary angiogram.

UK, January 24, 1959 John Cockcroft and Lewis Strauss succeed in creating nuclear fusion.

USA, March 9, 1959 Radar contact is made with the planet Venus.

California, USA, March 24, 1959 Charles Townes is granted a patent for the maser, a precursor to the laser, which is used to amplify radio signals.

World, December 1, 1959 The Antarctic Treaty is signed by 12 nations, agreeing to keep the continent free from military use and to use it for scientific research.

Pacific Ocean, January 23, 1960 The US Navy bathyscaphe *Trieste* descends 35,814 ft (10,916 m) below sea level to the bottom of Challenger Deep. Fish are observed at these great depths.

Jodrell Bank, England, March 14, 1960 British radio telescope makes contact with US *Pioneer V* satellite at a record distance of 409,000 miles (658,200 km).

California, USA, May 16, 1960 Physicist Theodore Maiman builds the world's first laser.

Sweden, December 10, 1960 US chemist Willard Libby is awarded the Nobel Prize in Chemistry for use of carbon-14 in determining age in archaeology and other sciences.

Olduvai Gorge, Tanganyika, February 24, 1961 Louis Leakey finds a partial skull of *Homo erectus*, pushing the origin of hominid species back 400,000 years to one million years.

Space, April 12, 1961 Soviet cosmonaut Yuri Gagarin becomes the first human space traveler, launched into a single Earth orbit aboard his *Vostok* spacecraft.

USSR, August 7, 1961 Russia's second cosmonaut, Gherman Titov spends 24 hours orbiting Earth.

Boston, USA, July 11, 1962 A new era of communications is promised as a live television picture is generated in the US and relayed to France and Britain by the Telstar communications satellite.

USA, September, 1962 *Silent Spring* by Rachel Carson is published. The book is credited with helping to launch the environmental movement, and facilitated the ban on the pesticide DDT.

Sweden, December 11, 1962 The Nobel Prize in Medicine is awarded to Crick, Watson, and Wilkins for their discoveries of DNA structure.

Leeds, England, February 14, 1963 Surgeons at Leeds General Infirmary transplant a kidney taken from a dead man into a living patient.

Stockholm, Sweden, March 18, 1964 US mathematician Norbert Wiener, father of the science of cybernetics, dies, aged 69.

Copenhagen, Denmark, November 18, 1962 Physicist Niels Bohr, a pioneer of quantum mechanics, dies, aged 77. He won the Nobel Prize for Physics in 1922.

USA, April 1964 IBM introduces a typewriter with a magnetic tape drive; "word processing" is born.

Holmdel, USA, 1964 Astronomers Arno Allan Penzias and Robert Woodrow Wilson find evidence of cosmic background radiation, which served as important confirmation of the Big Bang theory.

USA, February 19, 1965 The measles vaccine is introduced.

Space, March 18, 1965 Soviet cosmonaut Aleksei Leonov becomes the first man to engage in a space walk from his *Voskhod 2* craft.

Atlantic Ocean, August 29, 1965 American astronauts Charles Conrad and Gordon Cooper splash down near Bermuda after orbiting Earth for just under eight days in the craft *Gemini 5*.

The Moon, February 3, 1966 The Soviet space probe *Luna 9* is the first earthly item to touch down on the lunar surface, beaming pictures back to Earth.

Cape Canaveral, USA, June 6, 1966 Spaceship *Gemini 9* splashes down safely after failing to dock with an unmanned spacecraft.

Cape Town, South Africa, December 3, 1967 The first human heart transplant is carried out by a team led by Dr Christiaan Barnard.

UK and USA, 1967 UK geophysicists McKenzie and Parker and American W.J. Morgan develop the theory of plate tectonics.

England, 1967 Jocelyn Bell Burnell and Antony Hewish discover pulsars.

Sweden, December 12, 1968 Chemists Holley, Khorana, and Nirenberg win the Nobel Prize in Medicine for their work on the genetic code and protein synthesis.

Pacific Ocean, December 27, 1968 *Apollo 8* splashes down after a successful mission, orbiting the Moon ten times. Its occupants are the first humans to see the lunar surface first hand.

Cambridge, England, February 13, 1969 At Cambridge Physiological Laboratory, human eggs taken from female volunteers are fertilized in test tubes outside the body for the first time.

The Moon, July 21, 1969 US astronaut Neil Armstrong becomes the first man to walk on the surface of the Moon.

Western world, April 1970 Western countries begin to introduce mass vaccination for children.

California, USA, June 1970 Computer "floppy disks" are developed by IBM to store data.

Space, June 1–19, 1970 Vitaly Sevastyanov and Andriyan Nikolayev conduct the longest manned space flight aboard *Soyuz 9*.

UK, July 1970 Traces of PCBs (polychlorinated biphenyls) are found in British birds, with dramatic implications for the food chain.

South Africa, July 25, 1971 Dr Christiaan Barnard performs the first combined heart–lung transplant.

The Moon, July 30, 1971 *Apollo 15* lands a four-wheeled "moon buggy"; astronauts David Scott and Jim Irwin drive 17 miles (27 km) in it.

Left: NASA's space shuttle, the orbiter *Enterprise* disengages from the plane that has transported it, in a free-flying mission in 1977. NASA saw the future of the American space program in a reusable winged space shuttle. The *Enterprise* was named after the telegenic star ship.

World, March 1972 The El Niño weather pattern is seen to reverse trade winds on the Equator.

California, USA, July 1972 Robert Metcalf, a scientist at the Xerox company, lays the foundations for computer networks when he develops the "Ethernet."

Heimaey, Iceland, January 23, 1973 Helgafell volcano erupts after being dormant for 7,000 years. The island's 5,000 inhabitants flee to the mainland.

Florida, USA, May 14, 1973 NASA launches its first manned space station, *Skylab*.

London, England, August 25, 1973 The first use of a "CAT" scan heralds a breakthrough in medical imaging.

California, USA, September 1974 Scientists Frank Sherwood Rowland and Mario Molina warn that continued use of aerosol sprays propelled by chlorofluorocarbon gases (or CFCs) will cause ozone depletion, leading to climate change and skin cancer.

Hadar, Ethiopia, November 24, 1974 The "Lucy" hominid skeleton is found in East Africa.

USA, 1975 The Altair 8800, a simple mail-order microcomputer, goes on sale. The designers were surprised when they sold thousands in the first month. It is the spark that led to the personal computer revolution.

Boston, USA, January 1975 From his complex computer models, Syukoro Manabe shows how the atmospheric temperature rises when carbon dioxide levels are doubled.

Venus, October 1975 Soviet space probes land on the surface of Venus and send back the first pictures of the planet's surface.

USA, April 1, 1976 Inspired by the Altair, Steve Jobs and Steve Wozniak develop the Apple I personal computer kit, founding Apple Inc.

San Francisco, USA, April 7, 1976 Genentech, the first commercial company engaged in genetic engineering, is established.

Mars, July 20, 1976 *Viking 1* lands and transmits photographs of the Martian landscape.

California, USA, February 18, 1977 The *Enterprise* space shuttle makes its first flight atop a Boeing 747 jumbo jet.

Washington DC, USA, May 11, 1977 The US government announces that chlorofluorocarbons (CFCs) will be outlawed as propellants in aerosol cans in two years' time.

South Africa, October 23, 1977 Single-celled fossils that have been dated at 3.4 billion years old are discovered, pushing back the evidence of the first life on Earth by 100 million years.

Sweden, January 1978 Sweden becomes the first nation to legislate to ban aerosol sprays using chlorofluorocarbon gases as propellants.

Ethiopia, February 24, 1978 Mary Leakey finds footprints dated at 3.5 million years old. They are thought to belong to a bipedal hominid.

USA, June 22, 1978 The only moon of Pluto, Charon, is discovered.

Manchester, England, July 25, 1978 Louise Brown, the first baby to be conceived using *in vitro* fertilization, is born.

USSR, August 19, 1979 Soviet cosmonauts Lyakhov and Ryumin return home from 175 days in space.

Sweden, October 11, 1979 Godfrey Hounsfield is awarded the Nobel Prize for Medicine for the invention of the full body (computed axial tomography or "CAT") scanner.

Geneva, Switzerland, December 9, 1979 The World Health Organization (WHO) declares that smallpox has been globally eradicated.

Boston, USA, January 16, 1980 Scientists synthesize Interferon—a natural virus-fighting substance—using genetic engineering.

Washington DC, USA, June, 1980 The US Supreme Court hands down a decision that General Electric can patent a microbe developed to clean up oil spills.

USSR, October 11, 1980 Two cosmonauts return to Earth after 185 days in space aboard the *Salyut* space station, a new record.

Space, October 11, 1980 The US space probe *Voyager I* passes close to Saturn.

California, USA, April 14, 1981 The space shuttle *Columbia*, the first reusable space craft, returns after orbiting Earth for two days.

England, July 7, 1981 The first solar-powered aircraft crosses the English Channel.

New York, USA, August 1981 IBM releases the personal computer, combining the functionality of earlier machines in a sleeker package.

USA, December 1981 Doctors identify a new serious disease: Acquired Immune Deficiency Syndrome (AIDS).

Venus, March 1, 1982 The Soviet space probe *Venera 13*, lands on the surface of Venus, sending back the first pictures of the planet.

Salt Lake City, USA, December 1, 1982 Barney Clark, 61, is the first recipient of an artificial heart. Doctors replace two diseased ventricles.

Utah, USA, March 23, 1983 Barney Clark, the first recipient of an artificial heart, dies from failure of all his vital organs, except his heart.

USA, October 17, 1983 The EPA and National Academy of Sciences publish reports predicting catastrophic effects from global warming.

Space, February 7, 1984 The first untethered space walk takes place from the space shuttle *Columbia*, with the aid of a jet-propelled backpack.

France/USA, April 23, 1984 Researchers discover a virus that results in Acquired Immune Deficiency Syndrome (AIDS).

England, September 15, 1984 Alec Jeffreys stumbles upon genetic "fingerprinting"—DNA sequences unique to individuals that can be used for identification purposes.

Melbourne, Australia, February 10, 1985 A medical research team identifies the cause of Down Syndrome.

Antarctica, September 1985 A hole in the ozone layer is discovered.

Cape Canaveral, USA, January 28, 1986 The space shuttle *Challenger* explodes one minute after take off.

Space, March 6, 1986 The Soviet *Vega 1* space probe comes within 5,500 miles (8,800 km) of Halley's Comet.

Montreal, Canada, September 16, 1987 More than 70 nations pledge to limit their use of CFCs, which have been linked to the depletion of the ozone layer.

Arkalyk, USSR, December 29, 1987 Cosmonaut Yuri Romanenko returns to Earth after 327 days aboard the space station *Mir*, orbiting the planet.

USA, June 23, 1988 James Hanson, head of the Institute of Space Studies, warns Congress that the summer's high temperatures and record-breaking drought are caused by the "Greenhouse Effect."

England, May 1989 White blood cells produced by genetic engineering are transferred into cancer patients in order to attack tumors.

Space, August 25, 1989 *Voyager II* sends the first pictures of Neptune's surface back to Earth.

USA, November 1989 Pediatric surgeon Michael Harrison successfully carries out the world's first operation on a fetus, removed from the womb for lung surgery and then returned.

Space, April 25, 1990 The Hubble Space Telescope is launched from the space shuttle *Discovery*.

Bergen, Norway, May 10, 1990 A Conference on Sustainable Development is attended by 34 countries. The central issue is climate change, with a recognition that developed countries need to take a lead in reducing emissions.

London, England, November 1990 Tim Berners-Lee publishes a proposal for the World Wide Web and follows this by implementing it the next day, writing the first web page.

Worldwide, December 1991 The World Health Organization estimates that about 40 million people will be infected with HIV/AIDS by the year 2000.

Vatican City, October 31, 1992 The 1633 Inquisition had found Galileo Galilei guilty of heresy for suggesting that the Solar System did not revolve around Earth. Today, the Vatican rather tardily admits that he was right.

Cape Canaveral, USA, August 24, 1993 Contact has been lost with the Mars Observer, launched at a cost of $1 billion 11 months ago by NASA.

Antarctica, October 1993 British scientists observe the largest ozone hole ever over the southern continent. Decreases in the ozone level were first observed in the early 1970s and the hole has become progressively larger.

Illinois, USA, April 26, 1994 The missing link of the atom, the top quark, has been located after a 20-year project. Quarks are the building blocks of matter.

Jupiter, July 22, 1994 Fragments of the Shoemaker-Levy comet crash into Jupiter, leaving prominent scars from the impacts that could be seen on Jupiter for many months after the event.

Ethiopia, August 1994 Researchers find fossil remains of *Australopithecus ramidus*, thought to be a missing link, dating to 4.5 million years.

New South Wales, Australia, 1994 A new genus and species of coniferous tree, *Wollemia nobilis*, is discovered.

Space, February 6, 1995 The US space shuttle *Discovery* completes its rendezvous with the Russian *Mir* space station.

Kazakhstan, March 22, 1995 Russian cosmonaut Valeri Polyakov returns to Earth after a record 437 days in space.

California, USA, June 23, 1995 Jonas Salk dies, aged 79.

Edinburgh, Scotland, July 5, 1996 Dolly the sheep is successfully cloned.

USA, August 17, 1996 An Antarctic meteorite, ALH84001, is found to contain microscopic fossils thought to be ancient life from Mars.

Florida, USA, November 7, 1996 NASA launches the Mars Global Surveyor.

Nairobi, Kenya, December 9, 1996 Mary Leakey dies, aged 83.

Mars, July 4, 1997 The *Pathfinder* probe touches down on Mars. Onboard is the *Sojouner* rover that will make a three-month survey of the Martian surface.

UK, September 29, 1997 British scientists have conclusively discovered a link between mad cow disease (BSE) and the human variant, Creutzfeldt-JacobDisease (CJD).

USA, March 27, 1998 Pfizer's Viagra (Sildenafil citrate), used to treat impotence and angina, is approved by the Federal Drug Administration.

Space, February 11, 1999 Pluto becomes the furthest planet from the Sun again. Its eccentric orbit means that it does not revolve evenly around the Sun, unlike the other eight planets.

Worldwide, February 12, 1999 Scientists speak publicly on the health effects of genetically modified foods.

Europe, Asia, and Africa, August 11, 1999 The longest total solar eclipse since 1927 is seen by up to 350 million people.

2000s

Dassen Island, South Africa, July 5, 2000 An oil slick threatening a penguin rookery during breeding season leads conservationists to mount an airlift for 18,000 Jackass penguins.

Punta Arenas, Chile, September 9, 2000 The ozone hole over the Antarctic is at its largest extent ever, stretching over the city of Punta Arenas for the first time.

Stockholm, Sweden, October 10, 2000 The Nobel Prize in Physics is awarded to Jack Kilby (USA) and jointly to Zhores Alferov (Russia) and Herbert Kroemer (USA), for their work on the foundation of modern information technology.

USA and England, February 12, 2001 Two separate groups of researchers publish their results in mapping the sequence of the human genome in *Science & Nature*.

Stanford University, California, USA, July 9, 2001 Scientists at Stanford's particle accelerator shed more light on the mechanism of the Big Bang, proving that matter and anti-matter decay at different rates.

Stockholm, Sweden, October 8, 2001 The Nobel Prize in Medicine is awarded to cancer researchers Leland Hartwell, Tim Hunt, and Paul Nurse for their discoveries of the key regulators of the cell cycle.

Antarctic Peninsula, January 31, 2002 The Larsen B ice shelf begins to break up—eventually 1,255 square miles (3,250 square km) of shelf disintegrates.

Brandberg Mountains, Namibia, March 28, 2002 Entomologist Oliver Zomporo discovers an insect constituting a new order—the first for 87 years.

Chad, July 11, 2002 A skull dated to between six or seven million years old is discovered. The remains are the earliest to be found with distinctly human characteristics.

Ceduna, Australia, December 4, 2002 Thousands of people descend on this tiny town to view a total solar eclipse.

Bristol, England and Brisbane, Australia, May 29, 2003 Scientists discover an entirely new class of galaxy named ultra-compact dwarfs.

Ethiopia, June 12, 2003 The oldest remains of *Homo sapiens* are found, dating back some 160,000 years.

USA, September 1, 2003 The space shuttle *Columbia* disintegrates on re-entry after on its 28th mission killing all seven crew members.

Gobi Desert, China, October 15, 2003 Yang Liwei, onboard the *Shenzhou 5*, is the first man sent into space by the space program of China.

Worldwide, November 19, 2003 The World Conservation Union releases its Red List cataloguing endangered species.

USA, 2003 The Human Genome Project, established to map the sequence of the human genome, is completed.

Flores, Indonesia, 2003 Australian and Indonesian paleoanthropologists discover the remains of very small people, dubbed "Indonesian hobbits."

Mars, January 4, 2004 The first of two rovers, the *Spirit*, lands on Mars. The second, *Opportunity*, lands three weeks later. Even by 2008, both rovers are still working and continuing their mission of exploring the red planet.

Worldwide, June 2004 The Census of Marine Life is initiated to explore the mid-water and deep-sea regions.

Mojave Desert, California, USA, October 4, 2004 *SpaceShipOne* reaches a peak altitude of 70 miles (112 km) to win the Ansari X prize, set up to initiate private space travel.

Stockholm, Sweden, October 4, 2004 Americans Richard Axel and Linda Buck are awarded the Nobel Prize in Physiology for clarification of how the olfactory system works.

Flores, Indonesia, October 27, 2004 Australian archaeologists publish their discovery of *Homo floresiensis*, thought to have lived until 12,000 years ago.

Sumatra, Indonesia, December 26, 2004 An earthquake triggers a devastating tsunami that kills over 275,000 people across Asia. Measuring 9.3 on the Richter scale, it is the deadliest in recorded history.

Kyoto, Japan, February 16, 2005 The Kyoto Protocol, an international convention on climate change, comes into force.

Salt Lake City, USA, February 16, 2005 The Omo skulls, discovered by Richard Leakey in Ethiopia in 1967 are re-dated, pushing back the dawn of *Homo sapiens* to 195,000 years.

Seoul, South Korea, August 3, 2005 Scientists successfully produce the first cloned dog, named Snuppy.

New York, USA, December, 2005 NASA's Goddard Institute for Space Studies estimates that 2005 is the warmest year on record.

Utah, January 15, 2006 The *Stardust* spacecraft successfully returns samples of a comet to Earth.

France, August 24, 2006 The International Astronomical Union (IAU) formally downgrades Pluto from an official planet to a dwarf planet, joining Eris and Ceres.

UK, October 2006 Scientists create the first ever artificial liver cells using umbilical cord blood stem cells.

Sweden, October 3, 2006 John Mather and George Smoot win the Nobel Prize for Physics for their work on cosmic microwave background radiation. Roger Kornberg's studies of the molecular basis of eukaryotic transcription earn him the Chemistry prize.

World, January 12, 2007 Comet McNaught, the brightest comet in 40 years, becomes visible during daylight.

Switzerland, April 25, 2007 Gliese 581c is discovered in the constellation Libra, and identified as a potentially habitable Earth-like extrasolar planet.

USA, September 13, 2007 The Google Luna X PRIZE is announced. The challenge is to successfully launch, land, and operate a rover on the lunar surface.

USA, October 4, 2007 Celebrations mark the 50th anniversary of the Space Age, which began on this date in 1957 with the launch into orbit of *Sputnik*, by the Soviet Union, the world's first artificial satellite.

Sweden, October 9, 2007 The Nobel Prize for Physics is awarded to Albert Fert and Peter Grünberg for their discovery of giant magneto-resistance. The Nobel Prize for Chemistry is won by Gerhard Ertl for his studies of chemical processes on solid surfaces.

World, January 2008 Human embryonic stem cell lines are generated without the destruction of the embryo.

UK, April 2, 2008 A cross human–cow embryo survives for three days after fertilization. Scientists think that the 99 percent human embryo could improve research within the field of human diseases. The Catholic Church in England said that the creation was "monstrous" and that the destruction of it was unethical.

The chemical elements

The following table lists the chemical elements in alphabetical order.
To see the arrangement of the elements in the periodic table, see page 138.

Element	Symbol	Atomic Number
Actinium	Ac	89
Aluminium (Aluminum)	Al	13
Americium	Am	95
Antimony (Stibium)	Sb	51
Argon	Ar	18
Arsenic	As	33
Astatine	At	85
Barium	Ba	56
Berkelium	Bk	97
Beryllium	Be	4
Bismuth	Bi	83
Bohrium	Bh	107
Boron	B	5
Bromine	Br	35
Cadmium	Cd	48
Caesium (Cesium)	Cs	55
Calcium	Ca	20
Californium	Cf	98
Carbon	C	6
Cerium	Ce	58
Chlorine	Cl	17
Chromium	Cr	24
Cobalt	Co	27
Copper (Cuprum)	Cu	29
Curium	Cm	96
Darmstadtium	Ds	110
Dubnium	Db	105
Dysprosium	Dy	66
Einsteinium	Es	99

Element	Symbol	Atomic Number
Erbium	Er	68
Europium	Eu	63
Fermium	Fm	100
Fluorine	F	9
Francium	Fr	87
Gadolinium	Gd	64
Gallium	Ga	31
Germanium	Ge	32
Gold (Aurum)	Au	79
Hafnium	Hf	72
Hassium	Hs	108
Helium	He	2
Holmium	Ho	67
Hydrogen	H	1
Indium	In	49
Iodine	I	53
Iridium	Ir	77
Iron (Ferrum)	Fe	26
Krypton	Kr	36
Lanthanum	La	57
Lawrencium	Lr	103
Lead (Plumbum)	Pb	82
Lithium	Li	3
Lutetium	Lu	71
Magnesium	Mg	12
Manganese	Mn	25
Meitnerium	Mt	109
Mendelevium	Md	101
Mercury (Hydrargyrum)	Hg	80

Element	Symbol	Atomic Number
Molybdenum	Mo	42
Neon	Ne	10
Neptunium	Np	93
Nickel	Ni	28
Niobium	Nb	41
Nitrogen	N	7
Nobelium	No	102
Osmium	Os	76
Oxygen	O	8
Palladium	Pd	46
Phosphorus	P	15
Platinum	Pt	78
Plutonium	Pu	94
Polonium	Po	84
Potassium (Kalium)	K	19
Praseodymium	Pr	59
Promethium	Pm	61
Protactinium	Pa	91
Radium	Ra	88
Radon	Rn	86
Rhenium	Re	75
Rhodium	Rh	45
Roentgenium	Rg	111
Rubidium	Rb	37
Ruthenium	Ru	44
Rutherfordium	Rf	104
Samarium	Sm	62
Scandium	Sc	21
Seaborgium	Sg	106

Element	Symbol	Atomic Number
Silicon	Si	14
Silver (Argentum)	Ag	47
Sodium (Natrium)	Na	11
Strontium	Sr	38
Sulfur	S	16
Tantalum	Ta	73
Technetium	Tc	43
Tellurium	Te	52
Terbium	Tb	65
Thallium	Tl	81
Thorium	Th	90
Thulium	Tm	69
Tin (Stannum)	Sn	50
Titanium	Ti	22
Tungsten (Wolfram)	W	74
Ununbium	Uub	112
Ununhexium	Uuh	116
Ununoctium	Uuo	118
Ununpentium	Uup	115
Ununquadium	Uuq	114
Ununtrium	Uut	113
Uranium	U	92
Vanadium	V	23
Xenon	Xe	54
Ytterbium	Yb	70
Yttrium	Y	39
Zinc	Zn	30
Zirconium	Zr	40

The Greek alphabet

Symbol	Name
α	Alpha
β	Beta
γ	Gamma
δ	Delta
ε	Epsilon
ζ	Zeta
η	Eta
θ	Theta
ι	Iota
κ	Kappa
λ	Lambda
μ	Mu
ν	Nu
ξ	Xi
ο	Omicron
π	Pi
ρ	Rho
σ	Sigma
τ	Tau
υ	Upsilon
φ	Phi
χ	Chi
ψ	Psi
ω	Omega

Basic mathematical symbols

Symbol	Name
=	is equal to
≠	is not equal to
+	plus
-	minus
±	plus or minus
√	square root
x	the product of (multiplication)
÷	division
≤	is less than
≥	is more than
«	much less than
»	much greater than
Δ	symmetric difference
Σ	summation sign
π	pi
∫	integral
%	percent

Measures

Measure	Symbol	Description
Angström unit	Å	A unit of length equal to one 10,000 millionth of a meter or 10^{-10} meters.
Arc, degree of	°	A measure of angular separation defined as 1⁄360 of the circumference of a circle; hence there are 360° in a circle and 1° equals 60 arc-minutes or 3,600 arc-seconds.
Arc-minutes (arcmin; minute of arc)	′	A measure of angular separation equal to 1⁄60 of a degree or 60 arc-seconds.
Arc-seconds (arcsec; second of arc)	″	A measure of angular separation equal to 1⁄3,600 of a degree or 1⁄60 of an arc-minute.
Astronomical Unit	AU	A unit of length equal to the average distance between Earth and the Sun. One Astronomical Unit equals 92,955,807 miles (149,597,870 km). Jupiter is a distance of about 5 AU from the Sun.
Declination		The angle between a star or other body and the celestial equator, and is equivalent to latitude on Earth. *See* Right ascension.
Earth mass unit		When the mass of a planet is expressed in terms of the Earth's mass (5.974×10^{24} kg). For example, Jupiter's mass is 317.8 the mass of Earth.
Flux		The amount of energy from a star or other body received at Earth in an area of 1m² in one second.
Jupiter mass unit		When the mass of a planet is expressed in terms of Jupiter's mass (1.899×10^{27} kg).
Kelvin scale	K	A temperature scale starting at absolute zero and commonly used in astrophysics.
Light year		A unit of length defined as the distance a beam of light travels in one year, equal to about 6 million million miles (10 million million km).
Magnitude		A measure of the brightness of stars and other objects. Apparent magnitude is the magnitude as measured by an observer on Earth.
Micrometer	μm	A prefix to a meter indicating 10^{-6} of a meter and often used to write down the wavelength of infrared radiation.
Nanometer	nm	A prefix to a meter indicating 10^{-9} of a meter and often used to write down the wavelength of visible light.
Parsec	pc	Abbreviation for parallax second. A unit of length equal to 3.26 light-years and defined as the distance at which the width of Earth's orbit (2 AU) would appear to cover an angle of one arc-second on the celestial sphere.
Right ascension		The angle measured along the celestial equator, starting at a zero point defined by the position of the Sun for the March 21 equinox and is equivalent to longitude on Earth. *See* Declination.
Solar luminosity unit		When the luminosity of a star is expressed as a fraction of the Sun's luminosity (3.826×10^{26} J s^{-1}).
Solar mass unit		When the mass of a star is expressed in terms of the Sun's mass (1.898×10^{30} kg). For example, a star of mass 7.8×10^{30} kg equals 4.1 solar masses.

GLOSSARY AND INDEX

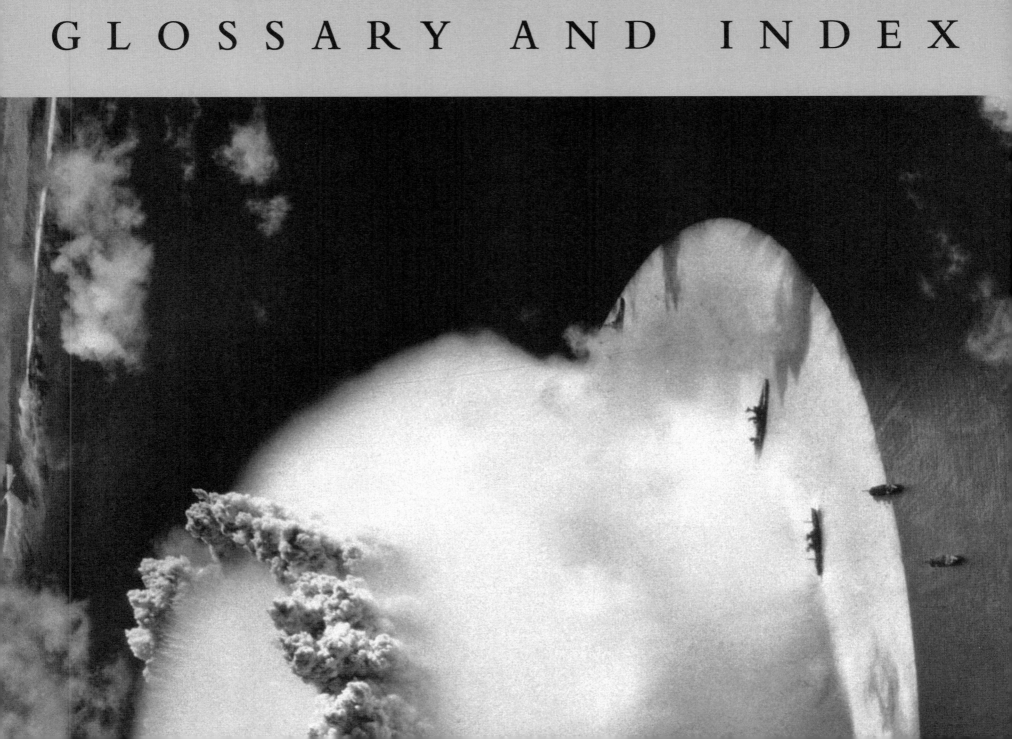

Glossary

Abacus A rectangular frame holding parallel rows of wires along which beads are moved in order to make calculations.

Absolute zero The coldest possible temperature an object can reach in the universe, occurring when all atoms in an object stop vibrating (that is, heat is caused by atoms vibrating). Absolute zero is equal to 0° Kelvin, -273.16° Celsius, or –459.69° Fahrenheit.

Absorption line A dark line or linear dip in a spectrum caused when electromagnetic radiation passing through a gas is selectively absorbed at a specific wavelength by a known type of atom. The presence of absorption lines in the spectrum of a star can tell the type of atoms in its atmosphere.

Acceleration The rate of change of velocity with respect to time.

Acid A substance that releases hydrogen ions when dissolved in water; a substance that neutralizes bases.

Acquired immunodeficiency syndrome (AIDS) A disease caused by the human immunodeficiency virus (HIV) in which the body loses its cellular immunity, thus making it highly susceptible to infection.

Active galaxy A galaxy with an abnormally bright central core. The power required to maintain the brightness of an active galaxy is thought to be provided by the infall of matter from an accretion disk into a massive black hole.

Active optics A mechanism that monitors and maintains the correct shape of a telescope's mirror when observations are being made.

Adaptive optics A mechanism that rapidly distorts the shape of a telescope's mirror based on corrections from a reference star, allowing it to produce the sharpest possible images by removing the blurring due to atmospheric distortions.

Left: A prism can create a dazzling rainbow from a single beam of light. A prism is a transparent object, usually made of glass with a triangular cross-section, which separates the white light into its spectrum of colors through a process known as refraction.

Previous spread: An aerial view of an atomic explosion during the 1940s. An atomic bomb gets its massive explosive force from the release of atomic energy through the splitting, or fission, of heavy nuclei.

Addiction The continued and uncontrolled use of a substance despite it causing severe physical, psychological, or social consequences.

Adrenal gland An endocrine gland near the front of each kidney. Secretes steroid hormones and adrenalin.

Adsorption The adhesion of a very thin layer of molecules of gases or liquid to the surface of a substance.

Afferent Carrying toward a central organ, as in a vein that carries blood toward the heart. Opposite to efferent.

Agar plate A sterile round plastic dish containing growth media used to grow organisms for the purpose of diagnostic testing.

AIDS See Acquired immunodeficiency syndrome (AIDS).

Algebra A system of mathematics for solving problems, in which variables, those numbers that can vary or are unknown, are given symbols or letters of the alphabet, such as x and y.

Algorithm A step-by-step set of rules, a systematic mathematical procedure, designed to solve a problem; used in computer programming.

Alkali Commonly used to mean a base, however can refer more specifically to a base containing elements from the first two columns (groups) in the periodic table.

Allele A variation of a specific gene, which is located at a specific site on a chromosome. Organisms contain two alleles for each gene, one from each parent.

Alloy A mixture of either two or more metals, or of metals and some other material.

Alluvial fans Depositions of river or stream sediments that have the shape of an open fan. They can be found at the mouths of rivers and streams that flow from a mountain valley out onto a flat plain.

Alveoli Small sacs in the lung that allow oxygen to enter the blood and carbon dioxide to be expelled into the air.

Ameba A single-celled organism that moves by extending projections called pseudopods. Amebas feed by engulfing their prey.

Amino acid An organic acid that is the chief component of proteins; particular proteins have a characteristic set of amino acids in a particular order. Twenty different amino acids are used in the proteins of living organisms.

Anabolism A process by which simpler substances are constructed into more complex substances that are responsible for growth.

Analgesic A drug that relieves pain.

Andesite A dark colored, fine-grained extrusive igneous rock of volcanic origin.

Anemia A condition where the blood is deficient in red blood cells, hemoglobin, or total volume. This results in pallor and a lack of energy.

Anesthesia The absence of sensitivity to pain, usually induced by the administration of certain drugs before surgery.

Anesthetic A drug that causes loss of sensation, with or without loss of consciousness.

Angina A disease marked by spasmodic attacks of chest pain or shortness of breath, occurring with arterial blockage.

Antenna A device used to transmit or receive electromagnetic radiation, like a radio telescope or a car aerial that receives radio waves.

Antibodies Proteins in the blood that respond to antigens. Antibodies protect the body by fighting foreign bodies such as viruses.

Antigen A foreign substance that prompts an immune response in the body, for example the production of antibodies.

Antimatter Matter made from particles with the same mass as normal matter but with opposite properties like charge. For a proton the antimatter particle is the antiproton and for the electron it's the positron.

Anxiolytic A drug for the treatment of anxiety, that is, a tranquilizer.

Aorta Large artery carrying blood away from the heart.

Aperture The width of a telescope's main mirror, lens, or dish antenna.

Aphelion The most distant point in the orbit of an object around the Sun.

Apoptosis A form of "programmed cell death," in which a cell undergoes an ordered sequence of events that lead to the death of the cell, such as occurs during growth and development of multicellular organisms as a part of normal cell ageing, or as a response to cellular injury.

Aquifer Any rock that contains a large quantity of saturated and permeable material so that it can conduct and hold groundwater and, if drilled, will yield significant quantities of water for human consumption.

Arteriole A small branch of an artery connecting it with capillaries.

Artery Any of the branching blood vessels that carry blood from the heart to the lungs and through the body.

Asteroid A small rocky body that orbits the Sun, ranging in size from hundreds of miles/kilometers to less than one mile/kilometer. Many are located in the asteroid belt and are also known as minor planets or planetoids.

Asteroid Belt A donut-shaped belt of asteroids located between the planets Mars and Jupiter.

Astrology The thoroughly discredited practice of predicting the future or character of a person based on the position of planets or the Sun against stars light years away, with the predictions being no more accurate than random change.

Astrophysics The study of the properties and behaviors of galaxies, stars, planets, and similar objects, and the intervening space between them.

Atherosclerosis A disease where arteries are clogged by hardened fatty deposits on their inner lining.

Atmosphere A gaseous envelope covering the surface of a star, planet, or moon and held in place by its gravity. Saturn's moon, Titan, and Earth both have an atmosphere.

Atom The basic building block of all matter. Most of the mass is in the central compact nucleus made of protons and neutrons, and surrounded by a cloud of electrons.

Atomic force microscopy (AFM) A high-resolution microscopic instrument, used to analyze and visualize surface features of materials, down to the atomic level.

Autoimmune disease A disease, such as multiple sclerosis, where the body's immune cells attack healthy tissue.

Autopsy A post-mortem examination of a body in order to determine the cause of death.

Auroras (polar lights) A phenomenon seen as dancing ribbons of colored lights high in Earth's atmosphere, occurring mostly within 20° of the North and South poles. To the north, it is called the Aurora Borealis and to the south, the Aurora Australis.

Avogadro constant The number of particles that defines one mole, approximately 6.0221 x 1023 mol–1. It is based on the number of carbon atoms in exactly 12 grams of pure carbon.

Axial In mathematics, along the same axis or centerline. In anatomy, belonging to the axis of the body.

Bacteria Unicellular microorganisms with a prokaryotic cell structure.

Barnard objects An object from the list of 349 dark objects or nebulae mostly located along the Milky Way and first published in 1927 by E. E. Barnard. For example, Barnard object 33 is the Horsehead Nebula.

Basalt A volcanic, dark-colored, mafic igneous rock composed primarily of clinopyroxene and calcic plagioclase feldspar, most commonly extrusive but also locally intrusive, as in the formation of dikes. Basalt is the fine-grained equivalent of the igneous rock gabbro.

Base A substance that can neutralize an acid by combining with hydrogen ions in water.

Batholith Large to very large, often discordant plutonic mass having a surface are of at least 39 square miles (100 km^2) and no known depth.

Beaufort scale A 12-point scale used to measure wind speed, based on observation of the wind's effects; below 1 is "calm" and 12 is a hurricane.

Big Bang A generally accepted theory explaining how the universe began and evolved to its current state. In this theory all the matter and energy in the universe originated in an explosion about 15 billion years ago.

Binary code A computer code using two discrete characters, 0 and 1.

Binary star A pair of stars in orbit around each other and held together by their mutual gravitational attraction.

Binomial In mathematics, particularly algebra, a term consisting of the sum or difference of two monomials. See also Monomial.

Biochemistry A branch of science that examines the chemical processes of cellular components of living organisms, such as proteins.

Biodiversity The variability among living things, including diversity within species, between species, and between ecosystems themselves.

Biogenesis The process of life forms producing life forms of the same type.

Biogenic Said of anything formed by biological processes. Rocks such as coal and shelly limestone are said to be biogenic rocks.

Bioluminescence The chemical emission of light from living creatures, such as some marine animals living in the ocean's depths.

Biome A climatically bounded type of vegetation that dominates and characterizes a particular geographic area.

Biopsy The removal of a small sample of cells or tissues (incisional) for testing or entire masses of suspicious tissues (excisional).

Biosphere Regions of land, air, and water in which life exists.

Bit Short for "binary digit," a single digit of binary notation, either a 0 or a 1. Also the smallest unit of information used by a computer.

Black dwarf A cold dead star that emits no radiation, making it appear black, and which forms when a white dwarf leaks all of its heat into space.

Black hole Formed when a massive star collapses, concentrating most of the mass into a compact object with a gravity that's so strong light cannot escape, making it look like a black ball in space.

Black smokers Hydrothermal vents that appear as chimneys along the deep sea floor near spreading zones or centers that eject hot water, hydrogen sulfide, and other gases. The hydrogen sulfide forms sulfide precipitates, making the vents appear to exhale black smoke.

Blastocyst Early form in the development of an embryo; a human blastocyst contains 70 to 100 cells. It is preceded by a zygote and suceeded by an embryo.

Bolide A bright meteor that looks like a ball of fire and can appear to explode.

Botany The scientific study of plants.

Botulism Serious and sometimes fatal food poisoning caused by botulin in food, usually from incompletely sterilized food preserved in an airtight container. Botulin is a toxin produced by the bacteria *Clostridium botulinum*.

Bronchus One of the two main branches of the windpipe that lead to the lungs.

Bronchiole A minute, thin-walled branch of a bronchus.

Brønsted-Lowry acid A compound that donates a proton.

Brønsted-Lowry base A compound that accepts a proton.

Brown dwarfs A type of small faint star that glows due to its own internal heat and has a mass that is to small to allow nuclear fusion to occur in its core. Unlike a normal star, where nuclear fusion maintains the heat, a brown dwarf is not able to compensate for losing its heat into space.

Buttes Obvious, often isolated, and generally flat-topped hills or small mountains, buttes stand out from their surrounding topography. They have steep sides and top layer capped by erosion-resistant rock.

Calcite A trigonal mineral with the chemical formula $CaCO_3$. Calcite, the carbonate, is a primary constituent of limestone and marble.

Calculus A branch of mathematics that looks at change and motion and how alterations in one quantity can affect another, related quantity. It is divided into differential calculus, which is applied to functions and phenomena in order to determine rates of change and maximum and minimum values; and integral calculus, which primarily deals with areas and volumes.

Caldera A more-or-less circular, large basin-shaped volcanic depression containing one or more volcanic vents.

Caldera lake A crater lake formed in a caldera.

Capillary A small blood vessel connecting small arteries with small veins. Capillaries enable the interchange of water, oxygen, carbon dioxide, and many other nutrient and waste substances between blood and surrounding tissues.

Catalysis The acceleration of a chemical reaction by means of a substance called a catalyst, which is added to the reaction but remains unchanged after the reaction has occurred.

Catalyst A substance that is added to a chemical reaction in order increase the rate of a reaction, but which is not itself consumed during the process.

CAT scan See Computed tomography (CT).

Catadioptric telescope A type of reflecting telescope that places a correcting lens in front of the primary mirror to achieve a sharper focus.

Celestial navigation Using the stars and other celestial bodies to determine one's position. Used by sailors in ancient and medieval times to cross the seas.

Cell The smallest biological unit of an organism, consisting of a cell nucleus and cytoplasm contained in a membrane. All known organisms are made up of one or more cells.

Cell division The division of one cell into two daughter cells, containing the same genetic material. Also known as mitosis.

Cell nucleus The central controlling body within a living cell that contains the genetic material and commands for maintaining life systems, growth, and reproduction of the organism.

Cenotes Natural, steep-walled water wells that extend below the water table, and are often the result of a collapsed cave ceiling.

Central nervous system The part of the nervous system that in vertebrates consists of the brain and spinal cord, and coordinates the activity of the entire nervous system.

Cerebrovascular disease A group of brain dysfunctions related to diseases of blood vessels supplying the brain.

Chaos theory A branch of mathematics concerned with systems sensitive to minor alterations, such that small changes can have great consequences.

Chemical bond A force of attraction that holds atoms together in a molecule or crystal.

Chlorophyll Green pigment molecules present in green plants and cyanobacteria used for absorption of light to provide energy.

Chromosome A threadlike structure containing DNA and carrying genetic, therefore hereditary, information.

Chyle A milky fluid that contains lymph and emulsified fats, formed during intestinal absorption of fats.

Chyme A semi-fluid mass of partly digested food expelled by the stomach into the intestines, as part of the digestive process.

Cilia Tiny, hair-like projections from a cell. In some cases they are capable of a lashing movement that produces locomotion in some single-celled organisms.

Circum-Pacific geothermal zone More commonly know as the Pacific Ring of Fire, the Circum-Pacific geothermal zone is a large area known for volcanic activity that manifests itself in the form of frequent earthquakes and volcanic eruptions encircling the basin of the Pacific Ocean. It is also sometimes called the Circum-Pacific seismic belt or more simply the Circum-Pacific belt.

Cirques Resulting from erosion by glaciers, cirques are steep-walled, deep, semi-circular or crescent-shaped depressions found most commonly at the head of mountainous glacial valleys.

Coefficient A number placed before a variable (for example, x) in an algebraic statement, thus multiplying it.

Cold-water geysers Due to a build up of gas pressure, primarily from carbon dioxide dissolved in the water, drilled cold-water wells sometimes erupt like geysers. Although these are not true geysers, they are often called cold-water geysers. The best known cold water geyser is Crystal Geyser near Green River in the State of Utah, United States.

Coma 1. The bright ball-shaped envelope of gas, typically about 93,200 miles (150,000 km) wide, that surrounds a comet's nucleus. 2. A state of deep unconsciousness that lasts for a long period of time.

Comet A small icy object made of frozen gas and dust that orbits the Sun. As a comet approaches the Sun it starts to melt, producing a long tail. The three main parts of a comet are the nucleus, coma, and tail.

Compound A substance formed by combining two (or more) different elements.

Computed tomography (CT) A specialized imaging method used commonly in medicine to create a three-dimensional image of inside the human body. Also known as CAT scan.

Cone geysers Any geyser that erupts from a mound or cone of siliceous sinter known as geyserite is a cone geyser. Cone geysers are one of two distinct types of geysers, the other being fountain geysers. Cone geysers usually erupt in steady jets that can last from only a few seconds to several minutes. Old Faithful in Yellowstone National Park in the United States is probably the best-known cone geyser.

Cone karst landscapes A type of karst topography or landscape, identifiable by the presence of stellate depressions at the bases of many steep-sided, cone-shaped hills. Cone karst landscapes are most common in the tropics.

Conjunction When the angle between a planet and the Sun, Moon, or another planet on the celestial sphere is at its smallest. A common conjunction is between the Moon and Venus.

Constellation A grouping of stars seen in the night sky that resemble an animal, person, or other object. There are 88 recognized constellations.

Continental crust The part of Earth's crust that underlies the continents and the continental shelves. It ranges in thickness from about 15 to 43 miles (25 to 70 km), with an average thickness of about 25 miles (40 km).

Convection The transfer of heat through a fluid (gas or liquid) by means of motion of the fluid resulting from the change in density with temperature.

Convection currents The transportation of heat energy by flowing groundwaters or by vertical movements within Earth's atmosphere.

Core The innermost region or nucleus of Earth's interior, which is thought to consist of two concentric spheres known as the inner core and the outer core. The outer core, which is approximately 1,350 miles (2,200 km) thick, is thought to consist of molten metal (possibly iron and nickel, as well as others); while the inner core, which is about 1,600 miles (2,600 km) in diameter, is thought to be solid and composed primarily of iron.

Corona The outermost part of a star's—including the Sun—gaseous atmosphere that extends for millions of miles/kilometers into space and reaches temperatures of millions of degrees.

Coronal mass ejection An energetic eruption of material from the solar corona that travels outward into space at high speed.

Coronary artery Either of the two arteries that arise from the aorta and supply blood to the tissues of the heart itself.

Cosmic background radiation A faint glow of electromagnetic radiation that fills the universe and has an almost uniform intensity over the whole celestial sphere. The source of the emission is thought to be the leftover glow of radiation from the Big Bang. As the radiation is strongest at microwave wavelength it is also known as the cosmic microwave background radiation.

Cosmic microwave background radiation see Cosmic background radiation.

Cosmic rays Atomic nuclei, mostly protons, moving at almost the speed of light through space and continually bombarding Earth's atmosphere from all directions.

Cosmology The study of our universe that seeks to understand its origin, how it has evolved to the present day, how it will change in the future, and how matter is distributed through it.

Covalent bond A chemical bond between two atoms that share one or more electrons.

Crater A bowl-shaped depression in the surface of a planet or moon caused by the impact of an asteroid or meteoroid.

Crescent A phase of Mercury, Venus, or the Moon with a banana-like appearance.

Crust The brittle, outermost layer or shell of Earth, from which the ocean floor and continents are formed. The crust ranges in thickness from about 3 to 6 miles (5 to 10 km) on the deep ocean floor to about 25 to 45 miles (40 to 70 km) under the surface-exposed continents.

Crustal deformation The warping, cracking, uplifting, twisting, crushing, melting, and bending of the various rocks forming Earth's crust, as demonstrated in the processes of mountain building, caused by tectonic pressures.

Cuesta Any hill or ridge having a gentle slope on one side and a steep slope on the other is known as a cuesta.

Cryptography The study of keeping data secret, by which plain information is transformed into an unintelligible form, or unintelligible data is decoded into plain language.

Crystal A solid substance made of atoms or molecules with a geometrically regular atomic structure.

Cyanobacteria Simple prokaryote cells with no nucleus.

Cytology The study of the structure, function, and life cycles of cells.

Cytoplasm The contents of a cell excluding the nucleus.

Dark energy A repulsive force that works against gravity to make the universe expand at a faster rate.

Dark matter Matter in the universe that is too faint to be seen by a telescope. Astronomers infer its existence by the way its gravity attracts visible matter.

Dark nebula Appears as a dark cloudy region against the bright Milky Way glow and is simply a large cloud of dust and gas blocking out the light from the distant Milky Way.

Deoxyribonucleic acid See DNA.

Dependent variable A mathematical variable whose value is determined by another.

Right: A high-angle view of the John Day Hydroelectric Dam, spanning the Columbia River in Oregon, USA. Hydroelectric dams make use of falling water to turn large turbines, which convert the water's energy into mechanical energy. A generator then converts this mechanical energy into electricity.

Depression A clinical psychology term used to summarize the general feelings of lowered mood, sadness, or loss of pleasure.

Dermis Skin layer below the epidermis. It contains blood vessels, nerves, sweat glands and the hair roots.

Desalination The chemical process of removing salt from seawater.

Diabetes A collection of disorders in which the metabolism is altered, and sugar levels are far higher than acceptable, resulting in unpleasant side effects, such as dizziness, for life.

Differential equation A mathematical equation that deals with how quantities vary; often quite difficult to solve.

Differentiate Process where a cell changes into a more specialized cell type with a different function and appearance.

Diffraction The interference of a wave with itself, which causes the wave to bend or spread out, especially after it encounters a small obstacle.

Digestion The process by which food is broken down into substances that can be utilized by the body.

Dike An igneous intrusion of generally tabular form that cuts across bedding planes and other recognizable zones of surrounding country rock.

Disk herniation A rupture of the fibrocartilage surrounding an intervertebral disk. This releases the nucleus pulposus that cushions the vertebrae, and presses on the spinal nerve roots.

Disinfection The removal of large numbers or specific types of microorganisms from an object. Unlike sterilization, disinfection does not guarantee their complete removal, and is a much less certain process.

DNA Deoxyribonucleic acid. The main constituent of chromosomes, it is the molecule containing genetic information.

DNA sequencing The method for determining the order of nucleotide bases in a DNA molecule.

Doline or sinkhole A solution-formed hollow in limestone that is funnel- or basin-shaped. Dolines range in depth from a few to several hundred feet (meters), and in width from a few to a thousand feet (meters) or more.

Doppler effect The increase or decrease in the wavelength of light from a star caused by the speed it moves at away from or toward Earth.

Double star
A pair of stars on the celestial sphere located within a few arc seconds of each other. The two types are 1. Optical double;
two stars appear to be close together but are at different distances. 2. Binary star; See binary star.

Ductile Able to stretched into various shapes.

Dunes A dune is for all practical purposes a hill made of sand formed by aeolian processes. Dunes can exhibit different sizes and forms as a direct result of their interaction with the wind. Most dunes are longer on the windward side where the sand is pushed up the dune by the wind, and are shorter on the opposing face since the crest of the dune blocks the wind.

Dwarf planet Defined by the IAU in 2006 as any solar system body that orbits the Sun, has enough mass to give it a nearly round shape, is not a moon of another object, and has not "cleared the neighborhood" of its own orbit. The three recognized dwarf planets are Ceres, Pluto, and Eris.

Dwarf star Any star located along the main sequence of the Hertzsprung-Russell diagram, including the Sun.

Dynamics The study of the motion and other possible changes of any physical system, which it seeks to understand in terms of the effects of other objects it interacts with. Its goal is to be able to make testable predictions about the behavior of the system in any given situation.

Eclipse When one body is partly or totally obscured by another body passing in front of it, like a solar eclipse.

Ecliptic A line drawn on the celestial sphere that is equal to the yearly path the Sun takes across the sky or where the plane of Earth's orbit is extended to meet the celestial sphere.

Ecology The study of the relationships and interactions between living organisms and the environment.

Ecosystem A complex system of interactions between a community of living organisms and their non-living environment within a given area.

Edema An increase of fluid in an organ causing swelling.

Efferent Carrying away from a central organ. Opposite to afferent.

Electrocardiogram A graphic record of the electrical activity of the myocardium. It detects the transmission of cardiac impulse through the muscle tissues of the heart.

Electrolyte An electrically charged atom or compound necessary for normal functioning of cells, for example sodium.

Electromagnetic radiation Energy that moves through space in the form of a wave with electric and magnetic components, and can also be described as packets or particles of energy, called photons.

Electromagnetic spectrum The range of wavelengths of electromagnetic radiation that is emitted by stars and other bodies. The spectrum is divided into radio waves, infrared radiation, visible light, ultraviolet radiation, X-rays, and gamma rays.

Electromagnetism The energy produced from a coupled electrical and magnetic field.

Electron A sub-atomic particle, negative in charge.

Element A substance that cannot be broken down into simpler materials by means of chemical reactions.

Elliptical galaxy A type of galaxy defined by an elliptical or oval bulge of stars and absence of a flat disk.

Embryo An unborn animal less developed than a fetus. In humans, an embryo is from the moment of fertilization until the end of the eighth week, when it is instead called a fetus.

Embryology The study of the development of embryos.

Endocrine gland
Secretes hormones directly into the bloodstream. These glands are part of the endocrine system, such as the adrenal glands, the pituitary gland, the pancreas, the testes, and the thyroid gland.

Endocrinology The study of the endocrine gland and system.

Endometrium The mucus membrane lining the uterus.

Endothermic A chemical reaction that absorbs heat.

Entomology The scientific study of insects.

Entropy In a closed system, a measure of the amount of thermal energy not available to do work. Can be thought of as a measure of the disorder or randomness in a closed system.

Enzyme A substance that acts as a catalyst for specific biochemical reactions.

Epidemiology A branch of medicine concerned with the causes, distribution, and control of diseases.

Epidermis The outermost skin layer.

Equinox When the Sun is directly above Earth's equator, occurring around March 21 and September 23. On these days, the length of day and night is nearly equal and the Sun crosses over the celestial equator.

Eukaryotes Describes cells that have their DNA enclosed in a nucleus and contain other membrane-bound organelles. Includes animals, plants, and fungi.

Eustatic changes Worldwide changes in sea-level that affect all the oceans. Generally caused in more recent times by the removal and melting of ice from the frozen polar regions.

Exothermic A chemical reaction marked by the release of heat.

Exponential In mathematics, the functions of the type $f(x) = a^x$. The indice x is the exponent. The function e^x (e is approximately equal to 2.7183) is known as the exponential function—a powerful tool in situations involving natural growth and decay.

Extrapolation Finding an approximation that lies outside the range of known values; relying upon continuing the trend of the known values.

Fenglin or "forest of stone" peaks Karst towers rising above an intervening alluvial plane and standing out prominently on the topography or terrain.

Ferrofluid A type of liquid that can take on solid-like properties by placing it near a strong magnetic field.

Fibrocartilage Cartilage that consists of a dense matrix of white collagenous fibres, such as disks in the spinal cord. It has the greatest tensile strength of all cartilage in the body.

Fjord A long, narrow, and winding, glacially-eroded inlet or arm of the sea. A fjord generally represents the seaward end of a deeply excavated glacial trough that becomes partially submerged after the melting of the glacial ice. It is also spelled fiord.

Flagellum A thread-like projection from a cell membrane. The function is usually to propel single cell or small multicellular organisms, such as spermatozoa and bacteria.

Flood plain The relatively smooth strip or surface of land adjacent to a river or stream channel, which has been naturally constructed by the present river or stream, and becomes covered with water when the river or stream overflows its banks during a flood.

Fluorescence The emission of electromagnetic radiation, usually as visible light, as a result of simultaneous absorption of shorter-wavelength radiation.

Fetus An unborn vertebrate; in humans usually from the eighth week after conception until before birth.

Fomite An inanimate object capable of carrying infectious organisms and which also serves in their transmission.

Food chain The linear representation of the feeding order in a community of organisms, with each organism gaining energy from the organism preceeding it.

Forensic science The application of science, including medicine, to legal problems.

Fractal An object that displays self-similarity at various scales.

Free radical A molecule with at least one unpaired electron. Electron-seeking molecules can break down other compounds as they snatch electrons from other molecules' atomic structures.

Functional analysis A branch of mathematics concerned with the study of spaces of functions.

Functional group Groups of atoms within molecules that are responsible for the characteristic chemical reactions of those molecules. The functional group will perform the same chemical reaction regardless of the size of the rest of the molecule.

Fungi Unicellular or multicellular organisms that lack chlorophyll. Examples include mushrooms, yeasts, and molds.

Galaxies A large grouping of slowly rotating stars, gas, and dust held together by gravity. A typical galaxy contains about 150,000 million stars. Common types of galaxies are named based on their shape—spiral galaxy, elliptical galaxy, and lenticular galaxy.

Galaxy, the Commonly called the Milky Way Galaxy, it is a spiral galaxy, about 100,000 light years wide, is home to the Sun, and contains about 200,000 million other stars.

Galilean satellites A term for the four large moons of Jupiter first seen through a small telescope by Galileo. They are name Io, Europa, Callisto, and Ganymede.

Gamete A germ cell with a single set of chromosomes that is capable of fusing with another gamete of the opposite sex to form a zygote from which a new organism will develop, for instance, a sperm and egg.

Gamma ray A form of high electromagenetic energy emitted in the radioactive decay of some unstable atomic nuclei.

Gamma-ray burst A short-lived flash of intense gamma rays.

Gas giant A term for any large planet made from gas like Jupiter, Saturn, Uranus, and Neptune.

Geiger counter A machine to measure radiation.

Gene expression The process by which information from a gene is made into RNA or a protein.

Gene pool The collective genetic variation within a population.

Gene splicing When a single gene is cut at a specific site and pasted together with another part of that gene. This process can occur multiple times and at different places in a gene. It allows multiple proteins to be made from the one gene.

General relativity A theory of gravitation developed by Albert Einstein. According to general relativity, the observed gravitational attraction between objects results from the masses of those objects warping nearby space–time; the bigger the mass, the larger the warping.

Genetic disorder A medical condition caused by abnormalities in DNA.

Genetic engineering A broad term that refers to a variety of techniques that manipulate the genes of a living cell or organism.

Genetics The discipline in biology dealing with the processes and structures involved in the heredity and variation in living organisms.

Genome The complete set of genes contained in a human cell's DNA. The genome consists of 46 chromosomes: 22 pairs plus the sex-determining chromosomes X and Y. One of each pair of chromosomes comes from an individual's mother, the other comes from the father.

Genotype The genetic configuration of an individual organism.

Geophysics The study of how Earth works, using physics.

Geothermal activity Geothermal activity is caused by the transfer of heat from deep within Earth, through the crust, to Earth's surface. Some surface manifestations of geothermal activity are hot springs, fumaroles, geysers, and volcanoes.

Geothermal gradient The rate of increase in temperature per depth of the earth. The geothermal gradient is not uniform but varies from region to region depending on the thermal conductivity of the rocks and the heat flow.

Geothermal steam vents Vents or openings in the surface of Earth's crust that allow steam to escape.

Geriatrics A branch of medicine concerned with the health of old people.

Gestation The carrying of young in the womb from conception to birth.

Gibbous A phase of Mercury, Venus, or the Moon where the visible disk is more than half, but less than fully illuminated.

Glacial epoch Any geologic time period from the Precambrian to the present day, in which Earth's climate was notably cold in both the northern and southern hemispheres.

Left: Children at a Dr Barnado's home in Hawkhurst, Kent, England, try on experimental gas masks *c.* 1940. Early gas masks were usually connected to a metal canister containing charcoal, which would absorb chemicals such as mustard gas. During World War II, the British government issued almost 40 million gas masks to the population.

Glacier A large to very large mass of ice surviving year to year, and formed over time through the compaction and recrystallization of snow. A glacier moves down-slope very slowly by creep, as a result of the stress of its own immense weight.

Gland An organ that makes substances such as hormones, often to secrete into the bloodstream.

Global warming The gradual increase in Earth's atmosphere's temperature caused by greenhouse gases.

Globular cluster A compact group of up to a few million stars that resemble a pile of sand when seen through a telescope. See star cluster.

Gluteus maximus The largest muscle of each buttock making up their shape.

Gneiss A foliated rock formed by regional metamorphism in which fine zones of granular minerals alternate with zones of minerals displaying a flaky habit.

Goethite An orthorhombic mineral with the chemical formula FeO(OH).

Gorge A deep and narrow valley with very steep rock walls, enclosed by mountains. Gorges are physically smaller than canyons.

Graben German for grave, a graben is an elongated trough or basin with both sides composed of high-angle normal faults.

Gravitation The ability of objects made of matter to attract all other objects. The strength of gravity increases with (1) increasing mass and (2) as the distance between objects decreases.

Gravity The force of attraction between two masses. The bigger the mass of an object, the stronger the force of attraction. Gravity attracts all objects to Earth, thus giving weight.

Greenhouse gas concentration The concentration of carbon dioxide and water vapor in the lower atmosphere that does not allow long-wavelength terrestrial radiant energy (heat) to escape the surrounding atmosphere.

Heisenberg's Uncertainty Principle The notion proposed by German physicist Werner Heisenberg that it is impossible to precisely measure physical attributes at the quantum scale (the scale of an atom and below). It states that an observer can know either the position and momentum of a particle at any given instant, but not both.

Helictite Usually composed of calcite or aragonite, a helictite is a thin, twig-like, curved, cave deposit with a small central canal that allows water to flow through to the end, thereby elongating the formation.

Helminth A type of parasitic worm, usually living in the digestive tract of an animal.

Herbivore An animal that eats only plants.

Heredity The genetic passing on of characteristics, either physical or mental, from generation to another.

Hertzsprung-Russell diagram (H-R diagram) A graph showing the relationship between spectral type or temperature and the absolute magnitude of stars. Stars form well-defined groups or bands, with 90 percent occupying a diagonal line called the main sequence. The diagram is of immense value in understanding the evolution of stars.

Homeostasis A state where the body's internal environment is maintained and balanced.

Hooke's Law The amount that an elastic material stretches out of shape is directly proportional to the force acting on it. Named for Robert Hooke (1635–1702) who discovered the science behind the stretching of elastic.

Hormone A substance produced by living cells that usually circulates in blood or sap and carries a signal from one cell to another.

Human papillomavirus (HPV) A virus that causes warts and tumurs in humans. Thought to cause cervical cancer.

Human immunodeficiency virus (HIV) A virus that breaks down the human body's immune system and can lead to AIDS.

Hydrothermal vents Vents or similar openings in the surface of Earth's crust through which hot water and the constituents it carries are expelled.

Hydroxyapatite An inorganic compound composed of calcium, phosphate, and hydroxide. Found in bones and teeth; gives them rigidity.

Hypothalamus Part of the brain that regulates body temperature, appetite, and other autonomic functions.

Ice cores Drill cores taken from ice sheets or glaciers that allow climatologists to study past climates and compare them to present-day conditions. They are also useful for studying the crystalline structure of the ice itself.

Igneous Any mineral or rock that solidified (crystallized) from molten or partially molten rock-forming material (magma) is referred to as igneous.

Immunization The administration of a vaccine to a person that results in resistance to a specific disease.

Immunology The biological study of the interaction of the body's immune cells with antigens; a broad term defined as anything the immune system can react against.

Immunosuppression Suppression of body's immunity, usually with drugs. Sometimes deliberate as in the case of preventing the rejection of a transplanted organ.

Independent variable In an equation of the type y = f(x), x is the independent variable (and y is the dependent variable) as the value for y depends upon the independent value assigned to x.

Inertia The property of a body that resists any change in its momentum. Equivalent to the mass of the body.

Infection An infection occurs when a species foreign to its host invades, multiplies causing illness or disease, for example, skin infection, bacterial in origin (acne).

Inflation A possible phase in the early universe after the Big Bang, when its size increased by an enormous factor over a very short period.

Influenza A highly infectious disease affecting the upper respiratory system, more serious than the common cold resulting in many unpleasant and debilitating side effects such as headaches and muscular pain, viral in origin. Potentially fatal in children if progresses to pneumonia.

Infrared
A type of electromagnetic radiation that can be felt as heat and is located between radio waves and visible light, covering a wave-length range from about one micrometer to about 1,000 micrometers.

Inner Planets A term referring to the four inner planets Mercury, Venus, Earth, and Mars.

Inorganic chemistry The branch of chemistry that deals with the chemical properties and behaviors of inorganic compounds.

Inorganic compounds Compounds excluded from the definition of organic compounds. Often containing metallic and non-metallic elements.

Intercostal muscles The muscles that run between the ribs and help form and move the chest wall.

Interferometer A method of building a large telescope by combining the signals from two or more smaller telescopes. The sharpness of the resulting image is determined by the largest separation between the smaller telescopes. Examples include the Very Large Telescope in Chile, the twin Keck Telescopes in Hawaii and the Very Large Array radio telescope in the United States.

International Astronomical Union (IAU) Founded in 1919, the organization's aims are to "promote and safeguard the science of astronomy in all its aspects through international cooperation." Its 10,000 members, from 48 countries, are professional astronomers involved in varied scientific and educational activities.

Interpolation Finding an approximation that lies between two or more known values.

Interstitial fluid The fluid that bathes and surrounds the internal cells of the body. Composed of water, salts, amino acids, proteins, sugars, and fatty acids, among other things, it is the medium through which nutrients and cellular wastes travel.

Invertebral disk A cartilaginous joint between vertebrae in the spine. These disks allow slight movement of the vertebrae and also act as a ligament to hold the vertebrae together.

***In vitro* fertilization** A medical technique involving the removal of ripe eggs from a woman's ovaries, combining the egg(s) with sperm from the father or from a donor, and allowing fertilization to take place. The fertilized embryo is then transferred back into the woman's uterus where, with luck, the embryo will implant into the wall of the uterus resulting in pregnancy.

Ion An atom missing one or more electrons, giving it a positive charge.

Ionization The conversion of an atom or molecule into ions by removing one or more electrons.

Ionizing radiation Highly energetic particles or waves, such as alpha or beta particles, that can knock free (ionize) electrons from an atom or molecule.

Irregular galaxy A type of galaxy with a distorted shape that lacks the symmetry of spiral and elliptical galaxies.

Isobars The lines on a weather map connecting points of equal atmospheric pressure.

Isomers Compounds that share the same formula yet have a different arrangement of atoms by virtue of different atoms being bonded to each other, or different spatial arrangements of the atoms.

Isotope An atom with the same number of protons in the nucleus of the atom but a different number of neutrons; a radioactive form of an element.

Joule A unit of energy, equal to the work done by a force of one newton for a distance of one meter. Named after the English physicist James Prescott Joule.

Kaposi's sarcoma A tumor caused by Human herpes virus 8. It became a defining illness for AIDS during the 1980s.

Karst A type of topography visually characterized by the presence of caves, sinkholes, and underground drainage systems that are formed in gypsum, limestone, and other rocks, primarily through dissolution.

Karstification The formation of karst topography through the chemical and mechanical actions of water on suitable rocks and minerals.

Karst tower An isolated hill surrounded by an alleviated plain in a karst region.

Kelvin scale (symbol K) A temperature scale starting at absolute zero and commonly used in astrophysics.

Kepler's Laws of Planetary Motion The astonomer Johannes Kepler's three laws explaining how the planets orbit the Sun.

Kepler's first law The orbit of each planet is elliptical or oval, with the Sun at one focus of the ellipse.

Kepler's second law A line connecting a planet to the Sun will sweep out equal areas in equal intervals of time as the planet orbits the Sun.

Kepler's third law The square of the time it takes a planet to orbit the Sun is proportional to the cube of its mean distance from the Sun.

Kinetic energy The energy possessed by an object in motion.

Kinetics A branch of chemistry concerned with measuring the rates of chemical reactions and studying the factors that affect those rates.

Kuiper Belt A belt of icy asteroids discovered in 1992 that orbit the Sun at distances between 35 AU, just beyond the orbit of Neptune, out to about 1,000 AU. Sometimes called the second asteroid belt of the Solar System.

Laccolith A concordant igneous intrusion bounded on its ends by a convex roof and a floor that is assumed, or preferably known, to be flat.

Large Magellanic Cloud (LMC) A nearby irregularly shaped galaxy in the constellation of Dorado that is 160,000 light years away and spans 8° of sky, making it the largest and brightest galaxy visible to the human eye.

Larynx The upper part of the trachea of air-breathing vertebrates, which contains the vocal cords in humans and other mammals.

Lenticular galaxy A galaxy with a large bulge and small disk, giving it the appearance of both a spiral and elliptical galaxy.

Leukocytes White blood cells. There are five types, which are lymphocytes, monocytes, neutrophils, basophils, and eosinophils.

Light microscope Optical microscope that uses visible light and lenses to magnify images of small samples.

Limestone An important sedimentary rock consisting mainly of calcium carbonate in the form of the mineral calcite. Other carbonate minerals such as magnesium carbonate (magnesite) may also be present. When limestone is subjected to metamorphism it forms another important rock, marble.

Lithium A silver-white soft metallic element, which is the lightest metal known.

Lithosphere The lithosphere is the layer or shell of Earth immediately above the asthenosphere. It is composed of the crust and part of the upper mantle and is approximately 62 miles (100 km) in thickness.

Local Group of Galaxies A group of 54 nearby galaxies that includes our Milky Way Galaxy, the Andromeda Galaxy, and the two Magellanic clouds.

Logarithm The logarithm of a number to a given base is the power to which that base is raised in order to create the number. That is the logarithm (base 10) of 100 is 2 as $10^2 = 100$.

Lopolith A large, concordant, typically layered igneous intrusion. Its floor is convex-down and the roof may be convex-down or flat.

Luminosity The total about of energy radiated by a star, galaxy, or other body, per second.

Lunar eclipse Occurs when Earth's shadow fully or partial covers the Moon.

Lymphocyte A white blood cell in lymph and blood that defends the body by immunological responses to invading matter, for instance, by producing antibodies.

Macrophage A white blood cell that ingests foreign matter and debris, and stimulates lymphocytes and other immune cells to respond to the pathogen.

Mafic Term used to describe an igneous rock that is composed primarily of one or more ferromagnesian (iron and magnesium) dark-colored minerals, such as olivine.

Magma Molten, mobile rock material below or within Earth's crust that is capable of intrusion into surrounding rocks or extrusion onto the surface. It may also contain suspended solids and gases.

Magmatism The formation of magma, its movement within Earth's mantle and crust, and its subsequent solidification to igneous rock.

Magnetism A physical phenomenon in which materials exert a force field that may attract or repulse other materials.

Magnetic Resonance Imaging (MRI) Used primarily in medicine to create a visual image of the structure of function of the body; differs from CAT scans as it is particularly useful for soft tissue visualization.

Malleable A material capable of pressed or rolled into a desired shape.

Mandelbrot Set A series of points in the complex plane, each of which is generated by an iterative process using a relatively simple equation. The Mandelbrot Set is named for its discoverer, Benoit Mandelbrot.

Mantle The layer or shell of Earth that extends from beneath the crust (Mohorovicic discontinuity) to the core. Approximately 1,800 miles (2,900 km) thick, it is divided into the upper mantle (asthenosphere and lower portion of the lithosphere) and the lower mantle (mesosphere).

Maser Abbreviation for Microwave Amplification by Stimulated Emission of Radiation. Masers operate on the same principle as lasers except they emit microwaves, a type of radio wave. Maser emission has been detected from comets, young stars, pulsars, and distant galaxies.

Mass A measure of the total amount of matter an object is made from. The unit of mass is the pound/kilogram or lb/kg. Large masses are often expressed in terms of the Sun's mass.

Mechanical energy The energy possessed by an object due to its motion and position. It is equal to the sum of the object's potential energy and kinetic energy.

Medical shock A serious medical condition where the vital organs are not receiving enough blood which could cause the patient to faint.

Melanin Naturally occurring dark pigment which gives skin and hair its color.

Mesa Bounded by steep erosion scarps on all sides, a mesa is an isolated nearly level-topped landmass standing distinctly above surrounding topography.

Mesosphere The lower mantle, or the layer or shell of Earth below the asthenosphere and next to Earth's core. The mesosphere is approximately 1,550 miles (2,500 km) thick. Very little is known about the mesosphere, but it is generally thought that it is probably not involved in Earth's tectonic processes.

Messier numbers The number of an object in a catalog of about 100 nebulae and star clusters first published in 1774, by Charles Messier. A famous object is Messier object 42, written M42, the Orion Nebula.

Metal A shiny, somewhat malleable and hard substance, often a conductor of heat and electricity. Metals form a large number of chemical elements in the periodic table; can make a metallic bond with other metal atoms.

Metamorphic rocks Rocks derived by alteration of pre-existing rocks, generally at depth in Earth's crust with very minimal or no melting involved in the transformation. The alteration is caused by dramatic increases in pressure and/or temperature, and often by changes in chemical environment as well. The process is described as contact or regional.

Metastasis A spreading of a malignant tumor to a site distant from the primary growth.

Meteor Also known as a falling star or shooting star. A streak of light seen in the night sky when a meteoroid entering Earth's upper atmosphere is heated by friction with the air causing it to completely burn up.

Meteorite A meteor that is not completely burnt up and crashes into the surface of Earth.

Meteoroids Small pieces of ice and dust from comet tails and debris from asteroid collisions. The majority of pieces are tiny particles with a mass under one ounce/gram, but can have a mass up to a few thousand pounds/kilograms.

Meteorology The scientific study of the atmosphere and the weather.

Microbial culture A method that allows microbial organisms such as bacteria, to reproduce on a petri dish, which mimics the conditions of the living organism.

Microbiology A branch of biology that focuses on the study of microorganisms.

Microwaves Part of the electromagnetic spectrum at frequencies between 1 and 300 GHz, placing them between regular radio waves and infrared waves; used in radar, industry, and communication.

Milankovitch cycle Solar energy received by any particular latitude over geologic time.

Milky Way A term describing how our Milky Way Galaxy looks in the night sky, appearing as a patchy diffuse belt of light that extends around the celestial sphere. See Galaxy, the.

Mitochondria A small rod-like organelle within the cell cytoplasm that functions in cellular metabolism and respiration. They provide cells with energy.

Mitosis The division of one cell into two daughter cells, containing the same genetic material. Also known as cell division.

Molar concentration The number of moles of solute per liter of a solution.

Mole A measure of the amount of a substance obtained by dividing the number of particles by the Avogadro constant. It is similar to counting objects, for example, by the dozen, however the Avogadro constant is very large to allow for the huge number of particles in even small amounts of any substance.

Molecule An arrangement of different types of atoms to form a single grouping with unique chemical properties.

Momentum A property of an object, equal to the product of its mass and velocity.

Monomer A molecule that joins with others to form a polymer.

Monomial In mathematics, particularly algebra, an expression that consists of just one term. See also Binomial.

Moraine Any accumulation, such as in a mound or ridge, of unsorted and un-stratified glacial till that is deposited and formed by movement of glacial ice.

Motility The ability to move spontaneously and independently. It can apply to either single-celled or multicellular organisms, such as sperm, or protozoa.

Motor neurons Neurons located in the central nervous system that directly or indirectly control the muscles.

MPEG A series of audio-visual encoding standards, developed by the Moving Picture Experts Group.

MP3 A popular encoding standard for producing compressed audio files, more formally known as MPEG-1 Layer 3.

MRI see Magnetic resonance imaging.

Multiple star A group of two or more stars held together by their mutual gravity and in orbital motion around each other.

Mutation The process of being changed; in particular a change in the chromosomes of an organism resulting in different appearance or behavior.

Mutually exclusive events Events are mutually exclusive if they share no common outcomes, for example, rolling two dice and coming up with a double (1, 1; 2, 2; 3, 3; 4, 4; 5, 5 and 6, 6) and rolling two dice with the sum of the numbers being 7, (1, 6; 2, 5; 3, 4; 4, 3; 5, 2; and 6, 1) are mutually exclusive events.

Myocardial infarction A medical condition where blood flow is interrupted to part of the heart; commonly known as a heart attack.

Myocardium A thick contractile middle layer of uniquely constructed and arranged muscle cells that forms the bulk of the heart wall.

Nanomaterials New materials, developed from the science of nanotechnology, which have properties that rely on the fact they exist on the nano-scale, nominally from 1 to 100 nm (nanometers).

Nanometers (nm) One-billionth of a meter.

Nanotechnology The science and technology of systems at a molecular level, nominally on a scale of 1 to 100 nm (nanometers) where a nanometer is one billionth of a meter (10^{-9} m).

Nebula A cloud of gas and dust in a galaxy that can be seen as (1) a dark patch against the Milky Way glow, a dark nebula, or as (2) a glowing patch of light.

Nematode A phylum of round worms, often parasitic in animals or plants, but also found freely living in soil or water.

Neurology The branch of medicine concerned with the nervous system.

Neutrino A type of elementary particle with no mass, zero charge, travels at the speed of light, and is produced at the center of all stars as a by-product of nuclear fusion.

Neutron A type of elementary particle with no charge that is found in the nucleus of all atoms except hydrogen.

Neutron star A type of compact star, about 6 miles (10 km) wide, formed when fusion stops allowing its own gravity to compress all the material in the star into neutrons.

Right: A view of Earth from the Moon. Our planet is almost 5 billion years old, having formed from the pre-solar nebula just like the other bodies in the Solar System. Earth goes round the Sun in an almost circular orbit at an average distance of 93 million miles (150 million km).

Newton's Laws of Motion Three laws developed by Isaac Newton that describe the relationship between forces acting on a body and the resulting motion of the body. The laws formed the basis for classical mechanics.

Newton A unit of force, equal to the force needed to accelerate a one kilogram mass to 1 m/s^2. Named after Isaac Newton.

Noble gases Helium, argon, neon, krypton, xenon, and radon. Unreactive elements in the right-most column of the periodic table.

Nova A class of binary star where the brightness can increase in a sudden unpredictable way by up to several magnitudes.

Nuclear fission A process where a heavy atomic nucleus splits into two lighter atomic nuclei, releasing large amounts of energy. An example of nuclear fission is an atomic bomb.

Nuclear fusion A process that powers the stars where the nuclei of two light atoms join together to form a single heavier nucleus, such as two hydrogen atoms joining to form a helium atom. This onlys happen at extremely high temperatures and pressures, such as at the center of a star.

Nuclear medicine A branch of radiology in which small amounts of radioactive substances are used to identify and treat diseases.

Nucleotide A basic structural unit of RNA and DNA, made up of a chemical base, a sugar molecule, and a phosphate group. DNA is a long stretch of nucleotides.

Nucleus (plural, nuclei) 1. Atom The central core of an atom made of protons and neutrons bound together by nuclear forces. 2. Comet A dark icy object made from ice and dust, and typically no larger than about 20 miles (30 km) wide. Material that evaporates off the nucleus forms a coma around it and a long tail.

Numerical analysis A branch of mathematics that studies algorithms (systematic mathematical procedures) for generating approximations.

Optics The science that describes the behavior, properties, and nature of light and particles that behave like light.

Orbital The region of space over which an electron is spread.

Organelles Membrane-bound structures found in eukaryotic cells.

Organic chemistry The chemistry of compounds containing carbon–hydrogen bonds and the study of properties of these compounds.

Organic compounds Compounds containing carbon, chiefly of biological origin.

Organism A living animal, plant, or single-celled life form.

Osmosis The movement of water or another solvent through a membrane, which only allows only certain substances to pass through it, into a solution of higher concentration. The process tends to equilize the concentrations on the two sides of the membrane.

Osteoarthritis A non-inflammatory form of arthritis in which joints undergo degeneration.

Osteoblasts A cell that resides in bones that creates bone tissue.

Orbit The path any object follows as it moves around the Sun, a planet, or other body.

Organic chemistry The branch of chemistry that deals with carbon compounds.

Outer planets A term referring to the four outer planets Jupiter, Saturn, Uranus, and Neptune.

Oxidation reaction A reaction in which an atom or ion loses electrons, for example an atom of sodium loses electrons to become a positive sodium ion when sodium reacts with chlorine to form sodium chloride (common salt).

Oxidized Mineral deposits modified by surface and groundwater, and gas mixtures such as air, are said to be oxidized and are found in oxidized zones. A good example would be when primary sulfides are altered to oxide and carbonate minerals.

Pandemic A disease that is widespread across the world.

Pangea Also spelled Pangaea, is the so-called supercontinent that existed during the Paleozoic and Mesozoic eras approximately 300 to 200 million years ago before each of the component continents was separated through the process of continental drift and formed their current configuration.

Parabola A two-dimensional shape in which each point is equidistant from a given point called the focus and a straight line called the directrix.

Parabolic dunes Sand dunes that are convex in the downwind direction with long scoop-shaped forms.

Paralysis The loss or partial loss of movement or sensation in a part of a body.

Parasite An organism that lives in, breeds in, and feeds from, another organism called the host, often to the host's detriment.

Particle accelerator A device that uses magnetic and electric fields to move charged particles, such as electrons. In research these devices can be used to accelerate one particle to collide with another.

Pathogen A microorganism that can cause disease.

Pathology The scientific study of the causes and effects of diseases. The word also covers the examination of body tissue samples to assist in medical diagnosis.

Periodic table A tabular arrangement of the chemical elements in order of their atomic numbers that shows a periodic variation in their properties. First developed by Dmitri Mendeleev.

Periodontal disease (periodontitis) A pathologic condition of the tissues around teeth. Usually causes inflammation to these tissues.

Peristalsis Waves of contractions along the walls of muscular structures such as the intestines, which force the contents to flow in a particular direction.

pH The concentration of hydrogen ions in a solution, determining acidity or alkinity. The scale is from 0–14, with 7 as neutral. Acids have a pH of below 7; bases have a pH above 7.

Pharynx The part of the throat that lies between the mouth cavity and the beginning of the esophagus in vertebrates.

Phase 1. Lunar Phases The appearance of the illuminated fraction of the Moon's disk as viewed from Earth and constantly changes as the Moon orbits Earth. The phase cycle starts when the Moon is at conjunction—a New Moon. As the Moon begins to move in its orbit, a small part of the disk is illuminated to produce a Waxing Crescent Moon. The illuminated section gets larger until half the Moon is illuminated, a First Quarter Moon, then increases through to a Waxing Gibbous Moon and then a Full Moon, occurring when the Moon reaches opposition and the visible disk is fully illuminated. The illuminated disk then gets smaller, going from a Waning Gibbous Moon, to a Third—or Last—Quarter Moon, to a Waning Crescent Moon and then to a New Moon to complete one cycle. 2. Mercury and Venus Like the Moon, the phase of Mercury and Venus changes as they orbit around the Sun.

Phenotype Observable characteristics of an organism, such as hair color, that result from the relationship between of its genotype with the environment.

Photoelectric effect The emission of electrons from materials when irradiated with light of an appropriate wavelength.

Photon A particle of electromagnetic radiation traveling at the speed of light. A beam of light can be described as a beam of photons, with an intensity proportional to the number of photons.

Photosphere The visible surface of a star or the Sun.

Photosynthesis The process by which green plants use the Sun's energy to make food from water and carbon dioxide.

Pigment A mineral based coloring agent that adds color to rocks and minerals. Examples include the iron oxide hematite to produce red; limonite for yellow; or the copper carbonate malachite for green. A pigment can also be referred to as a chromophore.

Placebo A medication that has no active ingredient and hence no physical effect on a patient's body. Any perceived health gain is through a positive psychological response from the patient, who does not know the medicine is a placebo.

Plain Any essentially flat surface area of a large or small size, situated at a low elevation, and having few or no prominent topographic surface features or irregularities.

Planet Defined by the International Astronomical Union in 2006 as any Solar System body that orbits the Sun, has enough mass to give it a nearly round shape and has "cleared the neighborhood" around its own orbit. The eight recognized planets are Mercury, Venus, Earth, Mars, Jupiter, Saturn, Uranus, and Neptune.

Planetary nebula An expanding cloud of gas ejected by a dying star that often looks like a circular cloudy disk.

Planktivorous Feeding on plankton.

Plaque In dentistry, an accumulation of bacteria on teeth. In medicine, accumulation of fat, usually cholesterol, on the inside walls of blood vessels.

Plasma A highly ionized gas of ions and electrons formed at high temperatures in stars or by intense ultraviolet light emitted by stars. The Sun and stars are made of plasma.

Plateau Any comparatively great flat surface area and elevation that obviously rises above the surrounding topography.

Platelets Disk-shaped cells that are essential for coagulation of blood.

Pneumonia Inflammatory condition of the lungs usually resulting from bacterial infection with side effects ranging from coughs and chest pain leading to difficulty breathing.

Polarization The production of polarized light, which are light waves moving in a restricted direction or pattern. Also can refer to the separation of positive and negative charges in a system.

Polar molecule A molecule that has slightly positively- and negatively-charged ends. Polar molecules are generally able to dissolve in water due to the polar nature of water.

Poles, celestial Two points where Earth's axis of rotation intercepts the celestial sphere.

Polymer A compound formed from a large number of small molecules (monomers) which join together.

Polymerase Chain Reaction (PCR) A rapid diagnostic test used commonly in medical Laboratories. It involves taking a small volume of biological material and replicating the DNA to multiple copies which are then analyzed looking for presence or absence of disease.

Polymorphism Variations in DNA sequence that occur in at least 1 percent of a population. These genetic variations may help predict a person's risk to a particular disease and their response to certain medication.

Positron The opposite on an electron. It has the same mass as an electron, but a positive instead of a negative one.

Positron emission tomography (PET) A nuclear medicine imaging technique in which radioactive material administered to a person releases positively charged particles (positrons). These particles bump into electrons, annihilating each other producing gamma rays. The PET camera uses sensitive crystals to pick up gamma rays and then a computer assembles the signals into a visual representation of positron activity.

Potential energy The energy stored in a physical system; for instance, an object held above the ground has potential energy.

Prefrontal cortex The part of the brain lying in the front part of the frontal lobes. This brain region is thought to be involved with personality expression and moderating correct social behavior.

Prime number A number that is not divisible by any number other than itself or 1.

Prokaryote A single-celled organism with no distinct nucleus or organelles.

Prominence An eruption of gas above the Sun's visible surface—photosphere—seen as bright arcs of gas at the edge of the Sun's disk.

Proper motion The apparent angular motion of a star on the celestial sphere specified as the number of arc seconds moved per year.

Prostaglandin A group of compounds derived from fatty acids that are found in body tissues and control various bodily processes, like the induction of labor and abortion.

Prosthesis An artificial device designed to replace a missing body part.

Protease inhibitor A type of drug used to treat or prevent infection by viruses such as HIV.

Proteasome Sub-cellular structure that digests unwanted proteins and resembles a "trashcan."

Protein Extremely complex chemical compounds found in nature which are essential components of all living cells. Proteins form an important part of the diet of humans and other animals.

Protista A biological group made up of microscopic single-celled organisms.

Proton A type of elementary particle with a positive charge that is found in the nucleus of all atoms.

DANGER EWD

Protozoan A single-celled animal. There are many different kinds, featuring different structures, habitats, and means of propulsion.

Psychoanalysis A method of treating and analyzing mental disorders through a process of communication between a patient and an analyst.

Pulsar A rapidly rotating neutron star that produces a regular pulsing emission of radiation. While the majority are radio pulsars, a smaller number of optical, X-ray, and gamma-ray pulsars have been discovered.

Pythagoras's Theorem For any right-angled triangle, the square of the hypotenuse (the longest side) is equal to the sum of the squares of the two shorter sides.

Quantum mechanics An important branch of physics that describes how the universe behaves by considering the smallest aspects of it, such as atoms and photons of light.

Quantum physics See quantum mechanics.

Quasar (quasi-stellar radio source) The most energetic type of active galaxy, with a distinct star-like appearance.

Qubit A unit of information in a quantum computer. Equivalent to a bit, except has the quantum property of superposition and so can be 0 and 1 at the same time.

Radiation Energy emitted in the form of waves or particles.

Radioactivity The emanation of radiation or particles from atoms when the nuclei fragment or drop to a lower energy state.

Radio astronomy The use of specialized radio receivers and antennas to study radio waves emitted by objects like galaxies, planets, and stars.

Radio galaxies A galaxy emitting radio waves from its central core, which can have long jets of radio emission rising above and below it.

Radiography The use of X-rays to view objects.

Left: The Deep Flight submersible, an exciting new submarine craft, piloted by its creator, inventor Graham Hawke, off the coast of Monterey, California, USA. Designed to reach the deepest parts of the ocean, Deep Flight carries only one passenger who must be in a prone position to navigate the craft.

Radioisotope A radioactive isotope.

Radiology A specialist field of medicine that uses X-rays for diagnostic purposes.

Radiotherapy Measured doses of radiation given to treat aggressive diseases such as cancer.

Reactant A material consumed during a chemical reaction.

Red dwarf A group of stars located at the lower end of the main sequence, that are smaller, less massive, and cooler than the Sun.

Red giant A star that has evolved away from the main sequence resulting in the surface temperature dropping to approximately 3,000° Kelvin as the star expands to 10 or more times its original size. In 5.5 billion years, the Sun will become a red giant.

Red supergiant A star that has evolved away from the main sequence after only a few million years, resulting in the surface temperature dropping to about 5,000° Kelvin as the star expands to 100 or more times it original size. Examples include Antares in the constellation Scorpius, and Betelgeuse in the constellation Orion.

Redshift When radiation emitted by an object is shifted toward the red end of the spectrum due to it receding from Earth. 1. Stars receding; for stars in our galaxy the redshift is a Doppler shift for stars moving away from Earth. 2. Galaxies receding; for galaxies the redshift increases with distance from Earth, a result of the Big Bang causing the universe to expand.

Reduction reaction A reaction in which an atom or ion gains electrons, for example, when positive copper ions gain electrons to become copper metal.

Reflection nebula When a nebula is visible due to light from a nearby star reflecting off it.

Refraction The bending of light as it passes between two different media, like air and glass.

Reservoir A place where anything is kept in store, like water collected for use. In microbiology, it is a site where microorganisms persist, thus being a continual source of infection.

Ribonucleic acid See RNA.

River delta Generally fan-shaped with multiple streams known as distributaries, a river delta is a landform where the mouth of a river flows into an ocean, sea, lake, estuary, or desert. The river transports built-up sediment outward into the flat area and then deposits those sediments as the current slows and loses energy.

RNA Ribonucleic acid. Nucleic acid related to protein synthesis. It carries messages from DNA in the cell's nucleus to the body of the cell.

Rotavirus A group of wheel-shaped viruses, which can cause gastroenteritis and diarrhea in children and animals.

Salinity The concentration of dissolved salts in a given amount of water.

Salt A compound composed of positive and negative ions formed by the reaction of an acid with a base.

Satellite 1. Natural satellite, also called a moon, that orbits around a planet. 2. Artificial satellite or man-made satellite launched by a rocket into orbit around Earth.

Scanning electron microscope A type of electron microscope that creates an image of a sample's surface by scanning it back and forth with a high-energy beam of electrons. The microscope can produce very detailed images of a sample surface, revealing features about 1 to 5 nanometers in size.

Scanning tunnel microscopy (STM) A device that employs a tunneling current from a conducting tip, through a sample on a conducting substrate. The electrical current change is used to visualize samples.

Scraping A process involving gently scratching the surface of the skin or nail to obtain dry biological material for testing.

Schist A group of medium-grade metamorphic rocks, chiefly notable for their contents of lamellar minerals, such as chlorite group minerals, graphite, micas, and talc. As they are defined, schists contain more than 50 percent platy and elongated minerals, sometimes finely inter-layered with feldspar and quartz.

Seyfert galaxy A type of active galaxy with strong emission lines that was discovered by Carl Seyfert in 1943.

Shield volcanoes Broad and low volcanoes in the shape of flattened domes, built by rhyolitic ash flows or basaltic lavas of low viscosity.

Shooting star Another name for a meteor.

Sidereal period The time it takes an object to complete one orbit around another, relative to the fixed stars. The sidereal period of Earth is 365.256 days.

Sidereal time A method of keeping time using the movement of the stars to measure the rotation of Earth.

Sign An objective indication of the presence of illness or disease, such as elevated blood pressure.

Small Magellanic Cloud (SMC) A nearby irregularly shaped galaxy in the constellation of Tucana that is about 190,000 light years away and spans 4° of sky—making it visible to the human eye as a cloudy patch.

Solar A term meaning of the Sun or caused by the Sun.

Solar activity A term for the various types of phenomena that change with time on the Sun and includes flares, sunspots, filaments, and prominence. The level of activity, termed the solar cycle, varies over about 11 years.

Solar eclipse When the Moon passes between Earth and Sun causing the visible disk of the Sun to be partially or totally covered by the Moon.

Solar System 1. A term describing the Sun and all the matter held in orbit by its gravitational influence, including eight planets and their moons, the dwarf planets, comets, asteroids, meteoroids, and other objects. 2. A general term referring to any star as a solar system.

Solar wind A wind-like stream of ions that flows outward from the Sun in all directions, reaching speeds of up to 560 miles (900 km) per second.

Solstices When the Sun reaches its highest and lowest point in the sky, occurring around June 21 and December 21. The longest day of the year occurs when the Sun is highest—Summer Solstice, while the shortest day is when the Sun is lowest—Winter Solstice.

Solute A substance that dissolves in a solvent to form a solution. The solute is often present in smaller quantity than the solvent.

Space–time Four-dimensional space, with three coordinates referring to physical three-dimensional space, and one coordinate referring to time.

Special Relativity A theory derived by Albert Einstein, which combines the principle of relativity (that all motion is relative with regard to an observer) with the observation that the speed of light is constant in all frames of reference. General relativity extends this theory by combining with it the effects of gravity.

Spectral type A method of classifying stars using their visible emission lines to determine the star's surface temperature or color. There are seven major types (O, B, A, F, G, K, and M), each divided into 10 numerical classes.

Spectrograph An optical instrument used to separate light into a spectrum and record it.

Spectroscopy A branch of science dealing with the measurement of the light emitted or absorbed by atoms and molecules. Spectroscopy can be used to obtain information about the structures of molecules or the amounts present.

Spectrum A display or plot of the variations in intensity of electromagnetic radiation with wavelength or frequence.

Speleothem Any secondary mineral deposit that is naturally formed in a cave, usually through the action of running or dripping mineralized water.

Spiral galaxy A type of galaxy defined by a central bulge of stars and flat disk with spiral arms. Some spiral galaxies have a barred center.

Sputum A biological sample from the upper respiratory tract, for example phlegm.

Stalactite A conical to cylindrical speleotherm or classic cave formation that develops downward from the ceiling or roof of a cave through the constant action of dripping, mineralized water.

Stalagmite A conical to cylindrical speleotherm or classic cave formation that develops upward from the floor of a cave as a result of the constant action of dripping, mineralized water.

Stapedius The smallest muscle attached to a bone in the human body. Its purpose is to stabilize the stapes, smallest bone in the body.

Stem cells Cells found in all multicellular organisms characterized by their fundamental ability to self-renew by indefinite, yet tightly controlled, divisions, and their capacity to differentiate into a variety of new cell types. Broadly categorized as embryonic stem cells or adult stem cells. Used in medical therapy to regenerate tissue.

Steppe A treeless grassland of extensive area in the semi-arid regions of southeastern Europe and Asia.

Star A massive ball of hot gas that glows brightly and is held together by its own gravity. A star generates enormous amounts of energy at its center by the conversion of hydrogen to helium, a process called nuclear fusion.

Star cloud When a region of the night sky glows due to an extremely high density of stars.

Star cluster A group of stars bound together by gravity. The two most common types are open clusters and globular clusters.

Star trails They are curved trails on photographs of the night sky. A long time exposure with a fixed camera allows Earth's rotation to move a star's image, leaving trails on the photograph.

Starburst galaxy A sudden intense burst of star formation triggered by a violent event, such as an encounter or collision with another galaxy.

Static electricity A stationary electrical charge that has built up on an insulated body, usually by friction.

Stellar A term meaning "of stars" or "composed of stars."

Sterilization The complete removal of all viruses, bacteria, and all other living organisms from an object.

Stereochemistry An area of science interested in the shapes and structures of molecules.

Stoichiometric coefficient The number that comes before a substance in a chemical equation with the convention that these values are positive for products of the reaction and negative for the reactants.

Stratosphere A layer of Earth's atmosphere that contains the ozone layer and is positioned between the troposphere and mesosphere.

Stratovolcanoes Volcanoes constructed of alternating layers of lava and pyroclastic debris, and intruded with numerous dikes and sills.

String theory The theory that says each fundamental particle is a closed loop of vibrating string. The difference between particles, such as mass and charge, are caused by the strings vibrating at different rates.

Subcutaneous tissue The layer of flesh just below the dermis containing fat, among other types of cells.

Subduction The geotectonic process of one lithospheric plate forcefully descending beneath another. Subduction drags crustal material downward into the upper mantle where it is recycled.

Sun A G5 dwarf star located at the center of our Solar System.

Sunspot A dark spot on the Sun's visible surface.

Supergiant stars See red supergiant.

Supernova The violent explosion of a star with a mass greater than about eight times that of the Sun.

Supernova remnant The expanding glowing cloud of gas left after a supernova explosion.

Super volcano A volcano of massive size and power that produces the largest and most voluminous kind of eruption on Earth. The actual potential explosivity of super volcano eruptions would vary, but the sheer volume of ejected tephra would be enough to radically alter the topography of a landscape and severely impact the global climate for years.

Surface brightness The brightness per unit area of sky of an object like a planet and often written as magnitudes per square arc-second.

Surface science The study of physical and chemical phenomena that occur at the interface (common boundary) between two phases of matter.

Surfactant A substance that can reduce surface tension of a liquid and thus is found at higher concentration at the surface of the solution than elsewhere in the solution.

Symbiotic relationship An interaction between two organisms, living in close association, usually to the benefit of both.

Synchrotron radiation Electromagnetic radiation emitted by high energy electrons moving through a magnetic field.

Syzygy When three objects are lined up in a straight line, such as the Sun, Moon, and Earth for a solar eclipse, or the Sun, Venus, and Earth for a transit.

Taxonomy The identification and naming of organisms.

Tectonic plates Sections of Earth's crust and uppermost mantle, together composing the lithosphere. Tectonic plates are approximately 60 miles (100 km) in thickness and consist mainly of oceanic crust and continental crust. The oceanic crust is mainly composed of basaltic rocks while continental crust consists principally of lower density rocks of granitic composition.

Terrestrial planets The four rocky planets Mercury, Venus, Earth, and Mars, located in the inner part of the Solar System.

Testes Male reproductive glands that produce sperm and male hormones.

Tetanus An infectious disease caused by the bacterium *Clostridium tetani* that is found in soil and the intestines and feces of animals. The bacteria produces a toxin that causes spasms of voluntary muscles, especially those of the neck and jaw.

Theory of Relativity Refers to two theories in physics developed by Albert Einstein—special relativity and general relativity.

Thermodynamics A branch of science concerned with heat and the conversion of heat to other forms of energy. Thermodynamics imposes fundamental limitations on all chemical and biological processes.

Thermophilic bacteria Bacteria that thrive at elevated temperatures.

Total eclipse When an object fully covers another object that emits light. Examples include a total solar eclipse, when the Moon covers the Sun as seen from Earth, or a total lunar eclipse, when Earth covers the Sun as seen from the Moon.

Tors High, isolated crags, peaks, pinnacles, or piles of heavily jointed rocks, exposed to intense weathering, and often having very odd shapes.

Tower karst Tropical karst characterized by steep-sided, isolated limestone hills surrounded by a flat alluvial plain.

Trachea The windpipe; a pipe that connects the larynx to the bronchi.

Transit 1. When the planets, Venus or Mercury, are observed from Earth to move across the face of the Sun. 2. When a moon or its shadow are observed to move across the face of the planet it orbits. 3. A star, the Sun, or Moon, is said to transit when it reaches the halfway point between rising and setting for any location on Earth.

Transpiration The process by which water absorbed by plants, most commonly through their root systems, is evaporated back into the atmosphere from plant surfaces.

Trigonometry A branch of mathematics concerned with the relationships between the sides and angles of triangles and their functions.

Tsunami Also called a seismic sea wave, a tsunami is a gravitational sea wave produced by any large-scale, short-duration disturbance of the ocean floor. The two main causes of tsunamis are shallow submarine earthquakes and undersea volcanic eruptions.

Ultrasound A pressure created to penetrate a medium which produces an image; medical uses.

Ultraviolet radiation Electromagnetic radiation located between visible light and X-rays, covering a wavelength range from 400 nanometers to 2 nanometers.

Umbra 1. The dark central region of a sunspot. 2. The darkest part of a shadow that is cast by an object illuminated by a light covering an area, like the Sun.

Universal Time The standard used for all time-keeping throughout the world, with all times referenced to the time at zero longitude Greenwich. Prior to 1928 it was known as Greenwich Mean Time.

Universe A word describing everything in the world that has been discovered and is waiting to be discovered.

Upwelling A process in which dense, cold deep water moves up toward the surface replacing the warm water.

Urethra The canal that carries urine from the bladder out of the body. In the male it also conveys semen.

U-shaped valleys Usually carved by glacial erosion and characterized by steep sides and a broad floor that is almost flat, U-shaped valleys have a pronounced cross profiles suggesting the shape of a broad letter "U."

Van der Waals forces A broad term for any force of attraction between molecules.

Variable star A star whose brightness varies with time. The three basic types are eclipsing, pulsating, and cataclysmic variables.

Variolation Deliberately infecting a person with smallpox (variola) in a controlled manner so as induce immunity against further infection.

Velocity The speed of an object in a particular direction.

Vein Any of the converging blood vessels that carry blood from the rest of the body back to the heart.

Vesicant A chemical that causes blisters.

Virus A submicroscopic infectious organism that often causes disease in a host organism. A virus consists of a DNA or RNA surrounded by a protein coat, and requires a living cell in order to replicate.

Volcanic Explosivity Index (VEI) Devised by the US Geological Survey and the University of Hawaii in 1982 to provide a relative measure of the explosiveness of volcanic eruptions, the VEI assigns magnitude numbers from 0 to 8 with corresponding terms from "gentle" to "mega-colossal." The 0 numerical designation is given to non-explosive eruptions, while 8 would be given to the full explosion of a super volcano.

Volcanic plug Also called a volcanic neck or lava neck, a volcanic plug is a volcanically generated landform that develops when lava or magma hardens within the vent of an active volcano.

V-shaped valley Characterized by steep sides and short tributaries, a V-shaped valley is a valley having a pronounced cross profile suggesting the shape of the letter "V."

Wavelength The distance over which a wave repeats itself. Electromagnetic radiation is a sine wave, like a wave on the ocean, making the wavelength the distance between two peaks or troughs. It is related to a wave's speed and frequency by wavelength × frequency = speed.

White dwarf A type of star that is the final step in the evolution of stars with a mass similar to the Sun's. When fusion stops in a Sun-like star it collapses under its own gravity to become a white dwarf about the size of Earth.

Wolf-Rayet star A group of unusual stars discovered in 1867 by C. J. Wolf and G. Rayet. They are extremely hot massive stars that continually generate bubbles of gas by ejecting the outer layers of their atmosphere. The star Gamma (γ) Velorum in the constellation Vela is a Wolf-Rayet star.

Xerophytes Plants that have adapted to arid climates and very dry conditions.

X-ray astronomy The study of X-rays emitted by galaxies, stars, comets, and other objects using X-ray telescopes placed in space to avoid Earth's atmosphere that blocks all X-rays.

X-ray background radiation A faint even glow over the whole celestial sphere thought to be caused by the combined X-ray emission of distant active galaxies.

X-rays High-energy electromagnetic radiation located between ultraviolet radiation and gamma rays, and covering a wavelength range from 10 nanometers to 0.005 nanometers.

Zygote A cell formed by the union of two gametes (conception).

Right: A selection of model dinosaur eyes. Paleontology is the science that concentrates on the study of fossils, providing us with a picture of the plants and animals that existed millions of years before humans evolved. Science uses observation and experimentation to make sense of our world. Indeed, the word "science" comes from the Latin word meaning "knowledge."

Index

N

O

P

Q

R

S

Credits and Acknowledgements

The Publisher would like to thank the following picture libraries and other copyright owners for permission to reproduce these images. Every attempt has been made to obtain permission for use of all images from the copyright owners, however Millennium House would be pleased to hear from copyright owners.

KEY—(t) top of page; (b) bottom of page; (l) left side of page; (r) right side of page; (c) center of page

Front cover: Getty Images, Nicholas Veasey

1 Getty Images: Dorling Kindersley, Clive Streeter; 2–3 Getty Images: UpperCut Images, Frank Bean; 4 Getty Images: Reportage, Philippe Bourseiller; 7 Getty Images: Science Faction, NASA-digital copyright Science Faction; 8–9 Getty Images: Science Faction, Jim Reed; 10–11 Getty Images: Glowimages; 12 Getty Images: Minden Pictures, Christian Ziegler; 14 Getty Images: Bridgeman Art Library, William Fettes Douglas; 16–17 Getty Images: Ian Waldie/Getty Images; 18 (t) Getty Images: Hulton Archive; 18 (l) Getty Images: Bridgeman Art Library, Joseph Wright of Derby; 18 (r) Getty Images: Photodisc, Nick Rowe; 19 (t) NASA,ESA, M. Robberto (Space Telescope Science Institute/ESA) and the Hubble Space Telescope Orion Treasury Project Team; 19 (b) Getty Images: Hulton Archive; 20 (tl) Getty Images: Bridgeman Art Library, Islamic School; 20 (tr) Getty Images: Bridgeman Art Library, Pierre Joseph Redoute; 20 (b) Getty Images: Time & Life Pictures/Getty Images; 21 (t) Getty Images: Science Faction, Dan McCoy–Rainbow; 21 (b) Getty Images: Science Faction, David Scharf; 22 Getty Images: Lonely Planet Images, Brent Winebrenner; 23 (t) Getty Images: Taxi, Kathy Collins; 23 (b) Getty Images: Stone, Mark Harmel; 24–25 Getty Images: Science Faction, Earl Zubkoff-Stock Connection; 26 (t) The Art Archive/Observatory Academy Florence Italy/Alfredo Dagli Orti; 26 (bl) Getty Images: Bridgeman Art Library, Greek; 26 (br) Getty Images: Hulton Archive; 27 (t) Getty Images: Hulton Archive; 27 (l) Getty Images: Photodisc, Don Farrall; 27 (c) Getty Images: Photonica, VEER George Diebold; 28 Getty Images: Nordic Photos, Bjorn Wiklander; 28–29 Getty Images: Photographer's Choice RF, Bryan Mullennix; 29 (t) Getty Images; 29 (c) Getty Images: Photographer's Choice, Brock Hanson; 30 (t) Getty Images: Stockbyte, John Foxx; 30 (b) Getty Images: Riser, Erik Von Weber; 31 (b) Getty Images: Roger Viollet/Getty Images; 32 (t) Getty Images: The Image Bank, Philippe Bourseiller; 32 (b) Getty Images: Hulton Archive; 33 (t) Getty Images: National Geographic/Getty Images; 33 (b) Getty Images: Science Faction, Peter Ginter; 34 (t) Getty Images: Digital Vision; 34 (l) Getty Images: Bridgeman Art Library; Enoch Seeman; 34 (b) Getty Images: Stone, David Madison; 35 Getty Images: Science Faction, Peter Ginter; 36 (c) Getty Images: Digital Vision, Yasuhide Fumoto; 36 (r) Getty Images: Hulton Archive; 36 (b) Getty Images: Reportage, Simon Roberts; 37 (t) Getty Images: Photodisc, PhotoLink; 37 (b) K. Sharon (Tel Aviv U) and E. Ofek (Caltech), ESA, NASA; 38 (t) Getty Images: Taxi, VCL/ Spencer Rowell; 38 (b) Getty Images: Time & Life Pictures/Getty Images; 39 Getty Images: Taxi, Space Frontiers; 40 (t) Getty Images; 40 (b) Getty Images: Stone, Jamey Stillings; 41 (t) Getty Images: Stone, Lester Lefkowitz; 41 (c) Getty Images: Gallo Images, Clinton Friedman; 42 (b) Getty Images: UpperCut Images, David J Turner; 42 (t) Getty Images: Digital Vision, Andre Kudyusov; 42 (b) Getty Images: Hulton Archive; 43 (c) Getty Images: Photo-grapher's Choice, Kevin Summers; 43 (b) Getty Images: Hulton Archive; 44 (t) Getty Images: PhotoAlto Agency RF Collections, Benoit Jeanneton; 44–45 Getty Images: Stone, Zane Williams; 46 (t) G. De Marchi (STScI and Univ. of Florence, Italy) and F. Paresce (STScI)/NASA, ESA; 46 (b) Getty Images: Photographer's Choice RF, Piccell; 47 Getty Images: Digital Vision, Pete Gardner; 47 (c) Getty Images; 49 Getty Images: Photographer's Choice RF, Nicholas Rigg; 50 (b) Getty Images: Gallo Images ROOTS RF collection, Steve Corner; 51 (tl) Stan Lamond; 51 (c) Getty Images: De Agostini Picture Library, DEA/I. Taborri; 52 (t) Getty Images: Taxi, Klauss; 52 (c) Getty Images: Popperfoto/Getty Images; 52 (bl) Getty Images: Time & Life Pictures/ Getty Images; 53 (t) Getty Images: Photodisc, Grant V. Faint;

53 (b) Getty Images: Riser, Brian Bailey; 54 (t) Getty Images: Photodisc, Arthur S. Aubry; 54 (b) Getty Images: Stockbyte; 55 (t) Getty Images: The Image Bank, Paul Souders; 55 (l) Getty Images: Hulton Archive, Hulton Collection; 55 (c) Getty Images: Science Faction, CMSP; 56 (t) Getty Images: Hulton Archive; 56 (b) Getty Images: Collection Mix: Subjects, Stocktrek Images; 56–57 Getty Images: Stocktrek Images; 57 (t) Getty Images: National Geographic/Getty Images; 57 (b) Getty Images: Photodisc, David Madison; 58 (b) Getty Images: Image Source, Image Source Black; 58–59 Getty Images: Science Faction, Peter Ginter; 59 (t) Getty Images: Library of Congress, digital version copyright Science Faction; 59 (c) Getty Images; 59 (b) Getty Images: Photonica, Steven Puetzer; 60 (t) Getty Images: Yoshikazu Tsuno/ AFP/Getty Images; 60 (br) Getty Images: Time & Life Pictures, Roy Stevens; 61 Getty Images: Photodisc, Stefano Stefani; 62 (t) Getty Images: Digital Vision; 62 (b) Getty Images: AFP/Getty Images; 62–63 Hulton Archive; 63 (l) Getty Images: Hulton Archive; 63 (tr) Getty Images: Time & Life Pictures/Getty Images; 64 (t) Getty Images: Hulton Archive; 64 (b) Getty Images: Science Faction, Peter Ginter; 64–65 Getty Images: Stone, Howard Kingsnorth; 65 (t) Getty Images; 65 (c) Getty Images: Time & Life Pictures/Getty Images; 65 (b) Getty Images: Science Faction, Peter Ginter; 66 (t) The Art Archive/Royal Astronomical Society/ Eileen Tweedy; 66 (b) Getty Images: fStop, Christian Thomas; 66–67 Getty Images: Stone, Gabriela Hasbun; 67 (l) Getty Images; 67 (r) Getty Images: Time & Life Pictures/Getty Images; 68 (t) Getty Images: Hulton Archive; 68 (b) Getty Images: Science Faction, Peter Ginter; 69 (tl) Getty Images; 69 (tr) Getty Images; 69 (b) Getty Images: Digital Vision; 70 (t) Getty Images: Hulton Archive; 70 (b) Getty Images: The Image Bank, Stephen Marks; 72 (t) Getty Images: Hulton Archive; 72 (c) Getty Images: Stone, Ray Massey; 72 (bl) AFP/Getty Images; 73 (t) Getty Images: Hulton Archive; 73 (b) Getty Images: Science Faction, Karen Kasmauski; 74 (c) Walter Jaffe/Leiden Observatory, Holland Ford/JHU/STScI, and NASA; 75 European Space Agency, NASA and Felix Mirabel (the French Atomic Energy Commission & the Institute for Astronomy and Space Physics/Conicet of Argentina); 76 (l) Getty Images: Hulton Archive; 76 (r) Getty Images: Hulton Archive; 77 (l) Getty Images: Science Faction, Library of Congress, digital version copyright Science Faction; 77 (b) Getty Images: Time & Life Pictures/Getty Images; 77 (r) Getty Images: Digital Vision, Roz Woodward; 78 (t) Getty Images: Hulton Archive; 78 (b) Getty Images: AFP/Getty Images; 79 (t) akg images; 79 (b) Getty Images: AFP/Getty Images; 80 (t) Getty Images: Hulton Archive; 80 (b) Getty Images: Science Faction, Peter Ginter; 81 (t) Getty Images: Science Faction, Peter Ginter; 81 (c) Getty Images: Fabrice Coffrini/AFP/Getty Images; 81 (b) Getty Images: AFP/Getty Images; 82 (t) Getty Images: Hulton Archive; 82 (b) Getty Images: Purestock; 83 (t) Getty Images: Hulton Archive; 83 Getty Images: Digital Vision; 84 Getty Images: AFP/Getty Images; 85 (c) Getty Images: O. Louis Mazzatenta/National Geographic/Getty Images; 85 (b) Getty Images: Science Faction, Louie Psihoyos; 86 (t) Getty Images: Photodisc, M. Freeman/PhotoLink; 86 (c) Getty Images: Science Faction, Peter Ginter; 86 (bl) Getty Images: Photodisc, Lawrence Lawry; 86 (2nd from l) Getty Images: PhotoAlto Agency RF Collections, PhotoAlto/Laurence Mouton; 86 (bc) Getty Images: Photodisc, StockTrek; 86 (2nd from

r) Getty Images: Stocktrek Images; 86 (br) NASA, ESA, S. Beckwith (STScI), and The Hubble Heritage Team (STScI/AURA); 88 (l) Getty Images: Photographer's Choice, Andrew Holt; 88 (br) Getty Images: Photodisc, Stuart Gregory; 89 (bl) Getty Images: Hulton Archive; 89 (c) Getty Images: Hulton Archive; 89 (b) Getty Images: Photonica, Steven Puetzer; 90–91 Getty Images: Aurora, Robb Kendrick; 92 Getty Images: Time & Life Pictures, Walter Sanders; 93 (t) Getty Images: Hulton Archive; 93 (c) Getty Images: Bridgeman Art Library; 93 (r) Getty Images: Hulton Archive; 93 (b) Getty Images: Hulton Archive; 94 (t) Getty Images: Digital Vision, Noel Hendrickson; 94 (b) Getty Images: Stockbyte; 94 (c) Getty Images: The Image Bank, Antonio M. Rosario; 95 Getty Images: Photographer's Choice RF, Tom Grill; 95 (b) Getty Images: Photodisc, Dick Luria; 96 (t) Getty Images: Bridgeman Art Library; 96 (c), (b) Getty Images: Time & Life Pictures/Getty Images; 96–97 Getty Images: Time & Life Pictures/Getty Images; 97 (t) Getty Images; 97 (b) The Art Archive; 98 (c) Getty Images: Time & Life Pictures/Getty Images; 98 (b) Getty Images: Hulton Archive; 99 (b) Getty Images: LOOK, Frank van Groen; 100 (c), (b) Getty Images: Bridgeman Art Library; 100 (r) Getty Images: Minden Pictures, Gerry Ellis; 101 Getty Images: Stockbyte, Jason Reed; 102 (t) Getty Images: Photographer's Choice, Ian Mckinnell; 102 (b) Getty Images: Digital Vision; 102 (c) Getty Images: Photodisc, Daryl Benson; 103 (tl) Getty Images: Photographer's Choice, Michael Dunning; 103 (tr) Getty Images: Stockbyte; 104 (b) Getty Images: Sean Gallup/Staff; 104 (c) Getty Images: Altrendo images; 104–105 Getty Images: Stone, Steven Peters; 105 (c) Getty Images: Photodisc, Nick Koudis; 105 (b) Getty Images: Photographer's Choice, Steven Hunt; 106 (b) Getty Images: Photographer's Choice, David Gould; 106 (c) Getty Images: Hulton Archive; 106–107 Getty Images: Bridgeman Art Library; 107 (l) Getty Images: Science Faction, Dan McCoy–Rainbow; 107 (r) Getty Images: Digital Vision, Jeffrey Coolidge; 108 (c) Roger Viollet/ Getty Images; 108 (b) Getty Images: Getty Images: Photodisc, Mark Downey; 109 Getty Images: China Span, Keren Su; 110 (t) Popperfoto/Getty Images; 110 (c) Getty Images: Science Faction, Caren Brinkema; 110 (b) Getty Images: Photographer's Choice RR, Chris Thomaidis; 111 Getty Images: Visuals Unlimited, Dr. David M. Phillips; 112 (t) Getty Images: Riser, Bruce Laurance; 112 (b) Getty Images: Altrendo Images; 113 (t) Getty Images: Bridgeman Art Library, Lucas van Valckenborch; 113 (c) National Geographic/Getty Images; 113 (b) Getty Images: Bridgeman Art Library, Sir Godfrey Kneller; 114 (b) Getty Images: Science Faction, Ed Darack; 114–115 Getty Images: Panoramic Images; 115 (t) Roger Viollet/Getty Images; 115 (c) Getty Images: Visuals Unlimited, Ken Lucas; 115 (b) Getty Images: Stone, Mark Harwood; 116 (t) Getty Images: Time & Life Pictures/Getty Images; 116 (b, t) Getty Images: Stone, Stephen Johnson; 117 Professor David E. Joyce, Mathematics and Computer Science, Clark University, Massachusetts, USA; 118–119 (c) Getty Images; 119 (t) Getty Images: Photographer's Choice, Michael Rosenfeld; 120 (t) AFP/Getty Images; 120 (c) Getty Images: The Image Bank, William King; 120 (b) Getty Images: Photodisc, Steve Cole; 121 Getty Images: Time & Life Pictures, Francis Miller; 122 (t) Getty Images; 122 (c) Getty Images: Photodisc, GeoStock; 122 (bl) Getty Images: National Geographic, Sisse Brimberg/Cotton Coulson/Keenpress; 122 (br) Getty Images: Hulton Archive; 123 Getty Images: Tetra Images, fotog; 123 (r) Getty Images; 124 (t) Getty Image: Photodisc, Steve Cole; 124 (b) Getty Images: Photographer's Choice RF, Justin Lightley; 125 (l) Getty Images: Bridgeman Art Library, A. Jamieson; 125 (r) Getty Images: Stone, Christian Lagereek; 124–125 (c) Getty Images: Dorling Kindersley, Dorota and Mariusz Jarymowicz; 126 (c) Getty Images: Dorling Kindersley, Brian Cosgrove; 126 (t) Getty Images: Reportage, Per-Anders Pettersson; 126–127 (c) Getty Images: Photo-grapher's Choice RR, R and R Images; 127 (t) Getty Images: Lonely Planet Images, Richard Cummins; 127 (b) Getty Images: Digital Vision, Alex Cao; 128 (t) Getty Images: Stock4B, T-Pool; (bl) AKG Images; 128 (br) Getty Images: Digital Vision, Will & Deni McIntyre; 128–129 (c) Getty Images: Photographer's Choice RR, Tom Mareschal; 129 (t) Getty Images: PhotoAlto Agency RF Collections, ZenShui/Michele Constantini; 130 (t) Getty Images: Digital Vision, John Lamb; 130 (b) Getty Images: The Image Bank, Pete Starman; 130–131 Getty Images: Photodisc, Don Smith; 131 (t) Aurora: Getty Images; 131 (b) Getty Images: The Image Bank, Luis Castaneda Inc; 132–133 Getty Images: Science Faction, William Radcliffe; 134 Getty Images: (l) Bridgeman Art Library; 134 (c) Science Museum/AFP/Getty Images; 134 (b) Getty Images: National Geographic, Justin Guariglia; 134–135 Getty Images: Bridgeman Art Library, Mehdi; 135 (t) Getty Images: Hulton Archive/Stringer; 135 (b) Getty Images: Hulton Archive; 136 (t) Getty Images: Time & Life Pictures, Fritz Goro; 136 (c) Getty Images: Photographer's Choice RF, Sam Clemens; 136 (b) Getty Images: Photodisc, Nick Koudis; 137 Getty Images: National Geographic, Rich Reid; 138 (c) Getty Images: Hulton Archive; 138 (t) Getty Images: Time & Life Pictures/Getty Images; 139 Getty Images; 140 (tl) Getty Images: Visuals Unlimited; 140 (tr) Getty Images: Dorling Kindersley, Harry Taylor; 140 (b) Getty Images: Iconica, Don Klumpp; 140 (c) Getty Images: Taxi, Ken Ross; 141 (r) Getty Images: De Agostini Picture Library, DEA/A.Rizzi; 142 Getty Images: Taxi, Mason Morfit; 143 (t) Getty Images: The Image Bank, Eightfish; 143 (cl) Getty Images: Stockbyte; 143 (cr) Getty Images: Dorling Kindersley; 143 (b) Getty Images: Stockbyte; 144 (t) Getty Images: Photographer's Choice, Images Etc Ltd; 144 (b) Getty Images: WireImage; 144 (c) Getty Images: Photodisc, Medioimages/Photodisc; 145 (t) Getty Images: The Image Bank, Jeffrey Coolidge; 145 (b) Getty Images; 146 (b) Getty Images: Dorling Kindersley, Andy Crawford & Tim Ridley; 147 Getty Images: ScienceFoto, nanolytics Austria; 148 Getty Images: Photodisc, Lawrence Lawry; 149 (t) Getty Images; 149 (c) Getty Images: Digital Vision, Yasuhide Fumoto; 149 (b) Getty Images: Time & Life Pictures/Getty Images; 150 (t) Getty Images: Photodisc, TRBfoto; 150 (b) Getty Images: Photodisc, PhotoLink; 151 (c) Getty Images: Visuals Unlimited, Mark Schneider; 151 (tr) Getty Images: Digital Vision, Chad Baker; 151 (b) Getty Images: Hulton Archive; 152 Getty Images; 153 (c) Getty Images; 153 (b) Getty Images; 154 (c) iStock-photo; 155 (t) Getty Images: All Canada Photos, Chris Cheadle; 155 (b) National Geographic/Getty Images; 156 (c) Getty Images: Photodisc, Don Farrall; 156 (r) Getty Images: Photonica, Steven Puetzer; 156 (b) Getty Images: Stone, Paul Sisul; 157 Getty Images: Minden Pictures, Carr Clifton; 158 (b) Getty Images: ScienceFoto, U. Bellhsuser; 158 (c) Getty Images: Time & Life Pictures/Getty Images; 158–159 Getty Images: ScienceFoto, nano-lytics Austria; 159 Getty Images: ScienceFoto, nano-lytics Austria; 160 (l) Getty Images: Science Faction, David Scharf; 160 (r) Getty Images: The Image Bank, Fredrik Skold; 161 (l) Getty Images: Peter Ginter; 161 (tl) Getty Images: J. Chech; 161 (b) Getty Images: Science Faction, Peter Ginter; 162 (t) Getty Images: Photographer's Choice RR, Kathy Collins; 162 (b) Getty Images: Science Faction, David Malin; 162 (r) Getty Images: Dorling Kindersley, Dave King; 163 Getty Images: Photo-grapher's Choice, Derek Croucher; 163 (b) Getty Images: Dorling Kindersley, Frank Greenaway; 164 (b) Getty Images: National Geographic, Michael Nichols; 164 (t) Getty Images: Photographer's Choice RR, Roger Holmes; 165 (t) Getty Images: Photographer's Choice, Guy Edwardes; 165 (b) Getty Images: Photodisc, Medioimages/Photodisc; 166 (t) Getty Images: Digital Vision, Ryan McVay; 166 (b) Getty Images: Science Faction, Richard Pasley–Doctor Stock; 166–167 National Geographic/Getty Images; 167 (t) Getty Images: Photodisc; 167 (b) Getty Images: Photodisc, pulp; 168 Getty Images: Time & Life Pictures/Getty Images; 169 (t) Getty Images: Aurora; 169 (b) Getty Images: The Image Bank, Jeffrey Coolidge; 170 (t) Getty Images: Stockbyte; 170 (b) Getty Images: 3D4Medical.com; 171 (t) Getty Images; 171 (b) Getty Images: Visuals Unlimited, Dr. John D. Cunningham; 172 (l) Getty Images: Hulton Archive; 172–173 Getty Images: National Geographic, Jodi Cobb; 173 (c) Getty Images: Science Faction, David Malin; 173 (b) Getty Images: Time & Life Pictures/Getty Image; 174 (c)

Getty Images: The Image Bank, SMC Images; 174–175 Getty Images: The Image Bank, Christoph Burki; 175 (t) Getty Images: Time & Life Pictures/Getty Images; 175 (b) Getty Images: Hulton Archive; 176 (c) Getty Images: The Image Bank, SMC Images; 176 (b) Getty Images: Science Faction, Bannor–CMSP; 176–177 Getty Images: Science Faction, David Scharf; 177 (b) Getty Images: Visuals Unlimited, Dr Kessel & Dr Kardon/Tissues & Organs; 178 (t) Getty Images: Stone, Kai Weichmann; 178 (b) Getty Images; 179 Getty Images: GAP Photos, Jo Whitworth; 180 (t) Getty Images: Photodisc, Andy Sotiriou; 180 (b) Getty Images: Photodisc, Andy Sotiriou; 181 (t) AFP/Getty Images; 181 (b) Getty Images: Stone, Mark Viker; 182 (b) Getty Images: Science Faction, Jay M. Pasachoff; 182 (c) Getty Images: The Image Bank, Kevin Arnold; 183 (t) Getty Images: Photographer's Choice, Hiroyuki Matsumoto; 183 (b) Getty Images; 184 Getty Images: Science Faction, Louie Psihoyos; 185 (t) Getty Images: Science Faction, Ed Darack; 185 (b) Getty Images: Stockbyte, George Doyle; 186 (c) Getty Images: Dorling Kindersley, Andy Crawford; 186 (b) Getty Images: StockFood Creative, Kroeger/Gross; 187 (t) Getty Images; 187 (b) Getty Images; 188 (t) Getty Images: fStop, Caspar Benson; 188 (b) Getty Images: The Image Bank, Pete Starman; 188–189 Getty Images: Stockbyte; 189 (t) Getty Images: Photodisc, PhotoLink; 189 (c) Getty Images: Collection Mix: Subjects, Car Culture; 189 (b) Getty Images: Time & Life Pictures/Getty Images; 190 (b) Getty Images: Digital Vision, Artifacts Images; 190–191 Getty Images; 191 (t) Getty Images; 191 (l) AFP/Getty Images; 192–193 Getty Images: Stocktrek; 194 Getty Images; 194–195 Getty Images: Time & Life Pictures/Getty Images; 195 Getty Images: Time & Life Pictures/Getty Images; 196 (t) Getty Images: Time & Life Pictures/Getty Images; 196 (b) Robert Williams and the Hubble Deep Field Team (STScI) and NASA; 197 (t) Getty Images; 197 (b) Getty Images: Digital Vision; 198 Getty Images: Bridgeman Art Library, after Domenicus van Wijnen; 199 (tl) Getty Images: Time & Life Pictures/Getty Images; 199 (tr) Getty Images: Collection Mix: Subjects, Stocktrek Images; 199 (b) NASA, ESA, R. Windhorst (Arizona State University) and H. Yan (Spitzer Science Center, Caltech); 200 (t) Getty Images: Science Faction, NASA Jet Propulsion Laboratory (NASA-JPL)–digital version copyright Science Faction; 200 (b) Getty Images; 200–201 Getty Images; 201 Getty Images: Digital Vision; 202 (c) Walter Jaffe/Leiden Observatory, Holland Ford/JHU/STScI, and NASA; 202 (b) Getty Images: Time & Life Pictures/Getty Images; 203 (t) Getty Images; 203 (c) NASA/HST/ASU/J. Hester et al; 203 (b) Getty Images: NASA-ESA–Hubble Heritage–digital version by Science Faction; 204 Getty Images: Time & Life Pictures/Getty Images; 205 (t) Getty Images: Science Faction, Fred Hirschmann; 205 (b) Getty Images: Aurora, John Lee; 206 Getty Images: Taxi, Space Frontiers; 207 (l) Getty Images: Time & Life Pictures/Getty Images; 207 (r) Getty Images; 207 (b) International Astronomical Union; 208 (t), (b) Getty Images: Bridgeman Art Library; 209 Getty Images: StockTrek; 210 (t) Getty Images: Stockbyte, StockTrek; 210 (b) Getty Images: Stone, Joel Simon; 211 (c) Getty Images: Time & Life Pictures/Getty Images; 211 (b) NASA Great Images in NASA; 212 (c) Getty Images; 212 (b) Getty Images: Photodisc, StockTrek; 213 (tl) Getty Images: Time & Life Pictures/Getty Images; 213 (tr) Getty Images: AFP/Getty Images; 213 (b) Getty Images: amana images, Yoshinori Watabe; 214 (t) NASA via Getty Images; 214 (b) Getty Images: America 24-7, Jim Hardy; 214–215 Getty Images: BLOOMimage; 215 (t) Getty Images: Purestock; 215 (c) Getty Images: Stockbyte; 215 (b) Getty Images: Stone, Pete Turner; 216 (t) NASA, ESA, and The Hubble Heritage Team (STScI/AURA); 216 (c) Getty Images; 216 (b) Getty Images: NASA-JPL-Caltech—Mars Rover/digital version by Science Faction; 217 (t) Getty Images; 217 (b) NASA, Hubblesite; 218 (t) Getty Images; 218 (b) Getty Images: AFP/Getty Images; 218–219 NASA, ESA, J. Clarke (Boston University), and Z. Levay (STScI); 219 (t) Getty Images; 219 (b) Getty Images: Science Faction, William Radcliffe; 220 (t) Getty Images: NASA-JPL-Caltech–Voyager/digital version by Science Faction; 220 (b) Getty Images: Getty Images: Time & Life Pictures/Getty Images; 221 (t) NASA, ESA and G. Bacon; 221 (b) Getty Images: Getty Images: Time & Life Pictures/Getty Images; 222 (c) Collection Mix: Subjects; Stocktrek Images; 222 (b) NASA, ESA, and Y. Momany (University of Padua); 222–223 Getty Images: Aurora, James Balog, 223 (t) Science Faction, Ctein; 224 (t) AFP/Getty Images; 224 (b) Getty Images: Roger Viollet/Getty Images; 225 (t) Getty Images: National Geographic, Jonathan Blair; 225 (b) Getty Images: Time & Life Pictures/Getty Images; 226 (b) Getty Images: Bridgeman Art Library, Andreas Cellarius; 226–227 Getty Images: Collection Mix: Subjects, Stocktrek Images; 227 (t) Getty Images; 227 (b) Getty Images: Stockbyte; 228 Getty Images: Hulton Archive; 229 (tl) NASA,ESA, M. Robberto (Space Telescope Science Institute/ESA) and the Hubble Space Telescope Orion Treasury Project Team; 229 (tr) NASA, ESA, and the Hubble Heritage Team (STScI/AURA)–ESA/Hubble Collaboration; 229 (c) Getty Images: Riser, Adrian Neal; 229 (b) Getty Images: HO/AFP/Getty Images; 230 (l) NASA and The Hubble Heritage Team (AURA/STScI); 230 (br) Getty Images: Purestock; 231 (t) NASA, ESA, N. Smith (University of California, Berkeley), and The Hubble Heritage Team (STScI/AURA); N. Smith (University of California, Berkeley) and NOAO/AURA/NSF; 231 (b) NASA, ESA and H.E. Bond (STScI); 232 (t) Getty Images: Collection Mix: Subjects, Stocktrek Images; 232 (b) Getty Images; 233 (c) Getty Images: Collection Mix: Subjects, Stocktrek Images; 233 (b) Getty Images: Time & Life Pictures/Getty Images; 234–235 Getty Images: Science Faction, Tony Hallis; 235 (t) NASA, ESA, and The Hubble Heritage Team (STScI/AURA), P. Goudfrooij (STScI); 235 (c) Getty Images: Science Faction, NASA–Hubble Heritage Team–digital version copyright Science Faction; 235 (b) Getty Images: Time & Life Pictures/Getty Images; 236 (t) Getty Images: Reportage, Joe McNally; 236 (b) Getty Images: Collection Mix: Subjects, Stocktrek Images; 237 (r) Getty Images: Aurora, Darron R. Silva; 238 (t) Getty Images: Photodisc, StockTrek; 238 (c) Getty Images: Collection Mix: Subjects, Stocktrek Images; 238 (b) Getty Images: The Image Bank, Kevin Kelley; 238–239 Getty Images: Collection Mix: Subjects, Stocktrek Images; 239 (t) NASA, Hubblesite; 239 (b) Getty Images: Science Faction, NASA-ESA–Hubble Heritage–digital version by Science Faction; 240 Getty Images; 240–241 Getty Images; 241 (t) NASA, J. English (U. Manitoba), S. Hunsberger, S. Zonak, J. Charlton, S. Gallagher (PSU), and L. Frattare (STScI), NASA, C. Palma, S. Zonak, S. Hunsberger, J. Charlton, S. Gallagher, P. Durrell (The Pennsylvania State University) and J. English (University of Manitoba); 241 (b) Getty Images; 242 (t) Getty Images: Hulton Archive; 242 (bl) Getty Images: Hulton Archive; 242 (br) Getty Images: Time & Life Pictures/Getty Images; 243 (t) Getty Images: Hulton Archive; 243 (b) Getty Images: Time & Life Pictures/Getty Images; 244 (t) Getty Images: Time & Life Pictures/Getty Images; 244 (b) Getty Images: Hulton Archive; 245 NASA Apollo, Great Images in NASA; 246 (t) Getty Images; 246 (b) Getty Images: Time & Life Pictures/Getty Images; 247 (t) NASA, Neil A. Armstrong, NASA, Great Images in NASA; 247 (bl) NASA, Great Images in NASA; 247 (br) Getty Images: Time & Life Pictures/Getty Images; 248 NASA, Great Images in NASA; 248–249 Getty Images: Time & Life Pictures/Getty Images; 249 (bl) NASA, Great Images in NASA; 249 (t) Getty Images: Time & Life Pictures/Getty Images; 249 (c) NASA, Great Images in NASA; 249 (br) Getty Images: Time & Life Pictures/Getty Images; 250 NASA, Mir-Crew, Great Images in NASA; 250–251 NASA, Great Images in NASA; 251 (t) Getty Images: Time & Life Pictures/Getty Images; 251 (b) NASA, Great Images in NASA; 252 (t) NASA, Great Images in NASA; 252 (b) Getty Images: AFP/Getty Images; 253 (t) Getty Images: Time & Life Pictures/Getty Images; 253 (c) Getty Images; 253 (b) Getty Images: AFP/Getty Images; 254–255 Getty Images: Science Faction, Norbert Wu; 256 (t) Getty Images: Popperfoto/Getty Images; 256 (b) Getty Images: Science Faction, David Scharf; 256 (tr) Getty Images: Photodisc, Adam Jones; 257 (t) Getty Images; 257 (bl) Getty Images:

ScienceFoto, G. Wanner; 257 (br) Getty Images: Taxi, Ed White; 258 (t) Getty Images: Time & Life Pictures/Getty Images; 258 (bl) Getty Images: The Image Bank, Andy Rouse; 258 (br) Getty Images: Photographer's Choice, John Warburton-Lee; 259 Getty Images: Stone, Pete Turner; 260 (t) Getty Images: National Geographic/Getty Images; 260 (c) Getty Images: MedicalRF.com; 260 (b) Getty Images: Photonica, EschCollection; 260–261 Getty Images: Photographer's Choice, Jeff Hunter; 261 (b) Getty Images: Gallo Images ROOTS RF collection, Shaen Adey; 262 (bl) Getty Images: Hulton Archive; 262 (br) Getty Images: The Bridgeman Art Library, George Richmond, 262–263 (t) Getty Images: Photographer's Choice RR, Frank Chmura; 263 (t) Getty Images: Minden Pictures, Martin Withers/ FLPA; 263 (bl) Getty Images: Stockbyte; 263 (br) Getty Images: Stockbyte; 264 (cl) Getty Images: AFP/Getty Images; 264 (cr) Getty Images: National Geographic, Kenneth Garrett; 264 (bl) Getty Images: National Geographic, George Grall; 264 (br) Getty Images: Photodisc, Medioimages/ Photodisc; 265 Getty Images: Tim Graham Photo Library, Tim Graham/Getty Images; 266 (t) Getty Images: Photodisc, ICHIRO; 266 (b) Getty Images: The Image Bank, Andy Rouse; 266–267 Getty Images: The Image Bank, Harald Sund; 267 (t) Getty Images: National Geographic, Jason Edwards; 267 (b) Getty Images: Discovery Channel Images, Jeff Foott; 268 (c) Getty Images Photographer's Choice, Ryan/Beyer; 268 (b) Getty Images: The Image Bank, Steve Allen; 268–269 Getty Images: Minden Pictures, Jan Van Arkel/Foto Natura; 269 (t) Getty Images: Hulton Archive, Getty Images; 269 (bl) Getty Images: Stone, Spike Walker; 269 (br) Getty Images: Visuals Unlimited, Dr Richard Kessel; 270 (b) Getty Images: Time & Life Pictures/Getty Images; 271 (t) Getty Images: Science Faction, Karen Kasmauski; 271 (l) Getty Images: Visuals Unlimited, Science VU/CDC; 271 (c) Getty Images: Stone, Hans Gelderblom; 271 (b) Getty Images: Visuals Unlimited, Dr F.A. Murphy; 272 (t) Getty Images: Dorling Kindersley, M.I. Walker; 272 (b) Getty Images: Science Faction, David Scharf; 273 (t) Getty Images: Science Faction, Karen Kasmauski; 273 (l) Getty Images: Visuals Unlimited, Wim van Egmond; 273 (br) Getty Images: StockFood Creative, Elisabeth Coelfen; 274 (t) Getty Images: Visuals Unlimited, Dr Gopal Murti; 274 (b) Getty Images: Stone, Ron Boardman; 275 (t) Getty Images: Stone, David Becker; 275 (c) Getty Images: Visuals Unlimited, Dr Gopal Murti; 275 (b) Getty Images: 3D4Medical.com; 276 (bl) Getty Images: Visuals Unlimited, Dr Terrence Beveridge; 276 (br) Getty Images: Visuals Unlimited, Dr Stanley Flegler; 276–277 Getty Images: Science Faction, David Scharf; 277 (t) Getty Images: ScienceFoto, G. Wanner; 277 (b) Getty Images: Visuals Unlimited, Dr Richard Kessel & Dr Gene Shih; 278 (c) Getty Images: Science Faction, Nancy Kedersha; 278–279 Getty Images: MedicalRF.com; 278 (b) Getty Images: Science Faction, Nancy Kedersha; 279 (b) Getty Images: 3D4Medical.com; 280 (t) Getty Images: Science Faction, O'Donnell–CMSP; 280 (b) Getty Images: Stone, Adrian T Sumner; 281 (t) Getty Images: Hulton Archive, Getty Images; 281 (b) Getty Images: Science Faction, Karen Kasmauski; 282 (t) Getty Images: AFP/Getty Images; 282 (c) Getty Images: Stock4B, Sabine Fritsch; 282 (b) Getty Images: Science Faction, Peter Ginter; 283 Getty Images: Stone, Andy Sacks; 284 (b) Getty Images; 283–284 Getty Images: Digital Vision, Monty Rakusen; 285 (bl) Getty Images: The Image Bank, Rob Atkins; 285 (br) Getty Images: Visuals Unlimited, Dr Dennis Kunkel; 286 (t) Getty Images: Visuals Unlimited, Dr Dennis Kunkel; 286 (b) Getty Images; 287 (t) Getty Images: Science Faction, Rawlins–CMSP; 287 (b) Getty Images; 288 (t) Getty Images; 288 (b) Getty Images; 289 (t) Getty Images: 3D4Medical.com; 289 (b) Getty Images: Getty Images; 290 (t) Getty Images: 3D4Medical.com; 290 (bl) Getty Images: Science Faction, Rawlins–CMSP; 290 (br) Getty Images: Stone, Yorgos Nikas; 291 Getty Images: The Image Bank, Anup Shah; 292 (t) Getty Images: 3D4Medical.com; 292 (b) Getty Images; 293 (t) Getty Images; 293 (b) Getty Images; 294 (t) Getty Images: The Image Bank, Tom Schierlitz; 294 (b) Getty Images: Visuals Unlimited, Dr Dennis Kunkel; 295 Getty Images: Stone, Ellen Martorelli; 296 Getty Images: Visuals Unlimited, Dr Fred Hossler; 297 (t) Getty Images: MedicalRF.com; 297 (c) Getty Images: LOOK, Harald Eisenberger; 297 (b) Getty Images: Visuals Unlimited, Veronika Burmeister; 298 (t) Getty Images: Rubberball Productions, Nicole Hill; 298 (b) Getty Images: Visuals Unlimited, Dr Gladden Willis; 299 Getty Images: 3D4Medical.com; 300 (t) Getty Images: PhotoAlto Agency RF Collections, ZenShui/Laurence Mouton; 300 (b) Getty Images; 300–301 Getty Images: Aurora, Henry Georgi; 301 (t) Getty Images: De Agostini Picture Library, DEA/C.Dani; 301 (b) Getty Images; Photographer's Choice, Cesar Lucas Abreu; 302 (t) Getty Images: Photodisc, PhotoLink; 302 (c) Getty Images: Image Source; 302 (b) Getty Images: Photographer's Choice, Derek Croucher; 303 Getty Images: Purestock; 304 Getty Images: Photo-grapher's Choice, David Maitland; 305 (t) Getty Images: Visuals Unlimited, Dr Richard Kessel & Dr Gene Shih; 305 (c) Getty Images: Stone, Dougal Waters; 305 (bl) Getty Images: 304 Getty Images: Minden Pictures, Andrea Denotti/Foto Natura; 305 (br) Getty Images: Digital Vision, Martin Ruegner; 306 (t) Getty Images: Minden Pictures, Birgitte Wilms; 306 (c) Getty Images: Taxi, Gary Bell; 306 (b) Getty Images: National Geographic, George Grall; 307 (t) Getty Images: Iconica, Jeff Rotman; 307 (b) Getty Images: Science Faction, Norbert Wu; 308 (t) Getty Images: Minden Pictures, Heidi & Hans-Jurgen Koch; 308 (c) Getty Images: Stockbyte, John Foxx; 308 (b) Getty Images: National Geographic, Tim Laman; 309 (l) Getty Images: National Geographic, George Grall; 309 (r) Getty Images: Taxi, Georgette Douwma; 309 (b) Getty Images: Visuals Unlimited, Joe McDonald; 310 (t) Getty Images: Photographer's Choice, Mitchell Funk; 310 (b) Getty Images: Photo-grapher's Choice, Ronald Wittek; 311 (t) Getty Images: National Geographic/Getty Images; 311 (c) Getty Images: The Image Bank, Steve Allen; 312 (r) Getty Images: Photographer's Choice, Georgette Douwma; 312 (c) Getty Images: Purestock; 312 (b) Getty Images: Taxi, Jeff Divine; 313 (t) Getty Images: The Image Bank, Mike Kelly, 313 (bl) Getty Images: Taxi, Peter David; 313 (br) Getty Images: Hulton Archive, Getty Images; 314 (c) Getty Images: Photographer's Choice RF, Grant Faint; 314 (b) Getty Images: The Image Bank, Steve Casimiro; 314–315 Getty Images: The Image Bank, Chip Porter; 315 (t) Getty Images: Visuals Unlimited, Wolf Fahrenbach; 315 (c) Getty Images: GAP Photos, Paul Debois; 315 (b) Getty Images: Photonica, Roy Mehta; 316–317 Getty Images: Photographer's Choice, Douglas Armand; 318 (t) Getty Images: Bridgeman Art Library, Leonardo da Vinci; 318 (b) Getty Images: Bridgeman Art Library, Gaston Melingue; 319 (tl) Getty Images: Visuals Unlimited, Dr Dennis Kunkel; 319 (tr) Getty Images: Hulton Archive, Getty Images; 319 (b) Getty Images: Photographer's Choice, Lester Lefkowitz; 320 (t) Getty Images: Hulton Archive, Getty Images; 320 (b) Getty Images: Hulton Archive, Getty Images; 320–321 Getty Images: Getty Images: Roger Viollet/Getty Images; 321 (t) The Art Archive/Musée Condé Chantilly/Gianni Dagli Ort; 321 (b) The Art Archive/Museo della Civilta Romana Rome/Alfredo Dagli Orti; 322 (t) Getty Images: Hulton Archive, Getty Images; 322 (b) Getty Images: Bridgeman Art Library, Egyptian; 322 (r) Getty Images: Time & Life Pictures/Getty Images; 323 (t) Getty Images: Bridgeman Art Library, French School; 323 (b) Getty Images: Bridgeman Art Library, Italian School; 324 (t) Getty Images: Bridgeman Art Library, English School; 324 (bl) Getty Images: De Agostini Picture Library, DEA/A.Dagli Orti; 324 (br) Getty Images: Bridgeman Art Library, After Frank Hancox; 325 (t) Getty Images: Hulton Archive, Getty Images; 325 (b) Getty Images: Stone, Hans Neleman; 326 (t) Getty Images: Photodisc, Siede Preis; 326 (b) Getty Images: Visuals Unlimited, Dr John D. Cunningham; 326–327 Getty Images: MedicalRF.com; 327 (t) Getty Images: Collection Mix: Subjects, Untitled X-Ray/Nick Veasey; 327 (b) Getty Images: Stone, Vince Michaels; 328 (t) Getty Images: 3D4Medical.com; 328 (bl) Getty Images: Image Source Black; 328 (br) Getty Images: Purestock; 329 Getty Images: Visuals Unlimited, George Musil; 329 (inset, l) Getty Images: Visuals Unlimited, Dr Gladden Willis;

329 (inset, r) Getty Images: Visuals Unlimited, Dr Gladden Willis; 330 Getty Images: MedicalRF.com; 330–331 Getty Images: MedicalRF.com; 331 (t) Getty Images: Visuals Unlimited, Dr Frederick Skvara; 331 (c) Getty Images: Science Faction, CMSP; 331 (b) Getty Images: Time & Life Pictures/Getty Images; 332 (t) Getty Images: Visuals Unlimited, Dr Dennis Kunkel; 332 (b) Getty Images: Visuals Unlimited, Dr Kessel & Dr Kardon/ Tissues & Organs; 333 Getty Images: 3D4Medical.com; 334 (l) Getty Images: Collection Mix: Subjects, 3D Clinic; 334 (c) Getty Images: Science Faction, David Scharf; 334 (br) Getty Images: Science Faction, Karen Kasmauski; 335 Collection Mix: Subjects, 3D Clinic; 335 (inset) Getty Images: 3D4Medical.com; 336 (l) Getty Images: 3D4Medical.com; 336 (b) Getty Images: Science Faction, CMSP; 336–337 Getty Images: Visuals Unlimited, Dr Kessel & Dr Kardon/Tissues & Organs; 337 (t) Getty Images: Glowimages; 337 (c) Getty Images: Visuals Unlimited, Craig Zuckerman; 337 (b) Getty Images: Science Faction, CMSP; 338 (t) Getty Images: Visuals Unlimited, Dr John D. Cunningham; 338 (c) Getty Images: ScienceFoto, G. Wanner; 338 (b) Getty Images: Visuals Unlimited, Dr Dennis Kunkel; 339 Getty Images: Visuals Unlimited, Dr David Phillips; 339 (inset) Getty Images: ScienceFoto, G. Wanner; 340 (t) Getty Images: Visuals Unlimited, Dr Don Fawcett; 340 (cl) Getty Images: Collection Mix: Subjects, 3D Clinic; 340 (b) Getty Images: Collection Mix: Subjects, 3D Clinic; 340–341 Getty Images; 341 Getty Images: Visuals Unlimited, Dr David M. Phillips; 342 (t) Getty Images: MedicalRF.com; 342 (b) Getty Images: PhotoAlto Agency RF Collections, Isabelle Rozenbaum; 343 (tl) Getty Images: Visuals Unlimited, Dr Michael Webb; 343 (r) Getty Images: Photodisc, Jim Wehtje; 343 (b) Getty Images: Visuals Unlimited, Dr Kessel & Dr Kardon/Tissues & Organs; 344 (t) Getty Images: Visuals Unlimited, SIU; 344 (l) Getty Images: Visuals Unlimited, Dr David M. Phillips; 344 (br) Getty Images: Visuals Unlimited, Dr Richard Kessel & Dr Gene Shih; 345 (t) Getty Images: Science Faction, David Scharf; 345 (c) Getty Images: Stone, Ben Edwards; 345 (b) Getty Images: Visuals Unlimited, Dr John D. Cunningham; 346 (t) Getty Images: Popperfoto/ Getty Images; 346 (b) Getty Images: 3D4Medical.com; 346–347 Getty Images: 3D4Medical.com; 347 (tl) Getty Images: Photographer's Choice, Craig van der Lende; 347 (tr) Getty Images: Iconica, Peter Cade; 347 (b) Getty Images: 3D4Medical.com; 348 (t) Getty Images: Iconica, Trevor Lush; 348 (b) Getty Images: Photodisc, Don Farrall; 349 (l) Getty Images: Stone, David Becker; 349 (r) Getty Images: The Image Bank, Hans Neleman; 349 (b) Getty Images: Stone, Mark Harmel; 350 (t) Getty Images: Hulton Archive, Getty Images; 350 (b) Getty Images: Visuals Unlimited, Dr F.A. Murphy; 350–351 Getty Images: The Image Bank, SMC Images; 351 (t) Getty Images: Blend Images, John Lund; 351 (c) Getty Images: 3D4Medical.com;
351 (bl) Getty Images: Photographer's Choice, Andersen Ross; 351 (br) Getty Images: Visuals Unlimited, Dr Kenneth Greer; 352 (c) Getty Images: Visuals Unlimited, Bill Beatty; 352 (tr) Getty Images; 352 (bl) Getty Images: Science Faction, Doctor Stock; 352 (br) Getty Images: The Image Bank, Claudia Uribe; 353 Getty Images: Taxi, Dana Neely; 354 (t) Getty Images: Visuals Unlimited, Dr Dennis Kunkel; 354 (b) Getty Images: Science Faction, Karen Kasmauski; 354–355 Getty Images: Science Faction, David Scharf; 355 (t) Getty Images: Visuals Unlimited, Dr John D. Cunningham; 355 (c) Getty Images: Science Faction, Karen Kasmauski; 355 (b) Getty Images: Visuals Unlimited, Boston Museum of Science; 356 (t) Getty Images: Visuals Unlimited, Dr Gladden Willis; 356 (b) Getty Images: Science Faction, J. Carson; 357 (l) Getty Images: Hulton Archive, Getty Images; 357 (r) Getty Images: Digital Vision; 357 (b) Getty Images: Science Faction, David Scharf; 358 (tl) Getty Images: Bridgeman Art Library, French School; 358 (c) Getty Images: Photonica, Charles Gullung; 358 (bl) Getty Images: Photographer's Choice, Bernard Van Berg; 358 (br) Getty Images: Photodisc, Keith Brofsky; 359 Getty Images: Roger Viollet/ Getty Images; 360 (c) Getty Images: Visuals Unlimited, Science VU/CDC; 360 (b) Getty Images: AFP/Getty Images; 360–361 Getty Images: The Image Bank, A. T. Willett; 361 (t) Getty Images: Visuals Unlimited, Dr Gopal Murti; 361 (c) Getty Images: Visuals Unlimited, Inga Spence; 361 (b) Getty Images: Axiom Photographic Agency, William Shaw; 362 Getty Images: Science Faction, Dan McCoy–Rainbow; 363 (tl) Getty Images: Stone, S. Lowry/Univ Ulster; 363 (tr) Getty Images: Visuals Unlimited, Dr K.G. Murti; 363 (c) Getty Images: Stone, Leland Bobbe; 363 (b) Getty Images; 364 (tl) Getty Images: Hulton Archive; 364 (bl) Getty Images: Time & Life Pictures/Getty Images; 364 (br) Getty Images; 365 (t) Getty Images: Science Faction, Rawlins-CMSP; 365 (b) Getty Images; 366 (t) Getty Images: Hulton Archive; 366 (l) Getty Images: 3D4Medical.com; 366 (cr) Getty Images: Digital Vision, Peter Dazeley; 366 (br) Getty Images: Stone, David Becker; 367 Getty Images: The Image Bank, SMC Images; 368 (c) Getty Images; 368 (b) Getty Images: The Image Bank, SMC Images; 368–369 Getty Images: Visuals Unlimited, Dr Fred Hossler; 369 (t) Getty Images: Science Faction, Karen Kasmauski; 369 (l) Getty Images: Visuals Unlimited, Dr David M. Phillips; 370 (t) Getty Images: Time & Life Pictures/Getty Images; 370 (b) Getty Images; 370–371 Getty Images: Stockbyte; 371 (t) Getty Images: Time & Life Pictures/Getty Images; 371 (b) Getty Images: Bridgeman Art Library, James Gillray; 372 (t) Getty Images: Hulton Archive; 372 (b) Getty Images: Time & Life Pictures/Getty Images; 372–373 Getty Images, Hulton Archive; 373 (t) Getty Images: Time & Life Pictures/Getty Images; 373 (c) Getty Images, Hulton Archive; 373 (b) Getty Images: Science Faction, Dan McCoy–Rainbow; 374 (t) Getty Images: Stockbyte, John Foxx; 374 (b) Getty Images: 3D4Medical.com; 374–375 Getty Images: AFP/ Getty Images; 375 (t) Getty Images: Image Source, Image Source Black; 375 (b) Getty Images: Visuals Unlimited, Dr Dennis Kunkel; 376 (t) Getty Images: Riser, A Bello; 376 (bl) Getty Images: Stone, RNHRD NHS Trust; 376 (br) Getty Images: Bridgeman Art Library, French School; 377 Getty Images: The Image Bank, Win Initiative; 378 Getty Images: Science Faction, Louie Psihoyos; 379 (t) Getty Images; 379 (c) Getty Images: The Image Bank, Jonathan Nourok; 379 (b) Getty Images: AFP/Getty Images; 380 (t) Getty Images: Stone, Hans Gelderblom; 380 (bl) Getty Images: Bridgeman Art Library, Italian School; 380 (br) Getty Images: Time & Life Pictures/Getty Images; 381 (t) Getty Images: Bridgeman Art Library; Pieter van Miereveld; 381 (b) Getty Images: Stone, S. Lowry/Univ Ulster; 382 (t) Getty Images: Digital Vision, Darrin Klimek; 382 (b) Getty Images: National Geographic/Getty Images; 383 (t) Getty Images; 383 (r) Getty Images; Photodisc, Keith Brofsky; 383 (b) Getty Images: Scoopt/Getty Images; 384–385 Getty Images: The Image Bank, Frank Krahmer; 386 (t) Getty Images: Dorling Kindersley, Gary Ombler; 386 (b) Getty Images: Discovery Channel Images, Jeff Foott; 387 Getty Images: Digital Vision, Tim Hibo; 388 (t) Getty Images: De Agostini Picture Library, DEA/R. Appiani; 388 (l) Getty Images: Stone, Wilfried Krecichwost; 389 (t) Getty Images: Visuals Unlimited, Dr Robert Calentine; 389 (b) Getty Images: Photographer's Choice, Kerrick James Photog; 390 (l) Getty Images: Science Faction, Louie Psihoyos; 390 (r) Getty Images: Discovery Channel Images, Jeff Foott; 391 Getty Images: National Geographic; 392 (b) Getty Images: Lonely Planet Images, Grant Dixon; 392 (t) Getty Images: National Geographic, Bobby Haas; 393 Getty Images: Science Faction, Ed Darack; 394 (t) Getty Images: Science Faction, Seth Resnick; 394 (b) Getty Images: Aurora, Dan Rafla; 395 (t) Getty Images: National Geographic, Ralph Lee Hopkins; 395 (b) Getty Images: Riser, Nevada Wier; 396 (l) Getty Images: The Image Bank, Andy Rouse; 396–397 Getty Images: Minden Pictures, Michael & Patricia Fogden; 397 (l) AFP/Getty Images; 397 (t) Getty Images: The Image Bank, Anup Shah; 398 Getty Images: National Geographic, Karen Kasmauski; 399 (t) AFP/Getty Images; 399 (b) Getty Images: Time & Life Pictures/Getty Images; 400 (tl) Getty Images: Photographer's Choice RF, Diane Macdonald; 400 (b) Getty Images: Science Faction, Fred Hirschmann; 400 (c) Getty Images: amana productions inc; 401 (t) Getty Images: Minden Pictures, Tim Fitzharris; 401 (b) Getty

Images: Robert Harding World Imagery, J P De Manne; 402 (l) Getty Images: The Image Bank, Wilfried Krecichwost; 402 (c) Getty Images: De Agostini Picture Library, DEA/N.Cirani; 402–403 Getty Images: Gallo Images, Stuart Fox; 403 (t) Getty Images: The Image Bank, G. Brad Lewis; 404 (t) Getty Images; 404 (b) Getty Images: MIXA, Makoto Watanabe; 405 (t) Getty Images: Science Faction, Jim Wark–Stock Connection; 405 (b) Getty Images: Stone, Claire Hayden; 406 (t) Getty Images: The Image Bank, Luis Veiga; 406 (b) Aurora/Getty Images; 407 (t) Getty Images: Minden Pictures, Rinie Van Meurs/ Foto Natura; 407 (bl) Getty Images: Stone, John Lund; 408 (t) Getty Images; 408 (b, from left) Getty Images: Photographer's Choice, Peter Hendrie; Digital Vision, Frank Krahmer; Photographer's Choice, Simeone Huber; Photodisc, John Wang; Photographer's Choice, Per Breiehagen; 409 Getty Images: Photographer's Choice RR, Frank Lukasseck; 410 (t) AFP/Getty Images; 410 (b) Getty Images: Minden Pictures. Tui De Roy; 411 (t) Getty Images; 411 (br) Getty Images; 412 (t) Getty Images: Panoramic Images; 412 (b) Getty Images: Stone, David Rosenberg; 413 (t) Getty Images: Photographer's Choice, Gary S Chapman; 413 (b) Getty Images: Stone, Burton McNeely; 414 Getty Images: Sebun Photo, Tomonari Tsuji; 415 (t) Getty Images: Photodisc, StockTrek; 415 (c) Getty Images; 415 (bl) Getty Images: Photographer's Choice, Hiroyuki Matsumoto; 416 (t) Getty Images: Tim Graham/ Getty Images; 416 (l) Getty Images: Dorling Kindersley, Irv Beckman; 416 (br) Getty Images: Stone, Oliver Strewe; 417 Getty Images: Minden Pictures, Konrad Wothe; 418 (t) Getty Images: Dorling Kindersley, Ken Findlay; 418 (bl) Getty Images: The Image Bank, David Madison; 418 (br) Getty Images: The Image Bank, Michele Westmorland; 419 Getty Images: Minden Pictures, Norbert Wu; 420 (l) Getty Images: Stone, Ted Mead; 420 (c) Getty Images: Frans Lemmens; 421 (t) Getty Images: Richard Packwood; 421 (b) Getty Images: Stone, Frans Lemmens; 422 (t) Getty Images: Gallo Images, Christopher Allan; 422 (b) Getty Images: Aurora, Alexander Nesbitt; 423 (t) Getty Images: National Geographic/Getty Images; 423 (c) Getty Images: Photodisc, Uyen Le; 423 (br) Getty Images: Discovery Channel Images, Jeff Foott; 424 Getty Images: The Image Bank, Adam Jones; 425 (tl) Getty Images: Gallo Images ROOTS RF collection, Hein von Horsten; 425 (tr) Getty Images: Stone, John William Banagan; 425 (b) Getty Images: Stockbyte, Adalberto Rios Szalay/Sexto Sol; 426 Getty Images: Robert Harding World Imagery, DH Webster; 427 (t) Getty Images: Stone, Joseph Van Os; 427 (b) Getty Images: Aurora; 428 (t) Getty Images: Time & Life Pictures/Getty Images; 428 (c) Getty Images: Photographer's Choice, Harald Sund; 428 (b) Getty Images: The Image Bank, Jochem D Wijnands; 429 Getty Images: The Image Bank, Michael Melford; 430 (l) Getty Images: Riser, Harald Sund; 430 (r) Getty Images: Photodisc, Albert J. Copley; 431 (t) Getty Images: Stone, Ed Freeman; 431 (b) Getty Images: Yoray Liberman; 432 Getty Images: Stone, David Schultz; 433 (tl) Getty Images; 433 (tr) Getty Images: Science Faction, Louie Psihoyos; 433 (b) Getty Images; 434 (t) Getty Images: National Geographic, Bobby Haas; 434 (b) Getty Images: Minden Pictures, Colin Monteath/Hedgehog House; 435 (t) Getty Images: Minden Pictures, Colin Monteath/Hedgehog House; (435 (b) Getty Images: Riser, Astromujoff; 436 (t) Getty Images: AFP/Getty Images; 436 (b) Getty Images: AFP/ Getty Images; 437 (tl) Getty Images: National Geographic/Getty Images; 437 (tc) Getti Images: Science Faction, NASA-JSC/digital version by Science Faction; 437 (b) Getty Images: Photographer's Choice RR, Sami Sarkis; 438 (t) Getty Images: Science Faction, NASA-JPL-SRTM/digital version by Science Faction; 438 (b) Getty Images: Stone, James Balog; 439 (tl) Getty Images: Minden Pictures, Tim Fitzharris; 439 (tr) Getty Images: National Geographic, Carsten Peter; 439 (b) Getty Images: Time & Life Pictures/Getty Images; 440 (t) Getty Images: National Geographic, Bobby Haas; 440 (b) Getty Images: Photographer's Choice, Ty Allison; 441 (c) Getty Images: Minden Pictures, Konstantin Mikhailov/Foto Natura; 441 (r) Getty Images: Hulton Archive; 442 (b) Getty Images: National Geographic, Kate Thompson; 442 (c) Getty Images: Lonely Planet Images, Bethune Carmichael; 442–443 Getty Images: National Geographic, Michael Nichols; 443 (t) Getty Images: Lonely Planet Images, Ross Barnett; 443 (b) Getty Images: Taxi, Jon Arnold; 444 Getty Images: Photodisc, Robert Glusic; 445 (t) Getty Images: Gallo Images, Travel Ink; 445 (b) Getty Images: LOOK, Juergen Richter; 446 (t) Getty Images: National Geographic, Stephen Alvarez; 446 (b) Getty Images: National Geographic, Stephen Alvarez; 447 (t) Getty Images: Stone, Schafer & Hill; 447 (b) Getty Images: Minden Pictures, Norbert Wu; 448 (t) Getty Images: National Geographic, Stephen Alvarez; 448 (b) Getty Images: National Geographic/Getty Images; 448–449 Getty Images: Stone, Chad Ehlers; 449 (t) Getty Images: The Image Bank, Kevin Cooley; 449 (c) Getty Images: Taxi, Richard Dobson; 450–451 Getty Images: Hulton Archive; 452 Getty Images: Boyer/Roger Viollet/Getty Images; 463 Getty Images: Time & Life Pictures/Getty Images; 468 Getty Images: Hulton Archive; 476–477 Getty Images: The Image Bank, Archive Holdings Inc.; 478 Getty Images: Altrendo, Altrendo Nature; 482 Getty Images: Photographer's Choice, Harald Sund; 486 Getty Images: Popperfoto/Getty Images; 491 Getty Images: Photographer's Choice, David Sutherland; 494 Getty Images: Science Faction, Norbert Wu; 499 Getty Images: Science Faction, Louie Psihoyos

Images used on the cover are from Getty Images.

Chapter opener images

pp 16–17: A student in Ceramic Engineering at the University of New South Wales, Australia, views a tiO2 photo-electrode used in an experiment to produce hydrogen from water and light.

pp 24–25: Fiber optic strands glow with light.

pp 90–91: Antique iron scale.

pp 132–133: Soap bubble bursting.

pp 192–193: The Alnitak region in Orion (Flame Nebula NGC 2024, Horsehead Nebula IC434).

pp 254–255: A pair of harlequin crabs (*Lissocarcinus orbicularis*) on a sea cucumber.

pp 316–317: X-ray effect image of a male skeleton, close-up.

pp 384–385: The wave rock formation reflected in water at Coyote Buttes North, Paria Canyon-Vermilion Cliffs Wilderness Area, Utah, USA.